Food Sciences

Wheat:
Chemistry
and
Technology

Volume II

Edited by
Y. Pomeranz

Washington State University
Pullman, Washington, USA

Published by the
American Association of Cereal Chemists, Inc.
St. Paul, Minnesota, USA

Front cover photograph courtesy of The National Association
  of Wheat Growers, Washington, DC.
Back cover photograph by M. V. Wiese, College of Agriculture,
  University of Idaho, Moscow, ID

Library of Congress Catalog Card Number: 88-71636
International Standard Book Number: 0-913250-73-2 (Vol. II)
                                    0-913250-65-1 (Vol. I)

Reference in this volume to a company or product name by personnel of the U.S.
Department of Agriculture or anyone else is intended for explicit description
only and does not imply approval or recommendation of the product to the
exclusion of others that may be suitable.

Printed in the United States of America

American Association of Cereal Chemists
3340 Pilot Knob Road
St. Paul, Minnesota 55121, USA

CONTRIBUTORS TO VOLUME II

E. J. Bass, Canadian International Grains Institute, Winnipeg, Manitoba, Canada

Antoinette A. Betschart, U.S. Department of Agriculture, Agricultural Research Service, Western Regional Research Center, Albany, California

A. H. Bloksma, Institute for Cereals, Flour and Bread TNO, Wageningen, The Netherlands

W. Bushuk, Food Science Department, University of Manitoba, Winnipeg, Manitoba, Canada

J. W. Dick, Department of Cereal Science and Food Technology, North Dakota State University, Fargo, North Dakota

Hamed Faridi, Nabisco Brands, Inc., RMS Technology Center, East Hanover, New Jersey

John W. Finley, Nabisco Brands, Inc., East Hanover, New Jersey

R. Carl Hoseney, Department of Grain Science and Industry, Kansas State University, Manhattan, Kansas

Karel Kulp, Cereal Science Research, American Institute of Baking, Manhattan, Kansas

William C. Mailhot (*retired*), Quality Control Department, General Mills, Inc., Minneapolis, Minnesota

R. R. Matsuo, Grain Research Laboratory, Agriculture Canada, Winnipeg, Manitoba, Canada

James C. Patton, Quality Control and Assurance, Campbell-Taggart, Inc., Dallas, Texas

Y. Pomeranz, Department of Food Science and Human Nutrition, Washington State University, Pullman, Washington

Peter Wade (*retired*), United Biscuits (U.K.) Ltd., Stoke Poges, Bucks., England

CONTRIBUTORS TO VOLUME I

D. B. Bechtel, U.S. Department of Agriculture, Agricultural Research Service, Grain Marketing Research Laboratory, Manhattan, Kansas

J. A. Bietz, U.S. Department of Agriculture, Agricultural Research Service, Northern Regional Research Center, Peoria, Illinois

A. D. Evers, Flour Milling and Baking Research Association, Chorleywood, Rickmansworth, Hertfordshire, England

John Halverson, U.S. Department of Agriculture, Federal Grain Inspection Service, Kansas City, Missouri

J. E. Kruger, Grain Research Laboratory, Canadian Grain Commission, Winnipeg, Manitoba, Canada

D. R. Lineback, Department of Food Science, North Carolina State University, Raleigh, North Carolina

William R. Morrison, Food Science Division, Department of Bioscience and Biotechnology, University of Strathclyde, Glasgow, Scotland

R. A. Orth, Bread Research Institute of Australia, North Ryde, New South Wales, Australia

Y. Pomeranz, Department of Food Science and Human Nutrition, Washington State University, Pullman, Washington

V. F. Rasper, Department of Food Science, University of Guelph, Guelph, Ontario, Canada

G. Reed, Universal Foods Corporation, Milwaukee, Wisconsin

J. A. Shellenberger (*deceased*), Department of Grain Science and Industry, Kansas State University, Manhattan, Kansas.

C. W. Wrigley, CSIRO Wheat Research Unit, North Ryde, New South Wales 2113, Australia

Lawrence Zeleny (*retired*), U.S. Department of Agriculture, Hyattsville, Maryland

PREFACE TO THE THIRD EDITION

But the images of men's wit and knowledge remain in books, exempted from the wrong of time and capable of perpetual renovation. Neither are they fitly to be called images because they generate still, and cast their seeds in the minds of others, provoking and causing infinite actions and opinions in succeeding ages.

Francis Bacon

Almost a quarter of a century has passed since the publication of the first edition of this monograph and almost two decades since the second edition. During those years, the book became accepted worldwide as the "wheat bible." I have seen it in practically every place that I have traveled around the world, and it invariably made me feel "at home" even in the most distant and isolated places.

The task of revising a "bible," even in the field of wheat chemistry and technology, is not a simple one. In conversations with the 27 coauthors (22 of whom are new ones) from six countries, I have been asked repeatedly whether the book should be a scientific primer, a textbook, a reference book, an overview of applications, a review of the state of the art, or a review of potential developments and projections of the future. My answer was "Yes" to all of these. The coauthors were asked to review critically the important material in the previous edition(s); to update and emphasize the new in information and interpretation, while making sure that the "classical" was retained, not merely out of respect but as landmarks in development; and to include new frontiers in biotechnology and genetic engineering while covering well-established and mature technologies.

For decades, much of what was accomplished in our field was based on rules of thumb. This has provided the basis for a tradition sanctified by time and strengthened by much reluctance and opposition to change. The motto has been "If it works, don't fix it." We have seldom asked how well it works, whether it works consistently or merely sometimes, whether it can be made to work better. To answer those questions, it is not enough to have know-how; we must also have know-why. Just as there is a need to make sure that students gain in college a wide range of technical skills and a broad spectrum of knowledge, there is a need to provide cereal chemists and technologists with a good data base with which to understand, interpret, and perform well in their field of expertise.

To provide that data base, the third edition of *Wheat: Chemistry and Technology* is a set of two books; the first one covers mainly the science (chemistry) and the second one mainly the technology. Every chapter has been revised extensively and most are completely new. The new edition has separate chapters on the microscopic structures and the composition of the wheat kernel. Two new chapters were added: one on nutritional aspects of wheat and wheat products and one on flat breads (including those produced in the Far and Middle East). Much space is devoted to, and information provided on, the theory and practice of evaluating wheat and wheat products around the world. All this required a considerable expansion (almost doubling) of the text and publication of the book in two volumes.

I would like to end this preface on a somewhat personal note. In discussions with the authors, many have expressed the view that a chapter should be either applied-pragmatic or basic-theoretical. I have always believed that the two are intertwined, that they reinforce and drive each other. In some of the positions I have held, there has been much pressure to complete the routine work first and engage in research only after that. Somehow the time that all routine work is done never comes and, consequently, some people never get to engage in research. This is unfortunate, because you really do not do satisfactory routine work if you do not, in a limited way at least, do some of your own exploratory work—and that is, after all, research. And finally, we often rely too much on machines, store reams of data, and do not take enough time to find out about their meaning and consequences. Let us let the modern automated instruments and computers do what they do best, generate and compute data, and let us see what they cannot see and, based on information in good books, interpret the uniqueness and research dimension of those data. If we do, we will be happier, more qualified, more equal and respected partners, and we can provide better services and cooperation in evaluating wheat in plant breeding, genetics, production, storage, marketing, processing, and utilization.

It is my hope that the new, revised edition of *Wheat: Chemistry and Technology* will help you attain that objective.

Y. Pomeranz

PREFACE TO THE SECOND EDITION

The present edition of the Wheat Monograph contains material that has been updated to include developments and advances in the last decade. This has resulted in a substantial increase in the size of the book. The main objective of the revision has been to describe, in detail, the present status of our knowledge in wheat chemistry and technology and to discuss critically significance and relevance of recent biochemical findings or new technological processes. Extensive changes were made in Part Three of the Monograph, on the composition and role of principal chemical components and elucidation of their structures by modern biochemical techniques. Part Four has been enlarged by addition of a chapter on relation between composition and functionality, and the chapter on dough properties has been revised extensively. The three chapters in Part Five have undergone major revisions to reflect the great changes that have taken place in recent years in processing of end products of wheat. The net result is a new book. It is based on the general outline and objectives of the previous edition, but some of the older material has been deleted and almost half has been revised.

Editors of multi-authored monographs are faced with two major problems: delays in submission of manuscripts and differences in scope and depth of presentation. It is, therefore, a pleasure to acknowledge the excellent cooperation of all contributors to the monograph. The contributors have prepared their manuscripts in a way that did not detract from their creative originality and individuality; yet, they conformed to a general basic approach; they sent in the material promptly so that all contributions could be processed for publication at approximately the same time.

It is a particular pleasure to acknowledge the generous help and useful suggestions of Dr. B. S. Miller, Past President of the American Association of Cereal Chemists. Mrs. Eunice R. Brown who had given nearly twenty years of devoted editorial service to the Association and was in charge of the technical editing for this edition, passed away before it was completed. Her work was carried on by Mrs. Joy McComb. They, with the aid of the Association's staff, have made every effort to produce an attractive and useful book.

Mrs. Eleanore V. Neu is thanked for excellent secretarial help.

Y. Pomeranz

PREFACE TO THE FIRST EDITION

The plant breeder who develops new varieties of wheat, the producer who grows the wheat, and grain inspector who grades it, the miller who mills it into flour, the baker who bakes the flour into bread, and the cereal chemist who studies the chemical composition and properties of the substances that make up the wheat kernel—all are preoccupied with some aspect of the same question: what constitutes quality in wheat. There is, if course, no simple answer to this question. Nevertheless, considerable progress has been made by cereal chemists over the years and recorded in widely scattered literature.

The present volume, the third in the Monograph series sponsored by the American Association of Cereal Chemists, is an attempt to provide a coherent set of reviews of our present knowledge on the cereal chemistry of wheat. It is a multiauthor work, with the advantage that each subject is dealt with by an expert in that field, but also with some disadvantages inherent in this type of book.

The quality of wheat is a function of the composition and properties of the basic material of wheat. But it is also a function of the technological methods by which wheat is transformed into intermediate, and ultimately into consumer, products. Accordingly, in the organization of this monograph both these aspects were included. The chapters on wheat, flour, dough, and end-use products emphasize the technological aspects, while the chapters on proteins, carbohydrates, lipids, etc., stress the chemical aspects. It is hoped that this book will serve as a useful reference for the cereal chemist and, in a secondary way, as a guide to original literature.

For the appearance of this book the credit must go to the individual authors. They gave generously of their time and knowledge both within and beyond their normally busy schedule of duties. Co-operation of the Monograph Committee helped to guide the Monograph especially through its earlier stages. The Committee comprised Dr. W. B. Bradley, the late Dr. W. F. Geddes, Dr. Lawrence Atkin, Mr. W. G. Bechtel, Drs. B. M. Dirks, J. W. Pence, J. A. Shellenberger, and Majel M. MacMasters. Thanks are also due to R. J. Tarleton, Executive Secretary of the Association, who was in charge of production, and to his assistant, Mrs. Eunice R. Brown, who was responsible for the technical editing. Finally, acknowledgment goes to F. D. Kuzina and Mrs. Mary Kilborn for technical assistance, and to others not mentioned but who assisted in many ways to make this Monograph a reality.

I. Hlynka

CONTENTS

VOLUME II

VOLUME I

CHAPTER 1

WHEAT FLOUR MILLING

E. J. BASS
Canadian International Grains Institute
Winnipeg, Manitoba, Canada

I. INTRODUCTION

This chapter is intended as a practical overview of flour milling. After digesting this chapter, the cereal scientist who began it with little or no prior knowledge of milling should be able to enter virtually any wheat flour mill in the world, understand its operation, and be equipped to study further the intricacies and refinements practiced by a skillful miller. It is assumed, of course, that the reader already has some fundamental training in cereal science. This chapter, then, may be regarded as a first step for a cereal scientist into the field of wheat flour milling.

The basic principles of modern wheat milling technology are practiced universally, and only slight variations occur regionally. In this chapter, therefore, the author has attempted to record a consensus among milling specialists in Canada, the United States, the United Kingdom, and continental Europe and to note briefly regional distinctions.

The principal portion of this chapter deals with the milling process in a logical sequence, beginning with a brief history of milling. The reader is next guided from the receiving elevator, through the screen room for wheat cleaning and tempering, and then into the mill. The milling process is next presented as a series of systems: break, grading, purification, sizings, reduction, flour dressing, and millfeed systems. The chief pieces of milling machinery (roller mill, sifter, and purifier) are described separately, whereas other pieces of milling machinery, such as destoner, impact machines, and powder feeders, are described in context. Other subjects inherent in the practice of milling are then described in the following order: flour dividing, flour extraction, mill stream analysis, flour treatment, milling by-products, and wheat germ recovery.

All of the preceding subjects having been elaborated with reference to hard wheat milling, the next three parts of the main section briefly describe soft wheat milling, durum milling, and air classification. The section then concludes with a very brief review of the role of starch damage in flour milling.

The final sections of the chapter provide reviews of recent developments in commercial milling and of experimental milling and milling research. An appendix lists milling terms used in the United Kingdom and gives their equivalent terms in the United States and Canada.

II. THE MILLING PROCESS

·A. History

Wheat has been a staple food for thousands of years, since people first began to settle in permanent communities. Wild cereal grasses were the early food grains. As civilization progressed, people learned to select seeds of superior plants, thereby improving the production of one of their primary food sources— a process that is still continuing today.

Primitive people used stones to pound grain and release the edible seeds from their hulls. The mortar and pestle (Fig. 1a) and the saddlestone (Fig. 1b) were later developed to improve this process. The ground material was winnowed in the air to remove the hulls, and the coarse meal was made into porridge or flat cakes. In all probability, each household owned its own mortar and pestle or saddlestone and used it to prepare its daily supply of crushed grain.

As methods of cultivation improved, and with them the yield of the harvested grain, so did the primitive milling process. Common wheats first appeared about 5,000 years ago in Asia Minor. For the next 2,000 years, milling remained a family process, until invention of the saddlestone paved the way for further progress, such as the production of a more palatable meal or flour, larger-scale technology, and the development of mills, which inevitably led to the specialized

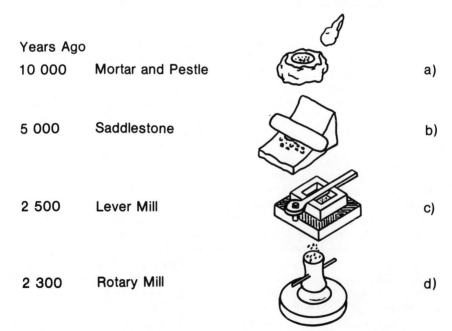

Years Ago		
10 000	Mortar and Pestle	a)
5 000	Saddlestone	b)
2 500	Lever Mill	c)
2 300	Rotary Mill	d)

Fig. 1. Ancient mills. a, Mortar and pestle; b, saddlestone; c, lever mill; d, rotary mill.

progression of milling. By ancient Egyptian times, the single-stage mortar and pestle grinding had evolved into a multistep process that added sieving, further grinding, and a final sieving to the grinding and winnowing process. The final sieving yielded a finer, whiter flour than could be produced by the simpler process. Subsequent advances, which began about 3,000 years ago (about the time leavened bread began to appear), involved the manufacture of larger saddlestones, grooved rubbing surfaces, and hoppers to feed grain to the lower stone.

The lever mill (Fig. 1c) was a logical improvement over the saddlestone process. Here, grain was subjected to both shearing and grinding by a flat, furrowed stone moving back and forth in a horizontal arc over a fixed lower stone, also furrowed.

Once the idea of the lever mill had evolved, it was not long before the Romans invented the rotary mill (Fig. 1d). Rotary mills, both animal-driven and hand-powered, came into use about 2,300 years ago and represented a major advance in the development of milling machinery. Hand-powered rotary mills, known as querns, soon superseded saddlestones and were widely used until the end of the 19th century. To this day, the quern is a popular piece of household equipment in many developing nations, where home milling is still practiced.

Animal power gave way to water power about 2,000 years ago. The dome-shaped stones of the earlier rotary mills gradually evolved into flatter, horizontal millstones (Fig. 2). Numerous water mills were built throughout Europe. Where water was not available, wind was harnessed. The first windmills were built about 1,000 years ago. Water- and wind-driven mills remained until the late 19th century and were eventually replaced by steam-powered units and subsequently by electrically driven mills.

By the early 19th century, bucket elevators and screw conveyors had been

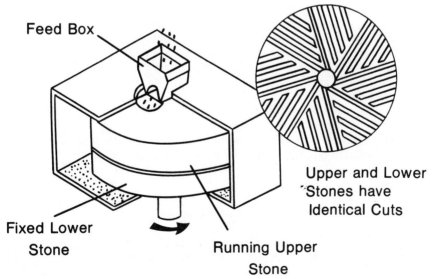

Fig. 2. Millstones.

introduced into mills in the United States, and flour milling had turned into a continuous process. Grinding and sifting, or bolting, were now highly mechanized processes. These developments paved the way for larger-scale milling with lower labor requirements.

The developments that led to the present concept of a gradual reduction process in milling occurred during the latter half of the 19th century. The middlings purifier, invented in France in 1860, was adapted for use in other European mills and in North America. This machine facilitated removal of much of the branny material that characterized flours of that time. This period also saw the introduction of the roller mill. As the advantages of this machine over millstones became apparent, there was a rapid transition to iron rolls. The gradual reduction system, adopted by most millers, needed higher capacities and greater area of sifting surface. At this time, then, the plansifter, another product of European milling engineers, became widely accepted for the grading and sifting functions of the mill. The plansifter rapidly replaced the earlier horizontal separating reels and centrifugal bolting (sifting) machines.

The gradual reduction system of milling was widely practiced in France and Hungary in the 19th century. Budapest became an important milling center, and many innovations in process and machines originated there. The new milling system was quickly accepted by the major milling countries—at that time, the United States, the United Kingdom, and Hungary. Each country's millers, using much the same basic principles, adopted slightly different means to achieve their objectives.

Thus, the milling process was an integral part of the Industrial Revolution. Some mills built in that era still continue to grind thousands of tonnes of wheat into flour for human consumption and millfeeds for animal consumption. The roller mill, sifter, and purifier have undergone many improvements since they were first introduced to the milling process over a century ago, but their basic principles remain unchanged, and most of today's modern mills incorporate these three machines into their design.

The changes in milling resulting from the introduction of these machines increased the need for separate and rapid conveying of the numerous mill stocks created during gradual reduction. Consequently, marked improvements were also necessary in mechanical handling of mill stocks. Pneumatic conveying, which originated in the 20th century, may well be the most significant development in this area.

It is perhaps safe to say that, subsequently in the 20th century, the most notable achievements in milling have taken place in materials handling, further refinement and improvement of existing milling machinery, remote control of machines and conveying systems, and finally, automation and computerization of the milling process. These developments are described in ensuing sections of this chapter.

B. Receiving and Storage of Wheat

Wheat is received and stored in that part of the mill known as the elevator. The elevator consists of unloading facilities, scales, storage bins, conveying systems, preliminary wheat cleaning equipment, and associated equipment such as exhaust systems, wheat dryers, and wheat turning and blending facilities. A typical mill elevator with maximum facilities is shown schematically in Fig. 3.

In general, mills are equipped to receive wheat by road, rail, water, or combinations of all three. The flow of wheat through the elevator is as follows. Incoming wheat is weighed, sampled, and immediately analyzed. Analyses are for foreign material such as other seeds, sand, straw, earth, or stones; insects; damaged kernels such as burnt, shriveled, immature, or sprouted kernels; moisture and protein content; and usually also α-amylase activity by some mechanical means, such as falling number. The wheat is weighed as received, dumped through a grate to remove coarse foreign material, passed over a magnet, then usually passed through a preliminary cleaner known as a "receiving separator" on its way to a storage bin, where it is stored according to class, grade, and protein content.

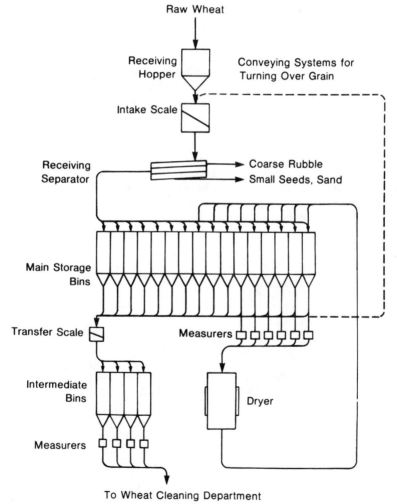

Fig. 3. Schematic diagram of a modern mill elevator.

A typical receiving separator is shown in Fig. 4. This is a cleaning machine of high capacity that does not unduly delay the unloading vehicle or vessel but provides sufficient preliminary clean-out of trash, which might otherwise block bin outlets or other equipment. The receiving separator removes material either much larger or much smaller than wheat kernels by means of inclined oscillating sieves. In addition, exhaust systems (either built-in or central) remove some of the light impurities such as shriveled kernels, wheat chaff, and dust.

The cleaned wheat is then conveyed to the main storage bins, which are usually hopper-bottomed, with either single or multiple bin outlets. A conveying system is used to transfer the wheat to intermediate storage bins, from which different grades may be blended to give a desired milling grist. The conveying system may be arranged to feed backward, to permit the "turning" of wheat from one bin to another during long storage periods. This turning over is sometimes necessary to keep the wheat in sound condition. Provision is often also made for drying wheat. This requires a separate conveying system to move the wheat through an automatic dryer and back to the dried-wheat bins and main storage block.

Usually, transfer systems that turn grain include facilities to fumigate the grain. The fumigant can be in liquid form but more commonly now is in the form of pellets that are dispensed as the grain moves along a conveyor.

Whenever dry wheat is moved, dust is produced. Accordingly, a dust control system is required to keep the atmosphere in the elevator relatively dust free. This is accomplished by using a central exhaust system with suction inlets to collect dust wherever it is generated on open conveyors and to keep a negative pressure in closed conveying systems and in the bins and all machinery. Figure 5 shows a typical dust collector installation.

In the United States, a light coat of mineral oil may be applied to grain entering grain elevators to suppress dust in the reception area and in subsequent handling. U.S. millers conservatively estimate that well over half of all milling

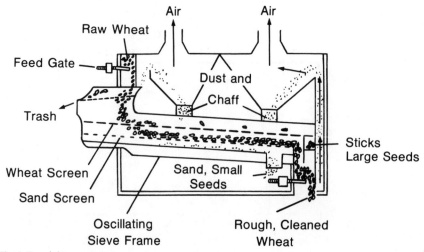

Fig. 4. Receiving separator.

wheat is treated, possibly several times, before arriving at a mill. Oil application rates may vary from 100 to 200 ppm, with approval by such agencies as the Food and Drug Administration, U.S. Department of Agriculture, and Federal Grain Inspection Service. It would be speculative at present to predict the future course of this practice.

C. Preparation of Wheat for Milling

Wheat delivered to a mill usually requires much more extensive cleaning than is provided by the precleaning machine to remove foreign or hazardous materials such as stones, mud, ergot, metals, straw, and other seeds, which might adversely affect the appearance or functionality of the milled product and might

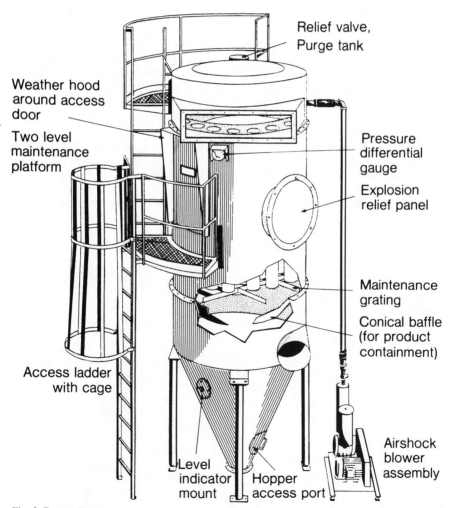

Fig. 5. Dust collector.

even damage the mill itself. Hence, before milling, wheat is transferred from the elevator to the wheat cleaning section, often referred to as the "screen room." The cleaned wheat is then prepared for milling by being "conditioned" (tempered) with an appropriate addition of water, as described in detail later.

The extent and sophistication of wheat cleaning equipment used in a mill depends largely on the range of wheat classes used in the milling grist(s) and on the miller's philosophy. For minimum loss of sound wheat, wheats of different classes or sizes should be cleaned separately, or grouped according to similar size, shape, and hardness before cleaning. Ideally, therefore, it would be desirable to have more than one cleaning line, each equipped with machinery best suited to a specific type of wheat. In practice, the general tendency is toward simplicity consistent with prevailing economics. This section describes a single cleaning passage that is commonly found in Canadian and U.S. mills. More complex cleaning systems can be readily visualized from this example.

Figure 6 shows a typical wheat cleaning flow. Wheat first passes through a milling separator, similar to the receiving separator used in the elevator but more discriminating in removing impurities that are marginally larger or smaller than wheat. As the grain passes through the machine, large impurities such as corn, soybeans, and unthreshed wheat are removed by the top screen, while the finer bottom sieve allows small impurities such as mustard seed, rapeseed, flax, and sand to pass through to be collected separately. The partially cleaned wheat "tails" over the finer sieve. As it falls off the end of the sieve, it is aspirated by air currents that remove shriveled kernels, chaff, and dust.

From the milling separator, the wheat passes to a destoner (Fig. 7). This machine separates materials that may be similar to wheat in shape and size but that differ in specific gravity—materials such as mudballs, small stones, glass,

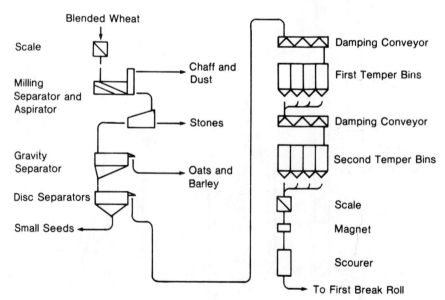

Fig. 6. Typical hard wheat cleaning flow.

and nonferrous metal. Uncleaned wheat is "floated," in a current of air, over an inclined metal screen that is oscillating rapidly. The wheat moves down the sloping screen to the outlet, while the denser stones are propelled upward toward the head of the screen by the oscillating movement and are collected separately.

The wheat from the destoner may still contain impurities either slightly longer or shorter than the wheat grains but of similar cross section. Oats and barley are examples of longer impurities, and cockle and wild buckwheat seeds, of shorter impurities. Either disc separators (Fig. 8) or cylinder separators are used to remove these types of impurities. A disc separator has a number of discs revolving in a vertical plane through the wheat. Each disc is indented, on both sides, with numerous pockets. As the discs rotate through a bed of wheat at a level slightly below the open center of the disc, they pick up particles short enough to lodge in the pockets on the upward part of the rotation and lift them out of the mixture. The lodged particles are discharged on the other side into a separate outlet on the downward part of the rotation. The grain is conveyed from the head to the tail of the machine by conveyor blades attached to the disc on the spokes near the hub. The conveyor blades convey the grain that is rejected by the pockets to a discharge point.

Figure 9 shows cross sections of a wheat disc and a seed disc. The pockets of a wheat disc are sized to lift the kernels of wheat and smaller-sized particles but to reject kernels longer than wheat, such as oats and barley. In a seed disc, the

Fig. 7. Destoner.

pockets are smaller, thereby lifting out the small seeds and rejecting wheat kernels. Disc separators are usually arranged in pairs, one machine being fitted with wheat discs and the other with seed discs.

Cylinder separators operate on the same principle as disc separators. Instead of rotating discs, the separator contains a revolving cylinder with surface indentations that pick up shorter seeds from a mass of wheat and other seeds. A second machine then picks up wheat from the residual mass of wheat and longer seeds. The liftings are dropped into an internal collecting trough, which can be adjusted in inclination to optimize the degree of length separation. Material not lifted is discharged through a separate outlet.

Ferrous metals are removed by magnets at various locations through the mill,

Fig. 8. Disc separator.

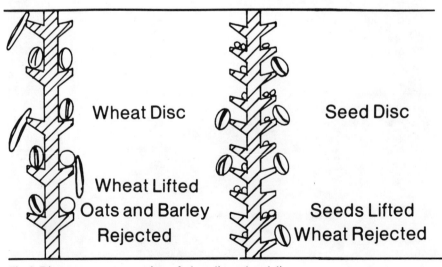

Wheat Disc

Wheat Lifted
Oats and Barley
Rejected

Seed Disc

Seeds Lifted
Wheat Rejected

Fig. 9. Disc separator: cross sections of wheat disc and seed disc.

usually at the wheat intake point at the start of the screen room cleaning and invariably preceding high-speed frictional cleaning machines and first break rolls.

In the not-too-distant past, the cleaned wheat was next washed and partially dried by passage through a "washer" and a "whizzer." This wet-cleaning system may have created many more problems than it solved, and its final demise was assured when sewage charges were levied on mills in proportion to the biological load in plant effluents. Today, only an occasional washer-whizzer may be found in some parts of the world. For all practical purposes, wheat washing has been replaced by "dry cleaning."

Most mills use a dry scourer to clean wheat before tempering. The dry scourer dislodges dirt and other impurities before water is applied to the surface of the wheat. A scourer cleans by means of rotating blades that impel the wheat against a cylindrical wire screen or perforated screen to impact and abrade the surface of the kernel. After scouring, the wheat must be aspirated to remove light impurities.

After cleaning, the wheat is ready for conditioning before milling. Conditioning (or tempering) is the controlled addition of moisture to wheat to achieve the following objectives: 1) to toughen the skin so that it will resist powdering during the milling process (powdered bran cannot be separated from flour at any stage in the milling process); 2) to facilitate the physical separation of endosperm from bran; 3) to "mellow" the endosperm so that it may be easily reduced to flour; 4) to ensure that all materials leaving the grinding rolls are in optimum condition for sifting; and 5) to ensure that grinding produces the optimum level of damaged starch consistent with the hardness of the wheat and the end use of the flour.

Correct tempering of cleaned wheat is essential to ensure maximum milling efficiency and optimum performance in the final product. The adverse effects of incorrect tempering can seldom, if ever, be reversed later in the milling process or in the bakery.

In tempering, a controlled amount of water is added, usually through spray nozzles, to a measured quantity of cleaned wheat in motion through a screw conveyor that is (usually) enclosed. At present, the tempering screw conveyor is being superseded by a high-speed mixing device that imparts a more vigorous mixing action and, therefore, a more uniform distribution of water. The damp wheat is then conveyed to a tempering bin, where it is retained for several hours to permit optimum moisture distribution throughout the wheat kernels. Usually, the miller applies a second tempering; the wheat is again dampened in another damping conveyor and conveyed to a second tempering bin, where it is usually retained for a shorter time. The tempered wheat is then sent to the first break rolls in the mill, sometimes receiving en route a final light spray of water to prevent bran shattering.

The extent and frequency of tempering depends on the initial moisture content, the class of wheat being milled, and the desired specifications of the milled products. In general, harder, more vitreous wheats require more time and can tolerate higher levels of moisture than softer wheats. Hard wheats may be tempered as long as 10–36 hr, whereas soft wheats may require as little as 4–6 hr. Durum wheat to be milled to semolina for pasta production is exceptional for a hard vitreous wheat, as it may be tempered most effectively in stages for as little as 6 hr. Obviously, then, for most effective milling, wheats should be tempered separately or only in combination with other wheats of similar tempering requirements.

The last step in the preparation of wheat for milling is scouring. After wheat from the final tempering bin has been passed over a magnet to remove ferrous metal contaminants, it is passed through a scourer to remove bran (beeswing) loosened by the tempering process and dust and dirt trapped in the crease. Scourers subject the grain to very vigorous frictional action. One type uses centrifugal movement in which steel beaters throw the grain against a woven metal screen. The dirt and beeswing come through the screen, whereas the cleaned grain passes through the grain outlet.

A vertical scourer of this type is shown in Fig. 10. Wheat enters the machine at the top and is distributed over an inverted cone to a perforated circular wire cover. It then travels in a spiral pattern down the interior of the cover, receiving a light scouring and polishing between the revolving beaters and perforated cover as it descends through the machine. The wheat is then ready to be milled.

In mills that receive infested grain, an impact scourer, such as the Entoleter, is used to break infested kernels and remove them, along with insect fragments and other impurities, by aspiration. The sound wheat is then directed to tempering.

D. Milling of Wheat into Flour

Attention can now be directed to the production of flour by milling. Milling is essentially a process of grinding and separating. Grinding is done on break rolls, sizing rolls, and reduction rolls. Separation is made using machines called sifters and purifiers.

The objective in milling is to break open the grain, scrape off as much endosperm from the bran skin as possible, consistent with obtaining flour of the required ash or color specification, and gradually grind (reduce) the practically pure endosperm into flour. After each grinding, the stock (material going to the sieve) is sifted to remove the flour. The material remaining on the sieve may then be classified into three types of particles larger than flour: 1) pure, or relatively pure, endosperm; 2) composites of endosperm and bran varying in size, shape, and proportion of endosperm to bran; and 3) pure, or relatively pure, bran.

The effectiveness of the milling depends on how well the miller is able to process the composite particles so as to recover endosperm essentially uncontaminated with bran. If the wheat has been correctly tempered, the miller may take advantage of the following three phenomena to achieve the milling objectives. First, in general, bran tends to remain in larger and coarser particles than endosperm. Second, as bran is much lighter than endosperm, composite particles of the two constituents are separable, according to their relative composition, in a machine such as the purifier, which segregates on the basis of density differences. Third, when a mixture of pure bran and pure endosperm particles of equal size (which could therefore not be separated by sifting) are ground with a slight shearing action through smooth rolls, bran tends to flatten, whereas endosperm tends to fragment into smaller particles, thereby providing a means of separation by subsequent sifting.

Thus, by judiciously sequencing grinding with corrugated and smooth rolls, sifting, and purification, a skilled miller can achieve an optimum separation of endosperm and bran.

Figure 11 shows schematically a simple milling process with four break passages, grading, purification, and eight reduction passages. Many mills are more complicated, the general rule being the larger the mill, the greater the number of grindings and separations. A large mill, however, is essentially a multiple of a small mill. Its physical appearance is extremely impressive, and there are many more connecting spouts among the machines, making the

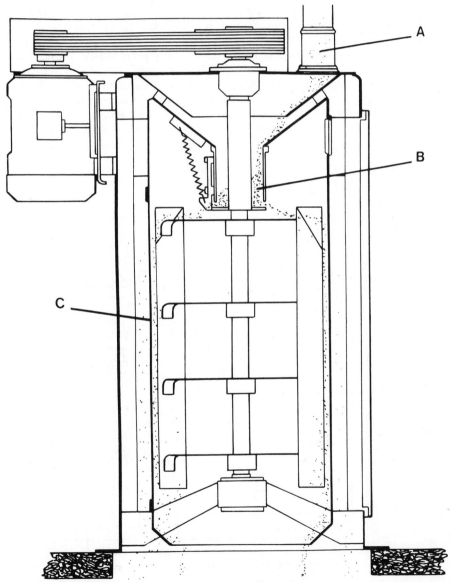

Fig. 10. Vertical scourer. A, top; B, inverted cone; C, wire cover.

process seem very complicated. However, all mills use the same basic principles of grinding and separating. Some may have five or even six break passages and up to six additional reduction passages, but the recent trend has been toward simpler systems with fewer break and reduction passages.

In principle, the milling process may be divided into the following seven distinct operating systems.

BREAK SYSTEM

Wheat is "opened up" on the first break rolls, which are grooved (corrugated) and impart a shearing action to the grain. The ground material leaving a break roll, known as "break chop," now passes to the grading (sieving) system. Here, sifting machines separate the mixture of particles according to size. The largest particles, referred to as "scalp" (*scalping* is a term applied to the specific removal of the coarsest fraction separated by a sifter), consist of the tough wheat bran and adhering endosperm. The scalp proceeds to the second break rolls, where it is again ground on corrugated rolls and is sent to the grading system (second break sifter) for further sifting. At each successive break stage, more endosperm is scraped away from the bran and separated in the grading system, until finally

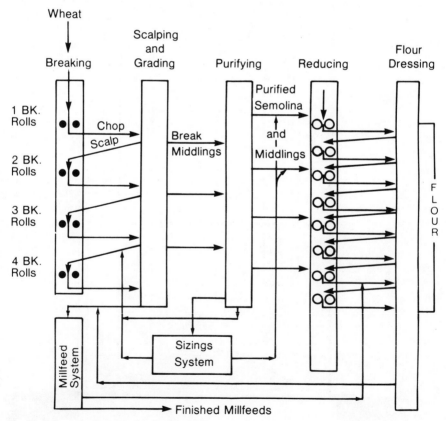

Fig. 11. Schematic diagram of a simple mill flow.

only the flat bran particles remain for final processing as a by-product. At each stage in the process, the miller tries to avoid excessive reduction of the bran, thus controlling the amount of endosperm removed at any one stage.

GRADING SYSTEM

Meanwhile, in the grading system, the smaller particles, called "break middlings," are separated according to size on sieves (i.e., graded) before being passed to the purifiers. Inevitably, as the wheat is broken open in the break system, a small amount of endosperm is reduced to flour particle size. This flour, called "break flour," is sifted out in the grading system. By far the larger proportion of endosperm passes to the purification system in the form of semolina and middlings (particles of endosperm larger than flour).

PURIFICATION SYSTEM

Break middlings are a mixture of pure endosperm, endosperm with particles of bran attached, and small particles of bran. The objective of purification is to separate this mixture into homogeneous fractions so that as much of the branny material as possible is removed before the endosperm particles (which at this stage are referred to as "farina" or "semolina") are further reduced to flour. The purer the endosperm being passed to the reduction system, the brighter and whiter the resulting flour. By selecting controlled air currents and appropriate sieves, the miller can "tune" a modern purifier effectively to maximize elimination of bran from middlings. In fact, purified semolina from the first and second break stages passing into the head of the reduction system is almost pure endosperm. In contrast, stocks originating from the third and possibly also fourth break passages always contain a higher proportion of branny material, which may be largely, but not completely, removed by efficient purification. The resulting relatively pure middlings, when ground on middle and lower reduction passages, will yield good-quality flour that is, however, darker than the flour from upper reduction passages.

SIZINGS (SCRATCH) SYSTEM

This system is best understood as a mini-break or rebreak process for the larger granules of semolina with adhering bran that come from the purification system. The main objective in the sizings or scratch system is to "scratch" or scrape this stock to release endosperm from adhering bran with minimal production of flour. This objective is achieved by grinding the branny stock lightly on corrugated ("scratch") rolls to facilitate separation of bran from endosperm. The ground stock is then sifted and the larger bran particles are separated first. The endosperm fraction may still contain smaller bran particles, which can then be removed by purification. If the purified stock is sufficiently pure, it may be directed to the reduction system; if not, it may be returned to the sizings system. In either event, light grinding is recommended in the sizings system, although the production of small quantities of middlings and flour is unavoidable.

REDUCTION SYSTEM

In the reduction system, purified semolina and middlings are gradually ground through smooth rolls to flour-particle size in six to eight successive stages of grinding and sifting. Flour is removed after each grinding stage. A small proportion of fine material remains on the sieve after the last reduction passage. This fine material passes to the by-products system along with branny material from the purifiers. The coarse by-product from the break system is called "bran," and the finer branny material from the break, purification, and reduction systems is known as "shorts."

At each grinding step, some particles may be only partly cracked, whereas others may be flaked. The extent to which each type of action occurs depends on roll spacings and load on the rolls. In some mills, particularly in western Europe, detachers or impact mills are installed between individual rollstands and the sifter to disintegrate the agglomerates and flakes, thereby increasing the yield of flour from each reduction step and reducing the load on the subsequent step.

A detacher consists of a horizontal beater-rotor, rotating in a drum. Flour is fed through an inlet situated on the upper part of one side of the drum and leaves the detacher on the opposite side. Flakes and agglomerates are reduced by the rotating beaters.

In an impact mill, or disintegrator, a rotor is formed by two parallel steel discs, separated from each other by pins of hardened steel. The rotor rotates at a speed of about 3,000 rpm inside a hoppered chamber. As flour is fed into the center of the rotor, the flour is thrown outward in a thin spray. Agglomerates and flakes are broken by impact against the rotor pins and the wall of the housing. These machines are reputedly more effective than detachers.

Impact mills make it possible to reduce the number of roller mills in the reduction system, thereby reducing the total roll length of the mill.

FLOUR DRESSING SYSTEM

Some flour is produced at each grinding stage. In the simple flow shown in Fig. 11, 12 different "machine" flours are produced, four from the break passages and eight from the reduction passages. Each machine flour has different characteristics. All machine flours may be combined to produce "straight run flour" or blended in various combinations to produce "split run flours," which are discussed later.

In the flour dressing system, all machine flours are sifted, mainly on very finely woven silk- or nylon-clothed sieves that are systematically stacked by sections in sifters. Any material sufficiently small in particle size to pass through the very fine apertures of these sieves is, by definition, flour. Any material that is not fine enough to sift through as flour is graded by size and returned to the most appropriate point in the system for further processing.

The whitest flours are made from the purest stocks of the early passages of the reduction system. At each succeeding grinding stage, stocks become less and less "pure," as the white flour has already been extracted. Consequently, on lower or "tail-end" passages of both break and reduction systems, the flours become progressively darker, higher in nonfunctional protein, and therefore inferior in baking performance. This suggests that the milling process tends to extract endosperm progressively from the center of the wheat kernel toward the bran layer as flour is produced from the head toward the tail end of the mill.

MILLFEED SYSTEM

The by-products of wheat flour milling, known as millfeeds, are bran, shorts,

and germ. Bran is the coarse, flaky product that has been passed through the break system after adhering endosperm has been removed. After it emerges from the final break passage sifter, it is passed through a bran duster (finisher) to recover the final vestiges of adhering endosperm. Shorts are the finer branny material emerging from the last break sifter, along with the final overtails of the lower passages of purification, and possibly also from the final reduction sifter. Shorts also are passed through a finisher (shorts duster) to strip the remaining endosperm. The bran and shorts dusters are similar in design, consisting of a brush or beaters rotating vertically or horizontally inside a cylinder of fine screen, through which the endosperm removed from the millfeed is passed.

In a conventional milling system, germ is recovered from the "overtails" of coarse reduction passages. Germ recovery is discussed in more detail later in this chapter.

The millfeed system also includes collection of foreign material recovered from the cleaning house, generally referred to as "screenings." Screenings contain all foreign materials except metals, stones, mudballs, and hazardous or toxic seeds as well as ergot. Screenings are usually ground in hammer mills and added to the millfeed. In many cases it is profitable to pelletize millfeeds for shipment to feed millers.

E. Milling Machinery

An understanding of the principles and operation of the major mill machines is essential to a practical knowledge of the milling process. The machines described here are the roller mill, the sifter, and the purifier.

ROLLER MILL

A modern roller mill is shown in Fig. 12. It consists of two pairs of cast-iron rolls, "chilled" on the outside to give a hard, wear-resistant surface and mounted in a heavy cast-iron frame. Standard rolls are available in two standard-diameter sizes, 225 mm (9 in.) and 250 mm (10 in.), and are manufactured in various lengths from 61 to 125 cm. Two pairs are normally mounted back to back in a single frame, each pair being driven separately and having separate feeding and adjustment mechanisms. Thus, a roller mill is really two pairs of rolls in one frame. The upper roll of a pair is carried in fixed bearings, but the lower roll is supported by pivot arms, which may be adjusted to vary the distance between the rolls. Adjusting the gap gives different degrees of grinding. Two smaller-diameter rolls, the feed rolls, are set above the grinding rolls. The purpose of the feed rolls is to regulate the stream of material to the "nip" of the grinding rolls. A balanced feed gate ensures even distribution of stock along the full width of the rolls. Cleaning brushes are mounted in the frame of the roller mill to remove any material that adheres to the surface of the break rolls. For reduction rolls, which are smooth, steel scrapers or hard felt strips are used instead of brushes. However, brushes are also used in the United Kingdom, as they last longer than felt strips.

Since about 1950, several shifts in the configuration of the grinding rolls have taken place. For example, Fig. 12 shows the pair mounted at an angle of approximately 45°. This configuration, known as the "diagonal roll," had traditionally been European, whereas U.S. manufacturers had traditionally produced horizontal rolls. In the intervening years, new or remodeled mills in the United States installed diagonal rolls, whereas in recent years most European manufacturers have turned to horizontal or nearly horizontal rolls.

Whatever the configuration, the two rolls turn in opposite directions at different speeds. The roll differential can be provided by gears, chain and sprocket, or v-belts and sheaves. The upper roll turns at a higher speed than the lower roll, thereby producing a shearing or tearing action, rather than the squeezing or flattening action that would be produced if the rolls turned at equal speeds.

The surface of the grinding rolls may be either corrugated or smooth. Break rolls are always corrugated, with "flutes" cut in a spiral pattern along the length of the roll. Figure 13A shows a pair of break rolls and illustrates the spiral pattern of the corrugations. When the rolls are turning (Fig. 13B), the corrugations cross each other. This action imparts a cutting and shearing effect on the material between them, the effect being greater the steeper the spiral (Fig. 13C). The spiral shape also prevents the rolls from accidentally "locking." The

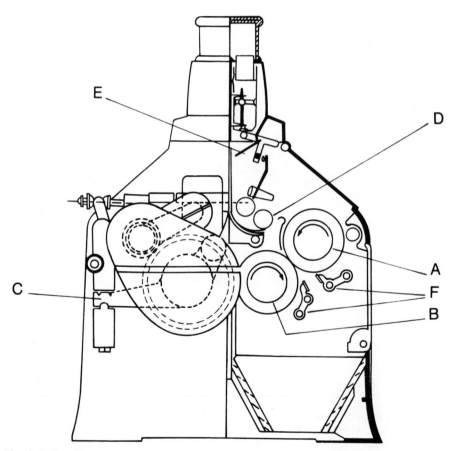

Fig. 12. Roller mill. A, upper roll; B, lower roll; C, pivot arms; D, feed rolls; E, feed gate; F, brushes.

slower roll tends to hold the stock (consisting of whole or partly broken kernels), while the speedier roll scrapes endosperm from it. The number of corrugations increases with successive breaking stages. For example, the first break rolls may have four to five corrugations per centimeter of roll circumference, whereas the last break passage may have 10–11. Roll speeds are usually 500–550 rpm for the fast rolls. Break roll differentials are 2.5:1 giving a slow-roll speed of about 200–220 rpm.

Numerous configurations of roll corrugations are available to the miller, who may therefore choose the best arrangement. Different wheats require different roll corrugations for best milling results. For example, first break rolls in U.K. mills may have only three and three-quarters or four corrugations per centimeter of roll circumference, to facilitate the grinding of the larger kernel characteristic of native wheats included in the grist. Most U.S. and Canadian millers prefer to have five corrugations on their first break rolls to accommodate the smaller-kernel domestic wheats.

The corrugations vary not only in the number per centimeter but also in the "profile," or shape of the tooth. A sharp angle is used primarily for hard wheats to make semolina or large middlings, whereas flatter angles are used for softer wheats. The corrugations may also be oriented to be either sharp or dull, as determined by the direction of the sharp angles as the rolls rotate. This is another variable used to optimize the cutting and scraping action of the break rolls.

Reduction rolls are usually smooth (frosted), but on occasion the head reduction surfaces may be finely corrugated. The two rolls of each pair also run at different speeds. The normal speed differential for reduction rolls is 5:4, with

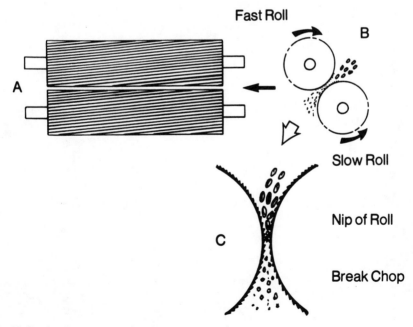

Fig. 13. Break rolls. A, corrugations; B and C, operation.

the fast roll running at 500–550 rpm. Some reduction rolls are cooled by circulating water through the center hollow part of the roll to disperse heat generated by grinding. This reduces the roll temperature and possible heat damage to the endosperm as it passes through the nip of the roll. Cool rolls also minimize evaporative loss and improve general milling performance.

It is very important to maintain all rolls in good condition to ensure that both extraction and quality of the flour are maintained. Break rolls must be recorrugated periodically to correct for wear. First and second break rolls may give about six to 12 months of continuous service, the exact interval depending on factors such as surface allocation or feed rate per unit of roll length, severity of grinding, hardness of wheat kernels, and presence of stones or other damaging impurities. Other break and reduction rolls may grind satisfactorily for several years before resurfacing becomes necessary.

SIFTER

A typical sifter is shown in Figure 14. It consists of several layers of sieves, rotating in a horizontal plane. The sieves are arranged in sections, each section

Fig. 14. Sifter.

containing as many as 30 layers of sieves. The complete machine may have two, four, six, or eight sections. A sifter provides a much larger area of sifting surface in much less space than was provided by the reels and centrifugals of yesterday's mills. A single eight-section sifter can be arranged to handle as many as 16 different mill stocks, and it is this versatility, combined with high capacity, that has made it the preferred machine for grading mill stocks and sifting flour.

The sieves of a sifter are clothed (covered) with coarse wire mesh for coarse separations, or fine wire mesh, nylon, or silk for finer separations of flour and middlings. Sieve covers are drawn tightly over the frames, which fit into flat metal trays. The sifted material (or "throughs") of the sieve fall into the tray to be directed out of the machine or to a finer sieve lower down in the sifter for further separation. Overtails leave the machine through an internal channel or may be directed to another sieve in the same section for further sifting. A 9.5-mm wire mesh fastened to the bottom of the sieve frame carries a number of hard rubber balls, cotton pads, or brushes. As the entire sifter gyrates, the pads are agitated to maintain a constant tapping action on the underside of the sieve, thereby preventing the apertures from being blocked.

Many different flow arrangements are possible in a sifter. For example, Fig. 15 shows schematically one section of a sifter with 26 sieves, arranged for grading the chop received from a break roll into five fractions. The numbers shown on the sieve covers indicate the mesh size in micrometers. In this example, the coarsest wire cloth at the top of the stack has an aperture size of 1,000 μm (1 mm), whereas the finest sieves used for "dressing out" the flour have openings of only 112 μm.

The operation of a sifter is relatively simple. Material enters the top of the machine, where, if the load is heavy, it may be split into two equal streams by a plate located just below the inlet. This provides easy passage of material over the head sieves, where the load is greatest, and thereby increases the capacity of the sifter section. The rotary motion of the sieves causes the finer, heavier particles to work their way down through the layer of ground material, leaving the coarser, lighter particles on the surface of the layer. The coarsest material that does not pass through the 1,000-μm mesh tails over the sieve into a vertical channel and then out of the sifter through a flexible sleeve. Meanwhile, the material passing through the 1,000-μm mesh is collected from the trays of several sieves of the same mesh and falls into a finer mesh sieve (e.g., 475 μm). The process of grading the material by size is repeated as it flows through the machine, each classification having its own outlet. An alternative arrangement is to remove the smallest particles first by inserting the finest meshes at the top of the sifter. The larger particles then tail over to a coarser mesh and are separated into two different sizes. The sieves in a section of a sifter may be arranged to suit almost any purpose and to make from two to seven separations from the stock feeding the machine.

The sieves in a sifter should be inspected and cleaned at regular intervals. Despite the cleaning action of the bouncing balls or pads, the sieve apertures can become clogged with floury or fibrous particles after prolonged operation, reducing the efficiency of the machine. Routine brushing of the cover and repairs to holes caused by wear and tear are essential to keep flour production and quality at the required levels. All covers are normally replaced when they become worn or torn beyond reasonable tolerance.

PURIFIER

The purifier is perhaps the least understood of the machines in a modern mill. Its purpose is to separate bran particles, and pieces of endosperm with adhering bran, from particles of pure endosperm. The feed to a purifier comes from the grading system and is a mixture of particles of similar size but of varying density. By subjecting the mixture to a controlled flow of air as it passes over a sieve, the purifier can separate the essentially pure endosperm from the branny particles.

Figure 16 illustrates the operation of a purifier. It consists of a frame about 2 m long with a slight downward slope. The frame carries a number of sieves, usually four per deck, that are clothed with progressively coarser material from

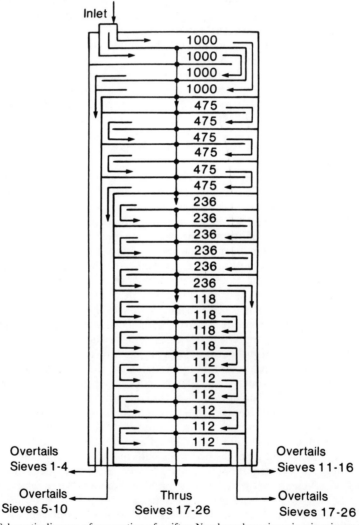

Fig. 15. Schematic diagram of one section of a sifter. Numbers show sieve sizes in micrometers.

head to tail end. The compartment above the sieves is connected to an air trunk. The bottom of the sieves is open to the atmosphere. Air currents are drawn through the total area. Valves, set in sections of the compartment and the main air trunk, allow adjustments to be made to the air volume passing through the machine. An eccentric drive oscillates the frame longitudinally. This motion agitates and stratifies the stock on the sieve into layers, the heavier endosperm particles riding closest to the sieve surface. The combination of oscillation and slope of the sieve causes the stock to travel slowly along its length. Carefully controlled air currents drawn from the underside of the sieve float the lighter branny particles to the surface of the layer of material on the sieve. Heavier particles of endosperm fall to the lowest strata of material and ride on the surface of the sieve until they reach the part of the sieve that has openings big enough for them to fall through. As the particles fall through, they are collected in a series of graded fractions. Thus, at the head end of the purifier, the purest, most dense particles of endosperm pass first through the mesh. As the mixture travels along the sieve, the composite particles of endosperm with adhering bran are next to

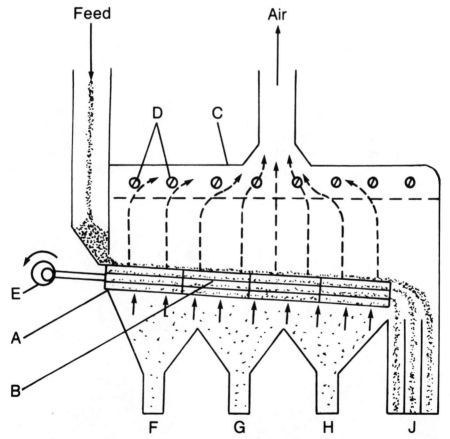

Fig. 16. Purifier. A, frame; B, sieves; C, air trunk; D, valves; E, drive; F, exit for dense endosperm particles; G and H, exit for composite particles; J, exit for branny particles.

pass through the mesh, and finally the light, coarse, branny particles tail over the sieve into separate collecting spouts. The air currents lift out the lightest branny material, which is drawn into the trunking above the compartment and conveyed in the air stream to a dust collector.

A modern purifier, such as is shown in Fig. 17, is equipped with two, and sometimes three, decks of sieves. When the air valves are correctly adjusted, feed stock may be sharply separated into pure endosperm, composite pieces, and branny particles. The pure endosperm is conveyed to reduction rolls for grinding into flour; composite particles are returned to the sizings system where finely corrugated rolls scrape the endosperm from the bran skins; and the bran is directed to a separate section of the break system. The sieves in a purifier are kept clean by moving brushes or rubber balls located immediately under the sieves.

Optimum purification is achieved when the graded semolina and middlings feeding the machine are classified into particle size ranges as narrow as is practical. At the same time, enough material must be available to keep the sieves covered to prevent air currents from becoming unbalanced, which causes imperfect separation of the branny particles. Efficient purification, followed by accurate grinding, increases the proportion of high-quality flour sifted out by the reduction sifters and increases flour extraction.

Purification is particularly important in production of granular products such as semolina (from durum wheat) and farina (from common wheat). These granular products are used for pasta, baby foods, and other specialty foods.

Fig. 17. Purifier.

Most mills repurify some of the cleanest middlings streams and package them separately. Sometimes the farinas (or semolinas) and other granular materials are passed over a gravity table as a final cleaning stage to remove black specks that may be residues of stones, dirt, or seed hulls.

As shown in Fig. 18, a gravity table operates similarly to a destoner. The feed is fluidized by air currents on an inclined deck, and oscillating action propels the desired material toward the discharge section. The heavier particles travel farther up the deck and discharge at the high point, whereas the lighter material, which is held back by the air currents, discharges at the low point of the deck. The premium product is obtained from the middle cut, which is usually the cleanest fraction.

F. Milled Products

Now that the fundamental principles of milling have been explained, attention may next be directed toward the main procedures practiced to produce economically viable end products. Procedures described include flour dividing, flour treatment, and by-product recovery. This section also deals with the concepts of flour extraction and mill stream analysis, which provide the major analytical tools used to control milling efficiency and end-product specifications.

FLOUR DIVIDING

As was mentioned earlier, the miller has two choices about how to utilize the machine flours produced at various stages in the milling process. The simplest,

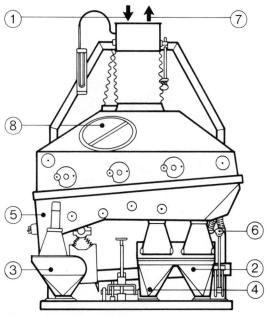

1. Product inlet
2. Outlet for heavy product
3. Outlet for light product
4. Outlet for mixed product
5. Pneumatic re-lift
6. Adjustment of table angle
7. Aspiration connection
8. Lighting

Fig. 18. Specific gravity separator.

and most obvious, choice is to combine all streams to produce one flour known as "straight run" or "straight grade" flour. The second, more complex and potentially more profitable, choice is to practice a technique referred to as "split-run milling" or "divide milling." Here, the miller combines flour streams to produce three or four grades with different properties, thus generating, by

Ash Content

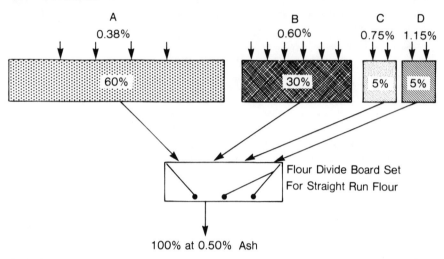

Ash Content

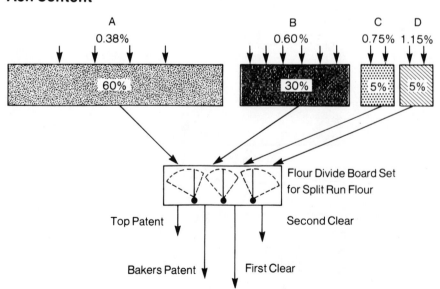

Fig. 19. Schematic diagram of a flour divide. Top, straight-run flour; bottom, split-run flour. Letters indicate flour streams.

judicious stream blending, a wide range of products. From a single type of wheat or appropriate grists, the miller can produce the wide variety of specialty flours demanded by the mill's customers. Obviously, also, flour dividing gives the miller considerable flexibility in selecting wheats and therefore in minimizing production costs.

The practice of flour dividing can best be explained by example. Figure 19 shows schematically a typical flour divide in a Canadian or U.S. mill using a good grade of hard red spring wheat. Similar machine flours are selected, primarily by specified ranges in ash content, and combined into four streams designated A to D. The final ash content and proportion of each stream (expressed as percent of the total flour produced) are shown at the top of the illustration. In this example, stream A contains four machine flours from the head of the reduction system, probably passages 1M (first middlings roll) to 4M (fourth middlings roll). Stream A represents 60% of the total flour being produced and has an ash content of 0.38%. Stream B contains six machine flours originating in the upper break and middle reduction passages. It represents 30% of the total flour and has an ash content of 0.60%. Stream C, representing 5% of the total at an ash content of 0.75%, contains two machine flours, possibly 3B and a low (tail end) reduction passage. Stream D, also 5% of the total, with an ash content of 1.15%, probably originates from the lowest break passages and the lowest reduction passages. In all probability, this flour stream would adversely affect the quality of any higher grade flour to which it might be added. The four streams flow into the top of a divide board (or box), which is installed in a central location. The "gates" in the box control the proportion of each stream that enters the blend. By setting the gates in different positions, the miller produces the various commercial flours.

In western Europe and the United States, split-run milling is practiced much less extensively than in Canada. Many other variations are possible, of course, but the intricacies of this technology are clearly beyond the scope of this review.

FLOUR EXTRACTION

Reference has already been made to flour extraction. Although the term appears to be self-explanatory, the many pitfalls in its use can, if not avoided, lead to confusion and misunderstanding. The two major pitfalls lie in the choice of method used to calculate extraction and the use of the term *flour yield* to denote *flour extraction*.

In general, flour extraction (perhaps a more descriptive term would be *flour extraction rate*) may be defined as the proportion of the wheat recovered as flour by the process of milling. It may be expressed as a percentage of either the raw material used or the products obtained. Thus, flour extraction rate may be calculated in at least the following five different ways, and the basis for the calculation must therefore be stated along with the rate to avoid ambiguity.

1. Basis: Wheat as received (sometimes referred to as "dirty wheat")

$$\% = \frac{\text{Weight of flour} \times 100}{\text{Weight of dirty wheat}}$$

2. Basis: Clean dry wheat to temper

$$\% = \frac{\text{Weight of flour} \times 100}{\text{Weight of clean dry wheat}}$$

3. Basis: Clean, tempered wheat to first break rolls

$$\% = \frac{\text{Weight of flour} \times 100}{\text{Weight of clean tempered wheat}}$$

4. Basis: Total products obtained in the milling process

$$\% = \frac{\text{Weight of flour} \times 100}{\text{Weight of flour} + \text{weight of millfeeds}}$$

5. Basis: Total products of the mill

$$\% = \frac{\text{Weight of flour} \times 100}{\text{Weight of flour} + \text{weight of millfeeds} + \text{weight of screenings}}$$

The two most important ways of expressing flour extraction rate are the first (basis wheat as received) and the third (basis clean, tempered wheat). The former provides a measure of the cost relationship between raw materials and flour produced, and the latter gives the efficiency of the milling process itself.

Flour yield, contrary to the conventional use of the word *yield*, does not represent the amount of flour obtained from a given amount of wheat. Instead, flour yield may be defined as the amount of wheat required to yield a given amount of flour, usually 100 lb. The term *flour yield* is most commonly used in the United States, and it too may be calculated differently for each way the "amount of wheat" is expressed. In the United States, flour yield is usually expressed as the number of bushels of wheat required to yield 100 lb of flour. Thus, the higher the flour yield, the lower the milling efficiency. This is the reverse of the meaning of flour extraction, where the higher the extraction, the higher the milling efficiency. Obviously, then, although flour yield and flour extraction are related, the relationship is inverse, and the two terms should not be used interchangeably. Usually, the miller finds flour yield a useful measure of the mill's wheat costs, thereby facilitating a response to daily changes in wheat price. On the other hand, flour extraction provides a useful measure of milling efficiency.

MILL STREAM ANALYSIS

Regular mill stream analysis is essential for maintaining accurate control over both flour divides and milling efficiency. As the name implies, mill stream analysis is merely analysis of each machine flour for moisture, ash, protein, and possibly color during normal operation of the mill. These analyses, along with a measure of the proportion of total flour represented by each machine flour, provide the miller with the information needed to make minor adjustments to the mill and to select flour divides for optimum financial return.

In practice, a complete mill stream analysis is conducted over an appropriate period, during which a uniform wheat mix is being milled and the mill is operating at normal efficiency. The data are collected and processed as shown in Table I, to provide a "cumulative ash" value (in percent). The cumulative ash represents the ash content of the flour that would be obtained if the machine flours of Table I were progressively blended. Thus, if all machine flours were blended, the straight grade flour in Table I would have an ash content of 0.45%. Of course, any combination of streams may be selected for blending, and the cumulative ash content of the blend may be calculated from columns 2, 4, and 5 of Table I, as shown in the example in Table II. Here, the ash content of a blend of the four machine flours from the first and second middling rolls and the first and second break rolls would be 20.29/60.2 = 0.34.

Cumulated ash data, such as those shown in Table I, are usually depicted graphically in an "ash curve," as shown in Fig. 20. Ash curves help the miller evaluate the efficiency of the process at a glance. Efficient milling is recognizable

TABLE I
Typical Mill Stream Analysis

Flour Stream[a]	Flour Production		Ash Content			
			Individual Streams		Cumulative	
	Percent of Total	Cumulative Percent	Percent[b]	Ash Units[c]	Ash Units	Percent[d]
1 M	29.3	29.3	0.30	8.79	8.79	0.30
2 M	14.6	43.9	0.33	4.82	13.61	0.31
3 M	5.2	49.1	0.36	1.87	15.48	0.31
1 + 2 B	16.3	65.4	0.41	6.68	22.16	0.34
3 B	6.5	71.9	0.45	2.92	25.08	0.35
2 Q	10.1	82.0	0.47	4.75	29.83	0.36
4 M	2.1	84.1	0.58	1.22	31.05	0.37
Tailings	6.0	90.1	0.67	4.02	35.07	0.39
5 M	4.7	94.8	0.79	3.71	38.78	0.41
4 B	3.8	98.6	0.92	3.50	42.28	0.43
LG	1.4	100	1.66	2.32	44.60	0.45

[a] M = middlings, B = break, Q = quality, LG = low grade.
[b] 14% moisture basis.
[c] Column 2 multiplied by column 4.
[d] Column 6 divided by column 3.

TABLE II
Typical Calculation for Split-Run Milling, Using Streams in Table I

Flour Stream[a]	Flour Production (% of total)	Ash Content (%)	Ash Units
1 M	29.3	0.30	8.79
2 M	14.6	0.33	4.82
1 B + 2 B	16.3	0.41	6.68
Total	60.2		20.29

[a] M = middlings, B = break.

by a curve that remains as nearly horizontal and low in ash content as possible up to the point represented by 80% of the total flour.

Cumulative flour protein may be calculated and utilized in a similar way to meet protein specifications of flour blends.

FLOUR TREATMENT

Flours may be treated with a variety of additives to achieve any desired combination of the following objectives: 1) to bleach the flour, 2) to improve the bread-baking quality of the flour, 3) to modify the gluten characteristics, 4) to supplement the natural amylase activity of the flour, or 5) to supplement the natural vitamin and mineral content of the flour.

Because many of the available chemicals may have multiple functions, it is important to select only those chemicals necessary to ensure a final flour with the desired functional properties. Correct choice of chemicals depends on their physicochemical properties in flour, a subject elaborated in considerable detail elsewhere in this book and therefore discussed here only in the context of milling technology.

Bleaching agents are added to flour primarily to eliminate the need for prolonged flour storage. Freshly milled flour has a slightly creamy color due to the small quantities of natural pigments in wheat. If the flour is stored for a period of about three weeks, most of the pigment is bleached by natural oxidation. However, millers find it impractical to store large quantities of flour for prolonged periods. Therefore, accelerated bleaching is usually achieved by the addition of chemicals that bleach the flour pigments by oxidation. Benzoyl peroxide is the most common bleaching agent used. At a feed rate of 50 ppm, bleaching occurs within 24 hr after addition. Chemical bleaching of flour is not permitted in the European Economic Community (EEC).

The baking performance of flour is also improved by prolonged storage. A

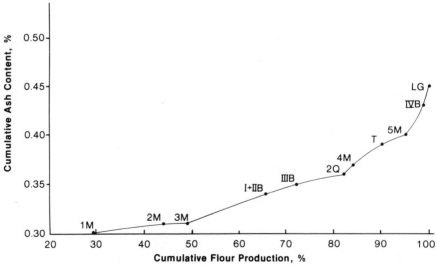

Fig. 20. Typical ash curve. B = break; LG = low grade; M = middlings; Q = quality; T = tailings.

similar, presumably oxidative, effect can be realized by treatment with chemicals referred to as flour improvers. Many improvers are available today, but the most common in Canada and the United States are potassium bromate and azodicarbonamide. The latter also acts as a bleaching agent. In the EEC, only ascorbic acid is permitted. In addition, numerous proprietary mixtures are available commercially, particularly for addition in the bakery.

Modification (usually weakening) of gluten characteristics for cake flours is frequently achieved by gaseous treatment with either chlorine or chlorine dioxide. For bread-baking flours, there are many proprietary products containing proteolytic enzymes or reducing agents. Again, most of these are intended for addition in the bakery. In the EEC, protease, cystine, and cysteine are additives permitted for flour.

Amylase activity is inversely related to the soundness of the wheat from which the flour is milled. Thus, when the wheat is very sound, supplementary amylase must usually be added to the flour to generate sufficient fermentable sugar during baking for conversion by yeast to carbon dioxide. Amylase may be added in the form of malted barley flour, malted wheat flour, or commercial enzyme preparations. Commercial preparations appear to be the most uniform and easily controlled, particularly for flours intended for specialty baked products.

Addition of supplementary vitamins and minerals is now widely practiced. In most cases, the vitamin-mineral supplement is available as a single additive, known as *enrichment* in the United States and Canada. Additional calcium and iron supplementation is becoming popular as well.

Overall, the type and amount of additives permitted in flour may vary widely from country to country, and regulations regarding their use are usually under close surveillance by government regulatory agencies. Consequently, current regulations should always be consulted for detailed information.

Needless to say, because additives are required in very small amounts (usually in parts per million), which must be uniformly distributed throughout the flour to be uniformly effective, the process by which flour is chemically treated is highly critical. The best method of distribution appears to be to convey the additive by high-speed air current into flour agitators. By this method, powder addition may be achieved simultaneously with flour stream blending of final products, thereby facilitating homogeneity in the final product.

In practice, it is preferable to have each powder (which may contain several different additives) metered separately into the appropriate flour stream. A wide range of accurate and reliable powder feeders is now available. The most recent ones are self-adjusting to compensate for variations in the flour flow rate, fully automated, and computer compatible. However, most mills today still use powder feeders that are manually adjustable and therefore require daily attention. A typical powder feeder is shown in Fig. 21.

Final sifting of flour immediately before packing may be considered a flour treatment designed to help ensure uniformity and freedom from impurities in the final product. This treatment is usually referred to as "redressing" or "rebolting." Rebolting may occur as the flour leaves the mill on its way to flour storage, or immediately before the flour is packed or conveyed to bulk loaders, or both. The type and location of the sifting equipment used for rebolting depend on several factors such as mill layout, available space, and proximity to agitators.

The final mechanical control of insects in flour is achieved by use of a centrifugal impact machine through which flour may be passed just before it leaves the mill and again before it is either packaged or shipped in bulk. Such impactors are now produced by several manufacturers. As with rebolt systems, the location of the impactors may vary from mill to mill.

A typical centrifugal impact machine is shown in Fig. 22. Two parallel, horizontal steel discs form a rotor. The discs are separated from each other by pins of hardened steel called impactors. This rotor is directly driven by an electric motor running at about 3,000 rpm in a hoppered chamber of special design. As flour is fed into the center of the high-speed rotor, it is flung outward in a thin spray through the space between the two steel discs. Live insects and insect eggs that have survived the entire milling process are killed by severe impact, against both the rotor impactors and the peripheral wall of the chamber.

Few, if any, millers consider impaction an alternative to good housekeeping and other traditional forms of insect control. However, it is effective and a valuable component in a comprehensive program of insect control.

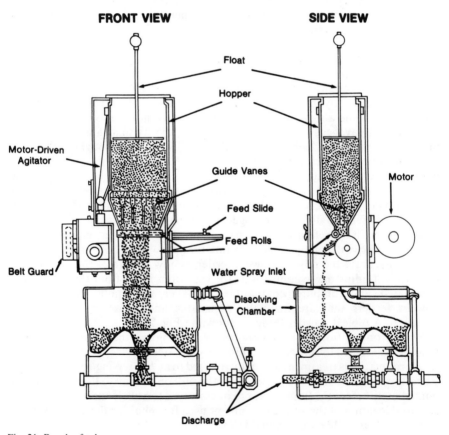

Fig. 21. Powder feeder.

G. Milling By-Products

Little attention has so far been given to the by-products of flour milling, which include the various forms of bran, wheat germ, and the "clean-out" of the cleaning house (or screen room). As these products represent about 25% of the original grain, they are of considerable economic significance to the miller.

SCREENINGS

Impurities removed in the cleaning house, called "screenings," are sold for

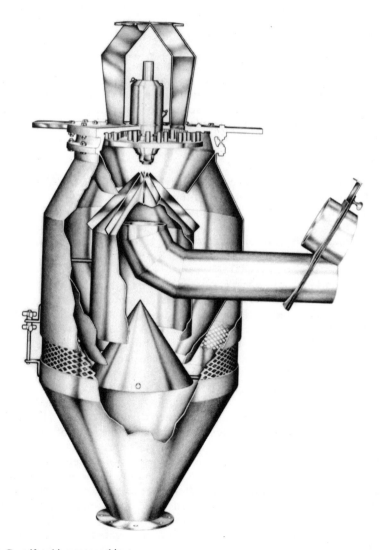

Fig. 22. Centrifugal impact machine.

processing into animal feeds. By-products of the milling process, or millfeeds, range in size from large bran flakes separated at the end of the break system to fine fibrous material tailing over the last flour cloths in the later stages of the reduction system. The by-products usually are collected in three different grades, having different chemical composition and commercial value. Conditions in the market determine the extent to which these by-products are segregated and sold separately by a mill. The three fibrous by-products vary widely in name and composition from country to country. In Canada, common names are bran, shorts, and wheat middlings. In the United States, comparable terms might be bran (coarse and fine), shorts, and red dog. Sometimes the three streams are combined to form a product known as millrun millfeeds in Canada and mixed millfeeds in the United States. Many mills in both Canada and the United States are now pelleting millfeeds for easier handling. Pelleting is also believed to increase the digestibility of millfeeds in animal feeds, the principal end use of millfeeds.

A new by-product in both the United States and Canada is high-fiber bran for human consumption. This is produced by sifting the pure bran following grinding by either roller mill or hammer mill to obtain the desired granulation and fiber specifications. This product can be provided in a wide variety of granulations, as some bakers demand large flakes of bran whereas others prefer small flakes of very high fiber content.

WHEAT GERM RECOVERY

Wheat germ, which theoretically constitutes 2–3% of the whole wheat kernel, is highly regarded in food and pharmaceuticals. Wheat germ is therefore a valuable, although minor, milling by-product. In mills with no special equipment for germ recovery, yields are low, rarely more than 0.5% for food-grade germ. In mills with sophisticated germ recovery systems, yields as high as 1.5–2.0% may be realized. Procedures for germ recovery in each instance are described briefly next.

In a conventional mill, wheat germ stock occurs primarily in the chop from the first two break rolls. The subsequent separation and purification of the germ is achieved by taking advantage of two physical characteristics of germ that distinguish it from the rest of the kernel. The first distinctive characteristic is pliability or plasticity, due no doubt to its high oil content. Because of its plasticity, germ tends to fragment into particles larger or coarser than bran or endosperm particles. Moreover, when germ particles pass through rolls, particularly smooth rolls, they tend to flatten, rather than crush like endosperm particles, thereby facilitating separation by size. Their second distinctive characteristic is that their density is greater than the density of other wheat particles. Thus wheat germ can also be isolated on the basis of specific gravity differences.

Separation of wheat germ begins in the purification system, where the germ stock from the first two breaks, along with branny particles of endosperm of similar size, arrives from the break sifters. Because of its size, the germ stock passes either through or over the third and fourth sheets of the semolina purifiers. These fractions are designated scratch, or sizings, stock. As described earlier, these stocks are ground lightly on corrugated rolls to ensure that: 1) endosperm particles are not excessively reduced, as excessive reduction at this stage prevents their subsequent purification and optimum reduction; 2) bran particles are not reduced in size, thereby causing an increase in the ash content of the flour; and 3) wheat germ is maintained as intact as possible to permit its subsequent separation.

When the sizings stock is sifted, the overtails of the appropriate coarse sieve (wire number[1] approximately 36 w) consists of wheat germ particles highly contaminated with relatively fine branny stock and a very small amount of endosperm. As wheat germ is denser than the other two constituents of similar size, it is readily separated from them on a gravity table known as a germ separator, which operates similarly to a destoner. The germ stock is dropped to an oscillating porous deck through which air is drawn upward to fluidize the stock. When both the magnitude of the airflow and the inclination of the deck are correctly adjusted, the lighter branny particles fall downward to be discharged at the lower end of the deck, whereas the heavier germ particles are propelled upward to emerge from the upper end.

The purified germ particles are now ready for the final processing steps, flaking and sifting. They are passed through smooth reduction rolls, which produce the large, flat flakes characterizing good-quality wheat germ. The stock is then given a final sifting on coarse sieves (between 16 w and 28 w) to remove finer particles that are destined for millfeeds. The overtails are the finished wheat germ, which must subsequently be conveyed gently to preserve the fragile flakes.

Although most of the germ stock is recovered from the scratch, or sizings, system, it is also possible to similarly recover some lower-quality germ from stocks reaching the reduction system. In general, the better food-grade germ is recovered from the third or fourth reduction passage, whereas germ recovered from later reduction passages is usually designated for industrial (nonfood) use.

In many mills adapted to maximize wheat germ recovery, the process actually begins before the first break. The incoming wheat stream is first passed through an impact machine, which releases nearly all of the wheat germ, along with a small quantity of fines, as the wheat kernels are cracked. Sifting then separates the fines from the broken wheat, which is sent to the first break rolls. The residual coarse mixture of branny endosperm, wheat germ, and bran is sent to a germ separator, where, as before, the purified germ emerges from the upper end of the inclined deck. From this point, the germ is flattened and sifted as described earlier.

Wheat germ recovered in the prebreak process is usually characterized by larger flakes with lower bran content and more golden color than germ recovered in a conventional mill. Moreover, the higher yield of germ realized in this process can be further augmented by residual germ recovered from the sizings and reduction systems, as in conventional milling. As with any other processing change requiring capital equipment, economics determine whether the benefits of improved germ recovery warrant the additional capital costs and other consequences of such a process change.

---

[1] Coarse metal sieves have traditionally been characterized by "wire number," e.g., 36 w, where the number refers to the number of meshes per linear inch of weave. Thus, 36 w refers to 36 wires per inch of screen. The gage or thickness of the wire, and therefore the size of the aperture in the sieve, are inversely related to the magnitude of the number. Thus, the thickness of the wire increases as the number decreases. In other words, the lower the number, the coarser the screen. Internationally approved nomenclature now refers exclusively to aperture size. Thus: 36 w = 500 μm aperture size, 28 w = 716 μm aperture size, and 16 w = 1,190 μm aperture size.

H. Soft Wheat Milling

Because of intrinsic differences in the structure of hard and soft wheats, they must be milled differently. For example, hard wheats usually have a vitreous, translucent endosperm that appears almost waxy, whereas the endosperm of soft wheats is white and completely opaque. The cell structure of soft wheat is very weak and readily broken. Under a microscope, hard wheat flour appears to be quite hard and crystalline, whereas soft wheat flour is quite "woolly" and not at all crystalline. In addition, the endosperm of soft wheat appears to adhere more strongly to the bran. Consequently, for efficient milling, soft wheats generally require less moisture addition and shorter tempering times, more grinding surface in the break system and less in the reduction system, less purification, and more sifting surface. If more sifting surface cannot be provided, the quantity of soft wheat going to the mill must be reduced, thereby reducing throughput and adversely affecting the economics of the milling process. Otherwise, the kind and sequence of operations are similar in the milling of both hard and soft wheats.

The differences between hard wheat and soft wheat milling are perhaps best illustrated by the differences in the milling surface commonly used to process each class effectively. Milling surface may be defined as the total amount of surface on milling equipment (rolls, sifters, purifiers) available to process a specific quantity of wheat within a specific time to meet specific production objectives. Although milling surface may be expressed in various ways, the most common is as follows: for roller mills: total grinding length per 100 kg of wheat per 24 hr; for sifters: area of total sieving surface per 100 kg of wheat per 24 hr; and for three-deck purifiers: total net sieve width per 100 kg of wheat per 24 hr.

Typical, although by no means exclusive, ranges of milling surface are shown in Table III.

Soft wheat mills are generally clothed in one of two ways. To produce a general-purpose, standard-granulation type of soft flour for crackers or cookies, the miller uses clothing of approximately the same aperture as for hard wheats. This means that the total flour is expected to pass through an aperture of 132 μm. To produce high-sugar-ratio cake flours and other specialty flours, the miller requires a flour of very fine granulation, most of which will pass through a 93-μm aperture. This latter requirement obviously can only be realized if more

TABLE III
Range of Milling Surfaces Allocated for Different Types of Wheat

Surface and Units Allocated per 24 hr	Wheat Type		
	Hard	Soft	Durum
Rolls			
mm/100 kg of wheat	10–15	10–13	16–20
inches/100 lb of flour	0.238–0.357	0.238–0.309	0.381–0.476
Sifters			
m²/100 kg of wheat	0.055–0.081	0.083–0.088	0.086–0.093
ft²/100 lb of flour	0.48–0.53	0.54–0.58	0.56–0.60
Purifiers			
mm of sieve width/100 g of wheat	3–7	0–3	8–12
ft²/100 lb of flour	0.04–0.093	0–0.04	0.106–0.159

roll and sifter surface is provided.

More specific details of soft wheat milling cannot be provided without reference to specific grades of soft wheat and the required flour specifications. However, it is safe to say that soft wheat is usually milled at a moisture content of 1/2 to 1% lower than that for hard wheat. The physical condition of the mill stocks is then optimal for conveying and sifting, and separation of bran from endosperm is facilitated. Nevertheless, soft wheat milling requires close attention to detail to prevent "choking" (higher feed rate than a machine or spout can handle, resulting in accumulation of stocks at bottlenecks) and irregular performance, with an accompanying significant reduction in the flour extraction rate. Even under the best conditions, soft wheat may be expected to provide an extraction rate up to 2% lower than that of hard wheat.

I. Durum Wheat Milling

Durum wheat milling is designed to produce semolina, a granular product analogous to the farina or flour middlings of the hard wheat milling process. Flour is actually regarded as an undesirable by-product of durum semolina milling. Thus the technologies of durum wheat milling and hard or soft wheat milling differ significantly.

The technology of durum wheat milling has evolved as a result of some of the characteristics of durum wheat and of its milled product, semolina. Sound durum wheat is very hard and flinty (vitreous), and is therefore readily reduced to the granular product, semolina, with minimal production of flour. Semolina is used in the production of premium pasta products the world over and is also used in the preparation of couscous, a staple food throughout much of North Africa. Moreover, as durum wheat contains relatively large amounts of yellow pigment, semolina is clear, bright, and golden in color. This desirable appearance is not appreciably diminished by the presence of small quantities of golden brown bran specks. However, dark specks originating from impurities in the wheat or blemishes on the skin of wheat kernels adversely affect the appearance ("dress") of semolina and the bright golden color of the finished pasta product. Because the pasta manufacturer requires, in addition to good color, a strong and somewhat elastic gluten at a high enough level (a minimum of 11% wheat protein is usually recommended) to ensure the desired degree of liveliness (al dente) in the cooked product, effective milling of durum wheat into semolina involves careful selection of high-quality grain, meticulous cleaning, a highly specialized mill flow, and skill in milling.

First, then, durum milling requires extensive cleaning to ensure that only uniform and unblemished wheat proceeds to the mill. Thus, particularly close attention is paid to cleaning by specific gravity, a principle that is used in the gravity table and destoner. Gravity retreatment is used primarily to remove ergot,[2] which would otherwise contaminate semolina with unsightly, if not hazardous, specks. In addition, batteries of disc separators or indent cylinders are used to eliminate other seeds that may differ only slightly in size and shape from the durum wheat itself.

---

[2] Ergot is a dark fungal body that grows in place of the grain kernel in grasses and cereals.

A typical cleaning and tempering flow for durum wheat is shown schematically in Fig. 23. Tempering may be done in two, or even three, stages. Tempering time is short, 4 hr or less. It is kept short because it is necessary only to "toughen" the bran, not to mellow the endosperm. A mellow endosperm favors flour production, which is not desired in durum milling. A final water addition of as much as 2% may be provided as little as 20 min before milling, and this has the desired effect of toughening the bran as required to produce clean semolina with a minimum of bran specks.

In the mill itself, durum milling differs markedly from hard or soft wheat milling at every stage. Figure 24 shows a typical block diagram of a durum semolina flow, clearly indicating the three major differences from hard or soft wheat milling discussed below. Table III shows typical milling surfaces for durum wheat milling.

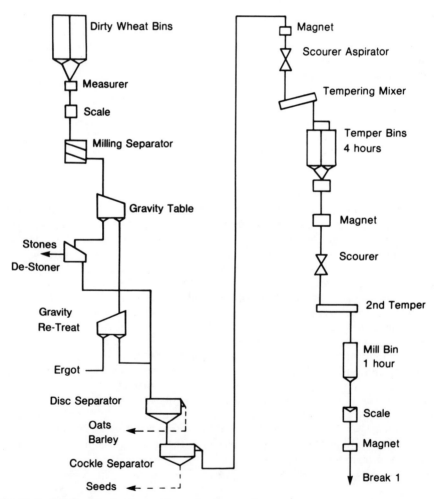

Fig. 23. Typical flow for cleaning and tempering durum wheat.

BREAK SYSTEM

Durum milling contains more break passages, frequently five and as many as six. The extra break passages enable the miller to grind more gently and gradually at each stage, thereby facilitating the release of endosperm in particles as large as possible to ensure maximum production of semolina. Moreover, the profile of the corrugations used in the durum mill is generally sharp, to minimize the amount of fine material produced during the breaks and sizings of semolina. Corrugations are usually set "sharp to sharp," which means that the sharpest angle cuts into the wheat kernel to produce larger chunks of endosperm with a minimum amount of fines.

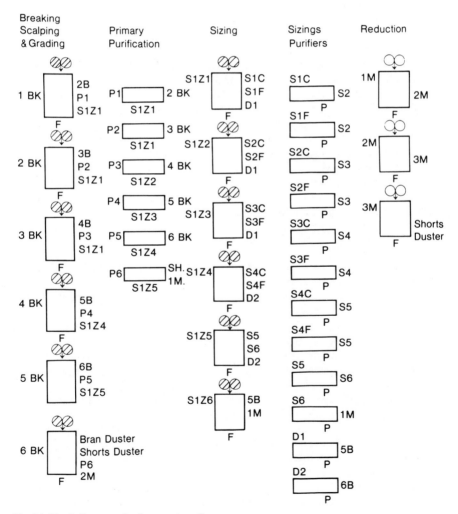

Fig. 24. Block diagram of a durum wheat flow.

PURIFICATION SYSTEM

After the break chop is sifted and purified, purified coarse semolina is sent to corrugated sizing rolls, instead of to smooth reduction rolls. The object here is to gradually reduce the coarse semolina, or to "size it down." The sized semolina is then again graded and purified. The major portion of the finished product, semolina, emerges from the second purification. Thus, a durum mill is easily recognized by the large number of purifiers in the process.

REDUCTION SYSTEM

Because the emphasis in traditional durum milling is on production of semolina, not flour, the process requires only a short reduction system with smooth rolls to grind the very fine fractions of semolina, which cannot be included with the granular product.

Many millers grind both durum and hard wheats on the same flow, but this always necessitates compromise in the mill design and flow. It is very difficult to produce the best yield of durum semolina on a mill that is also intended for milling other types of wheat.

In recent years, in the United States at least, durum flour has been increasingly demanded by the pasta and soup industries. A number of semolina mills are therefore equipped with a "grind-down" system to enable the miller to reduce semolina to flour. In fact, some pasta plants now use durum patent flour because flour hydrates more rapidly than semolina, thereby increasing production. To meet the increased demand for durum flour, some traditional semolina millers have installed additional rolls and sifters to produce flour instead of semolina. Others have modified standard hard wheat mills to produce durum flour. In addition to the extra power that must be provided to grind durum wheat in a modified hard wheat mill, a loss of 25–30% in capacity is experienced when durum wheat is milled on a hard wheat unit.

J. Air Classification

Air classification is a process developed commercially in the United States about 1960 to classify flour in the subsieve range by means of air currents. An air classifier looks like an air fan except that the air is drawn from the center of the rotor (Fig. 25C). The classification of flour is achieved under the influence of two opposing forces (Fig. 25A). The first, air traction, tends to pull small particles toward the center of the rotor. The second, centrifugal force, tends to propel particles toward the circumference of the housing. Finer particles of flour therefore tend to be carried along with the air current, whereas larger particles of greater mass tend to collect along the circumference (Fig. 25B). Adjustable baffles rotating in the walls along the circumference form a barrier, which permits particles of a selected size range to pass into the outer chamber, where they may be collected according to particle size range. The baffles may be adjusted to provide "cuts" of specific particle size ranging between 15 and 80 μm. Usually, two cuts are made, one at about 17 μm and the second at 35–40 μm, thereby providing three fractions:

1. Size 0–17 μm. This is the fine fraction consisting of essentially free protein particles and small starch granules. The protein content of this fraction can be as high as 25%.

2. Size 17–35 or 17–40 μm. This is the medium fraction consisting primarily of free starch granules. The protein content of this fraction can be as low as 5%.

3. Size 40–80 μm. This is the coarse fraction consisting primarily of endosperm particles of approximately the same composition as the parent flour.

To make the process commercially feasible, the miller first grinds the parent flour by impact in an attrition mill or pin mill to reduce endosperm particles sufficiently, thereby optimizing the the release of granules of essentially free starch.

Table IV shows a comparison of the three fractions obtained by air classifying conventional and impact-milled soft wheat flour. Each of the three fractions has different functional properties. The fine (high-protein) fraction is used primarily

TABLE IV
Comparison of Air-Classified Fractions from Conventional and Impact-Milled Soft Wheat Flours[a]

Size Range (µm)	Conventional Flour		Impact-Milled Flour	
	Proportion of Total (%)	Protein Content (%)	Proportion of Total (%)	Protein Content (%)
0–17	7.0	14.5	20.0	15.7
18–35	45.0	5.3	71.0	5.0
>35	48.0	8.9	9.0	9.5
Original flour	100.0	7.6	100.0	7.6

[a] Source: CIGI (1983); used by permission.

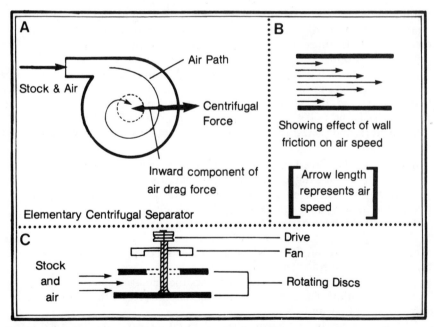

Fig. 25. Principles of air classification. A, opposing forces; B, separation by mass; C, means of circulation.

as a blending flour to increase protein content for breakmaking flours. Thus, it is blended with either the coarse fraction or some other lower-protein flour to achieve a final product of desired protein level. The medium (low-protein) fraction is used for cookies and cake flours and is particularly suited to angel cake production. The coarse fraction is used in much the same way as the parent flour, or it may be packaged as an "instant" flour because of its free-flowing properties. It may also be reprocessed to recover more of the medium (cake flour) fraction. The yield of each fraction depends on the hardness of the wheat source, the origin of the parent machine flour, and the setting of the air classifier.

Air classification is not practiced widely at this time because of its high energy requirement and other associated unfavorable processing costs. For example, high-horsepower motors are required to operate the grinding and air-classifying equipment. Moreover, heat generated in the processing causes moisture losses, which can result in products with less than 10% moisture. Consequently, it is usually less costly for the miller to select low-protein wheats for cake flours, high-protein spring wheats for bread flours, and granular flour streams for instant flour. Only a very few mills in the United States, France, and Germany practice air classification.

K. Starch Damage

It has been suggested that the term *damaged starch* may be misleading, as damage infers deterioration. Instead, the term *activated starch* more accurately reflects the functional difference between sound starch in an intact wheat kernel and starch in flour that has been subjected to the shearing and abrasive effects of milling the wheat. Be that as it may, the nature, origin, and functionality of damaged starch are discussed in detail elsewhere in this book. This section is confined only to those aspects of damaged starch that might directly affect milling operations.

Damaged starch is an important flour specification because it affects water absorption and gas production in fermenting doughs. Consequently, the miller must understand the factors affecting the generation of damaged starch, particularly if the mill's grists consist of wheats of different hardness levels. These factors, and their qualitative effects on damaged starch, are summarized briefly below.

Wheat Type. Damaged starch is directly related to wheat hardness. Thus, the harder the wheat, the higher the level of damaged starch.

Tempering Moisture. A moisture level below the optimum for milling will generate a higher than normal damaged starch level.

Feed Rate. Decreasing the feed rate increases the damaged starch level.

Roll Surface. In general, the damaged starch level increases with increasing degrees of corrugation. Thus, smooth rolls produce the least starch damage, heavily corrugated rolls the most.

Roll Speed. Increased roll speed increases the damaged starch level.

Roll Differential. The greater the differential, the greater the degree of starch damage, except on smooth rolls.

Corrugation Spiral. Increasing the roll spiral increases the damaged starch level.

Grinding Pressure. Increasing roll pressure increases the damaged starch level, except on smooth rolls.

Roll Temperature. Increasing roll temperature increases the damaged starch level. In general, smooth rolls tend to heat more readily than corrugated rolls.

As starch damage is still a subject of considerable research interest, the miller can only cope with the damaged starch specification by experience and trial and error, taking into consideration the role played by each of the above factors in a given mill.

III. RECENT DEVELOPMENTS IN COMMERCIAL MILLING

No revolutionary development has occurred in milling technology, either in terms of process or of basic equipment design, in the recent past. For example, most conventional milling machines have been modified from time to time to improve reliability and sturdiness, thereby also reducing their maintenance requirements, and to improve safety, sanitation, efficiency in operation, and economy in power consumption. Similarly, the trend toward shorter surface in milling, prevalent throughout the 1960s and early part of the 1970s, cannot be considered a major development and, in any event, seems to have leveled off since about the mid-1970s. Finally, the concept of a mill without purifiers, which became a contentious topic among millers, cannot now be considered a recent development. Although some movement away from purifiers has occurred in some mills that are exclusively designed for soft wheat milling, this does not represent a major recent development. It would appear, therefore, that significant changes in milling procedures in recent years have occurred only in wheat cleaning, tempering, and automation.

A. Wheat Cleaning

Since about 1970, there has been a growing trend toward a somewhat unconventional approach to wheat cleaning. The overall objective in newer wheat cleaning processes is to concentrate most of the impurities with the lighter wheat kernels. This, then, constitutes a relatively small fraction of the whole, which can be more effectively cleaned separately, whereas the main fraction, containing stones and heavier sound wheat, may then be passed through just one cleaning machine, a destoner. The heavier and denser wheat, which could comprise 70% or more of the whole, is then ready for tempering. This routing prevents good, sound wheat from going to other (unnecessary) cleaning machinery. The main advantage is that, because only a small fraction of the wheat requires extensive cleaning, smaller cleaners may be used. Alternatively, in large cleaning houses, one machine might take the place of two in parallel. Other advantages include lower air requirements and higher efficiency, as only a relatively small, concentrated fraction needs major attention.

New cleaning equipment has been designed exclusively to handle the relatively small fraction in which wheat contaminants are concentrated. Typical machines in this category include the Buhler concentrator, the Buhler combinator, and the Simon gravity selector. The concentrator, however, must be used in tandem with a separate destoner.

B. Tempering

Conventional tempering of the past failed to provide adequately uniform mixing of grain and water. Thus, it was difficult to add more than 3% water in one step, as excess surface moisture then interfered with the flow of grain through the hopper outlet. Moreover, as moisture distribution itself was not uniform, a longer-than-necessary tempering time was required to ensure both adequate moisture penetration within kernels and adequate moisture uniformity among kernels.

The Intensive Dampener and the Technovator now provide the technology to ensure both adequate moisture penetration and adequate moisture uniformity. These are achieved by a vigorous mixing action with provision for steam addition. These systems have been widely accepted, and temper times have been reduced accordingly.

The effectiveness of the new tempering equipment has been greatly increased by the advent of on-line automatic moisture control equipment such as the Simon H_2O-Kay and the Buhler Aquatron automatic dampening control system.

The H_2O-Kay moisture sensor uses a noncontacting, on-line, "transmission-through-the-product" technique. With its microcomputer, it gives a continuous indication of the instantaneous moisture content of the wheat as it flows past a sensor located in a special chute. Microwave energy is specifically absorbed by water in the grain, and the surviving microwave signal is sent to the microprocessor as a measure of the quantity of water present. Bulk density of the grain is similarly measured with transmitted γ-rays. The H_2O-Kay microcomputer continuously calculates the percentage of moisture in the grain, with automatic compensation for grain density and temperature. This permits accurate moisture measurement and positive feedback control during the tempering process. The microcomputer processes the measurement signals from the chute sensor, then displays and compares the grain moisture level with the preset target. Correcting signals are sent to the operator panel, where transducers convert electrical signals to pneumatic control signals. The moisture of the wheat entering the tempering mixer is measured, not deduced, and accuracy is thereby assured. Feedback control eliminates the need for accurate grain weighing equipment and flowmeters for water measurement, and it ensures consistency in the moisture content of the wheat being sent to the tempering bins.

The Buhler Aquatron automatic dampening control system also features on-line determination of moisture. Wheat moisture content is calculated continuously from an "electric capacity value" of the wheat as it moves through the system. The initial moisture content is measured together with the simultaneously measured bulk weight and temperature of the stock. The values of the initial moisture content and the required final moisture content, as well as the current throughput capacity, are all transmitted to the electronic system of the water-proportioning unit, which calculates the required water quantity and adds it to the flowing stock by means of a control valve.

C. Automation

Perhaps the most significant developments in recent years have occurred in the areas of automation and computerization of flour milling operations. The need for precise and simple monitoring of plant operation and flour yields, and for rapid generation of production reports, stimulated these developments, which have been achieved largely through the use of microprocessor-based controllers.

Initially, automation and computerization were addressed by two separate systems and therefore two separate packages were provided. The first was an equipment function control, performed primarily by programmable logic controllers, which are microprocessor devices. The second was a yield-management system package used to calculate flour extraction and generate production reports by means of microprocessor controllers and a microcomputer to provide operator interface with the system. However, more advanced versions of either system, with increased capability, can now perform most of the functions common to both. These systems perform the following important functions: 1) flour extraction monitoring, permitting corrective action if there is any deviation from target; 2) displaying the current status of production; 3) generating daily, weekly, monthly and yearly reports on production; 4) setting audiovisual alarms to warn of emergenices; 5) monitoring digital and analog sensors placed at critical points in the process; 6) routing of finished products through valve controls to desired locations; 7) activating start-up sequences for equipment in the mill; and 8) controlling the flow of additives to flour.

The total potential of such systems has not yet been fully realized. Many millers have already installed, or are planning to install, such yield-management systems. Some of these systems could be expanded into fully integrated automated systems.

IV. EXPERIMENTAL MILLING

Experimental or laboratory milling is a general term encompassing all milling procedures conducted on less than a commercial scale. Major applications of experimental milling include evaluation of lines from wheat breeding programs, prediction of the end-use quality of commercial samples, and research investigations. Many commercial flour mills use experimental milling to monitor mill performance and to obtain advance information on the properties of incoming wheat shipments. This section does not, however, deal with the commercial applications of experimental milling, which are usually proprietary. Instead, the subject is elaborated from a research point of view. Of course, the principles, equipment, and procedures used to generate milling information in a research environment are equally applicable to commercial practice.

The equipment and procedures described in this section are intended only as an introduction to the theory and practice of experimental milling. Standard procedures developed in one laboratory can rarely be directly applied in another laboratory, where objectives, wheat classes, and equipment may be different.

A. Ensuring Accuracy

The first requirement of an effective experimental milling program, as in any other analytical program, is accuracy and reproducibility of results. The first step in ensuring accuracy and reproducibility is minimizing sources of error, thereby reducing experimental error sufficiently to detect significant differences. In experimental milling, the major sources of error are sampling, inconsistent or inappropriate tempering, uncontrolled variation in ambient conditions, condition of equipment, and inadequate standardization of procedures. A homogeneous sample, representative of the whole, is essential if the experimental milling results are to be significant.

Just as tempering has a profound effect on the commercial milling properties of wheat, so too does it affect experimental milling. As the tempering moisture increases, the flour extraction rate decreases, flour color improves, and flour ash decreases (Butcher and Stenvert, 1973a; Hook et al, 1982b, 1982c, 1982d). Hard wheats will tolerate added water better than soft wheats (Hook et al, 1982c). For example, Table V shows the effect of tempering moisture on some Canadian wheat classes. The relatively soft Canada Prairie Spring wheat suffers a significant loss of extraction when tempering moisture is increased from 14.5 to 15.5%. Over the same range, the harder Canada western red winter and Canada western red spring wheats show improved flour quality with no significant loss of extraction.

Several factors affect optimum tempering time. Kernel hardness is perhaps the most significant factor here, as tempering time may range from 6 hr to more than 24 hr (Stenvert and Kingswood, 1977). If uncertain, the miller should determine the hardness of the wheat by an appropriate test, such as the particle size index (Williams and Sobering, 1986). Wheat protein content and initial wheat moisture content also affect tempering time (Moss, 1977a). Temperature is another significant variable. As temperature increases, the rate of moisture penetration increases (Campbell and Jones, 1955). These, and perhaps other, factors must be controlled to establish correct and reproducible tempering conditions.

The need for consistent tempering may be obvious to most, but the effect of ambient temperature and relative humidity on experimental milling performance is often overlooked. For example, Bayfield et al (1943) showed that

TABLE V
Effect of Tempering Moisture on Allis-Chalmers Milling Results for Wheats of Varying Hardness[a]

Wheat Class	Tempering Moisture (%)	Flour Extraction[b] (%)	Flour Ash[c] (%)	Flour Color (Kent-Jones units)
No. 1 Canada Western Red Spring	14.5	75.6	0.52	1.3
	15.5	75.3	0.50	0.8
	16.5	74.8	0.49	0.5
No. 1 Canada Western Red Winter	14.5	76.2	0.50	0.4
	15.5	76.3	0.49	−0.2
	16.5	75.9	0.49	−0.3
No. 1 Canada Prairie Spring	14.5	75.1	0.55	1.5
	15.0	74.9	0.55	1.0
	15.5	74.3	0.54	0.8

[a] Source: J. E. Dexter and D. G. Martin (unpublished data). Milled by method of Black et al (1980a).
[b] Proportion of clean wheat on constant moisture basis.
[c] 14% moisture basis.

flour yield, granulation, moisture, ash, maltose value, and baking quality are all affected by mill room environment. They concluded that changes in relative humidity have a greater effect on milling results than have temperature changes, a result confirmed recently by Hook et al (1984c). Overall, the effect of mill room environment on flour ash and color is not simply a consequence of its effect on flour extraction. As shown in Fig. 26, when wheat is milled under dry conditions, the ash contents of all flour streams increase, thereby reducing the apparent milling value of the wheat. Similarly, results of milling durum wheat semolina are particularly sensitive to fluctuations in relative humidity, because stocks are exposed to considerably more air during purification than are stocks in common wheat milling. When relative humidity declines, semolina quality (ash, color, and speckiness) deteriorates, leading to spaghetti that is duller and browner (has longer dominant wavelength), with reduced pigment intensity (lower purity), as shown in Table VI. Thus, control of relative humidity is essential to obtaining precisely reproducible results in experimental milling. If the mill room environment cannot be controlled, relative results can be obtained only if a standard reference sample is milled on the same day under the same environmental conditions as the test sample.

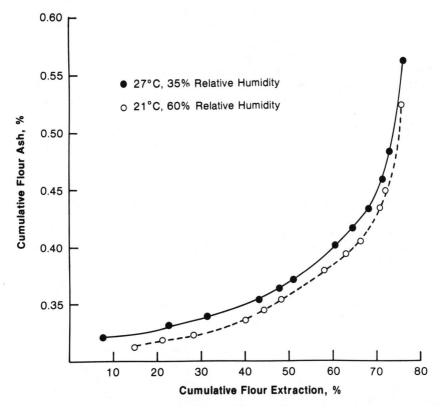

Fig. 26. Effect of mill room relative humidity on milling performance. Wheat milled in pilot mill described by Black (1980). Source: J. E. Dexter and R. J. Desjardins (unpublished data).

Turning next to the equipment, the major factors affecting consistency of results include roll grinding conditions, wheat feed rate, and, of course, equipment maintenance.

Incorrect or poorly controlled roll grinding can influence flour properties markedly. For example, heavy grinding in the break system adversely affects ash content and color of all flour streams (Dexter et al, 1984). Similarly, reduction roll grinding conditions affect hard wheat flour ash, flour color, and starch damage (Dexter et al, 1985b). Starch damage variations, in turn, affect physical dough properties, baking properties, and susceptibility of flour to the effects of α-amylase (Farrand, 1964). An example of the magnitude of these effects is shown in Table VII.

More subtle, often undetected, changes than those illustrated in Table VII occur as the mill "warms up." For example, Seeborg and Barmore (1951) observed a gradual increase in the grinding roll temperature of the Buhler laboratory mill during the first several hours of operation. They suggested prewarming the mill overnight with a burning light bulb placed under the rolls, then milling a "warm up" sample at the beginning of the day. At any rate, as the mill warms, the extraction rate increases (Hook et al, 1982a), and a corresponding increase may be observed in flour ash and flour color (darker). More significantly, however, the rolls expand as they warm, thereby increasing the grinding action which, in turn, increases the production of damaged starch (Ziegler, 1940) and alters the physical dough properties and baking performance of the flour. On the other hand, Jeffers and Rubenthaler (1977) observed the opposite effect of warming on Brabender Quadrumat mills. Here, reduced extraction was the result of expansion of the aluminum mill housing. They reported that permanent installation of thermostatically controlled heaters eliminated variations in flour extraction that might otherwise occur due to variations in mill temperature.

Performance of mill equipment, sifters as well as grinding rolls, is also affected by the feed rate of wheat to the mill. Efficiency of grinding and sifting decreases

TABLE VI
Effect of Mill Room Relative Humidity at Constant Temperature (21°C)
on Canada Western Amber Durum Wheat Semolina Properties
and Spaghetti Color[a]

Property	Relative Humidity		
	30%	45%	60%
Semolina			
Yield,[b] %	64.8	64.7	64.5
Ash, %	0.66	0.65	0.64
Agtron color, %	76	77	79
Specks per 50 cm^2	30	29	23
Spaghetti color			
Brightness, %	44.8	44.9	47.3
Purity, %	56.0	57.4	59.6
Dominant wavelength, nm	577.7	577.8	577.4

[a] Source: J. E. Dexter and J. J. Lachance (unpublished data). Milled on Bühler mill method of Dexter et al (1982a).
[b] Proportion of clean wheat on constant moisture basis.

as feed rate increases (Wingfield and Ferrer, 1984; Scanlon and Dexter, 1986). In an automatic mill, as feed rate increases, sifting efficiency decreases, resulting in movement of flour farther down the mill. Flour extraction declines and starch damage increases due to regrinding of unsifted flour (Okada et al, 1986). Because soft wheats sift less readily than hard wheats, soft wheats should be fed into the mill more slowly than hard wheats. Consequently, feed rate may have to be adjusted each time a different class of wheat is introduced into the mill.

Finally, consistency of experimental milling can be achieved only if all equipment is correctly maintained. A maintenance program is essential to prevent malfunctioning and unnecessary breakdown of equipment. Care must be taken to ensure that rolls are correctly aligned, that bearings, belts, and drives are in good working order, and that roll surfaces are in good condition. Sieves must be inspected and cleaned regularly. In addition, standard control samples should be milled each day to provide a regular test of mill performance and laboratory analyses.

B. Procedures and Equipment

Now that precision and accuracy have been assured, the reader may begin to consider the principal procedures and equipment used for experimental milling. These depend largely on the types of wheat to be milled and the objectives of the experimental milling. In this section, then, attention is focused briefly on equipment and procedures used in wheat preparation and, subsequently, in experimental mills ranging from micromills to large-scale pilot mills.

As in commercial milling, experimental milling begins with wheat

TABLE VII
Effect of Rate of Reduction of Farina into Flour
on Canada Western Red Spring (CWRS) Wheat Flour Properties[a,b]

Quality Factor	Gradual Reduction into Flour	Rapid Reduction into Flour
Flour		
Yield,[c] %	74.4	73.8
Ash,[d] %	0.47	0.50
Color, Kent-Jones units	0.2	1.1
Starch damage,[d] Farrand units	33	57
Farinograph absorption,[d] %	64.0	69.2
Alveograph		
Height, mm	98	156
Length, mm	123	78
Work, × ergs	385	512
Pan bread		
Baking absorption,[d] %	62	64
Loaf volume per 100 g of flour, cc	850	810
Hearth bread		
Specific volume, cc/g	6.4	5.6
Total score, %	72	55

[a] Source: Dexter et al (1985a); used by permission.
[b] Mean values for five CWRS wheats milled by a 254-mm Ross batch mill.
[c] Proportion of cleaned wheat on constant moisture basis.
[d] 14% moisture basis.

preparation, which includes both cleaning and conditioning (tempering) of wheat. Cleaning devices, such as the Carter dockage tester, which simulate the removal of foreign material and damaged wheat kernels by commercial cleaning houses, are available commercially. In the absence of such equipment, wheat samples can be cleaned using a series of sieves of sizes appropriate to retain sound wheat kernels but to reject foreign material and damaged wheat kernels. Although not essential, an experimental scourer is useful for removing surface dirt and insect-damaged kernels from cleaned wheat. A number of workers have described equipment designed to facilitate rapid, uniform tempering of wheat samples for experimental milling. Equipment includes rotating tubs or drums (Hook et al, 1982a), a rotary shaker (Sibbitt et al, 1960), and a wheat washer (Dexter et al, 1982b). Several methods designed to reduce tempering time for experimental milling have been reported in the scientific literature from time to time. Sullivan (1941) found that tempering time can be reduced from overnight to two to three hours if a wetting agent is added to the tempering water, with no apparent effect on flour extraction or flour quality. McNeill (1964) described a 20-min vacuum tempering procedure. More recently, it has been reported that total tempering time for hard and soft wheats can be reduced to 30 min by using a 15-min preliminary temper, a prebreak grinding pass, and a 15-min final temper (Finney and Bolte, 1985; Finney and Andrews, 1986a).

Experimental milling equipment is available commercially in numerous shapes and sizes, and customized equipment is limited only by the imagination and skill of the user. Some of the equipment in wide use, or reported in the literature, is described next.

Numerous micromilling procedures have been described for use primarily in conjunction with plant breeding programs. The most commonly used commercially available micromills are the Brabender Quadrumat Junior and Senior mills described in detail by Shellenberger and Ward (1967). The Brabender Quadrumat Junior mill is an automatic compact unit that makes three grinding passes using four 3-in. diameter rolls. The mill (Fig. 27) produces low- and high-ash flours and bran from as little as 100 g of wheat. The Brabender Quadrumat Senior mill has two four-roll units, one for break passes and one for reduction passes. The Brabender Quadrumat Junior mill provides a reliable method of producing flour of sufficiently good quality for large-scale wheat quality testing programs.

Commercially available equipment is often modified to achieve the desired combination of small sample size, reliability, and sample throughput for early-generation wheat screening. For example, Seeborg et al (1951) modified the Bühler laboratory mill to give reliable milling quality data for 100-g samples, and other workers (Yamazaki and Andrews, 1982b; Finney and Andrews, 1986b) have described modifications to the Brabender Quadrumat Junior mill that allow an estimate of soft wheat milling quality for 20-g samples. Finney and Yamazaki (1946) described a micromilling procedure for 100-g samples incorporating preliminary breaks through the rolls of a Tag-Heppenstall moisture meter, followed by one break and two reductions on a Hobart grinder. A similar procedure using break and reduction heads of the Brabender Quadrumat Senior mill instead of the Hobart grinder is still in use for early-generation screening of U.S. hard red winter wheats (Finney and Bolte, 1985). Custom-designed micromills for 100-g samples have been reported by many workers, including Geddes and Frisell (1935) and Kemp et al (1961). A two-break milling machine for the rapid milling quality evaluation of 5-g wheat samples has been described (Kitterman et al, 1960).

Larger scale experimental milling equipment is conveniently classified as either continuous or batch, as suggested by Shellenberger and Ward (1967).

Continuous mills are automatic and therefore provide reliable results, even when used by relatively inexperienced operators. The most commonly used continuous experimental mills are the previously described Brabender Quadrumat Junior experimental mill and the Buhler laboratory mill.

Fig. 27. Brabender Quadrumat Jr. experimental mill.

The Buhler laboratory mill originally used belt conveyors, but more modern versions use pneumatic conveyors (Black et al, 1983). The Buhler laboratory mill (Fig. 28) has a simple flow comprising only three breaks and three reductions. The three-break passages are built on one roll pair and the three reduction passes are built on a second roll pair. The reduction rolls may be smooth or corrugated. The capacity of the mill is in excess of 100 g/min. Many workers, including Finney et al (1949), have shown that the Buhler laboratory mill can produce a straight-grade flour comparable in quality to commercial flour. The flour extraction rate of the Buhler laboratory mill, however, falls short of commercial

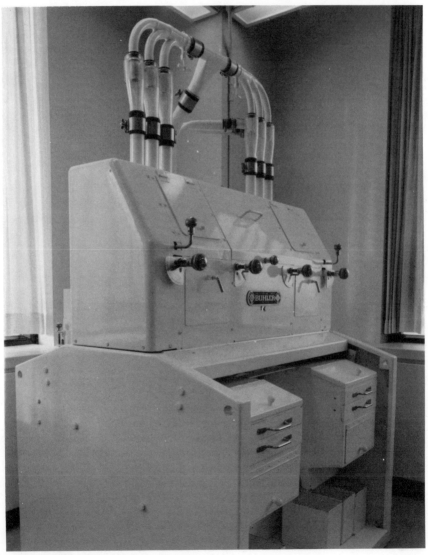

Fig. 28. Buhler laboratory mill.

extraction rates. As shown by Orth and Mander (1975), commercial flour extraction rates should be achieved to correctly evaluate the commercial potential of wheat. With minor modifications to the Buhler mill (Butcher and Stenvert, 1972; Black et al, 1983) and a bran finisher such as that described by Black et al (1961) in the flow, commercial flour extraction rates are readily achieved (Black et al, 1983).

Batch mills consist of at least one roll stand and sifter. Stock must be conveyed manually. Batch mills have the advantage of flexibility, but they require skilled operators to achieve consistent results. The two most widely used batch mills are the Allis-Chalmers experimental mill (Black et al, 1980b; Yamazaki and Andrews, 1982b) and the Ross experimental mill (Black et al, 1980a; Posner and Deyoe, 1986). A multistand Allis-Chalmers mill is shown in Fig. 29. Batch mill roll gaps and speeds are adjustable, but better precision of milling results can be achieved by using several roll stands with fixed settings. Sieving is generally performed on a multisieve gyrating laboratory box sifter, sieves being removed and changed as the milling progresses. Black et al (1980b) described a custom-built sifter for improving reproducibility of results.

Purifiers are usually not used in experimental flour milling. Ziegler (1938) showed that a purifier is simply another variable in experimental flour milling that offers no additional improvement in straight-grade flour quality.

There are several advantages to using a batch system for experimental milling.

Fig. 29. Allis-Chalmers experimental mill. (Courtesy Grain Research Laboratory, Canadian Grain Commission)

First, the miller has an opportunity to examine the distribution and visual quality of stock throughout the millflow, thereby gaining considerable insight into wheat milling properties. Next, as observed by Yamazaki and Andrews (1982a), a batch milling system allows the miller to vary the milling procedure according to the quantity and quality of flour required and the nature of the wheat, by adjusting roll settings and shortening or lengthening millflow. Moreover, individually and collectively, batch mill roll stands are valuable research tools for examining the influence of grinding conditions on wheat milling performance and flour quality (Hsieh et al, 1980; Dexter and Martin, 1986). For example, the Buhler-Miag "Vario" rollstand features easily removed commercial-diameter rolls and a wide range of roll speeds and feed rates, which make it particularly effective for research on grinding under simulated commercial conditions (Gehle, 1965). Finally, batch mills are also particularly useful for durum wheat semolina milling. In this application, a laboratory purifier, such as that described by Black (1966) or Sibbitt et al (1964), is required to produce a high yield of clean, granular semolina. Black and Bushuk (1967) described how the Buhler laboratory mill can be modified for use as a batch mill for durum wheat semolina milling. More recently, Dexter et al (1982a) described a system for producing 65% extraction semolina using the modified Buhler laboratory mill in conjunction with a laboratory purifier.

Although the experimental mills described to this point provide valuable milling information, they cannot satisfactorily simulate commercial milling. Consequently, pilot mills have been designed to bridge this gap. Ideally, pilot mills retain sufficient flexibility to permit large-scale experimentation with a variety of commercial millflows, thereby providing a vehicle for testing experimental results obtained with small-scale laboratory mills. The Buhler-Miag Multomat mill (Auer, 1962) is one example of a large-scale automatic experimental mill with a capacity of about 1 kg/min. Black (1980) and Shuey and Gilles (1969) described laboratory-scale pilot mills that made use of the Multomat rolls, but they incorporated more sophisticated millflows and ancillary equipment such as purifiers, impact finishers, and redesigned sifters. These mills provide complete commercial milling evaluations on wheat samples weighing less than 30 kg (Shuey and Gilles, 1969; Preston et al, 1983). Several other examples of larger-scale pilot milling systems using commercial equipment are the Kansas State University pilot mill (Farrell and Ward, 1965), the Canadian International Grains Institute pilot mill (Fig. 30) (Holas and Tipples, 1978), and the recently completed Bread Research Institute of Australia pilot mill (Orth, 1985).

C. Interpretation of Results

Whatever the objectives, hardware, and techniques of experimental milling may be, they can only be useful if the data they generate can be interpreted meaningfully. Often, of course, interpretation of data is confined to the solution of a single problem. Usually, however, the problem is complex, and various attempts have been made to find general solutions. Some of the approaches that have been used to evaluate experimental milling data are described next to conclude this section.

In wheat breeding, an estimate of the overall milling value of a wheat sample may provide a useful guide to the future conduct of a breeder's program. Many factors in addition to flour extraction determine the milling value of wheat. For example, the lower the flour ash and the brighter the flour color, the more desirable is the wheat for milling. Shellenberger and Ward (1967) developed the following formula for calculating wheat milling value on the basis of straight-grade flour ash:

$$\text{Milling rating} = \% \text{ extraction} - (\text{ash} \times 100)$$

In the United Kingdom., where flour color is an important measure of flour quality, the following formula is used (Shellenberger and Ward, 1967):

$$\text{Milling value} = \% \text{ extraction} - \text{Kent-Jones flour color}$$

Another widely used technique for comparing milling quality of different wheats is to compare their cumulative ash curves (Shellenberger and Ward, 1967; Lillard and Hertsgaard, 1983). Cumulative ash curves are generated, as in commercial milling, by arranging millstreams in ascending order of ash on a constant moisture basis and tabulating cumulative ash and cumulative extraction for each successive millstream. Wheats that exhibit the lowest initial flour ash and the slowest rate of ash increase with increasing flour extraction are

Fig. 30. Pilot mill of Canadian International Grains Institute.

preferred because they produce the highest proportion of patent flour. Cumulative ash curves provide a basis to calculate the proportion of patent flour, other flour divides, and millfeed. In fact, if a realistic commercial value is assigned to each product, a monetary milling value may be calculated for the wheat (Shellenberger and Ward, 1967; Posner and Deyoe, 1986).

The behavior of wheat during milling and stock distribution is also important in appraising wheat milling performance. The proportion of sizings stock produced is associated with hardness in common and durum wheats and influences flour mill balance (Posner and Deyoe, 1986) and durum wheat semolina yield (Dexter and Matsuo, 1981). A high yield of break flour and fine flour particle size are associated with better cake and cookie quality for soft wheats (Gaines, 1985). The ease with which endosperm can be separated from bran flakes, the "milling separation index" (Yamazaki and Andrews, 1982a), is also associated with soft wheat milling quality.

Other factors that have a direct bearing on the milling value of wheat include cleanliness, wheat moisture, ability to withstand increasing levels of tempering water, and maximum practical feed rate. Maximum feed rate can be estimated by assessing the ease with which stock flows and the grittiness and sifting efficiency of reduction stock.

V. MILLING RESEARCH

Milling research encompasses a broad spectrum of topics, including studies aimed at developing predictive tests for milling performance, the implications of wheat structure on milling performance, and the effect of milling conditions on flour extraction and quality. An exhaustive survey of milling research literature would obviously be beyond the scope of this chapter. The discussion below is therefore intended only as a general introductory survey of recent milling research.

Researchers continue their pursuit of a simple, reliable predictor of milling performance. Specific, or test, weight is still a primary international grading standard. However, many workers, most recently Hook (1984), have shown that test weight by itself is a poor predictor of wheat milling potential. Shuey (1960) found that kernel size was more reliable than test weight for predicting flour extraction. Marshall et al (1984) predicted on theoretical grounds that increasing the grain size of Australian wheats by 50–100% should improve potential milling yields by 2–3%. However, in a later experimental study, they concluded that other factors including the amount of germ tissue, thickness of bran, and depth of crease were more important than grain size in determining milling yield (Marshall et al, 1986). Baker and Golumbic (1970) found that bran content was related to flour extraction for all wheat classes. This relationship was used by Australian researchers to develop a modified fiber test that predicts wheat milling extraction (Stenvert and Moss, 1974).

The implications of kernel morphology on milling properties have been examined by many workers. Using light microscopy, Larkin et al (1951) and Bradbury et al (1956) found evidence that aleurone cell thickness may play a role in determining flour extraction. Crewe and Jones (1951) showed that marked irregularity in aleurone cell thickness appeared to be associated with difficult bran separation during milling. It has been suggested that bran thickness is, in fact, related to flour extraction (Larkin et al, 1951; Lineback et al, 1978), but conclusive evidence is lacking. Larkin et al (1952) presented evidence that endosperm cell wall thickness may be related to soft wheat millability.

Wheat hardness or texture plays an important role in flour milling, influencing the pattern of endosperm fracture, ease of separation of bran from endosperm, flour particle size, break flour release, and flour starch damage (Stenvert, 1972). Blakeney et al (1979) used high-speed photography to show that hard grains shattered into large pieces, whereas soft grains released fine particles when milled by the Brabender Quadrumat mill. They found a close relationship between break flour release and grain hardness. Kilborn et al (1982) showed (Table VIII) that wheat hardness is also closely related to milling energy consumption, break release, and flour starch damage.

Neel and Hoseney (1984a, 1984b) investigated the well-known tendency for soft wheats to flow poorly and sift inefficiently during milling. They concluded that flour flowability and sifting efficiency were related to moisture content, presence or absence of fat, and particle size independently of wheat hardness. Soft wheat flour was more sensitive to changes in these flour properties because of rough particle surface characteristics.

Scanning electron microscopy has been used to advantage by Moss et al (1980) and Davis and Eustace (1984) to compare the structures of wheats of varying hardness and to relate structural differences to milling behavior. Moss (1985) reviewed how both light microscopy and scanning electron microscopy can be used in conjunction with other techniques for flour milling research and trouble-shooting.

Tempering, or conditioning, continues to be a popular research topic. Recently Hsu (1984) used mathematical models based on Fick's law of diffusion in spherical bodies to predict that tempering time should increase with increase in kernel size and should decrease with increase in initial moisture content and water diffusivity, the latter being dependent on kernel (endosperm) structure. Moss (1973) used a number of staining techniques to show how kernel morphology affects rate of moisture penetration and milling response to conditioning and temperature. Autoradiography has proven to be a very useful technique for following the rate of penetration of water into wheat during conditioning. Butcher and Stenvert (1973b) used autoradiography to demonstrate that water penetrated into selected wheat varieties at different rates. Stenvert and Kingswood (1976) used it to show rapid entry of water into

TABLE VIII
Comparison of Milling Performance of Wheats of Different Hardness Levels[a]

Property	ESW[b] (Soft)	CWRS[b] (Hard Red Spring)	CWAD[b] (Durum)
Hectoliter weight, kg/hl	77.8	80.9	82.7
Flour extraction, %	74.0	76.5	75.8
Break flour release, %	32.9	24.2	12.7
Energy consumption, $J \times 10^5$/kg	4.6	6.7	11.9
Flour starch damage, Farrand units	8	19	35

[a] Source: Kilborn et al (1982); used by permission.
[b] ESW = English soft wheat; CWRS = Canada Western Red Spring, 13.5% protein content; CWAD = Canada Western Amber Durum.

the starchy endosperm where the germ met the bran on the dorsal side of the grain. Using an improved autoradiographic technique, Moss (1977b) showed that, within 1 hr, labeled water had penetrated into the aleurone cells, and in many cases, into the starchy endosperm. Stenvert and Kingswood (1977) used autoradiography to demonstrate the importance of endosperm structure, protein content, and protein distribution on rate of moisture penetration. They found that the subaleurone region appeared to be limiting to the rate of moisture penetration. Moss (1977a) showed that protein content, endosperm structure, and grain moisture all influenced the rate of moisture penetration. Pomeranz et al (1984) concluded that, based on conductance measurements, water distribution in the wheat kernel was stabilized in 2-4 hr after water addition at 21°C.

Some attention has been given to two special conditioning cases: milling high-moisture wheat and conditioning wheat blends. Hook et al (1984b) concluded that flour extraction and flour color of wheat with a natural moisture content suitable for milling could be improved by drying the wheat and conditioning back to optimum moisture before milling, although the benefits must be weighed against increased cost. In a separate study, Hook et al (1984a) concluded that separate optimum tempering of dissimilar wheats before milling offered no milling advantage over single tempering of the blended wheats to an intermediate moisture level. Posner et al (1974) came to the same conclusion, although they point out that the best results are achieved by milling the wheats separately and blending flours.

The composition of the various constituents of the wheat kernel has been well documented by MacMasters et al (1971). The nonuniform distribution of components within the wheat kernel gives rise to the variability in composition and functional quality of the various millstreams. Ziegler and Greer (1971) described the properties of various millstreams in detail. Recently, Jensen et al (1982) showed how autofluorescence can be used to determine quantitatively the botanical constituents (pericarp, aleurone, testa, and endosperm) of wheat flour streams, thereby providing a means to pursue optimum mechanical separation in the milling process. Variations in constituents such as α-amylase (Kruger, 1981), lipids (Morrison and Hargin, 1981), pentosans (Ciacco and D'Appolonia, 1982), proteins (Nelson and McDonald, 1977), and individual amino acids (Jensen and Martens, 1983) among individual millstreams have been recently documented.

The work of Hinton (1959) and Morris et al (1945, 1946) demonstrated not only that there is greater concentration of ash in the bran, aleurone, and germ than in the endosperm, but also that ash content diminishes toward the center of the starchy endosperm. This has made ash a widely used index of flour purity. However, in countries like the United Kingdom, where flour must be enriched with nutrients, including some minerals, ash content loses much of its former significance. Here, light reflectance from flour-water pastes is used instead to differentiate between flours. In theory, the amount of light reflected from flour-water pastes should diminish with increasing bran contaminants. Recently, Barnes (1986) showed that pure endosperm from different wheats varied widely in reflectance, and differences could not be eliminated by bleaching of carotenoid pigments. He concluded that flour-water paste reflectance cannot be regarded as an accurate indicator of bran content when comparing flours from different wheats. Similarly, of course, variations in endosperm ash content between wheats of different class or origin limits the usefulness of flour ash as an index of flour quality when different wheats are being compared.

A less time-consuming alternative to determining flour ash or flour-water paste reflectance that is rapidly gaining acceptance in the milling industry is near-infrared reflectance (NIR). Williams et al (1981) showed that by monitoring an absorbance band of cellulose, flour ash and bran content could be predicted. Bolling and Zwingelberg (1984) and Posner and Wetzel (1986) have described how on-line monitoring of key flour mill streams by NIR can improve flour mill performance.

Increasing awareness of the dangers of naturally occurring toxins and chemical residues in food has prompted numerous studies on the fate of toxic compounds during milling. A number of authors have reported on the partitioning of chemical residues in milled products (Kadoum and Lahue, 1977; Alnaji and Kadoum, 1979; Mensah et al, 1979). Chemical residues generally concentrate in the millfeed. Mycotoxins in cereal products have also prompted numerous studies. For example, recent concern in Canada and the United States over the occurrence in wheat of a mycotoxin, deoxynivalenol, a metabolite of *Fusarium graminearum*, led to detailed milling studies (Hart and Braselton, 1983; Scott et al, 1983; Young et al, 1984; Seitz et al, 1986). Deoxynivalenol is not destroyed in milling but partitions itself within milled products. As shown in Table IX, the lowest ash flour streams contain the lowest concentrations, whereas bran and shorts contain the highest.

Needless to say, physical state of the wheat is of primary importance in determining milling performance. For example, Posner and Deyoe (1986) showed that the milling performance of newly harvested wheat is erratic for at least two months after harvest. They recommended blending freshly harvested wheat with older wheat to avoid fluctuations in the mill.

As might be expected, environmental damage incurred by before harvest can have major effects on milling quality. Dexter et al (1985a) showed (Table X) that as the degree of visual frost damage increased for hard red spring wheat, flour extraction decreased, flour ash increased, and flour color darkened. The

TABLE IX
Distribution of Milling Products and Corresponding Concentrations
of Deoxynivalenol (DON)[a]

Product	Product Yield (%)	DON (μg/g)
Clean wheat[b]	100	4.6
Dockage	0.9	16.7
Bran	16.6	4.6
Shorts	4.6	6.9
Feed flour	2.9	8.0
Straight-grade flour	75.6	4.1
Bread[c]	...	4.2

[a] Source: Scott et al (1983); used by permission.
[b] Canadian Western Red Spring.
[c] On equivalent flour weight basis.

extreme hardness of badly frosted wheat necessitated high energy during milling and resulted in poor mill balance and excessive flour starch damage. Simmons and Meredith (1979) showed that frosting of wheat grains prevented or slowed pericarp degradation, resulting in grains with lighter relative amounts of bran and inherently lower flour extraction potential. Many of the effects of frost damage can be attributed to immaturity. Tipples (1980) showed that when hard red spring wheat was cut at a moisture content in excess of 35%, the resultant harvested wheat began to show reduced flour extraction potential, increased hardness, and darker flour color.

Rees et al (1984) found that black point in hard wheat had little influence on flour milling properties, although flour color was slightly darker for black-pointed wheat. Dexter and Matsuo (1982) reached a similar conclusion when investigating the effect of black point on durum wheat semolina milling. Dexter and Matsuo (1981, 1982) also documented the adverse effects of starchy kernels, immaturity, shrunken kernels, mildewed kernels, and smudge and black point on durum wheat semolina milling.

Preharvest sprouting can have serious effects on flour baking quality. Recently Liu et al (1986) described how pearling sprouted wheat before conventional roller milling can greatly reduce the α-amylase content of flour, thereby improving flour baking quality.

Mill settings greatly influence mill performance. The rate of stock movement over a sieve, sieve length, mesh irregularities in sieve covers, and particle shape influence sifter performance, according to Nuret et al (1979). Wingfield and Ferrer (1984) showed that the rate at which stock travels over sieves decreased with decreasing stock particle size and was independent of stock feed rate. Sifter efficiency was influenced by feed rate, sifter rotation, and sifter speed.

Many workers have examined the effect of roller-mill settings on milling performance and flour properties. Niernberger and Farrell (1970) found that, at constant differential, as the speed of the fast first break roll increased, the proportion of first break flour produced and flour ash both increased. Cleve and Will (1966) showed that mill performance was highly dependent on the condition and profile of the break roll corrugations, particularly the first break roll corrugations. Hsieh et al (1980) found for hard red spring wheat that first

TABLE X
**Comparison of Milling Performance of Canadian Red Spring Wheats
with Varying Degrees of Frost Damage[a]**

Property	No. 3 Canada Western Red Spring	Canada Feed	
		A	B
Severely frozen kernels, %	2	33	55
Test weight, kg/hl	80.1	76.7	77.5
Particle size index, %	15.1	12.9	10.8
Flour extraction, %	73.7	69.3	68.4
First break release, % through 24 w	20.0	15.7	13.4
Energy consumption, $J \times 10^5$/kg	8.5	10.5	12.0
Flour ash	0.53	0.60	0.60
Flour color, Kent-Jones units	2.0	4.7	5.6
Flour starch damage, Farrand units	36	52	59

[a] Source: Dexter et al (1985a); used by permission.

break release increased with increase in tempering moisture and roll differential and decrease in roll gap, whereas feed rate and roll speed had no effect. First break flour ash decreased with increasing temper moisture, decreasing roll differential, and decreasing roll gap. Starch damage in first break flour was primarily influenced by roll differential, increasing with increasing differential.

Many workers (e.g., Dexter et al, 1984) have shown that heavy grinding by the break system causes a deterioration in flour quality. The first break roll gap is particularly important in establishing flour quality. Recently, Dexter and Martin (1986) showed that a very small first-break release (prebreak) led to improved flour ash throughout the milling process, as shown in Table XI. Zwingelberg (1972) reported a similar beneficial effect of prebreaking on rye flour ash.

An aspect of roller milling that continues to attract attention is the relationship between reduction-roll grinding conditions and flour starch damage. Flour starch damage greatly influences flour rheological properties (Farrand, 1964). For pan bread products, relatively high starch damage is desirable because flour water absorption is increased, thereby increasing the yield of pan bread per unit of flour weight (Jones et al, 1961). Research in this field was pioneered by Jones (1940) and Ziegler (1940). Jones (1940) described how roller milling induces two types of starch damage in flour starch granules. Compressive forces cause internal starch damage, whereas shearing forces effect surface granule damage from the breakdown of endosperm particles. He identified wheat hardness, reduction roll surface, roll pressure, and roll differential as the primary factors influencing starch damage. Ziegler (1940) independently came to similar conclusions. Recently, Scanlon and Dexter (1986) demonstrated the effect of smooth roll grinding conditions on the reduction of hard red spring wheat farina. Starch damage in reduction flour was highly significantly related to milling energy consumption and primarily influenced by roll differential. Vorwerck (1977) demonstrated the effect of roll surface—smooth or sandblasted—and roll differential on flour starch damage and flour water absorption. Zwingelberg et al (1983) found that the best results for farina reduction were obtained for matted reduction rolls of low hardness at a differential of 1.5 to 1. They found that lower differentials produced too little starch damage, whereas higher differentials had too drastic an effect. Rolls of low hardness were better than harder rolls because the surface roughness became

TABLE XI
Effect of Prebreak on Flour Properties of Some Canadian Wheats[a,b]

Property	Milled without Prebreak			Milled with Prebreak[c]		
	CF	No. 1 CWRS	No. 1 CPS	CF	No. 1 CWRS	No. 1 CPS
Milling energy, $J \times 10^5$/kg	9.5	7.5	5.4	9.3	7.2	5.7
Flour extraction, %	69.8	75.0	72.7	69.7	74.7	74.1
Flour ash, %	0.56	0.48	0.47	0.51	0.45	0.46
Flour color, Kent-Jones units	4.5	1.0	0.4	3.9	0.7	−0.2
Starch damage, Farrand units	51	31	18	49	30	16

[a] Source: Dexter and Martin (1986); used by permission.
[b] CF = Canada Feed; CWRS = Canada Western Red Spring; CPS = Canada Prairie Spring.
[c] Prebreak of 2% release using smooth rolls at 1:1 differential.

self-adjusting in use and they operated at a lower equilibrium temperature than harder rolls. A number of workers, including Willm (1977) and Evers et al (1984), have shown that finely fluted reduction rolls yield lower flour starch damage than smooth matted rolls.

The use of increased differential to increase starch damage has the disadvantage of producing darker, higher-ash flour (Scanlon and Dexter, 1986). Increasing roll pressure, which also imparts increased starch damage, results in lower flour ash and improved flour color by increasing bran flattening (Ward and Shellenberger, 1951). The potential deleterious effects of extreme roller mill reduction grinding conditions on protein properties must also be considered (Holas and Tipples, 1978).

ACKNOWLEDGMENTS

The following are gratefully acknowledged for their contribution to the text: P. W. Brennan and A. K. Sarkar, Canadian International Grains Institute, Winnipeg; J. E. Dexter and P. C. Williams, Grain Research Laboratory, Canadian Grain Commission, Winnipeg; A. N. Hibbs, International Multifoods, Minneapolis; C. A. Rogers, formerly Canadian International Grains Institute, Winnipeg; and K. Vorwerck, Buhler-Miag, Braunschweig. Special acknowledgments are due to the author's secretary, S. L. Miller, and the communications staff of the Canadian International Grains Institute, particularly D. J. Bride and V. S. Sloan.

LITERATURE CITED

ALNAJI, L. K., and KADOUM, A. M. 1979. Residue of methyl phoxin in wheat and milling fractions. J. Agric. Food Chem. 27:583-584.

AUER, E. 1962. The Miag Multomat. Oper. Millers Tech. Bull. (April) pp. 2643-2645.

BAKER, D., and GOLUMBIC, C. 1970. The estimation of the flour yielding capacity of wheat. Miller 277(Feb.):8-11.

BARNES, P. J. 1986. The influence of wheat endosperm on flour colour grade. J. Cereal Sci. 4:143-155.

BAYFIELD, E. G., ANDERSON, J. E., GEDDES, W. F., and HILDEBRAND, F. C. 1943. The effect of millroom temperature and relative humidity on experimental flour yields and flour properties. Cereal Chem. 20:149-171.

BLACK, H. C. 1966. Laboratory purifier for durum semolina. Cereal Sci. Today 11:533-534, 542.

BLACK, H. C. 1980. The GRL Pilot Mill. Oper. Millers Tech. Bull. (Sept.) pp. 3834-3837.

BLACK, H. C., and BUSHUK, W. 1967. Modification of the Buhler laboratory mill for milling semolina. Cereal Sci. Today 12:164, 166-167.

BLACK, H. C., FISHER, M. H., and IRVINE, G. N. 1961. Laboratory milling. I. A small-scale bran finisher. Cereal Chem. 38:97-101.

BLACK, H. C., HSIEH, F. H., MARTIN, D. G., and TIPPLES, K. H. 1980a. Two Grain Research Laboratory research mills and a comparison with the Allis-Chalmers mill. Cereal Chem. 57:402-406.

BLACK, H. C., HSIEH, F. H., TIPPLES, K. H., and IRVINE, G. N. 1980b. The GRL sifter for laboratory flour milling. Cereal Foods World 25:757-760.

BLACK, H. C., PRESTON, K. R., and DEXTER, J. E. 1983. Modifications to the Bühler laboratory mill to produce a flour comparable in yield and quality to the Allis-Chalmers laboratory mill. Can. Inst. Food Sci. Tech. J. 16:191-195.

BLAKENEY, A. B., ALMGREN, G., and JACOB, E. H. 1979. Analysis of first break milling of hard and soft wheats. Milling Feed Fert. 162(9):22, 24-25, 28.

BOLLING, H., and ZWINGELBERG, H. 1984. Kontinuierliche Erfassung von Mehlenhaltsstoffen durch NIR. Getreide Mehl Brot 38:3-5.

BRADBURY, D., MacMASTERS, M. M., and CULL, I. M. 1956. Structure of the mature wheat kernel. III. Microscopic structure of the endosperm of hard red winter wheat. Cereal Chem. 33:361-373.

BUTCHER, J., and STENVERT, N. L. 1972. An Entoleter for the Buhler laboratory mill. Milling (July) 154:27-29.

BUTCHER, J., and STENVERT, N. L. 1973a. Conditioning studies on Australian wheat. I. The effect of conditioning on milling behavior. J. Sci. Food Agric. 24:1055-1066.

BUTCHER, J., and STENVERT, N. L. 1973b. Conditioning studies on Australian wheat. III. The role of the rate of water penetration into the wheat grain. J. Sci. Food Agric. 24:1077-1084.

CAMPBELL, J. D., and JONES, C. R. 1955. The effect of temperature on the rate of penetration of moisture within damped wheat grains. Cereal Chem. 32:132-139.

CIACCO, C. F., and D'APPOLONIA, B. L. 1982. Characterization and gelling capacity of water-soluble pentosans isolated from different mill streams. Cereal Chem. 59:163-166.

CIGI. 1983. Grains and Oilseeds, 3rd ed. Canadian Int. Grains Inst., Winnipeg.

CLEVE, H., and WILL, F. 1966. Research with the help of the Vario Roll. Cereal Sci. Today 11:128-130, 132.

CREWE, J., and JONES, C. R. 1951. The thickness of wheat bran. Cereal Chem. 28:40-49.

DAVIS, A. B., and EUSTACE, W. D. 1984. Scanning electron microscope views of material from various stages in the milling of hard red winter, soft red winter, and durum wheat. Cereal Chem. 61:182-186.

DEXTER, J. E., and MARTIN, D. G. 1986. The effect of prebreak conditions on the milling performance of some Canadian wheats. J. Cereal Sci. 4:157-169.

DEXTER, J. E., and MATSUO, R. R. 1981. Effect of starchy kernels, immaturity, and shrunken kernels on durum wheat quality. Cereal Chem. 58:395-400.

DEXTER, J. E., and MATSUO, R. R. 1982. Effect of smudge and blackpoint, mildewed kernels, and ergot on durum wheat quality. Cereal Chem. 59:63-69.

DEXTER, J. E., BLACK, H. C., and MATSUO, R. R. 1982a. An improved method for milling semolina in the Bühler laboratory mill and a comparison to the Allis-Chalmers mill. Can. Inst. Food Sci. Tech. J. 15:225-228.

DEXTER, J. E., MATSUO, R. R., MARTIN, D. G., and TIPPLES, K. H. 1982b. A history of laboratory milling at the Grain Research Laboratory. 1982 Annual Report. Canadian Grain Commission, Grain Research Laboratory, Winnipeg.

DEXTER, J. E., PRESTON, K. R., MATSUO, R. R., and TIPPLES, K. H. 1984. Development of a high extraction flow for the GRL pilot mill to evaluate Canadian wheat potential for the Chinese market. Can. Inst. Food Sci. Tech. J. 17:253-259.

DEXTER, J. E., MARTIN, D. G., PRESTON, K. R., TIPPLES, K. H., and MacGREGOR, A. W. 1985a. The effect of frost damage on the milling and baking quality of red spring wheat. Cereal Chem. 62:75-80.

DEXTER, J. E., PRESTON, K. R., TWEED, A. R., KILBORN, R. H., and TIPPLES, K. H. 1985b. Relationship of flour starch damage and flour protein to the quality of Brazilian-style hearth bread and remix pan bread produced from hard red spring wheat. Cereal Foods World 30:511-514.

EVERS, A. D., BAKER, G. J., and STEVENS, D. J. 1984. Production and measurement of starch damage in flour. I. Damage due to roller milling of semolina. Staerke 36:309-312.

FARRAND, E. A. 1964. Flour properties in relation to the modern bread processes in the United Kingdom, with special reference to alpha-amylase and starch damage. Cereal Chem. 41:98-111.

FARRELL, E. P., and WARD, A. B. 1965. Flow rates and analyses for ash and protein of all streams in the Kansas State University pilot mill. Oper. Millers Tech. Bull. (March) pp. 2842-2847.

FINNEY, K. F., and BOLTE, L. C. 1985. Experimental micromilling: Reduction of tempering time of wheat from 18–24 hours to 30 minutes. Cereal Chem. 62:454-458.

FINNEY, K. F., and YAMAZAKI, W. T. 1946. A micro-milling technique using the Hobart grinder. Cereal Chem. 23:484-492.

FINNEY, K. F., HEIZER, H. K., SHELLENBERGER, J. A., BODE, C. E., and YAMAZAKI, W. T. 1949. Comparison of certain chemical, physical, and baking properties of commercial, Buhler and Hobart-milled flours. Cereal Chem. 26:72-80.

FINNEY, P. L., and ANDREWS, L. 1986a. A 30-minute conditioning method for micro-, intermediate-, and large-scale experimental milling of soft red winter wheat. Cereal Chem. 63:18-21.

FINNEY, P. L., and ANDREWS, L. C. 1986b. Revised microtesting for soft wheat quality evaluation. Cereal Chem. 63: 177-182.

GAINES, C. S. 1985. Associations among soft wheat flour particle size, protein content, chlorine response, kernel hardness, milling quality, white layer cake volume, and sugar-snap cookie spread. Cereal Chem. 62:290-292.

GEDDES, W. F., and FRISELL, B. 1935. An experimental flour mill for 100-gram wheat samples. Cereal Chem. 12:691-695.

GEHLE, H. 1965. The Miag "Vario" Rollstand—Design and purpose. Oper. Millers Tech. Bull. (May) pp. 2861-2862.

HART, L. P., and BRASELTON, W. E. 1983. Distribution of vomitoxin in dry milled fractions of wheat infected with *Gibberella zeae*. J. Agric. Food Chem. 31:657-659.

HINTON, J. J. C. 1959. The distribution of ash

in the wheat kernel. Cereal Chem. 36:19-31.

HOLAS, J., and TIPPLES, K. H. 1978. Factors affecting farinograph and baking absorption. I. Quality characteristics of flour streams. Cereal Chem. 55:637-652.

HOOK, S. C. W. 1984. Specific weight and wheat quality. J. Sci. Food Agric. 35:1136-1141.

HOOK, S. C. W., BONE, G. T., and FEARN, T. 1982a. The conditioning of wheat. The influence of roll temperature in the Buhler laboratory mill on milling parameters. J. Sci. Food Agric. 33:639-644.

HOOK, S. C. W., BONE, G. T., and FEARN, T. 1982b. The conditioning of wheat. The influence of varying levels of water addition to UK wheats on flour extraction, moisture and colour. J. Sci. Food Agric. 33:645-654.

HOOK, S. C. W., BONE, G. T., and FEARN, T. 1982c. The conditioning of wheat. The effect of increasing wheat moisture content on the milling performance of UK wheats with reference to wheat texture. J. Sci. Food Agric. 33:655-662.

HOOK, S. C. W., BONE, G. T., and FEARN, T. 1982d. The conditioning of wheat. An investigation into the conditioning requirements of Canadian Western Red Spring No. 1. J. Sci. Food Agric. 33:663-670.

HOOK, S. C. W., BONE, G. T., and FEARN, T. 1984a. The conditioning of wheat. Moisture migration between the components of a mixed grist, and its effect on milling performance. J. Sci. Food Agric. 35:584-590.

HOOK, S. C. W., BONE, G. T., and FEARN, T. 1984b. The conditioning of wheat. A comparison of UK wheats milled at natural moisture content and after drying and conditioning to the same moisture content. J. Sci. Food Agric. 35:591-596.

HOOK, S. C. W., BONE, G. T., and FEARN, T. 1984c. The influence of air temperature and relative humidity on milling performance and flour properties. J. Sci. Food Agric. 35:597-600.

HSIEH, F. H., MARTIN, D. G., BLACK, H. C., and TIPPLES, K. H. 1980. Some factors affecting the first break grinding of Canadian wheat. Cereal Chem. 57:217-223.

HSU, K. H. 1984. A theoretical approach to the tempering of grains. Cereal Chem. 61:466-470.

JEFFERS, H. C., and RUBENTHALER, G. L. 1977. Effect of roll temperature on flour yield with the Brabender Quadrumat experimental mills. Cereal Chem. 54:1018-1025.

JENSEN, S. A., and MARTENS, H. 1983. The botanical constituents of wheat and wheat milling fractions. II. Quantification by amino acids. Cereal Chem. 60:172-177.

JENSEN, S. A., MUNCK, L., and MARTENS, H. 1982. The botanical constituents of wheat

and wheat milling fractions. I. Quantification by autofluorescence. Cereal Chem. 59:477-484.

JONES, C. R. 1940. The production of mechanically damaged starch in milling as a governing factor in the diastatic activity of flour. Cereal Chem. 17:133-169.

JONES, C. R., GREER, E. N., THOMLINSON, J., and BAKER, G. J. 1961. Technology of the production of increased starch damage in flour milling. Factors in individual reductions. Northwest. Miller 265:34, 36-38, 45, 46, 48.

KADOUM, A. M., and LAHUE, D. W. 1977. Degradation of malathion in wheat and milling fractions. J. Econ. Entom. 70:109-110.

KEMP, J. G., WHITESIDE, A. G. O., MacDONALD, D. C., and MILLER, H. 1961. Ottawa micro flour mill. Cereal Chem. 38:50-59.

KILBORN, R. H., BLACK, H. C., DEXTER, J. E., and MARTIN, D. G. 1982. Energy consumption during flour milling: Description of two measuring systems and the influence of wheat hardness on energy requirements. Cereal Chem. 59:284-288.

KITTERMAN, S. J., SEEBORG, E. F., and BARMORE, M. A. 1960. A note on the modification of the five-gram milling-quality test and the five-gram micromill. Cereal Chem. 37:762-764.

KRUGER, J. E. 1981. Severity of sprouting as a factor influencing the distribution of amylase levels in pilot mill streams of Canadian wheats. Can. J. Plant Sci. 61:817-828.

LARKIN, R. A., MacMASTERS, M. M., WOLF, M. J., and RIST, C. E. 1951. Studies on the relation of bran thickness to millability of some Pacific Northwest wheats. Cereal Chem. 28:247-258.

LARKIN, R. A., MacMASTERS, M. M., and RIST, C. E. 1952. Relation of endosperm cell wall thickness to the milling quality of seven Pacific Northwest wheats. Cereal Chem. 29:407-413.

LILLARD, D. W., Jr., and HERTSGAARD, D. M. 1983. Computer analysis and plotting of milling data: HRS wheat cumulative ash curves. Cereal Chem. 60:42-46.

LINEBACK, D. R., CASHMAN, W. E., HOSENEY, R. C., and WARD, A. B. 1978. Note on measuring thickness of wheat bran by scanning electron microscopy. Cereal Chem. 55:415-419.

LIU, R., LIANG, Z., POSNER, E. S., and PONTE, J. G., Jr. 1986. A technique to improve functionality of flour from sprouted wheat. Cereal Foods World 31:471-476.

LOCKWOOD, J. F. 1960. Flour Milling, 4th ed. Henry Simon Ltd., Stockport, Cheshire, England.

MacMASTERS, M. M., HINTON, J. J. C., and

BRADBURY, D. 1971. Microscopic structure and composition of the wheat kernel. Pages 51-113 in: Wheat: Chemistry and Technology, 2nd ed. Y. Pomeranz, ed. Am. Assoc. Cereal Chem., St. Paul, MN.

MARSHALL, D. R., ELLISON, F. W., and MARES, D. J. 1984. Effects of grain shape and size on milling yields in wheat. I. Theoretical analysis based on simple geometric models. Aust. J. Agric. Res. 35:619-630.

MARSHALL, D. R., MARES, D. J., MOSS, H. J., and ELLISON, F. W. 1986. Effects of grain shape and size on milling yields in wheat. II. Experimental studies. Aust. J. Agric. Res. 37:331-342.

McNEILL, F. J. 1964. A quick temper for experimental milling of wheat. Cereal Sci. Today 9:408-409.

MENSAH, G. W. K., WATTER, F. L., and WEBSTER, G. R. B. 1979. Insecticide residues in milled fractions of dry or tough wheat treated with malathion, bromophos, iodofenphos, and pirimiphos-methyl. J. Econ. Entom. 72:728-731.

MORRIS, V. H., ALEXANDER, T. L., and PASCOE, E. D. 1945. Studies of the composition of the wheat kernel. I. Distribution of ash and protein in center sections. Cereal Chem. 22:351-361.

MORRIS, V. H., ALEXANDER, T. L., and PASCOE, E. D. 1946. Studies of the composition of the wheat kernel. III. Distribution of ash and protein in central and peripheral zones of whole kernels. Cereal Chem. 23:540-547.

MORRISON, W. R., and HARGIN, K. D. 1981. Distribution of soft wheat kernel lipids into flour milling fractions. J. Sci. Food Agric. 32:579-587.

MOSS, R. 1973. Conditioning studies on Australian wheat. II. Morphology of wheat and its relationship to conditioning. J. Sci. Food Agric. 24:1067-1076.

MOSS, R. 1977a. The influence of endosperm structure, protein content and grain moisture on the rate of water penetration into wheat during conditioning. J. Food Technol. 12:275-283.

MOSS, R. 1977b. An autoradiographic technique for the location of conditioning water in wheat at the cellular level. J. Sci. Food Agric. 28:23-33.

MOSS, R. 1985. The application of light and scanning electron microscopy during flour milling and wheat processing. Food Microstruct. 4:135-141.

MOSS, R., STENVERT, N. L., KINGSWOOD, K., and POINTING, G. 1980. The relationship between wheat microstructure and flour milling. Scanning Electron Microscopy III:613-620.

NEEL, D. V., and HOSENEY, R. C. 1984a. Sieving characteristics of soft and hard wheat flours. Cereal Chem. 61:259-261.

NEEL, D. V., and HOSENEY, R. C. 1984b. Factors affecting flowability of hard and soft wheat flours. Cereal Chem. 61:262-266.

NELSON, P. N., and McDONALD, C. E. 1977. Properties of wheat flour protein in flour from selected mill streams. Cereal Chem. 54:1182-1191.

NIERNBERGER, F. F., and FARRELL, E. P. 1970. Effects of roll diameter and speed on first break grinding of wheats. Oper. Millers Tech. Bull. (Jan.) pp. 3154-3158.

NURET, H., HERSANT, P., and ENSMIC, T. S. 1979. Considerations nouvelles sur le blutage. Bull. Anc. Eleves Ec. Fr. Meun. 294:342-359.

OKADA, K., NEGISHI, Y., and NAGAO, S. 1986. Studies on heavily ground flour using roller mills. I. Alteration in flour characteristics through overgrinding. Cereal Chem. 63:187-193.

ORTH, R. A. 1985. Pilot flour mill for Australia. Food Technol. Aust. 37:318-319.

ORTH, R. A., and MANDER, K. C. 1975. Effect of milling yield on flour composition and breadmaking quality. Cereal Chem. 52:305-314.

POMERANZ, Y., BOLTE, L. C., and AFEWORK, S. 1984. Time-dependent moisture gradients in conditioned wheat, determined by electrical methods. Cereal Chem. 61:559-561.

POSNER, E. S., and DEYOE, C. W. 1986. Changes in milling properties of newly harvested hard wheat during storage. Cereal Chem. 63:451-456.

POSNER, E. S., and WETZEL, D. L. 1986. Control of flour mills by NIR on-line monitoring. Oper. Millers Tech. Bull. (April) pp. 4711-4722.

POSNER, E. S., WARD, A. B., and NIERNBERGER, F. F. 1974. Evaluation of wheat tempering and blending methods of hard winter wheats under experimental conditions. Oper. Millers Tech. Bull. (Jan.) pp. 3425-3428.

PRESTON, K. R., BLACK, H. C., TIPPLES, K. H., DEXTER, J. E., SADARANGANEY, G. T., and TKAC, J. J. 1983. The GRL pilot mill. III. Comparison with two Canadian commercial hard red spring wheat mills. Can. Inst. Food Sci. Tech. J. 16:97-103.

REES, R. G., MARTIN, D. J., and LAW, D. P. 1984. Black point in bread wheat: Effects on quality and germination, and fungal associations. Aust. J. Exp. Agric. Anim. Husb.

24:601-605.

SCANLON, M. G., and DEXTER, J. E. 1986. Effect of smooth roll grinding conditions on reduction of hard red spring wheat farina. Cereal Chem. 63:431-435.

SCOTT, P. M., KANHERE, S. R., LAU, P.-Y., DEXTER, J. E., and GREENHALGH, R. 1983. Effects of experimental flour milling and breadbaking on retention of deoxynivalenol (vomitoxin) in hard red spring wheat. Cereal Chem. 60:421-424.

SEEBORG, E. F., and BARMORE, M. A. 1951. Suggestions for improving the uniformity of Buhler experimental milling. Trans. AACC 9:25-30.

SEEBORG, E. F., SHOUP, N. H., and BARMORE, M. A. 1951. Modification of the Buhler mill for micro milling. Cereal Chem. 28:299-308.

SEITZ, L. M., EUSTACE, W. D., MOHR, H. E., SHOGREN, M. D., and YAMAZAKI, W. T. 1986. Cleaning, milling, and baking tests with hard red winter wheat containing deoxynivalenol. Cereal Chem. 63:146-150.

SHELLENBERGER, J. A., and WARD, A. B. 1967. Experimental milling. Pages 445-469 in: Wheat and Wheat Improvement. K. S. Quisenberry, ed. Am. Soc. Agron., Inc., Madison, WI.

SHUEY, W. C. 1960. A wheat sizing technique for predicting flour milling yield. Cereal Sci. Today 5:71-72, 75.

SHUEY, W. C., and GILLES, K. A. 1969. Laboratory scale commercial mill. Oper. Millers Tech. Bull. (May) pp. 3100-3105.

SIBBITT, L. D., CLASSON, D. H., and HARRIS, R. H. 1960. Improved tempering and modified milling techniques for small samples of wheat. Cereal Chem. 37:398-404.

SIBBITT, L. D., SHUEY, W. C., and GILLES, K. A. 1964. A small laboratory purifier. Cereal Sci. Today 9:436, 438-439.

SIMMONS, L., and MEREDITH, P. 1979. Width, height, endosperm and bran of the wheat grain as determinants of flour milling yield in normal and shrivelled wheats. N.Z. J. Sci. 22:1-10.

STENVERT, N. L. 1972. The measurement of wheat hardness and its effect on milling characteristics. Aust. J. Exp. Agric. Anim. Husb. 12:159-164.

STENVERT, N. L., and KINGSWOOD, K. 1976. An autoradiographic demonstration of the penetration of water into wheat during tempering. Cereal Chem. 53:141-149.

STENVERT, N. L., and KINGSWOOD, K. 1977. Factors influencing the rate of moisture penetration into wheat during tempering. Cereal Chem. 54:627-637.

STENVERT, N. L., and MOSS, R. 1974. The

separation and technological significance of the outer layers of the wheat grain. J. Sci. Food Agric. 25:629-635.

SULLIVAN, B. 1941. Quick tempering of wheat for experimental milling. Cereal Chem. 18:695-696.

TIPPLES, K. H. 1980. Effect of immaturity on the milling and baking quality of red spring wheat. Can. J. Plant Sci. 60:357-369.

VORWERCK, K. 1977. Mechanisch beschädigte Stärke in Weizenmehlen-Beziehungen zu Analytik und Verarbeitungseigenschaften. Getreide Mehl Brot 31:234-239.

WARD, A. B., and SHELLENBERGER, J. A. 1951. Grinding with controlled roll pressure. Oper. Millers Tech. Bull. (Aug.) pp. 1907-1912.

WILLIAMS, P. C., and SOBERING, D. C. 1986. Attempts at standardization of hardness testing of wheat. I. The grinding/sieving (particle size index) method. Cereal Foods World 31:359, 362-364.

WILLIAMS, P. C., THOMPSON, B. N., WETZEL, D., McLAY, G. W., and LOEWEN, D. 1981. Near-infrared instruments in flour mill quality control. Cereal Foods World 26:234-237.

WILLM, C. 1977. Contribution à l'étude de l'endommagement de l'amidon en mouture de blé tendre. Bull. Anc. Eleves Ec. Fr. Meun. 277:13-26.

WINGFIELD, J., and FERRER, A. 1984. Multiple sieve sifter performance using various combinations of feed rates, circles and speeds. J. Food Process. Eng. 7:91-110.

YAMAZAKI, W. T., and ANDREWS, L. C. 1982a. Experimental milling of soft wheat cultivars and breeding lines. Cereal Chem. 59:41-45.

YAMAZAKI, W. T., and ANDREWS, L. C. 1982b. Small-scale milling to estimate the milling quality of soft wheat cultivars and breeding lines. Cereal Chem. 59:270-272.

YOUNG, J. C., FULCHER, R. G., HAYHOE, J. H., SCOTT, P. M., and DEXTER, J. E. 1984. Effect of milling and baking on deoxynivalenol (vomitoxin) content of Eastern Canadian wheats. J. Agric. Food Chem. 32:659-664.

ZIEGLER, E. 1938. The use of a purifier in experimental milling. Cereal Chem. 15:663-671.

ZIEGLER, E. 1940. The effect on diastatic activity of surface, pressure, differential, and temperature of the reduction rolls in milling. Cereal Chem. 17:668-679.

ZIEGLER, E., and GREER, E. N. 1971. Principles of milling. Pages 115-199 in: Wheat: Chemistry and Technology, 2nd ed.

Y. Pomeranz, ed. Am. Assoc. Cereal Chem., St. Paul, MN.

ZWINGELBERG, H. 1972. Vorzerkleinerung und Mahleigenschaften des Brotgetreides. Getreide Mehl Brot 31:66-69.

ZWINGELBERG, H., MAYER, D., and GERSTENKRON, P. 1983. Beeinflussung der Mehlausbeute und Mehlqualität von Weizen durch Glattwalzen unterschiedlicher Beschaffenheit. Getreide Mehl Brot 37:112-117.

APPENDIX

Milling Terms Used in the United Kingdom and in the United States and Canada[a]

United Kingdom	United States/Canada
I Bk.	I Bk. or IB
II Bk. coarse	II Bk/B. coarse
II Bk. fine	II Bk/B. fine
III Bk. coarse	III Bk/B. coarse
III Bk. fine	III Bk/B. fine
IV Bk. coarse	IV Bk/B. coarse
IV Bk. fine	IV Bk/B. fine
V Bk. coarse	V Bk/B. coarse
V Bk. fine	V Bk/B. fine
X^1	Chunks 1
X^2	Chunks 2
Y	Chunks 3
Z	Chunks 4
A	1 Sizings
B	1st Middlings
C	2nd Middlings
B^2	2nd Quality
D	3rd Middlings
E	4th Middlings
F	1st Tailings
G	5th Middlings
H	6th Middlings
J	2nd Tailings
K	7th Middlings
L	8th Middlings
M	3rd Tailings
N	1st Low grade
O	2nd Low grade
P	3rd Low grade
Germ I	Germ I
Germ II	Germ II
Pollard I	Shorts 1
Pollard II	Shorts 2
Bran flaking	Bran flaking
Coarse semolina (CS)	1st and 2nd Coarse middlings
Medium semolina (MS)	1st and 2nd Medium middlings
Fine semolina (FS)	1st and 2nd Fine middlings
Coarse middlings (CM)	
Medium middlings (MM)	
Fine middlings (FM)	
Coarse middlings duster (CMD)	Midds Re
Medium middlings duster (MMD)	
Fine middlings duster (FMD)	
Semolina grader (SG)	Coarse middlings grader
Dunst	Dunst

(continued on next page)

Milling Terms (continued)

United Kingdom	United States/Canada
Screenings	Screenings
Bran	Bran
Broad bran	...
Coarse weatings	Coarse middlings
Fine weatings	Fine middlings
Bakers' grade flour	Bakers' grade flour
Patent flour	Patent flour
Straight-run flour	Straight-run flour
Low-grade flour	Clears
Breaks	Breaks
Scalping	Scalping
Grading	Grading
Dusting	Dusting
Purification	Purification
Dressing	Bolting
Reductions	Reductions
Offal grading	Feed grading
Redressing	Rebolting

[a] Source: Lockwood (1960); used by permission.

CHAPTER 2

CRITERIA OF FLOUR QUALITY

WILLIAM C. MAILHOT (*retired*)
Quality Control Department
General Mills, Inc.
Minneapolis, Minnesota

JAMES C. PATTON
Quality Control and Assurance
Campbell Taggart, Inc.
Dallas, Texas

I. INTRODUCTION

Quality, in the broadest definition, is conformance to requirements: requirements are established, and the supplier satisfies the customer by conformance to those requirements. Flour quality represents conformance to several measurable characteristics that are significant in terms of end use. Flour quality can be defined as the ability of the flour to produce a uniformly good end product under conditions agreed to by the supplier and the customer. Pratt (1971) previously reviewed flour quality criteria. In this chapter, only flours of common wheats are discussed; durum wheat products are discussed in Chapter 8 of this volume.

In a broad sense, flour strength has been synonymous with flour quality. The presence or absence of strength factors governs the suitability of a flour for a specific end use. Strength is associated with wheat or flour protein and encompasses both quantity and quality measurements. There are many methods for evaluating flour strength, both objective and subjective; however, these tests must be related to other characteristics of the flour.

II. COMPONENTS OF QUALITY

Several chapters of this monograph deal specifically with some of the quality components that are related to end use applications, as well as those that are related to the evaluation and selection of the raw material. In this chapter, the testing of flour components that are recognized as contributing to quality is discussed in a general manner.

Flour quality factors can be divided into at least two basic groups: those that are inherent in the wheat as a result of the genetic makeup of the particular class and variety plus changes brought about by growing conditions, including fertilization, weather (drought, heat, frost, rainfall, etc.), and disease; and those that might be altered during the process of converting the wheat into flour. One might subdivide the latter group of quality factors, some of which are controllable within reasonable limits, to separate out those that may be intentionally modified to improve the performance of the end product: bleaching and maturing, addition of enzyme supplements, fine grinding, and air classification. In any discussion of quality components, a different weight must be assigned to each component, depending on the end use requirements.

A. Protein—Quantity and Quality

Protein quantity and quality are both considered primary factors in measuring the potential of a flour in relation to its end use. The quantitative expression of crude protein is related to total organic nitrogen in the flour, whereas quality evaluations relate specifically to physicochemical characteristics of the gluten-forming component.

QUANTITY

Protein quantity has been measured by the classic Kjeldahl nitrogen analysis, which assumes a constant relation between total nitrogen and the array of amino acids that link together to form proteins. In wheat flour, this relation is expressed by multiplying the nitrogen content by 5.7. For many years, protein quantity as determined by this measurement has been designated "crude protein."

Another method sometimes employed for determining protein quantity is to determine the wet and dry gluten content of a flour by washing techniques. This method is somewhat crude as well as tedious and does not lend itself to the multiple determinations necessary to control a manufacturing process. In addition, gluten is very difficult to wash out from ground wheat. The experimental error of this test is large. Gluten washing does, however, allow the chemist to visually observe some physical characteristics of the gluten, such as elasticity and color. Automated instruments have been developed to shorten the time required to complete testing for gluten.

Udy (1956) developed a method using a dye-binding technique to determine crude protein of wheat and wheat flour. More recently, methods have been developed to determine protein by near-infrared reflectance (NIR). Williams (1979) and Norris (1978) were involved in establishing these methods for determining wheat and wheat flour protein.

U.S. Department of Agriculture researchers at the Grain Marketing Research Center, Manhattan, Kansas, reported on the quantity of protein in wheat or flour as a criterion of breadmaking quality (Pomeranz, 1980; Finney, 1985). Webb et al (1971) conducted a study on the rheological assessment of protein quality and quantity for breadmaking.

QUALITY

Protein quality criteria are related to the gluten portion of the flour protein.

Quality can be measured by subjecting the flour to several physical testing devices that measure rheological characteristics (see section II-L). Branlard and Dardevet (1985) discussed the diversity of grain proteins and bread wheat quality.

Protein quality is much more difficult to evaluate than protein quantity; therefore, several test methods are used, with the bake test as the final performance test. Most quality evaluations of wheat flours are based on the flours' unique property of contributing viscoelastic characteristics to doughs. Rye and triticale proteins show this property to a much lesser degree. In both rye and triticale, breadmaking potential may be affected by the presence of large amounts of gums.

Variations in the physical properties of doughs from different wheats are of major concern to bakers, depending on their products and processes. Consistent dough properties are necessary for quality control of finished baked goods.

For bread-baking purposes, the proteins in hard red winter and spring wheats are considered superior to those from soft wheats. Soft wheat flours are most often chosen for production of cookies, crackers, and cakes. However, fractions mechanically selected from hard red wheat flours by a process known as "air classification" have been used quite satisfactorily for production of a wide variety of baked products.

Relatively broad ranges for energy used in dough development and for dough strength, elasticity, and extensibility are needed in baking, depending on the particular products and/or processes involved. Therefore, a choice of wheats to produce flours of various strengths and mix requirements is needed.

MacRitchie (1984) and Pyler (1983) have presented comprehensive summaries of current knowledge relative to quality in wheat protein and flour. Gluten proteins are essential for bread production because elasticity and extensibility are considered most important in breadmaking. Glutenins appear to contribute to mixing time, strength, and elasticity, whereas gliadins contribute to extensibility and stickiness. The molecular structures and weights of these proteins are being carefully studied in attempts to clarify their relationships to flour quality.

Research efforts to identify and measure quality factors of wheat proteins have continued for many years. Much progress has been made in developing tests that can be used to evaluate various characteristics of flour. However, more research is needed to identify the specific quality factors inherent in wheat and flour protein that are necessary for production of acceptable end products.

B. Minerals

For many years, the mineral (ash) content of flour has been considered an important measure of flour quality. The mineral content of a flour per se is not related to final performance, but it gives some indication of the miller's skill and of the degree of refinement in processing.

The minerals in the wheat kernel are concentrated in areas adjacent to the bran coat and in the bran itself. Flour products that contain more ash are darker in color and may be assumed to contain more fine bran particles or more of the portion of the endosperm adjacent to the bran. Unfortunately, some wheats naturally have a higher endosperm ash because of soil conditions and, possibly, genetic factors. Such wheats are sometimes rejected as undesirable simply because they cannot produce flour with an ash content within the limitations placed by end users. Ash content is closely related to the flour components that influence color. Morris et al (1949) published studies on the distribution of minerals in the wheat kernel.

C. Color

Color measurement may be approached in two ways. The first approach is to measure whiteness, which primarily determines the extent of color removal by bleaching compounds; however, the presence of finely divided bran has some influence on available instruments. Measurements of this kind can be made with the Agtron (Blue) instrument, which measures light reflectance of the product within a relatively narrow range of light source (blue spectrum). Such measurements do not correlate well with ash because the extent of color removal by chemical bleaching or natural oxidation of the carotenoid pigments of the flour is a function of time, and therefore the values for flours vary with the age and treatment level of the flour.

The second approach to color measurement largely ignores whiteness and concentrates on the influence of the branny material in the flour by measuring reflectance with a light source in the green band of the light spectrum. Either of two instruments can be used: the Kent-Jones and Martin color grader or the Agtron (Green). Gillis (1963) established that Agtron (Green) measurements are related to ash content, as long as wheat class in the mill mix remains reasonably uniform. When the mill mix is changed, the instrument must be restandardized to restore its effective performance. The changes involved relate somewhat to particle size. Shuey (1975) further discussed flour color (as measured on the Agtron) as a measurement of flour quality.

D. Moisture

Wheat moisture variables that affect milling and baking properties have been studied by Nagi and Bains (1983) and by the Association of Operative Millers in technical bulletins in the United States.

E. Absorption

Dough absorption can be related to flour moisture; however, it also depends on mix formulation and other factors. Flour absorption is an important consideration in the production of all types of baked goods. Usually, high absorption values are desirable.

Absorption is measured as the amount of water required to yield a batter or dough of predetermined consistency. In the laboratory, the farinograph is used to determine this factor, and the value thus obtained can be used as a guide in the bakery, provided variations among instruments are recognized. Other instruments used to determine flour absorption are considered in section V.

The water absorption level of a flour is influenced by several factors. The gluten portion of flour has a relatively constant water-imbibing capacity (about 2.8 times the dry gluten content). The water-soluble portion of the total protein content of the flour has no water-imbibing properties. Other constituents, such as dextrins, pentosans, and cellulose, have a minor influence on absorption because they are present in small quantities. Sandstedt (1955) pointed out that starch damaged during milling takes up additional water. Particle size has been proposed as a possible factor involved in determining the water-holding capacity of doughs or batters.

F. Viscosity

Measurements of the viscosity of flour-water mixtures as criteria of flour quality take two forms. The viscosity of an acidulated flour-water suspension at room temperature can be measured by the MacMichael or the Brookfield viscometer.

In a dilute lactic acid solution, flour gluten swells considerably, as shown by the MacMichael instrument, and starch swells to a limited extent, increasing the viscosity of the flour-water suspension. The increase is in direct relation to the swelling properties and quantity of gluten present in the test flour, assuming that changes in viscosity due to starch are relatively constant. If damaged starch changes, the total value measurable by the MacMichael viscometer changes. Thus, in some instances, the value reflects both a protein and a starch variable. This test is generally used to evaluate soft wheat flours in which the range in protein content is rather narrow, and the measurement reflects the condition of the starch.

In the second form of viscosity testing, wheat protein is held constant and flour protein remains constant, so that any differences in observed viscosity are measurements of the type and severity of grinding applied by the miller in the manufacturing process—that is, starch damage changes due to the severity of grinding. The Brookfield viscometer is currently the alternate choice because MacMichael equipment parts are not available.

The amylograph can be used to measure viscosity (Shuey and Tipples, 1982). A standard quantity of flour solids is placed in a buffered water suspension, and the viscosity of this uniform suspension is measured and charted throughout a standardized heating cycle. The amylograph value indicates the rate and extent to which the viscosity of the suspension changes during the controlled cycle. Swelling and gelatinization of the starch thicken the suspension and thus raise its viscosity. Under the test conditions, as temperature is increased, the activity of thermostable starch-liquefying enzymes increases and part of the total starch is hydrolyzed, thus reducing viscosity. The recorded maximum viscosity or peak measurement can be used to estimate enzyme activity as a means of determining the quantity of enzyme that might be added to the flour. The enzyme active in this test is predominantly α-amylase.

Meredith and Pomeranz (1982) used the amylograph in studies of U.S. wheat flours and starches. Additional instruments used to determine viscosity are listed in the table in section V.

G. Enzymatic Activity

The general subject of enzymes in wheat and flour is discussed in Volume I, Chapter 8; however, because enzymes influence flour behavior, enzymes related to flour quality are discussed in this section.

AMYLASE

In the manufacture of yeast-leavened products such as bread, rolls, and soda crackers, carbon dioxide is the gaseous agent that causes the product to rise during both fermentation and baking. This gas is produced by yeast cells from simple sugars in the dough that were present in the flour or were added as an ingredient of the process formula or produced during fermentation. To regulate the production of carbon dioxide at a rate that does not exceed the ability of the gluten network of the dough to stretch and retain the gas, the extent of enzyme modification must be controlled.

The two main types of amylases present in wheat flour are α-amylase and β-amylase. Most cereal chemists agree that β-amylase, a saccharogenic enzyme, is present in adequate quantity in flour milled from sound (unsprouted) wheat. α-Amylase is not present in adequate quantity. Wheat flour that is to be used in yeast-fermented products must be supplemented with malted wheat, malted barley flour, or fungal enzymes.

The following measurements of amylase activity in flour are made. Maltose value is a quantitative measurement of maltose produced enzymatically from the flour substrate under controlled conditions. The gassing-power (pressuremeter) method measures the quantity of carbon dioxide produced by a fixed quantity of yeast from 10 g of a flour dough under controlled time and temperature conditions. Because the carbon dioxide is produced by the yeast primarily from simple flour sugars or fermentable sugars made available by enzyme action on the starch, the pressuremeter value basically reflects the enzyme activity of the flour. The amylograph and the falling number test are currently used rather than the maltose value test or pressuremeter method. If fungal enzymes are added to the flour, then the grain amylase analyzer method is used.

Because α-amylase may be added to wheat flour to achieve any desired level of enzyme activity, the response to this additive is an important criterion of flour quality. This response can be controlled by careful grinding to maintain a desirable and uniform level of damaged starch to serve as the substrate for amylase action. The optimum level of enzyme activity is ultimately governed by the end use of the flour and the type of processing involved in this end use. Warchalewski and Klockiewicz-Kaminska (1983) reported studies involving the influence of α-amylase supplementation.

PROTEASE

At present, no attempt is made to control or alter the activity of proteolytic enzymes present in wheat flour apart from adding amylase supplements that contain quantities of proteolytic enzymes. Bread bakers have employed proteolytic enzymes to reduce the mechanical dough development requirements of unusually strong or tough glutens. In this instance, the proteases mellow the gluten enzymatically rather than mechanically; the mechanical process might be limited by mixing capacity or schedule. In general, wheat gluten tolerates a fairly broad range of proteolytic supplementation.

LIPASE

The presence of fat-splitting enzymes or lipases in wheat flour has been recognized for a number of years. The activity of lipase is important only insofar as stability of the flour under prolonged storage or adverse conditions is concerned. Lipase converts the flour fat into fatty acids, which are associated with "off" or rancid odors in the flour. Lipase activity is generally high in flours that are highest in fat content and is normally of little concern in flour produced from sound wheat. Measurement of fat acidity of the extracted fat from wheat or flour can reveal incipient deterioration before organoleptically evident rancid odors develop. There are indications that excessive free fatty acids in a flour adversely affect its baking properties.

H. Granulation (Particle Size) and Grain Hardness

Granulation, or particle size, may be considered a quality component of flour only if one recognizes that a particular degree of fineness may have resulted from other changes. As mentioned previously, particle size is related to water absorption of flour. Finer average particle size accelerates the rate of flour hydration, thus bringing the plastic dough mass more rapidly to a stage where work can be applied to it to develop gluten.

Particle size in a finished, conventionally milled flour is a measurement of the friability of the wheat endosperm under the conditions of milling. Some mellow wheats tend to "shell out" their endosperm readily, whereas the friability of more vitreous wheats requires greater work input. Measurement of particle size distribution reveals these characteristics quite well and can be applied to compare friability of different wheats milled under identical conditions. Particle size and shape also affect bulk density.

Air classification of flour requires tests for particle size distribution. With the use of these tests, it is possible to predict classification of the conventionally milled parent flour. One test for determining particle size distribution is the Whitby sedimentation test. Devices such as the Coulter Counter, Fischer subsieve analyzer, Micromerograph, Cintel counter, Andreasen pipet, and Microtrac have been used. The Fisher (average particle size) analyzer is also used.

Particle size distribution tests confirm that flour is not a homogeneous material. Particles falling in the range of 0–20 μm in Stokes equivalent diameter (SED) are largely free protein material mixed with small starch particles, cell wall material, and in some instances, severely damaged starch. Particles falling between 20 and 35 μm in SED are predominantly free starch particles with little adhering protein. Flour particles above 35 μm in SED are a mixture of chunk material composed of starch with adhering protein. If air classification is used to separate various groupings in terms of their size in micrometers, the severity of grinding can be judged by measuring the starch damage in the various fractions.

Use of the Microtrac (a light-diffraction technique) has been reviewed (Anonymous, 1978). Donelson and Yamazaki (1972) reported particle size analysis in soft wheat flour, and Yamazaki and Donelson (1972) analyzed the significance of particle size in soft wheat flours. Chaudhary (1975) investigated the relation between particle size and white layer cake quality.

Research is continuing on grain hardness. Stenvert (1972) compared Australian wheat varieties. Pomeranz et al (1984) reported on wheat hardness and the baking properties of wheat flours.

Hardness tests are of two types, those for single kernels and those for bulk samples. The single-kernel tests can be made on the whole grain or on sections. They include penetration, abrasion, crushing, cutting, and related tests. Tests on bulk samples measure the power or time required to grind, resistance to grinding, percentage of abraded material, and particle size of ground material (by actual sieving, as in the particle size index test, or by NIR).

I. Starch Damage

The influence of starch damage on the rehydration rate of doughs and batters and on the rate of enzyme activity and its effect on viscosity measurements has been mentioned previously. Yamazaki (1955) observed an inverse relation between the starch tailings fraction and cookie spread factor; the starch tailings fraction contained damaged starch, whereas the prime starch fraction did not. Work indicates that starch damage plays an important role in governing the performance of bread flours. Further, under a fixed set of milling conditions, the level of starch damage is related to kernel hardness or vitreousness of the parent wheat. These studies have also revealed that starch damage can be modified to some extent by the tempering process and grinding practices (roll cuts and speed ratios). Normally, break flours have the least and middlings flours the most starch damage; these values increase as the product moves down through the reduction system.

There is some relation between starch damage and insoluble (gluten) protein content in terms of performance; that is, flours with higher gluten protein seem to tolerate higher levels of starch damage and still perform satisfactorily.

Several methods for measuring damaged starch have been studied by committees of the American Association of Cereal Chemists (AACC), and a standard procedure has been published (AACC, 1983, Method 76-30A). Barnes (1978) developed a rapid enzymatic determination of starch damage in flour, and McDermott (1980) reported a rapid nonenzymatic determination of damaged starch. Farrand (1972a, 1972b) observed the effect of levels of starch damage on bread quality and also the influence of particle size and starch damage on bread flours.

J. Response to Additives

Flours respond differently to additives such as bleaching compounds, maturing agents, and potassium bromate and to enzyme supplementation. Because these additives influence flour performance, how the flour responds to them is important from a quality standpoint. The additives permitted by the U.S. Standards of Identity (Code of Federal Regulations, 1985, Title 21, Para. 137) fall within four basic groups: color-removing agents, gaseous or powdered bleaching or maturing agents; enzymes in the form of malted wheat flour, malted barley flour, or fungal amylase; and oxidizers (potassium bromate). Internationally, the Codex Committee on Cereals, Pulses and Legumes, Codex Alimentarius Commission (Food and Agriculture Organization, World Health Organization) has completed a draft standard for wheat flour and has recommended optional ingredients and food additives for wheat flour (Alinorm 85/29, 1985). The AACC book of approved methods and its supplements

(AACC, 1983) provide methods to determine additives to wheat flour, as described in this section.

COLOR-REMOVING AGENTS

Only one chemical agent with color-removing properties, exclusively, is permitted in the United States at present. Benzoyl peroxide reacts with the carotenoid pigments in flour to render them white, whereas their natural color is creamy yellow to yellow. If flour is aged under ideal storage conditions, the same chemical change takes place through natural oxidation. Although the presence of the natural yellow pigments has no influence on flour performance, some end products are considered more desirable if they are white rather than creamy yellow.

The optimum level of treatment can be determined by treating a flour at various levels of the bleaching powder and extracting the treated flours with an appropriate solvent. After centrifugation or filtration, the presence of residual unoxidized pigments can be determined visually or with a colorimeter.

MATURING AGENTS

Cereal chemists have recognized for many years that the performance of a flour improves with age. This natural aging process requires considerable time, which may vary with storage temperature and wheat type. To rely on natural aging would place heavy economic demands on both the miller and the user of the flour.

The maturing process is one of oxidation and, as it relates to flour performance, is specifically modification of the protein. Researchers suggest that the modification involves sulfhydryl and disulfide groups of the protein molecule, resulting in cross-linkage of the protein chains. Application of a maturing compound decreases extensibility of the dough and increases dough resistance to extension.

Flours milled from different types of wheat require different amounts of maturing compounds; because overtreatment with these compounds can result in undesirable performance, the rate of treatment must be controlled carefully and the optimum level determined accurately. The required treatment level is customarily determined by an oxidation series in which the flour is treated at several levels; a standard bake test indicates which level gives the best bread.

Until 1962, chlorine dioxide gas was the principal maturing agent used in the United States, although other agents have been used at various times in the history of milling, for example, ozone (Alsop process), Agene (nitrogen trichloride), and persulfates. Persulfates were abandoned when more effective agents became available or were prohibited by law because of possible harmful physiological side effects.

During 1962, two new maturing compounds were approved for use in flour in the United States. Azodicarbonamide and acetone peroxides have an effect similar to that of chlorine dioxide. Azodicarbonamide does not react until the flour is made into a dough, which may indicate that the reaction sites differ from those of chlorine dioxide. The reaction of acetone peroxides is apparently completed in the flour within 24 hr. There is some evidence that the acetone peroxides may not be entirely specific or restricted to the flour protein, inasmuch as some color-removing effects have been noted.

Chlorine dioxide and acetone peroxides are seldom used in the United States at present; ascorbic acid is used in flours designed for specific end uses. Because azodicarbonamide is applied to flour in powder form, feed rates are somewhat easier to control for it than for a gas such as chlorine dioxide, which requires complex metering and agitator equipment. In general, azodicarbonamide is used only for treatment of bread flours.

Another gaseous maturing agent, chlorine, is widely used in the treatment of soft wheat flours and low-protein, air-classified flours when the flour is to be used for cakes, certain types of nonspread cookies, and certain pastries. It may also be used on all-purpose family flours.

Sollars (1958) and others have studied the effects of chlorine on soft wheat flour fractions—prime starch, tailings starch, fat, and protein. Although chlorine residues can be found in all fractions after treatment, indicating chemical reaction with part of the involved fraction, the action of chlorine on prime starch appears to have the greatest effect on performance. However, although the major improvement from chlorine treatment is due to action on the starch, excessive treatment has a mellowing effect on the protein, which is desirable in strong or harsh glutens but undesirable if the gluten is mellow or weak.

Optimum application rates for chlorine are established by cake-baking tests of the flour after treatment to several pH levels. Chlorine has some color-removing effect, indicating action on flour pigments, but the creamy yellow pigment cannot be removed completely at the treatment levels used to achieve optimum performance.

Although the use of chlorine is primarily confined to cake flours, very light chlorine treatment is sometimes applied to all-purpose family flours to improve their cake-baking performance. This usually requires some compromise, because the chlorine adversely affects bread-baking properties.

POTASSIUM BROMATE

Bread flours are often treated with potassium bromate in addition to maturing compounds. Although the end result of bromate addition in bread baking is similar to that of the maturing agents, bromate can further improve the performance of flours that have been optimally treated with other maturing compounds. Bromate response, or the response of a flour to the addition of successively greater increments of potassium bromate, is sometimes called "bromate tolerance." Bromate treatment is considered optimum at the level after which further addition results in minimal baking improvement.

ENZYMES

Modern methods of producing bread and other yeast-raised baked goods require the fermentation processes to proceed at a uniform rate. For this to occur, the yeast must be supplied with fermentable sugars, which are converted to carbon dioxide and thus cause the dough to rise, and to other intermediate products such as alcohols and aldehydes, which are associated with flavor. Although some of the necessary sugar may be supplied as sucrose or dextrose as an ingredient of the dough, this procedure does not assure a continuing supply that would sustain uniform fermentation. A relatively continuous supply of sugar at an ample level can be achieved by the action of amylase. Another desirable effect of added malt or fungal amylases is the limited modification of starch. Wheat flour deficient in α-amylase is fortified by adding to it an ingredient such as malted wheat flour or malted barley flour or fungal enzymes to produce the desired effect. As mentioned in section II-G, several methods measure the effect the enzyme has on the flour substrate.

K. Nutritional Fortification

Several countries regulate the nutritional fortification of wheat flour. Kulp et al (1980) reported on the natural levels of nutrients in wheat flours. U.S. requirements for enriched wheat flour are listed in the Code of Federal Regulations (1985, Title 21, Para. 137). These regulations provide for addition of niacin, riboflavin, thiamine, and iron. Calcium addition is optional for enriched flour. The U.S. National Academy of Sciences has recommended additional nutritional fortification of wheat flour in the United States; however, this recommendation has not yet been implemented by regulation. Nutritional aspects of wheat flour are elaborated on in Chapter 3 of this volume.

L. Rheological Properties

Several physical testing devices measure various rheological properties of wheat flour doughs. The tests are usually performed on flour-water doughs and are widely employed in quality testing. They characterize the gluten portion of the protein by measuring such factors as extensibility and resistance to extension of doughs at rest, hydration time, maximum development time, and tolerance or resistance to breakdown at a predetermined consistency during mechanical mixing. Dick et al (1977) and Bloksma (1972) reviewed the rheological properties of flour doughs.

Recording dough mixers such as the farinograph, mixograph, and rheograph evaluate the mixing characteristics of gluten development in a dough. The principles upon which these instruments are based are discussed in detail in Chapter 4 of this volume.

Because the mixing characteristics of a flour are usually related to the gluten quality measurements, they can be defined by the use of recording dough mixers in the selection and evaluation of experimentally milled wheat. Mixing requirements of bread doughs can be correlated with the measurements obtained with recording dough mixers. These mixers are not infallible, however, despite efforts of cereal chemists to standardize equipment and procedures.

Dough extensibility can be measured using the extensigraph or the Chopin Alveograph. These instruments measure the extent to which a properly developed gluten can be extended or sheeted and the ability of that gluten to retain gas (resistance to extension). Basically, the extensibility of gluten is related to wheat variety, and careful wheat selection can result in conformance to predetermined standards insofar as this characteristic is concerned.

Both extensibility and resistance to extension can be modified to some degree by maturing treatment. These two factors, then, can be effectively adjusted or controlled through proper addition of maturing agents during the milling process.

M. Bake Tests for Quality Attributes

None of the rheological and chemical measurements is capable of fully predicting the end performance of a flour. These measurements serve as indexes that, when properly interpreted, increase the probability of satisfactory performance. The ultimate criteria of quality in flour are its conformance to chemical and physical requirements plus its adherence to certain standards as established by a performance or bake test.

For many years, cereal scientists have worked to develop one or more laboratory tests that would give an objective measurement of the potential baking quality of wheat and/or flour. Although other research work has yielded much valuable information, no single available laboratory test or series of tests to evaluate baking performance can replace properly designed and implemented bake tests. Bake tests and the interpretation of their results are subjective in nature, and well-trained persons are required to properly perform the tests and subsequently evaluate the final product.

Bake tests are often perceived as being universal in application regardless of the manufacturing process and final product requirements. Such a perception does not take into account the significant differences in objectives between research and practical applications.

Cereal chemists involved in research and wheat quality evaluation traditionally perform bake tests that can be done with small amounts of flour. The results are primarily used to compare different wheats or flours by indicating the relative merits of samples. Without these small baking evaluations, the potential baking quality of new wheat varieties in early stages of development could not be estimated. Bakes of larger loaves are also in relatively widespread use. Examples of small-scale tests are described by Finney (1984), Uchida (1982), and Bond and Moss (1983) and in the standard AACC-approved methods (AACC, 1983). Many variations of these methods can be used, depending on the worker's concept of what information is needed.

On the other hand, scientists employed by baking companies are primarily interested in applying bake tests to predict the performance of a flour in the various processes used by their companies. These scientists must deal with day-to-day problems that involve formulations and conditions associated with commercial production of their products. Bakers started a project at the American Institute of Baking (1985) to standardize various bake tests for predicting flour performance in different production processes. The tests are designed to incorporate formulations and mechanical conditions that approximate those encountered during production. Breeders and millers may also use the results from these tests to help them understand more fully the qualities of flour necessary for different production procedures.

When enough flour is available, fermentation tolerance and "baking strength" can be evaluated in commercial-type bake tests. Fermentation tolerance refers to the ability of a particular flour to allow production of high-quality products over a relatively wide range of fermentation conditions. Baking strength relates to the ability of a flour used for yeast-raised products to retain appropriate amounts of leavening gas while producing a desirable grain and texture within a well-structured product. Whether it be used in whole-grain breads or a rich formulation, a flour with good baking strength will carry relatively heavy loads of inert ingredients and yield a final product with good expansion and desirable crumb grain and texture. Long mixing time is not a measure of baking strength. Many flours classified as "strong" by mixing requirements do not exhibit the strength to maintain the gas cells in their appropriate configuration even when the dough is developed to optimum conditions.

III. IMPORTANCE OF QUALITY COMPONENTS

Tables I–III indicate criteria for measuring quality of flour for specific end uses. Minor (1984) further discussed soft wheat flour characteristics. Schiller (1984) reported on hard wheat bakery flour specifications. Spicher and Pomeranz (1985) compared the quality specifications for wheat flours used in Germany, France, and the United States.

IV. MICROBIOLOGY

Wheat flour is used in the manufacture of a variety of products. Utilization of flour products in convenience foods (cereal-based mixes, frozen or refrigerated doughs and batters, cream sauces, soups, candies, etc.) emphasizes the need for audit in the area of microbiological concerns.

The total microbial population is lower in flour than in parent wheat because much of the microbial population of wheat is on the outer surfaces. Available data indicate that the microbial population in wheat is related to growth, harvest, transport, and storage conditions.

Patent flours are generally lower in total microbial population than clears and second-clears. The number of microorganisms in millfeed or offal need not be high in relation to that in flour, depending on the wheat cleaning and conditioning process. Adding chlorine to wheat wash water and tempering water reduces the microbial population of the wheat and flour, but the effectiveness of this treatment is somewhat limited.

In published data concerning the identification of fungal and bacterial species, no currently recognized pathogenic species have been found with sufficient frequency to indicate a potentially hazardous situation. This does not rule out the possibility of contamination occurring as a result of poor manufacturing practices or environment. Three major concerns of the cereal microbiologist are public health, spoilage, and good manufacturing practices.

Hobbs and Greene (1984) reported on the microbiology of cereals and cereal products. Routine analyses used to determine the microbiological condition of wheat and wheat flour include mold and yeast counts, aerobic plate count, and counts of coliform organisms and *Escherichia coli*, staphylococci, and salmonellae. Special analyses of wheat flour could include testing for "rope" spores, certain mycotoxins, and other microorganisms, depending on the end use of the product.

Daftary et al (1970) reported changes in wheat flours damaged by mold during storage. Bothast et al (1981) described the effects of moisture and temperature on microbiological properties of wheat flour.

With proper growing conditions, handling and processing controls, and storage practices, there should be no fear of microbiological contamination of wheat and wheat flours.

TABLE I
Generally Used Criteria for Measuring Quality of Flour for Yeast Breads

Quality Criterion	Pan Bread			Hearth Breads	Variety Breads, Cracked Wheat		
	Wholesale, Conventional	Continuous Dough	Retail Shop		Bread, Rye Bread	Soft Rolls	Sweet Goods
Wheat class[a]	HS, HW, or blend	HS, HW, or blend	HS, HW, or blend	HS	HS	HS, HW, or blend	HS, HW or blend
Flour absorption	Medium to high	Medium	High	High	High	Medium	Medium to high
(% by farinograph)	60–64	59–64	60–65	63–68	65–75	59–64	60–64
Ash (%)	0.44–0.50	0.44–0.50	0.42–0.50	0.44–0.55	0.45–0.55	0.44–0.50	0.45–0.50
Alk. water retention	NS[b]	NS	NS	NS	NS	NS	NS
Baking test: bread; dough	Pan; straight or sponge	Pan; sponge	Pan; straight or sponge	Hearth; straight	Pan; straight	Pan; sponge	Pan; straight
Bleaching, maturing agents	Azodicarbonamide, acetone peroxide, ascorbic acid, benzoyl peroxide, and/or $KBrO_3$ as required to optimize quality of flour for end use						
Color	Creamy white	Creamy white	Creamy white	Creamy white	NS	Creamy white to white	Creamy white
Enzyme activity							
Amylase							
Amylograph (BU)	475–625	525–600	450–600	400–600	350–550	475–625	475–625
Maltose (mg)	290–350	280–330	300–360	300–360	300–375	290–350	290–350
Pressuremeter,							
5th-hr (mm)	500–550	475–550	500–550	500–550	500–550	500–550	500–550
Falling number	200–300	200–300	200–300	175–275	200–300	200–300	200–300
Amylase analyzer	300–600	300–600	300–600	350–600	300–600	300–600	300–600
Lipase	NS	NS	NS	NS	NS	NS	Low for prepared mixes
Protease	NS	NS	NS	NS	NS	NS	NS

Particle size							
Range (SED)[c] (μm)	0–150	0–150	0–150	0–150	0–150	0–150	0–150
Light diffraction	0–150	0–150	0–150	0–150	0–150	0–150	0–150
Fisher (average particle size)	16–22	16–22	16–22	18–22	18–22	16–22	16–22
Protein							
Quantity (%)	11.0–13.0	11.0–12.5	11.0–13.0	13.5–14.5	13.5–16.0	11.5–12.5	11.5–13.0
Quality (recording mixer values)							
Hydration	Medium	Short to medium	Medium	Medium to long	Medium to long	Medium	Short to medium
Peak development (min)	6–8	5–7	6–8	7–9	7–9	6–8	5–8
Stability (min)	7.5 minimum	8 minimum	9 minimum	10 minimum	10 minimum	8 minimum	8 minimum
Extensibility[d]	Medium	Medium	Medium	Medium to long	Medium to long	Medium to long	Long
Resistance to extension[c]	Medium	Medium	Medium	Maximum	Maximum	Medium to low	Medium to low
Starch damage[d]	5.5–7.8	5.5–7.8	6.0–8.0	7.0–8.5	Low as possible	5.5–7.8	5.5–7.8
Viscosity							
MacMichael	NS	NS	NS	NS	NS	NS	NS
Amylograph (BU)	475–625	525–600	450–600	400–600	350–550	475–625	475–625
Falling number	200–300	200–300	175–275	200–300	200–300	200–300	200–300

[a] HS = hard spring, HW = hard winter.
[b] NS = not significant at present, or significance not known.
[c] Stokes equivalent diameter.
[d] Descriptive terms in each instance refer to relative values for wheat variety involved.

TABLE II
Criteria for Evaluating Flours for Home Baking, Cake Doughnuts, Biscuits, Crackers, and Pastry

Quality Criterion	Home Baking	Cake Doughnuts	Biscuits	Crackers Sponge	Crackers Dough	Pastry
Wheat class[a]	HW or HW-SW blend	SW-HW blend	HW-SW blend or SW	SW or SW-HW blend	SW	SW
Flour absorption (% by farinograph)	Medium low 52–58	Medium low 52–58	Low 50–54	Low 48–52	Low 48–52	Low 48–52
Ash (%)	0.44–0.48	0.44–0.48	0.44–0.48	0.40–0.48	0.40–0.48	0.40–0.48
Alk. water retention	NS[b]	64–68	NS	58–68	50–56	50–56
Baking test: bread; dough	Biscuits; kitchen layer cakes; straight-dough bread	Doughnuts	Biscuits	AACC cookie spread; W/T 6.0–7.5	AACC cookie; W/T 7.5–8.5	AACC cookie; W/T 7.5–8.5
Bleaching, maturing agents	Azodicarbonamide, acetone peroxide, ascorbic acid, benzoyl peroxide, Cl$_2$ (light) as required to optimize quality of flour for end use			None	None	None
Color	Creamy white to white	Creamy	White	Creamy	Creamy	Creamy
Enzyme activity Amylase						
Amylograph (BU)	450–600	NS	NS	700–800	NS	NS
Maltose (mg)	290–320	NS	NS	200–250	NS	NS
Pressuremeter, 5th-hr (mm)	400–450	NS	NS	300–350	NS	NS
Falling number	200–300	250 minimum	250 minimum	250 minimum	250 minimum	250 minimum
Lipase	NS	Low for prepared mixes	NS	NS	NS	NS
Protease	NS	NS	NS	NS	NS	NS

Particle size						
Range (SED)[c] (μm)	0–125	0–125	0–90	0–125	0–125	0–125
Light diffraction	0–125	0–125	0–90	0–125	0–125	0–125
Fisher (average particle size)	12–18	11–16	10–15	11–14	10.5–13	10.5–13
Protein						
Quantity (%)	9.5–11.0	9.5–10.0	9.0–10.0	9.0–10.0	8.0–9.0	8.5–10.5
Quality (recording mixer values)						
Hydration	Short	Short	Short	Short	Short	Short
Peak development (min)	3–5	2–3	2–3	1–3	1–3	1–3
Stability (min)	3–6	2–5	1–3	1–3	1–2	1–2
Extensibility[d]	Medium	Maximum	Medium short	Medium	Medium	Medium
Resistance to extension[c]	Medium to low	Medium low	Low	Low	Low	Low
Starch damage[d]	Low as possible	Moderate	Low as possible	Moderate	Low as possible	Low as possible
Viscosity						
MacMichael	90–120°	60–90°	60–90°	65–90°	45–60°	45–60°
Amylograph (BU)	450–600	NS	NS	NS	NS	NS
Falling number	200–300	250 minimum	250 minimum	250 minimum	250 minimum	250 minimum

[a] HW = hard winter, SW = soft winter.
[b] NS = not significant at present, or significance not known.
[c] Stokes equivalent diameter.
[d] Descriptive terms in each instance refer to relative values for wheat variety involved.

TABLE III
Criteria for Evaluating Flours for Cookies, Cakes, and Soups and Gravies (Thickeners)

Quality Criterion	Cookies	Cakes		Soups and Gravies (Thickeners)
		Layer	Foam	
Wheat class[a]	SW	SW	SW	SW
Flour absorption (% by farinograph)	Low; 48−52	Low; 48−52	Very low; 44−48	NS[b]
Ash (%)	0.42−0.50	0.34−0.40	0.29−0.33	0.40−0.58
Alk. water retention	50−54	54−62	52−56	48−55
Baking test: bread; dough	AACC cookie; W/T 8.0−9.5	White layer cake	Angel food cake	None
Bleaching, maturing agents	None or light Cl_2	Cl_2 to pH 4.6−5.0; benzoyl peroxide	Cl_2 to pH 4.8−5.1; benzoyl peroxide	None
Color	Creamy	White	White	NS
Enzyme activity				
Amylase				
Amylograph (BU)	NS	NS	NS	700+, high as possible
Maltose (mg)	NS	NS	NS	200 mg, low as possible
Pressuremeter, 5th-hr (mm)	NS	NS	NS	NS
Falling number	250 minimum	250 minimum	250 minimum	250 minimum
Lipase	NS	Low for prepared mixes	Low for prepared mixes	NS
Protease	NS	NS	NS	NS
Particle size				
Range (SED)[c] (μm)	0−125	0−125	20−60	0−150
Light diffraction	10.5−13.5	10−12	10.5−12	11−14
Protein				
Quantity (%)	7.0−9.5	7.0−9.5	5.5−7.5	9.0−10.5
Quality (recording mixer values)				
Hydration	Short	Short	Short	NS
Peak development (min)	1−3	1−2	1−1.5	NS
Stability (min)	1−2	1−1.5	1−1.5	NS
Extensibility[d]	Medium	Short	Short	NS
	Low	Low	Low	NS
Starch damage[d]	Low as possible	High	Low	Low as possible
Viscosity				
MacMichael	40−65°	35−65°	30−45°	NS
Amylograph (BU)	NS	NS	NS	NS
Falling number	250 minimum	250 minimum	250 minimum	250 minimum

[a] SW = soft winter.
[b] NS = not significant at present, or significance not known.
[c] Stokes equivalent diameter.
[d] Descriptive terms in each instance refer to relative values for wheat variety involved.

V. METHODOLOGY AND INSTRUMENTATION

This monograph is not a laboratory manual but is intended as a reference source. Table IV is a reference list of detailed analytical procedures for the evaluation of flour quality.

TABLE IV
AACC Approved Methods (AACC, 1983) Used in Evaluating Flour Quality

Test Factor	Method Number
Acidity	
Fat acidity	02-01A, 02-02A, 02-31
pH	02-52
Ash	08-01, 08-02, 08-03, 08-12
Baking tests	
Bread, straight dough	10-10B
Bread, sponge dough	10-11
Sweet yeast products	10-20
Biscuits, self-rising flour	10-31A
Cookie flour	10-50D
Pie flour	10-60
Rye bread flour	10-70
Cake flour	10-90
Angel food cake flour	10-15
Color and pigments	
Pekar color test	14-10
Color of macaroni products	14-21
Pigments	14-50
Agtron color test	14-30
Enzymes	
Diastatic activity (amylograph)	22-10, 22-12
Diastatic activity (pressuremeter)	22-11
Diastatic activity (falling number)	56-81B
Diastatic activity of flour and semolina	22-15
Lipoxygenase activity of flour and semolina	22-40
Proteolytic activity of flour	22-60
Experimental milling	26-10, 26-20, 26-21, 26-30, 26-95
Extraneous matter	
Sand and cinders	28-06, 28-07
Insect, rodent contamination	28-20, 28-21, 28-22, 28-31, 28-32, 28-40, 28-41A, 28-43, 28-44, 28-50, 28-51, 28-60, 28-70, 28-85
Crude fat in flour	30-10
Crude fiber	32-10, 32-20
Gluten	38-10, 38-11, 38-20
Microorganisms	42-10, 42-11, 42-15, 42-17, 42-20, 42-25A, 42-26, 42-30B, 42-35, 42-40, 42-45, 42-50
Minerals	
Calcium	40-20
Iron	40-40, 40-41A
Moisture	44-10, 44-15A, 44-16, 44-18, 44-19, 44-20, 44-32, 44-40
Nitrogen	
Crude protein	39-10, 46-08, 46-09, 46-10, 46-11A, 46-12, 46-13, 46-14A, 46-19

(*continued on next page*)

TABLE IV (continued)

Test Factor	Method Number
Oxidizing, bleaching, and maturing agents	
Oxidizing agents	48-02
Acetone peroxides	48-05
Ascorbic acid	86-10
Benzoyl peroxide	48-06B, 48-07
Chlorine dioxide	48-30
Azodicarbonamide	48-71A
Potassium bromate	48-42
Particle size	50-10
Physical tests	
Extensigraph	54-10
Farinograph	54-20, 54-21, 54-28A, 54-29
Mixograph	54-40
Physicochemical tests	
Sedimentation	56-60
Viscosity	56-80
Falling number	56-81B
Sampling	64-60
Starch, damaged	76-30A
Tables	82-21, 82-22, 82-23, 82-24
Viscosity (MacMichael)	56-80
Vitamins	
Vitamin A	86-01A, 86-02, 86-03, 86-05
Ascorbic acid	86-10
Niacin	86-49, 86-50, 86-51, 86-52
Riboflavin	86-70, 86-72, 86-73
Thiamine	86-80

VI. SUMMARY

This discussion of flour quality criteria can only provide an overview of the subject and a source reference for more detailed information. It is evident that more work must be completed and more objective test methods developed to predict flour performance. Each end use for flour requires different tests to evaluate quality. Some of the more subjective test methods must be replaced with more exacting, objective ones.

More knowledge is required to explain the recognized differences in physical behavior of wheat varieties, such as mixing time, dough extensibility, and gas retention. The response of different flours to oxidizing agents such as bromate, gaseous maturing agents, and the maturing compounds such as acetone peroxides, azodicarbonamide, and ascorbic acid is well recognized, but the mechanics of the activity of these agents have not been completely explained.

At present, research indicates that the amino acid composition of wheat proteins is similar, if not identical, but we do not yet know whether they are identically bonded in forming the gluten protein. This may partly explain functional differences among wheat varieties.

Again, flour quality requirements must be established by the supplier and the end user. Tests to determine flour quality must be specific to end use requirements.

LITERATURE CITED

AACC. 1983. Approved Methods of the American Association of Cereal Chemists, 8th ed. The Association, St. Paul, MN.

ALINORM 85/29. 1985. Report of the fourth session of the Codex Committee on Cereals, Pulses and Legumes (FAO/WHO). Appendix II, p. 33. Food Agric. Org., Rome.

AMERICAN INSTITUTE OF BAKING. 1985. AIB Addendum to the 36th Annual Report of the Wheat Quality Council. Am. Inst. of Baking, Manhattan, KS.

ANONYMOUS. 1978. Faster, better size distribution analysis of dry flour particles achieved in Microtrac instrument. Food Process. 10:88.

BARNES, W. C. 1978. Rapid enzymic determination of starch damage in flours from sound and rain damaged wheat. Staerke 30:114-119.

BLOKSMA, A. H. 1972. Flour composition, dough rheology, and baking quality. Cereal Sci. Today 17:380-386.

BOND, E. E., and MOSS, H. J. 1983. Test baking from breeder to baker. Dev. Food Sci. 5A:649-654.

BOTHAST, R. J., ANDERSON, R. A., WARNER, K., and KWOLEK, W. F. 1981. Effects of moisture and temperature on microbiological and sensory properties of wheat flour and corn meal during storage. Cereal Chem. 58:309-311.

BRANLARD, G., and DARDEVET, M. 1985. Diversity of grain proteins and bread wheat quality. J. Cereal Sci. 3:329-343, 345-354.

CHAUDHARY, V. 1975. An investigation of the relation between particle size and white layer cake quality. Diss. Abstr. Int. B 36:3, 1122-1123.

CODE OF FEDERAL REGULATIONS. 1985. Title 21, Food and Drugs, Parts 100 to 169, Para. 137, Cereal Flours and Related Products, pp. 266-282.

DAFTARY, R. D., POMERANZ, Y., HOSENEY, R. C., SHOGREN, M. D., and FINNEY, K. F. 1970. Changes in wheat flours damaged by mould during storage. Effects in breadmaking. Agric. Food Chem. 18:617-619.

DICK, J. W., SHUEY, W. C., and BANASIK, O. J. 1977. Adjustment of rheological properties of flours by fine grinding and air classification. Cereal Chem. 54:246-255.

DONELSON, D. H., and YAMAZAKI, W. T. 1972. Soft wheat flour particle-size analysis by integrated sieve and Coulter Counter procedures. Cereal Chem. 49:641-648.

FARRAND, E. A. 1972a. Controlled levels of starch damage in a commercial United Kingdom bread flour and effects on absorption, sedimentation value, and loaf quality. Cereal Chem. 49:479-488.

FARRAND, E. A. 1972b. The influence of particle size and starch damage on the characteristics of bread flours. Bakers Dig. 46(1):22-24, 26, 55.

FINNEY, K. F. 1984. An optimized, straight-dough, bread-making method after 44 years. Cereal Chem. 61:20-27.

FINNEY, K. F. 1985. Experimental bread-making studies, functional (breadmaking) properties, and related gluten protein fractions. Cereal Foods World 30:794-801.

GILLIS, J. A. 1963. The Agtron; Photoelectric method of determining flour color. Cereal Sci. Today 8:40-42, 44, 46, 55.

HOBBS, W. E., and GREENE, V. W. 1984. Cereals and cereal products. Pages 690-699 in: Compendium of Methods for the Microbiological Examination of Foods, 2nd ed. Am. Pub. Health Assoc., Washington, DC.

KULP, K., RANUM, P. M., WILLIAMS, P. C., and YAMAZAKI, W. T. 1980. Natural levels of nutrients in commercially milled wheat flours. I. Description of samples and proximate analysis. Cereal Chem. 57:54-58.

MacRITCHIE, F. 1984. Baking quality of wheat flours. Adv. Food Res. 29:201-277.

McDERMOTT, E. E. 1980. The rapid non-enzymic determination of damaged starch in flour. J. Sci. Food Agric. 31:405-413.

MEREDITH, P., and POMERANZ, Y. 1982. Inherent amylograph pasting ability of U.S. wheat flours and starches. Cereal Chem. 59:355-360.

MINOR, G. K. 1984. Soft wheat flour characteristics. Cereal Foods World 29:659-660.

MORRIS, V. H., ALEXANDER, T. L., and PASCOE, E. D. 1946. Studies of the composition of the wheat kernel. III. Distribution of ash and protein in central and peripheral zones of whole kernels. Cereal Chem. 23:540-547.

NAGI, H. P. S., and BAINS, G. S. 1983. Effect of scouring and conditioning variables on milling, rheological and baking properties of Indian wheats. J. Food Sci. Technol. 20:122-124.

NORRIS, K. H. 1978. Near infrared reflectance spectroscopy—The present and future. Pages 245-251 in: Cereals '78: Better Nutrition for the World's Millions. Y. Pomeranz, ed. Am. Assoc. Cereal Chem., St. Paul, MN.

POMERANZ, Y. 1980. What? How much? Where? What function? in bread making. Cereal Foods World 25:656-662.

POMERANZ, Y., BOLLING, H., and ZWINGELBERG, H. 1984. Wheat hardness and baking properties of wheat flours. J. Cereal Sci. 2:137-143.

PRATT, D. B., Jr. 1971. Criteria of flour quality. Pages 201-226 in: Wheat: Chemistry and Technology, 2nd ed. Y. Pomeranz, ed. Am. Assoc. Cereal Chem., St. Paul, MN.

PYLER, E. J. 1983. Flour protein's role in baking performance. Bakers Dig. 57(3):24-33.

SANDSTEDT, R. M. 1955. Photomicrographic studies of wheat starch. III. Enzymatic digestion and granule structure. Cereal Chem. 32(Suppl.):17.

SCHILLER, G. W. 1984. Bakery flour specifications. Cereal Foods World 29:647-651.

SHUEY, W. C. 1975. Flour color as a measurement of flour quality. Bakers Dig. 49(5):18-19, 22-23, 26.

SHUEY, W. C., and TIPPLES, K. H., eds. 1982. The Amylograph Handbook. Am. Assoc. Cereal Chem., St. Paul, MN.

SOLLARS, W. F. 1958. Cake and cookie flour fractions affected by chlorine bleaching. Cereal Chem. 35:100-110.

SPICHER, G., and POMERANZ, Y. 1985. Bread and other baked products. Pages 331-389 in: Ullmann's Encyclopedia of Industrial Chemistry, 5th ed., Vol. A4. VCH Publ. Co., Weinheim, Federal Republic of Germany.

STENVERT, N. L. 1972. The measurement of wheat hardness and its effect on milling

characteristics. Aust. J. Exp. Agric. Anim. Husb. 12:159-164.

UCHIDA, M. 1982. Test baking methods and their applications in Japan. Cereal Foods World 27:597-598.

UDY, D. C. 1956. Estimation of protein in wheat and flour by ion-binding. Cereal Chem. 33:190-197.

WARCHALEWSKI, J. R., and KLOCKIEWICZ-KAMINSKA, E. 1983. The influence of alpha-amylase and supplementation on dough rheological properties and bread making performance. Dev. Food Sci. 5A:361-365.

WEBB, T., HEAPS, P. W., and COPPOCK, J. B. M. 1971. Protein quality and quantity: A rheological assessment of their relative importance in breadmaking. J. Food Technol. 6:47-62.

WILLIAMS, P. C. 1979. Screening wheat for protein and hardness by near infrared reflectance spectroscopy. Cereal Chem. 56:169-172.

YAMAZAKI, W. T. 1955. The concentration of a factor in soft wheat flours affecting cookie quality. Cereal Chem. 32:26-37.

YAMAZAKI, W. T., and DONELSON, D. H. 1972. The relationship between flour particle size and cake-volume potential among Eastern soft wheats. Cereal Chem. 49:649-653.

CHAPTER 3

NUTRITIONAL QUALITY OF WHEAT AND WHEAT FOODS

ANTOINETTE A. BETSCHART
Western Regional Research Center
Agricultural Research Service
U.S. Department of Agriculture
Albany, California

I. INTRODUCTION

Bread has often been described as the staff of life. Wheat and wheat foods, long recognized as a major staple and source of calories in the diets of people in many cultures, also contribute significant quantities of other nutrients to the diet.

Evaluations of the nutritional quality of wheat and wheat foods range from the simplest data—i.e., nutrient content determined mainly by chemical methods—to complex assays for bioavailability of nutrients, determined by in vivo animal and human studies. Recently, research and consumer interests have focused on a more integrated approach to the nutritional value and health implications of wheat foods. Proceedings of symposia and documents supportive of national policy (National Research Council, 1974; Spicer, 1975; Birdsall, 1985) reflect the interest in the nutritional quality of wheat foods and their role in the human diet.

This chapter presents basic information, for purposes of reference, and discusses recent findings and issues in selected areas of nutritional quality. This is not an exhaustive review, for such reviews already exist for wheat as well as for specific classes of nutrients in wheat. Rather, the objective of this chapter is to provide the reader with a basic understanding of the nutritive aspects of wheat as well as an exposure to state-of-the-art knowledge in several major areas of nutritional quality of wheat foods. Nutritional quality is addressed from the perspectives of classes of nutrients, factors that affect nutritional quality and the availability of nutrients, ways to improve nutritional quality, and future research needs.

II. ROLE OF WHEAT IN HUMAN NUTRITION

Wheat and wheat foods are a major source of nutrients for people in many regions of the world. Although often seen mainly as a source of carbohydrate, wheat foods are also a substantive source of protein, vitamins, and minerals when consumed as a major component of the diet.

In the United States, during a recent 15-year period, the consumption of wheat flour remained relatively constant; from 1970 to 1984, per capita consumption ranged from 138 to 147 g/day, and total per capita consumption of wheat products, including cereals, ranged from 141 to 151 g/day (USDA, 1985). Per capita consumption exceeded 200 g/day in 43% of the 63 wheat-consuming countries reported by the Food and Agriculture Organization (FAO) and exceeded 100 g/day in 68% (FAO, 1970). The populations of more than 50% of these countries obtained 20% or more of their calories from wheat foods; people in about 25% of these countries obtained 10% or fewer of their dietary calories from wheat.

Wheat consumption has declined in Western Europe and North America during the 20th century. In the United States, consumption fell until about 1970 and has stabilized somewhat since then; per capita consumption decreased from 312 g/day during 1924–1928 to 203 g/day during 1959–1961. Although it continued to decrease somewhat through 1970, to 188 g/day, it increased slightly, to 201 g/day, by 1984 (FAO, 1970; USDA, 1985).

The distribution of nutrients within the wheat kernel (Figs. 1 and 2) is typical of that of many cereals. Although the endosperm consists mainly of starch, since it makes up the major portion of the wheat kernel, a significant proportion of many minerals and vitamins in wheat is located in the endosperm. Nutrients are generally found in the highest concentrations in the germ or embryo and in the aleurone cells surrounding the starchy endosperm. Aleurone cells are rich sources of minerals, many of the B vitamins, and protein. Significant quantities of the minerals and vitamins are lost when whole wheat is milled to produce white, endosperm flour because the outer layers of bran are removed along with aleurone cells and germ. Data compiled by FAO showed that 72–75% extraction flour, typical of white flour in the United States, contains from as little as 20% to about 60% of the B vitamins originally present in whole wheat flour (FAO, 1970).

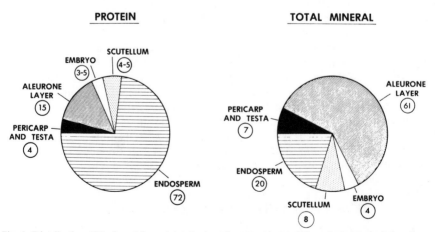

Fig. 1. Distribution (%) of protein and total mineral content located in various parts of the wheat kernel. Reprinted, with modifications, from MacMasters et al (1971).

The general composition of wheat is described in Volume I of this monograph, with chapters on the composition of the kernel (Chapter 4), and on proteins and amino acids (Chapter 5), carbohydrates (Chapter 6), and lipids (Chapter 7). This chapter emphasizes the nutritional quality and its implications rather than the composition of wheat and wheat foods.

III. PROTEIN

The contributions of cereal grains in general and wheat in particular to total calories and protein are shown in Table I (FAO, 1980). Cereal grains contribute 50 and 45% of the world's dietary calories and protein, respectively; wheat provides slightly less than 20% of total calories and protein. In general, people in the developed countries obtain slightly more protein from wheat (23%) than do those in the developing countries (18%). Wheat provides 25% or more of dietary protein in Western Europe (25%), Eastern Europe and the Soviet Union (29%), and the Near East (43%). The caloric contribution of wheat to the diets in these areas parallels the protein data somewhat: 22, 29, and 40% of dietary calories are provided by wheat in these three regions, respectively.

A. Protein Content and Amino Acid Composition

The nutritional quality of protein depends on the digestibility and availability of amino acids. Protein content and amino acid composition, however, can be used to assess potential protein quality. The protein content and amino acid composition of various wheat fractions and flours of several extraction rates have been extensively reviewed (Millers' National Federation, 1972; Betschart, 1978; Kulp et al, 1980; Young and Pellett, 1985). In addition, Pedersen and

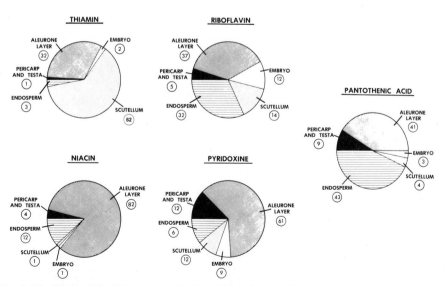

Fig. 2. Distribution (%) of selected vitamins located in various parts of the wheat kernel. Reprinted, with modifications, from MacMasters et al (1971).

Eggum (1983) reported systematic studies of protein and lysine content, percentage of nitrogen digestibility, and net protein utilization (NPU) of hard winter wheat flours ranging in extraction rate from 66 to 100%. Relevant data from these and other sources are presented in Table II to illustrate both the range of values obtained for specific samples and the relative differences in types of wheat and wheat fractions.

In general, the protein content of wheat fractions is greatest in germ, followed by middlings, bran, whole wheat flour, and white flour, in decreasing order. Within types of wheat, the hard wheats are generally higher in protein than the soft wheats; for example, hard red spring wheat contains more protein than hard red winter wheat, which contains more than soft wheat (Kulp et al, 1980). When wheat is milled, the protein content decreases as the extraction rate decreases; the drop in protein from 14.2% for 100% extraction flour to 12.7% for 66% extraction flour reflects the removal of germ and the aleurone-containing bran, which are relatively rich in protein (Pedersen and Eggum, 1983). In general, bread contains less protein than the flours from which it is made. Table II includes data for flours and breads from Denmark, the United States, Canada, Africa, and Sri Lanka and reflects variability in flour composition and bread formulations.

Protein digestibility, as indicated by nitrogen digestibility, is generally highest for milled white flour and lowest for bran (Miladi et al, 1972). Since whole wheat flour includes bran, its protein digestibility is invariably lower than that of white flour.

The limited data in Table II illustrate that different biological indexes produce different data and conclusions. Relative nutritional value (RNV) clearly differentiates protein quality, rating germ the highest, followed by middlings, bran, whole wheat, and white flour protein (Miladi et al, 1972). However, if

TABLE I
Contribution of Cereal Grains and Wheat to Human Diets by Region[a]

Region	Population (millions)	Food Energy			Protein		
		Daily Intake (kcal)	Percent from Cereal Grains	Percent from Wheat	Daily Intake (g)	Percent from Cereal Grains	Percent from Wheat
Western Europe	367	3,376	26	22	94.8	29	25
Eastern Europe and the Soviet Union	369	3,481	38	29	103.3	37	29
North America	240	3,557	17	13	105.7	18	15
Africa	344	2,205	47	10	55.0	51	13
Latin America	336	2,557	39	15	65.5	38	16
Near East	196	2,620	61	40	73.5	62	43
Far East	1,159	2,029	67	15	48.7	63	18
Developed countries	1,137	3,395	31	21	99.1	30	23
Developing countries	3,017	2,260	61	16	57.3	55	18
World total	4,154	2,571	50	18	68.8	45	19

[a] Adapted from FAO (1980), with permission from the Food and Agriculture Organization of the United Nations.

TABLE II
Protein Content and Biological Data—Wheat, Wheat Fractions, Flours, and Breads

Wheat Sample	Protein (%)	Nitrogen (%)	Biological Data			Reference
			N Digestibility (%)	RNV[a]	NPU[b]	
Wheat	8.0–21.9					Millers'
Durum wheat	11.5–21.5					National
Bran	11.9–22.9					Federation
Durum bran	17.6					(1972)
Hard wheat bran	16.7–21.5					
Soft wheat bran	16.2–16.4					
Shorts	18.5–20.3					
Red dog	17.5–25.0					
Whole wheat	16.7		91	39 (63)[c]		Miladi et al
White flour	14.4		99	24 (62)		(1972)
Middlings	19.7		84	57 (69)		
Bran	17.2		69	51 (59)		
Germ	25.2		92	79 (80)		
Whole wheat	12.2	2.09				FAO (1970)
Germ	22.9	3.95				
Bran	13.6	2.16				
Bulgur (parboiled)	11.2	1.92				
Flour[d]						
80–90%	11.7	2.05				
70–80%	10.9	1.91				
60–70%	9.2	1.61				
Hard winter wheat[d]						Pedersen and
100% whole						Eggum (1983)
wheat	14.2		91.6		56.2	
95%	13.9		93.9		58.2	
91%	13.8		...		...	
87%	13.8		94.9		54.4	
80%	13.4		94.2		53.8	
75%	13.3		96.6		54.0	
66%	12.7		95.7		56.4	
Hard red winter						Kulp et al
Wheat	11.8 ± 0.7					(1980)
Flour	10.6 ± 0.6					
Hard red spring						
Wheat	13.6 ± 0.6					
Flour	12.8 ± 0.4					
Soft wheat						
Wheat	10.2 ± 0.4					
Flour	8.5 ± 0.6					
Hard-soft blends						
Wheat	10.6 ± 0.5					
Flour	9.3 ± 0.7					
Flour						Betschart
whole wheat	11.9–13.3					(1978)
white	10.4–12.2					
Bread						
whole wheat	8.4–10.8					
white	7.8– 8.9					

[a] Relative nutritional value.
[b] Net protein utilization.
[c] Values in parentheses obtained with added lysine.
[d] % = extraction rate of flour.

NPU is used as the criterion, protein quality is similar in flours with extraction rates of 66–100% (Pedersen and Eggum, 1983).

Amino acid composition indicates protein quality when compared with a given amino acid profile. Patterns such as the FAO provisional amino acid scoring pattern and suggested patterns of amino acid requirements for infants, children, and adults (FAO, 1973) have been used to evaluate wheat protein quality (Betschart, 1978; Bodwell, 1985; Young and Pellett, 1985). Table III briefly summarizes the amino acid composition of various wheats and wheat fractions, together with their amino acid scores, and illustrates the high quality of germ, bran, and other milling by-products (red dog, shorts, middlings) and the relatively lower quality of whole wheat and white flours.

Lysine is clearly the first limiting amino acid in all wheat flours and wheat fractions except germ. The RNV of lysine-supplemented fractions, with the exception of germ, also indicates that lysine is the first limiting amino acid (Table II). The lack of improvement in lysine-supplemented germ protein is explained by its amino acid pattern, in which isoleucine is limiting. Although wheat protein is limited in lysine when compared to the FAO reference pattern and suggested requirements for infants and children, it is adequate for adults, who require less lysine (Betschart, 1978; Young and Pellett, 1985).

Some of the recent reviews on wheat protein place great emphasis on amino acid data while often overlooking or only discussing briefly in vivo animal and human studies (Bodwell, 1985; Young and Pellett, 1985). Bodwell (1985) did argue for some form of correction or adjustment using nitrogen digestibility or other data. The limitations of relying solely on amino acid data are illustrated in the study by Pedersen and Eggum (1983): although lysine decreased with increased milling, the flours did not differ in NPU. These results emphasize the importance of biological data in the evaluation of wheat and wheat products.

B. Protein Quality—Animal and Human Studies

The many indexes used to evaluate protein quality may be divided into the percentage of nitrogen digestibility [(N intake − fecal N)/(N intake) × 100], which reflects protein (N) digestion and absorption, and other methods that also evaluate retention or availability and protein utilization, such as nitrogen balance, protein efficiency ratio (PER), RNV, and NPU. Several authors have reviewed the protein quality of wheat from the perspective of amino acid availability, digestibility, and several other criteria for protein utilization (Carpenter, 1975; Betschart, 1978; Finley and Hopkins, 1985; Young and Pellett, 1985).

Protein digestibility tends to decrease as fiber content increases in wheat fractions, flours, and selected breads. Nitrogen digestibility is lowest for fractions containing the most fiber and highest for low-fiber fractions, such as white flour. Although whole wheat bread has a more favorable amino acid profile, the diminished protein digestibility of the bran protein is reflected in the fact that its protein quality (PER) is not significantly different from that of white bread (Table IV). A review of several rat studies showed, however, that the protein quality of white bread improves when the bread is supplemented with amino acids such as lysine and threonine (Betschart, 1978).

An assessment of several studies of children showed that results vary

TABLE III
Amino Acid Composition of Wheat and Wheat Fractions[a]

Wheat Sample	Number of Samples	Isoleucine	Leucine	Lysine	Methionine + Cystine	Phenylalanine + Tyrosine	Threonine	Tryptophan	Valine	Amino Acid Score
HRW wheat[b]		233–239	383–413	155–176*	174–194	431–472	164–173	62–124	291–319	46–52
HRS wheat[b]		224–249	384–391	156–166*	194–219	433–451	173–183	64–79	289–314	46–49
HRS-HRW blend[b]										
Red dog		214	361	258*	194	400	194	78	307	76
Shorts		207	353	261	186	391	189*	81	303	76
Bran		206	344	236*	188	400	179	99	293	69
Germ		218	259	330	184*	389	214	61	306	84
Whole wheat[c]				156*	269		175			46
Whole flour[c]				119*	289		163			35
Middlings				263*	256		206			77
Bran				256*	250		206			75
Germ				306*	238		219			90
Whole wheat[d]	≥69	204	417	179*	253	469	183		276	53
Bulgur	2	203	399	161*	219	436	177	66	244	47
Germ	9	225*	433	407	252	451	265		314	90
Gluten	2	258	433	89*	231	553	159	63[f]	266	26
Bran	9	209	415	270*	270	460	223		315	79
Flour[d]										
80–90%	20	232	379	159*	224	462	192	68	270	47
70–80%	12	228	440	130*	250	449	168	67[g]	258	38
60–70%	15	217	400	113*	229	423	153	58[g]	240	33
FAO PSP[e]		250	440	340	220	380	250	60	310	

[a] Expressed in milligrams per gram of nitrogen. An asterisk indicates limiting amino acid. HRW = hard red winter, HRS = hard red spring.
[b] Data from Millers' National Federation (1972).
[c] Data from Miladi et al (1972).
[d] Data from FAO (1970). % = extraction rate of flour.
[e] PSP = provisional scoring pattern (FAO, 1973).
[f] Chemical method.
[g] Microbiological method.

somewhat with experimental conditions (Betschart, 1978). Laboratory studies of infants and small children, often hospitalized and recovering from severe malnutrition, indicated that lysine supplementation of white flour and gluten diets improved nitrogen retention and weight gain (Graham et al, 1971). No effect was observed when whole wheat diets were supplemented with lysine. In several field studies of children, the addition of lysine, either to supplemental breads given at school or to all wheat products consumed, did not produce any beneficial effects. Other field studies reported an increase in either weight or height of children when lysine was added to wheat breads. Betschart (1978) summarized the specific conditions of these studies and some of the reasons for the variation in observations.

Differences between children and adults in protein and amino acid requirements are reflected in studies showing that adults can be maintained in nitrogen equilibrium or slightly positive nitrogen balance when consuming bread diets (Bolourchi et al, 1968; Betschart et al, 1985). The importance of sufficiently long control or adjustment periods was apparent in both of these studies. Protein digestibility was significantly greater in white bread diets (92.1%) than in whole wheat bread diets (85.0%) when normal young men (21–35 years old) consumed 1 lb (454 g) of bread per day. The difference in nitrogen digestibility was significant irrespective of the number of six-day periods (one to four) over which data were gathered, but this was not the case for nitrogen balance data. Not until the third six-day period was the nitrogen balance of those on white bread diets significantly higher than that of those on whole wheat bread diets (Betschart et al, 1985). These results illustrate the critical importance of designing human studies with sufficiently long treatment periods, including the preliminary control or adjustment period. In addition, the importance of energy intake and the dependence of protein requirements on energy consumption must not be overlooked, as was stressed in a recent FAO publication on energy and protein requirements (FAO/WHO, 1985).

TABLE IV
Protein Content and Quality of Breads and Breakfast Foods[a]

Wheat Food	Nitrogen[b] (%)	Protein[b] (%)	Protein Efficiency Ratio[c] (adjusted)
Breads			
Enriched white	2.5	14.5	0.78 d
Dark style	3.2	18.2	0.53 d
Whole wheat	2.9	16.3	0.72 d
Wheat berry	2.8	15.7	0.59 d
Seven grains	2.6	14.8	0.98 cd
Wheat germ	2.8	16.0	0.96 cd
Breakfast foods			
Whole wheat, processed	2.0	12.4	0.18 e
Natural no. 1	2.2	13.6	1.19 c
Natural no. 2	2.3	14.3	0.92 cd
ANRC[d] casein	···	···	2.50 a

[a] Reprinted with modifications, with permission, from Betschart (1982).
[b] Percent moisture-free basis.
[c] Values not followed by the same letter are significantly different (P <0.05). Values relative to casein adjusted to 2.50.
[d] Animal Nutrition Research Council.

C. Processing and Preparation

Carpenter (1975) reviewed the effects of processing and extrusion on wheat protein quality. In general, baking has a moderate effect on protein quality. Some rat studies have reported a small decrease in protein quality, mainly in the crust, but others have not (Carpenter, 1975). Small decreases in the crust are presumed to be a result of the Maillard reaction. Data reported by Milner and Carpenter (1969) suggest that the apparent improvement in protein quality (PER) when wheat is steeped may be mainly a result of improved palatability because NPU (i.e., net protein gain divided by grams of protein consumed) for steeped and dried wheat flour is not significantly different from that of raw wheat flour.

Linko et al (1981) reviewed the effect of extrusion cooking on protein digestibility and nutritional quality. Although moderate heat treatment improves enzymatic hydrolysis of protein, extremely high temperatures are likely to decrease cystine and diminish the availability of lysine. For example, pepsin hydrolysis of extruded corn and durum wheat semolina and of wheat flour increased with temperatures up to 225° C at 14% moisture and decreased at higher initial moisture levels. These data emphasize the importance of moisture-temperature interactions. In general, protein quality (PER) is not adversely affected by conventional extrusion cooking.

IV. CARBOHYDRATES AND DIETARY FIBER

Carbohydrates are discussed here in the broad context of starch, dietary fiber, and wheat foods containing various proportions of starch and dietary fiber. Studies of the physiological, nutritional, and metabolic effects of dietary fiber are reviewed in this section. The discussion of the nutritional influence of wheat fiber is confined to the effects on digestion and metabolism of macronutrients. The effects on micronutrients are considered in the sections on vitamins and minerals. This section focuses on wheat fiber, with examples of other fibers for purposes of comparison only.

Several extensive reviews have appeared recently, either focusing specifically on dietary fiber in cereals (Wisker et al, 1985) or taking a broader view of the nutritional implications of consuming wheat foods (Birdsall, 1985). Leeds (1985) compiled and edited a comprehensive review of dietary fiber, which includes an extensive bibliography. Other relevant publications include that of Vahouny (1985) on nutritional, pharmacological, and pathological aspects of dietary fiber and the report of the Canadian Expert Advisory Committee on Dietary Fiber (Health and Welfare Canada, 1985).

Much has been written and many claims continue to be made regarding the effects of dietary fiber and, specifically, wheat fiber. Anderson (1985) described the state of the art as follows: "While wheat fiber might have important health advantages, almost every area of potential benefit needs to be rigorously examined with well designed studies." Data from various studies do, of course, support several major effects or functions of fiber and of wheat bran.

The Canadian report, for example, concluded that "there are three and possibly four specific physiological effects for different types of dietary fiber: 1) regulating colonic function, 2) normalizing serum lipid levels, 3) attenuating the postprandial glucose response and perhaps 4) suppressing appetite" (Health and Welfare Canada, 1985). This report also stressed that the effects of dietary fiber may not all be positive. Wheat fiber is but one type of dietary fiber, and its effects on physiological function and nutritional status are somewhat specific to wheat and wheat bran. Although various authors hold divergent views, there is general agreement that the most consistent physiological effects of wheat fiber involve alterations in intestinal function, which include an increase in fecal bulk and weight and a decrease in intestinal transit time. Wheat fiber is especially effective in the treatment of constipation and diverticular disease. The effects of wheat bran or fiber on cancer of the colon are, however, less clear, and data are inconclusive (Wisker et al, 1985).

A. Physiological Effects—Intestinal Function

The physiological effects of wheat fiber that most directly influence nutritional quality and health are discussed in this section. Fiber in general (Leeds, 1985), as well as starch (Shetty and Kurpad, 1986), can increase fecal bulk, and wheat bran is one of the most effective fecal-bulking agents (Anderson, 1985; Health and Welfare Canada, 1985). The results of several studies evaluating the relationship between fecal weight and fecal bulk (volume) suggest that fecal volume is the more sensitive indicator (Health and Welfare Canada, 1985). However, because of the inadequacy of techniques for measuring volume, fecal weight is most often measured.

In human studies comparing the effectiveness of comparable quantities of fiber from various fruits, vegetables, and cereal grains, Wisker and co-workers (1985) showed that cereal grains were most effective in increasing fecal bulk. The cereals studied included whole wheat, wheat bran, and rye.

The selected data summarized in Table V show clearly that, in human studies with controlled diets, whole wheat bread and wheat bran markedly increased fecal weight relative to white bread, and coarse bran had a much greater effect than did fine bran. Factors that increased the effectiveness of wheat bran in increasing stool weight were coarse versus fine bran and raw versus cooked bran (Vahouny, 1985). The decrease in bulking capacity of cooked wheat bran may be

TABLE V
Effect of Wheat Fractions on Fecal Weight[a]

Source of Fiber	Number of Subjects	Intake of Dietary Fiber[b] (g/day)	Fecal Weight (g/day)
White bread[c]	8 females	···	101
Whole wheat bread[c]		···	176
White bread[d]	12 males	9	77
White bread + coarse bran[d]		22	140
White bread + fine bran[d]	4 males	22	102
Whole wheat bread[d]		22	143
White bread + coarse bran[d]		35	202

[a] Adapted from Wisker et al (1985).
[b] Neutral detergent fiber.
[c] Data from Feldheim (1980).
[d] Data from Van Dokkum et al (1983).

due to structural changes that lead to increased bacterial degradation and, subsequently, less fecal bulk. The particle size of wheat bran also influences degradation in the gut. Heller et al (1980) reported that about half of the hemicellulose in both coarse and fine wheat bran was degraded by humans, whereas 77 and 94% of the cellulose in fine and coarse wheat bran, respectively, was recovered in the feces. Thus, smaller particle size enhanced the degradation of cellulose in the gut and was also associated with decreased fecal bulk.

The mechanisms responsible for increased fecal weight and bulk are not entirely clear. Although water-holding capacity is often suggested, not all agree on the importance of in vitro water-holding capacity relative to fecal bulk (Wisker et al, 1985). Decreased intestinal transit time would, however, be expected to decrease water absorption by the intestinal lumen, resulting in higher moisture in stools. Other possible mechanisms include susceptibility to bacterial digestion and the effect on bacterial growth, and the production of osmotically active bacterial metabolites (Vahouny, 1985). The production of short-chain fatty acids may, through their osmotic action, affect fecal weight and transit time. Although these fatty acids may be mainly absorbed in the rectum, their influence could be significant in the other portions of the large intestine before they are absorbed (Wisker et al, 1985).

In contrast to water-soluble fibers, such as pectin, which slow gastric emptying and increase small intestine transit time, foods containing more insoluble fiber, such as wheat bran, decrease transit time. Wheat bran mainly affects events in the small intestine and colon (Anderson, 1985). Colon disorders, especially constipation, are effectively and consistently treated with wheat fiber or bran. Consumption of wheat bran or whole wheat foods improves bowel function and relieves constipation (Health and Welfare Canada, 1985; Vahouny, 1985). Cereal fibers, including not only wheat but also maize and rye, with their relatively insoluble dietary fiber, are more effective than fruits and vegetables in increasing fecal bulk, decreasing intestinal transit time, and improving bowel function. In addition, wheat fiber may possibly reduce the incidence of diverticular disease (Health and Welfare Canada, 1985).

Evidence for the protective effect of wheat fiber against colon cancer is less clear (Anderson, 1985). Although colon and breast cancer are more closely linked to dietary factors than are most other forms of cancer, the data are mainly epidemiological and, at best, suggestive. Since carcinoma of the colon is the third most common form of fatal cancer in the United States, clearer data leading to an understanding of the etiology of this disease are of utmost importance. Of interest is the observed negative correlation between intake of pentose-containing fiber (pentosans) and colon cancer reported by Bingham et al (1979).

In light of the complexity of the issue, Wisker et al (1985) have proposed three hypotheses whereby dietary fiber could function as a protective factor against colon cancer. First, contact of carcinogens with the gut wall diminishes and the concentration of carcinogens decreases because of shortened transit time and increased fecal bulk. Second, dietary fiber reduces the conversion of bile acids into potential carcinogens in the large intestine. Third, fiber alters the ratio of anaerobic to aerobic bacteria, thereby affecting the degradation of bile acids. Additional research is needed to clarify the role of fiber in the etiology of colon cancer.

B. Digestion and Absorption

Digestion and fermentation of wheat bran by monogastrics are slight or minimal (Anderson, 1985) and are influenced by factors that alter its physical state, such as particle size. Ehle and co-workers (1982) increased the degradation of wheat bran by reducing particle size. In general, more lignified cell walls, such as those of wheat bran, are less susceptible to fermentation in the gut.

A well-designed study by Sandberg et al (1982) investigated the effects of passage through the stomach and small intestine on the digestion of wheat fiber and its influence on other nutrients. Otherwise healthy subjects who had undergone an ileostomy were fed 15 g of wheat bran. Without the benefit of colonic bacteria, 75–100% of the hemicellulose and the components, arabinose and xylose, was recovered in the ileostomy. The degradation that does occur appears to take place in the colon. In contrast, between about 40 and 75% of the phytic acid consumed was degraded in the small intestine. Thus, the major sites of digestion or degradation of wheat fiber and phytic acid are different. Phytic acid becomes hydrolyzed sufficiently early in the digestive process to allow for the absorption of at least some of the minerals with which it is most likely to form complexes. Experiments using synthetic phytic acid should always be interpreted with some caution because phytic acid in situ is part of a more complex biological system and would not necessarily be expected to undergo hydrolysis and function in the same way as the synthetic preparation.

Wheat fiber has various effects on the digestion and absorption of starch, protein, and lipids. Of the several possible mechanisms cited by Vahouny (1985) to explain the effects of fiber on nutrient availability, the following are relevant to wheat fiber: effects on intestinal motility and intestinal transit, altered pH in the stomach and intestine, effects of entrapment of digestive enzymes and food nutrients, and accessibility of absorbable nutrients to intestinal mucosa.

Wheat fiber often decreases the digestibility of macronutrients, especially starch, and increases fecal excretion of nitrogen and fat (Anderson, 1985; Leeds, 1985). Vahouny (1985) evaluated the decrease in digestibility in the broader context of increased fecal energy loss, from about 60 to more than 300 kcal/day, associated with an increased consumption of cereal fiber. The energy loss was mainly protein and fat.

Several authors have stressed the difficulty of interpreting the increased fecal excretion of protein (N) and fat associated with consumption of cereal fiber (Anderson, 1985; Vahouny, 1985). Specifically, Vahouny (1985) suggested that it is unclear whether fecal nitrogen and fat are of bacterial origin or whether they reflect altered nutrient metabolism and absorption. He suggested that the fecal energy losses are mainly due to bacterial proliferation. Proposed sources of fecal nitrogen include bacterial products, mucosal cell debris, and intestinal secretions. Some of the unmetabolized nitrogen would be available for bacterial growth and development. Vahouny also stated that fiber sources may contain enzyme inhibitors that decrease protein utilization through their effect on proteases.

The recognition of unmetabolizable nitrogen and the presence of protease inhibitors suggest that some portion of the excreted nitrogen resulting from the consumption of wheat bran or whole wheat foods was originally wheat protein. Excreted nitrogen has been shown to be mainly amino acid nitrogen (Saunders and Betschart, 1980). A study of normal and gnotobiotic rats (Betschart et al, 1983) attempted to determine the contributions of wheat bran and microbial protein to excreted fecal nitrogen. Neither apparent nor true nitrogen digestibility of the AACC soft white or hard red wheat brans differed significantly between normal and germ-free rats; true nitrogen digestibility ranged from 71 to 75%. Since no microbial nitrogen was available to and thus excreted by the gnotobiotic rats, and since nitrogen digestibility of bran in normal and gnotobiotic rats was similar, the nitrogen excreted in the feces in this study appears to represent unavailable wheat protein. Experiments with intrinsically labeled wheat protein could yield definitive data regarding this question.

In addition to affecting protein and fat excretion, wheat bran has been reported to decrease starch digestibility. This decrease in digestibility may be due to the inhibition of enzyme activity as well as to a decrease in the time during which the surface area of starch is exposed to hydrolytic enzymes (Anderson, 1985). From a broader perspective, many factors play a major role in the rate, site, and quantity of carbohydrate digested and absorbed. These factors include the food form (physical state, processing), fiber, starch-nutrient (protein or fat) interactions, and the presence of antinutrients such as phytates, tannins, and other enzyme inhibitors (Jenkins et al, 1985). Other factors, such as the physical and chemical state or nature of the starch and the quantity of starch consumed, could also be included. This section reviews selected studies relating to the digestion of wheat starch and then discusses the more complex topic of carbohydrate metabolism, including glucose tolerance, glycemic indexes of wheat foods and diets, and the role of carbohydrates in the diets of diabetics.

Recently, much discussion and debate has centered on the presence of some indigestible starch in cereal grains, including wheat. Although it is not a previously unknown phenomenon, some authors have recently argued that the portion of indigestible starch that reaches the large intestine or hindgut and is subsequently fermented and metabolized by microorganisms has physiological properties similar to those of dietary fiber (Berry, 1986). In most cereals, as in wheat, indigestible starch constitutes a minor portion of the starch ingested. A well-designed rat study by Björck et al (1986) investigated the effect of digestion of wheat starch, wheat flour, and white wheat bread in the small intestine. An antibiotic, nebacitin, was used to significantly reduce microbial activity in the hindgut. Pure wheat starch, starch in wheat flour milled from a Danish winter wheat (69.5% extraction, 1.8% N, dry-weight basis), and bread baked from this flour were almost completely digested and absorbed in the rat small intestine. As a result of baking, however, a small amount of the starch (0.6–0.9%) was resistant to in vitro enzymatic digestion unless it was initially solubilized with potassium hydroxide. In vivo data also indicated that this portion of the starch was not digested in the small intestine but was readily degraded by the hindgut microorganisms.

The physical nature and composition of the food, including the effects of processing and preparation, can influence starch digestibility. Spaghetti starch is digested more slowly than that of white bread. Both processing and type or source of wheat may be involved. The effect of heat processing on starch through gelatinization and retrogradation can both improve and diminish digestibility (Berry, 1986). Swedish workers reported that thermal processing such as extrusion cooking, drum drying, autoclaving, popping, and steam flaking of whole wheat and white flour did not significantly affect the in vitro digestion of starch (Siljestroem et al, 1986). Particle size also appears to affect starch digestion. Brown rice is more slowly digested than white rice, and pumpernickel bread made with whole rye kernels produced a flatter glucose response curve than did rye bread made with flour (Jenkins et al, 1985).

The nature of the starch—that is, the amylose or amylopectin content—influences starch digestion. The higher the amylose content, the more slowly the starch is digested. Amylopectin, with its open, branched structure, appears to be more susceptible to enzymatic degradation than the more compact amylose. Jenkins and co-workers (1982) reported that the starch in lentils, with a higher amylose content, was digested more slowly than that in white bread. Such data support an "ease of enzymatic accessibility" hypothesis for high-amylopectin starches.

Both fiber and phytic acid influence differences in the rate and extent of starch digestion in whole grain and milled white flours. Fiber diminishes in vitro activity of pancreatic enzymes (Roksson et al, 1982). Other in vitro studies showed that phytic acid decreases α-amylase digestion of starch (Knuckles and Betschart, 1987). Yoon et al (1983) added phytic acid to wheat bread and found that in vitro digestibility of starch decreased. Thus, reported differences in digestibility of whole wheat and milled wheat fractions and foods involve more than the single factor of relative fiber content.

The quantity of wheat food consumed also appeared to influence the quantity of undigested carbohydrate in infants (about six to 18 months old). A linear relationship was reported between fecal carbohydrate content and consumption of pasta at levels of 25, 50, and 75% of calories consumed (MacLean et al, 1979). Although the linear relationship was significant, the increases in energy loss were relatively small; for example, the incremental energy loss on the 75% pasta diet was about 3% greater than that of the 50% pasta diet.

Modification of the morphology of intestinal epithelial cell membranes depends on the source of dietary fiber. Scanning electron micrographs showed that a 10% wheat bran diet had little, if any, effect on the number of epithelial cells in weanling rats, in contrast to bile-sequestering fiber, which did have an effect (Vahouny, 1985).

C. Carbohydrate Metabolism—Glycemic Index

Although closely related to digestion and the factors that affect digestion, glucose tolerance and sensitivity to insulin warrant additional discussion because of recent research. Jenkins and co-workers (1980) developed the concept of the glycemic index (GI) to evaluate the influence of foods and diets on postprandial glycemia. The GI is the area under the blood glucose curve 2 hr after ingestion of a 50-g carbohydrate sample of food, expressed as a percentage of the response to 50 g of a reference carbohydrate such as glucose (Jenkins et al, 1981).

Factors that affect digestion of carbohydrate are suggested to be the key factors influencing the GI. In addition to the factors cited by Jenkins et al (1985)—food form, fiber content, starch-nutrient interactions, presence of antinutrients, and nature of the starch—cell walls, accessibility of starch to enzymatic digestion, and the physical mix of carbohydrates with other components in the diet should be considered. Given the number, complexity, and potential interaction of these factors affecting postprandial glycemic and insulin response, it is not surprising that earlier hypotheses, which proposed that dietary fiber content was the critical or determining factor, have come into question (Jenkins et al, 1983; Leeds, 1985). From a more integrated perspective, the inconsistencies in data from attempts to correlate the GI solely with dietary fiber would be expected in the evaluation of such a complex process. For example, GIs of pulses and legumes have generally ranged from 20 to 50% of that of wheat bread; the GI of all-bran breakfast cereal is similar to that of sucrose, whereas the GI of shredded wheat is similar to that of white bread. The GI of spaghetti is lower than that of white bread, and the GI of white bread is similar to that of whole wheat bread (Jenkins et al, 1981, 1983). These data have been interpreted as supporting the absence of any effect of fiber (Anderson, 1985; Leeds, 1985), when it may be that the fiber effect is one of several variables involved.

Although the effects of the more water-soluble fibers are more pronounced than those of wheat bran and whole wheat products, wheat fractions and foods are the focus of the selected data presented here. The data are mixed on the effects of wheat fiber. In several studies, wheat fiber or bran influenced GI to a very limited extent (Jenkins et al, 1978; Bosello et al, 1980) or not at all in healthy subjects (Wahlqvist et al, 1979) as well as diabetics (Cohen et al, 1980). Although the long-term effects of wheat bran on GI were variable (Cohen et al, 1980), others reported a progressive reduction of meal-induced glucose response (Villaume et al, 1984). Vahouny (1982) suggested that wheat bran may bring about some adaptive changes in the intestine because improved glucose metabolism and increased apparent insulin sensitivity were observed in subjects one to two days after cessation of a diet containing wheat bran.

Some of the differences in the GIs of healthy and diabetic subjects are shown in Table VI. Among 62 commonly consumed foods and sugars, vegetables produced the largest increase in blood glucose ($70 \pm 5\%$), followed in decreasing order by breakfast cereals ($65 \pm 5\%$), cereals and cookies ($60 \pm 3\%$), fruit ($50 \pm 5\%$), dairy products ($35 \pm 1\%$), and dried legumes ($31 \pm 3\%$) (Jenkins et al, 1981).

The data in Table VI suggest that several factors influence the GI. The differences between whole wheat and white flour products were small relative to the lower indexes associated with spaghetti and pastry. Other factors, such as type of wheat flour (durum), drying to low moisture content, and the presence of fat (pastry), seem to decrease the GI. Corn-based cereal produced the highest, oat porridge the lowest, and wheat cereals intermediate GIs. The GI of potatoes varied with composition and processing; that of dried legumes was consistently low. Instantizing improved starch digestibility of potatoes, whereas several factors, including the presence of phytic acid, amylose, and dietary fiber, and possibly the nature of the cell walls surrounding legume carbohydrate, were associated with decreased apparent digestion. Of interest are the relatively high GI of honey, the intermediate values for sucrose, and the low value for fructose. In general, the relatively high GIs of wheat and corn products may be modified by consumption of legumes and guar crispbread (Jenkins et al, 1980, 1982).

Results from the studies of healthy subjects should be considered in light of

TABLE VI
Glycemic Index of Foods and Meals[a]

Food or Meal	Number of Subjects	Glycemic Index[b]
Cereal products		
Bread, white	10	69 ± 5
Bread, whole wheat	10	72 ± 6
Pastry	5	59 ± 6
Rice, white	7	72 ± 9
Rice, brown	7	66 ± 5
Spaghetti, white	6	50 ± 8
Spaghetti, whole wheat	6	42 ± 4
Sponge cake	5	46 ± 6
Breakfast cereals		
All-bran	6	51 ± 5
Cornflakes	6	80 ± 6
Porridge oats	6	49 ± 8
Shredded wheat	6	67 ± 10
Vegetables		
Frozen peas	6	51 ± 6
Potato, instant	8	80 ± 13
Potato, new	8	70 ± 8
Potato, sweet	5	48 ± 6
Dried legumes		
Beans, baked	7	40 ± 3
Beans, kidney	6	29 ± 8
Beans, soy	7	15 ± 5
Peas, blackeye	6	33 ± 4
Peas, chick	6	36 ± 5
Lentils	7	29 ± 3
Fruit		
Apple, Delicious	6	39 ± 3
Banana	6	62 ± 9
Orange	6	40 ± 3
Sugars		
Fructose	5	20 ± 5
Glucose	35	100
Maltose	6	105 ± 12
Sucrose	5	59 ± 10
Miscellaneous		
Honey	6	87 ± 8
Candy (Mars bars)	6	68 ± 12
Potato chips	6	51 ± 7
Test meals		
Whole wheat bread and cheese	6	100
Whole wheat bread and guar crispbread	6	51
Whole wheat bread and soybeans	6	65
Guar crispbread and soybeans	6	25
Soybeans and lentils	6	29
Cornflakes and toast	6	108

[a] Data on single foods from Jenkins et al (1981); data on test meals from Jenkins et al (1980). Subjects given single foods were healthy; subjects given test meals were diabetic. Food portions had 50 g of carbohydrate; test meals included 42–44 g of carbohydrate.
[b] Mean ± standard error of the mean.

the tea, milk, and tomatoes consumed at the same time as the selected foods tested (Jenkins et al, 1981). Some or all of these other foods may have affected the rate and extent of carbohydrate digestion. The authors reported a significant negative correlation between GI and the fat content ($r = -0.386$, $P < 0.01$) and between GI and the protein content ($r = -0.523$, $P < 0.001$) of the food. No relationship between GI and fiber or sugar content was detected. Given the influence of fat and protein, it would be of interest to evaluate these data in terms of the relationship of total fat and protein intake to the GI, especially when milk intake was high.

In contrast to the data in Table VI, Crapo et al (1980) found that 11 subjects with impaired glucose tolerance had a similar glucose response to white potatoes and dextrose, but the GIs for rice and white bread were somewhat lower. Neither the composition of food nor dietary fiber content alone effectively predicts GI because this response seems to be the result of the interaction of many factors that affect the rate of carbohydrate digestion and absorption.

Collier et al (1986) investigated the possibility of predicting glycemic response of non-insulin-dependent diabetics to mixed meals from the GIs obtained for individual foods. With diets containing 50% carbohydrate, 30% fat, and 20% protein, the GI of meals based on potatoes, white bread, white rice, spaghetti, or barley with lentils was highly correlated ($r = 0.988$) with calculated GI.

As with most acute tests, data for GI must eventually be validated in more long-term studies. In addition, quicker in vitro methods would provide useful data for screening foods. Using a method in which digestive juices are incubated with the substrate inside a dialysis bag, Jenkins et al (1982) showed a strong correlation between glucose collected from the medium surrounding the dialysis bag and GI for various foods.

On the basis of research data, several major diabetes associations from the United States, Canada, and England have suggested that "there may be definite benefits from increasing the consumption of high fiber, carbohydrate foods" (Health and Welfare Canada, 1985). Although this encompasses both soluble and insoluble fiber, it does indicate the positive effect of high-carbohydrate, high-fiber diets on glycemic response.

Bierman (1985) summarized the trend toward recommending higher-carbohydrate and lower-fat diets within a fixed caloric content. During the 50-year period from 1930 to 1980, the recommended carbohydrate intake for diabetics quadrupled, with a ratio of simple to complex carbohydrates of 1.0:1.8, whereas the recommended fat intake decreased to 40% of the level recommended in 1930. Wheat fractions and foods appear to be beneficial in the diet of diabetics.

D. Lipid Metabolism

Interest in the role of wheat and wheat foods in lipid metabolism has increased because of the observation that some forms of dietary fiber are associated with a reduction in serum lipid levels. Because serum lipids, especially cholesterol, have been identified as one of several risk factors in coronary heart disease, dietary alterations that reduce circulating levels of serum lipids are of continuing interest to researchers.

The influence of dietary fiber, including wheat fiber, on serum lipids is the subject of several recent reviews (Kies and Fox, 1977; Anderson, 1985; Brown and Karmall, 1985; Judd and Truswell, 1985; Vahouny, 1985; Wisker et al, 1985). In addition, professional organizations and governmental agencies have developed recommendations on the role of dietary fiber in lipid metabolism (IFT, 1979; Vahouny, 1982; Health and Welfare Canada, 1985).

The source and nature of dietary fiber are critical determinants of its efficacy in reducing serum lipid levels. The reported effects of wheat bran and whole wheat on blood lipid patterns are variable. Wheat bran does not appear to alter serum lipids. Cholesterol-lowering effects are most often associated with gelling, mucilaginous, viscous fibers, such as pectin and guar gum, as well as legumes and oat products.

The importance of pectin in altering blood lipids is clearly exemplified in a recent review by Judd and Truswell (1985). In an extensive table in this review summarizing the effects of fiber on plasma and fecal lipids in humans, pectin from various sources was the only source of dietary fiber listed. In the 12 studies cited with pectin, the decrease in plasma cholesterol was, in general, greater with hyperlipidemic than with healthy subjects and greater in shorter studies, that is, those lasting two to three weeks rather than four to eight weeks. These observations suggest that hyperlipidemic subjects are more responsive, that studies should be of sufficient length, and that data obtained from healthy and hyperlipidemic subjects should be reported as such and not combined.

The variable influence of wheat bran and whole wheat bread on serum cholesterol levels relative to pectin and oat products is shown in Table VII. In the studies summarized, which lasted four or five weeks, the effects of wheat bran and whole wheat bread on total cholesterol and low-density lipoprotein (LDL) were not encouraging. In contrast, pectin (consumed in relatively small doses) and oat products consistently decreased total cholesterol without decreasing high-density lipoprotein (HDL). Thus, the ratio of HDL to LDL increased, which is a favorable effect because LDL has been identified as atherogenic.

A critical review of 32 human studies that used wheat bran as the source of dietary fiber indicated that mean plasma total cholesterol was reduced in only eight of the studies (Judd and Truswell, 1985). In the majority of the studies, plasma cholesterol either rose or was unchanged. Most of the studies that reported no change or an increase used more rigorous experimental conditions, such as crossover design, control-test-control, long experimental periods, and frequent measurements of serum cholesterol. In contrast, among the eight studies that reported a cholesterol-lowering effect of wheat bran, only one was a significant result (Avgerinos et al, 1977), one reported that hard wheat but not soft wheat bran had an effect (Munoz et al, 1979), and others used the more tenuous one-way design, that is, control followed by the test period. Finally, these studies made relatively few measurements of plasma cholesterol. The merits of developing rigorous, standardized protocols and the difficulties of comparing data obtained from studies with different protocols are obvious.

Several mechanisms have been proposed for the effect of viscous polysaccharides on serum cholesterol. The increase in fecal excretion of bile acids is one explanation. Data for the gel-forming, viscous sources of dietary fiber show an association between bile acid excretion and lower serum cholesterol levels. Difficulty occurs, however, when data for wheat bran are also

TABLE VII
Effect of Dietary Fiber on Serum Lipids[a]

Source of Dietary Fiber	Amount of fiber (g/day)	Number of Subjects[b]	Length of Study (wk)	Cholesterol[c]			TG[c,f]	Reference
				Total	HDL[d]	LDL[e]		
Pectin	15	9	3	190 (224)	...	...	95 (94)	Kay and Truswell (1977)
	9	15	5	158 (171)	63 (62)	...	...	Stasse-Wolthuis et al (1980)
Wheat bran	24	27	5	197 (194)	...	...	97 (97)	Weinreich et al (1977)
	40	20	4	147 (189)	...	...	...	Mathur et al (1977)
	50	9	4	168 (161)	53 (52)	...	112 (113)	McDougall et al (1978)
Soft wheat	26	6	4	163 (166)	...	101 (103)	...	Munoz et al (1979)
Hard wheat	26	6	4	149 (168)	...	83 (105)	...	Munoz et al (1979)
	35	7	4	169 (187)	51 (59)	104 (111)	82 (106)	Van Berge-Henegouwen et al (1979)
	45	20	...	No change	...	...	...	Liebman et al (1983)
Hard wheat bread	12[g]	8	3	+15	...	...	+4	Van Dokkum (1978)
Oat bran	100	8	1.5	234 (269)	48 (49)	159 (184)	...	Kirby et al (1981)
Oat flakes	125	10	3	187 (204)	60 (57)	...	74 (78)	Judd and Truswell (1981)

[a] Adapted from Wisker et al (1985).
[b] All subjects were healthy.
[c] Expressed in milligrams per 100 ml. Control values are in parentheses.
[d] High-density lipoprotein.
[e] Low-density lipoprotein.
[f] Triglycerides.
[g] Neutral detergent fiber.

examined (Table VIII); increases in bile acid excretion are not associated with commensurate decreases in serum cholesterol. Thus, bile acid excretion may, in part, explain decreases in serum cholesterol but does not appear to be the controlling mechanism. Wisker et al (1985) summarized other possible mechanisms, including reduced absorption of cholesterol and reduced postprandial levels of insulin. As Judd and Truswell (1985) suggested, "it would appear that diets high in fiber-containing foods—such as fruits, vegetables and legumes—may be of benefit to the general population, irrespective of whether the effects are due to replacement of saturated fat and sugar in the diet or to modification in rate of absorption." Since viscous polysaccharides are most effective in lowering levels of blood lipids, various new bread products containing guar have been formulated (Judd et al, 1983; Burley et al, 1985).

E. Weight Loss

Dietary fiber, including wheat bran, may have some potential in the management of weight loss. This proposed effect is derived from the potential influence of fiber on several aspects of food intake and nutrient availability (Vahouny, 1982), including texture and palatability of food, energy density of food (diet), rate of satiation, extent of bulking, interference with nutrient availability, altered hormonal response, and altered thermogenesis.

The role of wheat and wheat foods in weight loss is, however, ill defined. Data on weight loss are conflicting and often obtained from short-term studies. The suggested effects on weight loss are often deduced from effects on satiety, decreased caloric intake, and increased fecal excretion of energy in the form of fat and nitrogen (Health and Welfare Canada, 1985; Leeds, 1985; Vahouny, 1985; Wisker et al, 1985). More research effort seems to have been devoted to indirect indexes of weight loss and the mechanisms involved than to actual studies of weight loss. The tenuous nature of the data in this area is reflected in the conclusion of the Canadian Expert Advisory Committee on Dietary Fiber

TABLE VIII
Influence of Dietary Fiber on Steroid Excretion and Serum Cholesterol[a]

Source of Dietary Fiber	Change in Fecal Excretion of Bile Acids (%)	Change in Total Serum Cholesterol (%)	Reference
Pectin	+34	−15	Jenkins et al (1976)
	+33	−13	Kay and Truswell (1977)
	+51 (male) +53 (female) }	−8	Stasse-Wolthuis et al (1980)
Wheat bran	+90	...	Cummings et al (1976)
	−49	0	Tarpila et al (1978)
	−23 (male) +41 (female) }	+8	Stasse-Wolthuis et al (1980)
Oat bran	+51	−13	Kirby et al (1981)
Oat flakes	+35	−8	Judd and Truswell (1981)

[a] Adapted from Wisker et al (1985).

(Health and Welfare Canada, 1985) that different types of dietary fiber have three and possibly four specific physiological effects. The possible fourth mechanism was appetite suppression.

Few long-term, well-controlled studies have been conducted to evaluate the effects of wheat fiber on weight loss. A high-fiber bread was associated with weight loss on an energy-restricted diet (Mickelsen et al, 1979). Caloric intake was also reduced slightly in subjects consuming high-fiber bread, compared with those consuming white bread. Grimes and Gordon (1978) reported that 10 of 12 normal-weight subjects consumed less energy from whole meal than from white bread, even though whole meal bread leaves the stomach sooner than white bread.

Energy density seems to be important in influencing the number of calories consumed. Duncan et al (1983) found that normal and obese individuals on low-energy-density diets (unrefined carbohydrates) consumed about half the calories consumed by those on high-energy-density diets. In general, subjects consuming whole wheat or high-fiber breads feel fuller than those consuming white bread, and satiety value is greater for the former group (Grimes and Gordon, 1978). Thus, the effects of whole wheat breads are due to the lower caloric density, the bulking effect (including distension of the stomach), and the increased time of chewing, all of which decrease the quantity of food needed to satisfy the appetite. These factors are viewed as regulating food intake and inducing satiety (Heaton, 1980).

A recent review of studies conducted at weight-loss clinics summarized the effects of various forms of dietary fiber on weight loss (Health and Welfare Canada, 1985). The results of these studies, in which caloric intake was restricted to 1,000–1,200 kcal/day, were conflicting. Wheat bran had no effect on weight loss, whereas guar gum was more effective than oat bran. In reviewing the literature, Krothiewski and Smith (1985) also concluded that gel-forming fibers, such as guar gum and pectin, are more effective in promoting weight reduction than are non-gel-forming fibers, such as wheat bran. If one of the major modes of action of wheat bran is to decrease total consumption of calories by suppressing appetite, the effects of wheat bran would be more evident in diets fed ad libitum than in calorie-restricted diets.

The decrease in apparent digestibility of fat and protein associated with an increased intake of whole wheat or wheat bran has been the rationale for presuming that fewer calories are available. However, increased fecal excretion does not necessarily imply decreased availability of calories, for several reasons, and furthermore, the decrease in apparent digestibility is too small to have a significant impact on the total number of available calories (Health and Welfare Canada, 1985; Krothiewski and Smith, 1985; Vahouny, 1985).

In summary, most of the studies to date used to support the benefits of wheat fiber in weight loss have reported indirect evidence, such as decreased consumption or increased excretion of calories. Thus, because of lack of reasonable evidence that wheat fiber induces weight loss, it would appear advisable to use dietary fibers "as supportive measures rather than as pharmacological agents in weight reducing programs" (Health and Welfare Canada, 1985).

V. MINERALS

A. Content

Many reports in the literature on the mineral content of wheat and wheat fractions present data for only one or a limited selection of wheat types or varieties. The data summarized in Table IX were taken from studies or reports that examined a broader spectrum of wheat samples. The concentration in the bran and germ of the trace minerals reported in this table ranges from about two to more than five times that in the wheat kernel. In contrast, the milled flour contains from one third to one tenth the mineral content of the wheat kernel. The aleurone cells, which are a component of bran, contain about 60% of the total minerals in the wheat kernel (Fig. 1). The data in Table IX also illustrate the broad range of values found in wheat (Millers' National Federation, 1972). Data reported by Lorenz and Loewe (1977) show that although the mineral contents of hard and soft wheat are similar, hard wheat generally contains at least 10% more than soft wheat.

Data on U.S. and Canadian flour samples (Table X) illustrate the differences

TABLE IX
Mineral Content (mg/kg) of Wheat and Wheat Fractions[a]

Wheat Sample	Fe	Zn	Cu	Mn	Se
Wheat[b]	18–31	21–63	1.8–6.2	24–37	0.04–0.71
Bran	74–103	56–141	8.4–16.2	72–144	0.10–0.75
Germ	41–58	<100–144	7.2–11.8	101–129	0.01–0.77
Flour	3.5–9.1	3.4–10.5	0.62–0.63	2.1–3.5	0.01–0.45
Hard wheat[c]	33 ± 6	26 ± 3	4.7 ± 1.2	42 ± 3	···
Soft wheat[c]	29 ± 4	23 ± 8	4.4 ± 1.0	37 ± 6	···

[a] Adapted from Burk and Solomons (1985) and Turnlund (1982).
[b] Ranges for several varieties of wheat. Data from Miller's National Federation (1972).
[c] Mean ± standard deviation. Data from Lorenz and Loewe (1977).

TABLE X
Mineral Content (mg/100 g) of Commercially Milled U.S. and Canadian Wheat Flours[a]

Laboratory	Flour Type	Ca	Mg	Fe	Zn
1	All[b]	14.2 ± 2.6	22.8 ± 7.9	0.99 ± 0.36	0.74 ± 0.21
	Hard	14.3 ± 2.8	26.1 ± 6.5	1.10 ± 0.33	0.80 ± 0.18
	Soft	14.0 ± 2.2	15.8 ± 6.0	0.75 ± 0.29	0.60 ± 0.22
2	All[b]	15.6 ± 2.9	26.0 ± 7.8	1.88 ± 0.37	1.23 ± 0.33
	Hard	15.6 ± 3.1	28.8 ± 6.7	1.94 ± 0.35	1.28 ± 0.34
	Soft	15.5 ± 2.5	20.2 ± 6.9	1.77 ± 0.41	1.13 ± 0.26
3	All[b]	17.7 ± 3.0	28.6 ± 8.5	1.29 ± 0.42	0.83 ± 0.26
	Hard	17.7 ± 3.2	32.2 ± 7.0	1.40 ± 0.42	0.88 ± 0.20
	Soft	17.5 ± 2.7	20.8 ± 6.1	1.06 ± 0.35	0.72 ± 0.36

[a] Adapted from Lorenz et al (1980). Data expressed on moisture-free basis. Data shown are means ± standard deviations.
[b] Includes hard, soft, baker's bread, family/all-purpose, hearth, cake, and cookie-cracker flours.

reported among laboratories. The values reported by laboratory 3 are consistently higher than those of laboratory 1. This study also confirmed that, with the exception of calcium, the mineral content was higher in hard wheat than in soft wheat. The distribution of calcium is different from that of other minerals. Only about 25% of calcium is in the aleurone cell layer, and 50% is in the endosperm. In contrast, less than 10% of magnesium and zinc is located in the endosperm, and about 70 and 50%, respectively, are present in the aleurone layer (MacMasters et al, 1971).

Davis et al (1984) reported on the mineral content of one variety, Century, grown in 13 locations. The ranges for mineral levels were as follows (mg/kg): iron, 51–133; zinc, 24–57; copper, 1–8; and manganese, 35–65. In general, the range of values is broader for Century grown at different locations than for the various wheat samples (Table IX) reported by the Millers' National Federation (1972).

These data on the mineral content of wheat emphasize the many factors that can influence mineral concentration. Type and variety of wheat, field location, milling methods, and analytical methods may all affect trace mineral content of wheat and wheat foods. Mahoney (1982) presented a comprehensive listing of the iron, zinc, copper, and manganese content of various wheat foods.

B. Bioavailability

Information on the bioavailability of minerals in diets containing wheat foods is often conflicting and contradictory. The data obtained from studies on bioavailability of minerals are influenced by many variables. Some of these factors are the nutritional status and requirements of the subjects; the presence of other foods in the diet, which may improve (protein, peptides, ascorbic acid) or impair (phytate, fiber, oxalate) bioavailability of minerals; production and milling effects; form of mineral consumed; level of minerals of interest as well as other minerals in the diet; length of experimental treatment period; and adaptation by subjects (Erdman, 1981; INACG, 1982; Harland and Morris, 1985; Health and Welfare Canada, 1985; Vahouny, 1985; Wisker et al, 1985). Many of these factors have been summarized on the bases of effects on intestinal absorption and various mechanisms involved (Burk and Solomons, 1985).

The difficulty of interpreting data on bioavailability of minerals in wheat foods is further complicated by differences of opinion on the role of fiber-phytate interactions (Hallberg, 1981; Kelsey, 1982; Forbes and Erdman, 1983; Harland and Morris, 1985). Data presently available have not clarified the relative impact of fiber and phytate on the bioavailability of calcium, magnesium, iron, and zinc.

Because of the many sources of variability, data from different studies are seldom compared in this section; rather, relative differences in bioavailability of minerals within studies are emphasized. Although much useful information may be gleaned from animal and in vitro studies, the discussion focuses mainly on human studies.

Bioavailability of minerals in wheat food diets may be approached from at least two perspectives. First, there is the question of whether whole wheat or wheat bran impairs the availability of minerals in the diet. Second, there is the question of the availability of minerals contained in the wheat kernel and contributed by wheat foods. The latter question is often difficult to answer directly and is usually addressed indirectly. For example, if bioavailability of minerals decreases when whole wheat or bran is added to the diet, the conclusion is that minerals in the aleurone cell layer are not readily available. However, if the total quantity of minerals absorbed (not excreted) is greater, the presumption is that at least some of the minerals contributed by wheat are available. Intrinsically labeled wheat would provide clearer evidence of the true bioavailability of minerals in wheat.

In general, the minerals in cereal grains are not readily absorbed (INACG, 1982). The iron in wheat flour of 60–80% extraction is an exception. The problem is that little iron is present in milled flour, and therefore the total quantity of iron absorbed from flour is low.

Turnlund (1982) discussed the importance of examining data in terms of quantity absorbed as well as the percent absorbed (Table XI). The percent zinc absorbed from white bread was more than twice that from whole wheat bread, but since whole wheat bread contained more than three times as much zinc as white bread, the absolute quantity of zinc absorbed was greater from the whole wheat bread (Sandstrom et al, 1980).

The addition of $ZnCl_2$ to both bread diets illustrates several difficulties (Sandstrom et al, 1980). Inorganic salts such as $ZnCl_2$ are absorbed differently than are minerals that are an integral part of the wheat kernel. Also, the level of zinc intake was much higher when $ZnCl_2$ was added (Table XI). Thus, although more zinc was absorbed with both breads when $ZnCl_2$ was added, the percent absorbed fell dramatically, to approximately one third and one half the values from the unsupplemented white and whole wheat breads, respectively. The absorption of added $ZnCl_2$ was quite different with the two breads; about 10% was absorbed from white bread, compared to less than 3% from whole wheat bread. This observation supports the relative efficacy of fortifying white bread rather than whole wheat bread.

Van Dokkum et al (1982) reported on the differences in absorption of several minerals from a broader perspective. Their data showed that the effect of wheat bran on mineral balance varied with the mineral and the particle size of the bran. The apparent balances for calcium and copper decreased markedly when coarse bran was added to white bread, in spite of the higher intake with the added bran. When the bran was milled to a finer particle size, the apparent balance for all minerals was higher than with the coarse bran, and for all but copper the balance

TABLE XI
Zinc Absorption in Humans from White and Whole Wheat Bread[a]

Bread	Source of Zinc Added as $ZnCl_2$ (mg/meal)	Total Zinc Content of Meal (mg)	Absorption Percent	Quantity
White	0	0.4	38.2	0.15
Whole wheat	0	1.3	16.6	0.22
White	3.1	3.6	13.2	0.48
Whole wheat	2.2	3.5	8.2	0.29

[a] Adapted from Sandstrom et al (1980) as cited by Turnlund (1982).

was higher than with the white bread alone. Data on mineral balance with fine bran suggest that the bran contributed significant quantities of calcium, magnesium, iron, and zinc. In addition, apparent balance for all minerals except calcium was markedly improved when whole wheat rather than white bread was consumed (Table XII).

The increased mineral absorption associated with fine wheat bran and whole wheat bread, as opposed to the negative effects of coarse bran, emphasizes the importance of particle size. The bran component of whole wheat flour is also finer than coarse bran. Increased milling to produce smaller particles would be expected to increase the disruption of aleurone cell walls. This could increase the accessibility of nutrients contained in aleurone cells as they pass through the gastrointestinal tract and improve their potential for being absorbed.

Although some animal experiments confirmed the conclusions of Van Dokkum et al (1982), a comprehensive study of commercial variety breads using the hemoglobin repletion technique indicated that wheat bran interfered with the bioavailability of iron (Ranhotra et al, 1979). In contrast, the human study of Van Dokkum et al (1982) clearly showed that iron balance was only slightly decreased with coarse bran and markedly increased with fine bran and whole wheat. Although the specific objectives of these two studies were different, they illustrate the limitations of applying conclusions from small animal studies to humans without some qualifications.

Turnlund recently reported the effects of white and whole wheat breads on the bioavailability of several minerals in humans (Turnlund, 1987; Turnlund et al, 1987). Through the use of stable isotopes, absorption of several minerals was investigated. Data were obtained in an 84-day metabolic ward study with young men, using a crossover design. Results indicated that retention of magnesium, zinc, and copper was only slightly lower with the whole wheat bread diet. Iron retention appeared to improve with whole wheat bread, whereas calcium retention fell significantly, from +151 to −36 mg/day. These data on calcium agree with those reported by Van Dokkum et al (1982) and suggest that the consumption of large quantities of whole wheat bread may adversely affect calcium nutriture (Turnlund, 1987).

Results from the studies of Sandstrom et al (1980), Turnlund et al (1987), and Van Dokkum et al (1982) are not in complete agreement on the change in absorption of minerals in the presence of wheat bran or whole wheat. They do, however, conclude that there is selected interference in the percent of mineral absorption in the presence of whole wheat, although the absolute amount

TABLE XII
Effect of Various Wheat Fractions on Mineral Balance[a]

Bread	Number of Subjects	Length of Study (wk)	Fiber[b] (g/day)	Apparent Balance (mg/day)				
				Ca	Mg	Fe	Zn	Cu
White	12	3	9	+14	−8	+0.8	−0.6	+0.2
White + coarse bran	12	3	22	−42	−3	+0.7	−0.4	−0.3
White + fine bran	4	3	22	+48	+10	+1.3	+0.2	+0.1
Whole wheat	4	3	22	−10	+12	+2.2	+0.6	+0.4

[a] Adapted from Wisker et al (1985); data from Van Dokkum et al (1982).
[b] Neutral detergent fiber.

absorbed may be higher than in white bread. In circumstances in which wheat bran or fiber diminishes the bioavailability of minerals, there is considerable interest in the mechanisms involved. Two factors, dietary fiber and phytate, have received much attention and considerable support for research.

C. Phytate and Fiber

Several authors have reviewed phytate-fiber-mineral interactions (Kelsey, 1982; Forbes and Erdman, 1983; Frohlich, 1984; Harland and Morris, 1985; Morris, 1986; Turnlund, 1987). The early hypothesis of McCance and Widdowson (1942) that phytate was mainly responsible for the effects of whole wheat bread on calcium and iron balance was reinforced by the initial studies of Reinhold et al (1973). In subsequent studies, when enzymatic hydrolysis of phytate during baking was considered along with the determination that fiber had the ability to bind some cations, Reinhold et al (1976, 1981) suggested that fiber per se could be responsible for mineral imbalances. This conclusion was supported by Kies and co-workers, who reported that cellulose and hemi-cellulose increased fecal excretion of divalent cations in adolescent boys (Drews et al, 1979).

Frohlich (1984) provided a broader perspective through a comprehensive review of the effects of cereals on the bioavailability of minerals in normal humans. The review included studies of chlorine, sodium, potassium, phosphorus, magnesium, iron, zinc, and copper. Cereal grain fiber negatively affected mineral balance in 35 of the 55 human studies reported between 1942 and 1984, no change was observed in 18, and the balance improved in two. Notwithstanding a critical interpretation of these studies, it seems clear that cereals impaired mineral balance in a significant number of human studies.

Harland and Morris (1985) and Morris (1986) thoroughly reviewed selected studies on the influence of phytate and fiber on mineral bioavailability. Anderson et al (1983) conducted a human study to determine the effect of fiber in whole wheat bread. The intake of minerals and phytate was kept constant, and they were added to white bread to match the levels in whole wheat bread. Six subjects consumed 200 g of the respective breads for 24 consecutive days; data were gathered for the last 12 days of the period. Neither fecal excretion nor balance of calcium, iron, or zinc was affected by the type of bread consumed. These data suggest that fiber does not play a major role in decreasing the availability of minerals in the presence of whole wheat bread and imply that differences reported by others are due to the level of phytate.

Morris and co-workers conducted several studies to determine the influence of phytate on mineral bioavailability (Morris and Ellis, 1982, 1985). Wheat bran (untreated) or dephytinized wheat bran was incorporated into muffins, which were consumed by humans for 15-day periods. Apparent absorption of iron, zinc, and magnesium with the wheat bran muffins was significantly lower for the first five days than for the last 10 days. However, the apparent balance reported for the last 10 days of the study was not significantly affected by dephytinizing wheat bran. The apparent balance for calcium, iron, zinc, copper, and manganese, in general, was positive for both types of bran.

Most of the phytic acid in yeast-leavened whole grain wheat and rye breads is hydrolyzed during the fermentation of the dough (Wisker et al, 1987). Adding wheat bran to the bread produced higher levels of phytic acid than did brewers' spent grains. The effects of phytase during fermentation and the effect of adding wheat bran to whole grain breads were clearly demonstrated in this study.

The effect of phytate was also studied by adding sodium phytate at two levels (1.7 and 2.9 g/day) to muffins in addition to the natural level present in muffins (0.5 g/day) (Morris et al, 1984, 1985). Increasing the level of phytate had no effect on the balance of iron, copper, and manganese. However, absorption of calcium, magnesium, and zinc tended to decrease. Turnlund et al (1984, 1985) also reported mixed effects of sodium phytate on the balance of various minerals. The effect of α-cellulose and sodium phytate on zinc and copper absorption was studied in normal humans with the use of stable isotopes. As reported by Morris et al (1984, 1985), sodium phytate (2.3 g, consumed with the diet) did not influence copper absorption but decreased zinc absorption by about 50%. In contrast, α-cellulose had no effect on zinc absorption. Data from these studies suggest that the effect of sodium phytate on mineral balance varies, depending on the element.

Information reported to date indicates the difficulty of separating the effects of natural products such as wheat bran or whole wheat into major components, such as fiber and phytate. Caution should be exercised when comparing data on the effects of sodium phytate with those of phytate present as a component of the wheat kernel. When both fiber and phytate are present, a negative effect on mineral balance is more likely. The influence of wheat bran on availability of minerals may be due to a synergistic effect of fiber and phytate. The degree to which aleurone cell walls are disrupted may also be important in determining availability of minerals in whole wheat and wheat bran. Regrinding wheat bran or whole wheat to a smaller particle size may have several effects. With a smaller particle size, minerals present in the aleurone cells may be more available for absorption. Phytate, also present in the aleurone cells, may be more susceptible to hydrolysis by phytase in yeast-leavened breads and under conditions in the gastrointestinal tract. Smaller particle size is also associated with a diminished physiological effect of fiber and its potential for binding or occluding minerals. Other factors in wheat, such as starch, have also been reported to influence bioavailability of minerals (Turnlund, 1987). Thus, the availability of minerals from wheat depends on several factors.

The phenomenon of adaptation may well be yet another factor that has led to conflicting reports and conclusions in the literature. The data of Morris and co-workers (Morris and Ellis, 1982; Morris et al, 1984) strongly suggest that adaptation occurred in their studies. Absorption was lower in the first five days of the 15-day periods for calcium, iron, zinc, and manganese with both untreated and dephytinized wheat bran. Longer treatment periods, an initial adaptation period, and crossover designs are approaches that help to minimize the influence of adaptation on experimental data and results (Anderson et al, 1983; Turnlund et al, 1987).

General recommendations for the consumption of whole wheat and wheat bran products usually suggest that these foods be consumed in moderation. A modest intake of whole wheat and bran products would not be expected to adversely affect the mineral status of an adequately nourished population. There is concern, however, for populations that are clearly or marginally deficient, consume large quantities of high-fiber foods, and are among the high-risk groups, i.e., children, pregnant and lactating women, and the elderly, especially postmenopausal women (Harland and Morris, 1985; Health and Welfare Canada, 1985; Vahouny, 1985). Suggested ways to improve absorption of minerals in the diet include the consumption of meat, sufficient protein, and ascorbic acid.

VI. VITAMINS

Interest in the influence of fiber, especially wheat bran, on bioavailability of vitamins has been somewhat less than the corresponding interest in minerals. The more complex question of the availability of vitamins in bran and whole wheat products has been addressed even less frequently. Conclusions have often been based on indirect evidence (Kahlon et al, 1986; Keagy and Oace, 1986; Hudson et al, 1988). Methodological problems often complicate the investigations, especially in the area of biochemical and analytical methods required to assess nutritional status and vitamin content.

A. Content

Vitamins are not distributed uniformly throughout the wheat kernel. Vitamin distribution in the endosperm, scutellum, embryo, aleurone layer, and other fractions differs more than the corresponding distribution of minerals. As illustrated in Fig. 2, the endosperm contains less than 5% of the thiamine and more than 40% of the pantothenic acid. The aleurone layer contains 32% of the thiamine and more than 80% of the niacin. Data reported by Keagy et al (1980) showed that commercially milled flour contained 32, 35–42, and 15% of the whole grain thiamine, riboflavin, and pyridoxine, respectively. Although these values are higher than those previously reported, they illustrate the differences in distribution of the various vitamins.

In addition, vitamins vary tremendously in their molecular structure, stability, and sensitivity to light, temperature, and moisture. Therefore, altering the conditions of milling, processing, or storage of wheat and wheat foods can have profoundly different effects on the stability and bioavailability of specific vitamins.

The mean values of 63 unenriched commercially milled flours from the United States and Canada provide a meaningful base with which to compare other values (Keagy et al, 1980). In these samples, all vitamins except thiamine were present at slightly higher levels in the hard than in the soft wheats (Table XIII). Commercial milling removed about 68% of the thiamine, 58–65% of the riboflavin, and 85% of the pyridoxine contained in whole wheat. Data such as these are used along with other information to develop guidelines for the fortification of flour or replacement of nutrients based on the original content of the whole wheat kernel.

B. Bioavailability

Fat-soluble vitamins are influenced by some of the same factors and mechanisms of absorption as is dietary fat. Interest in fat-soluble vitamins is due, in part, to their potential interference with bile salt reabsorption as well as the effect of various forms of fiber on bioavailability (Kasper et al, 1979; Kelsey, 1982).

Reports on the effects of fiber on bioavailability of fat-soluble vitamins are varied, and conclusions are often contradictory. Kasper and co-workers (1979) reported an increase in postprandial uptake of vitamin A into the serum of men who consumed a vitamin A-rich meal containing large quantities of wheat bran, compared to those consuming a control meal. There have been reports that 10% levels of various dietary fibers caused a decrease in liver and serum levels of vitamin A in rats, whereas others have reported no effect of 22% coarse or fine wheat bran on plasma levels of vitamin A in rats (Kahlon et al, 1986a). Results and conclusions may vary for several reasons. Serum lipid levels alone often provide tenuous data and are not necessarily correlated with other indexes such as liver stores. Different assays or indexes are often used to evaluate vitamin A status, and various sources of dietary fiber have different effects. For example, viscous fibers interfere with normal digestibility and absorption of lipids and, thus, often also influence fat-soluble vitamins (Vahouny, 1982).

In contrast to the findings reported for vitamin A, wheat bran does affect the bioavailability of vitamin E. Coarse wheat bran was reported to significantly decrease liver stores of vitamin E in rats (Kahlon et al, 1986a). Thus, consumption of large quantities of wheat bran (22%) appears to have a detrimental effect on vitamin E status.

The effect of wheat bran on lipid digestibility and absorption has been examined by Canadian and U.S. scientists (Mongeau et al, 1986; Kahlon et al, 1986b). Initial studies concluded that the addition of 16 and 20% wheat bran to rat diets significantly increased fecal fat excretion compared to a diet containing 8% bran (Mongeau et al, 1986). Kahlon et al (1986b) reported a small increase in the absolute quantity of fat excreted and a lack of a significant increase in fecal fat in rats fed 22% coarse or fine wheat bran. If fat-soluble vitamins are excreted in proportion to total fat, these data suggest that wheat bran does not have a major deleterious effect on bioavailability of fat-soluble vitamins.

Data on the influence of wheat bran or whole wheat products on bioavailability of water-soluble vitamins are somewhat limited. Available studies indicate that bioavailability of riboflavin and nicotinic acid is diminished in diets containing cereal brans or whole wheat bread (Hollman, 1954; Carter

TABLE XIII
Vitamin Content of Commercially Milled U.S. and Canadian Flours[a]

Nutrient	All Flours	Wheat Type	
		Hard	Soft
Protein (%)	12.6	13.8	9.9
Ash (%)	0.52	0.54	0.47.
Vitamins (mg/100 g)			
Thiamine	0.146	0.144	0.150
Riboflavin	0.040	0.041	0.038
Niacin	1.42	1.48	1.31
Pyridoxine	0.046	0.052	0.034
Folacin	0.019	0.021	0.016
Pantothenic acid	0.37	0.39	0.34

[a] Adapted from Keagy et al (1980). Values are given on a moisture-free basis.

and Carpenter, 1982). Niacin occurs mainly in the "bound" form in whole wheat and wheat bran. The lack of availability of bound niacin is often little affected by processing such as toasting. In a study of selected wheat products, Hepburn (1971) found that most of the niacin in the enriched products was in the free form, compared to only about half of the niacin in the whole wheat and unenriched products. The nutritional advantages of niacin enrichment are clear.

There has been somewhat more interest in vitamin B_6, or pyridoxine, as reflected in both human and animal studies (Leklem et al, 1980; Lindberg et al, 1983; Kies et al, 1984; Betschart et al, 1986; Hudson et al, 1988). Wheat bran decreases the bioavailability of vitamin B_6 in humans (Lindberg et al, 1983; Kies et al, 1984). Leklem et al (1980) also reported that B_6 is less available from whole wheat bread than from white bread enriched with B_6. Recent data obtained with rats did not show a significant decrease in B_6 bioavailability when 20% wheat bran was included in the diet (Hudson et al, 1988). Species differences, the presence of the cecum in rats, and the level of wheat bran consumed may partly explain differences in the results.

Vitamin B_6 in enriched white bread is less available than in nonfat dry milk (Gregory, 1980). Other workers found that B_6 is more available from animal tissues (beef and tuna) than from plant-based foods (peanut butter and soybeans) (Leklem et al, 1980; Kabir et al, 1983a; Miller et al, 1985). The lower availability of vitamin B_6 in plant foods may be a result of fiber or glycosylated B_6, both of which are present in plant but not in animal foods (Kabir et al, 1983b, 1983c).

Wheat bran, high-extraction flours, and dietary fiber, in general, do not interfere with the bioavailability of folate in humans (Babu and SriKantia, 1976; Russell et al, 1976; Ristow et al, 1982; Keagy et al, 1987). No interference with folate absorption was observed in Iranian men consuming traditional high-extraction of flat breads (Russell et al, 1976). In vitro studies showed no binding effects of bran with folate, and increased levels of bran in chick diets did not reduce plasma or liver folate concentrations (Ristow et al, 1982). A rat study showed no detrimental effect of cellulose, xylan, pectin, or wheat bran on the utilization of folic acid added to the diet (Keagy and Oace, 1986). Thus, wheat bran and whole wheat products do not diminish the bioavailability of folate and in some instances appear to contribute additional available folate (Keagy and Oace, 1986).

Based on the evidence to date, there need be little concern about the negative effects of wheat bran or whole wheat foods on the bioavailability of vitamins in well-nourished populations. In populations with marginal intakes, especially of niacin and vitamin B_6, it would seem advisable not to consume excessive quantities of wheat bran.

VII. IMPROVING THE NUTRITIONAL QUALITY OF WHEAT

Nutritional quality depends on both the quantity and the quality of nutrients present. The quantity of nutrients in wheat is affected by wheat variety, growing location, cultural and agronomic practices employed, and milling or extraction rate. In addition, fortification of wheat products effectively increases nutrient concentration. The availability of nutrients depends on many factors, including the form of the nutrient, extraction rate, fiber, phytate, and the nature of cell walls surrounding the nutrients. The presence of other components in wheat or in the diet that improve or impair availability, the level of the nutrient consumed, the age and physiological status of the subject, and processing methods also influence nutrient availability.

The number and complexity of the variables that affect nutritional quality of wheat and wheat foods offer many possible routes for improving nutritional quality. For example, selection or modification of wheat varieties, alteration of agronomic practices or degree of milling, addition of specific ingredients or nutrients to wheat flour or wheat products, or any combination of these approaches could be explored.

The critical question, however, is "for whom and for what specific purpose should nutritional quality be improved?" Once the specific objective has been defined, it is essential to critically evaluate the various means of improving nutritional quality. It is important to assess the relative efficacy in achieving the nutritional goal as well as the cost-effectiveness of alternative approaches. Only then can the appropriate specific solution be selected. Too often specific approaches are selected before alternative approaches have been evaluated, and research and development continue long after other solutions have proved to be more effective. In addition, the limits of some approaches often dictate that the nutritional impact, as evaluated by other methods, would not be significant. An example has been the continuation of plant breeding and genetic engineering programs designed to increase the lysine level in wheat in light of the limited impact of lysine fortification at significantly higher levels. Thus, workers in various disciplines should continue to evaluate the potential of their approach in relation to results obtainable by alternative approaches.

It is, of course, important that the perceived nutritional deficiency of wheat be real. Several studies have shown that when wheat protein from white bread is the major source of protein, adults are maintained in positive nitrogen balance. Thus, food products and programs designed to improve the protein quality of wheat should focus on subpopulations for which wheat protein, consumed as part of a total diet, would be inadequate for growth and development. Focusing on specific target groups is also more cost-effective.

The data in this chapter indicate that one of the simplest and more cost-effective ways to improve the nutritional quality of wheat and wheat foods is to increase the extraction rate and/or use whole wheat flour. Widdowson (1975) described the nutritional implications of higher extraction rates. As shown in Table XIV, nutrients are generally present in equal or higher concentrations in whole wheat than in enriched white flours and breads. Although the percentage of absorption of minerals is lower in whole wheat flours, the total quantity of minerals available increases because of the higher absolute concentration in whole wheat. In general, although protein digestibility is decreased, wheat fiber does not have major deleterious effects on either mineral or vitamin bioavailability. Calcium is an exception, in that calcium balance either decreased or was negative when subjects consumed whole wheat bread. Vitamin B_6 is somewhat less available from whole wheat bread, but most of the water-soluble vitamins are not negatively affected by the presence of wheat fiber. Likewise, the fat-soluble vitamins A and E are relatively unaffected, with the exception of vitamin E in the presence of coarse wheat bran. Thus, the additional nutrients present in whole wheat products and the physiological effect of the fiber on fecal bulk and transit time suggest that Western industrialized populations would continue to benefit from the consumption of more whole wheat foods.

Fortification is an effective means of increasing the level of nutrients in white flour. Hepburn and McQuilkin (1981) summarized data on white enriched and whole wheat bread (Table XIV). The white breads were enriched with calcium, iron, thiamine, riboflavin, and niacin. With the exception of these nutrients, the remaining minerals and vitamins were two to six times more concentrated in whole wheat bread. Sodium is mainly contributed by the salt in bread formulations.

Interest in the fortification of white flour has centered on the nutritional impact, the technical feasibility, and the health, social, and economic ramifications of fortification programs, as well as on efforts to increase the number and level of nutrients used in fortification (National Research Council, 1974, 1978; Elton, 1975; Ranum, 1980; Vetter, 1982; Cook and Welsh, 1987). Cook and Welsh (1987) summarized the impact of fortification on subjects two years old or older who were part of the 1977–1978 National Food Consumption Surveys. Enrichment and fortification of cereal grain products provided 18–20% of the total intake of riboflavin, niacin, and iron and 32% of the thiamine intake of these subjects. These nutrients contributed both a larger portion of the

TABLE XIV
Nutrient Content of White and Whole Wheat Bread[a]

Nutrient	White (Enriched)		Whole Wheat	
	Number of Samples	Mean ± SD	Number of Samples	Mean ± SD
Proximate composition (g/100 g)				
Water	713	36.9 ± 1.7	96	38.5 ± 2.3
Protein	788	8.3 ± 0.6	117	9.6 ± 0.7
Total lipid	787	3.9 ± 0.9	117	4.2 ± 0.9
Total carbohydrate	⋯	48.4	⋯	45.4
Crude fiber	11	0.3 ± 0.2	29	1.5 ± 0.4
Neutral detergent fiber	1	0.6	7	3.0 ± 1.4
Ash	720	2.1 ± 0.3	76	2.3 ± 0.5
Minerals (mg/100 g)				
Calcium	717	126.0 ± 49.2	59	72.0 ± 24.8
Iron	703	2.8 ± 0.6	65	3.4 ± 0.8
Magnesium	446	22.0 ± 3.8	23	89.0 ± 29.6
Phosphorus	22	100.0 ± 20.4	2	200.0 ± 14.1
Potassium	30	110.0 ± 23.2	4	210.0 ± 28.5
Sodium	48	510.0 ± 157.4	9	640.0 ± 143.1
Zinc	432	0.6 ± 0.2	29	1.5 ± 0.4
Copper	48	0.14 ± 0.02	18	0.30 ± 0.06
Manganese	49	0.31 ± 0.07	17	1.90 ± 0.79
Vitamins (mg/100 g)				
Thiamine	641	0.47 ± 0.09	79	0.35 ± 0.11
Riboflavin	671	0.31 ± 0.10	55	0.21 ± 0.15
Niacin	638	3.75 ± 0.60	56	3.83 ± 0.80
Pantothenic acid	17	0.40 ± 0.06	8	0.80 ± 0.10
Vitamin B-6	33	0.04 ± 0.02	11	0.19 ± 0.03
Folacin	5	0.035 ± 0.005	5	0.052 ± 0.004

[a] Reprinted, with permission, from Hepburn and McQuilkin (1981).

recommended daily allowance and a higher proportion of the total intake in the diets of children than in adults. Wheat flour is, thus, an effective vehicle for fortification.

The technical feasibility of adding several other nutrients (vitamins A and B_6, folic acid, magnesium, and zinc) has been documented (National Research Council, 1978; Vetter, 1982). Other considerations, however, such as the effect of some of these nutrients on vulnerable or sensitive individuals, have curtailed initial efforts to expand the fortification of cereal grains.

Dietary guidelines have been developed in an effort to reflect the impact of food on general health and to reduce the risk of diseases that are associated with diet and nutrition. The U.S. Dietary Guidelines (USDA/USDHHS, 1985) recommend decreasing consumption of fat (especially saturated fat) and cholesterol and increasing consumption of starch and fiber. These guidelines are similar to those proposed in other national nutrition policies, including those of Norway (Royal Ministry of Health and Social Affairs, 1981–1982). An increase in consumption of foods high in complex carbohydrates and low in fat, such as wheat and wheat foods, would bring the diets of the populations of Western industrialized countries closer to the dietary guidelines.

In summary, whole wheat and its various fractions, white flour, bran, and germ are all nutritious foods that help to meet the nutritional and health needs of many. The relatively stable level of consumption of wheat foods in the United States during the past few decades suggests a need to increase consumer interest in and desire for wheat foods. Awareness of the healthful qualities of wheat foods, along with the development of new and attractive wheat food products, could contribute to increased consumption of wheat foods.

VIII. RESEARCH NEEDS

Future research on wheat would benefit if at least as much effort were devoted to health-related properties as to specific nutritional inadequacies; to availability of nutrients as to nutrient content; and to nutritional quality as to functional quality. Research and development should be directed to the use of new principles and procedures for fractionating and stabilizing wheat. In addition to traditional whole wheat and white flour, components such as the aleurone cell layer and germ would seem to have good potential marketability if they could be clearly separated. Research should also address the need to improve the stability of wheat fractions and products through storage at optimum conditions and new stabilization methods, especially for germ, bran, and whole wheat flour.

Research on the nutritional quality of wheat would also benefit from a more holistic approach to nutritional quality and health benefits. Rather than focus on one nutrient or class of nutrients, researchers should at least consider the impact of variables on total nutritional quality. Studies could also be improved by the development of more standardized methods and by the use of the dose-response approach rather than feeding selected but varying levels in different studies. A good portion of seemingly contradictory data in the literature may well be the result of differences in experimental diet, experimental design, length of treatment periods, and amount of nutrients fed.

Studies of wheat would also benefit from a more integrated approach between those interested in nutritional and those interested in functional quality. Greater coordination is also needed among the various disciplines investigating wheat quality. As with any commodity in which diverse disciplines are involved and differing objectives are operative, research on wheat would benefit from a clearer understanding of the impact of specific objectives on the overall marketability of wheat, which is both a nutritious food and a cereal with unique functional properties. That is, those working to improve nutritional quality should keep functional qualities in mind and those attempting to improve functional quality should not overlook nutritional qualities.

There is also a need for new and creative approaches to long-standing problems. For example, the work of Stone and co-workers (Stone, 1985; Fincher and Stone, 1986; Cheng et al, 1987) on the nature of cell walls in wheat could provide some new insights and avenues for research on improving nutritional quality. Characterization of cell walls would help to more effectively define methods of breaking down the cell walls to provide greater accessibility for digestion and absorption of nutrients.

Clearly defined, relevant objectives would greatly improve research on the nutritional quality of wheat. Although seemingly self-evident, deciding which aspect of nutritional quality to improve and for whom could assist in a more efficient use of resources. Studies driven by scientific interest in a specific nutrient or nutrients would be more appropriately based on a real nutritional need of a specific population group.

When relevant research needs have been defined, a systematic evaluation of the alternative means of obtaining the objective must follow. Considering questions such as whether certain nutrients might best be added, be obtained by blending wheat with another component, be increased by altering wheat genetically or culturally, or be increased by different milling and other processing practices is essential. Ease of approach and cost-effectiveness factor heavily in the selection process. As a result of this process, one or more viable approaches would be defined and pursued.

Through the development and use of new techniques, including stable isotope methodology, much progress has been made in expanding our understanding of the nutritional quality of wheat in the past decade. From another perspective, we may have just begun.

LITERATURE CITED

ANDERSON, H., NAVERT, B., BINGHAM, S. A., ENGLYST, H. N., and CUMMINGS, J. H. 1983. The effects of breads containing similar amounts of phytate but different amounts of wheat bran on calcium, zinc and iron balance in men. Br. J. Nutr. 50:503-510.

ANDERSON, J. W. 1985. Health implications of wheat fiber. Am. J. Clin. Nutr. 41:1103-1112.

AVGERINOS, G. C., FUCHS, H. M., and FLOCH, M. H. 1977. Increased cholesterol and bile acid excretion during a high fiber diet. Gastroenterology 72:1026-1030.

BABU, S., and SRIKANTIA, S. G. 1976. Availability of folate from some foods. Am. J.

Clin. Nutr. 29:376-379.

BERRY, C. 1986. Production of resistant starch in food by heat processing: The role of amylose retrogradation and its implications for dietary fibre determination. Pages 37-49 in: Dietary Fibers. R. Anado and T. F. Schweitzer, eds. Academic Press, London.

BETSCHART, A. A. 1978. Improving protein quality in bread—Nutritional benefits and realities. Adv. Exp. Med. Biol. 105:703-734.

BETSCHART, A. A. 1982. Protein content and quality of cereal grains and selected cereal foods. Cereal Foods World 27:395-401.

BETSCHART, A. A., HUDSON, C. A., and

IRVING, D. W. 1983. Nutritional quality of wheat bran—In vivo and histochemical studies. Pages 1109-1114 in: Progress in Cereal Chemistry and Technology, Vol. 5B. J. Holas and J. Kratochvil, eds. Elsevier, Amsterdam.

BETSCHART, A. A., HUDSON, C. A., and TURNLUND, J. R. 1985. Protein digestibility and nitrogen balance of white and whole wheat breads in humans. (Abstr.) Cereal Foods World 30:538-539.

BETSCHART, A. A., HUDSON, C. A., TURNLUND, J. R., KRETSCH, M. J., and SAUBERLICH, H. E. 1986. Protein utilization by young women on animal or plant protein diets at various levels of vitamin B-6 intake. (Abstr.) Cereal Foods World 31:590-591.

BIERMAN, E. L. 1985. Diet and diabetes. Am. J. Clin. Nutr. 41:1113-1116.

BINGHAM, S., WILLIAMS, D. R. R., COLE, T. J., and JAMES, W. P. T. 1979. Dietary fiber and regional large bowel cancer mortality in Britain. Br. J. Cancer 40:456-463.

BIRDSALL, J. J. 1985. Wheat foods: Nutritional implications in health and disease. Summary and areas for future research. Am. J. Clin. Nutr. 41:1172-1176.

BJÖRCK, I., NYMAN, M., PEDERSEN, B., SILJESTRÖM, M., ASP, N.-G., and EGGUM, B. O. 1986. On the digestibility of starch in wheat bread—Studies in vitro and in vivo. J. Cereal Sci. 4:1-11.

BODWELL, C. E. 1985. Amino acid content as an estimate of protein quality for humans. Pages 1-14 in: Digestibility and Amino Acid Availability in Cereals and Oilseeds. J. W. Finley and D. T. Hopkins, eds. Am. Assoc. Cereal Chem., St. Paul, MN.

BOLOURCHI, S., FRIEDMAN, M., and MICKELSEN, O. 1968. Wheat flour as a source of protein for adult human subjects. Am. J. Clin. Nutr. 21:827-834.

BOSELLO, O., OSTUZZI, R., ARMELLINI, F., MICCIOLO, R., and SCURO, L. A. 1980. Glucose tolerance and blood lipids in bran-fed patients with impaired glucose tolerance. Diabetes Care 3:46-49.

BROWN, W. V., and KARMALLY, W. 1985. Coronary heart disease and the consumption of diets high in wheat and other grains. Am. J. Clin. Nutr. 41:1163-1171.

BURK, R. F., and SOLOMONS, N. W. 1985. Trace elements and vitamins and bioavailability as related to wheat and wheat foods. Am. J. Clin. Nutr. 41:1091-1102.

BURLEY, V., LEEDS, A. R., ELLIS, P. R., and PETERSON, D. B. 1985. Wholemeal guar bread: Acceptability and efficacy combined. A study of blood glucose, plasma insulin and palatability in normal subjects. (Abstr.) Proc. Nutr. Soc. 43:48A.

CARPENTER, K. J. 1975. The nutritional value of wheat proteins. Pages 93-114 in: Bread—Social, Nutritional and Agricultural Aspects of Wheaten Bread. A. Spicer, ed. Applied Science Publ., London.

CARTER, E. G. A., and CARPENTER, K. J. 1982. The bioavailability for humans of bound niacin from wheat bran. Am. J. Clin. Nutr. 36:855-861.

CHENG, B.-Q., TRIMBLER, R. P., ILLMAN, R. T., STONE, B. A., and TOPPING, D. L. 1987. Comparative effects of dietary wheat bran and its morphological components (aleurone and pericarp-seed coat) on volatile fatty acid concentrations in the rat. Br. J. Nutr. 57:69-76.

COHEN, M., WU LEONG, V., SALMON, E., and MARTIN, F. I. R. 1980. Role of guar and dietary fiber in the management of diabetes mellitus. Med. J. Aust. 1:59-61.

COLLIER, G. R., WOLEVER, T. M. S., WONG, G. S., and JOSSE, R. G. 1986. Prediction of glycemic response to mixed meals in noninsulin-dependent diabetic subjects. Am. J. Clin. Nutr. 44:349-352.

COOK, D. A., and WELSH, S. O. 1987. The effect of enriched and fortified grain products on nutrient intake. Cereal Foods World 32:191-196.

CRAPO, P. A., KOLTERMAN, O. G., WALDECK, N., REAVER, G. M., and OLEFSKY, J. M. 1980. Postprandial hormonal responses to different types of complex carbohydrate in individuals with impaired glucose tolerance. Am. J. Clin. Nutr. 33:1723-1728.

CUMMINGS, J. H., HILL, M. J., JENKINS, D. J. A., PEARSON, J. R., and WIGGINS, H. S. 1976. Changes in fecal composition and colonic function due to cereal fiber. Am. J. Clin. Nutr. 29:1468-1473.

DAVIS, K. R., PETERS, L. J., CAIN, R. F., LeTOURNEAU, D., and McGINNIS, J. 1984. Evaluation of the nutrient composition of wheat. III. Minerals. Cereal Foods World 29:246-248.

DREWS, L. M., KIES, C., and FOX, H. M. 1979. Effect of dietary fiber on copper, zinc and magnesium utilization by adolescent boys. Am. J. Clin. Nutr. 32:1893-1897.

DUNCAN, K. H., BACON, J. A., and WEINSIER, R. L. 1983. The effects of high and low energy density diets on satiety, energy intake and eating time of obese and nonobese subjects. Am. J. Clin. Nutr. 37:763-767.

EHLE, F. R., ROBERTSON, J. B., and VAN SOEST, P. J. 1982. Influence of dietary fibers on fermentation in the human large intestine. J. Nutr. 112:158-166.

ELTON, G. A. H. 1975. The fortification of flour. Pages 249-258 in: Bread—Social, Nutritional and Agricultural Aspects of Wheaten Bread. A. Spicer, ed. Applied Science Publ., London.

ERDMAN, J. W., Jr. 1981. Bioavailability of trace minerals from cereals and legumes. Cereal Chem. 58:21-26.

FAO. 1970. Amino acid content and biological data on proteins. FAO Nutr. Stud. No. 24. Food Agric. Org. U.N., Rome.

FAO. 1973. Energy and protein requirements. FAO Nutr. Meet. Rep. Ser. No. 52. Food Agric. Org. U.N., Rome.

FAO. 1980. Food balance sheets, 1975–1977. Food Agric. Org. U.N., Rome.

FAO/WHO. 1985. Energy and protein requirements. WHO Tech. Rep. Ser. 724. World Health Org., Geneva.

FELDHEIM, W. 1980. Stuhlgewicht und transitzeit bei ballaststoffreicher und ballaststoffarmer kost. Pages 82-87 in: Pflanzenfasern-Ballaststoffe. H. Rottka, ed. Thieme Verlag, Stuttgart.

FINCHER, G. B., and STONE, B. A. 1986. Cell walls and their components in cereal grain technology. Pages 207-295 in: Advances in Cereal Science and Technology, Vol. VIII. Y. Pomeranz, ed. Am. Assoc. Cereal Chem., St. Paul, MN.

FINLEY, J. W., and HOPKINS, D. T., eds. 1985. Digestibility and Amino Acid Availability in Cereals and Oilseeds. Am. Assoc. Cereal Chem., St. Paul, MN.

FORBES, R. M., and ERDMAN, J. W., Jr. 1983. Bioavailability of trace mineral elements. Annu. Rev. Nutr. 3:213-231.

FROHLICH, W. 1984. Bioavailability of minerals from unrefined cereal products: In vitro and in vivo studies. Department of Food Chemistry, Chemical Center, University of Lund, Lund, Sweden.

GRAHAM, G. G., MORALES, E., CORDANO, A., and PLACKO, R. P. 1971. Lysine enrichment of wheat flour: Prolonged feeding of infants. Am. J. Clin. Nutr. 24:200-206.

GREGORY, J. F., III. 1980. Bioavailability of vitamin B-6 in nonfat dry milk and a fortified rice breakfast cereal product. J. Food Sci. 45:84-86.

GRIMES, D. S., and GORDON, C. 1978. Satiety value of wholemeal and white bread. Lancet 2:106-108.

HALLBERG, L. 1981. Bioavailability of dietary iron in man. Annu. Rev. Nutr. 1:123-147.

HARLAND, B. F., and MORRIS, E. R. 1985. Fibre and mineral absorption. Pages 72-82 in: Dietary Fibre Perspectives. A. R. Leeds, ed.

John Libbey, London.

HEALTH AND WELFARE CANADA. 1985. Report of the Expert Advisory Committee on Dietary Fibre. Minister of National Health and Welfare, Ottawa.

HEATON, K. W. 1980. Food intake regulation and fibre. Pages 223-238 in: Medical Aspects of Dietary Fiber. G. A. Spiller and R. M. Kay, eds. Plenum Medical, New York.

HELLER, S. N., HACKLER, L. R., RIVERS, J. M., VAN SOEST, P. J., ROE, D. A., LEWIS, B. A., and ROBERTSON, J. 1980. Dietary fiber. The effect of particle size of wheat bran on colonic function in young adult men. Am. J. Clin. Nutr. 33:1734-1744.

HEPBURN, F. N. 1971. Nutrient composition of selected wheats and wheat products. VII. Total and free niacin. Cereal Chem. 48:369-372.

HEPBURN, F. N., and McQUILKIN, C. 1981. Nutritional profile of variety breads. Pages 43-54 in: Variety Breads in the United States. B. S. Miller, ed. Am. Assoc. Cereal Chem., St. Paul, MN.

HOLLMAN, W. I. M. 1954. Pages 92-98 in: Med. Res. Council Spec. Rep. Ser. No. 287. Her Majesty's Stationery Office, London.

HUDSON, C. A., BETSCHART, A. A., and OACE, S. M. 1988. Bioavailability of vitamin B-6 from rat diets containing wheat bran or cellulose. J. Nutr. 118:65-71.

IFT. 1979. Dietary fiber. Scientific status summary of the Institute of Food Technologists' expert panel on food safety and nutrition. Food Technol. (Chicago) 33:1-5.

INACG. 1982. The Effects of Cereals and Legumes on Iron Availability. Report of the International Nutritional Anemia Consultative Group. The Nutrition Foundation, Washington, DC.

JENKINS, D. J. A., LEEDS, A. R., GASSULL, M. A., HOUSTON, H., GOFF, D. V., and HILL, M. J. 1976. The cholesterol-lowering properties of guar and pectin. Clin. Sci. Mol. Med. 51:8-9.

JENKINS, D. J. A., WOLEVER, T. M. S., LEEDS, A. R., GASSULL, M. A., HAISMAN, P., DILAWARI, J., GOFF, D. V., METZ, G. L., and ALBERTI, K. G. M. M. 1978. Dietary fibers, fiber analogues, and glucose tolerance: Importance of viscosity. Br. Med. J. 1:1392-1394.

JENKINS, D. J. A., WOLEVER, T. M. S., TAYLOR, R. H., BARKER, H. M., FIELDEN, H., and JENKINS, A. L. 1980. Effect of guar crisp-bread with cereal products and leguminous seeds on blood glucose concentration of diabetics. Br. Med. J. 281:1248-1250.

JENKINS, D. J. A., THOMAS, D. M.,

WOLEVER, M. S., TAYLOR, R. H., BARKER, H., FIELDEN, H., BALDWIN, J. M., BOWLING, A. C., NEWMAN, H. C., JENKINS, A. L., and GOFF, D. V. 1981. Glycemic index of foods: A physiological basis for carbohydrate exchange. Am. J. Clin. Nutr. 34:362-366.

JENKINS, D. J. A., THORNE, M. J., CAMELON, K., JENKINS, A. L., RAO, A. V., KALMUSKY, J., REICHERT, R., and FRANCIS, T. 1982. Effect of processing on digestibility and the blood glucose response: A study of lentils. Am. J. Clin. Nutr. 36:1093-1101.

JENKINS, D. J. A., WOLEVER, T. M. S., JENKINS, A., LEE, R., WONG, G. S., and JOSSE, R. 1983. Glycemic response to wheat products: Reduced response to pasta but no effect of fiber. Diabetes Care 6:155-159.

JENKINS, D. J. A., JENKINS, A. L., WOLEVER, T. M. S., THOMPSON, L. U., and RAO, A. V. 1985. Variations in the availability of carbohydrates. Pages 242-246 in: Proc. Int. Congr. Nutr., 13th. T. G. Taylor and N. K. Jenkins, eds. John Libbey, London.

JUDD, P. A., and TRUSWELL, A. S. 1981. The effect of rolled oats on blood lipids and fecal steroid excretion in man. Am. J. Clin. Nutr. 34:2061-2067.

JUDD, P. A., and TRUSWELL, A. S. 1985. Dietary fibre and blood lipids in man. Pages 23-39 in: Dietary Fibre Perspectives. A. R. Leeds, ed. John Libbey, London.

JUDD, P. A., LEEDS, A. R., ELLIS, P. R., APLING, C., and JEPSON, E. 1983. A new guar bread: An effective hypocholesterolaemic agent in subjects with hypercholesterolaemia. (Abstr.) Proc. Nutr. Soc. 42:120A.

KABIR, H., LEKLEM, J. E., and MILLER, L. T. 1983a. Measurement of glycosylated vitamin B-6 in foods. J. Food Sci. 48:1422-1425.

KABIR, H., LEKLEM, J. E., and MILLER, L. T. 1983b. Comparative vitamin B-6 bioavailability from tuna, whole wheat bread and peanut butter in humans. J. Nutr. 113:2412-2420.

KABIR, H., LEKLEM, J. E., and MILLER, L. T. 1983c. Relationship of the glycosylated vitamin B-6 content of food to vitamin B-6 bioavailability in humans. Nutr. Rep. Int. 28:709-716.

KAHLON, T. S., CHOW, F. I., HOEFER, J. L., and BETSCHART, A. A. 1986a. Bioavailability of vitamins A and E as influenced by wheat bran and bran particle size. Cereal Chem. 63:490-493.

KAHLON, T. S., CHOW, F. I., HOEFER, J. L., and BETSCHART, A. A. 1986b. Effect of dietary fiber and wheat bran particle size on digestibility and GI tract development.

(Abstr.) Cereal Foods World 31:590.

KASPER, H., RABAST, V., FASSL, H., and FEHLE, F. 1979. The effect of dietary fiber on postprandial serum vitamin A concentration in man. Am. J. Clin. Nutr. 32:147-149.

KAY, R. M., and TRUSWELL, A. S. 1977. Effect of wheat fibre on gastrointestinal function, plasma lipids and steroid excretion in man. Br. J. Nutr. 37:227-235.

KEAGY, P. M., and OACE, S. M. 1986. Folic acid utilization from high fiber diets in rats. J. Nutr. 114:1252-1259.

KEAGY, P. M., BORENSTEIN, B., RANUM, P., CONNOR, M. A., LORENZ, K., HOBBS, W. E., HILL, G., BACHMAN, A. L., BOYD, W. A., and KULP, K. 1980. Natural levels of nutrients in commercially milled wheat flours. II. Vitamin analysis. Cereal Chem. 57:59-65.

KEAGY, P. M., SHANE, B., and OACE, S. M. 1988. Folate bioavailability in humans: Effects of wheat bran and beans. Am. J. Clin. Nutr. 47:80-88.

KELSEY, J. L. 1982. Effects of fiber on mineral and vitamin bioavailability. Pages 91-103 in: Dietary Fiber in Health and Disease. G. V. Vahouny and D. Kritchevsky, eds. Plenum Press, New York.

KIES, C., and FOX, H. M. 1977. Dietary hemicellulose interactions influencing serum lipid patterns and protein nutritional status of adult men. J. Food Sci. 42:440-443.

KIES, C., KAN, S., and FOX, H. M. 1984. Vitamin B-6 bioavailability from wheat, rice and corn brans for humans. Nutr. Rep. Int. 30:483-496.

KIRBY, R. W., ANDERSON, J. W., SIELING, R., ROSE, E. D., CHEN, W. J. L., MILLER, R. E., and KAY, R. M. 1981. Oat bran intake selectively lowers serum low density lipoprotein cholesterol concentration. Am. J. Clin. Nutr. 34:824-829.

KNUCKLES, B. E., and BETSCHART, A. A. 1987. The effect of phytate and other myoinositol phosphate esters on α-amylase digestion of starch. J. Food Sci. 52:719-721.

KROTHIEWSKI, M., and SMITH, U. 1985. Dietary fibre in obesity. Pages 61-67 in: Dietary Fibre Perspectives. A. R. Leeds, ed. John Libbey, London.

KULP, K., RANUM, P. M., WILLIAMS, P. C., and YAMAZAKI, W. T. 1980. Natural levels of nutrients in commercially milled wheat flours. I. Description of samples and proximate analysis. Cereal Chem. 57:54-58.

LEEDS, A. R., ed. 1985. Dietary Fibre Perspectives—Reviews and Bibliography. John Libbey, London.

LEKLEM, J. E., MILLER, L. T., PERERA, A. D., and PEFFERS, D. E. 1980. Bioavailability of vitamin B-6 from wheat bread in humans. J. Nutr. 110:1819-1828.

LIEBMAN, M., SMITH, M. C., IVERSON, J., THYE, F. W., HINKLE, D. E., HERBERT, W. G., RITCHEY, S. J., and BRISKELL, J. A. 1983. Effects of coarse wheat bran fiber and exercise on plasma lipids and lipoproteins in moderately overweight men. Am. J. Clin. Nutr. 37:71-81.

LINDBERG, A. S., LEKLEM, J. E., and MILLER, L. T. 1983. The effect of wheat bran on bioavailability of vitamin B-6 in young men. J. Nutr. 113:2578-2586.

LINKO, P., COLONNA, P., and MERCIER, C. 1981. High-temperature, short-time extrusion cooking. Pages 145-235 in: Advances in Cereal Science and Technology, Vol. IV. Y. Pomeranz, ed. Am. Assoc. Cereal Chem., St. Paul, MN.

LORENZ, K., and LOEWE, R. 1977. Mineral composition of U.S. and Canadian wheats and wheat blends. Agric. Food Chem. 25:806-809.

LORENZ, K., LOEWE, R., WEADON, D., and WOLF, W. 1980. Natural levels of nutrients in commercially milled wheat flours. III. Mineral analysis. Cereal Chem. 57:65-69.

MacLEAN, W. C., Jr., DeROMANA, G. L., KLEIN, G. L., MASSA, E., MELLITS, E. D., and GRAHAM, G. G. 1979. Digestibility and utilization of the energy and protein of wheat by infants. J. Nutr. 109:1290-1298.

MacMASTERS, M. M., HINTON, J. J. C., and BRADBURY, D. 1971. Microscopic structure and composition of the wheat kernel. Pages 51-113 in: Wheat: Chemistry and Technology, 2nd ed. Y. Pomeranz, ed. Am. Assoc. Cereal Chem., St. Paul, MN.

MAHONEY, A. W. 1982. Mineral contents of selected cereals and baked products. Cereal Foods World 27:147-150.

MATHUR, M. S., SINGH, F., and CHADDA, V. S. 1977. Effect of bran on blood lipids. J. Assoc. Physicians India 25:275-278.

McCANCE, R. A., and WIDDOWSON, E. M. 1942. Mineral absorption of healthy adults on white and brown bread dietaries. J. Physiol. (London) 101:44-85.

McDOUGALL, R. M., YAKYMYSHYN, L., WALKER, K., and THURSTON, O. G. 1978. Effect of wheat bran on serum lipoproteins and biliary lipids. Can. J. Surg. 5:433-435.

MICKELSEN, O., MAKDANI, D. P., COTTON, R. H., TITCOMB, S. T., COLMEY, J. C., and GATTY, R. 1979. Effects of a high fiber bread on weight loss in college-age males. Am. J. Clin. Nutr. 32:1703-1709.

MILADI, S., HEGSTED, D. M., SAUNDERS, R. M., and KOHLER, G. O. 1972. The relative nutritive value, amino acid content, and digestibility of the proteins of wheat mill fractions. Cereal Chem. 49:119-127.

MILLER, L. T., LEKLEM, J. E., and SHULTZ, T. D. 1985. The effect of dietary protein on the metabolism of vitamin B-6 in humans. J. Nutr. 115:1663-1672.

MILLERS' NATIONAL FEDERATION. 1972. Millfeeds Manual. The Federation, Chicago.

MILNER, C. K., and CARPENTER, K. J. 1969. Effect of wet heat-processing on the nutritive value of whole-wheat protein. Cereal Chem. 46:425-434.

MONGEAU, R., BEHRENS, W. A., MADERA, R., and BRASSARD, R. 1986. Effect of dietary fiber on vitamin E status in rats: Dose-response to wheat bran. Nutr. Res. 6:215-224.

MORRIS, E. R. 1986. Phytate and dietary mineral bioavailability. Pages 57-76 in: Phytic Acid: Chemistry and Applications. E. Graf, ed. Pilatus Press, Minneapolis, MN.

MORRIS, E. R., and ELLIS, R. 1982. Phytate, wheat bran, and bioavailability of iron. Pages 121-141 in: Nutritional Bioavailability of Iron. C. Kies, ed. Am. Chem. Soc., Washington, DC.

MORRIS, E. R., and ELLIS, R. 1985. Bioavailability of dietary calcium, effect of phytate on adult men consuming non-vegetarian diets. Pages 63-72 in: Nutritional Bioavailability of Calcium. C. Kies, ed. Am. Chem. Soc., Washington, DC.

MORRIS, E. R., ELLIS, R., HILL, A. D., COTTRELL, S., STEELE, P., MOY, T., and MOSER, P. 1984. Trace element nutriture of adult men consuming three levels of phytate. Fed. Proc. Fed. Am. Soc. Exp. Biol. 43:846.

MORRIS, E. R., ELLIS, R., HILL, A. D., STEELE, P., COTTRELL, S., MOY, T., and MOSER, P. 1985. Magnesium and calcium nutriture of adult men consuming omnivore diets with three levels of phytate. Fed. Proc. Fed. Am. Soc. Exp. Biol. 44:1675.

MUNOZ, J. M., SANDSTEAD, H. H., JACOB, R. A., LOGAN, G. M., RECK, S. J., KLEVAY, L. M., DINTZIS, F. R., INGLETT, G. E., and SHUEY, W. C. 1979. Effects of some cereal brans and textured vegetable protein on plasma lipids. Am. J. Clin. Nutr. 32:580-592.

NATIONAL RESEARCH COUNCIL. 1974. Proposed fortification policy for cereal-grain products. Natl. Acad. Sci., Washington, DC.

NATIONAL RESEARCH COUNCIL. 1978. Proceedings of a workshop on technology of fortification of cereal-grain products, Washington, DC, May 16-17, 1977. U.S.

Dep. Commer. Natl. Tech. Inf. Serv., Washington, DC.

PEDERSEN, B., and EGGUM, B. O. 1983. The influence of milling on the nutritive value of flour from cereal grains. 2. Wheat. Plant Foods Hum. Nutr. 33:51-61.

RANHOTRA, G. S., LEE, C., and GELROTH, J. A. 1979. Bioavailability of iron in some commercial variety breads. Nutr. Rep. Int. 19:851-857.

RANUM, P. M. 1980. Note on levels of nutrients to add under expanded wheat flour fortification/enrichment programs. Cereal Chem. 57:70-72.

REINHOLD, J. G., HEDAYATI, H., LAHIMGARZADEH, A., and NASR, K. 1973. Zinc, calcium, phosphorus, and nitrogen balances of Iranian villagers following change from phytate-rich to phytate-poor diets. Ecol. Food Nutr. 2:157-162.

REINHOLD, J. G., FARADJI, B., ABADI, P., and ISMAIL-BEIGI, F. 1976. Decreased absorption of calcium, magnesium, zinc and phosphorus by humans due to increased fiber and phosphorus consumption as wheat bread. J. Nutr. 106:493-503.

REINHOLD, J. G., GARCIA, J. S., and GARZON, P. 1981. Binding of iron by fiber of wheat and maize. Am. J. Clin. Nutr. 34:1384-1391.

RISTOW, K. A., GREGORY, J. F., III, and DAMRON, B. L. 1982. Effects of dietary fiber on the bioavailability of folic acid monoglutamate. J. Nutr. 112:750-758.

ROKSSON, G., LUNDQUIST, I., and IHSE, I. 1982. Effect of dietary fiber on pancreatic enzyme activity in vitro: The comparison of viscosity, pH, ionic strength, absorption and time of incubation. Gastroenterology 82:918-924.

ROYAL MINISTRY OF HEALTH AND SOCIAL AFFAIRS. 1981–1982. Rep. No. 11 to the Storting. On the follow-up of Norwegian nutrition policy. Norwegian Government, Oslo.

RUSSELL, R. M., ISMAIL-BEIGI, F., and REINHOLD, J. G. 1976. Folate content of Iranian breads with the effect of their fiber content on the intestinal absorption of folic acid. Am. J. Clin. Nutr. 29:799-802.

SANDBERG, A.-S., HASSELBLAD, C., and HASSELBLAD, K. 1982. The effects of wheat bran on the absorption of minerals in the small intestine. Br. J. Nutr. 48:185-191.

SANDSTROM, B., ARVIDSON, B., CEDERBLAND, A., and BJORN-RASMUSSEN, E. 1980. Zinc absorption from composite meals. I. The significance of wheat extraction rate, zinc, calcium, and

protein content in meals based on bread. Am. J. Clin. Nutr. 33:739-745.

SAUNDERS, R. M., and BETSCHART, A. A. 1980. The significance of protein as a component of dietary fiber. Am. J. Clin. Nutr. 33:960-961.

SHETTY, P. S., and KURPAD, A. V. 1986. Increasing starch intake in the human diet increases fecal bulking. Am. J. Clin. Nutr. 43:210-212.

SILJESTRÖM, M., WESTERLUND, E., BJÖRCK, I., HOLM, J., ASP, N. G., and THEANDER, O. 1986. The effect of various thermal processes on dietary fiber and starch content of whole grain wheat and white flour. J. Cereal Sci. 4:315-323.

SPICER, A., ed. 1975. Bread—Social, Nutritional and Agricultural Aspects of Wheaten Bread. Applied Science Publ., London.

STASSE-WOLTHUIS, M., ALBERS, H. F. F., VAN JEVEREN, J. G. C., DEJONG, J. W., HAUTVAST, J. G. A. J., HERMUS, R. J. J., KATAN, M. B., BRYDON, G., and EASTWOOD, M. A. 1980. Influence of dietary fiber from vegetables and fruits, bran or citrus pectin on serum lipids, fecal lipids, and colonic function. Am. J. Clin. Nutr. 33:1745-1756.

STONE, B. A. 1985. Aleurone cell walls— Structure and nutritional significance. Pages 349-354 in: New Approaches to Research on Cereal Carbohydrates. R. D. Hill and L. M. Munck, eds. Elsevier Sci. Publ., Amsterdam.

TARPILA, S., MIETTINEN, T. A., and METASARANTA, L. 1978. Effects of bran on serum cholesterol, faecal mass, fat, bile acids and neutral steroids and biliary lipids in patients with diverticular disease of the colon. Gut 19:137-145.

TURNLUND, J. R. 1982. Bioavailability of selected minerals in cereal products. Cereal Foods World 27:152-157.

TURNLUND, J. R. 1987. Carbohydrate, fiber and mineral interactions. Pages 37-40 in: Nutrition 1987. AIN Symp. Proc. Am. Inst. Nutr., Bethesda, MD.

TURNLUND, J. R., KING, J. C., KEYES, W. R., GONG, B., and MICHAEL, M. C. 1984. A stable isotope study of zinc absorption in young men: Effects of phytate and α-cellulose. Am. J. Clin. Nutr. 40:1071-1077.

TURNLUND, J. R., KING, J. C., GONG, B., KEYES, W. C., and MICHAEL, M. C. 1985. A stable isotope study of copper absorption in young men: Effect of phytate and α-cellulose. Am. J. Clin. Nutr. 42:18-23.

TURNLUND, J. R., BETSCHART, A. A., KEYES, W. R., and ACORD, L. L. 1987. A stable isotope study of zinc bioavailability in young men from diets with white vs. whole wheat bread or beef vs. soy. Fed. Proc. Fed. Am. Soc. Exp. Biol. 46:879.

USDA. 1985. Agricultural Statistics. U.S. Dep. Agric., Washington, DC.

USDA/USDHHS. 1985. Dietary guidelines for Americans. Home Gard. Bull. 232. U.S. Dep. Agric., U.S. Dep. Health Hum. Serv., Washington, DC.

VAHOUNY, G. V. 1982. Conclusions and recommendations of the symposium on dietary fiber in health and disease, Washington, DC, 1981. Am. J. Clin. Nutr. 35:152-156.

VAHOUNY, G. V. 1985. Dietary fibers—Aspects of nutrition, pharmacology, and pathology. Pages 207-277 in: Nutritional Pathology—Pathobiochemistry of Dietary Imbalances. H. Sidransky, ed. Marcel Dekker, New York.

VAN BERGE-HENEGOUWEN, G. P., HUUBREGTS, A. W., VAN DE WERF, S., DEMACKER, P., and SCHADE, R. W. 1979. Effect of standardized wheat bran preparation on serum lipids in young healthy males. Am. J. Clin. Nutr. 32:794-798.

VAN DOKKUM, W. 1978. Zemelin in brood: Verteerbaarheid en invloed op het defecatie-patroon, de mineralenbalans en de serum lipid-concentraties bij de mens. Voeding-smiddelentechnologie 41(Nov.):18-21.

VAN DOKKUM, W., WESSTRA, A., and SCHIPPERS, F. A. 1982. Physiological effects of fibre-rich types of bread. I. The effect of dietary fibre from bread on the mineral balance of young men. Br. J. Nutr. 47:451-460.

VAN DOKKUM, W., PIKAAR, N. A., and THISSEN, J. T. N. M. 1983. Physiological effects of fibre-rich types of bread. II. Dietary fibre from bread: Digestibility by the intestinal microflora and water holding capacity in the colon of human subjects. Br. J. Nutr. 50:61-74.

VETTER, J. L., ed. 1982. Adding Nutrients to Foods: Where Do We Go from Here? Am. Assoc. Cereal Chem., St. Paul, MN.

VILLAUME, C., BECK, B., GARIOT, P., DESALME, A., and DEBRY, G. 1984. Long-term evolution of the effect of bran ingestion on meal-induced glucose and insulin responses in healthy men. Am. J. Clin. Nutr. 40:1023-1026.

WAHLQVIST, M. L., MORRIS, M. J., LITTLEJOHN, G. O., BOND, A., and JACKSON, R. V. J. 1979. The effects of dietary fibre on glucose tolerance in healthy males. Aust. N.Z. J. Med. 9:154-158.

WEINREICH, J., PEDERSEN, O., and DINESEN, K. 1977. Role of bran in normals. Acta Med. Scand. 202:125-130.

WIDDOWSON, E. M. 1975. Extraction rates—Nutritional implications. Pages 235-248 in: Bread—Social, Nutritional and Agricultural Aspects of Wheaten Bread. A. Spicer, ed. Applied Science Publ., London.

WISKER, E., FELDHEIM, W., POMERANZ, Y., and MEUSER, F. 1985. Dietary fiber in cereals. Pages 169-238 in: Advances in Cereal Science and Technology, Vol. VII. Y. Pomeranz, ed. Am. Assoc. Cereal Chem., St. Paul, MN.

WISKER, E., HUDTWALCKER, G., RAMMINGER, B., and FELDHEIM, W. 1987. Phytic acid content of high-fibre breads. Dtsch. Lebensm. Rundsch. 83:13-15.

YOON, J. H., THOMPSON, L. U., and JENKINS, D. J. A. 1983. The effect of phytic acid on in vitro rate of starch digestibility and blood glucose response. Am. J. Clin. Nutr. 38:835-842.

YOUNG, V. R., and PELLETT, P. L. 1985. Wheat protein in relation to protein requirements and availability of amino acids. Am. J. Clin. Nutr. 41:1077-1090.

CHAPTER 4

RHEOLOGY AND CHEMISTRY OF DOUGH

A. H. BLOKSMA
Institute for Cereals, Flour and Bread TNO
Wageningen, The Netherlands

W. BUSHUK
Grain Industry Research Group
Food Science Department
University of Manitoba
Winnipeg, Manitoba, Canada

I. INTRODUCTION

Dough is a complex mixture of flour constituents, water, yeast, salt, and other ingredients. It is a principal intermediate stage in the transformation of wheat, through flour, into bread. Rheological properties of dough are important to the baker for two reasons. First, they determine the behavior of dough pieces during mechanical handling, such as dividing, rounding, and molding. Second, they affect the quality of the finished loaf of bread. By choosing the formula and process, the baker controls dough properties in order to produce bread that meets the specific quality requirements for its type. The relation between the rheological properties of dough and its behavior in the bakery, including their effect on the quality of the baked product, is an important theme of this chapter. This relation is becoming even more important as mechanization and automation expand in the baking industry.

Another, equally important theme is the relationship between dough structure and dough rheology. Rheological properties of materials depend on their structure—the spatial arrangement of the constituents and the forces acting between them. Therefore, this chapter discusses the chemistry of dough in relation to its rheology. This relation is an essential link in the understanding of what constitutes quality in wheat and flour.

Section II is an introduction to dough structure. The molecular structure of dough is discussed in sections VII and VIII; section IX-A discusses further the microscopic structure and its technological implications.

Gas retention, a property typical of wheat flour doughs, is introduced in section III; its technological aspects are further discussed in section IX-B.

Section IV describes what has been learned about the rheological behavior of dough by the application of basic experimental techniques. Section V discusses the various physical dough-testing techniques that breeders, millers, and bakers apply for quality control and for the adaptation of processes. Sections VI, VII, and VIII discuss how physical and chemical factors influence dough properties.

Section IX describes the breadmaking process from the rheological and chemical points of view. Because other doughs and batters made from flour have been investigated less extensively than bread doughs (and simpler unleavened doughs), a similar description of the preparation of products other than bread is not yet possible.

Because of the increased interest in bread with a higher fiber content, the rheology of doughs from high-extraction flours is discussed in section X. This section also discusses the rheology of composite flour doughs.

II. DOUGH STRUCTURE

A. Constituents

Before one can discuss the structure of a material, one must know its constituents. The constituents of dough are discussed in other chapters of this monograph; therefore, we will merely recapitulate them here.

The main constituents are those derived from the flour, most importantly the proteins, both soluble and insoluble. The carbohydrates, which include starch, sugars, and soluble and other insoluble polysaccharides, are the most abundant. The lipids form a small but significant part of the flour. The molecules of some minor constituents are composed of a lipid part and either a carbohydrate part (glycolipids) or a protein part (lipoproteins); these are interesting because of their possible role in the interactions between more abundant constituents. Finally, flour enzymes may affect dough properties.

Water, another constituent, plays a key role in the formation of dough.

During dough mixing, some air is occluded. The air forms the nuclei of the gas cells, which expand during fermentation and oven rise. Atmospheric oxygen is involved directly in oxidative reactions that affect dough properties (see section VIII-D).

The remaining constituents include all other ingredients that are added to dough, including yeast, salt, malt, enzymes, flour improvers, sugar, fats, emulsifiers, milk or soy solids, and mold inhibitors.

B. Structure

Dough is composed of a continuous phase in which gas cells are dispersed; this continuous phase is called "dough phase." The microscopic structure of this dough phase has been studied by light microscopy, transmission electron microscopy, and scanning electron microscopy (Moss, R., 1972, 1974; Simmonds, 1972; Bechtel et al, 1978; Fretzdorff et al, 1982; Pomeranz et al, 1984a, 1984b). The following constituents can be distinguished:

1. Starch granules, occupying about 60% of the volume of the dough phase. Large elliptical granules with a large diameter of 30–40 μm occur side by side with small spherical granules with a diameter of less than 10 μm.

2. Swollen protein. Changes in the organization of the protein phase on a microscopic scale that occur during mixing are discussed in section IX-A. Its structure on a molecular level is an important subject of sections VII and VIII.

3. Yeast cells, with a diameter of about 2 μm.

4. Lipids. In doughs both without and with added shortening, droplets less than 1 μm up to 3 μm in diameter are observed (Simmonds, 1972; Bechtel et al, 1978).

5. Irregularly shaped remnants of cell organelles and wheat grain tissues.

Immediately after mixing, the gas cells form spherical holes with diameters between 10 and 100 μm (Carlson and Bohlin, 1978; Junge et al, 1981; Sluimer, 1986). Their number at that stage is estimated to be between 10^{11} and 10^{13} m^{-3} (Bloksma, 1981). At an advanced stage of fermentation, they lose their spherical shape, and dough approaches a foam with angular gas cells.

III. GAS RETENTION AND ITS MEASUREMENT

To produce a loaf of bread with a light and even crumb, the yeast must produce sufficient carbon dioxide, and the dough must retain the gas produced over a sufficiently long period (see section IX-B). Insufficient gas production can easily be remedied by the addition of sugars, enzymes, or more yeast, or by an increase in dough temperature. Therefore, gas retention, embracing many relevant factors, is the most comprehensive quality requirement for doughs for breadmaking; wheat's unique position among cereals is due to the ability of doughs from wheat flour to retain gas during fermentation and oven rise. This ability, in turn, depends on the ability of dough to be stretched into thin membranes. This dough property is attributed to its protein phase. This is plausible because thin sheets can also be obtained by stretching isolated gluten.

Technological aspects of gas production and gas retention are discussed in section IX. Table I summarizes methods and instruments for measuring gas production and/or retention and is arranged according to the principles behind the various methods. Shuey (1975) gives a more extensive description of some instruments. The oven-rise recorder is the only instrument that measures gas retention under oven conditions.

Most methods do not distinguish between the gas that is retained in the dough and the gas that escapes from it; therefore, they measure gas production. Exceptions are methods that measure the height of a fermenting dough, such as the maturograph (Fig. 1), or the buoyant force on a test piece of dough (oven-rise recorder). Measurements of gas production can be converted into measurements of gas retention if the method is modified to absorb, in a solution of alkali, the escaped carbon dioxide.

IV. BASIC APPROACHES TO DOUGH RHEOLOGY

This section summarizes the results of rheological measurements with wheat flour doughs and addresses the extent to which results from different authors agree quantitatively. The results are based on the application of basic experimental techniques. These techniques differ from the routine physical dough-testing techniques discussed in section V in that the testing conditions are not only well defined but also physically simple; the results are expressed in terms with simple physical definitions and in SI units, independent of the particular instrument used and of the shape and dimensions of the test piece. This simplicity and independence make the results suitable as a starting point for engineering calculations, including attempts to understand the effect of the rheological properties of dough on its behavior in the breadmaking process and on the qualities of the resulting loaf. These results also can be used to validate hypotheses on the molecular structure of dough by comparing the properties predicted by a hypothesis with the experimentally determined properties. The knowledge of dough structure is, however, not yet sufficiently advanced for a quantitative comparison.

The following discussion assumes that the reader is familiar with the elementary concepts of rheology. If necessary, the reader should consult

TABLE I
Methods and Instruments for Measuring Gas Production and/or Gas Retention

Increase in volume at atmospheric pressure
  Measurement of volume by means of gas buret or similar device
    AACC Approved Method 22-14 (AACC, 1983)
    Fermentometer (Burrows and Harrison, 1959; Parisi, 1981, method 1)
    SJA Fermentograph,[a] gas collected under a bell jar floating in an oil bath, Naessjoe
      Mettallverkstadt, Naessjoe, Sweden
  Measurement of volume by means of buoyant force on test piece
    Oven-rise recorder,[a,b] Brabender OHG, Duisburg, Federal Republic of Germany and
      C. W. Brabender Instruments, South-Hackensack, NJ (Seibel, 1968)
  Measurement of height of fermenting test piece
    Maturograph,[a,c] Brabender OHG, Duisburg, Federal Republic of Germany and
      C. W. Brabender Instruments, South-Hackensack, NJ (Seibel, 1968)
    Rheofermentometer,[a,d] Tripette & Renaud, Paris
    Synthetic dough in graduated cylinders[e] (Schulz, 1966; Parisi, 1981, method 2)
Increase in pressure at constant volume
  Without release of pressure
    AACC Approved Method 22-11[f] (AACC, 1983)
    Gasograph,[a] D & S Instruments, Pullman, WA (Rubenthaler et al, 1980)
  With periodic release of pressure
    Zymotachygraph,[a,g] Tripette & Renaud, Paris (Chopin, 1953, 1973)

[a] Recording instrument.
[b] The instrument is designed to measure gas retention under baking conditions. Therefore, the temperature of the oil bath in which the test piece is immersed is gradually increased from 30 to 100°C. The instrument has also been used at a constant temperature (Marek and Bushuk, 1967; Marek et al, 1968).
[c] Test piece height is sensed by a stamp; the weight of the stamp is normally compensated for by a counterweight. At intervals of 2 min, weight compensation is temporarily removed and the sensing stamp sinks, under the load of its own weight, into the test piece. This periodic loading makes the recorded curve a sawtooth line (Fig. 1).
[d] To be used in conjunction with the zymotachygraph.
[e] Designed for the determination of yeast activity. Therefore, a "synthetic dough," composed of corn starch, carob flour, sucrose or maltose, and water is used as substrate.
[f] To distinguish between the activities of different lots of yeast, the same procedure with a richer formula has been developed (Shogren et al, 1977).
[g] The curve recorded, falling to zero at each release of pressure, consists of a series of peaks; peak height indicates the rate of gas production. By connecting the manometer with the vessel containing the fermenting dough alternately directly and via a bottle with alkali, the instrument records simultaneously the rate of gas production and changes in the quantity of gas retained. An instrument based on the same principle, but without measurement of gas retained and with readout of the quantity of gas produced, has been described by Voisey et al (1964).

introductory texts on rheology (Reiner, 1949; Scott Blair, 1969; Muller, 1973).

The rheological properties of a material are derived from the relationship between the stress on a material, the corresponding deformation or strain of it, and time. For simplicity, in most rheological tests the deformation or the stress is held constant or is made to vary as a simple function of time. This leads to the following classification of test programs: 1) stress constant: creep (see section IV-A); 2) deformation increases linearly with time: viscometry (see section IV-B); 3) deformation constant: stress relaxation (see section IV-C); 4) deformation and/or stress varies sinusoidally with time: oscillatory or dynamic measurements (see section IV-D).

In addition to choosing among these programs, the experimenter must choose between experiments with shear or extension as the type of deformation. Methods in which test pieces are stretched are technically more difficult than shearing experiments. On the other hand, they provide more relevant information. During fermentation and oven rise, dough membranes between gas cells are stretched rather than sheared, and their behavior under these conditions determines gas retention (see sections IX-B and IX-C). Because dough is a concentrated dispersion of nonspherical starch granules in the protein phase, it cannot be taken for granted that the (Trouton) coefficient of viscous traction is three times the coefficient of viscosity, as it is for rheologically simple materials (Giesekus, 1983). Consequently, the behavior of dough in extension cannot be predicted reliably on the basis of its behavior in shear, and vice versa.

The above definition of rheological properties implies that experiments covering infinitely wide ranges of deformation, stress, and time are required to fully describe rheological properties. This is, of course, impossible and also unnecessary. For engineering purposes, it is sufficient to cover the range of deformations or rates of deformation to which the material is subjected in the process being considered (Prins and Bloksma, 1983). For the breadmaking process, this encompasses still a wide range (Fig. 2). From the engineering point of view, the whole range of shear rates from 3×10^{-4} to 5×10^{1} sec^{-1} is of interest.

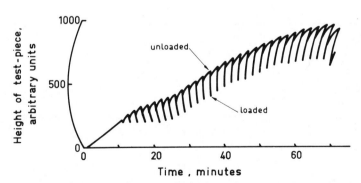

Fig. 1. A representative maturogram from a maturograph. The top envelope of the sawtooth line describes the net result of gas production and gas loss under the condition of periodic loading. In the ascending part of the envelope, gas loss is slower than production; in the descending part, it is more rapid. A late and high maximum indicates that gas retention can stand mechanical abuse. The difference between top and bottom envelopes reflects the periodic loading; it is often referred to as elasticity.

Measurements over a wide range of deformation rates show that the rheological behavior of dough is nonlinear; that is, stress and (rate of) deformation are not proportional. Dough has this property in common with most colloidal systems and foods. In section IV-B, we show that over the range mentioned in the preceding paragraph, the apparent viscosity (i.e., the ratio of stress and rate of deformation) varies by a factor of as much as 2,000.

In some studies, the rheological behavior of a material is presented in the form of a mechanical model, that is, an array of springs and dashpots representing elastic and viscous behavior, respectively. For a quantitative description, numerical values must be assigned to the various elements. Figure 3 is a crude model of dough behavior. It does not account for nonlinear behavior, nor does it describe effects of time and deformation. Such models are helpful to visualize qualitatively the behavior of a material in various testing programs. More precise and complete descriptions require highly complicated models, including other elements besides springs and dashpots (Lerchenthal and Funt, 1967; Lerchenthal and Muller, 1967). These models are too complicated for easy visualization.

A. Creep and Recovery

Loading a test piece by a constant stress is a logical and simple approach to the rheological characterization of materials; such tests are called creep tests. They can be performed in shear and in extension. The latter type is technically more

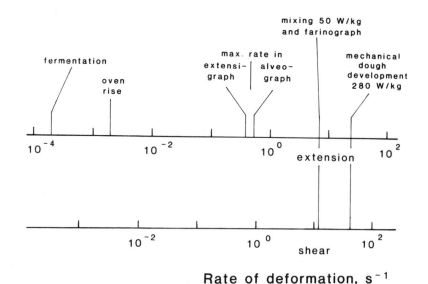

Rate of deformation, s^{-1}

Fig. 2. Rates of deformation of dough under various processing and testing conditions. Except for the extensigraph and alveograph, the indicated values are mean values; during mixing in particular, much larger rates of deformation occur locally and temporarily. Scales for the rates of shear and of extension are positioned relative to each other so that they have the second invariants of the rate of strain tensors in common; for rheologically simple materials, this invariant determines the flow properties. Rates are based on estimates by Bloksma (1957, 1962, 1981, 1984, 1986).

difficult to perform, and its interpretation is less straightforward. Figure 4, curve a, shows an example of results obtained in shear.

If, after some time, the load is removed, the material recovers partly or completely. Curves b–e in Fig. 4 are examples of recovery curves, recorded after various durations of creep. The permanent part of the deformation is ascribed to viscous flow during the creep test, the temporary part to elasticity. The power of creep and recovery experiments lies in this distinction between elastic and viscous deformations.

Hibberd and Parker (1979) found a measurable permanent deformation after loading with a shear stress as low as 10 Pa. It follows from this observation that, if dough has a yield value, it is smaller than 10 Pa. Therefore, no element accounting for a yield value is incorporated in the model in Fig. 3.

The beginnings of both the creep curve and the recovery curves in Fig. 4 show an instantaneous elastic deformation or recovery, which is accounted for by spring A in Fig. 3. It is followed by a delayed elastic reaction, accounted for by

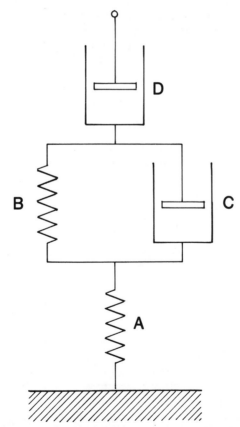

Fig. 3. Mechanical model for the rheological behavior of dough, showing instantaneous (A) and retarded (B and C) elasticity and viscous flow (D). In the rheological literature, this model is called the Burgers body (Scott Blair, 1969; Muller, 1973).

the combination of elements B and C in Fig. 3. It is impossible to pinpoint the borderline between instantaneous and delayed elasticity. The sum of the instantaneous and delayed elastic deformations can be described by a Young modulus (for extension) of (1 to 3) $\times 10^4$ Pa (Glucklich and Shelef, 1962a, 1962b; Lerchenthal and Muller, 1967; Lerchenthal, 1969) or by a shear modulus of (1 to 2) $\times 10^3$ Pa (Hibberd and Parker, 1979).

After completion of the delayed elastic deformation, further deformation is due to viscous flow, accounted for by element D in Fig. 3. If the viscosity is constant, the creep curve continues as a straight line whose slope is equal to the stress divided by the coefficient of viscosity. Then a steady state is attained that cannot be distinguished from the steady flow discussed in section IV-B. However, Hibberd and Parker (1979), applying a shear stress of 50 Pa, did not attain a steady state within 10,000 sec (3 hr).

Figure 5 shows results of creep and recovery tests reported by various authors (Table II). Despite differences in dough composition, temperature, and type of deformation, results for the total deformation agree remarkably well. However, the elastic parts of the deformations are about five times larger in the shear experiments of Hibberd and Parker (curves HP) than in the extension experiments of the other investigators; a small part of this difference can be explained by the longer recovery time in the experiments of Hibberd and Parker.

Linear behavior would show in a graph like Fig. 5 as straight lines with slopes equal to unity. In fact, with increasing stress, the deformation increases more than proportionally to stress.

B. Viscometry

SHEAR FLOW

The aim of viscometry is to determine the relation between the rate of shear and shear stress in a material in a steady state. If the stress is proportional to the

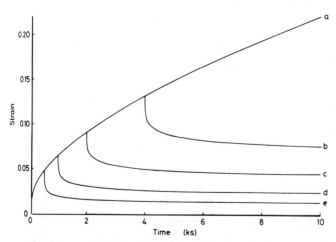

Fig. 4. Creep and recovery curves. Curve a shows the course of creep under a shear stress of 50 Pa. If, after 500 sec, the stress is removed, recovery proceeds along curve e. Curves d, c, and b show the course of recovery if the stress is removed after 1,000, 2,000, and 4,000 sec, respectively. (Reprinted, by permission, from Hibberd and Parker, 1979)

rate of shear, the material is called a Newtonian liquid; the quotient stress over the rate of shear is its (coefficient of) viscosity. For non-Newtonian liquids, among them dough, this ratio is called the apparent viscosity; it is a function of the rate of shear or the shear stress. Interpretation of experiments with these materials is greatly simplified if the rate of shear is homogeneous, that is, the same throughout the test piece. This condition is satisfied in flow between concentric cylinders with a narrow gap between them, or between a plate and a cone with a small free angle between them. Therefore, most of the experiments reported in this section were performed with instruments of these types.

In these instruments, shear stress can be recorded as a function of time. Upon the instantaneous imposition of a constant shear rate, the shear stress increases at a finite rate because of the elastic response of the dough. From the initial slopes of the curves of stress versus time, Bloksma and Nieman (1975) derived shear moduli of the order of 4×10^3 Pa, slightly increasing with increasing rate of shear; this value is in fair agreement with the values derived from creep tests

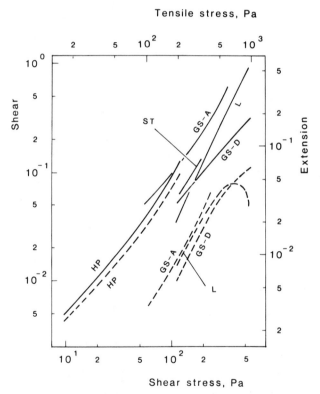

Fig. 5. Results of creep and recovery tests with dough (see also Table II). Solid lines show total deformation after 250 sec of creep; dashed lines show temporary or elastic parts of the deformation. Scales for shear stress and tensile stress are positioned relative to each other so that they have the second invariants of the stress tensors in common; the same applies to the scales of shear and extension and the second invariants of the strain tensors. By this positioning, curves for shear and extension of rheologically simple materials coincide. The Cauchy definition of extension (the increase in length divided by the original length) is used.

TABLE II
Experimental Conditions in Creep and Recovery Tests

Test Reference	Label in Fig. 5	Type and Protein Content of Flour(s) or Parent Wheat(s)[a]	Water Added (%)[b]	Temperature (°C)	Type of Deformation	Duration (sec)	
						Creep[c]	Recovery
Glucklich and Shelef (1962a, 1962b)	GS-A	80% Hard red spring wheat and 20% soft wheat, 13% protein	54	24	Extension	330	150–210
Hibberd and Parker (1979)	GS-D	Durum flour, 12% protein	64	24	Extension	330	150–210
	HP	Australian flour, 13% protein; 2% of salt added	60	27	Shear	250	5,000
Lerchenthal (1969)	L	...	...	...[d]	Extension	330	270
Smith, T. L., and Tschoegl (1970)	ST	Hard red winter wheat, 14% protein	46	...	Extension	...	...

[a] Protein contents are reported on dry basis.
[b] Water addition in milliliters per 100 g of flour with 14% moisture.
[c] Duration of the creep before unloading. This applies only to the dashed lines.
[d] Oscillatory measurements were made at 24°C in the same laboratory and with the same flour (Table V, label ST).

(section IV-A). With rates of shear up to about 2×10^{-2} sec^{-1}, shear stress gradually approaches a steady value; the shape of the curve is qualitatively similar to that of a Maxwell body (elements A and D in Fig. 3) but is quantitatively very different. With higher rates of shear, shear stress goes through a maximum and then decreases again, which indicates breakdown of structures as a result of continuous shearing. If this occurred, the maximum stress value was used in the construction of Fig. 6.

The classic way to determine the viscosity of a liquid is to have it flow through a cylindrical tube and to establish the relation between the pressure gradient and the volume rate of flow. This type of flow is nonhomogeneous; the rate of shear is zero in the axis of the tube and maximum near the wall. Nevertheless, for Newtonian liquids the interpretation is straightforward and a single measurement suffices. Flow through a cylindrical tube can also be applied to establish the relation between shear rate and shear stress in non-Newtonian liquids; then a series of measurements and elaborate calculations are required. Some of the results reported in Fig. 6 were obtained by such a procedure. Table III reports conditions used in shear flow tests.

Figure 6 is a collation of experimental results obtained with various experimental techniques and with very different ranges of shear rate. A straight line in this graph means that the relation between the rate of shear and shear stress is described by a power law:

$$\tau = \tau_0 \cdot (D/D_0)^n , \qquad (1)$$

in which D = the rate of shear; D_0 = an arbitrarily chosen reference rate of shear; τ = the shear stress; τ_0 = the shear stress if $D = D_0$ (its numerical value depends on the choice of D_0); and n = an exponent. For Newtonian liquids, $n = 1$. In graphs

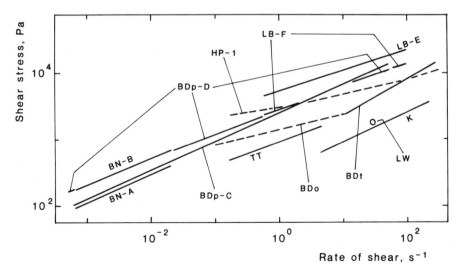

Fig. 6. Shear stress in dough as a function of the rate of shear (see also Table III). Solid lines represent steady shear flow experiments; dashed lines were derived from oscillatory measurements (section IV-D).

TABLE III
Experimental Conditions in Tests of Shear Flow

Test Reference	Label in Fig. 6	Type and Protein Content of Flour(s)[a]	Water Added (%)[b]	Temperature (°C)	Type of Measurement
A. H. Bloksma and J. A. Duiser (unpublished results, 1980)	BDp-C	Dutch bread flour, 14% protein	59[c]	30	Cone and plate
	BDp-D	Dutch biscuit flour, 12% protein	55[c]	30	Cone and plate
	BDt	Dutch bread flour, 14% protein	59[c]	30	Cylindrical tube
Bloksma and Nieman (1975)	BN-A	Dutch bread flour, 12% protein	56	30	Cone and plate
	BN-B	Dutch bread flour (protein content not determined; two extremes of three flours; 2% salt added to all three flours	54	30	Cone and plate
Kieffer et al (1982)	K	0.5% ash	58[d]	22	Cylindrical tube
Launay and Buré (1973)	LB-E	French flour, 13% protein	...	...	Cone and plate
	LB-F	French flour, 10% protein, two extremes of seven flours; about 1.4% salt added	...	...	Cone and plate
Lenz and Wutzel (1984)	LW	...	76	...	Cone and plate
Tscheuschner and Treiber (1978)	TT	...	...[e]	...	Concentric cylinders
Derived from oscillatory measurements; see section IV-D and Table V	BDo				
	HP-1				

[a] Protein contents are reported on dry basis.
[b] Water addition in milliliters per 100 g of flour with 14% moisture.
[c] Corresponding to farinograph water absorption (500 FU).
[d] Corresponding to maximum consistency of 430 FU in the farinograph.
[e] Corresponding to maximum consistency of 380 FU in the farinograph.

like Fig. 6, Newtonian behavior is represented by a straight line with unit slope.

The experiments depicted in Fig. 6 were made with widely different flours, without a uniform basis for the choice of the water addition, and at different temperatures; nevertheless, their results agree remarkably well. In the range $6 \times 10^{-4} < D < 3 \times 10^{2} \text{ sec}^{-1}$, results can be summarized by equation 1 with parameters $\tau_0 = 2 \times 10^{3}$ Pa, if $D_0 = 1 \text{ sec}^{-1}$, and $n = 0.37$. Because $n < 1$, it follows that the apparent viscosity decreases with increasing rate of shear, from a value of 1.6×10^{5} Pa·sec at $D = 10^{-3} \text{ sec}^{-1}$ to 1.1×10^{2} Pa·sec at $D = 10^{2} \text{ sec}^{-1}$.

EXTENSIONAL FLOW

Figure 7 shows some typical results of extensional flow experiments. In this section, extensions are expressed in the Hencky measure, that is, the natural logarithm of the ratio of the actual length to the initial length of the test piece. The tensile stress is proportional to this measure of extension over a wider range of extensions than if the Cauchy measure of extension (increase in length divided by initial length) is used. In the experiments discussed, the length of the test piece was increased at a constant rate. As a consequence, the Cauchy measure also increased at a constant rate, whereas the rate of increase of the Hencky measure gradually decreased with time; the rates of extension in Fig. 7 are the Cauchy rates at any time or the Hencky rates at zero time.

Tschoegl et al (1970a, 1970b) and Rasper et al (1974) separated the effects of extension and time on tensile stress up to extensions of between 0.5 and 1 (Hencky measure), as shown in the following equation:

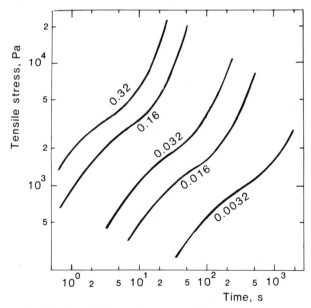

Fig. 7. Tensile stress as a function of time at various rates of extension. Labels indicate the (Hencky) rate of extension (the natural logarithm of the ratio of the actual length and the initial length) in sec^{-1}. Doughs from hard red spring wheat flour with 14% protein; water addition 52%; temperature 25° C. (Data from Tschoegl et al, 1970b)

$$\sigma = F(t_o) \cdot (t/t_o)^n \cdot \Gamma(\epsilon) , \tag{2}$$

where σ = the tensile stress (force divided by the actual cross-sectional area); $F(t_o)$ = the "isochronal modulus" at an arbitrarily chosen standard time t_o; t = time; ϵ = the Hencky measure of extension; $\Gamma(\epsilon)$ = a function of the extension ϵ that, for small values of ϵ, approaches ϵ; and n = an exponent. In nearly all experiments, the exponent n had a value between -0.2 and -0.4, indicating that the rheological behavior of dough is between that of an elastic ($n = 0$) and a viscous ($n = -1$) material. In these experiments, steady-state viscous flow was not attained before rupture occurred.

C. Stress Relaxation

If a test piece is deformed to a certain extent, after which the deformation is maintained at a constant level, the stress built up during deformation gradually relaxes. In the model in Fig. 3, this corresponds to a displacement of the plunger of dashpot D, which compensates for changes in the lengths of springs A and B. Stress relaxation is, also at the molecular level, due to the same mechanism as viscous flow.

Figure 8 shows some experimental results from the literature (Table IV). Stress relaxation already begins during the buildup of the stress. If the stress buildup takes considerable time, the initial, rapid part of the relaxation process is covered by the buildup of the stress; consequently, only the recorded

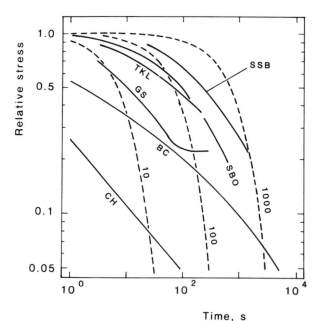

Fig. 8. Stress relaxation in dough (see also Table IV). Solid lines are experimental results; dashed lines are calculated curves for Maxwell bodies with single relaxation times of 10, 100, and 1,000 sec. Time is measured from the completion of the buildup of the stress. The relative stress is the stress at a certain time divided by the stress at zero time.

TABLE IV
Experimental Conditions in Stress-Relaxation Experiments

Test Reference	Label in Fig. 8	Type and Protein Content of Flour or Parent Wheat[a]	Water Added[b] (%)	Temperature (°C)	Type of Deformation	Magnitude of Deformation	Loading Time (sec)
Bohlin and Carlson (1981)	BC	Starke II, 9% protein	55	24	Shear	0.1	0.3
Cunningham, J. R., and Hlynka (1954)	CH	Red spring wheat, 13% protein	55	23	Extension	...	1
Glucklich and Shelef (1962b)	GS	80% Hard red spring wheat and 20% soft wheat	54	24	Extension	...	...
Shelef and Bousso (1964)	SBO	Manitoba, 13% protein	50[c]	30	Extension	0.01	10
Schofield, R. K., and Scott Blair (1932)	SSB	...			Extension	1.5	...
Telegdy Kováts and Lásztity (1966)	TKL	...	...	30	Extension	...	13

[a] Protein contents are reported on dry basis.
[b] Water addition in milliliters per 100 g of flour with 14% moisture.
[c] Corresponding to maximum consistency of about 850 FU in the farinograph.

relaxation appears slower than it really is. Therefore, Fig. 8 shows results of experiments in which the deformation was completed within a short time (up to 13 sec). Nevertheless, this figure still suggests that stress relaxation is more rapid after shorter loading times. Therefore, the curves recorded after the shortest loading times (BC and CH) are considered the more reliable ones.

None of the curves in Fig. 8 corresponds to the exponential decay characteristic of a linear Maxwell body with a single relaxation time. To describe them, a relaxation time spectrum can be used, which corresponds to a model in which a large or infinite number of Maxwell bodies with different relaxation times are placed parallel (Cunningham, J. R., and Hlynka, 1954). Bohlin and Carlson (1981) described their curves by means of an equation based on a theory of cooperative flow, in which the coordination number of the flow units is an important parameter.

A slow stress relaxation has been associated with good baking quality (Halton and Scott Blair, 1936, 1937; Launay and Buré, 1974; Launay, 1979); this could not be confirmed in the laboratory of one of the authors (A. H. Bloksma). Frazier and co-workers applied stress-relaxation measurements with dough balls after uniaxial compression to follow mechanical dough development in the presence of various quantities of flour improvers. With increasing mixing energy, relaxation time goes through a maximum; this maximum is considered to indicate optimum dough development (Frazier et al, 1975, 1985; Fitchett and Frazier, 1986; and papers cited therein).

Figure 8 shows that if dough is loaded during an interval of 1 sec or less, little stress relaxation occurs; after the stress is removed, the dough largely recovers elastically. In sheeting rolls and molders, dough therefore behaves predominantly as an elastic material. On the other hand, in processes that take 100 sec or more, dough behavior approximates that of a highly viscous liquid.

D. Oscillatory Measurements

Faubion et al (1985) reviewed the application of oscillatory or dynamic measurements to doughs. In these measurements, a strain that varies sinusoidally with time is imposed. After a transition period, during which the effect of the accidental initial conditions fades, a steady state is attained in which, if the material behaves linearly, the stress also varies sinusoidally with the same frequency as the strain. In general, there is a phase difference between stress and strain. To describe the behavior of the material, the stress can be resolved into a component that is in phase with the strain and one that is a quarter of a cycle ($\pi/2$ rad) ahead of it. The amplitudes of these two components, divided by the amplitude of strain, are called the storage modulus and the loss modulus, respectively. If these two moduli are known as a function of frequency, the behavior of the material is defined in the range of frequencies examined. The storage modulus describes the elastic properties; the viscous energy dissipation is, at constant strain amplitude, proportional to the loss modulus.

Figure 9 summarizes results of applications (Table V) of this method to wheat flour doughs. Wheat flour doughs do not satisfy the condition of linear behavior, even at strain amplitudes as low as 10^{-3}. As the strain amplitude increases, the storage and loss moduli decrease. With increasing strain amplitude, this effect is already observed before the stress signal deviates from the sinusoidal form. Parker and Hibberd found the relative decrease to be independent of frequency and water addition (Parker and Hibberd, 1974; Hibberd and Parker, 1975). The magnitude of the effect can be seen in Fig. 9; the curves HP-1, HP-2, and HP-3 refer to strain amplitudes of 7×10^{-4} (at which the effect of nonlinearity can hardly be observed), 10^{-2}, and 10^{-1}, respectively. Szczesniak et al (1983) reported even larger effects (see curves SL-1 and SL-2). Other authors have paid little attention to this phenomenon. Their results must be considered "apparent moduli," which are valid for the strain amplitudes in Table V, which were estimated assuming linear behavior and therefore indicate only an order of magnitude. This effect of strain amplitude largely explains why the results in Fig. 9 vary more than those in Fig. 6; variations in strain amplitude act in addition to the causes of variation mentioned in section IV-B.

To compare the results of oscillatory measurements with those of steady shear flow, one can make use of the absolute value of the complex dynamic modulus, G^*, defined by:

$$G^* = (G'^2 + G''^2)^{\frac{1}{2}}, \tag{3}$$

where $G' = $ the storage modulus and $G'' = $ the loss modulus. For many materials, G^* depends on the angular frequency in a way similar to the way the shear stress

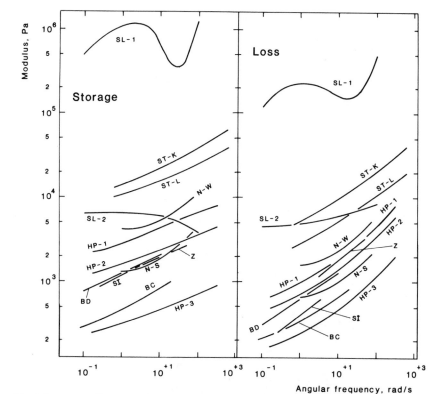

Fig. 9. Storage and loss moduli of doughs as a function of frequency (see also Table V).

TABLE V
Experimental Conditions in Oscillatory Measurements in Doughs

Test Reference	Label in Fig. 9	Type and Protein Content of Flour(s) or Parent Wheat(s)[a]	Water Added[b] (%)	Temperature (°C)	Shear Amplitude[c]
Bohlin and Carlson (1980)	BC	Starke wheat, 10% protein	55	25	0.37
A. H. Bloksma and J. A. Duiser (unpublished results, 1980)	BD	Dutch bread flour, 14% protein	59[d]	30	0.02–0.04
Hibberd and Parker, 1975; Parker and Hibberd, 1974	HP-1	Mendos wheat	62	25	7×10^{-4}
	HP-2	Mendos wheat	62	25	10^{-2}
	HP-3	Mendos wheat	62	25	10^{-1}
Navickis et al (1982)	N-S	Hard red spring wheat, 18% protein	69[d]	22	Linear region
	N-W	Soft red winter wheat, 11% protein	54[d]	22	Linear region
Shimizu and Ichiba (1958)	SI	Manitoba Northern No. 3 wheat, 13% protein; 2% salt added	63	28	0.25
Szczesniak et al (1983)	SL-1	U.S. flour, 16% protein	67[e]	22	5×10^{-3}
	SL-2	U.S. flour, 16% protein	67[e]	22	10^{-1}
Smith, J. R., et al (1970)	ST-K	Hard red winter wheat, 14% protein	46	24	7×10^{-4}
	ST-L	Weak flour, 9% protein	49	24	3×10^{-4}
Zangger (1979)	Z	Diplomat wheat	72	30	...

[a] Protein contents are reported on dry basis.
[b] Water addition in milliliters per 100 g of flour with 14% moisture.
[c] Center to peak value.
[d] Corresponding to farinograph water absorption (500 FU).
[e] Corresponding to maximum consistency of about 650 FU in the farinograph.

depends on the rate of shear (Cox and Merz, 1958). G^* was calculated from two sets of results in Fig. 9 and was plotted, as dashed lines, in Fig. 6. The agreement is quite satisfactory.

The results of Szczesniak et al (1983) (curves SL-1 in Fig. 9) deviate from the other results by their high values at a small shear amplitude and, in particular, by the minimum in the curves at a frequency of about 30 rad/sec. This minimum was not observed when testing gluten and could be eliminated by increasing the shear amplitude (curves SL-2), by increasing the time between mixing and measurement, by coating the free surface of the test piece with silicone oil, or by adding shortening. The authors suggested that the minimum is caused by a breakdown of aggregates of starch granules.

In oscillatory measurements, deformations remain small. In contrast with, for example, viscometry, existing structures are hardly damaged. This has advantages for studies of dough structure. On the other hand, this procedure does not yield information on the behavior of dough at large deformations, which is technologically important.

V. PHYSICAL DOUGH-TESTING INSTRUMENTS

Numerous instruments have been devised to obtain objective data on dough properties in order to predict its behavior in the bakery. The properties measured can be classified as physical, and in many cases this classification can be narrowed to rheological. To describe the results in simple rheological terms is, however, much more difficult than with the procedures discussed in the preceding section because of the complicated geometry of the deformation and its nonhomogeneity; moreover, elastic and viscous deformations are mixed together in such a way that they are difficult to separate.

Table VI lists the most important commercially available instruments, their manufacturers or suppliers, and references to more complete descriptions and to available standard procedures for using them. It is arranged to follow the text of this section.

A. Recording Dough Mixers

Recording dough mixers record the power that is required to mix a dough at constant speed or, what is essentially the same thing, the resistance to mixing. Provided that both scales of the record are linear, the area under the curve is proportional to mixing energy.

The recorded curves yield information about changes in rheological properties during mixing. They usually consist of a rising part, showing an increase in resistance with mixing time, followed by a more or less slow decrease in resistance. The maximum is either flat and broad or peaked, depending on the instrument and the type of flour. The records are usually interpreted in terms of dough development and subsequent breakdown.

BRABENDER FARINOGRAPH AND RELATED INSTRUMENTS

The farinograph (Fig. 10) was developed about 1930. In the mixer, two Z-shaped mixing blades rotate in opposite directions toward each other at different speeds. After the flour and optional other dry ingredients have been placed in the mixer, the instrument is started, and water is added rapidly from a buret. The torque on the driving shaft of the mixing blades causes a proportional rotation of the dynamometer (3–7 in Fig. 10); this rotation is transmitted to a pointer and recorder. The mixer wall is hollow, and water is circulated through it for temperature control. Mixers for doughs made from 300 and 50 g of flour can be placed on the standard base; there is also a microfarinograph with a 10-g mixer on a special base.

The resistograph is a variant of the farinograph with a more intensive mixing

TABLE VI
Physical Dough-Testing Instruments

Instrument	Manufacturer(s) or Supplier(s)	References to Descriptions	References to Standard Procedures
Recording dough mixers			
Farinograph	Brabender OHG, Duisburg, Federal Republic of Germany; C. W. Brabender Instruments, South-Hackensack, NJ, USA	Shuey (1984a)	AACC (1983, Method 54-21); ICC (1972b)
Resistograph	Brabender OHG; C. W. Brabender Instruments	Brabender, M. (1973a, 1973b)	···
Do-Corder	Brabender OHG; C. W. Brabender Instruments	Sietz (1984)	···
Valorigraph	Labor-MIM, Budapest, Hungary	Lüddeke (1969); Lásztity and Hegedüs (1975)	···
Mixograph	National Mfg. Corp., Lincoln, NE, USA	400 g: Swanson and Working (1933); 35 g: Larmour et al (1939); 10 g: Finney and Shogren (1972)	AACC (1983, Method 54-40)
Load-extension meters			
Brabender Extensigraph	Brabender OHG; C. W. Brabender Instruments	Munz and Brabender (1940)[a]	AACC (1983, Method 54-10); ICC (1972a)
"Research" extensometer	Henry Simon Ltd., Stockport, Cheshire, UK	Shuey (1975)	···
Chopin Alveograph	Tripette & Renaud, Paris, France	Chopin (1927, 1973;[b] Faridi and Rasper (1987)	ISO (1983); AACC (1983, Method 54-30)
Extrusion meter			
"Research" water absorption meter	Henry Simon Ltd.	Shuey (1975)	···

[a] The revised ICC standard and an ISO standard, which is being prepared, will contain more up-to-date descriptions.
[b] In the present models, the air for inflation of the dough bubble is pumped into the instrument at a constant rate (Launay and Buré, 1970). The manometer has also been changed.

action. Strong flours produce curves with one peak, whereas medium strong and weak flours produce curves with two peaks.

Another variant of the farinograph, the Brabender Do-Corder, has a nearly closed mixer for 700 or 1,000 g of dough, a variable speed, and a wider range of torque-to-deflection ratios. It was designed to simulate conditions during mechanical dough development (section IX-A).

The valorigraph is essentially similar to the farinograph. Only mixers for 50 g of flour are available. The dynamometer of the valorigraph is technically quite different from that of the farinograph. Temperature is controlled by air flowing around the mixer; this type of control is much slower than that of the farinograph (Bloksma, 1984).

Figure 11 shows a representative farinograph record, or farinogram. The height of the middle of the band is called the consistency and is expressed in instrument-related farinograph units (FU). The term *consistency* is also frequently used for the maximum consistency of a dough; this dual use is confusing.

One would expect the consistency of a dough to be closely related to its viscosity; however, doughs with the same maximum consistency can have different viscosities (Launay, 1979) and can behave differently in creep tests and in sensory assessments of stiffness (Bloksma and Meppelink, 1973). The measured consistency is partly due to the sticking of the dough to the mixer wall and blades (Bloksma, 1984 and references cited therein).

Maximum consistency decreases with increasing water content, on the average about 30 FU per milliliter of water per 100 g of flour (see also Table X); in addition, the time taken to reach the maximum in the curve becomes longer. The amount of water added to a flour with 14% moisture to produce a maximum consistency of 500 FU is called farinograph water absorption. Water absorption is an important flour property. In the literature, the term is often misused for water addition to a dough.

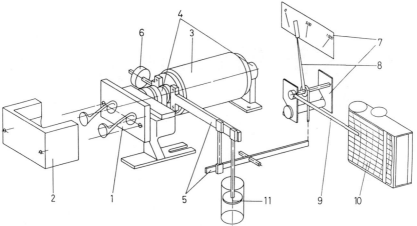

Fig. 10. Diagram showing the principle of the Brabender Farinograph. 1 = Back wall of mixer with mixing blades; 2 = remainder of mixer; 3 = housing of motor and gears; 4 = ball race bearings; 5 = levers; 6 = counterweight; 7 = scale head and scale; 8 = pointer; 9 = pen arm; 10 = recorder; 11 = dashpot damper. (Courtesy Brabender Instruments Co.)

In addition to the water absorption, the shape of a farinogram with a maximum consistency of 500 FU is used to characterize flour properties. Various indexes are used; indexes defined by the American Association of Cereal Chemists (AACC, 1983, Method 54-21) and the International Association for Cereal Chemistry and Technology (ICC, 1972b) are indicated in Fig. 11. Precise definitions are given in the standards referred to in Table VI.

For farinograms with a broad maximum, the AACC and ICC definitions of the dough development time lead to different numbers. The Brabender Valorimeter value attempts to characterize the shape of a farinogram by a single number (Brabender, C. W., 1937; Shuey, 1984b) that is higher the longer the dough development time and the smaller the degree of softening.

SWANSON AND WORKING MIXOGRAPH

The original Swanson and Working Mixograph, a recording dough mixer with a bowl for doughs made from 400 g of flour, was first described in 1933. Modified instruments with mixers for 35 and 10 g of flour followed in 1939 and 1972, respectively, and have superseded the original unit. The mixing bowl contains three or four vertical pins. Four other vertical pins, lowered into it, travel in a planetary motion around the stationary pins. The torque on the mixing bowl is recorded by means of its rotation against a spring.

Variation of the water addition affects the mixogram in the same way as it does the farinogram (see Table X). The shape of the mixogram can be characterized by indexes similar to those defined for the farinogram (AACC, 1983, Method 54-40).

B. Load-Extension Instruments

In several commercially available instruments, a test piece of dough is

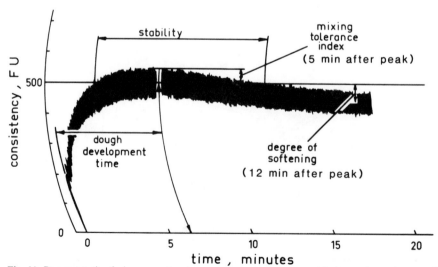

Fig. 11. Representative farinogram showing some commonly measured indexes. A consistency of 500 farinograph units (FU) corresponds to a power of 68 and 81 W per kilogram of dough in mixers for 300 and 50 g of flour, respectively.

stretched until it ruptures and a curve of force versus extension is recorded. From these records, resistance against deformation and extensibility can be read.

BRABENDER EXTENSIGRAPH OR EXTENSOGRAPH

The Brabender Extensigraph (Fig. 12) was introduced in 1936 to supplement the information supplied by the Brabender Farinograph. The extensigraph shows the effect of flour improvers such as bromate and iodate, which can scarcely be observed on the farinogram.

A dough is prepared from flour, 2% of sodium chloride, and a quantity of water based on farinograph data. A test piece is rounded into a ball, shaped into a cylinder, and clamped in a cradle; after a rest period, it is stretched in the instrument. The dough in the cradle is placed on the fork on the right-hand side of a balance beam. Then the motor of the stretching hook is started; the hook moves downward, engages the middle of the test piece, and extends it. The resulting force on the test piece is transmitted through the levers to another

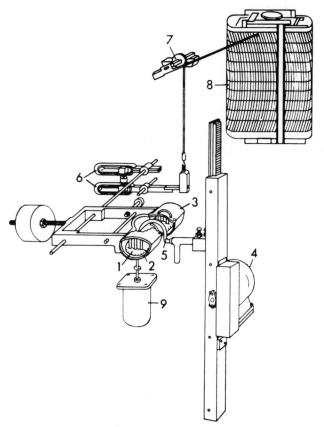

Fig. 12. Diagram showing the principle of the Brabender Extensigraph. 1 = Test piece; 2 = cradle; 3 = clamp; 4 = motor; 5 = stretching hook; 6 = levers; 7 = scale head; 8 = recorder; 9 = dashpot damper. (Courtesy Brabender Instruments Co.)

balance and the pen of a recorder. The chart paper in the recorder moves at a constant speed, and a curve of force versus time is recorded.

Doughs for testing in the extensigraph are usually prepared in the farinograph mixer. Water addition is adjusted to make the final consistency 500 FU. The standard mixing procedures of AACC (1983, Method 54-10) and ICC (1972a) are different. The AACC method calls for mixing for 1 min, a rest period of 5 min, and then mixing until the consistency reaches the maximum value. The ICC method stipulates uninterrupted mixing for a fixed period of 5 min. This difference in mixing procedures implies a difference in water addition. The use of the rest period in the AACC procedure is questionable (Muller and Hlynka, 1964). For research purposes, mixers, mixing times, and water additions can be varied.

According to both standard procedures, each test piece is stretched three times, namely at slightly more than 45, 90, and 135 min after the end of mixing. Two test pieces are rounded and shaped immediately after mixing; after a rest period of 45 min, they are stretched. After these stretchings, the pieces of dough are collected, rounded, and shaped, and after another 45 min, the second stretchings are made. This operation is repeated once more. This procedure was designed to simulate the fermentation period in conventional breadmaking, interrupted by punching and molding; it was devised to predict changes in dough properties during fermentation. However, if each test piece is stretched only once, results are more reproducible and easier to interpret (Muller and Hlynka, 1964). This result is realized in the structural-relaxation procedure (Hlynka, 1955b) and in any procedure in which the time intervals both between mixing and shaping and between shaping and stretching are varied (see also section VI-A). It is also realized in the procedure for the "Kurz-Extensogramm," wherein test pieces are stretched after a rest period of 10 min; the preparation of doughs and test pieces is also different (Brümmer, 1976, 1980).

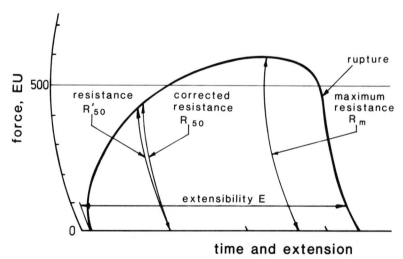

Fig. 13. Representative extensigram showing some commonly measured indexes. One extensigraph unit (EU) equals 1.00 gramforce (9.81 mN) in American instruments (AACC, 1983, Method 54-10) and 1.25 gramforce (12.3 mN) in German instruments (ICC, 1972a).

A representative extensigraph record, or extensigram, is shown in Fig. 13. In this schematic drawing, the curve falls from the rupture point immediately to zero, that is, following a circular curve parallel to the ordinate. Actually recorded curves fall less rapidly as a result of inertia of the instrument and the test pieces and also because the two dough strands do not break at exactly the same time.

Because the speed of the stretching hook is fixed, time is also an approximate measure of extension. For precise calculations, one must take into account the fact that the position of the cradle depends on the force (see also the discussion below of resistance at a fixed extension) and the fact that the mass of dough being stretched increases as stretching proceeds. In a semiempirical way, the curve in Fig. 13 can be transformed into a curve of tensile stress versus extension (Rasper, 1975 and references cited therein). Such a curve rises until the point of rupture. The maximum in Fig. 13 is explained by the fact that beyond it the increase in stress is slower than the decrease in cross-sectional area. After an extensigram has been transformed into a curve of stress versus extension, it can be compared with results of viscometry with extensional flow (discussed in section IV-B); the difference remains that in the extensigraph, the rate of extension is not constant.

The following characteristics of the extensigram are widely used for purposes of quality control (some of them are indicated in Fig. 13):

1. The maximum resistance, R_m;
2. The extensibility, E, expressed as the length of the curve until the point of rupture.
3. The area under the curve. This area is proportional to the energy that is required to bring about rupture of the test piece along the predetermined path; therefore, it is sometimes indicated by energy. The true reason for its importance is that it is a convenient single figure for characterizing flour strength (section V-F), which is associated with both a large resistance and a large extensibility.
4. The resistance at a fixed extension, usually corresponding to 50-mm transposition of the chart paper, R'_{50}. Its advantage over the maximum resistance is that all samples are compared at the same extension and cross-sectional area (Dempster et al, 1952). In a further refinement, one can allow for displacement of the cradle proportional to the force. This means that the points of constant extension in Fig. 13 shift more to the right with increasing force (Dempster et al, 1953); this is indicated by the curve labeled "corrected resistance R_{50}."

Additional extensigram characteristics are the ratio E/R_m, characterizing the shape of the extensigram, and the extension at the maximum force (Merritt and Bailey, 1945). To characterize the "Kurz-Extensogramm," three quantities have been defined that reflect the shape of the curve as well as the area under it (Brümmer, 1980).

It is remarkable that the standard procedures for the extensigraph allow a differentiation among flours. Doughs are prepared with water contents such that their resistance to mixing appears equal; nevertheless, their resistance to extension varies widely. One cannot exclude the possibility that extensigrams reflect differences among flours under mixing conditions rather than under stretching conditions. The low resistance of sticky doughs, for example, is explained by the contribution of stickiness to the measured farinograph consistency (section V-A). As a consequence, the internal resistance to deformation of a sticky dough is lower than that of a nonsticky dough with the same farinograph consistency; this difference will show up when the test pieces are stretched in the extensigraph. This has prompted some researchers to consider a constant water addition to doughs prepared for extensigraph tests (Muller and Hlynka, 1964). A meaningful report on extensigraph tests must indicate the quantities of water added.

HALTON EXTENSIGRAPH OR
SIMON "RESEARCH" EXTENSOMETER

This instrument is based on a design by Halton and Fisher in the late 1930s (Halton, 1949). In it a dough ball is impaled on two halves of a split pin. Then, one of the halves is moved at a constant speed, and the force on the other is recorded. The water content of the dough is based on extrusion measurements (see below). The apparatus, together with the interpretation of its curves, is similar to the Brabender Extensigraph, except that it was designed for measurements with yeasted doughs.

CHOPIN EXTENSIGRAPH OR ALVEOGRAPH

The principle of the Chopin Extensigraph, or Alveograph, dates from the early 1920s. A circular sheet of dough, clamped at its circumference, is inflated by air flowing through a hole in the base plate into an expanding, nearly spherical dough bubble; eventually, the bubble ruptures. The excess pressure of the air in the dough bubble is recorded as a function of time. An accessory, the "relaxomètre," makes it possible to stop the flow of air at a chosen volume of the bubble and to record the decay of pressure; this corresponds to the measurement of stress relaxation (Launay, 1979).

Whereas in the Brabender and Halton extensigraphs the extension of the test piece is in only one direction, or uniaxial, in the alveograph it is in all directions in the surface of the bubble, or biaxial. Time is a measure of the volume of the bubble and therefore of the extension of its wall. The excess air pressure in the bubble is a measure of the tensile stress in the wall. Both extension and tensile stress are nonhomogeneous; at any moment they increase from the base to the top of the bubble.

The alveograph requires doughs with a low water content. The standard procedure (ISO, 1983) specifies the addition of 51.8 ml of salt solution (25 g per liter of solution) to 100 g of flour with 14% moisture; this corresponds to a water addition of 51.4%, irrespective of the water absorption of the flour. Just as the use of a constant water addition to doughs has been considered for the Brabender Extensigraph, so has the use of doughs with a constant consistency been considered for the alveograph. The discussion on the advantages and disadvantages of constant water and constant consistency started in the 1950s and is still continuing (Rasper et al, 1985 and references cited therein; Faridi and Rasper, 1987).

In the standard procedure (ISO, 1983), the test pieces are shaped partly immediately after mixing, but mainly 20 min later. Stretching is done immediately after the final shaping.

A typical alveograph record, or alveogram, is shown in Fig. 14. The remarkable, sharp maximum in the curve does not reflect a change in dough properties; it is explained by the geometry of the test piece and its deformation (Bloksma, 1957, 1958; Launay et al, 1977). The usual interpretation of the alveogram is similar to that of the extensigram. The maximum height of the curve is a measure of resistance, and its length is a measure of extensibility. Because the doughs are made with a constant water addition, the resistance is strongly affected by the water absorption of the flour. Instead of using the area under the curve itself, this area is multiplied by a constant factor; the product is called the *W* of Chopin. It is widely used for quality assessment in France and countries trading with France. Its physical meaning and the reason for its use are the same as for the area under the extensigram.

Unlike the Brabender Extensigraph, the alveograph does not show the effect of flour improvers. This limitation can be overcome by applying special procedures that provide for longer reaction times and for structural activation (section VI-A) (Bennett and Coppock, 1952; Hlynka and Barth, 1955).

C. Extrusion Meter

In the normal operation of the Simon "Research" water absorption meter, the time required to extrude a fixed volume of dough through an orifice under a fixed force is measured. The rate of extrusion of doughs increases with increasing water content (see Table X). Assuming that the optimum water content of dough corresponds to a particular extrusion rate, one can use this instrument to determine the optimum water addition to an unknown flour. The method can be applied only if a suitable mixer is available.

D. Electronic Recording

In the physical dough-testing instruments discussed in the preceding sections, force, torque, and pressure are measured and recorded by mechanical means. When transducers became available, they made it possible to feed an electrical signal into electronic amplifiers and recorders. Laboratory engineers replaced mechanical devices in existing instruments with electrical ones; such

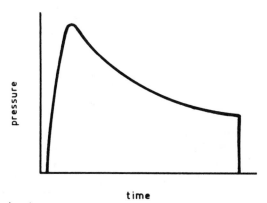

Fig. 14. Representative alveogram.

replacements have been described for the farinograph (Voisey et al, 1966, 1971), extensigraph (Evans et al, 1974), mixograph (Voisey and Kloek, 1980 and references cited therein), and alveograph (Launay et al, 1977). Voisey (1976) reviewed the technical aspects and developments. Some of the instruments discussed above are now commercially available with electronic recording (Sietz, 1983, 1984).

The main advantage of electronic recording is that it prepares the way for automatic processing and filing of data. It adds little to the accuracy of data from, for example, dynamometer deflections. Maintenance and adjustment of electronic instruments require additional special knowledge that may not yet be available in some laboratories.

For research purposes, electronic measurement has the advantage of speed. It gives more information on rapid fluctuations of force and torque than is contained in the bandwidth of conventional recording mixers. It also makes possible better control of mixer speed, as well as mixing at constant power consumption and mixing until a chosen energy consumption (Frazier et al, 1975; Kilborn and Preston, 1985).

E. Comparing Results from Different Instruments

Very high correlations among results of different instruments are not to be expected because the instruments measure different properties, or at least different mixtures of similar properties. In addition, the mechanical history of the test pieces differs among instruments, which also affects results (section VI-A). Experiments with the Brabender and Halton extensigraphs, the latter with various cradles for the test pieces, demonstrated that the history of the test piece is at least as important as the instrument used to make the test (Hlynka, 1955a).

In general, there are significant correlations between the resistances to mixing and between dough development times as determined with various recording mixers (Sibbitt et al, 1953; Miller et al, 1956; Saurer, 1976). Similarly, there are significant correlations between the resistances to deformation and between the extensibilities measured with various load-extension instruments (Aitken et al, 1944).

Correlations between farinogram or mixogram and extensigram character-istics have been established (Johnson et al, 1946; Saurer, 1976). The correlation between farinograph and extensigraph results is also obvious from an examination of curves (Aitken et al, 1944).

F. Application to Flour Quality Control and Research

Tests with the instruments described above provide information about flour strength, balance of properties, and consistency.

Strong wheats or flours usually have a high protein content. They are further characterized, in recording dough mixers, by a high water absorption, a long dough development time, and a small degree of softening or small mixing tolerance index and, in load-extension tests, by a large resistance and a large extensibility. The combination of the latter two characteristics leads to a large area under the curve. Flours lacking most of these properties are called weak.

In the bread bakery, doughs made from strong flours are easy to handle and yield loaves with a large volume and with a fine, resilient crumb. They usually have a high optimum mixing energy (section IX-A) and are also tolerant of overmixing and of fermentation times that are longer than optimum. Strong flours have carrying power; when added to a weak flour that, if used alone, would not produce an acceptable loaf, they can result in a mixture with satisfactory bread-baking properties (Tipples et al, 1982).

Strong flours are the preferred raw material for the production of bread and rusks. A flour may be too strong for breadmaking; such a flour can be used to improve weaker flours. Truly weak flours are the preferred raw material for biscuits, cookies, pastry, and cakes.

Oxidation increases the resistance of dough to deformation and decreases its extensibility (section VIII-D). Usually, this is accompanied by improved loaf properties. Too much oxidation, however, results in impairment of loaf properties; it is called overtreatment. The reversal of the effect in baking is not due to a reversal of changes in rheological properties, but rather to the imbalance of an excessive resistance and a deficient extensibility (Dempster et al, 1956; Bloksma, 1974). Load-extension tests with various quantities of flour improver are useful for obtaining a preliminary estimate of the optimum addition for a particular application. In the evaluation of such tests, it is logical to use one of the measures that characterize the shape of the curve (section V-B) (Chamberlain et al, 1962). For a successful application of such tests, it is necessary to consider the product to be made as well as the process by which it is made; the rheologically optimum dough properties depend on both (Schäfer, 1984 and references cited therein). It is also disputable whether the standard procedures for the extensigraph and, in particular, the alveograph are the most suitable ones for this purpose; different time schedules may give more useful information.

The discussion in the preceding paragraph is equally applicable to more complex cases in which combinations of oxidizing and reducing agents, enzymes, and emulsifiers are applied (Brümmer et al, 1980).

Load-extension tests with untreated flours give information on the inherent balance or imbalance of the dough properties. They guide the millers in determining how to compose a grist in which defects of one wheat lot are compensated for by the attributes of another lot and how to exploit fully the potential in the available raw materials (Bolling, 1980).

Recording dough mixers and extrusion meters give some measure of consistency. Penetrometers, though not specifically designed for measurements on dough, are also useful for this purpose (Auerman, 1962; Schmieder and Zabel, 1966 and references cited therein; Wassermann and Dörfner, 1971; Nieman, 1978c). If a target consistency value for these measurements is specified in terms of maximum deflection, extrusion time, or depth of penetration, the measurements can be used to predict the optimum water addition to an unknown flour, which has direct importance for the baker. The better temperature control and the more regular shape of the recorded band in the farinograph make it better suited and more widely used to determine water absorption than the mixograph. The mixograph is more widely used to evaluate flour strength. The optimum addition of water to a flour in a bakery is correlated with the farinograph water absorption. Their relation depends on the product to be made and the process applied; the same considerations apply here as apply to the balance between resistance and extensibility.

Physical dough tests are used also as a rapid substitute for baking tests. A condition for this application is that results of baking tests can be predicted on the basis of physical measurements. Loaf volume can be predicted more precisely if all flour samples have a similar origin and the same extraction rate. Load-extension tests in which the same flour improvers in the same quantities are used as in the baking tests, and from which both the area under the curve and its shape are used, give the most satisfactory predictions. Data of Brümmer (1978, 1981), Weipert (1981), and Bolling and Weipert (1984) indicate that under the most favorable conditions, the standard deviation of the difference between measured and predicted loaf volumes, divided by the loaf volume, is 0.05.

Physical dough tests are a useful complement of baking tests in that they pinpoint the causes of poor baking performance of a flour.

The application of physical dough tests for research purposes is restricted only by the imagination of the scientist and the fact that their results cannot be described in terms of clear-cut rheological properties, as mentioned in the introduction to section V. In combination with analytical techniques and baking tests, physical dough tests have played an important role in the elucidation of the action of flour improvers (Dempster et al, 1956).

VI. PHYSICAL FACTORS THAT INFLUENCE DOUGH PROPERTIES

A. Mechanical Work and Time

Some changes in dough properties that appear to be due to time are in fact consequences of chemical reactions slowly proceeding in a resting dough. For example, bromate increases the resistance of dough to deformation in the course of some hours (Dempster et al, 1956). The slow decrease in resistance of untreated doughs is ascribed to proteolytic and amylolytic reactions. The following discussion of time effects is restricted to changes that cannot easily be explained by chemical reactions.

The most obvious example of mechanical work affecting dough properties is the mixing process, which develops a dough from a mixture of flour, water, and other optional ingredients. Continuation of the same process causes structures to break down again; this is called overmixing. Mixing, overmixing, and "unmixing" or "undevelopment" are discussed in more detail in section IX-A.

Another effect of mechanical work is the maximum in the curve of stress versus time in viscometric experiments with a sufficiently high rate of shear (section IV-B); in these experiments, shear results in a reduction of the apparent viscosity. If this reduction depended on the quantity of shear only, the maximum stress would occur at the same shear, independent of the rate of shear. However, the shear at the maximum stress increases with increasing rate of shear (A. H. Bloksma and J. A. Duiser, unpublished results). Apparently, the onset of structure breakdown depends on both the quantity and the rate of shear.

Effects of combinations of mixing and time on dough properties can be demonstrated by recording load-extension curves with different rest periods between shaping the test piece and stretching it. Thus, after mixing with a small energy input, the resistance to extension gradually decreases as the rest period increases (Fig. 15, curve corresponding to mixing energy of 10 kJ/kg). Rounding and shaping the test piece change its structure so as to make it more resistant to deformation; this is called structural activation. The increased resistance to deformation may be due to a better alignment of glutenin molecules, permitting more physical cross-links between them. Structural activation is followed by structural relaxation, a spontaneous return to an equilibrium structure with fewer cross-links and less resistance to extension (Hlynka, 1955b).

In several cases, structural relaxation could be described by the hyperbolic equation (Dempster et al, 1955):

$$R = R_a + C/t , \qquad (4)$$

where R = the resistance measured at a fixed extension and corrected for the displacement of the cradle (section V-B), t = the rest period (i.e., the time interval between shaping and stretching), R_a = the asymptotic resistance or asymptotic load, and C = the (structural) relaxation constant. R_a and C are parameters that are adapted to fit a series of measurements after various rest periods. The rest period is to be distinguished from the reaction time, which is the time interval between mixing and shaping.

After mixing in air with larger energy inputs, the curves of resistance versus rest period have a very different shape (Fig. 15). With the highest energy level,

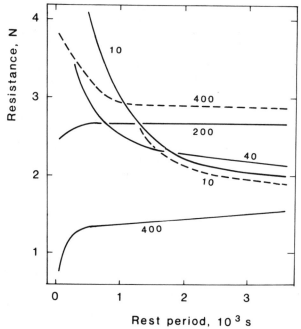

Fig. 15. Resistance to extension as a function of mixing energy and rest period (i.e., the time interval between shaping and stretching). Labels indicate mixing energy in kJ/kg; mixing power was 333 W/kg. Solid lines refer to mixing in air, dashed lines to mixing in nitrogen. Resistance was measured at a constant extension of the test pieces. (Data from Frazier et al, 1985)

the initial resistance is small, then increases with increasing rest period; the dough shows structural recovery instead of structural relaxation. For this phenomenon to occur, two conditions must be satisfied. First, mixing must be severe; the energy must exceed the optimum energy for breadmaking (section IX-A). Second, the mixing atmosphere must contain sufficient oxygen; the (dashed) nitrogen curves in Fig. 15 do not show small initial resistance, irrespective of the mixing energy. A possible explanation for this phenomenon is discussed in section VIII-D.

Sheeting is a practical application of mechanical work to change dough properties. Sheeting of a premixed dough can be used to achieve full dough development. Development by sheeting requires considerably less energy (3.1 kJ/kg) than development by mixing (22 kJ/kg) at roughly the same power consumption (Kilborn and Tipples, 1974).

These examples demonstrate that the application of mechanical work to dough can act in either direction: it can strengthen dough structure, increasing its resistance to deformation, or it can weaken it. The direction depends on the

TABLE VII
Effect of Temperature on Rheological Properties of Dough[a]

Procedure and Property	Factor[a]	Energy of Activation (MJ/kmol)	No.[b]	References
Viscometry, shear flow				
Apparent viscosity	0.953–0.967	26[c]–36	5	Launay and Buré (1973)
	0.949–0.956	34–40	3	Bloksma and Nieman (1975)
Modulus	0.953–0.967	···	3	Bloksma and Nieman (1975)
Viscometry, extensional flow				
Stress at constant time	0.960–0.974	30–80[d]	2	Tschoegl et al (1970b)
or extension	0.965–0.985	30–110[d]	6	Rasper et al (1974)
	0.974	20	1	Budicek and Bulena (1973)
Stress relaxation				
Rate	1.06–1.14	46–101	3[e]	Cunningham, J. R., and Hlynka (1954)
	1.17	120	1	Telegdy Kováts and Lásztity (1966)
	1.09–1.12	62–81	1[f]	Launay and Buré (1974)
Oscillatory measurements				
Storage modulus	0.964	130	1[g]	Shimizu and Ichiba (1958)
	0.964–0.973	76	1[h]	Hibberd and Wallace (1966)
Loss modulus	0.974	46	1[h]	Hibberd and Wallace (1966)
Brabender Farinograph Consistency				
At maximum	0.966	26	1	Bohn and Bailey (1936)
	0.96–0.99[i]		1	Bayfield and Stone (1960)
	0.973–0.984[j]	13–21	3	Hlynka (1962)
At constant time	0.964	28	1	Bayfield and Stone (1960)
Mixograph				
Height of curve	0.978–0.988	9–17	4	Harris and Sibbitt (1945)
Brabender Extensigraph Resistance at constant				
extension	0.94–0.96	···	1[k]	Dempster et al (1955)
Rate of structural				
relaxation	1.07–1.10	50–80[l]	1[k]	Dempster et al (1955)
				(continued on next page)

TABLE VII (continued)

Procedure and Property	Factor[a]	Energy of Activation (MJ/kmol)	No.[b]	References
Chopin Extensigraph				
Resistance at constant extension	0.969–0.987		1[m]	Jelaca and Dodds (1969)
Rate of structural relaxation	1.026–1.061	19–45	1[m]	Jelaca and Dodds (1969)
Extrusion meter				
Extrusion rate	1.06	⋯	1	Stamberg and Bailey (1940)
Laboratory mixer				
Torque on driving shaft	0.979	16	1	Budicek and Bulena (1973)
Commercial mixer				
Net power consumption	0.976	19	1	Bloksma (1985)

[a] For each K increase in temperature, the properties listed in column 1 change by the factors shown in column 2.
[b] Number of dough compositions examined.
[c] After excessively long mixing times, lower values were found.
[d] Energies of activation reported by these authors are based on a different physical model; therefore, they cannot be compared with the other values in this table.
[e] Three water levels in combination with four rest periods.
[f] Two values of initial stress.
[g] Shear amplitude 0.25; frequency range 0.3–4 rad/sec.
[h] Shear amplitude in the linear region; frequency range 0.1–260 rad/sec. The time-temperature superposition principle was applied separately to the storage and the loss moduli.
[i] Increasing with increasing temperature. Dough development time decreased with increasing temperature.
[j] Increasing with increasing water content.
[k] Three reaction times.
[l] Increasing with increasing addition of potassium bromate.
[m] Four series of measurements. Control doughs after 20 min of rest. Mixing, reaction time, and rest period at the same temperature as the stretchings.

history of the dough and the type of action. Consequently, mixing energy, and even the combination of power and energy, do not completely characterize the mixing process; their effect also depends on the geometry of the mixing action, which is much more difficult to specify.

Because of the effects of mechanical work and of time, results of physical dough tests cannot be compared unless the mechanical history of the test pieces is the same. In research reports, this history must be described precisely, and in routine tests, specified procedures and time intervals must be strictly adhered to.

B. Temperature

The rheological properties of most materials are highly sensitive to temperature. Dough is no exception. Table VII summarizes data from the literature on the effect of temperature on dough properties. The data are based on measurements over a narrow temperature range, namely roughly 15–40° C; freezing and starch gelatinization were its natural limitations. The table shows that the resistance of dough to deformation decreases relatively about 0.03 per Kelvin increase in temperature; the rates of stress relaxation, structural relaxation, and extrusion increase more strongly.

In the case of rate processes like viscous flow, the effect of temperature can be described by the so-called energy of activation, or the energy that a molecule or part of it must acquire before it can take part in the rate process. The relation is described by the Arrhenius equation:

$$\eta = \eta_o \cdot \exp = [(W/R) \cdot (1/T - 1/T_o)] \quad , \tag{5}$$

where η = the (apparent) viscosity at the absolute temperature T, η_o = the (apparent) viscosity at the reference (absolute) temperature T_o, W = the energy of activation, and R = the gas constant.

In the case of stress relaxation and oscillatory measurements, the energy of activation can be estimated on the basis of the time-temperature superposition principle. This principle is applicable if graphs of the property considered versus the logarithm of time or frequency at various temperatures are identical except for a shift parallel to the time or frequency axis. If the temperature changes from T_o to T, the shift of the curve along a time axis is:

$$\ln a = (W/R) \cdot (1/T - 1/T_o) , \tag{6}$$

analogous to equation 5. The shift along a frequency axis has the opposite sign.

Table VII shows that not only the viscosity of dough decreases with increasing temperature, but also its modulus. The modulus of rubber and similar polymers, the elastic resistance of which is caused by a decrease of entropy upon deformation, is proportional to the absolute temperature. Therefore, the cause of elasticity in dough is different from the cause of elasticity in rubber (Bloksma and Nieman, 1975).

If temperature is increased above 50° C, viscosity increases sharply (Fig. 16). This sharp increase is explained by the gelatinization of starch and, to a lesser extent, by the polymerization of glutenin molecules (section IX-C). That gelatinization proceeds in time at a finite rate explains why the minimum in the curves shifts to higher temperatures as the rate of heating is increased. The curve in Fig. 16 that corresponds to an average rate of heating of 0.10K/sec is representative of the course of viscosity in a dough baked in tins about 12 cm wide at an oven temperature of about 230° C.

Temperature can also have an indirect effect on dough properties. Chemical reactions proceed faster at a higher temperature. If the reaction results in a stiffer dough, the normal softening of a dough with increasing temperature can be reduced or even reversed to a stiffening. Such a reversal has in fact been observed with doughs containing bromate (Dempster et al, 1955).

VII. INFLUENCE OF FLOUR CONSTITUENTS ON DOUGH PROPERTIES

In this section, various flour constituents are discussed briefly. For more information on them, see Volume I, Chapters 5 (proteins), 6 (carbohydrates), 7 (lipids), and 8 (enzymes and pigments), and Volume II, Chapter 5 (composition and functionality).

A. Fractionation and Reconstitution Methods

The fractionation of flour and the study of doughs from reconstituted flours is a logical approach to the question of what contribution each of the dough constituents makes to the physical properties. Flours can be reconstituted from flour fractions in the same proportions as they occur in normal flour; or one or more components can be deleted or added in abnormal concentrations; or the naturally occurring components may be replaced by materials other than flour components. To reach valid conclusions from fractionation and reconstitution techniques, one must establish that a reconstituted dough of normal composition behaves like a dough from normal flour.

The isolation of gluten from wheat flour dough by kneading the dough in water or dilute salt solutions is the oldest fractionation technique (Bailey, 1941 [translation of a 1728 lecture by Beccari]). The method has been used extensively for quality testing as well as for isolating gluten for reconstitution experiments. Methods have been expanded and refined, leading to successful fractionation of wheat flour into its major constituents and fractions: starch, gluten, tailings, water-solubles, and lipids. Proteins and lipids have been extensively subfractionated.

MacRitchie (1985) made a comprehensive study of different steps in the separation, fractionation, and reconstitution of wheat flours. Important

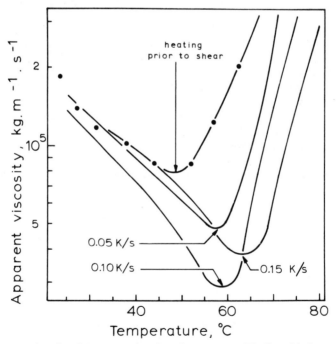

Fig. 16. Apparent viscosity of dough as a function of temperature. The line with dots represents a series of measurements in which the test piece was brought to the desired temperature in about 30 min and then sheared. The solid lines show the course of viscosity when the test piece was heated during shearing; labels indicate the average rate of heating between 30 and 70°C. Rate of shear was 6.5×10^{-4} sec^{-1}. (Reprinted, by permission, from Bloksma, 1980)

precautions that must be observed in order to retain the native functional properties of fractions are the "extraction of lipid with an appropriate solvent . . . , the washing out of gluten at an optimum temperature (15°C), minimization of contact time for gluten protein fractions with acid solution, neutralization of extracting solutions to the correct pH and grinding of freeze-dried fractions to a suitable particle size" (MacRitchie, 1985).

B. Proteins and Gluten Structure

COMPOSITION AND FUNCTIONAL PROPERTIES

The functional properties of wheat flour proteins in doughs at various stages of the breadmaking process depend on the composition of the protein constituent, the molecular properties of the proteins, and their interactions with themselves and other flour constituents (Bushuk and MacRitchie, 1988). In the context of functionality, flour proteins can be classified in two different ways, both based on differences in solubility.

A crude classification is a distinction between water-soluble proteins (mainly albumins and globulins) and water-insoluble proteins (gliadins and glutenins) or gluten proteins (MacRitchie, 1984). The quantity of the water-insoluble protein and its structure appear to constitute the important factor in functionality in breadmaking (Table VIII).

A more detailed classification results from the modified Osborne procedure, which divides flour proteins into five groups (Chen and Bushuk, 1970). It has proved quite useful in ranking flours according to breadmaking potential. Orth and Bushuk (1972) showed that loaf volume for a group of widely different flours varied inversely with the amount of glutenin that is soluble in $0.05N$ acetic acid solution and directly with the amount that is insoluble in that solvent (residue protein). The composition of flour proteins of hard red spring and durum wheats is given in Table IX. Note that durum wheat contains substantially more gliadin and less glutenin than the bread wheat; this explains in part the higher extensibility of durum flour doughs (and gluten). Albumin, globulin, and gliadin contents of widely different bread wheats are relatively constant (Orth and Bushuk, 1972). Hydrated glutenin is a tough, elastic material; hydrated gliadin is a viscous liquid. This suggests that a mixture of the

TABLE VIII
Properties of Gluten and Nongluten Proteins[a]

Property	Gluten Proteins	Nongluten Proteins
Amount in wheat flour	6–18%	0–2%
Chemical composition	High glutamic acid, mainly glutamine (to 40%)	Glutamic acid (to 12%)
	High proline (to 14%)	Proline (to 8%)
	Low lysine (to 1.5%)	Lysine (to 3%)
Solubility	Insoluble in neutral solutions, salts; soluble (partially) in acid, alkali, and urea; solubility increases with increasing temperature	Soluble in neutral solution; precipitated by heating
Physical properties	Responsible for viscoelasticity of dough and baking properties	Less important for dough properties or in baking

[a] Source: MacRitchie (1984); used by permission.

two is essential to impart to dough the viscoelastic properties that are associated with good breadmaking performance (Dimler, 1963; for reviews see Wall, 1979; Bushuk and MacRitchie, 1988).

MOLECULAR STRUCTURE OF GLUTEN PROTEINS

Gluten is the proteinaceous material that is recovered after washing out of flour-water doughs the soluble and occluded substances of flour. The resulting gluten contains approximately 65% water. For use as an additive to flour, it is usually dried. From the point of view of functionality in breadmaking, gluten of high water-binding capacity is preferred. Approximately 90% of flour protein can be recovered in the gluten. It therefore contains considerable soluble protein in addition to the so-called gluten proteins, the gliadins and glutenins. On a dry-matter basis, gluten contains 75–86% protein; the remainder is carbohydrate and lipid, held strongly within the gluten protein matrix.

The viscoelastic properties of dough are primarily due to its continuous protein phase. At the molecular level, the behavior of the continuous phase can be studied in relation to two relatively distinct aspects: the molecular properties of the constituent polymeric substances (mainly proteins), and the interactions among the proteins and between the proteins and nonprotein constituents. Rheological behavior of polymeric networks depends on the concentration and strength of cross-links, the molecular weight of intercross-link regions, and the average molecular weight and weight distribution of the constituent polymers (Bushuk and MacRitchie, 1987).

It is well established that the most important cross-links in gluten structure are the disulfide bonds, the hydrogen bonds, and the hydrophobic interactions. A separate subsection is devoted to each of these cross-links. Electrostatic bonds make a significant contribution, as can be inferred from the well-known effects of salt on the rheological properties of dough. However, the number of these bonds is considered to be quite low, as inferred from the relatively low content of amino acids with ionizable side groups.

Disulfide Cross-Links. The fact that mercaptans and sulfite improve the solubility of gluten (and especially of glutenin) is a first indication that disulfide bonds are important for its cohesion (De Deken and Mortier, 1955; Wren and Nutt, 1967, Bloksma, 1975). In glutenin, the disulfide bonds (of the interpolypeptide type) actually bind together protein chains with varying

TABLE IX
Distribution of Flour Proteins Among Modified
Osborne Solubility Fractions[a]

Protein Fraction	Solvent	Hard Red Spring Wheat (%)	Durum Wheat (%)
Nongluten proteins			
Albumin	Water	12.4	12.3
Globulin	0.5N NaCl	5.4	4.7
Gluten proteins			
Gliadin	70% Ethanol	29.6	41.1
Soluble glutenin	0.05N Acetic acid	17.3	18.5
Insoluble glutenin	···	35.3	23.4

[a]Source: Bushuk (1985); used by permission.

molecular weights in the range of 30,000–140,000 to form giant structures with molecular weights up to some millions (Ewart, 1968, 1972, 1977, 1978, 1979; Graveland et al, 1985). In contrast, the disulfide bonds in those gliadins that contain them are of the intrapolypeptide type (Wall, 1979).

As early as 1936, it was suggested that the stiffening of a dough induced by oxidative flour improvers might be due to the formation of disulfide cross-links (SS) from thiol groups (SH) (Balls and Hale, 1936). The reaction corresponds to the conversion of the amino acid cysteine into cystine. It can be written as follows:

$$2 \text{ Pr-CH}_2\text{-SH} \rightleftharpoons \text{Pr-CH}_2\text{-SS-CH}_2\text{-Pr} + 2\text{H}$$

  cysteine   cystine ,

where Pr represents a protein chain molecule. The double arrow indicates that disulfide bonds can also be reduced to thiol groups; it is not, however, a readily established equilibrium.

Only after methods for the quantitative determination of thiol and disulfide groups became available in the late 1950s was it possible to obtain a better understanding of the role of these groups. Flour contains on the order of 10 and 1 mmol of total disulfide and thiol groups, respectively, per kilogram, or 100 and 10 mmol per kilogram of protein (Mecham, 1968). Graveland et al (1978) reported 5–7 mmol of thiol groups and 11–18 mmol of disulfide per kilogram of flour, determined in the presence of urea and sodium dodecyl sulfate to unfold the proteins and make those groups accessible to analytical reagents. These higher values have not been replicated. When comparing literature data, one must keep in mind that the level of the results and their interpretation also depend on the method used (Bloksma, 1959). The distribution of disulfide and thiol groups in the various protein and nonprotein fractions is discussed later in this section.

In the last 20 years, many determinations of the thiol and disulfide contents of doughs have been reported. Many of these reports give information on the effect of oxidation or thiol-blocking on the thiol and disulfide contents of doughs. The effects on the thiol contents are reviewed in the discussion of the various reagents (section VIII-D). Disulfide contents are hardly changed in these experiments.

The contribution of disulfide bonds to the rheological properties of dough can be explained on the basis of either of two models of gluten structure. In the early literature, the continuous protein network model, with covalent interpolypeptide disulfide cross-links, was preferred. Results on the effect of potassium iodate and thioacetic acid on the birefringence of stretched gluten are consistent with this model (Frazier and Muller, 1967). However, the model has some serious limitations (Bloksma, 1975) and more recently has been superseded by a model in which noncovalent cross-links play a key role in the rheological properties of the gluten. Models based on this concept have been described by Ewart (1968, 1972, 1977, 1978, 1979) and MacRitchie (1980a and references therein). In the noncovalent cross-link model, the molecular weight distribution of the glutenins is the main factor that determines rheological properties. Thiol-disulfide interchange (see below) acts via this distribution. Elasticity is ascribed to deformation within glutenin subunits and viscosity to slip between the interacting molecules of the gluten complex.

We need to focus our attention not only on the disulfide groups but also on the thiol groups for two reasons. First, thiol reagents like *N*-ethylmaleimide (NEMI), which do not form disulfide cross-links, have effects on rheological properties similar to effects of oxidizing reagents. Second, a network with a number of permanent cross-links can only be deformed elastically; that is, the deformation disappears when the load is removed. From the data in section V, it can be concluded that, except under conditions of short duration of the load and a small stress, the deformation of dough is predominantly viscous or permanent. A permanent deformation can be attained only if the cross-links are not permanent or if they are broken under severe stress.

Goldstein (1957) was the first to suggest that thiol-disulfide interchange reactions can explain the occurrence of permanent deformations of a continuous protein network with disulfide cross-links. These reactions are shown in Fig. 17. The first stage shows two protein molecules cross-linked by two disulfide groups. Reaction 1 opens the 1,2- disulfide group. Reaction 2 forms a new disulfide cross-link between positions 2 and 3. Between reactions 1 and 2, the conformations of the protein molecules may change as a result of Brownian motion. In Fig. 17, it is arbitrarily assumed that the upper chain retains its conformation and that the conformation of the lower molecule is changed. Reaction 3 is the net result of reactions 1 and 2. The thiol compound XSH becomes available again for another cycle of interchange reactions. A small amount of a thiol compound, the nature of which is discussed later, permits any number of reaction cycles, however large, provided that sufficient time is available. Under stress, the direction of the Brownian motion is biased; as a consequence, the combined conformation changes result in measurable deformation.

McDermott and Pace (1961) demonstrated for the first time the actual occurrence of these interchange reactions in dough. They added thiolated gelatin to dough, isolated gluten from it, and found hydroxyproline in its hydrolysate; this amino acid occurs naturally in gelatin but not in gluten. Since then, similar experiments have been reported with disulfides from amino acids that do not naturally occur in proteins (Redman and Ewart, 1967a, 1967b; McDermott et al, 1969), radioactively labeled amino acids (Mauritzen and Stewart, 1963; Stewart and Mauritzen, 1966) or glutathione (Kuninori and Sullivan, 1968), and with radioactive flour and inactive thiols (Lee and Lai, 1968).

Thiol-disulfide interchange during mixing is a logical explanation for the formation of a protein network in dough, in which protein molecules originating from different flour particles are cross-linked to form a continuous and coherent phase. The observation that the interchange reaction is most rapid in wheat flour doughs, slower in doughs from rye, and still slower in doughs from other cereals may explain the differences between these doughs in gas retention (Redman and Ewart, 1967b). Why the rate differs among cereals is not known.

In the model with noncovalent cross-links, thiol-disulfide interchange reactions affect the rheological properties via the molecular weight distribution.

According to the continuous network model of protein structure, the elastic (temporary) deformation of gluten and dough is restricted by the number of disulfide bonds. On the other hand, the presence of thiol groups is required for a viscous (permanent) deformation. The model predicts that the modulus increases with increasing disulfide content and that the viscosity decreases with increasing thiol content. As a consequence, oxidation of thiol to disulfide groups will increase both the viscosity and the modulus. Qualitatively this has indeed been observed in experiments with various oxidizing, reducing, and thiol-blocking agents (Bloksma, 1972b, 1975). A quantitative explanation could not be attained, however. The change in rheological properties was greater (in proportion) than the change in thiol and/or disulfide contents. The relations between the disulfide content and the elastic deformation and between the thiol content and the viscous deformation were not unequivocal but depended on the reagent used to change the thiol and disulfide contents.

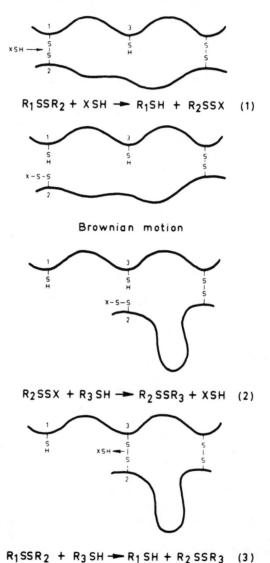

$$R_1SSR_2 + XSH \longrightarrow R_1SH + R_2SSX \quad (1)$$

Brownian motion

$$R_2SSX + R_3SH \longrightarrow R_2SSR_3 + XSH \quad (2)$$

$$R_1SSR_2 + R_3SH \longrightarrow R_1SH + R_2SSR_3 \quad (3)$$

Fig. 17. Viscous flow as a result of thiol-disulfide interchange reactions in the protein network.

That a small change in disulfide content results in a relatively large change in the modulus might be explained if only a small fraction of the total disulfide groups participates in the cross-linking of long-chain molecules; this fraction is called rheologically effective. Observations of the elastic deformation of isolated gluten are in accordance with this assumption. Estimates of the rheologically effective disulfide groups, based on application of the theory of rubber elasticity to these observations, are 3% (Bloksma, 1968) or even less (Muller, 1969) of the total disulfide content. A partial explanation of this low percentage is that in glutenin, about half of the disulfide groups are required for binding the subunits into long-chain molecules and do not contribute to the modulus (Ewart, 1968, 1972).

The factors that determine the rheological effectiveness of disulfide bonds are, in general, different from those that determine their chemical reactivity. Therefore, no correlation between these two properties is to be expected for disulfide bonds.

Unlike disulfide bonds, thiol groups are expected to show a parallel between their chemical reactivity and rheological effectiveness. Chemically reactive thiol groups will readily enter the interchange reactions in Fig. 17 and, consequently, will also be rheologically effective. Small thiol compounds, like cysteine and glutathione, diffuse much more rapidly in dough than do proteins. Therefore, they are expected to be chemically more reactive as well as rheologically more effective. In extensigraph measurements, added glutathione (mol wt 307) was indeed found to be more effective than thiolated gelatin (mol wt 10,000 or more) with the same amount of thiol groups (Villegas et al, 1963). Differences in rheological effectiveness are found not only between added reagents, but also between thiol groups from flour. Upon oxidation by iodate, the viscous compliance decreased more rapidly than did the total thiol content; this observation could be explained by the assumption that 30–40% of the thiol groups in dough are rheologically ineffective (Bloksma, 1968). Jones et al (1974) estimated, on the basis of farinograph measurements, that approximately 25–35% of the thiol groups and 4–13% of the disulfide bonds are rheologically effective.

Because the rheological effectiveness of thiol and disulfide groups depends on the compound or protein fraction of which they are a part, it may be useful to review our knowledge of the distribution of these groups in flour fractions.

By far the most disulfide groups are found in the proteins, and only less than 10% is found in smaller molecules (Proskuryakov and Zueva, 1964; Hird et al, 1968). Slightly reactive, basic peptides contribute a few percent, and the much more reactive oxidized glutathione, an acid peptide, contributes only 0.03 mmol per kilogram of flour, or 0.3% of total disulfide content (Hird et al, 1968; Jones and Carnegie, 1969a). Added disulfides with low molecular weight decrease the resistance of dough to extension in the same way as thiols do (section VIII-E). However, it has not been determined whether the effect on dough properties of this small amount of naturally present oxidized glutathione is sufficiently large to be detected.

The thiol groups were found to be less restricted to the proteins than the disulfide groups. About 10–30% of the total thiol groups is found in the nonprotein fraction (Proskuryakov and Zueva, 1964; Hird et al, 1968), about 5% is in glutathione (Kuninori and Matsumoto, 1964b; Hird et al, 1968), and about

10% is in the total of the highly reactive, small peptides (Tkachuk, 1969). Within the proteins, the soluble proteins contain relatively more thiol groups than the gluten proteins (Matsumoto and Hlynka, 1959; Sullivan et al, 1961a; Agatova and Proskuryakov, 1962). The differences in the rheological effectiveness of flour thiol groups, mentioned above, may be related to the distribution of these groups in the various protein fractions and nonprotein compounds.

Titration results support the conclusions in the preceding paragraphs. If urea is omitted from the solvent in which the material is dispersed for titration, only the reactive groups are titrated, presumably those in smaller molecules. It has been found that at least 60% of the thiol groups, but only a much lower percentage of the disulfide groups, is reactive (Stevens, 1966; Tsen and Bushuk, 1968). This is qualitatively in accordance with the fact that by far the most disulfide bonds are found in proteins, whereas a considerable fraction of the thiol groups forms part of compounds with lower molecular weights.

In the gliadin molecules that contain disulfide groups, these disulfide groups appear to be randomly distributed along the polypeptide chain (Shewry et al, 1984). In glutenin, SH groups in reduced subunits of most of the high-molecular-weight subunits are located near the ends of the polypeptide chains, thereby facilitating end-to-end linkage of subunits to form the large glutenin molecules postulated by Ewart (1979). The disulfide groups in some high-molecular-weight glutenin subunits are located at some distance from the ends of the polypeptide chains, because cleavage at these residues produced a marked reduction of subunit molecular weight (Moonen et al, 1985).

Hydrogen Bonds. Noncovalent interactions between side groups of proteins play an important role in the physical properties of concentrated protein systems such as dough. Two interactions, hydrogen bonds and hydrophobic interactions, appear to be extremely important to the functional properties of gluten proteins.

The presence of numerous hydrogen bonds in gluten can be inferred from its high content of glutamine and asparagine (about 35%). The direct contribution of hydrogen bonds to the viscoelastic properties of gluten has been demonstrated by two different experiments. First, desirable viscoelastic properties are impaired when the glutamine side chains are converted into carboxy groups by mild acid hydrolysis (Holme and Briggs, 1959) or into methyl esters by methanolysis (Beckwith et al, 1963; Mita and Matsumoto, 1984). Second, mixing flour with heavy water instead of ordinary water increased the resistance of the dough to deformation because deuterium bonds are stronger than hydrogen bonds (Kretovich and Vakar, 1964; Tkachuk and Hlynka, 1968).

Hydrophobic Interactions. Gluten proteins contain about 30% nonpolar amino acid residues. Accordingly, relatively high average hydrophobicity values can be estimated for gluten proteins (Belitz et al, 1982; McMaster, 1982) using the procedure of Bigelow (1967). Van der Waals forces between these nonpolar amino acid residues can form numerous though weak cross-links. The formation of such cross-links reduces the thermodynamically unfavorable contact between nonpolar groups and water, which strengthens the cross-link. This is why interactions between nonpolar groups of proteins in water are called hydrophobic bonds. They can result in the formation of areas from which water is excluded and which are occupied by the nonpolar amino acid residues of one or more protein molecules.

The presence of hydrophobic interactions in gluten proteins explains the solubility of gliadin in 70% ethanol, in which aggregates dissociate. Dissociation also occurs in the presence of detergents (Schofield, J. D., 1986), soaps (Kobrehel and Bushuk, 1977; Hamauzu et al, 1979), and chaotropic salts (Preston, 1981). There is evidence that the hydrophobic interactions in proteins of higher-quality flours are stronger (or greater in number) because they require a higher concentration of soap for equivalent dissolution (Kobrehel, 1984).

C. Carbohydrates

The last decade has produced numerous excellent reviews on the carbohydrate constituents of flour and their functionality in dough and bread (Greenwood, 1976; MacRitchie, 1984; Evers and Stevens, 1985; Manners, 1985a, 1985b; Shelton and D'Appolonia, 1985; Blanshard, 1986).

The most abundant carbohydrate in dough is starch. Its water absorption and gelatinization are essential phases in the transformation of flour into bread. In this respect, wheat starch cannot be replaced entirely by any other material or by starch from other sources (Sandstedt, 1961). Of the starches from closely related cereals, barley and rye starches are most similar to wheat starch in functionality in the breadmaking process (Hoseney et al, 1971).

Until the onset of gelatinization during baking, the starch granules are in the native state. Native starch granules have been considered an inert, rigid filler in the continuous protein phase. However, recent research (Shelton and D'Appolonia, 1985) suggests that wheat starch granules interact specifically with other dough constituents. This interaction may involve specific proteins present on the surface of some wheat starches (Greenwell and Schofield, 1986). In the context of basic dough rheology, two conclusions emerge from published data. First, with increasing starch content, the extensibility of dough decreases (Heaps and Coppock, 1968). Second, variations in water content affect dough properties more if the starch content is high (Hibberd, 1970b; Bloksma, 1972a). This latter observation can be explained if interactions between starch granules contribute to the rigidity of dough; hydration of the continuous phase reduces the number of contact points.

During milling, some of the starch granules (about 5–10%) are damaged, with the result that they absorb more water than native starch and become available to starch-degrading enzymes. Farinograph measurements (section VIII-A) indicate that water uptake by damaged starch can vary from one and a half to two times its dry mass, occasionally even more (Greer and Stewart, 1959; Moss, H. J., 1961; Meredith, P., 1966b; Belderok and Slager, 1983). Upon enzymatic breakdown, the water absorbed by the damaged starch is set free, which results in a softening of the dough. In addition, maltose is formed and can be utilized by the yeast. As a yeast substrate, available starch supplements the mono-saccharides and disaccharides that are present in the flour (Evers and Stevens, 1985).

Pentosans form a small fraction of flour (2–3%) but have a large effect on its water absorption (Meuser and Suckow, 1986). Farinograph measurements (section VIII-A) indicate that water uptake by pentosans is about 10 times their own dry mass; as a consequence, they hold one fourth of the dough water (Bushuk, 1966; Kulp, 1968). This high water-binding capacity is probably due to a generic effect because in baking tests with starch and gluten, various gums could replace the flour solubles (Cawley, 1964).

A remarkable property of the soluble pentosan fraction is its ability to form gels upon oxidation. The cross-linking process has been localized in a chelate bridge between oxidized ferulic acid groups in a glycoprotein (Painter and Neukom, 1968; Neukom and Markwalder, 1978). Other mechanisms for the gelation have been proposed recently (Hoseney and Faubion, 1981).

D. Lipids

The composition, structure, and functionality of flour lipids have been comprehensively reviewed in numerous recent publications (MacRitchie and Gras, 1973; MacRitchie, 1977, 1980a, 1980b, 1984; Morrison, 1978; Barnes, 1983; Békés et al, 1983a, 1983b; Pomeranz, 1985; Chung, 1986; Larsson, 1986).

The effect of flour lipids and added lipids on dough properties can result from various actions, as Bungenberg de Jong recognized in 1956 (Bungenberg de Jong, 1956). Here we shall make a distinction between chemical reactions in which lipids are involved on the one hand and physical and interface phenomena on the other hand. Because chemical reactions can interfere with experiments on physical actions, we shall discuss them first.

CHEMICAL REACTIONS

The polyunsaturated linoleic acid makes up more than half of the fatty acids in nearly all lipid compounds of flour. In the presence of oxygen, this fatty acid can be enzymatically oxidized. The oxygen consumption during mixing of dough depends on the presence of easily extractable lipids as well as on a heat-labile, water-soluble component, presumably lipoxygenase (Smith, D. E., and Andrews, 1957). Oxygen consumption is extremely rapid in doughs from high-extraction and whole wheat flours (Galliard, 1986; see also section X). Graveland (1968) found that when a dough is mixed in air or oxygen, linoleic and linolenic acid present as free fatty acids or in α-monoglycerides are oxidized; in other compounds, they remain unchanged. Later, Graveland (1970) demonstrated that the primary reaction products of linoleic acid are hydroperoxides along with hydroxy-epoxides. Experiments by Morrison with batters confirmed that free fatty acids are less stable against oxidation than are the esterified acids. In addition to the specific oxidation of polyunsaturated fatty acids, Morrison found another oxidation process that results in a proportional decrease of all fatty acids (Morrison, 1963; Morrison and Maneely, 1969).

The hydroperoxides oxidize thiol groups in the dough. This has been demonstrated for flour lipids that were extracted, oxidized, and added back and for oxidized methyl linoleate (Tsen and Hlynka, 1963). In accordance with this, a definite increase in extensigraph resistance has been reported after addition of peroxides of flour lipids, oxidized methyl linoleate (Tsen and Hlynka, 1963), or oxidized sodium linoleate (Dahle and Sullivan, 1963) in amounts of about 1 mmol per kilogram of flour. The fact that lipids and their peroxides contribute to the oxidation of thiol groups also explains why thiol groups are more slowly oxidized by oxygen in doughs from defatted flours than in doughs from the complete flours (Bloksma, 1963; Tsen and Hlynka, 1963). Whether oxidation of thiol groups by oxygen proceeds solely by way of peroxides or whether direct oxidation also occurs has not yet been established. The peroxides involved are not necessarily formed during dough mixing; even if dough is mixed in the absence of oxygen, peroxides formed during storage of the flour can oxidize thiol groups (Mecham and Knapp, 1966). As a consequence, excessive aging of flour results in a decrease in the extensibility of dough (Sullivan et al, 1936).

Our chemical understanding of the role of flour lipids in the oxidation of thiol groups is contradictory to Bungenberg de Jong's view that flour lipids, by consuming the available oxygen, protect proteins against oxidation (Bungenberg de Jong, 1956). His conclusion was based on the observation that the extensigraph resistance of doughs from defatted flours increased more by mixing in air than that of doughs from normal flour. Smith, D. E., et al (1957) and Narayanan and Hlynka (1962) confirmed these observations. In general, the effect was more obvious with low-grade flours with high lipid and enzyme contents (Bungenberg de Jong, 1956; Galliard, 1986). Also, it is in line with the observation that defatting the flour retards the consumption of bromate during mixing in air (Bushuk and Hlynka, 1961c). At present, the chemical evidence and the rheological evidence have not yet been brought into accordance.

In addition to the reactions discussed above, carotenoid pigments are destroyed through a coupled reaction involving peroxidation of unsaturated fatty acids under the influence of lipoxygenase (Irvine and Winkler, 1950). This bleaching is undesirable in the production of paste products (see Volume I, Chapter 8 and Volume II, Chapter 9). On the other hand, it can be useful to obtain a whiter bread crumb. For a review of patents based on this principle, see Hawthorn (1961).

PHYSICAL PHENOMENA

Having discussed the chemical reactions, we can turn our attention to the physical phenomena involving lipids. Experimenting with doughs from lipid-extracted flour is one approach. Adding other lipids, alone or in combination with emulsifiers, is another approach that is widely used in breadmaking. The effects observed can reliably be ascribed to physical phenomena if precautions are taken to exclude chemical effects, for example, by mixing in the absence of oxygen or by adding only saturated lipids.

Extraction and Reconstitution Experiments. For a comprehensive study of the extraction and reconstitution of flour lipids, see the series of excellent publications by MacRitchie (1977, 1980a, 1980b, 1985).

In experiments with defatted flours, one must keep in mind that nonpolar fat solvents like light petroleum or diethyl ether extract only a part of the lipids, which we call the free lipids; they usually form less than 1% of flour mass. A subsequent extraction with a more polar solvent, such as a chloroform-methanol, butanol-water, or benzene-ethanol-water mixture, removes the so-called bound lipids; they form about 0.5% of flour mass (Wootton, 1966; Graveland, 1968; Chung, 1986). Even after that, acid hydrolysis frees an additional small amount of lipids. Because some lipid fractions bind to flour proteins, they may be considered as part of lipoproteins. Triglycerides and fatty acids are the most abundant constituents of the free lipids; more polar lipids, like glycolipids and phospholipids, form the main part of the bound fraction (Graveland, 1968; Chung, 1986).

The phenomenon of nonpolar solvents extracting only part of the lipids is even more obvious in dough than in flour. During mixing, some of the free lipids are converted into bound ones. Both polar and nonpolar lipids are bound (see discussion of constituent interactions below).

Creep tests after mixing in nitrogen showed that the extraction of free lipids from flour reduced the deformation of dough at the same stress to roughly 0.4 of its original value (Bloksma, 1972a); recovery measurements showed that this was mainly the result of a decrease in the viscous component of the deformation (A. H. Bloksma, unpublished). In accordance with this, measurements with dough-testing machines showed an increase in the resistance against extension (Cookson and Coppock, 1956; Germain et al, 1968; Tao and Pomeranz, 1968). In the latter cases, however, doughs were mixed in air, and interference by oxidation reactions cannot be excluded. The increased resistance may explain why baking tests gave contradictory results on the effect of extraction on loaf quality (Cookson and Coppock, 1956; Pomeranz et al, 1968a). The increased resistance corresponds to the effect of oxidation. With some flours, this action is beneficial, but with others, which were already in a state of optimum oxidation, the extra action induces signs of overtreatment.

After extraction of free and bound lipids from the flour, creep tests after mixing in nitrogen resulted in a sevenfold increase in the total deformation at the same shear stress, primarily because of an increase in the viscous component of the deformation (A. H. Bloksma, unpublished). Dough and gluten have hardly any extensibility, gluten development is slow, and loaf quality is extremely poor (Mecham and Mohammad, 1955; Bloksma, 1966).

Recovery of the original properties after the extracted constituents are added back into the flour can serve as a check that the effect of the extraction is solely the result of the removal of particular constituents. Several instances of successful reconstitution after extraction of free lipids have been reported (Sullivan et al, 1936; Sullivan, 1940; Cole et al, 1960; Pomeranz et al, 1968a; MacRitchie, 1985). Reconstitution after extraction of free and bound lipids is much more difficult (Mecham and Mohammad, 1955; Pomeranz et al, 1968a). Nevertheless, successful reconstitutions after extraction of free and bound lipids were described by Hoseney et al (1969) and MacRitchie (1985).

Fractionation of the lipid extract and reconstitution with a specific fraction can yield information on the effect of the particular fraction. Using this technique, Tao and Pomeranz (1968) found that the nonpolar fraction of the free lipids was more effective in restoring the original mixogram and extensigram of the flour, but that the polar fraction (in particular the glycolipids) was essential for restoring the breadmaking properties (Daftary et al, 1968; MacRitchie, 1985).

Addition of Fats. Triglycerides in amounts of 1–5% of flour mass are commonly added in breadmaking. In general, the addition of triglycerides improves loaf quality considerably. Remarkably, the effect is hardly detected by farinograph measurements (Moore, C. L., and Herman, 1942; D'Appolonia, 1984) or extensigraph measurements (Merritt and Bailey, 1945) at 30°C. In creep tests at the same temperature, the addition of 1% of a triglyceride mixture with 15% solid material approximately doubled the deformation of dough at the same stress. With flour from which the free lipids had been extracted, the effect was even larger (Bloksma, 1972a); recovery measurements showed that this was mainly due to changes in the viscous component of the deformation

(A. H. Bloksma, unpublished). The effect of added fat becomes clearly visible only during oven rise (Baker and Mize, 1942; Elton and Fisher, 1966).

As early as 1942, Baker and Mize noted that only those fats that are at least partly solid at the temperature of mixing and fermentation are effective in breadmaking (Baker and Mize, 1942); others confirmed this observation (Baldwin et al, 1963; Bayfield and Young, 1963a, 1963b, 1964; Chamberlain et al, 1965). If the dough temperature is higher after mixing, as is often the case after mechanical development, a shortening with a higher solid fat content is required (Bayfield and Young, 1963b). Chamberlain et al (1965) concluded that the maximum improvement is attained if the content of solid fat at the fermentation temperature is at least 0.02% of flour mass. Similar observations were made with mineral fats (hydrocarbons); only mixtures with melting points above 61°C improved loaf quality (Pomeranz et al, 1966b; Elton and Fisher, 1968). Also in line with these observations is the fact that the polar fraction of flour lipids, if added to a complete flour, has a beneficial effect on loaf quality, whereas the nonpolar lipids have hardly any effect; their softening points are 65°C and below 25°C, respectively (Pomeranz et al, 1966a).

There are no comprehensive theories that explain the physical effects of flour lipids and added fats. The hypothesis that they act as lubricating agents goes as far back as a paper by Working published in 1924 (Working, 1924). The models of gluten structure by Bungenberg de Jong (1956) and Grosskreutz (1961) are also based on this hypothesis. There are indeed several arguments for a layerlike arrangement of lipids in dough. A layerlike structure can be observed by the eye when gluten is stretched. With X-ray diffraction, a reflection was found that corresponds with a distance of 45–48 Å and that is due to a lipid fraction; this distance corresponds to the thickness of a bimolecular layer of phosphatides containing fatty acids with 18 carbon atoms (Traub et al, 1957; Daniels, D. G. H., 1958; Carlisle et al, 1960; Grosskreutz, 1961). The process of lipid binding during mixing, discussed below, may be due to the formation of layers of absorbed or chemically bound lipids. Added lipids could then be absorbed between layers of bound lipids and thus facilitate shear between them.

This hypothesis explains neither the specific effects of solid fats on loaf quality, discussed above, nor the observation that the effect of added fats shows up only during oven rise. To explain these facts, Baker and Mize (1942) assumed that added fat plugs pores in dough membranes during oven rise; however, no other evidence supports this hypothesis.

The effect of lipids on the gelatinization of starch is discussed in section IX-C.

Addition of Emulsifiers. In practice, various emulsifiers derived from fatty acids are added in amounts up to 0.5% of flour mass and usually in combination with triglycerides. A multitude of compounds or mixtures of compounds are mentioned in technical publications, patents, and food legislation, and a comprehensive review is beyond the scope of this section. The compounds have in common that they contain fatty acids, usually palmitic or stearic acid. These are bound in various ways to one or more polyfunctional molecules with carboxylic, hydroxyl, and/or amino groups, e.g., glycerol, lactic acid, or tartaric acid. Sometimes the carboxylic group is converted into its sodium or calcium salt. In most of the compounds, one or more of these functional groups have remained free; they impart a polar character to part of the molecule. This, together with the nonpolar fatty acid residue in the molecule, is a reason to classify these compounds as emulsifiers, although their emulsifying action has not always been established. Only in a few cases are data on their mode of action available. If added to dough without shortening, they cause only minor changes in rheological properties. A better distribution of added fat as a consequence of the addition of glyceryl monostearate could not be detected. That it improved loaf quality is ascribed to its effect on the gelatinization and retrogradation of starch (Knightly, 1968).

E. Interconstituent Interactions

Interconstituent interactions have been implicated in the physical properties of dough throughout the breadmaking process. Much of the early evidence that such interactions do indeed occur was indirect, based on loss of extractability of lipids upon addition of water and upon mixing of dough. Compelling direct evidence has been published in the past decade (for reviews, see Frazier, 1983; Bushuk, 1986; Chung, 1986; Frazier and Daniels, 1986).

In the context of functionality in dough, protein-lipid interactions appear to be particularly important. Addition of water to flour induces gluten protein to "bind" certain of the flour lipids (Chung and Pomeranz, 1981; Frazier and Daniels, 1986). Subsequent dough mixing and development modify lipid binding, depending on the nature of the mixing action and on the atmosphere; lipid binding is stronger in the absence of oxygen than in its presence (Baldwin et al, 1963; Daniels, N. W. R., et al, 1966, 1967; Frazier et al, 1981; Frazier and Daniels, 1986). Addition of salt decreases lipid binding in dough (Mecham and Weinstein, 1952; Pomeranz et al, 1968b; Tao and Pomeranz, 1968). Specific binding of lipids by gluten proteins was first reported about 40 years ago (McCaig and McCalla, 1941; Olcott and Mecham, 1947). Olcott and Mecham (1947) proposed the name "lipoglutenin" for the highly stable lipid-protein complex that they isolated from gluten. Lipid-protein complexes have taken on new significance in dough rheology with the discovery that specific gluten proteins have a high lipid-binding capacity (Frazier et al, 1981; Békés et al, 1983a; Zawistowska et al, 1985a) and that the bound lipids appear to be involved in the aggregation of some gluten proteins (Békés et al, 1983b; Zawistowska et al, 1985b).

Good evidence supports the hypothesis that wheat starch is not just an inert filler in dough but rather contributes actively to dough development and structure. Studies of synthetic gluten-starch doughs have shown that wheat starch, and to a lesser degree barley and rye starches, have unique characteristics required for the formation of viscoelastic dough (Hoseney et al, 1971). The key to this interaction may be the unique surface of wheat starch granules (Shelton and D'Appolonia, 1985). Specific gluten proteins can interact with a low-molecular-weight carbohydrate to form large aggregates (McMaster and Bushuk, 1983; Zawistowska et al, 1985b). Such aggregates may be functionally important in dough.

Pentosans interact with gluten proteins with varying degrees of specificity (Shelton and D'Appolonia, 1985). It is common knowledge that addition of water-soluble pentosans in excess of the indigenous pentosan content of wheat flour can completely inhibit the formation of a viscoelastic gluten. Meuser and Suckow (1986) found that ferulic acid can serve as a link between protein and pentosan chains and that this link occurs side by side with the diferulic acid coupling in gelled glycoproteins mentioned above in section VII-C.

Interactions between starch and lipids contribute little to the physical properties of dough but are probably quite important in subsequent stages of the breadmaking process. Complexing of lipids by starch can modify starch gelatinization during baking and may thereby affect the rheology of doughs during oven spring (section IX-C).

F. Enzymes

PROTEASES

Jørgensen (1945) attempted to explain the effect of oxidizing flour improvers on the basis of inhibition of flour proteases. He assumed that these proteases are of the papain type and are therefore inactivated by oxidation. Without oxidation, proteolysis causes softening of the dough, which is reflected in poor baking properties; this softening can be prevented by the addition of oxidizing compounds that inhibit proteolysis. Jørgensen's theory raised a considerable controversy in cereal chemistry. At present, there is convincing evidence that oxidizing flour improvers act directly on thiol groups of the proteins (section VIII-D).

Proteases may be added intentionally to some doughs to prevent buckiness. Indigenous proteases are of no consequence in flours from sound wheat but do contribute significantly to rheological properties of doughs from flours milled from wheat damaged by preharvest sprouting (Bushuk and Lukow, 1987).

AMYLASES

Amylases in dough break down the starch to dextrins and fermentable sugars. During fermentation, only the small percentage of the starch granules that have been damaged in the milling process are available for attack by amylases. The susceptibility of the starch to amylolytic breakdown is greatly increased only during the time between the start of gelatinization of starch in the oven and the inactivation of amylases.

Amylases may affect dough properties in three ways: by the formation of sugars that can be fermented by yeast, by the removal of the damaged starch fraction, and by the formation of dextrins. The formation of fermentable sugars increases the gassing power in doughs that are deficient in sugar. In these doughs, the formation of fermentable sugars is the most important effect of added amylases. However, if 3–6% of the easily fermentable sucrose is included in the formula, an increase in gassing power can hardly be expected from the addition of amylases. Under these conditions, the beneficial effects of amylase additions must be due to the other two processes.

The disappearance of starch, and particularly of the damaged starch granules with a higher water-binding capacity, decreases dough consistency as measured by the farinograph. Because of the limited amount of this available starch, the consistency does not decrease beyond a limiting value, even with extremely large additions of α-amylase (Johnson and Miller, 1949). The addition of large quantities of α-amylase to dough from normal flours does not adversely affect the dough's rheological properties during the fermentation stage, provided that accompanying proteases are effectively inactivated (Miller and Johnson, 1948).

Under this condition, extremely large amounts of α-amylase increase the loaf volume, although the crumb qualities are affected adversely (see below). The gas retention of the dough is increased, particularly during proofing, more probably because of the formation of dextrins than because of the disappearance of the damaged starch fraction (Johnson and Miller, 1949).

The breakdown of the starch after it has gelatinized in the oven depends on the amount and the thermal stability of the amylases present. At about 65° C, the native starch gelatinizes and becomes available for amylases. At the same temperature, β-amylase is inactivated, but wheat α-amylase remains active for another 1 or 2 min, until the temperature reaches 75° C (Walden, 1955). An excessive amount of α-amylase, as in sprout-damaged flour, or the use of a thermostable α-amylase, such as bacterial amylase, causes excessive liquefaction and dextrinization and consequently results in a wet and sticky crumb that is characteristic of bread from sprouted wheat.

LIPASES AND LIPOXYGENASES

Under normal conditions, the effect of lipase in dough is negligible. The effect of lipoxygenase is discussed in section VII-D and section X.

VIII. OTHER CHEMICAL FACTORS
THAT INFLUENCE DOUGH PROPERTIES

A. Water

Increasing the water content makes a dough less and less stiff until eventually a batter or a flour suspension is obtained. Doughs with too little water lack cohesion.

Table X summarizes various experiments on the effect of water content on dough stiffness. In general, dough stiffness changes between 5 and 15% if the water content is changed by 1% of flour mass. Exceptions are measurements with extrusion meters and creep measurements, where much larger effects were found. In most cases, the viscosity is more sensitive to changes in water content than is the modulus.

Tests with various types of extensigraphs show that if the water content is increased, an increase in extensibility accompanies a decrease in resistance.

In oscillatory measurements with very small deformations, Hibberd (1970a) found that the storage and loss moduli depend in quantitatively the same way on the water addition. More remarkable is the fact that Hibberd was able to separate the effects of wheat variety, frequency of oscillations, and water addition. The physical consequence is that the rheological properties of an arbitrary dough can be reproduced in a dough from any flour by selecting the proper water addition. It also means that the rheological behavior of any flour can be characterized by a single figure. Hibberd calls this the "principle of corresponding water contents." However, the principle is valid only for very small deformations.

Various constituents absorb water in dough. Native starch is the only constituent whose water content in dough can be estimated with some precision. In equilibrium with water, it absorbs 0.45 kg of water per kilogram of dry matter (Dengate et al, 1978); because the water activity in dough is hardly less than that of pure water, native starch in dough will absorb nearly the same quantity.

It is questionable whether water contents of constituents that do not form separate phases have a physical meaning. Nevertheless, they have been estimated on the basis of (multiple) correlations of the faringraph water absorptions of flours with their content of various constituents, or on the basis of the change in water absorption when a constituent is added. In both cases, it is tacitly assumed that the properties of dough do not change upon the addition of a moderate quantity of a constituent with the same moisture content as it has in dough; this assumption cannot be true, among other reasons because it neglects the

TABLE X
Effect of Water Content on Rheological Properties of Dough[a]

Procedure and Property	Ratio		Number of Values Determined	References
	Mean	Range		
Creep, shear flow				
Commercial flour				
Total deformation	1.26[b]	1.18–1.41	6	Bloksma (1964b)
Elastic deformation	1.15	...	1	Bloksma (1964b)
Viscous deformation	1.33[b]	1.19–1.44	3	Bloksma (1964b)
Experimental flour (cultivar Felix)				
Total deformation	1.54[b]	1.23–2.8	9	Bloksma (1964b)
Elastic deformation	1.23	...	1	Bloksma (1964b)
Viscous deformation	1.66[b]	1.26–3.0	4	Bloksma (1964b)
Creep, extensional flow				
Elastic deformation	1.12	1.09–1.16	3	Halton and Scott Blair (1936, 1937)
Viscous deformation	1.16	1.13–1.20	3	Halton and Scott Blair (1936, 1937)
Viscometry, shear flow				
Stress	0.77	0.72–0.84[c]	5	Launay and Buré (1973)
	0.72	0.70–0.82[c]	3[d]	Quendt et al (1974)
	0.72	0.70–0.77	2[e]	Rudolph et al (1977)
Viscometry, extensional flow				
Stress	0.90	0.83–0.92	9[f]	Tschoegl et al (1970b)
Stress relaxation				
Stress	0.95	0.95–0.96	2	Bohn and Bailey (1936)
	0.94	...	1	Cunningham, J. R., and Hlynka (1954)
Rate of relaxation	1.21	1.20–1.23	2[g]	Launay and Buré (1974)
Oscillatory measurements[h]				
Storage modulus	0.87	0.81–0.90	4	Shimizu and Ichiba (1958)
	0.91	...	1	Hibberd (1970a)
	0.91	0.91–0.92	2	Smith, J. R., et al (1970)
	0.92	...	1	Hibberd and Parker (1975)
	0.87	0.85–0.89	5	Navickis et al (1982)
Loss modulus	0.85	0.81–0.95	4	Shimizu and Ichiba (1958)
	0.91	...	1	Hibberd (1970a)
	0.92	0.91–0.94	2	Smith, J. R., et al (1970)
	0.92	...	1	Hibberd and Parker (1975)
	0.88	0.87–0.89	5	Navickis et al (1982)

(continued on next page)

TABLE X (continued)

Procedure and Property	Ratio		Number of Values Determined	References
	Mean	Range		
Brabender Farinograph				
Maximum consistency	0.97	···	1	Bohn and Bailey (1936)
	0.95	0.94–0.96	4	Markley and Bailey (1938)
	0.94	0.93–0.94	7	Hlynka (1959, 1960)
	0.94	0.94–0.95	5	Louw and Krynauw (1961)
	0.93	0.92–0.94	3	Nieman (1978a)
Swanson and Working Mixograph				
Curve height	0.99	0.99–0.99	10[i]	Harris and Sibbitt (1945)
	0.97	···	1	Shuey and Gilles (1966)
Brabender Extensigraph				
Maximum resistance	0.93	0.90–0.95	4	Fisher et al (1949)
	0.90	0.89–0.90	3[j]	A. H. Bloksma (unpublished)
Extensibility	1.07	1.04–1.10	3[j]	A. H. Bloksma (unpublished)
Modulus	0.89	0.88–0.90	5	Bloksma (1967)
Viscosity	0.84	0.81–0.85	5	Bloksma (1967)
Chopin Extensigraph				
Maximum resistance	0.90	0.87–0.91	9	Tchetveroukhine (1947, 1948)
	0.88	0.85–0.91	7	Soenen (1949)
Extrusion meters				
Extrusion time	0.81	0.79–0.83	3	Stamberg and Bailey (1940)
	0.76	···	1	Bennett and Coppock (1953)
	0.63	···	1	Wensveen and De Miranda (1955)
	0.75	0.75–0.76	3	Nieman (1978b)
Laboratory mixer				
Torque on driving shaft	0.92	0.91–0.93	3[k]	Tscheuschner et al (1975)

[a] If the water addition is increased by 1 ml per 100 g of flour, the property in column 1 is multiplied by the factor in column 2.

[b] The high factor for the viscous and total deformations may be due to slip between the test piece and the smooth surfaces of plate and cone, increasing with increasing water addition.

[c] With a larger water content, the exponent n in equation 4 is larger, that is, the behavior is closer to Newtonian behavior; as a consequence, the factor in this table differs less from unity as the rate of shear is larger.

[d] One flour at three rates of shear between 1 and 30 sec^{-1}.

[e] One flour in two different mixers.

[f] Two flours at four or five temperatures.

[g] One flour at two different initial shear stresses.

[h] For conditions, see Table III.

[i] Five flours at two temperatures.

[j] One flour after 45, 90, and 135 min.

[k] One flour at three speeds.

accompanying changes in the volume fraction of the starch in the dough. Therefore, the following estimates, which are based on experimental data from Aitken and Geddes (1939), Belderok and Slager (1983), Bloksma (1972a), Greer and Stewart (1959), Kulp (1968), Meredith, P. (1966b), and Moss, H. J. (1961), have hardly more significance than a rank order: protein, 1–3 kg of water per kilogram of dry protein; damaged starch, 1.5–2 kg/kg; and pentosans, about 10 kg/kg.

The binding of water to dough constituents is reflected by observations by nuclear magnetic resonance (Toledo et al, 1968). Differential scanning calorimetry (Davis and Webb, 1969) and differential thermal analysis (Bushuk and Mehrotra, 1977) showed that at water contents below 0.3 kg per kilogram of dry flour, no water freezes upon cooling to about $-20°$ C.

B. Yeast

Yeast is added to doughs to produce carbon dioxide that is necessary to obtain a light crumb texture. At the same time, yeast ferments sucrose, glucose, fructose, and maltose.

The extensibility of unyeasted dough that contains no improvers increases with increasing reaction time, and the resistance to deformation decreases, according to measurements with the Halton Extensigraph. The extensigrams of yeasted doughs, however, become higher and much shorter with increasing reaction time. In both cases, the area under the extensigram and the extrusion time decrease (Halton, 1949, 1959).

Fermentation with 2% yeast lowers the pH of dough by about 0.5 unit or less in 3 hr (Blish and Hughes, 1932; Halton and Fisher, 1932; Dörner and Stephan, 1956). This increase in acidity cannot have a marked effect. It is not precisely known how the disappearance of fermentable sugars, the reduction of the oxygen pressure, the formation of fermentation by-products, and the stretching of the dough by expanding gas cells during fermentation affect dough properties; their effects are superimposed on the changes that proceed in unyeasted dough.

C. Electrolytes and pH

Common salt is generally one of the dough ingredients in breadmaking. A small percentage of sodium chloride stiffens the dough and makes it less sticky. The greater stiffness is confirmed by measurements with the Brabender Extensigraph (Fisher et al, 1949; Grogg and Melms, 1956) and the Chopin Extensigraph (Alveograph) (Margulis and Campagne, 1955; Calvel, 1969); the curves with added salt show a higher resistance. Measurements with the Swanson and Working Mixograph and the Simon "Research" Water Absorption Meter are in accordance with an increase in stiffness from the addition of salt (Bennett and Coppock, 1953). The only instrument that does not fit into these observations is the Brabender Farinograph, where a decrease in consistency is often reported (Bennett and Coppock, 1953; Hlynka, 1962; Tanaka et al, 1967), although under certain conditions, effects in the opposite direction have also been observed (Moore, C. L., and Herman, 1942; Fisher et al, 1949; D'Appolonia, 1984). The observed decrease in consistency has been explained as the result of the decrease in stickiness rather than of a decrease in stiffness (Bennett and Coppock, 1953).

Extensigraph measurements show that addition of salt increases not only the resistance but usually also the extensibility. By contrast, oxidizing reagents increase resistance but decrease extensibility (Fisher et al, 1949).

The possibility of dispersing gluten proteins in dilute acid solutions, and of their precipitation at an ionic strength of about $0.02M$ (Lusena, 1950; Cunningham, D. K., et al, 1955), can be explained by electrostatic effects. At an extreme pH, the protein molecules carry electrical charges all of the same sign. Their mutual repulsion prevents the precipitation of the proteins, unless sufficient electrolyte is present to shield the field of point charges. The latter is the case if the ionic strength is of the order of $0.02M$ or higher, the more so because wheat proteins carry only a few ionizable side chains. Differences between ions of the same charge at large electrolyte concentrations must be ascribed to specific adsorption, in particular of anions (Bennett and Ewart, 1965; Guy et al, 1967; Tanaka et al, 1967), or to changes in hydrophobic bonding (Preston, 1984 and references cited therein).

Between pH 4.2 and 7.3, the resistance of dough decreases with increasing pH (Tsen, 1966; Tanaka et al, 1967). This has been explained by the hypothesis that mercaptide ions, RS^-, rather than neutral thiol groups, RSH, participate in the interchange reactions in Fig. 17; at a higher pH, the ions are more abundant. This hypothesis does not explain why the decrease in resistance is accompanied by an increase in the maximum consistency in the farinograph, at least in the presence of salt (Gasiorowski, 1967; Tanaka et al, 1967).

The rate of the bromate reaction increases with decreasing pH (Bushuk and Hlynka, 1960b). This is reflected by the change in the extensigram and also by a decrease in the optimum dosage in baking tests (Hlynka and Chanin, 1957).

D. Oxidation

In countries where the law permits it, the treatment of flour with minute amounts of oxidizing reagents is an established practice. The flour may be blended with dry chemicals, such as potassium bromate or potassium persulfate, or it may be treated with gaseous reagents, such as chlorine or chlorine dioxide. The most important flour improvers are listed in Table XI (Fitchett and Frazier, 1986). Other compounds with flour improving and/or bleaching action are mentioned in the remainder of this section; additional compounds are to be found in earlier reviews (Jørgensen, 1945; Sullivan, 1954; Kent-Jones and Amos, 1957).

The amount of oxidizing reagent used varies with the type of oxidant and type of flour; usually it is about 25 mg per kilogram of flour. With mechanical dough development, much larger amounts, about 100 mg/kg, are used. Current improver application data are summarized in Table XI.

The effect of oxidizing compounds is twofold: flour improvement and bleaching. Flour improvement refers to changes in the rheological properties of the dough. Flour bleaching is due to the destruction of its yellow pigments and results in white flour and bread. Nitrogen dioxide and benzoyl peroxide bleach only. Bromate, iodate, ascorbic acid, and azodicarbonamide improve dough properties without bleaching. Chlorine, chlorine dioxide, and acetone peroxides exert both effects simultaneously.

Of the oxidizing improvers, only the fast-acting ones affect the farinogram (Meredith, P., and Bushuk, 1962). On the other hand, the effect of slow-acting improvers such as bromate can be demonstrated clearly by load-extension tests: an increase in resistance and a decrease in extensibility, depending on the type and amount of the agent and the reaction time. These phenomena are correlated

TABLE XI
Improver Application Data[a]

Agent	Action	Year Introduced	Accept-ability[b]	Equivalent Mass (kg/kmol)	Typical Usage (ppm)	U.K. Permitted Maximum (ppm)	Relative Rate of Reaction
Azodicarbonamide	Bread dough improver	1962	Limited	58	10–20	45	Fast
Halogenates							
Potassium bromate	Bread dough improver	1916	Moderate	28	20–50	50	Slow
Potassium iodate	Bread dough improver	1916	Limited	36	10–20	0	Fast
Chlorine dioxide	Bread flour improver and bleach	1948	Moderate	51	16	30	Fast
L-Ascorbic acid	Bread dough improver	1935	Wide	88	75	200	Slow
L-Cysteine hydrochloride	Bread/biscuit dough-softening agent	1962	Moderate	158	30–70	75 bread 300 biscuit	Fast
Acetone peroxide	Bread flour improver and bleach	1961	Limited	27	25	0	Fast
Sodium metabisulfite	Biscuit/pastry-dough softening agent	1950	Limited	95	200	300[c]	Fast
Chlorine	Cake flour improver and bleach	1900	Moderate	35	2,000	2,500	Fast

[a] Source: Fitchett and Frazier (1986); used by permission.
[b] In general, use of improvers is most permissive in the United States, more restrictive in the United Kingdom, and very restrictive in the rest of the European Economic Community.
[c] 200 ppm as SO_2.

with a decrease in thiol content. Even if disulfide groups were the only oxidation product, the increase in their contents would be only a few percent and would therefore be too small to be detected. There is evidence that, besides disulfide groups, higher oxidation products can be formed (Hird and Yates, 1961a, 1961b; Zentner, 1964).

With iodate, a reversal of the improver effect—that is, a decrease in extensigram height—has occasionally been found. In all cases reported, one or more of the following factors were involved: a high dosage of iodate, the simultaneous action of another fast-acting oxidizing agent (e.g., oxygen), long and intensive mixing, and a short time interval between mixing and shaping (Frater et al, 1960; Bushuk and Hlynka, 1962; Meredith, P., and Bushuk, 1962; Tsen and Bushuk, 1963). A possible explanation is that the oxidizing agents cause a considerable reduction in thiol content and consequently hamper thiol-disulfide interchange during mixing. This, combined with the high speed of mixing, leads to high stresses, which cause breakage of cross-links. The fact that the effect is observed only when the time interval between mixing and shaping is short suggests that the broken bonds are mainly reversible ones (e.g., hydrogen bonds) that are restored in time. However, Tsen and Bushuk (1963) also observed a decrease in disulfide content under oxidative conditions that led to a reduction in extensigram height. The decrease in extensigram height has also been observed with another fast-acting oxidizing reagent, butanone peroxide (Tsen and Hlynka, 1963), as well as with the thiol-blocking reagents N-ethyl-maleimide (NEMI) and p-chloromercury(II) benzoic acid (Frater et al, 1960; Sullivan et al, 1961b, 1963; Bushuk and Hlynka, 1962; Meredith, P., and Bushuk, 1962).

Similarly, the observations of Frazier et al (1985) show (Fig. 15) a reversal of the effect of atmospheric oxygen after intensive mixing. Bromate acts too slowly to cause a sufficiently stiff dough during mixing. The reversal of the improver effect should not be confused with reduction of gas retention and loaf quality if the amount of oxidizing improver exceeds the optimum level. In section V-F, such overtreatment was explained by an imbalance of resistance and extensibility.

In principle, it would be most logical to start the discussion of flour improvers with the one that is used, though unconsciously, by every baker, namely the oxygen of the air. However, we prefer to start with potassium bromate and iodate because the effect of these reagents is easier to demonstrate and has been investigated most extensively.

BROMATE AND IODATE

The effect of bromate on dough properties has been studied extensively with the Brabender Extensigraph (Dempster et al, 1955, 1956). Figure 18A shows that after the addition of a sufficient amount of bromate, the structural-relaxation curve shifts upward with increasing reaction time. Figures 18B and 18C show how the structural relaxation constant C and the asymptotic load R_a (section VI-B) vary with reaction time and the amount of bromate. In untreated dough, both C and R_a decrease with increasing reaction time. With short reaction times, bromate causes a small increase in R_a but a negligible decrease in C. Only after reaction times of more than 1 hr does the increase in C become evident; then both C and R_a increase with increasing amount of bromate as well as with reaction time. In general, these phenomena are more pronounced at higher temperatures (Dempster et al, 1955).

Apparently both 1) an interval between mixing and shaping and 2) structural activation are needed to show the effect of bromate, which suggests that bromate forms reactive groups on the protein chains that form cross-links only during structural activation. Possibly, these groups are too far removed from one another in a resting dough and react only when they are brought into juxtaposition by the work of rounding and shaping. The exact nature of the assumed reactive groups is still unknown; there is little doubt, however, that bromate and iodate act via the oxidation of thiol groups.

Analytical determinations of residual bromate have revealed that its rate of disappearance is higher during mixing than in a resting dough (Bushuk and Hlynka, 1961a; Bloksma, 1964a). Under normal conditions, about half of the added bromate is still present after mixing and a 4-hr rest (Bushuk and Hlynka,

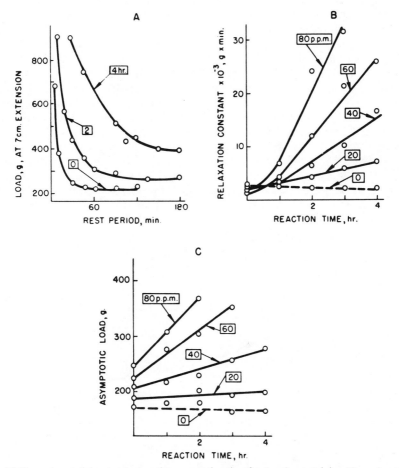

Fig. 18. Dependence of structural relaxation on reaction time for doughs containing 40 ppm bromate (A) and dependence of structural-relaxation parameters on reaction time and on amount of bromate (B and C). (Reprinted, by permission, from Dempster et al, 1955)

1960a). The activation energy of the reaction is about 47 MJ/kmol, with little difference between mixing and rest (Bushuk and Hlynka, 1961a); this means a tripling in the rate if the temperature is raised from 30 to 50°C. That the bromate reaction is slow except at elevated temperatures is confirmed by thiol determinations (Tsen, 1968) and by rheological measurements (Jelaca and Dodds, 1969; Nagao, 1986). The assumption that bromate reacts with thiol groups explains why the presence of other oxidizing agents and thiol-blocking agents reduces the rate of bromate consumption (Bushuk and Hlynka, 1960b), whereas added thiol compounds increase this rate (Bushuk and Hlynka, 1961b). Ethylenediamine tetraacetic acid (EDTA), which complexes bivalent cations and heavy metal ions, increases both the rate of bromate consumption (Bushuk and Hlynka, 1960b) and its effect on the extensigram (Hlynka, 1957); the mechanism of this effect is unknown. A decrease in pH causes more rapid bromate consumption (Bushuk and Hlynka, 1960b) and a larger effect on the extensigram (Hlynka and Chanin, 1957). In accordance with this, the amount of bromate that is optimum for breadmaking decreases.

The major difference between iodate and bromate is the fast action of iodate. Analyses made immediately after mixing revealed that additions of up to 0.1 mmol of iodate per kilogram of flour (21 ppm) were completely consumed under normal mixing conditions (Bushuk, 1961; Bloksma, 1964a). The difference in rate has also been demonstrated by thiol determinations (Hird and Yates, 1961b; Bloksma, 1964a). Figure 19 shows that, unlike bromate, iodate increases the structural-relaxation constant even after a short reaction time. The structural-relaxation constant of dough with iodate decreases with increasing reaction time. This behavior is similar to that of untreated dough; enzymatic breakdown of proteins or available starch may offer an explanation for it. Because of the faster action of iodate, fewer equivalents of iodate than of bromate are required to bring the rheological properties of dough to the level that is optimum for breadmaking (Dempster et al, 1956; see also Table XI).

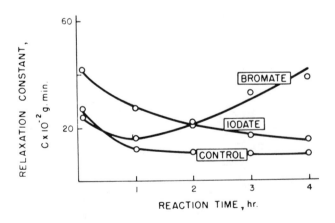

Fig. 19. Structural-relaxation constants for untreated dough and dough containing 15 ppm bromate or 9 ppm iodate after various reaction times. (Reprinted, by permission, from Dempster et al, 1956)

Potassium chlorate, although a powerful oxidizing agent, does not act as a flour improver (Jørgensen, 1945). Extrapolation of our knowledge from studies with iodate and bromate suggests that chlorate reacts too slowly for practical uses. Hypobromite, bromine, and the corresponding chlorine compounds act rapidly but are rather ineffective flour improvers. Chlorite is rapid and effective (Tkachuk and Hlynka, 1961).

Rheological measurements on mechanically developed doughs showed little effect of bromate alone except at elevated temperatures (Fitchett and Frazier, 1986). Consistent with its fast action in dough, iodate had a marked effect on mechanical dough development. At moderate levels of iodate addition, doughs showed a bimodal development profile; at higher levels (approximately equivalent to the total thiol content), rapid initial development was followed by very rapid breakdown (Fig. 20).

OXYGEN

Oxygen from the air affects flour during storage and is mainly responsible for its natural maturing. Apart from its effect on flour, which is claimed to be small (Bennett and Coppock, 1957), oxygen that is incorporated into dough during mixing can affect its properties considerably. Changes in the rheological properties of the dough can be observed immediately after mixing; nevertheless, extensigram height increases further after a sufficient reaction time (Smith, D. E., and Andrews, 1952; Dempster et al, 1954). Oxygen also bleaches the dough.

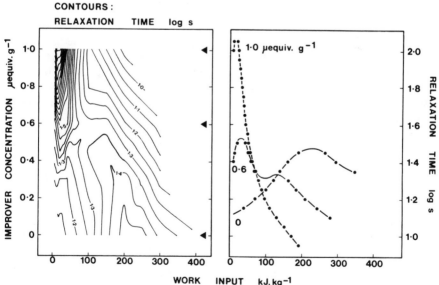

Fig. 20. Combined effect of potassium iodate concentration and mechanical work input on the development of dough structure as determined by compressive stress-relaxation measurement: three-dimensional plot representing the development surface by contour lines joining points of equal relaxation time (left); development profile cross sections taken at three levels of iodate addition (indicated by arrows on contour plot) (right). (Reprinted, by permission, from Fitchett and Frazier, 1986)

The mechanism of the action of oxygen and the interaction of oxygen with flour lipids were discussed in section VII-D and are mentioned again in section X. The disappearance of thiol groups when dough is mixed in air or oxygen has been well established (Sokol et al, 1960; Bloksma, 1963; Sullivan et al, 1963; Tsen and Bushuk, 1963; Tsen and Hlynka, 1963). During mixing in the absence of oxygen, the thiol content gradually increases or remains constant (Bloksma, 1963; Tsen and Hlynka, 1963; Mecham and Knapp, 1966).

CHLORINE DIOXIDE

Chlorine dioxide is a gaseous improving and bleaching reagent that is believed to react with flour instantaneously and quantitatively. The treatment results in an increase in extensigram height. This effect, like that of oxygen, can be observed immediately after mixing and increases as the time interval between mixing and testing is increased (Smith, D. E., and Andrews, 1952; Jaska, 1961). The effect of chlorine dioxide on the thiol content of dough and on dough properties appears to be through the chlorite ion formed by the reaction of chlorine dioxide with moisture (Fitchett and Frazier, 1986).

AZODICARBONAMIDE

Azodicarbonamide, $H_2N \cdot CO \cdot N \cdot N \cdot CO \cdot NH_2$, is a rapid-acting reagent that reduces the thiol content of dough and increases its resistance to extension. During the reaction, the oxidant is converted into biurea. It has no bleaching action (Joiner et al, 1963; Tsen, 1963, 1964; Fitchett and Frazier, 1986). Its effects in mechanically developed doughs are slightly different from those of iodate, probably because of its insolubility in water (Fitchett and Frazier, 1986).

ORGANIC PEROXIDES

Whereas benzoyl peroxide only bleaches, peroxides of acetone have both a bleaching and an improving effect. The commercially available product contains a mixture of peroxides in which 2,2-dihydroperoxypropane is dominant (Johnson et al, 1962; Ferrari et al, 1963). Acetone peroxide rapidly reduces the thiol content of dough and increases its resistance. To attain the same effect in dough, more equivalents of acetone peroxide than of iodate or azodicarbonamide are required (Tsen, 1964). Peroxides of butanone and 2-methyl propanone have similar effects on thiol groups and on the extensigram (Tsen and Hlynka, 1963).

L-ASCORBIC ACID

L-ascorbic acid, although itself a reducing agent, can exert an effect on dough properties similar to that of the oxidizing reagents mentioned above (Jørgensen, 1935). Actually, the effective reagent is not L-ascorbic acid but its oxidation product, namely, dehydro-L-ascorbic acid (Melville and Shattock, 1938).

L-ascorbic acid is oxidized by atmospheric oxygen under the influence of both a heat-labile enzyme (Grant and Sood, 1980; Grosch, 1986) and a heat-stable catalyst. D-ascorbic acid and a number of related compounds are oxidized in the same way and at comparable rates. The oxidation product, dehydro-L-ascorbic acid, can oxidize glutathione under the influence of a specific enzyme. This reaction proceeds at an appreciable speed only with dehydro-L-ascorbic acid and dehydroreductic acid but not with the D-compounds or other related compounds. The specificity of this second enzyme reaction confines the improving action to L-ascorbic acid, reductic acid, and their dehydro compounds (Sandstedt and Hites, 1945; Maltha, 1953). More recent studies on the fate of ascorbic acid and dehydroascorbic acid in doughs and flour suspensions confirmed this theory (Kuninori and Matsumoto, 1963, 1964a; Carter and Pace, 1965; Grosch, 1986).

Determinations of thiol groups in doughs and extensigrams support the conclusion that L-ascorbic acid is effective only if oxygen is present; the effect of dehydro-L-ascorbic acid depends much less on the oxygen pressure. Unlike the effects of bromate and iodate, the effect of L-ascorbic acid in a resting dough is greatly reduced by the presence of yeast, which makes the conditions more anaerobic (Meredith, P., 1965; Tsen, 1965). On the basis of thiol determinations and extensigrams, Tsen (1965) and Meredith, P., (1965, 1966a) suggested a synergistic action not only of ascorbic acid and oxygen, but also of ascorbic acid and bromate; the effect of the combination of reagents was larger than the sum of their separate effects. Determinations of ascorbic acid in dough by Chamberlain et al (1966b) did not support this conclusion.

The rheological behavior of mechanically developed doughs in the presence of L-ascorbic acid is quite different from that of iodated doughs under similar conditions (Fitchett and Frazier, 1986). The apparent relaxation time (section IV-C) increases linearly with the quantity of ascorbic acid added up to a certain limit and then remains rather constant. Overoxidation with L-ascorbic acid is not encountered in practice. This observed rheological effect is consistent with the relatively slow enzymatic conversion of L-ascorbic acid to dehydro-L-ascorbic acid. Load-extension measurements, however, suggest a rapid effect of L-ascorbic acid (Meredith, P., 1965; Tsen, 1965).

E. Compounds That React Specifically with Disulfide and Thiol Groups

This section deals with compounds that affect the disulfide and thiol groups in dough by other ways than oxidation or reduction. We shall discuss the effect of added thiol compounds and disulfides, of compounds like NEMI that block thiol groups from the flour, and of sulfite, which specifically splits disulfide bonds. Finally, chemical dough development is discussed; it is based on temporary splitting of disulfide bonds.

THIOL COMPOUNDS

Most of the experiments described in the literature were performed by adding the amino acid cysteine or the tripeptide glutathione to dough. Addition of glutathione causes a rapid decrease in consistency in the farinograph (Hird, 1966). In the extensigraph, cysteine decreased the resistance and increased the extensibility (Frater et al, 1961); glutathione decreased both (Bloksma, 1972b). Cysteine accelerated both stress relaxation and structural relaxation (Frater et al, 1961). These effects can easily be explained by an increase in the rate of thiol-disulfide interchange reactions. Mercaptoacetic acid (Coppock and Muller, 1969; Meredith, P., and Hlynka, 1964; Muller, 1968) and γ-glutamyl cysteine (Jones and Carnegie, 1969b), as well as thioctic acid and dithiothreitol in their reduced state (Dahle and Hinz, 1966), have also been used; they gave similar results.

If the consistency of a dough being mixed in the faringraph had been decreased by dithiol compounds, the consistency could be increased again by introducing oxygen into the mixing bowl. Monothiols cause a decrease that could not be neutralized by oxygen. Contrary to what has been said in the section on protein structure, in this case the effect of oxygen can better be explained by the (re-)formation of disulfide bonds than by the removal of thiol groups (Meredith, P., and Hlynka, 1964).

DISULFIDES

Oxidized glutathione (GSSG) has been reported to affect the farinogram and extensigram in the same direction as does the reduced form (GSH), although to a lesser extent (Frater and Hird, 1963; Meredith, P., and Hlynka, 1964; Hird, 1966; Jones and Carnegie, 1969b). This can be explained by a series of interchange reactions:

$$GSSG + XSH \rightarrow GSSX + GSH$$
$$PrSSPr + GSH \rightarrow PrSSG + PrSH$$
$$GSSX + PrSH \rightarrow PrSSG + XSH ,$$

where Pr = protein, GSH = glutathione, and XSH = an arbitrary thiol compound. In this series of interchange reactions, rheologically ineffective mixed disulfides PrSSG are formed at the expense of effective cross-links between protein chains. Similar effects have been recorded for β-dithiodiglycol (Meredith, P., and Hlynka, 1964) and bis-γ-glutamyl cystine (Jones and Carnegie, 1969b); many others were ineffective (Jones and Carnegie, 1969b). The reverse effect—that is, an increase in resistance—has been reported for formamidine disulfide (Sullivan and Dahle, 1966). Why the various disulfides behave differently has not yet been explained.

THIOL-BLOCKING AGENTS

NEMI is the thiol-blocking compound that has been studied most extensively. It reduces dough development time and increases peak height of the farinogram; after the peak, the farinogram shows a rapid decrease in consistency (Mecham, 1959; Sullivan et al, 1961b; Meredith, P., and Bushuk, 1962; Jankiewicz and Pomeranz, 1965; Hird, 1966). After the addition of small amounts (about 0.5 mmol per kilogram of flour) and gentle mixing, the resistance of the dough is increased; both stress relaxation and structural relaxation are retarded (Frater et al, 1960; Meredith, P., and Bushuk, 1962; Jankiewicz and Pomeranz, 1965).

An analysis of extensigraph results led to the conclusion that NEMI increases both the modulus and viscosity (Bloksma, 1967). With larger additions, longer mixing time, or both, the effect on the extensigram can easily be reversed (Frater et al, 1960; Meredith, P., and Bushuk, 1962; Sullivan et al, 1963). These effects are in accordance with the hypothesis that blocking of thiol groups hinders the interchange reactions, which results in a higher resistance. It has been previously explained (section VIII-D) why too high a resistance in combination with prolonged mixing causes breakdown of the dough. The reduction of the thiol content of dough by NEMI has indeed been demonstrated (Matsumoto and Hlynka, 1959; Sullivan et al, 1963).

Other thiol-blocking reagents with which similar results have been obtained are the inorganic salts silver nitrate and mercury(II) chloride (Resnitschenko and Popzowa, 1934), *p*-chloromercury(II) benzoic acid (Goldstein, 1957; Matsumoto and Hlynka, 1959; Mecham, 1959), and the alkylating reagent iodoacetamide (Mecham, 1959).

SULFITE

The treatment of flour with sulfur dioxide or the addition of sulfite or bisulfite to dough results in a reduction of dough development time; after the peak, the farinogram shows a rather rapid decrease in consistency. The resistance to extension is reduced, and the extensibility is increased. The increase in extensibility can be used to advantage in the preparation of biscuits. Gluten washing is made difficult or impossible (Hlynka, 1949; Halton, 1959; Morrison and Hawthorn, 1960). This is explained by the splitting of disulfide cross-links according to the reaction:

$$RSSR + NaHSO_3 \rightarrow RSH + RSSO_3Na .$$

Both in dough (Matsumoto and Hlynka, 1959) and in flour or gluten dispersions (Matsumoto et al, 1960), bisulfite causes a considerable increase in the amount of thiol groups that can be titrated. This reaction with excess sulfite is used extensively for the titration of the sum of disulfide and thiol groups. The reversibility of the reaction is confirmed by the observation that acetaldehyde, which forms an addition compound with sulfite, nullifies the effect of the latter on dough properties and the thiol titer of gluten dispersions (Hlynka, 1949; Matsumoto et al, 1960).

ACTIVATED OR CHEMICAL DOUGH DEVELOPMENT

In the process of activated or chemical dough development, both cysteine and bromate or ascorbic acid are added to the dough. As a result, a normal mixing and short fermentation process yields acceptable bread. Without cysteine, the same fermentation process gives satisfactory loaves only in combination with high-speed mixing (see also section IX-A). Cysteine is supposed to split disulfide bonds rapidly, which facilitates the unfolding of protein molecules during mixing and thereby aids dough development. As a consequence, a normal mixing process with cysteine can develop the protein network to the same extent as high-speed mixing without cysteine. Bromate and ascorbic acid oxidize the excess of thiol groups after dough development has been completed, to avoid too soft a dough during proof. In commercial applications, whey (Henika and Rodgers, 1965) or soy flour is also added.

In experimental variants of the process, bisulfite has been used as a disulfide-splitting reagent and a combination of ascorbic acid and bromate has been used as oxidant (Chamberlain et al, 1966a; Pace and Stewart, 1966).

IX. RHEOLOGY AND CHEMISTRY
OF THE BREADMAKING PROCESS

A. Mixing

The mixing of bread doughs has three functions. First, the ingredients—flour, water, yeast, salt, and minor constituents—are transformed into a mass that is homogeneous when examined with a resolution of 0.5 mm. Second, flour proteins swell and are organized in such a way that they impart to the dough the desired gas retention; this is called dough development. And third, air is occluded, forming the nuclei of the gas cells that grow in size (though not in number) during fermentation (Baker and Mize, 1941; see also section II-B). Upon continued mixing, the dough becomes extremely extensible and sticky and consequently difficult to handle; then it is overmixed.

Recent mixers have an energy input of about 25 kJ per kilogram of dough in five to 12 min. After such a mixing process, sufficient gas retention is developed only during a long fermentation process. In Europe some decades ago, mixers were even slower, imparting 10–15 kJ/kg in about 30 min. Dough can be developed mechanically if it contains an extra quantity of oxidizing flour improver and an emulsifier and if both mixing energy and mixing speed are above critical levels; after this type of mixing process, hardly any fermentation between mixing and molding is required. For the Chorleywood bread process, the optimum mixing energy is claimed to be 40 kJ/kg, to be imparted within a few minutes, and is largely independent of flour and mixer type (Axford and Elton, 1960; Chamberlain et al, 1962; Axford et al, 1963). Kilborn and Tipples (1972) found that the critical energy level varied with flour and mixer type between 20 and 75 kJ/kg but was usually between 45 and 55 kJ/kg. Oxidation reduces the optimum mixing energy (Kilborn and Tipples, 1979). Addition of cysteine, up to 100 ppm, reduces both the critical mixing speed and the optimum energy (Kilborn and Tipples, 1973). The application of cysteine in combination with an oxidizing agent is called activated dough development or chemical dough development (section VIII-E); a more correct term would be "chemically accelerated dough development" (Tipples and Kilborn, 1974).

A recording of mixing power versus time (at constant speed), as is made, for example, in recording dough mixers (section V-A), is a convenient way to visualize dough development and breakdown. Very often the time taken to reach maximum power or consistency is considered the optimum mixing time. Tipples and Kilborn (1974) claimed that this time, or a slightly longer one, is indeed the optimum mixing time for baking procedures with only short first and intermediate proof times. More reliable information on optimum mixing time can be obtained from rheological tests, if the rest period is long enough for structural relaxation or recovery to be completed (section VI-A). By means of stress-relaxation measurements after a rest period of 45 min (section V-C), Frazier and co-workers demonstrated that oxidation reduces the optimum mixing energy (Fig. 20). In contrast, the energy consumption during mixing until maximum consistency hardly changes (Frazier et al, 1985; Fitchett and Frazier, 1986; and references cited therein).

Moss, R. (1972, 1974) used light microscopy to study the changes in the structure of the protein phase in the dough during mixing. In an undermixed dough, the swollen protein forms compact masses in which starch granules occur; other starch granules lack a protein envelope. Upon continued mixing, the protein masses are stretched into a continuous network of sheets with starch granules embedded in them or on them. These sheets can be considered the precursors of the membranes in the fermented dough. Further mixing results in a finer network with more cross-links. Eventually, protein veils surrounding all starch granules are formed; then the dough is overmixed. Descriptions of other authors (Bechtel et al, 1978; Fretzdorff et al, 1982; Paredes-Lopez and Bushuk, 1983b), based on other microscopic techniques, are essentially similar. In these descriptions, overmixing shows up as rupture of the protein sheets and a loss of continuity of the network.

Oxidation retards and reduction accelerates the development of a continuous network of protein (Moss, R., 1974). Doughs from strong flours develop slowly (Moss, R., and Stenvert, 1979) and are also more resistant to overmixing (Bechtel et al, 1978; Paredes-Lopez and Bushuk, 1983b).

Mixing and overmixing of doughs are accompanied by an increase in the extractability of the proteins; this is explained by disaggregation and/or depolymerization of high-molecular-weight glutenin, the latter by breaking of disulfide bonds (Tanaka and Bushuk, 1973a, 1973b, 1973c; Danno and Hoseney, 1982 and references cited therein). This phenomenon was, however, less apparent in the presence of salt (Meredith, O. B., and Wren, 1969). In addition, lipids become less extractable (see section VII-E).

If a mechanically developed dough is mixed for a further period of time well below the minimum speed required for optimum development, "undevelopment" occurs. Dough properties become similar to those of undermixed doughs: upon baking, a loaf with a small volume and poor crumb texture is obtained, and proteins are less extractable. The "undevelopment" can be reversed by mixing for some time at high speed (Tipples and Kilborn, 1975; Paredes-Lopez and Bushuk, 1983a). Microscopic examination showed that "undevelopment" destroyed the continuous protein structure (Paredes-Lopez and Bushuk, 1983b).

B. Fermentation

The objectives of fermentation are to bring the dough to an optimum condition for baking. During fermentation, dough development is continued; that is, gas retention is improved, if this has not already been achieved by mixing. The number of gas cells is increased by punching and/or molding. At the end of the last proof, the dough must contain a large volume of gas and must yet have sufficient gas retention left for oven rise. If dough development requires a long fermentation process, which would result in an excessively large final volume, excess gas is expelled by punching or remixing.

During fermentation, the yeast converts fermentable sugars into carbon dioxide and ethanol. Not enough fermentable sugars are present in the flour to maintain gas production until baking. Supplements are formed during fermentation by the action of amylases on available starch or are added upon dough-making.

The carbon dioxide and ethanol produced dissolve in the dough phase; some evaporates from there into gas cells. At a concentration of 3.3×10^{-5} kmol of carbon dioxide per kilogram of dough, the dough phase is in equilibrium with this gas at $27°C$ at atmospheric pressure (Hibberd and Parker, 1976). As this concentration is approached, the rates of evaporation and of expansion of the gas cells, which in the beginning are small, gradually increase, approach the rate of production, and eventually keep pace with it. With a gas production of 2×10^{-8} kmol·kg^{-1}·sec^{-1} (5×10^{-4} L·kg^{-1}·sec^{-1}, corresponding to about 2% of compressed yeast), growth of gas cells changes from slow to rapid after about

25 min of fermentation. No new gas cells are formed by gas production (Baker and Mize, 1941). Ethanol, which is more soluble in water than is carbon dioxide, hardly evaporates during fermentation.

The total pressure in the gas cells is greater than atmospheric pressure because of the surface tension in the gas-dough interface and the viscous resistance of the dough to expansion. Measurements (Bailey, 1955) and calculations (Bloksma, 1981) have demonstrated that the excess pressure is relatively small—on the order of 0.01 of atmospheric pressure. As a consequence, the volume of a given mass of carbon dioxide is virtually independent of the excess pressure, and therefore of dough stiffness, except with extremely stiff doughs; other conditions being equal, the rate of expansion of doughs does not depend on their stiffness.

During fermentation, carbon dioxide is lost by diffusion to the external surface of the dough followed by evaporation from this surface, and by rupture of dough membranes at this surface. Self-diffusion experiments have shown that diffusion and evaporation are rapid enough to be effective within normal fermentation times (Burtle and Sullivan, 1955). Comparisons of the gas production in fermenting doughs and their volume expansion show that the loss is slow for about 1 or 2 hr and then accelerates rapidly (Bailey and Johnson, 1924; Elion, 1940; Chopin, 1953; Hagberg, 1957; Marek and Bushuk, 1967; Marek et al, 1968). The slow initial loss can be explained by diffusion and evaporation and the rapid increase by rupture of membranes. Under normal bakery conditions, the stage of rapid loss is not reached before baking. As a result, only small differences in gas retention are found among flours in the fermentation stage. This is not true for the maturograph, in which the loss of gas begins earlier as a result of the mechanical abuse of the test piece.

Apart from the observation that mechanical dough development can replace part of the fermentation process, there is remarkably little experimental evidence for the theory that fermentation continues dough development. The theory is supported by observations of Halton (1949, 1959) that the resistance of fermenting doughs in stretching tests increased during fermentation and that of nonfermenting ones decreased with time. Observations suggesting the opposite effect, however, are also found in the literature (Bohn and Bailey, 1937; Hoseney et al, 1979; Casutt et al, 1984). The decrease in the pH of dough as a result of fermentation is too small to have a marked effect on dough properties (section VIII-B).

In contrast to gas production, punching and molding operations may increase the number of gas cells by subdividing existing ones, in particular during the passage between sheeting rolls, or by the occlusion of new gas cells upon folding layers of dough (Baker and Mize, 1946). An increase in the number of gas cells leads, of course, to thinner membranes between them.

C. Baking

When the fermentation process is finished, the dough is baked in an oven. Bloksma (1986) reviewed changes in a dough during baking. Temperature increases, in the interior up to 100°C and in the crust above 100°C. During baking, three important changes in dough properties take place: First, the dough expands further; the volume of gas increases from three to four times the volume of the gas-free dough to four to six times this volume. This is called oven rise. Second, the predominantly fluid dough is transformed into a predominantly solid bread crumb or crust. And third, the foam structure of the dough with separate gas cells is transformed into a sponge structure with interconnected gas cells (Baker, 1939).

The expansion of dough in the oven is caused by the following processes, some of which occur simultaneously. In the first stage of baking, the yeast continues to produce carbon dioxide, even at an increased rate, until it is inactivated by heat at a temperature of about $50°C$ (Stokes, 1971). As a result of the increasing saturated vapor pressure, water evaporates into the gas cells. Carbon dioxide and ethanol, produced by the yeast and dissolved in the dough water, also evaporate. Thermal expansion of the gas in the cells according to the law of Gay-Lussac is of minor importance in the expansion of the dough volume during baking.

As a result of the temperature rise, dough viscosity first decreases, as in other materials in which no transformations take place (see Fig. 16). A temperature rise from 26 to $60°C$ reduces the viscosity by a factor of five. In the center of the dough piece, viscosity is further decreased by the transport of water during baking from the outer to the inner parts (Sluimer and Krist-Spit, 1987); at $60°C$, the increase in moisture content is of the order of 4% of the flour mass, resulting in a further reduction in the viscosity by a factor of about two. In this stage of baking, dough is a foam that is much more fluid than the fermenting doughs that are handled and can be observed in a bakery.

Above $60°C$, dough viscosity increases rapidly as a result of the swelling of the starch granules and the exudation of amylose from them, which are part of the gelatinization process (Fig. 16). Starch granules with widely different degrees of gelatinization, from swollen to disrupted granules, are found in bread crumb (Greenwood, 1976). The enormous increase in viscosity reflects the transformation of the predominantly viscous (fluid) dough into the predominantly elastic (solid) crumb and is mainly caused by starch gelatinization. The effect is increased by the polymerization of glutenins (LeGrys et al, 1981; Schofield, J. D., et al, 1984).

A consequence of gelatinization is that most of the starch is susceptible to amylases. If the enzyme activity is high, as in flour from sprouted wheat, enzymatic breakdown may be excessive, leading to a sticky bread crumb.

At the surface of the dough piece, starch gelatinization and evaporation of water cause the formation of a semirigid crust. Browning occurs as a result of the high temperature and may be caused by chemical reactions other than those of the Maillard type (Ziderman and Friedman, 1985). Flavor compounds are also formed.

At a given rate of expansion, excess pressure in the gas cells and the tensile stress in the membranes between them increase with increasing dough viscosity (Bloksma, 1981). Consequently, starch gelatinization is accompanied by a sharp increase in excess pressure and tensile stress. The latter can initiate rupture of membranes and the formation of holes in them, through which excess pressure is released to the outside. As a result, gas retention is lost; the foam structure is transformed into a sponge structure with interconnected gas cells.

This mechanism predicts that the end of oven rise will coincide with an increase in the rate of carbon dioxide release. In the first stage of baking, carbon dioxide is released by diffusion only; after rupture of membranes, it flows outward through the holes. Experimental verification of this prediction is complicated by the fact that oven rise of the outer parts of the dough is completed long before expansion of the center stops; the increase in temperature, expansion, and rupture of cell membranes proceed from the crust to the center of the dough piece (Sluimer and Krist-Spit, 1987). Only in a "resistance oven," in which the heat is generated by an electrical current through the dough itself (Junge and Hoseney, 1981), is dough temperature homogeneous. The experimental data (Elton and Fisher, 1966; Daniels, D. G. H., and Fisher, 1976; Junge and Hoseney, 1981; Moore, W. R., and Hoseney, 1986) contain some contradictions and some observations that are difficult to understand. Nevertheless, they confirm that a long time interval before the rapid release of carbon dioxide is positively correlated with a long duration of oven rise and with a large loaf volume. Similar phenomena were observed in the baking of cakes with various sugar contents (Mizukoshi et al, 1980). The proposed mechanism explains why additives that delay starch gelatinization, such as shortening and emulsifiers, increase loaf or cake volume. The mechanism described does not explain why proteins are important for baking quality. They probably determine the number and sizes of the gas cells in the foam before starch gelatinization.

The formation of a rigid crust does not explain the end of oven rise. If this were the cause, there would be no tensile stress in the cell membranes and, consequently, no cause for the transformation into the sponge structure.

During cooling, water vapor diffuses from the center of the loaf outward (Sluimer and Krist-Spit, 1987) and is replaced by air flowing inward. The latter flow is possible because of the transformation into a sponge structure.

X. RHEOLOGICAL PROPERTIES OF DOUGHS FROM HIGH-EXTRACTION, WHOLE WHEAT, AND COMPOSITE FLOURS

The last decade has witnessed rapid growth in the use of high-extraction and whole wheat flours for the production of bread. This growth has been fueled by documented evidence on the beneficial contribution of cereal fiber to human health and general well-being. In the United Kingdom, consumption of brown and whole wheat bread increased from 7–8% of total bread consumption in 1975 to 25% in 1985 (Galliard, 1986). Although most of the early processing problems have been overcome, it soon became apparent that published information on the rheological properties of doughs from low-extraction flours could not be applied directly to doughs from high-extraction flours.

Although published information on the rheological properties of doughs from high-extraction flours is limited, there is a clear indication that both chemical and physical factors are involved in the observed differences. The detrimental effect of an increased proportion of germ can be related to the introduction of excessive amounts of reduced glutathione. The main functional effect of this constituent appears to be the modification of gluten proteins resulting from the reduction of disulfide bonds (Sullivan et al, 1937). In baking practice, the detrimental effect can be minimized by addition of bromate (Moss, R., et al, 1984).

The negative effect of bran on bread texture and loaf volume cannot be fully accounted for on the basis of dilution of gluten proteins (Pomeranz et al, 1977; Rogers and Hoseney, 1982). Obviously, active constituents of bran and germ interact with gluten proteins to prevent the development of anticipated rheological properties.

The key constituents of bran and germ appear to be the lipid and the lipolytic enzymes (Galliard, 1986). Lipid hydrolysis, producing free fatty acids, proceeds rapidly during storage of high-extraction flours. During dough mixing, polyunsaturated free fatty acids (e.g., linoleic acid) are quickly oxidized by the increased levels of lipoxygenase to a variety of oxidation products. Modification of rheological properties of doughs can result by several mechanisms: inhibition of gluten formation by free fatty acids (Warwick and Shearer, 1982); rapid depletion of oxygen by the lipoxygenase-catalyzed oxidation of polyunsaturated fatty acids (Galliard, 1986); and oxidation of thiol groups by the lipid oxidation products (Tsen and Hlynka, 1963). Preliminary evidence suggests that several reactions are involved and that they are interdependent and probably synergistic (Galliard, 1986) (Fig. 21). Further research is needed to elucidate the interplay of what appear to be numerous chemical and physical effects.

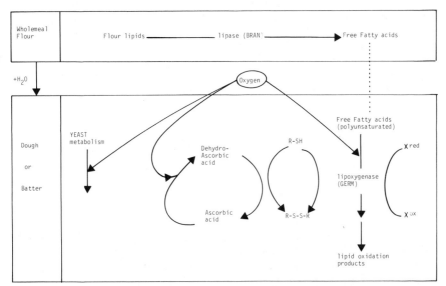

Fig. 21. Schematic representation of lipid degradation, oxygen utilization, and redox reactions in whole-meal flour, doughs, and batters. The lipase reaction, which occurs over a period of weeks in whole-meal flour (natural moisture basis), is distinguished from the rapid (seconds) reactions that occur when water is added to make a dough or batter. Also illustrated are possible co-oxidation reactions of lipoxygenase activity by which thiols (R-SH) may be converted into disulfides (RSSR) and, in general, oxidizable substrates (X_{red}) may be converted into oxidized forms (X_{ox}). (Reprinted, by permission, from Galliard, 1986)

The term *composite flour* is used to describe a flour made by blending nonwheat flour with wheat flour and used for the production of baked goods traditionally made from wheat only. The nonwheat component can be flour from other cereals, namely barley, maize, millet, oats, or sorghum; starches from roots or tubers; or high-protein flours, concentrates, or isolates from oilseeds and legumes (Civetta, 1974).

Interest in the commercial use of composite flours exists for two widely different reasons. First, in countries that do not grow wheat, it can be economically advantageous to replace some of the imported wheat by lower priced domestically grown crops, for example, cassava, millet, or sorghum. This interest in composite flours has waned in the mid-1980s, mainly because wheat is available from exporting countries at very low prices. Second, in some other countries where bread contributes a substantial proportion of dietary protein, the addition of high-protein flours from oilseeds and legumes increases the content and nutritional quality of the protein in bread.

The functional properties of doughs from composite flours containing high-starch cereal flours such as maize flour can be predicted from simple dilution of gluten proteins (Mukulumwa, 1976). In general, the breadmaking potential decreases linearly with the proportion of wheat flour. The proportion of nonwheat flour that the wheat flour can effectively "carry" depends on the inherent strength of the wheat flour (Youssef and Bushuk, 1986).

In the context of functionality, most additions have a direct detrimental effect arising from simple dilution of gluten proteins. This detrimental effect may be accentuated by the presence in the nonwheat product of undesirable enzymes or constituents that interact strongly with gluten proteins and thereby inhibit development of desirable rheological properties. The method of preparation can increase or diminish the functionality of protein isolates in bread doughs. Youssef and Bushuk (1986) reported that two high-protein products prepared from faba beans by different procedures had widely different functionality in breadmaking. As in the case of composite flours from starchy cereals, the basic strength of the wheat flour used to prepare the composite is important to the functionality of doughs from such flours (Youssef and Bushuk, 1986). Flours from stronger wheat cultivars can carry a higher percentage of the nonwheat products and still meet the specifications of the baked product. Successful utilization of composite flours in bread production requires ingredient and process optimization for each specific product. Economics and consumer acceptance of the baked product ultimately determine the extent of commercial use of any composite flour.

Results of physical dough tests applied to composite flours must be interpreted with caution; the correlations that are valid for wheat flour may not apply. For example, some additions that are detrimental to loaf quality nevertheless increased the farinograph water absorption (Khan et al, 1975; Campos and El-Dash, 1978; Olatunji and Akinrele, 1978).

XI. CONCLUSIONS AND OUTLOOK

A comparison of this chapter with the corresponding chapters in the first (Bloksma and Hlynka, 1964) and second (Bloksma, 1971) editions of this book shows that our knowledge of dough properties has increased enormously. Figure 6 is an example of this new knowledge; in the previous editions, it was not yet possible to present results of measurements on viscous flow by various authors and over five decades of the rate of shear. However, the expansion of our knowledge has been more quantitative than qualitative; it is more a matter of facts and precision than of understanding.

The two questions to be considered, as mentioned in section I, are first, what is the relationship between the molecular and microscopic structure of dough and its rheological behavior, and second, how do the rheological properties of dough affect its behavior in the bakery?

We understand that the cohesion and extensibility of dough depend on the presence of sufficient glutenins with a high molecular weight. Disulfide bonds link the glutenin subunits together, which explains the effect of oxidation and reduction on the distribution of molecular weights and, consequently, on the rheological properties of dough. There are some hypotheses on the relation between the primary structure of glutenin subunits and their ability to form long, unbranched molecules. Ideas on the role of gliadins and soluble protein fractions are still vague. Differences among cultivars in improver response have not yet been explained. Our knowledge of the role of polar and nonpolar lipids in the binding between the various protein fractions consists of bits and pieces. The role of polysaccharides other than starch likewise needs further exploration.

Once the rheological behavior of the protein phase of dough, including its nonprotein constituents, can be understood on the basis of its molecular structure, deriving from this understanding how the complete dough, with a large volume fraction of starch, will behave will still be a difficult task. Theories on the rheological behavior of dispersions fail under the conditions of nonlinear behavior of the continuous phase and a large volume fraction of nonspherical particles.

Perusal of section IX on the breadmaking process does not lead to an accurate specification of rheological properties that dough must have in order to be handled safely in a bakery and to yield a loaf of bread with a light and even crumb structure. The only exception is that it obviously must be extensible in order to avoid premature rupture of membranes between gas cells; extensibility must be maintained long enough under baking conditions to permit sufficient oven rise. Extensibility under baking conditions is more important because only under these conditions do the large differences between flours and the effects of fats and emulsifiers show up. That only the extensibility emerges as a potentially important property is a rather negative conclusion on the importance of the rheological properties of dough. A more penetrating analysis of the breadmaking process than has been given in section IX may, however, reveal rheological properties that are indeed essential for good baking quality.

Bread doughs in an advanced stage of fermentation and during baking are foams with a large area of gas-dough interface. Although this is a difficult subject for experimental studies, it will be important to learn more of the properties of this interface and their possible effect on gas retention. This conclusion applies both to the fermentation stage and to the baking stage, in which the foam structure is converted into a sponge structure.

LITERATURE CITED

AAAC. 1983. Approved Methods of the American Association of Cereal Chemists, 8th ed. Method 22-11, approved 1961; Method 22-14, approved 1961; Method 54-40, approved 1961; Method 54-10, approved 1961, reviewed 1982; Method 54-21, approved 1961, reviewed 1982; Method 54-30, approved 1984. The Association, St. Paul, MN.

AGATOVA, A. I., and PROSKURYAKOV, N. I. 1962. On sulphydryl groups and disulphide bonds in the proteins of wheat flour. Biochimia 27:88-93.

AITKEN, T. R., and GEDDES, W. F. 1939. The relation between protein content and strength of gluten-enriched flours. Cereal Chem. 16:223-231.

AITKEN, T. R., FISHER, M. H., and ANDERSON, J. A. 1944. Effect of protein content and grade on farinograms, extenso-grams, and alveograms. Cereal Chem. 21:465-488.

AUERMAN, L. J. 1962. Bestimmung der physikalischen Eigenschaften von Kleber, Teig und Brotkrume mit Hilfe des Pene-trometers. Nahrung 6:545-553.

AXFORD, D. W. E., and ELTON, G. A. H. 1960. The mechanical development of bread doughs. Chem. Ind. London 1960:1257-1258.

AXFORD, D. W. E., CHAMBERLAIN, N., COLLINS, T. H., and ELTON, G. A. H. 1963. Continuous breadmaking. The Chorley-wood process. Cereal Sci. Today 8:265-266, 268, 270.

BAILEY, C. H. 1941. A translation of Beccari's lecture "Concerning grain" (1728). Cereal Chem. 18:555-561.

BAILEY, C. H. 1955. Gas pressure in fermented doughs. Cereal Chem. 32:152-156.

BAILEY, C. H., and JOHNSON, A. H. 1924. Carbon dioxide diffusion ratio of wheat flour doughs as a measure of fermentation period. Cereal Chem. 1:293-304.

BAKER, J. C. 1939. The permeability of bread by air. Cereal Chem. 16:730-733.

BAKER, J. C., and MIZE, M. D. 1941. The origin of the gas cell in bread dough. Cereal Chem. 18:19-34.

BAKER, J. C., and MIZE, M. D. 1942. The relation of fats to texture, crumb, and volume of bread. Cereal Chem. 19:84-94.

BAKER, J. C., and MIZE, M. D. 1946. Gas occlusion during dough mixing. Cereal Chem. 23:39-51.

BALDWIN, R. R., JOHANSEN, R. G., KEOUGH, W. J., TITCOMB, S. T., and COTTON, R. H. 1963. Continuous bread-making. The role that fat plays. Cereal Sci.

Today 8:273-274, 276, 284, 296.

BALLS, A. K., and HALE, W. S. 1936. Further studies on the activity of proteinase in flour. Cereal Chem. 13:656-664.

BARNES, P. J. 1983. Lipids in Cereal Technology. Academic Press, New York.

BAYFIELD, E. G., and STONE, C. D. 1960. Effects of absorption and temperature upon flour-water farinograms. Cereal Chem. 37:233-240.

BAYFIELD, E. G., and YOUNG, W. E. 1963a. Liquid (oil) shortenings in white bread. Bakers Dig. 37(5):58-62.

BAYFIELD, E. G., and YOUNG, W. E. 1963b. Liquid (oil) shortenings in white bread. II. Effect of elevated dough temperatures. Bakers Dig. 37(6):59-60, 62-65.

BAYFIELD, E. G., and YOUNG, W. E. 1964. Fluid shortenings for white bread. Cereal Sci. Today 9:363-364, 366-368, 370-371, 381.

BECHTEL, D. B., POMERANZ, Y., and DE FRANCISCO, A. 1978. Breadmaking studied by light and transmission electron microscopy. Cereal Chem. 55:392-401.

BECKWITH, A. C., WALL, J. S., and DIMLER, R. J. 1963. Amide groups as interaction sites in wheat gluten proteins: Effects of amide-ester conversion. Arch. Biochem. Biophys. 103:319-330.

BÉKÉS, F., ZAWISTOWSKA, U., and BUSHUK, W. 1983a. Protein-lipid complexes in the gliadin fraction. Cereal Chem. 60:371-378.

BÉKÉS, F., ZAWISTOWSKA, U., and BUSHUK, W. 1983b. Lipid-mediated aggregation of gliadin. Cereal Chem. 60:379-380.

BELDEROK, B., and SLAGER, E. A. 1983. Effect van zetmeelbeschadiging in tarwebloem op deeg- en broodeigenschappen. Voedings-middelentechnologie 16(2):34-39.

BELITZ, H.-D., SEILMEIER, W., WIESER, H., KIEFFER, R., KIM, J.-J., and DIRNDORFER, M. 1982. Struktur und technologische Eigenschaften von Kleber-proteinen des Weizens. Getreide Mehl Brot 36:285-291.

BENNETT, R., and COPPOCK, J. B. M. 1952. Measuring the physical characteristics of flour: A method of using the Chopin Alveograph to detect the effect of flour improvers. J. Sci. Food Agric. 3:297-307.

BENNETT, R., and COPPOCK, J. B. M. 1953. Dough consistency and measurement of water absorption on the Brabender Farino-graph and Simon "Research" water absorption meter. Trans. Am. Assoc. Cereal Chem.

11:172-182.

BENNETT, R., and COPPOCK, J. B. M. 1957. The natural ageing of flour. J. Sci. Food Agric. 8:261-270.

BENNETT, R., and EWART, J. A. D. 1965. The effects of certain salts on doughs. J. Sci. Food Agric. 16:199-205.

BIGELOW, C. C. 1967. On the average hydrophobicity of proteins and the relation between it and protein structure. J. Theor. Biol. 16:187-211.

BLANSHARD, J. M. V. 1986. The significance of the structure and function of the starch granule in baked products. Pages 1-13 in: Chemistry and Physics of Baking: Materials, Processes and Products. J. M. V. Blanshard, P. J. Frazier, and T. Galliard, eds. R. Soc. Chem., London.

BLISH, M. J., and HUGHES, R. C. 1932. Some effects of varying sugar concentrations in bread dough on fermentation by-products and fermentation tolerance. Cereal Chem. 9:331-356.

BLOKSMA, A. H. 1957. A calculation of the shape of the alveograms of some rheological model substances. Cereal Chem. 34:126-136.

BLOKSMA, A. H. 1958. A calculation of the shape of the alveograms of materials showing structural viscosity. Cereal Chem. 35:323-330.

BLOKSMA, A. H. 1959. The amperometric titration of thiol groups in flour and gluten. Cereal Chem. 36:357-368.

BLOKSMA, A. H. 1962. Slow creep of wheat flour doughs. Rheol. Acta 2:217-230.

BLOKSMA, A. H. 1963. Oxidation by molecular oxygen of thiol groups in unleavened doughs from normal and defatted wheat flours. J. Sci. Food Agric. 14:529-535.

BLOKSMA, A. H. 1964a. Oxidation by potassium iodate of thiol groups in unleavened wheat flour doughs. J. Sci. Food Agric. 15:83-94.

BLOKSMA, A. H. 1964b. Rheologie von Brotteig bei langsamen Deformationen. Brot Gebaeck 18:173-181.

BLOKSMA, A. H. 1966. Extraction of flour by mixtures of butanol-1 and water. Cereal Chem. 43:602-622.

BLOKSMA, A. H. 1967. Detection of changes in modulus and viscosity of wheat flour doughs by the "work technique" of Muller et al. J. Sci. Food Agric. 18:49-51, 132.

BLOKSMA, A. H. 1968. Effect of potassium iodate on creep and recovery and on thiol and disulfide contents of wheat flour doughs. Pages 153-166 in: Rheology and Texture of Foodstuffs. SCI Monogr. 27. Soc. Chem. Ind., London.

BLOKSMA, A. H. 1971. Rheology and chemistry of dough. Pages 523-584 in: Wheat

Chemistry and Technology, 2nd ed. Y. Pomeranz, ed. Am. Assoc. Cereal Chem., St. Paul, MN.

BLOKSMA, A. H. 1972a. Flour composition, dough rheology, and baking quality. Cereal Sci. Today 17:380-386.

BLOKSMA, A. H. 1972b. The relation between the thiol and disulfide contents of dough and its rheological properties. Cereal Chem. 49:104-118.

BLOKSMA, A. H. 1974. The chemical and rheological basis of oxidative flour improvement. Pages 251-262 in: Lebensmittel—Einfluss der Rheologie auf die Lebensmittelverarbeitung und die Lebensmittelqualität. Dechema Monographien 77. Verlag Chemie, Weinheim, Federal Republic of Germany.

BLOKSMA, A. H. 1975. Thiol and disulfide groups in dough rheology. Cereal Chem. 52:170r-183r.

BLOKSMA, A. H. 1980. Effect of heating rate on viscosity of wheat flour doughs. J. Texture Stud. 10:261-269.

BLOKSMA, A. H. 1981. Effect of surface tension in the gas-dough interface on the rheological behavior of dough. Cereal Chem. 58:481-486.

BLOKSMA, A. H. 1984. Theoretical aspects of the farinograph. Pages 7-12 in: The Farinograph Handbook, 3rd ed. B. L. D'Appolonia and W. H. Kunerth, eds. Am. Assoc. Cereal Chem., St. Paul, MN.

BLOKSMA, A. H. 1985. The effect of temperature on dough viscosity, and its consequence for the control of dough temperature. J. Food Eng. 4:205-227.

BLOKSMA, A. H. 1986. Rheological aspects of structural changes during baking. Pages 170-178 in: Chemistry and Physics of Baking: Materials, Processes and Products. J. M. V. Blanshard, P. J. Frazier, and T. Galliard, eds. R. Soc. Chem., London.

BLOKSMA, A. H., and HLYNKA, I. 1964. Basic considerations of dough properties. Pages 465-526 in: Wheat Chemistry and Technology, 1st ed. I. Hlynka, ed. Am. Assoc. Cereal Chem., St. Paul, MN.

BLOKSMA, A. H., and MEPPELINK, E. K. 1973. Sensory assessment of stiffness and rheological measurements on doughs of identical farinograph consistencies. J. Texture Stud. 4:145-153.

BLOKSMA, A. H., and NIEMAN, W. 1975. The effect of temperature on some rheological properties of wheat flour doughs. J. Texture Stud. 6:343-361.

BOHLIN, L., and CARLSON, T. L.-G. 1980. Dynamic viscoelastic properties of wheat flour dough: Dependence on mixing time. Cereal Chem. 57:174-177.

BOHLIN, L., and CARLSON, T. L.-G. 1981. Shear stress relaxation of wheat flour dough and gluten. Colloids Surf. 2:59-69.

BOHN, L. J., and BAILEY, C. H. 1936. Effect of mixing on the physical properties of dough. Cereal Chem. 13:560-575.

BOHN, L. J., and BAILEY, C. H. 1937. Effect of fermentation, certain dough ingredients, and proteases upon the physical properties of flour doughs. Cereal Chem. 14:335-348.

BOLLING, H. 1980. Zur Optimierung der Backeigenschaften von Weizenmischungen unter besonderer Berücksichtigung spezifischer Rohstoffeigenschaften. Getreide Mehl Brot 34:310-314.

BOLLING, H., and WEIPERT, D. 1984. Zur Beurteilung der Eigenschaften von Weizenteigen mit Hilfe des Extensogramms. Getreide Mehl Brot 38:131-136.

BRABENDER, C. W. 1937. Beitrag zur Weizenstandardisierung. Ein Vorschlag für den Welt-Weizenstandard. Pages 152-163 in: Nordisk Cereal-Kjemiker-Forening og Nordisk Cerealistforbund Kongres, Oslo.

BRABENDER, M. 1973a. Resistography. A dynamic quick method to classify wheat and flour quality. Cereal Sci. Today 18:206-210.

BRABENDER, M. 1973b. Eine dynamische Schnellanalyse an Teigen für Weizen-und Mehlqualitätsbeurteilung mit dem Resistographen. Getreide Mehl Brot 27:258-262.

BRÜMMER, J.-M. 1976. Erfahrungen mit dem Kurzextensogramm zur Erkennung von Weizen, die nicht zur Brotherstellung geeignet sind. Getreide Mehl Brot 30:213-217.

BRÜMMER, J.-M. 1978. Untersuchungen von Weizenmehlen mit dem Kurzextensogramm. 1. Erkennung des Backverhaltens von Weizenmehlen der Type 550. Getreide Mehl Brot 32:220-224.

BRÜMMER, J.-M. 1980. Arbeitsanleitung für das Kurzextensogramm. Muehle Mischfuttertechnik 117:593.

BRÜMMER, J.-M. 1981. Einsatzmöglichkeiten des Kurzextensogramms im Mühlenlaboratorium. Muehle Mischfuttertechnik 118:213-215, 228-231.

BRÜMMER, J.-M., SEIBEL, W., and STEPHAN, H. 1980. Backtechnische Wirkung von L-Cystein-Hydrochlorid und L-Cystin bei der Herstellung von Brot und Kleingebäck. Getreide Mehl Brot 34:173-178.

BUDICEK, L., and BULENA, V. 1973. Kneten teigartiger Stoffe. Getreide Mehl Brot 27:325-328.

BUNGENBERG DE JONG, H. L. 1956. Gluten oxydation and fatty substances. Rev. Ferment. Ind. Aliment. 11:261-270.

BURROWS, S., and HARRISON, J. S. 1959. Routine method for determination of the activity of baker's yeast. J. Inst. Brew. London 65:39-45.

BURTLE, J. G., and SULLIVAN, B. 1955. The diffusion of carbon dioxide through fermenting bread sponges. Cereal Chem. 32:488-492.

BUSHUK, W. 1961. Accessible sulfhydryl groups in dough. Cereal Chem. 38:438-448.

BUSHUK, W. 1966. Distribution of water in dough and bread. Bakers Dig. 40(5):38-40.

BUSHUK, W. 1985. Flour proteins: Structure and functionality in dough and bread. Cereal Foods World 30:447-448, 450-451.

BUSHUK, W. 1986. Protein-lipid and protein-carbohydrate interactions in flour-water mixtures. Pages 147-154 in: Chemistry and Physics of Baking: Materials, Processes and Products. J. M. V. Blanshard, P. J. Frazier, and T. Galliard, eds. R. Soc. Chem., London.

BUSHUK, W., and HLYNKA, I. 1960a. The bromate reaction in dough. I. Kinetic studies. Cereal Chem. 37:141-150.

BUSHUK, W., and HLYNKA, I. 1960b. The bromate reaction in dough. II. Inhibition and activation studies. Cereal Chem. 37:343-351.

BUSHUK, W., and HLYNKA, I. 1961a. The bromate reaction in dough. III. Effect of continuous mixing and flour particle size. Cereal Chem. 38:178-186.

BUSHUK, W., and HLYNKA, I. 1961b. The bromate reaction in dough. IV. Effect of reducing agents. Cereal Chem. 38:309-316.

BUSHUK, W., and HLYNKA, I. 1961c. The bromate reaction in dough. V. Effect of flour components and some related compounds. Cereal Chem. 38:316-325.

BUSHUK, W., and HLYNKA, I. 1962. The effect of iodate and N-ethylmaleimide on extensigraph properties of dough. Cereal Chem. 39:189-195.

BUSHUK, W., and LUKOW, O. M. 1987. Effect of sprouting on wheat proteins and baking properties. Pages 188-196 in: Proc. 4th Int. Symp. Pre-Harvest Sprouting in Cereals. Daryl J. Mares, ed. Westview Press, Boulder, CO.

BUSHUK, W., and MacRITCHIE, F. 1988. Wheat proteins: Aspects of structure that are related to breadmaking quality. Pages 337-361 in: Protein Quality and the Effects of Processing. R. D. Phillips and J. W. Finley, eds. Marcel Dekker, New York.

BUSHUK, W., and MEHROTRA, V. K. 1977. Studies of water binding by differential thermal analysis. II. Dough studies using the melting mode. Cereal Chem. 54:320-325.

CALVEL, R. 1969. L'action du sel sur les pâtes fermentées et sur les caractéristiques du produit fabriqué. Bull. Anc. Eleves Ec. Fr. Meun. 40:241-246.

CAMPOS, J. E., and EL-DASH, A. A. 1978.

Effect of addition of full fat sweet lupine flour on rheological properties of dough and baking quality of bread. Cereal Chem. 55:619-627.

CARLISLE, C. H., HOLMES, K. C., and HUTCHINSON, J. B. 1960. X-ray diffraction patterns of doughs from strong and weak flours. Chem. Ind. London 1960:1216-1217.

CARLSON, T., and BOHLIN, L. 1978. Free surface energy in the elasticity of wheat flour dough. Cereal Chem. 55:539-544.

CARTER, J. E., and PACE, J. 1965. Some interrelationships of ascorbic acid and dehydroascorbic acid in the presence of flour suspensions and in dough. Cereal Chem. 42:201-208.

CASUTT, V., PRESTON, K. R., and KILBORN, R. H. 1984. Effects of fermentation time, inherent flour strength, and salt level on extensigraph properties of full-formula remix-to-peak processed doughs. Cereal Chem. 61:454-459.

CAWLEY, R. W. 1964. The rôle of wheat flour pentosans in baking. II. Effect of added flour pentosans and other gums on gluten-starch loaves. J. Sci. Food Agric. 15:834-838.

CHAMBERLAIN, N., COLLINS, T. H., and ELTON, G. A. H. 1962. The mechanical development of bread doughs. Pages 173-195 in: Recent Advances in Processing Cereals. SCI Monogr. 16. Soc. Chem. Ind., London.

CHAMBERLAIN, N., COLLINS, T. H., and ELTON, G. A. H. 1965. The Chorleywood bread process. Improving effects of fat. Cereal Sci. Today 10:415-416, 418-419, 490.

CHAMBERLAIN, N., COLLINS, T. H., DODDS, N. J. H., and ELTON, G. A. H. 1966a. Chemical development of bread doughs. Milling 146:319, 338.

CHAMBERLAIN, N., DODDS, N. J. H., and ELTON, G. A. H. 1966b. Interaction of ascorbic acid and potassium bromate in mechanically developed doughs. Chem. Ind. London 1966:1456-1457.

CHEN, C. H., and BUSHUK, W. 1970. Nature of proteins in *Triticale* and its parental species. I. Solubility characteristics and amino acid composition of endosperm proteins. Can. J. Plant Sci. 50:9-14.

CHOPIN, M. 1927. Determination of baking value of wheat by measure of specific energy of deformation of dough. Cereal Chem. 4:1-13.

CHOPIN, M. 1953. Le zymotachygraphe. Appareil de contrôle de la fermentation panaire. Bull. Anc. Eleves Ec. Fr. Meun. 24:70-76.

CHOPIN, M. 1973. Cinquante Années de Recherches Relatives aux Blés et à Leur Utilisation Industrielle. M. Chopin & Cie.,

Boulogne, France.

CHUNG, O. K. 1986. Lipid-protein interactions in wheat flour, dough, gluten, and protein fractions. Cereal Foods World 31:242-244, 246, 247, 249-252, 254-256.

CHUNG, O. K., and POMERANZ, Y. 1981. Recent research on wheat lipids. Bakers Dig. 5(5):38-50, 55.

CIVETTA, A. 1974. Developing a market for composite flour. Cereal Sci. Today 19:146-148.

COLE, E. W., MECHAM, D. K., and PENCE, J. W. 1960. Effect of flour lipids and some lipid derivatives on cookie-baking characteristics of lipid-free flours. Cereal Chem. 37:109-121.

COOKSON, M. A., and COPPOCK, J. B. M. 1956. The role of lipids in baking. III. Some breadmaking and other properties of defatted flours and of flour lipids. J. Sci. Food Agric. 7:72-87.

COPPOCK, J. B. M., and MULLER, H. G. 1969. Measurement of rheological properties of flour doughs subjected to large deformations: The effect of certain bonding reagents on the viscous and elastic components of such doughs. Pages 729-740 in: Proc. 1st Int. Congr. Food Sci. Technol. I. Chemical and Physical Aspects of Food. J. M. Leitch, ed. Gordon & Breach Sci. Publ., New York.

COX, W. P., and MERZ, E. H. 1958. Correlation of dynamic and steady flow viscosities. J. Polym. Sci. 28:619-622.

CUNNINGHAM, D. K., GEDDES, W. F., and ANDERSON, J. A. 1955. Precipitation by various salts of the proteins extracted by formic acid from wheat, barley, rye, and oat flours. Cereal Chem. 32:192-199.

CUNNINGHAM, J. R., and HLYNKA, I. 1954. Relaxation time spectrum of dough and the influence of temperature, rest, and water content. J. Appl. Phys. 25:1075-1081.

DAFTARY, R. D., POMERANZ, Y., SHOGREN, M., and FINNEY, K. F. 1968. Functional bread-making properties of wheat flour lipids. 2. The role of flour lipid fractions in bread-making. Food Technol. 22:327-330.

DAHLE, L. K., and HINZ, R. S. 1966. The weakening action of thioctic acid in unyeasted and yeasted doughs. Cereal Chem. 43:682-688.

DAHLE, L. K., and SULLIVAN, B. 1963. The oxidation of wheat flour. V. Effect of lipid peroxides and antioxidants. Cereal Chem. 40:372-384.

DANIELS, D. G. H. 1958. Polar lipids in wheat flour. Chem. Ind. London 1958:653-654.

DANIELS, D. G. H., and FISHER, N. 1976. Release of carbon dioxide from dough during baking. J. Sci. Food Agric. 27:351-357.

DANIELS, N. W. R., RICHMOND, J. W., RUSSELL EGGITT, P. W., and COPPOCK,

J. B. M. 1966. Studies on the lipids of flour. III. Lipid binding in breadmaking. J. Sci. Food Agric. 17:20-29.

DANIELS, N. W. R., RICHMOND, J. W., RUSSELL EGGITT, P. W., and COPPOCK, J. B. M. 1967. Effect of air on lipid binding in mechanically developed doughs. Chem. Ind. London 1967:955-956.

DANNO, G., and HOSENEY, R. C. 1982. Effects of dough mixing and rheologically active compounds on relative viscosity of wheat proteins. Cereal Chem. 59:196-198.

D'APPOLONIA, B. L. 1984. Types of farinograph curves and factors affecting them. Pages 13-23 in: The Farinograph Handbook, 3rd ed. B. L. D'Appolonia and W. H. Kunerth, eds. Am. Assoc. Cereal Chem., St. Paul, MN.

DAVIES, R. J., and WEBB, T. 1969. Calorimetric determination of freezable water in dough. Chem. Ind. London 1969:1138-1139.

DE DEKEN, R. H., and MORTIER, A. 1955. Étude du gluten de froment. I. Solubilité. Biochim. Biophys. Acta 16:354-360.

DEMPSTER, C. J., HLYNKA, I., and WINKLER, C. A. 1952. Quantitative extensograph studies of relaxation of internal stresses in non-fermenting bromated and unbromated doughs. Cereal Chem. 29:39-53.

DEMPSTER, C. J., HLYNKA, I., and ANDERSON, J. A. 1953. Extensograph studies of structural relaxation in bromated and unbromated doughs mixed in nitrogen. Cereal Chem. 30:492-503.

DEMPSTER, C. J., HLYNKA, I., and ANDERSON, J. A. 1954. Extensograph studies of the improving action of oxygen in dough. Cereal Chem. 31:240-249.

DEMPSTER, C. J., HLYNKA, I., and ANDERSON, J. A. 1955. Influence of temperature on structural relaxation in bromated and unbromated doughs mixed in nitrogen. Cereal Chem. 32:241-254.

DEMPSTER, C. J., CUNNINGHAM, D. K., FISHER, M. H., HLYNKA, I., and ANDERSON, J. A. 1956. Comparative study of the improving action of bromate and iodate by baking data, rheological measurements, and chemical analyses. Cereal Chem. 33:221-239.

DENGATE, H. N., BARUCH, D. W., and MEREDITH, P. 1978. The density of wheat starch granules: A tracer dilution procedure for determining the density of an immiscible dispersed phase. Staerke 30:80-84.

DIMLER, R. J. 1963. Gluten. The key to wheat's utility. Bakers Dig. 37(1):52-57.

DÖRNER, H., and STEPHAN, H. 1956. Über pH-Untersuchungen an Teigen und Broten. Brot Gebaeck 10:171-175.

ELION, E. 1940. The importance of gas-production and gas-retention measurements during the fermentation of dough. Cereal Chem. 17:573-581.

ELTON, G. A. H., and FISHER, N. 1966. A technique for the study of the baking process, and its application to the effect of fat on baking dough. J. Sci. Food Agric. 17:250-254.

ELTON, G. A. H., and FISHER, N. 1968. Effect of solid hydrocarbons as additives in breadmaking. J. Sci. Food Agric. 19:178-181.

EVANS, G. C., DE MAN, J. M., RASPER, V., and VOISEY, P. W. 1974. An improved dough extensigraph. J. Inst. Can. Sci. Technol. Aliment. 7:263-268.

EVERS, A. D., and STEVENS, D. J. 1985. Starch damage. Pages 321-349 in: Advances in Cereal Science and Technology, Vol. VII. Y. Pomeranz, ed. Am. Assoc. Cereal Chem., St. Paul, MN.

EWART, J. A. D. 1968. A hypothesis for the structure and rheology of glutenin. J. Sci. Food Agric. 23:617-623.

EWART, J. A. D. 1972. A modified hypothesis for the structure and rheology of glutenins. J. Sci. Food Agric. 23:687-699.

EWART, J. A. D. 1977. Re-examination of the linear glutenin hypothesis. J. Sci. Food Agric. 28:191-199.

EWART, J. A. D. 1978. Glutenin and dough tenacity. J. Sci. Food Agric. 29:551-556.

EWART, J. A. D. 1979. Glutenin structure. J. Sci. Food Agric. 30:482-492.

FARIDI, H., and RASPER, V. F. 1987. The Alveograph Handbook. Am. Assoc. Cereal Chem., St. Paul, MN. 56 pp.

FAUBION, J. M., DREESE, P. C., and DIEHL, K. C. 1985. Dynamic rheological testing of wheat flour doughs. Pages 91-116 in: Rheology of Wheat Products. H. Faridi, ed. Am. Assoc. Cereal Chem., St. Paul, MN.

FERRARI, C. G., HIGASHIUCHI, K., and PODLISKA, J. A. 1963. Flour maturing and bleaching with acyclic acetone peroxides. Cereal Chem. 40:89-101.

FINNEY, K. F., and SHOGREN, M. D. 1972. A ten-gram mixograph for determining and predicting functional properties of wheat flours. Bakers Dig. 46(2):32-35, 38-42, 77.

FISHER, M. H., AITKEN, T. R., and ANDERSON, J. A. 1949. Effects of mixing, salt, and consistency on extensograms. Cereal Chem. 26:81-97.

FITCHETT, C. S., and FRAZIER, P. J. 1986. Action of oxidants and other improvers. Pages 179-198 in: Chemistry and Physics of Baking: Materials, Processes and Products. J. M. V. Blanshard, P. J. Frazier, and T. Galliard, eds. R. Soc. Chem., London.

FRATER, R., and HIRD, F. J. R. 1963. The

reaction of glutathione with serum albumin, gluten and flour proteins. Biochem. J. 88:100-105.

FRATER, R., HIRD, F. J. R., MOSS, H. J., and YATES, J. R. 1960. A role for thiol and disulphide groups in determining the rheological properties of dough made from wheaten flour. Nature 186:451-454.

FRATER, R., HIRD, F. J. R., and MOSS, H. J. 1961. Rôle of disulphide exchange reactions in the relaxation of strains introduced in dough. J. Sci. Food Agric. 12:269-273.

FRAZIER, P. J. 1983. Lipid-protein interactions during dough development. Pages 189-212 in: Lipids in Cereal Technology. P. J. Barnes, ed. Academic Press, London.

FRAZIER, P. J., and DANIELS, N. W. R. 1986. Protein-lipid interactions in bread dough. Pages 299-326 in: Interactions of Food Components. G. G. Birch and M. G. Lindley, eds. Elsevier Appl. Sci. Publ., London.

FRAZIER, P. J., and MULLER, H. G. 1967. A note on the effect of potassium iodate and thioacetic acid on the birefringence of stretched gluten. Cereal Chem. 44:558-559.

FRAZIER, P. J., DANIELS, N. W. R., and RUSSELL EGGITT, P. W. 1975. Rheology and the continuous breadmaking process. Cereal Chem. 52:106r-130r.

FRAZIER, P. J., DANIELS, N. W. R., and RUSSELL EGGITT, P. W. 1981. Lipid-protein interactions during dough development. J. Sci. Food Agric. 32:877-897.

FRAZIER, P. J., FITCHETT, C. S., and RUSSELL EGGITT, P. W. 1985. Laboratory measurement of dough development. Pages 151-175 in: Rheology of Wheat Products. H. Faridi, ed. Am. Assoc. Cereal Chem., St. Paul, MN.

FRETZDORFF, B., BECHTEL, D. B., and POMERANZ, Y. 1982. Freeze-fracture ultrastructure of wheat flour ingredients, dough, and bread. Cereal Chem. 59:113-120.

GALLIARD, T. 1986. Whole meal flour and baked products: Chemical aspects of functional properties. Pages 199-215 in: Chemistry and Physics of Baking: Materials, Processes and Products. J. M. V. Blanshard, P. J. Frazier, and T. Galliard, eds. R. Soc. Chem., London.

GASIOROWSKI, H. 1967. Influence du pH sur les changements de certaines propriétés de la pâte pendant le pétrissage. Ind. Aliment. Agric. 84:1581-1586.

GERMAIN, B., PERRET, G., POMA, J., and BURÉ, J. 1968. Rôle des lipides et de leur évolution dans le comportement technologique des farines. Ind. Aliment. Agric. 85:803-810;

Meun. Fr. 250:29-36.

GIESEKUS, H. 1983. Disperse systems: Dependence of rheological properties on the type of flow with implications for food rheology. Pages 205-220 in: Physical Properties of Foods. R. Jowitt, F. Escher, B. Hallström, H. F. T. Meffert, W. E. L. Spiess, and G. Vos, eds. Appl. Sci. Publ., London.

GLUCKLICH, J., and SHELEF, L. 1962a. A model representation of the rheological behaviour of wheat-flour dough. Kolloid Z. Z. Polym. 181:29-33.

GLUCKLICH, J., and SHELEF, L. 1962b. An investigation into the rheological properties of flour dough. Studies in shear and compression. Cereal Chem. 39:242-255.

GOLDSTEIN, S. 1957. Sulfhydryl- und Disulfidgruppen der Klebereiweisse und ihre Beziehung zur Backfähigkeit der Brotmehle. Mitt. Geb. Lebensmittelunters. Hyg. 48:87-93.

GRANT, D. R., and SOOD, V. K. 1980. Studies of the role of ascorbic acid in chemical dough development. II. Partial purification and characterization of an enzyme oxidizing ascorbate in flour. Cereal Chem. 57:46-49.

GRAVELAND, A. 1968. Combination of thin layer chromatography and gas chromatography in the analysis on a microgram scale of lipids from wheat flour and wheat flour doughs. J. Am. Oil Chem. Soc. 45:834-840.

GRAVELAND, A. 1970. Enzymatic oxidations of linoleic acid and glycerol-1-monolinoleate in doughs and flour-water suspensions. J. Am. Oil Chem. Soc. 47:352-361.

GRAVELAND, A., BOSVELD, P., and MARSEILLE, J. P. 1978. Determination of thiol groups and disulphide bonds in wheat flour and dough. J. Sci. Food Agric. 29:53-61.

GRAVELAND, A., BOSVELD, P., LICHTENDONK, W. J., MARSEILLE, J. P., MOONEN, J. H. E., and SCHEEPSTRA, A. 1985. A model for the molecular structure of the glutenins from wheat flour. J. Cereal Sci. 3:1-16.

GREENWELL, P., and SCHOFIELD, J. D. 1986. A starch granule protein associated with endosperm softness in wheat. Cereal Chem. 63:379-380.

GREENWOOD, C. T. 1976. Starch. Pages 119-157 in: Advances in Cereal Science and Technology, Vol. I. Y. Pomeranz, ed. Am. Assoc. Cereal Chem., St. Paul, MN.

GREER, E. N., and STEWART, B. A. 1959. The water absorption of wheat flour: Relative effects of protein and starch. J. Sci. Food Agric. 10:248-252.

GROGG, B., and MELMS, D. 1956. A method of analyzing extensograms of dough. Cereal Chem. 33:310-314.

GROSCH, W. 1986. Redox systems in dough. Pages 155-169 in: Chemistry and Physics of Baking: Materials, Processes and Products. J. M. V. Blanshard, P. J. Frazier, and T. Galliard, eds. R. Soc. Chem., London.

GROSSKREUTZ, J. C. 1961. A lipoprotein model of wheat gluten structure. Cereal Chem. 38:336-349.

GUY, E. J., VETTEL, H. E., and PALLANSCH, M. J. 1967. Effect of the salts of the lyotropic series on the farinograph characteristics of milk-flour dough. Cereal Sci. Today 12:200-203.

HAGBERG, S. 1957. Evaluation of flour and dough performance and bread quality. Cereal Sci. Today 2:224-227.

HALTON, P. 1949. Significance of load-extension tests in assessing the baking quality of wheat flour doughs. Cereal Chem. 26:24-45.

HALTON, P. 1959. Physical properties of wheat protein systems. Pages 12-25 in: The Physico-chemical Properties of Proteins with Special Reference to Wheat Proteins. SCI Monogr. 6. Soc. Chem. Ind., London.

HALTON, P., and FISHER, E. A. 1932. The significance of hydrogen-ion concentration in panary fermentation. Cereal Chem. 9:34-44.

HALTON, P., and SCOTT BLAIR, G. W. 1936. A study of some physical properties of flour doughs in relation to their bread-making qualities. J. Phys. Chem. 40:561-580.

HALTON, P., and SCOTT BLAIR, G. W. 1937. A study of some physical properties of flour doughs in relation to their bread-making qualities. Cereal Chem. 14:201-219.

HAMAUZU, Z., KHAN, K., and BUSHUK, W. 1979. Studies of glutenin. XIV. Gel filtration and sodium dodecyl sulfate electrophoresis of glutenin solubilized in sodium stearate. Cereal Chem. 56:513-516.

HARRIS, R. H., and SIBBITT, L. D. 1945. Comparative effects of absorption and temperature on mixogram patterns of different wheat varieties. Bakers Dig. 19:5-9, 15.

HAWTHORN, J. 1961. Oxygen in the mixing of bread doughs. Bakers Dig. 35(4):34-35, 38, 40, 42-43.

HEAPS, P. W., and COPPOCK, J. B. M. 1968. Rheology of wheat flour doughs in relation to baking quality. Pages 168-180 in: Rheology and Texture of Foodstuffs. SCI Monogr. 27. Soc. Chem. Ind., London.

HENIKA, R. G., and RODGERS, N. E. 1965. Reactions of cysteine, bromate, and whey in a rapid breadmaking process. Cereal Chem. 42:397-408.

HIBBERD, G. E. 1970a. Dynamic viscoelastic behaviour of wheat flour doughs. II. Effects of water absorption in the linear region. Rheol. Acta 9:497-500.

HIBBERD, G. E. 1970b. Dynamic viscoelastic behaviour of wheat flour doughs. III. The influence of the starch granules. Rheol. Acta 9:501-505.

HIBBERD, G. E., and PARKER, N. S. 1975. Dynamic viscoelastic behaviour of wheat flour doughs. IV. Non-linear behaviour. Rheol. Acta 14:151-157.

HIBBERD, G. E., and PARKER, N. S. 1976. Gas pressure-volume-time relationships in fermenting doughs. I. Rate of production and solubility of carbon dioxide in dough. Cereal Chem. 53:338-346.

HIBBERD, G. E., and PARKER, N. S. 1979. Nonlinear creep and creep recovery of wheat-flour doughs. Cereal Chem. 56:232-236.

HIBBERD, G. E., and WALLACE, W. J. 1966. Dynamic viscoelastic behaviour of wheat flour doughs. I. Linear aspects. Rheol. Acta 5:193-198.

HIRD, F. J. R. 1966. Wheat quality in relation to chemical bonds. Hereditas 2(Suppl.):29-46.

HIRD, F. J. R., and YATES, J. R. 1961a. The oxidation of cysteine, glutathione and thioglycollate by iodate, bromate, persulphate and air. J. Sci. Food Agric. 12:89-95.

HIRD, F. J. R., and YATES, J. R. 1961b. The oxidation of protein thiol groups by iodate, bromate and persulphate. Biochem. J. 80:612-616.

HIRD, F. J. R., CROKER, I. W. D., and JONES, W. L. 1968. Low molecular weight thiols and disulphides in flour. J. Sci. Food Agric. 19:602-604.

HLYNKA, I. 1949. Effect of bisulfite, acetaldehyde, and similar reagents on the physical properties of dough and gluten. Cereal Chem. 26:307-316.

HLYNKA, I. 1955a. Some observations on the "research" extensometer. J. Sci. Food Agric. 6:763-767.

HLYNKA, I. 1955b. Structural relaxation in dough. Bakers Dig. 29(2):27-30, 51.

HLYNKA, I. 1957. The effect of complexing agents on the bromate reaction in dough. Cereal Chem. 34:1-15.

HLYNKA, I. 1959. Dough mobility and absorption. Cereal Chem. 36:378-385.

HLYNKA, I. 1960. Intercomparison of farinograph absorption obtained with different instruments and bowls. Cereal Chem. 37:67-70.

HLYNKA, I. 1962. Influence of temperature, speed of mixing, and salt on some rheological properties of dough in the farinograph. Cereal Chem. 39:286-303.

HLYNKA, I., and BARTH, F. W. 1955. Chopin Alveograph studies. II. Structural relaxation in dough. Cereal Chem. 32:472-480.

HLYNKA, I., and CHANIN, W. G. 1957. Effect of pH on bromated and unbromated doughs. Cereal Chem. 34:371-378.

HOLME, J., and BRIGGS, D. R. 1959. Studies on the physical nature of gliadin. Cereal Chem. 36:321-340.

HOSENEY, R. C., and FAUBION, J. M. 1981. A mechanism for the oxidative gelation of wheat flour water-soluble pentosans. Cereal Chem. 58:421-424.

HOSENEY, R. C., FINNEY, K. F., POMERANZ, Y., and SHOGREN, M. D. 1969. Functional (breadmaking) and biochemical properties of wheat flour components. V. Role of total extractable lipids. Cereal Chem. 46:606-613.

HOSENEY, R. C., FINNEY, K. F., POMERANZ, Y., and SHOGREN, M. D. 1971. Functional (breadmaking) and biochemical properties of wheat flour components. VIII. Starch. Cereal Chem. 48:191-201.

HOSENEY, R. C., HSU, K. H., and JUNGE, R. C. 1979. A simple spread test to measure the rheological properties of fermenting dough. Cereal Chem. 56:141-143.

ICC. 1972a. Method for using the Brabender Extensograph (under revision). Standard No. 114, Standard Methods of the International Association for Cereal Science and Technology. Moritz Schäfer, Detmold, Federal Republic of Germany.

ICC. 1972b. Method for using the Brabender Farinograph (under revision). Standard No. 115, Standard Methods of the International Association for Cereal Science and Technology. Moritz Schäfer, Detmold, Federal Republic of Germany.

IRVINE, G. N., and WINKLER, C. A. 1950. Factors affecting the color of macaroni. II. Kinetic studies of pigment destruction during mixing. Cereal Chem. 27:205-218.

ISO. 1983. Wheat flour—Physical characteristics of doughs—Part 4: Determination of rheological properties using an alveograph. International Standard 5530/4. Int. Organ. for Standardization, Geneva.

JANKIEWICZ, M., and POMERANZ, Y. 1965. Comparison of the effects of N-ethylmaleimide and urea on rheological properties of dough. Cereal Chem. 42:37-43.

JASKA, E. J. 1961. Effect of bleaching on flour as measured by structural relaxation of dough. Cereal Chem. 38:369-374.

JELACA, S., and DODDS, N. J. 1969. Studies of some improver effects at high dough temperatures. J. Sci. Food Agric. 20:540-545.

JOHNSON, J. A., and MILLER, B. S. 1949. Studies on the role of alpha-amylase and proteinase in breadmaking. Cereal Chem. 26:371-383.

JOHNSON, J. A., SHELLENBERGER, J. A., and SWANSON, C. O. 1946. Extensograph studies of commercial flours and their relation to certain other physical dough tests. Cereal Chem. 23:400-409.

JOHNSON, J. A., MILLER, D., and FRYER, H. C. 1962. A new bleaching and maturing agent for flour. Acetone peroxide. Bakers Dig. 36(6):50-52, 54-55.

JOINER, R. R., VIDAL, F. D., and MARKS, H. C. 1963. A new powdered agent for flour maturing. Cereal Chem. 40:539-553.

JONES, I. K., and CARNEGIE, P. R. 1969a. Isolation and characterisation of disulphide peptides from wheat flour. J. Sci. Food Agric. 20:54-60.

JONES, I. K., and CARNEGIE, P. R. 1969b. Rheological activity of peptides, simple disulphides and simple thiols in wheaten dough. J. Sci. Food Agric. 20:60-64.

JONES, I. K., PHILLIPS, J. W., and HIRD, F. J. R. 1974. The estimation of rheologically important thiol and disulphide groups in dough. J. Sci. Food Agric. 25:1-10.

JØRGENSON, H. 1935. Über die Natur der Bromatwirkung. Muehlenlaboratorium 5:113-125.

JØRGENSON, H. 1945. Studies on the Nature of the Bromate Effect. Oxford Univ. Press, London.

JUNGE, R. C., and HOSENEY, R. C. 1981. A mechanism by which shortening and certain surfactants improve loaf volume in bread. Cereal Chem. 58:408-412.

JUNGE, R. C., HOSENEY, R. C., and VARRIANO-MARSTON, E. 1981. Effect of surfactants on air incorporation in dough and the crumb grain of bread. Cereal Chem. 58:338-342.

KENT-JONES, D. W., and AMOS, A. J. 1967. Modern Cereal Chemistry, 6th ed. Northern Publ. Co., Liverpool.

KHAN, M. N., RHEE, K. C., ROONEY, L. W., and CATER, C. M. 1975. Bread baking properties of aqueous processed peanut protein concentrates. J. Food Sci. 40:580-583.

KIEFFER, R., KIM, J., KEMPF, M., BELITZ, H.-D., LEHMANN, J., SPRÖSSLER, B., and BEST, E. 1982. Untersuchung rheologischer Eigenschaften von Teig und Kleber aus Weizenmehl durch Capillarviscometrie. Z. Lebensm. Unters. Forsch. 174:216-221.

KILBORN, R. H., and PRESTON, K. R. 1985. Grain Research Laboratory instrumentation for studying the breadmaking process. Pages 133-149 in: Rheology of Wheat Products. H. Faridi, ed. Am. Assoc. Cereal Chem., St. Paul, MN.

KILBORN, R. H., and TIPPLES, K. H. 1972. Factors affecting mechanical dough development. I. Effect of mixing intensity and work input. Cereal Chem. 49:34-47.

KILBORN, R. H., and TIPPLES, K. H. 1973. Factors affecting mechanical dough development. IV. Effect of cysteine. Cereal Chem. 50:70-86.

KILBORN, R. H., and TIPPLES, K. H. 1974. Implications of the mechanical development of bread dough by means of sheeting rolls. Cereal Chem. 51:648-657.

KILBORN, R. H., and TIPPLES, K. H. 1979. The effect of oxidation and intermediate proof on work requirements for optimum short-process bread. Cereal Chem. 56:407-412.

KNIGHTLY, W. H. 1968. Rôle of surfactants in baked foods. Pages 131-157 in: Surface-Active Lipids in Foods. SCI Monogr. 32. Soc. Chem. Ind., London.

KOBREHEL, K. 1984. Sequencial solubilization of wheat proteins with soaps: Protein-protein and protein-lipid interactions. Pages 31-37 in: Proc. 2nd Int. Workshop on Gluten Proteins. A. Graveland and J. H. E. Moonen, eds. TNO Cereals, Flour and Bread Institute, Wageningen, The Netherlands.

KOBREHEL, K., and BUSHUK, W. 1977. Studies of glutenin. X. Effect of fatty acids and their sodium salts on solubility in water. Cereal Chem. 54:833-839.

KRETOVICH, V. L., and VAKAR, A. B. 1964. Effect of D_2O on the physical properties of wheat gluten. Proc. Acad. Sci. USSR Sect. Biochem. 155:71-72.

KULP, K. 1968. Pentosans of wheat endosperm. Cereal Sci. Today 13:414-417, 426.

KUNINORI, T., and MATSUMOTO, H. 1963. L-ascorbic acid oxidizing system in dough and dough improvement. Cereal Chem. 40:647-657.

KUNINORI, T., and MATSUMOTO, H. 1964a. Dehydro-L-ascorbic acid reducing system in flour. Cereal Chem. 41:39-46.

KUNINORI, T., and MATSUMOTO, H. 1964b. Glutathione in wheat and wheat flour. Cereal Chem. 41:252-259.

KUNINORI, T., and SULLIVAN, B. 1968. Disulfide-sulfhydryl interchange studies of wheat flour. II. Reaction of glutathione. Cereal Chem. 45:486-495.

LARMOUR, R. K., WORKING, E. B., and OFELT, C. W. 1939. Quality tests on hard red winter wheats. Cereal Chem. 16:733-752.

LARSSON, K. 1986. Functionality of wheat lipids in relation to gluten gel formation. Pages 62-74 in: Chemistry and Physics of Baking: Materials, Processes and Products. J. M. V. Blanshard, P. J. Frazier, and T. Galliard, eds. R. Soc. Chem., London.

LÁSZTITY, R., and HEGEDÜS, J. 1975. Ungarische Instrumente für die Laboratoriumskontrolle in der Bäckerei. Baecker Kond. 23:15-18.

LAUNAY, B. 1979. Propriétés rhéologiques des pâtes de farine: Quelques progrès récents. Ind. Aliment. Agric. 96:617-623.

LAUNAY, B., and BURÉ, J. 1970. Alvéographe Chopin et propriétés rhéologiques des pâtes. Lebensm. Wiss. Technol. 3:57-62.

LAUNAY, B., and BURÉ, J. 1973. Application of a viscometric method to the study of wheat-flour doughs. J. Texture Stud. 4:82-101.

LAUNAY, B., and BURÉ, J. 1974. Stress relaxation in wheat flour doughs following a finite period of shearing. I. Qualitative study. Cereal Chem. 51:151-162.

LAUNAY, B., BURÉ, J., and PRADEN, J. 1977. Use of the Chopin Alveographe as a rheological tool. I. Dough deformation measurements. Cereal Chem. 54:1042-1048.

LEE, C. C., and LAI, T.-S. 1968. Studies with radioactive tracers. XIV. A note on the disulfide-sulfhydryl interchange in doughs made with ^{35}S-labeled flour. Cereal Chem. 45:627-630.

LeGRYS, G. A., BOOTH, M. R., and AL-BAGHDADI, S. M. 1981. The physical properties of wheat proteins. Pages 243-264 in: Cereals: A Renewable Resource; Theory and Practice. Y. Pomeranz and L. Munck, eds. Am. Assoc. Cereal Chem., St. Paul, MN.

LENZ, J., and WUTZEL, H. 1984. Der Einfluss des Zusatzes von Hemicellulosen auf die Fliesseigenschaften von Weizenmehlteig. Rheol. Acta 23:570-572.

LERCHENTHAL, C. H. 1969. On decay of elastic properties in a viscoelastic material under tensile stress. Pages 361-378 in: Contributions to Mechanics, Marcus Reiner 80th Anniversary Volume. D. Abir, ed. Pergamon Press, Oxford.

LERCHENTHAL, C. H., and FUNT, C. 1967. The strength of wheat flour dough in uniaxial tension. Pages 203-224 in: Rheology and Texture of Foodstuffs. SCI Monogr. 27. Soc. Chem. Ind., London.

LERCHENTHAL, C. H., and MULLER, H. G. 1967. Research in dough rheology at the Israel Institute of Technology. Cereal Sci. Today 12:185-187, 190-192.

LOUW, J. B., and KRYNAUW, G. N. 1961. The relationship between farinograph mobility and absorption. Cereal Chem. 38:1-7.

LÜDDEKE, J. 1969. Valorigraph—Ein neues Konsistenzprüfgerät aus Ungarn. Baecker Kond. 17:233-235.

LUSENA, C. V. 1950. Preparation of dried native wheat gluten. Cereal Chem. 27:167-178.

MacRITCHIE, F. 1977. Flour lipids and their

effects in baking. J. Sci. Food Agric. 28:53-58.

MacRITCHIE, F. 1980a. Physicochemical aspects of some problems in wheat research. Pages 271-326 in: Advances in Cereal Science and Technology, Vol. III. Y. Pomeranz, ed. Am. Assoc. Cereal Chem., St. Paul, MN.

MacRITCHIE, F. 1980b. Protein-lipid interaction in baking quality of wheat flours. Bakers Dig. 54(5):10-13.

MacRITCHIE, F. 1984. Baking quality of wheat flours. Pages 201-277 in: Advances in Food Research, Vol. 29. C. O. Chichester, E. M. Mrak, and B. S. Schweigert, eds. Academic Press, New York.

MacRITCHIE, F. 1985. Studies of the methodology for fractionation and reconstitution of wheat flours. J. Cereal Sci. 3:221-230.

MacRITCHIE, F., and GRAS, P. W. 1973. The role of flour lipids in baking. Cereal Chem. 50:292-302.

MALTHA, P. 1953. Über den Einfluss von L-Askorbinsäure und Verbindungen mit verwandter Struktur auf die Backfähigkeit des Mehles. Getreide Mehl 3:65-69.

MANNERS, D. J. 1985a. Some aspects of the structure of starch. Cereal Foods World 30:461-467.

MANNERS, D. J. 1985b. Some aspects of the metabolism of starch. Cereal Foods World 30:722-727.

MAREK, C. J., and BUSHUK, W. 1967. Study of gas production and retention in doughs with a modified Brabender oven-rise recorder. Cereal Chem. 44:300-307.

MAREK, C. J., BUSHUK, W., and IRVINE, G. N. 1968. Gas production and retention during proofing of bread doughs. Cereal Sci. Today 13:4-6, 13.

MARGULIS, H., and CAMPAGNE, Y. 1955. Influence des acides et des sels minéraux sur les indices alvéographiques des farines. Ind. Aliment. Agric. 72:485-492.

MARKLEY, M. C., and BAILEY, C. H. 1938. The collodial behavior of flour doughs. II. A study of the effects of varying the flour-water ratio. Cereal Chem. 15:317-326.

MATSUMOTO, H., and HLYNKA, I. 1959. Some aspects of the sulfhydryl-disulfide system in flour and dough. Cereal Chem. 36:513-521.

MATSUMOTO, H., OSHIMA, I., and HLYNKA, I. 1960. Effect of bisulfite and acetaldehyde on the disulfide linkage in wheat protein. Cereal Chem. 37:710-720.

MAURITZEN, C. A. M., and STEWART, P. 1963. Disulphide-sulphydryl exchange in dough. Nature 197:48-49.

McCAIG, J. D., and McCALLA, A. G. 1941. Changes in the physical properties of gluten with aging of flour. Can. J. Res. C19:163-176.

McDERMOTT, E. E., and PACE, J. 1961. Modification of the properties of flour protein by thiolated gelatin. Nature 192:657.

McDERMOTT, E. E., STEVENS, D. J., and PACE, J. 1969. Modification of flour proteins by disulphide interchange reactions. J. Sci. Food Agric. 20:213-217.

McMASTER, G. J. 1982. Association of carbohydrate and protein in wheat gluten. Ph.D. thesis. Univ. of Manitoba, Winnipeg, Canada.

McMASTER, G. J., and BUSHUK, W. 1983. Protein-carbohydrate complexes in gluten: Fractionation and proximate composition. J. Cereal Sci. 1:171-184.

MECHAM, D. K. 1959. Effects of sulfhydryl-blocking reagents on the mixing characteristics of doughs. Cereal Chem. 36:134-145.

MECHAM, D. K. 1968. The sulfhydryl and disulfide contents of wheat flours, doughs, and proteins. Bakers Dig. 42(1):26-28, 30, 59.

MECHAM, D. K., and KNAPP, C. 1966. The sulfhydryl content of doughs mixed under nitrogen. Cereal Chem. 43:226-236.

MECHAM, D. K., and MOHAMMAD, A. 1955. Extraction of lipids from wheat products. Cereal Chem. 32:405-415.

MECHAM, D. K., and WEINSTEIN, N. E. 1952. Lipid binding in doughs. Effects of dough ingredients. Cereal Chem. 29:448-455.

MELVILLE, J., and SHATTOCK, H. T. 1938. The action of ascorbic acid as a bread improver. Cereal Chem. 15:201-205.

MEREDITH, O. B., and WREN, J. J. 1969. Stability of the molecular weight distribution in wheat flour proteins during dough making. J. Sci. Food Agric. 20:235-237.

MEREDITH, P. 1965. The oxidation of ascorbic acid and its improver effect in bread doughs. J. Sci. Food Agric. 16:474-480.

MEREDITH, P. 1966a. Combined action of ascorbic acid and potassium bromate as bread dough improvers. Chem. Ind. London 1966:948-949.

MEREDITH, P. 1966b. Dependence of water absorption of wheat flour on protein content and degree of starch granule damage. N.Z. J. Sci. 9:324-330.

MEREDITH, P., and BUSHUK, W. 1962. The effects of iodate, N-ethylmaleimide, and oxygen on the mixing tolerance of doughs. Cereal Chem. 39:411-426.

MEREDITH, P., and HLYNKA, I. 1964. The action of reducing agents on dough. Cereal Chem. 41:286-299.

MERRITT, P. P., and BAILEY, C. H. 1945. Preliminary studies with the extensograph. Cereal Chem. 22:372-391.

MEUSER, F., and SUCKOW, P. 1986. Non-starch polysaccharides. Pages 42-61 in:

Chemistry and Physics of Baking: Materials, Processes and Products. J. M. V. Blanshard, P. J. Frazier, and T. Galliard, eds. R. Soc. Chem., London.

MILLER, B. S., and JOHNSON, J. A. 1948. High levels of alpha-amylase in baking. II. Proteolysis in straight and sponge doughs. Cereal Chem. 25:178-190.

MILLER, B. S., HAYS, B., and JOHNSON, J. A. 1956. Correlation of farinograph, mixograph, sedimentation, and baking data for hard red winter wheat flour samples varying widely in quality. Cereal Chem. 33:277-290.

MITA, T., and MATSUMOTO, H. 1984. Dynamic viscoelastic properties of concentrated dispersions of gluten and gluten methyl ester: Contributions of glutamine side chain. Cereal Chem. 61:169-173.

MIZUKOSHI, M., MAEDA, H., and AMANO, H. 1980. Model studies of cake baking. II. Expansion and heat set of cake batter during baking. Cereal Chem. 57:352-355.

MOONEN, J. H. E., SCHEEPSTRA, A., and GRAVELAND, A. 1985. Biochemical properties of some high molecular weight subunits of wheat glutenin. J. Cereal Sci. 3:17-27.

MOORE, C. L., and HERMAN, R. S. 1942. The effect of certain ingredients and variations in manipulations on the farinograph curve. Cereal Chem. 19:568-587.

MOORE, W. R., and HOSENEY, R. C. 1986. Influence of shortening and surfactants on retention of carbon dioxide in bread dough. Cereal Chem. 63:67-70.

MORRISON, W. R. 1963. The free fatty acids of wheat flour: Their role in simple flour-water mixtures. J. Sci. Food Agric. 14:245-251.

MORRISON, W. R. 1978. Cereal lipids. Pages 221-348 in: Advances in Cereal Science and Technology, Vol. II. Y. Pomeranz, ed. Am. Assoc. Cereal Chem., St. Paul, MN.

MORRISON, W. R. 1979. Lipids in wheat and their importance in wheat products. Pages 313-335 in: Recent Advances in the Biochemistry of Cereals. D. L. Laidman and R. G. Wyn Jones, eds. Academic Press, London.

MORRISON, W. R., and HAWTHORN, J. 1960. Effect of atmospheric sulphur dioxide on wheat flour. Chem. Ind. London 1960:529-530.

MORRISON, W. R., and MANEELY, E. A. 1969. Importance of wheat lipoxidase in the oxidation of free fatty acids in flour-water systems. J. Sci. Food Agric. 20:379-381.

MOSS, H. J. 1961. Milling damage and quality evaluation of wheat. Aust. J. Exp. Agric.

Anim. Husb. 1:133-139.

MOSS, R. 1972. A study of the microstructure of bread doughs. CSIRO Food Res. Q. 32:50-56.

MOSS, R. 1974. Dough microstructure as affected by the addition of cysteine, potassium bromate, and ascorbic acid. Cereal Sci. Today 19:557-561.

MOSS, R., and STENVERT, N. L. 1979. The relationship between oxidising and reducing agents, dough microstructure and bread quality in two breadmaking systems. Pages 303-314 in: Food Texture and Rheology. P. Sherman, ed. Academic Press, London.

MOSS, R., MURRAY, L. F., and STENVERT, N. L. 1984. Wheat germ in bakers flour—its effect on oxidant requirements. Bakers Dig. 58(3):12-17.

MUKULUMWA, A. M. 1976. Effect of maize flour and its major components on bread-making properties of wheat flour. M.S. thesis. Univ. of Manitoba, Winnipeg, Canada.

MULLER, H. G. 1968. Aspects of dough rheology. Pages 181-196 in: Rheology and Texture of Foodstuffs. SCI Monogr. 27. Soc. Chem. Ind. London.

MULLER, H. G. 1969. Die Anwendung der statistischen Theorie der Gummielastizität auf Weizenkleber und Teig. Brot Gebaeck 23:153-155.

MULLER, H. G. 1973. An Introduction to Food Rheology. W. Heinemann, London.

MULLER, H. G., and HLYNKA, I. 1964. Brabender Extensigraph techniques. Cereal Sci. Today 9:422-424, 426, 430.

MUNZ, E., and BRABENDER, C. W. 1940. Prediction of baking value from measurements of plasticity and extensibility of dough. I. Influence of mixing and molding treatments upon physical dough properties of typical American wheat varieties. Cereal Chem. 17:78-100.

NAGAO, S. 1986. The Do-Corder and its application in dough rheology. Cereal Foods World 31:231-240.

NARAYANAN, K. M., and HLYNKA, I. 1962. Rheological studies of the role of lipids in dough. Cereal Chem. 39:351-363.

NAVICKIS, L. L., ANDERSON, R. A., BAGLEY, E. B., and JASBERG, B. K. 1982. Viscoelastic properties of wheat flour doughs: Variation of dynamic moduli with water and protein content. J. Texture Stud. 13:249-264.

NEUKOM, H., and MARKWALDER, H. U. 1978. Oxidative gelation of wheat flour pentosans: A new way of cross-linking polymers. Cereal Foods World 23:374-376.

NIEMAN, W. 1978a. Bepaling van de Wateropname van Bloem. II. Registrerende Kneders. Voedingsmiddelentechnologie

11(31/32):9-11.

NIEMAN, W. 1978b. Bepaling van de Wateropname van Bloem. III. De Simon Research water absorption meter. Voedingsmiddelentechnologie 11(36):10-11.

NIEMAN, W. 1978c. Bepaling van de Wateropname van Bloem. IV. De Penetrometer. Voedingsmiddelentechnologie 11(40):10-11.

OLATUNJI, O., and AKINRELE, I. A. 1978. Comparative rheological properties and bread qualities of wheat flour diluted with tropical tuber and breadfruit flours. Cereal Chem. 55:1-6.

OLCOTT, H. S., and MECHAM, D. K. 1947. Characterization of wheat gluten. I. Protein-lipid complex formation during doughing of flours. Lipoprotein nature of the glutenin fraction. Cereal Chem. 24:407-414.

ORTH, R. A., and BUSHUK, W. 1972. A comparative study of the proteins of wheats of diverse baking qualities. Cereal Chem. 49:268-275.

PACE, J., and STEWART, B. A. 1966. Dough development in relation to disulphide and thiol groups. Milling 146:317-318, 338.

PAINTER, T. J., and NEUKOM, H. 1968. The mechanism of oxidative gelation of a glycoprotein from wheat flour. Evidence from a model system based upon caffeic acid. Biochim. Biophys. Acta 158:363-381.

PAREDES-LOPEZ, O., and BUSHUK, W. 1983a. Development and "undevelopment" of wheat dough by mixing: Physiocochemical studies. Cereal Chem. 60:19-23.

PAREDES-LOPEZ, O., and BUSHUK, W. 1983b. Development and "undevelopment" of wheat dough by mixing: Microscopic structure and its relations to bread-making quality. Cereal Chem. 60:24-27.

PARISI, F. 1981. An evaluation of two simple methods for the measurement of fermenting power of baker's yeast. Pure Appl. Chem. 53:603-610.

PARKER, N. S., and HIBBERD, G. E. 1974. The interpretation of dynamic measurements on non-linear viscoelastic materials. Rheol. Acta 13:910-915.

POMERANZ, Y. 1985. Wheat flour lipids—What they can and cannot do in bread. Cereal Foods World 30:443-446.

POMERANZ, Y., RUBENTHALER, G. L., DAFTARY, R. D., and FINNEY, K. F. 1966a. Effects of lipids on bread baked from flours varying widely in bread-making potentialities. Food Technol. 20:1225-1228.

POMERANZ, Y., RUBENTHALER, G. L., and FINNEY, K. F. 1966b. Studies on the mechanisms of the bread-improving effects of lipids. Food Technol. 20:1485-1488.

POMERANZ, Y., SHOGREN, M., and FINNEY, K. F. 1968a. Functional bread-making properties of wheat flour lipids. 1. Reconstitution studies and properties of defatted flours. Food Technol. 22:324-327.

POMERANZ, Y., TAO, R. P.-C., HOSENEY, R. C., SHOGREN, M. D., and FINNEY, K. F. 1968b. Evaluation of factors affecting lipid binding in wheat flours. J. Agric. Food Chem. 16:974-978.

POMERANZ, Y., SHOGREN, M. D., FINNEY, K. F., and BECHTEL, D. B. 1977. Fiber in breadmaking—Effects on functional properties. Cereal Chem. 54:25-41.

POMERANZ, Y., MEYER, D., and SEIBEL, W. 1984a. Wheat, wheat-rye, and rye dough and bread studied by scanning electron microscopy. Cereal Chem. 61:53-59.

POMERANZ, Y., MEYER, D., and SEIBEL, W. 1984b. Rasterelektronenmikroskopische Untersuchungen über die Struktur von Weizen(mehl)-, Weizenmisch- und Roggenteigen und Broten. Getreide Mehl Brot 38:138-146.

PRESTON, K. R. 1981. Effects of neutral salts upon wheat gluten protein properties. I. Relationship between the hydrophobic properties of gluten proteins and their extractability and turbidity in neutral salts. Cereal Chem. 58:317-324.

PRESTON, K. R. 1984. Use of lyotropic salts to study the hydrophobic properties of wheat gluten proteins. Pages 207-215 in: Proc. 2nd Int. Workshop on Gluten Proteins. A. Graveland and J. H. E. Moonen, eds. TNO Cereals, Flour and Bread Institute, Wageningen, The Netherlands.

PRINS, A., and BLOKSMA, A. H. 1983. Guidelines for the measurement of rheological properties and the use of existing data. Pages 185-189 in: Physical Properties of Foods. R. Jowitt, F. Escher, B. Hallström, H. F. T. Meffert, W. E. L. Spiess, and G. Vos, eds. Appl. Sci. Publ., London.

PROSKURYAKOV, N. I., and ZUEVA, E. S. 1964. Enzymatic reduction of disulfide bonds of proteins and low-molecular components of wheat flour. Dokl. Akad. Nauk SSSR 158:232-234.

QUENDT, H., TSCHEUSCHNER, H.-D., and HEINICKEL, U. 1974. Rotationsviskometrische Untersuchungen an Weizenteig. Baecker Kond. 22:197-200.

RASPER, V. F. 1975. Dough rheology at large deformations in simple tensile mode. Cereal Chem. 52:24r-41r.

RASPER, V., RASPER, J., and DE MAN, J. 1974. Stress-strain relationships of chemically improved unfermented doughs. I. The evaluation of data obtained at large deformations in simple tensile mode. J. Texture Stud. 4:438-466.

RASPER, V. F., HARDY, K. M., and FULCHER, G. R. 1985. Constant water content vs. constant consistency techniques in alveography of soft wheat flours. Pages 51-73 in: Rheology of Wheat Products. H. Faridi, ed. Am. Assoc. Cereal Chem., St. Paul, MN.

REDMAN, D. G., and EWART, J. A. D. 1967a. Disulphide interchange in dough proteins. J. Sci. Food Agric. 18:15-18.

REDMAN, D. G., and EWART, J. A. D. 1967b. Disulphide interchange in cereal proteins. J. Sci. Food Agric. 18:520-523.

REINER, M. 1949. Twelve Lectures on Theoretical Rheology. North-Holland Publ. Co., Amsterdam.

RESNITSCHENKO, M. S., and POPZOWA, A. I. 1934. Über die Wirkung ungesättigter Fettsäuren auf die kolloide Beschaffenheit des Klebers. Muehlenlaboratorium 4:57-62.

ROGERS, D. E., and HOSENEY, R. C. 1982. Problems associated with producing whole wheat bread. Bakers Dig. 56(6):22.

RUBENTHALER, G. L., FINNEY, P. L., DEMARAY, D. E., and FINNEY, K. F. 1980. Gasograph: Design, construction, and reproducibility of a sensitive 12-channel gas recording instrument. Cereal Chem. 57:212-216.

RUDOLPH, H., TSCHEUSCHNER, H.-D., and QUENDT, H. 1977. Einfluss der Feuchtigkeit von Brötchenteigen auf ihre Verarbeitungseigenschaften und die Brötchenqualität. Baecker Kond. 25:100-105.

SANDSTEDT, R. M. 1961. The function of starch in the baking of bread. Bakers Dig. 35(3):36-42, 44.

SANDSTEDT, R. M., and HITES, B. D. 1945. Ascorbic acid and some related compounds as oxidizing agents in doughs. Cereal Chem. 22:161-187.

SAURER, W. 1976. Frühzeitige Erfassung von Qualitätsmerkmalen als Grundlage für eine wirkungsvolle Selektion in der Weizenzüchtung. II. Das Mixogramm als Mass für teigphysikalische Kriterien. Muehle Mischfuttertechnik 113:453-455.

SCHÄFER, W. 1984. Das rheologische Optimum—Erfahrungen und Aspekte. Muehle Mischfuttertechnik 121:251-255.

SCHMIEDER, W., and ZABEL, S. 1966. Möglichkeiten des Einsatzes des automatischen Penetrometers AP 4/1 für rheologische Teigmessungen. Nahrung 10:619-633.

SCHOFIELD, J. D. 1986. Flour proteins: Structure and functionality in baked products. Pages 14-29 in: Chemistry and Physics of Baking: Materials, Processes and Products. J. M. V. Blanshard, P. J. Frazier,

and T. Galliard, eds. R. Soc. Chem., London.

SCHOFIELD, J. D., BOTTOMLEY, R. C., LeGRYS, G. A., TIMMS, M. F., and BOOTH, M. R. 1984. Effects of heat on wheat gluten. Pages 81-90 in: Proc. 2nd Int. Workshop on Gluten Proteins. A. Graveland and J. H. E. Moonen, eds. TNO Cereals, Flour and Bread Institute, Wageningen, The Netherlands.

SCHOFIELD, R. K., and SCOTT BLAIR, G. W. 1932. The relationship between viscosity, elasticity and plastic strength of soft materials as illustrated by some mechanical properties of flour doughs. I. Proc. R. Soc. London A138:707-719.

SCHULZ, A. 1966. Methode zur Ermittlung der Triebkraft von Backhefen. Brot Gebaeck 20:92.

SCOTT BLAIR, G. W. 1969. Elementary Rheology. Academic Press, London.

SEIBEL, W. 1968. Experiences with the maturograph and oven-rise recorder. Bakers Dig. 42(1):44-46, 48.

SHELEF, L., and BOUSSO, D. 1964. A new instrument for measuring relaxation in flour dough. Rheol. Acta 3:168-172.

SHELTON, D. R., and D'APPOLONIA, B. L. 1985. Carbohydrate functionality in the baking process. Cereal Foods World 30:437-442.

SHEWRY, P. R., FIELD, J. M., FAULKS, A. J., PARMAR, S., MIFLIN, B. J., DIETTER, M. D., LAW, E. J. L., and KASARDA, D. D. 1984. The purification and N-terminal amino acid sequence analysis of the high molecular weight gluten polypeptides of wheat. Biochim. Biophys. Acta 788:23-24.

SHIMIZU, T., and ICHIBA, A. 1958. Rheological studies of wheat flour dough. I. Measurement of dynamic viscoelasticity. Bull. Agric. Chem. Soc. Jpn. 22:294-298.

SHOGREN, M. D., FINNEY, K. F., and RUBENTHALER, G. L. 1977. Note on determination of gas production. Cereal Chem. 54:665-668.

SHUEY, W. C. 1975. Practical instruments for rheological measurements on wheat products. Cereal Chem. 52:42r-81r.

SHUEY, W. C. 1984a. The farinograph. Pages 1-6 in: The Farinograph Handbook, 3rd ed. B. L. D'Appolonia and W. H. Kunerth, eds. Am. Assoc. Cereal Chem., St. Paul, MN.

SHUEY, W. C. 1984b. Interpretation of the farinogram. Pages 31-32 in: The Farinograph Handbook, 3rd ed. B. L. D'Appolonia and W. H. Kunerth, eds. Am. Assoc. Cereal Chem., St. Paul, MN.

SHUEY, W. C., and GILLES, K. A. 1966. Effect of spring settings and absorption on mixograms for measuring dough characteristics. Cereal Chem. 43:94-103.

SIBBITT, L. D., HARRIS, R. H., and CONLON, T. J. 1953. Some relations between farinogram and mixogram dimensions and baking quality. Bakers Dig. 27:76-79.

SIETZ, W. 1983. Computerauswertung bei teigrheologischen Messungen. Getreide Mehl Brot 37:39-43.

SIETZ, W. 1984. Modifications and new developments with the farinograph. Pages 44-47 in: The Farinograph Handbook, 3rd ed. B. L. D'Appolonia and W. H. Kunerth, eds. Am. Assoc. Cereal Chem., St. Paul, MN.

SIMMONDS, D. H. 1972. Wheat-grain morphology and its relationship to dough structure. Cereal Chem. 49:324-335.

SLUIMER, P. 1986. De rol van de gasfase in deeg en brood. Voedingsmiddelentechnologie 19(9):25-31.

SLUIMER, P., and KRIST-SPIT, C. E. 1987. Heat transport in dough during the baking of bread. Pages 355-363 in: 1st Eur. Congr. Food Sci. Technol., Cereals in a European Context. I. D. Morton, ed. Ellis Harwood, Chichester, West Sussex, England.

SMITH, D. E., and ANDREWS, J. S. 1952. Effect of oxidizing agents upon dough extensograms. Cereal Chem. 29:1-17.

SMITH, D. E., and ANDREWS, J. S. 1957. The uptake of oxygen by flour dough. Cereal Chem. 34:323-336.

SMITH, D. E., VAN BUREN, J. P., and ANDREWS, J. S. 1957. Some effects of oxygen and fat upon the physical and chemical properties of flour doughs. Cereal Chem. 34:337-349.

SMITH, J. R., SMITH, T. L., and TSCHOEGL, N. W. 1970. Rheological properties of wheat flour doughs. III. Dynamic shear modulus and its dependence on amplitude, frequency, and dough composition. Rheol. Acta 9:239-252.

SMITH, T. L., and TSCHOEGL, N. W. 1970. Rheological properties of wheat flour doughs. IV. Creep and recovery in simple tension. Rheol. Acta 9:339-344.

SOENEN, M. 1949. Kwaliteitsonderzoek van tarwe door meting van de elastische eigenschappen van deeg. Landbouwk. Tijdschr. 61:715-727.

SOKOL, H. A., MECHAM, D. K., and PENCE, J. W. 1960. Sulfhydryl losses during mixing of doughs: Comparison of flours having various mixing characteristics. Cereal Chem. 37:739-748.

STAMBERG, O. E., and BAILEY, C. H. 1940. Plasticity of doughs. Cereal Chem. 17:37-44.

STEVENS, D. J. 1966. The reaction of wheat proteins with sulphite. II. The accessibility of disulphide and thiol groups in flour. J. Sci. Food Agric. 17:202-204.

STEWART, P. R., and MAURITZEN, C. M. 1966. The incorporation of (35S)cysteine into the proteins of dough by disulphide-sulphydryl interchange. Aust. J. Biol. Sci. 19:1125-1137.

STOKES, J. L. 1971. Influence of temperature on the growth and metabolism of yeasts. Pages 119-134 in: The Yeasts, Vol. 2: Physiology and Biochemistry of Yeasts. A. H. Rose and J. S. Harrison, eds. Academic Press, New York.

SULLIVAN, B. 1940. The function of the lipids in milling and baking. Cereal Chem. 17:661-668.

SULLIVAN, B. 1954. Proteins in flour. Review of the physical characteristics of gluten and reactive groups involved in change in oxidation. J. Agric. Food Chem. 2:1231-1234.

SULLIVAN, B., and DAHLE, L. K. 1966. Disulfide-sulfhydryl interchange studies of wheat flour. I. The improving action of formamidine disulfide. Cereal Chem. 43:373-383.

SULLIVAN, B., NEAR, C., and FOLEY, G. H. 1936. The role of the lipids in relation to flour quality. Cereal Chem. 13:318-331.

SULLIVAN, B., HOWE, M., and SCHMALZ, F. 1937. An explanation of the effect of heat treatment on wheat germ. Cereal Chem. 14:489-495.

SULLIVAN, B., DAHLE, L., and LARSON, E. 1961a. The oxidation of wheat flour. I. Measurement of sulfhydryl groups. Cereal Chem. 38:272-280.

SULLIVAN, B., DAHLE, L., and NELSON, O. R. 1961b. The oxidation of wheat flour. II. Effect of sulfhydryl-blocking agents. Cereal Chem. 38:281-291.

SULLIVAN, B., DAHLE, L. K., and SCHIPKE, J. H. 1963. The oxidation of wheat flour. IV. Labile and nonlabile sulfhydryl groups. Cereal Chem. 40:515-531.

SWANSON, C. O., and WORKING, E. B. 1933. Testing the quality of flour by the recording dough mixer. Cereal Chem. 10:1-29.

SZCZESNIAK, A. S., LOH, J., and MANNELL, W. R. 1983. Effect of moisture transfer on dynamic viscoelastic parameters of wheat flour/water systems. J. Rheol. 27:537-556.

TANAKA, K., and BUSHUK, W. 1973a. Changes in flour proteins during dough-mixing. I. Solubility results. Cereal Chem. 50:590-596.

TANAKA, K., and BUSHUK, W. 1973b. Changes in flour proteins during dough-mixing. II. Gel filtration and electrophoresis results. Cereal Chem. 50:597-605.

TANAKA, K., and BUSHUK, W. 1973c. Changes in flour proteins during dough-mixing. III. Analytical results and mechanisms. Cereal Chem. 50:605-612.

TANAKA, K., FURUKAWA, K., and MATSUMOTO, H. 1967. The effect of acid and salt on the farinogram and extensigram of dough. Cereal Chem. 44:675-680.

TAO, R. P.-C., and POMERANZ, Y. 1968. Functional bread-making properties of wheat flour lipids. 3. Effects of lipids on rheological properties of wheat flour doughs. Food Technol. 22:1145-1149.

TCHETVEROUKHINE, B. 1947. Les alvéogrammes Chopin à matière sèche constante et à matière sèche variable. Bull. Anc. Eleves Ec. Fr. Meun. 18:217-225.

TCHETVEROUKHINE, B. 1948. Les alvéogrammes Chopin à matière sèche constante et à matière sèche variable. (Suite et fin) Bull. Anc. Eleves Ec. Fr. Meun. 19:25-29.

TELEGDY KOVÁTS, L., and LÁSZTITY, R. 1966. Neuere Ergebnisse in der Rheologie des Teiges. II. Spannungsrelaxation der Weizenmehl-Teige. Period. Polytech. Chem. Eng. 10:53-64.

TIPPLES, K. H., and KILBORN, R. H. 1974. Dough development for shorter breadmaking processes. Bakers Dig. 48(5):34-39.

TIPPLES, K. H., and KILBORN, R. H. 1975. "Unmixing"—The disorientation of developed bread doughs by slow speed mixing. Cereal Chem. 52:248-262.

TIPPLES, K. H., PRESTON, K. R., and KILBORN, R. H. 1982. Implications of the term "strength" as related to wheat and flour quality. Bakers Dig. 56(6):16-18, 20.

TKACHUK, R. 1969. Involvement of sulfhydryl peptides in the improver reaction. Cereal Chem. 46:203-205.

TKACHUK, R., and HLYNKA, I. 1961. Some improving effects of halogenates and their reduction intermediates in dough. Cereal Chem. 38:393-398.

TKACHUK, R., and HLYNKA, I. 1968. Some properties of dough and gluten in D_2O. Cereal Chem. 45:80-87.

TOLEDO, R., STEINBERG, M. P., and NELSON, A. I. 1968. Quantitative determination of bound water by NMR. J. Food Sci. 33:315-317.

TRAUB, W., HUTCHINSON, J. B., and DANIELS, D. G. H. 1957. X-ray studies of the wheat protein complex. Nature 179:769-770.

TSCHEUSCHNER, H.-D., and TREIBER, G. 1978. Untersuchungen zur rheologischen Beschreibung der Rohrförderung von Weizenteig. Baecker Kond. 26:260-262.

TSCHEUSCHNER, H.-D., QUENDT, H., and HEINICKEL, U. 1975. Zur Analyse des Prozesses der Weizenteigbereitung mit hoher Knetintensität. Baecker Kond. 23:231-233.

TSCHOEGL, N. W., RINDE, J. A., and SMITH, T. L. 1970a. Rheological properties of wheat flour doughs. I. Method for determining the large deformation and rupture properties in simple tension. J. Sci. Food Agric. 21:65-70.

TSCHOEGL, N. W., RINDE, J. A., and SMITH, T. L. 1970b. Rheological properties of wheat flour doughs. II. Dependence of large deformation and rupture properties in simple tension on time, temperature, and water absorption. Rheol. Acta 9:223-238.

TSEN, C. C. 1963. The reaction mechanism of azodicarbonamide in dough. Cereal Chem. 40:638-647.

TSEN, C. C. 1964. Comparative study on reactions of iodate, azodicarbonamide, and acetone peroxides in simple chemical systems and in dough. Cereal Chem. 41:22-31.

TSEN, C. C. 1965. The improving mechanism of ascorbic acid. Cereal Chem. 42:86-97.

TSEN, C. C. 1966. A note on effects of pH on sulfhydryl groups and rheological properties of dough and its implication with the sulfhydryl-disulfide interchange. Cereal Chem. 43:456-460.

TSEN, C. C. 1968. Oxidation of sulfhydryl groups of flour by bromate under various conditions and during the breadmaking process. Cereal Chem. 45:531-538.

TSEN, C. C., and BUSHUK, W. 1963. Changes in sulfhydryl and disulfide contents of doughs during mixing under various conditions. Cereal Chem. 40:399-408.

TSEN, C. C., and BUSHUK, W. 1968. Reactive and total sulfhydryl and disulfide contents of flours of different mixing properties. Cereal Chem. 45:58-62.

TSEN, C. C., and HLYNKA, I. 1963. Flour lipids and oxidation of sulfhydryl groups in dough. Cereal Chem. 40:145-153.

VILLEGAS, E., POMERANZ, Y., and SHELLENBERGER, J. A. 1963. Effects of thiolated gelatins and glutathione on rheological properties of wheat doughs. Cereal Chem. 40:694-703.

VOISEY, P. W. 1976. Instrumental measurement of food texture. Pages 79-141 in: Rheology and Texture in Food Quality. J. M. de Man, P. W. Voisey, V. F. Rasper, and D. W. Stanley, eds. Avi Publ. Co., Westport, CT.

VOISEY, P. W., and KLOEK, M. 1980. Note on methods of recording dough development curves from electronic recording mixers. Cereal Chem. 57:442-444.

VOISEY, P. W., MILLER, H., and KLOEK,

M. 1964. Apparatus for determining the gassing power of dough. Cereal Sci. Today 9:393-396, 412.

VOISEY, P. W., MILLER, H., and KLOEK, M. 1966. An electronic recording dough mixer. IV. Applications in farinography. Cereal Chem. 43:438-446.

VOISEY, P. W., MILLER, H., and BYRNE, P. L. 1971. The Ottawa electronic recording farinograph. Cereal Sci. Today 16:124-128, 130-131.

WALDEN, C. C. 1955. The action of wheat amylases on starch under conditions of time and temperature as they exist during baking. Cereal Chem. 32:421-431.

WALL, J. S. 1979. The role of wheat proteins in determining baking quality. Pages 275-311 in: Recent Advances in the Biochemistry of Cereals. D. L. Laidman and R. G. Wyn Jones, eds. Academic Press, London.

WARWICK, M. J., and SHEARER, G. 1982. The association of saturated and unsaturated triglycerides and oleic acid with wheat flour components during dough making. J. Sci. Food Agric. 33:918-924.

WASSERMANN, L., and DÖRFNER, H. H. 1971. Der Einfluss des Wasser-Mehl-Verhältnisses in Brotteigen auf die Zusammensetzung und Eigenschaften der daraus hergestellten Brote. I. Teigkonsistenz, pH-Wert und Zuckergehalt. Brot Gebaeck 25:148-153.

WEIPERT, D. 1981. Teigrheologische Untersuchungsmethoden—Ihre Einsatzmöglichkeiten im Mühlenlaboratorium. Getreide Mehl Brot 35:5-9; Muehle Mischfuttertechnik 118:350-353.

WENSVEEN, C. J., and DE MIRANDA, H. 1955. Eine einfache Methode zur Bestimmung der optimalen Wassermenge bei der Teigbereitung. Brot Gebaeck 9:141-144.

WOOTTON, M. 1966. Binding and extractability of wheat flour lipid after dough formation. J. Sci. Food Agric. 17:297-301.

WORKING, E. B. 1924. Lipoids, a factor influencing gluten quality. Cereal Chem. 1:153-158.

WREN, J. J., and NUTT, J. 1967. The effects of three thiols on the extractability of wheat-flour proteins. J. Sci. Food Agric. 18:119-123.

YOUSSEF, M. M., and BUSHUK, W. 1986. Breadmaking properties of composite flours of wheat and faba bean protein preparations. Cereal Chem. 63:357-361.

ZANGGER, R. R. 1979. Viscoelasticity of doughs. Pages 343-354 in: Food Texture and Rheology. P. Sherman, ed. Academic Press, London.

ZAWISTOWSKA, U., BEKES, F., and BUSHUK, W. 1985a. Gluten proteins with high affinity to flour lipids. Cereal Chem. 62:284-289.

ZAWISTOWSKA, U., BEKES, F., and BUSHUK, W. 1985b. Involvement of carbohydrates and lipids in aggregation of glutenin proteins. Cereal Chem. 62:340-345.

ZENTNER, H. 1964. The oxidation of mechanically developed doughs. J. Sci. Food Agric. 15:629-634.

ZIDERMAN, I. I., and FRIEDMAN, M. 1985. Thermal and compositional changes of dry wheat gluten-carbohydrate mixtures during simulated crust baking. J. Agric. Food Chem. 33:1096-1102.

CHAPTER 5

COMPOSITION AND FUNCTIONALITY OF WHEAT FLOUR COMPONENTS

Y. POMERANZ
Department of Food Science and Human Nutrition
Washington State University
Pullman, Washington

I. INTRODUCTION

The problem of relating the chemical composition and structure of wheat components to functional properties has kept more cereal chemists at work than any other single problem in the field (Pomeranz, 1968, 1973a, 1985, 1987; Pomeranz et al, 1970c; Finney et al, 1987). The functional properties of a flour depend on several factors, including wheat variety, environmental and soil conditions under which the wheat was grown, the process used to mill the wheat into flour, and the chemical composition of the flour. Wheat varies widely in chemical composition. Percentages of protein, minerals, vitamins, pigments, and enzymes show a range up to fivefold among cargoes of wheat. Such differences have far-reaching effects on processing and best use. Cereal chemists use the term *quality* to describe the suitability of a wheat flour for producing specific end products, such as bread, pastry, cakes, macaroni, or crackers. Quality of wheat cannot be expressed in terms of a single property; it depends on several milling, rheological, and processing characteristics of the raw materials and on their suitability and functionality in producing an acceptable food.

Durum wheat is the only tetraploid ($2n = 28$) wheat grown commercially in the United States. It is used, mainly, to make semolina flour for macaroni products. All hexaploid ($2n = 42$) wheats are capable of producing some type of leavened bakery product. It is generally accepted that the value of a wheat flour for a particular baking purpose depends on the protein content and the rheological properties of the gluten that can be made from the flour (Hehn and Barmore, 1965).

Endosperm proteins of wheat possess the unique and distinctive property of forming gluten when wetted and mixed with water. Gluten imparts physical properties that differ from those of doughs made from any other cereal grain. It is gluten formation, rather than any distinctive nutritive property, that gives wheat its prominence in the diet. When water is added to wheat flour and mixed,

219

the water-insoluble proteins hydrate and form gluten, a complex coherent mass, in which starch, added yeast, and other dough components are embedded (Beccari, 1745). Thus, the gluten is, in reality, the skeleton or framework of wheat flour dough and is responsible for gas retention, which makes the production of light, leavened products possible. The percentage of dry gluten is closely related to the percentage of protein. (See also Chapter 5 in Volume I of this monograph.)

On the basis of their suitability for the manufacture of yeast-leavened bread, common or vulgare wheats and the flours milled therefrom are classified broadly into two groups: strong and weak. Strong wheat flours contain a relatively high percentage of proteins, which form a tenacious, elastic gluten of good gas-retaining properties and are capable of being baked into well-risen, shapely loaves possessing good crumb grain and texture. They require considerable water to make a dough of proper consistency to give a high yield of bread. The doughs have excellent handling qualities and do not have critical mixing and fermentation requirements; for this reason, they yield good bread over a wide range of baking conditions and have good fermentation tolerance. In contrast, weak flours have a relatively low protein content and form a soft, weak, relatively nonelastic gluten of poor gas-retaining properties. They have relatively low water-absorbing capacity and yield doughs of inferior handling quality (which give trouble in machine baking), and the fulfillment of exact mixing and fermentation requirements is very critical, so that they are more likely to fail in baking. Weak flours require less mixing and fermentation than strong flours to give optimum baking results. The protein test, although not included as a grading factor in the U.S. standards for wheat, is accepted as a marketing factor. Customarily, data on protein content are made available to buyers of wheat and flour (Fig. 1). (See also Chapter 1 in Volume I of this monograph.)

American bakery practice and the quality demands of the U.S. consumer

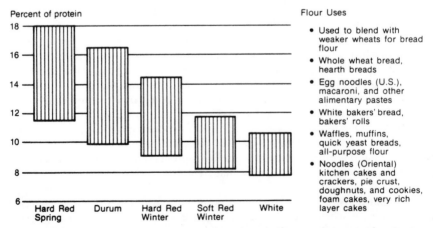

Fig. 1. Protein range and flour uses of major wheat classes. Flour uses are listed according to the approximate level of protein required for specified wheat products. Durum is not traded on the basis of protein content. (Reprinted, with permission, from Heid, 1979)

dictate the use of strong, relatively high-protein flours for breadmaking. Of the commercial market classes, the hard red spring (HRS) and hard red winter (HRW) wheats are excellent for yeast-leavened bread. Strong wheats surpass weak wheats for breadmaking, but the reverse is true for chemically leavened products. In the latter, there is no fermentation to "mellow" or "ripen" the gluten, and the low protein content and the soft, mellow characteristics of the gluten formed by weak flours produce a lighter, more tender product than do strong flours. Various degrees of strength or weakness are required by the bakery trade. Thus, the general strength of bread flours sold for commercial pan bread is greater than that of bread flour sold for the family trade, and flour for commercial hearth bread is even stronger. Home baking involves rather mild treatment (hand mixing or very slow-speed mixing and gentle fermentation), so that good results are obtained with a flour of lower protein content and more easily conditioned gluten than would be satisfactory for commercial bakeries. For the manufacture of pan bread, a medium-strong flour is required, to withstand high-speed mixing and produce a dough possessing the physical characteristics that permit machine manipulation. Bread baked on the hearth of the oven without pans requires a flour of still higher protein content to yield a strong dough that does not flatten unduly under its own weight. Other types of bread flours must be supplied by the flour miller to suit different markets.

With soft wheat flours, the protein content and the desirable gluten quality also vary widely (Yamazaki, 1953). For example, in soda cracker manufacture, flours varying in protein content from about 9 to 10% are required (depending upon whether they are used in the sponge or as doughing flours). For cake making, very weak flours, with protein contents ranging from about 7 to 9% and yielding batters that have a pH value of 5.1–5.3, give the best results. For the manufacture of alimentary pastes (macaroni, spaghetti, and noodles) of the desired brilliant yellow and cooking characteristics, semolina from durum is the preferred raw material. The wheat should be clean, large, vitreous, and high in protein and should have medium-strong gluten, a high concentration of carotenoid pigments, and low lipoxygenase activity. The particle size distribution and granulation of semolina are highly important in the manufacture of macaroni.

Since uses of soft wheats in the Japanese market are different from those in Western countries, quality requirements for them have peculiar aspects. Testing methods for sponge cake and Japanese-type noodles were most valuable in evaluating the processing quality of soft wheats (Nagao et al, 1976). General quality evaluation of soft wheats can be derived by combining the results of sponge-cake and Japanese-type noodle tests, as well as by the AACC cookie test and by consideration of test milling and analytical results.

Soft wheats, such as soft white, white club, and soft red winter wheats from the United States, Victoria soft, Victoria fair average quality (F.A.Q.), Western Australia F.A.Q., French, and domestic Japanese wheats were compared with respect to their utility for Japanese products (Nagao et al, 1977a). White club, soft white, and soft red winter wheats were superior in their suitability for making confectionery products. White club was better than soft white. Small quantities of Victoria soft and some types of French and Japanese wheats can be blended with soft wheats from the United States to produce confectionery flour. In spite of the low protein content, the kernel characteristics of both Australian

F. A. Q. wheats were rather hard, and they were least preferable in terms of their sponge-cake baking qualities. As the material for Japanese-type noodle flour, wheats similar to Japanese wheat were considered most desirable. Although Australian F.A.Q., soft white, and white club wheats were different from Japanese wheat, they possessed favorable characteristics for noodle flour.

In a subsequent study, 370 samples of soft white and white club wheats harvested in the United States Pacific Northwest in 1970–1973 and 33 samples of Western Australia and Victoria wheats harvested in 1970–1973 were tested to evaluate their utility for Japanese products (Nagao et al, 1977b). Sponge-cake and cookie-baking qualities of United States white club wheat were superior to those of soft white, even though the protein content of white club is higher than that of soft white. A combination of tests (protein content, maltose value, flour particle size, sponge cake, and cookie baking) confirmed that kernel softness is the most important factor for flour in cake, cookie, biscuit, cracker, and bun production in Japan. Australian F.A.Q., soft white, and white club wheats showed their unique and favorable characteristics for Japanese-type noodles. The best wheats for noodles were about 10% in protein content.

II. CHEMICAL BONDS IN DOUGH

Most cereal products require mixing flour with water to form a dough. Dough is a highly complex chemical system. The wheat flour components include proteins; mono-, oligo-, and polysaccharides; lipids; minerals; and enzymes. The additives yeast, milk solids, soy flour, eggs, and malt each contain several chemical moieties, whereas water, shortening, sugar, and salt are basically chemically pure. In addition, the dough contains a mixture of gases. The components interact and are constantly modified during mixing, fermentation (or chemical leavening), and baking. According to Holme (1966), foods containing flour as the main constituent can be best differentiated from one another on the basis of the ratio of flour to water used in their manufacture. We distinguish between dough systems (such as bread, pasta doughs, and cookie batter) and batter systems (mainly cake batters). In dough systems, the ratio of flour to water is about 1:0.6, and the gluten proteins provide the main structural matrix. In systems of higher ratios of water to flour, the importance of the gluten matrix decreases. Batter systems have free water and an aqueous phase containing a certain concentration of all the soluble components in the system. The aqueous phase forms the continuous phase in which the other ingredients are dispersed and into which the leavening agents release carbon dioxide and produce a foam. During baking, this foam rises, forms the typical foam structure, and is set in the later stages of heating (Handelman et al, 1961; Wilson and Donelson, 1963, 1965; Miller and Trimbo, 1965; Howard et al, 1968). To relate chemical composition to functional properties and explain chemical and physical modifications in dough processing, it is essential to understand the nature of the chemical bonds among dough components.

Wheat proteins are generally credited with attaining the final objectives of the baker—the production of a well-risen loaf of bread of satisfactory texture. In addition, wheat proteins largely govern the flour's water absorption, oxidation requirement, and mixing and fermentation tolerance. Groups of proteins in wheat flour are given in Fig. 2. Yet, it is well recognized that flour proteins are not the only wheat component uniquely suited for breadmaking and that practically all components are essential to produce an acceptable bread. Wheat proteins occupy a central position in the manufacture of macaroni and alimentary pastes. The role of wheat proteins in the manufacture of cookies and cakes is less prominent. However, studies by Donelson and Wilson (1960a, 1960b) and by Baxter and Hester (1958) pointed to the importance of the gluten fraction for optimum cake performance. The work of Miller and Trimbo (1965), Baldi et al (1965), and Sollars (1958a, 1958b) indicates an important role of starch in cake baking. Soft flours used in cake baking are generally bleached. The bleaching results in a drop in flour pH, and cakes with a higher ratio of sugar and shortening to flour can be produced. Cakes from bleached flours are more tender and less likely to collapse and have finer grain, improved color, and often greater volume. Starch, gluten proteins, and lipids were implicated in the improving effect.

A. Protein Structure and Types of Bonds

Proteins are generally considered to have at least three levels of structure—primary, secondary, and tertiary—although for some proteins no clear distinction can be made between the last two categories. The sequential arrangement of amino acids in a protein chain constitutes the first level of organization and is termed *primary structure*. The chains of amino acids are not straight but are folded to form characteristic three-dimensional structures,

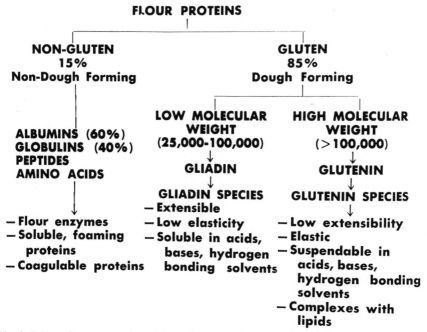

Fig. 2. Schematic representation of the main protein fractions of wheat flour. (Reprinted, with permission, from Holme, 1966)

called *secondary structure*. To account for the folding of the chains into relatively compact and rigid globular molecules, we visualize another level of organization, the *tertiary structure*. Finally, if the protein consists of two or more loosely linked globular subunits, we speak of its *quaternary* or *multimeric structure*. The native structures of most proteins are relatively compact and rigid; there is little internal space that is accessible to solvents. In general, the nonpolar amino acids (such as leucine and valine) are buried in the interior of the molecule, where they are shielded from contact with solvents. Finally, and most significantly, the complete molecular organization of a protein is essentially dictated by its amino acid sequence. The sequence governs the primary, secondary, and tertiary structures.

Chemical bonds in proteins can be divided into two major types: covalent and noncovalent. The covalent bonds include bonds within the amino acid, bonds between amino acids (peptide linkages), and disulfide (SS) links within or between peptide chains. The energy associated with (or required to break) those bonds is high. In addition, noncovalent bonds include ionic, hydrogen, and van der Waals bonds, which generally are much weaker than covalent bonds. It must be emphasized that covalent bonds and the three types of noncovalent bonds are theoretical models that are useful in describing the natural bonds, which are never purely covalent, ionic, hydrogen, or van der Waals bonds, but always a mixture of bonds.

In addition to the chemical bonds, there is the hydrophobic bond. Hydrophobic and van der Waals bonds have one thing in common: both are significant in interactions between nonpolar groups. However, the hydrophobic bond is not a chemical bond (i.e., due to the attraction between protons and electrons) but a thermodynamic phenomenon or entropy effect. A summary of functional groups involved in bond formation in wheat proteins is given in Table I.

B. Covalent Bonds

The only covalent bonds that are known to be significant in dough structure are the SS linkages between proteins. Such bonds have an energy of about 50 kcal per mole and probably are not broken at room temperature except by a chemical reaction.

TABLE I
Amounts[a] of Functional Groups in the Flour Proteins[b]

Group	Amino Acids	Albumins and Globulins	Gliadin	Glutenin
Acidic	Glutamic and aspartic acid	91	27	35
Basic	Lysine, arginine, histidine, tryptophan	100	39	52
Amide	Glutamine, asparagine	90	309	266
Sulfhydryl and disulfide	Cysteine, cystine	45	12	12
Total ionic	(Acidic + basic)	191	66	87
Total polar	(Hydroxy + amide)	190	381	365
Total nonpolar		372	390	301

[a] Millimoles per 100 g of protein.
[b] Source: Holme (1966); used by permission.

About 1.4% of the amino acids in gluten are either cystine or cysteine. The SS bonds in cystine can link together portions of the same polypeptide chain or different polypeptide chains and contribute to dough firmness. A network with permanent cross-links is not capable of viscous flow, which characterizes bread dough. Viscous flow requires opening of the rigid cross-links. If the dough is not to lose cohesion, the cross-links must reform at the same rate at other sites. By reacting with sulfhydryl (SH) groups (of a cysteine residue), SS groups can interchange and provide the required mobility. The involvement of such interchange reactions in rheological properties of dough was first suggested by Goldstein in 1957.

The effects of oxidizing agents on the rheological properties of dough may be quantitatively explained as the breaking of SS cross-links and their concurrent reformation by exchange reactions with SH groups. The velocity of exchange reactions would be expected to be proportional to the combined number of SH and SS groups; viscosity would be inversely proportional to the SH content. In practice, these effects are complicated by variations in SH and SS reactivity (Bushuk, 1961). That reactivity depends on the size of the molecules of which the groups are a part (Villegas et al, 1963), on the three-dimensional availability of the groups, and probably on interaction with proteins and nonprotein components.

According to Belderok (1967), baking quality of wheat is governed by protein content and the ratio of SS to SH. For a given protein level, optimum breadmaking results are obtained if the SS-SH ratio is around 15. The SS-SH ratio was above 7 in strong flours investigated by Kuchumova and Strelnikova (1968); the ratio increased during flour storage.

During dough mixing, only 1–2% of all gluten SS bonds can be broken by exchange with free SH groups, indicating that only a few reactive SS bonds are crucial to rheological properties. Total and reactive SH groups and the ratio of reactive to total SH groups increase with decreasing flour strength. Total SS contents decrease slightly, whereas reactive SS groups (or the ratio of reactive to total SS groups) increase, with decreasing mixing strength. Thus, mixing strength appears inversely related to reactive SH and SS contents (Tsen and Bushuk, 1968).

MacRitchie (1979) reported a highly significant correlation between gluten protein amide content and loaf volume index for 37 flours covering a wide range of baking quality. The percentage of glutamine plus asparagine in gluten protein ranged from 31.2 to 33.7% (the standard deviation for the amide determination was 0.3%). Loaf volume index and gluten protein amide content were affected both by cultivar and by growth conditions. MacRitchie (1980a) also calculated the total glutamic and aspartic acid content of different cereal grains, from the data of Tkachuk and Irvine (1969), and pointed to the higher percentage of amidation in wheat than in barley, triticale, and rye (but not in oats). MacRitchie (1980a) also pointed to the fact that according to Ronalds (1974), the ratio of amide nitrogen is slightly higher in bread wheats than in durum. The results were interpreted to mean that the primary amide content of gluten protein may be an important factor that confers baking quality to a flour. No significant correlation, however, was found between loaf volume per gram of flour protein and moles of amide nitrogen per 100 g of gluten protein in 36 English wheats studied by Ewart (1982).

C. Ionic Bonds

The importance of ionic bonds is demonstrated by several effects of adding salt to dough. Ions in a dough may complex with ionic groups of lipids (Fullington, 1969), proteins (Bennet and Ewart, 1965), and pentosans (Neukom et al, 1967). Theoretically, ions may improve both association and dissociation of dough components. In practice, the former prevails, since adding salts increases dough rigidity and reduces extensibility.

The ions occurring in wheat flour are summarized in Table II. About 7.3% of all amino acid residues in gliadin and 9.3% of those in glutenin are charged. In addition to these ions, about 2% sodium chloride and 0.5% other mineral salts (in the form of yeast food) are added in commercial baking (see also section VIII of this chapter).

D. Hydrogen Bonds

Hydrogen bonds result from the affinity of hydrogens of hydroxyl, amide, or carboxyl groups for the oxygen of carbonyl or carboxyl groups. Indirect evidence of the importance of hydrogen bonds in dough is twofold: 1) the presence of the necessary elements for their formation and 2) profound changes in rheological properties from the action of hydrogen-bond-disrupting agents. A large portion, about 90%, of dough components is highly polar. Over one third of the amino acid content in gluten is glutamine, which has an amide group that can participate in hydrogen bonding. Studies with water-insoluble and soluble synthetic polypeptides containing side-chain amide groups have indicated hydrogen-bonding phenomena (Krull et al, 1965; Krull and Wall, 1966). Acetylation of free amino acids destroys the cohesiveness of gluten (Barney et al, 1965). Conformation, surface properties, and functional properties of

TABLE II
Ionic Components in 100 g of Wheat Flour[a]

Component	Positive Charges (m val)	Negative Charges (m val)
K^+	2.87	
Mg^{2+}	1.16	
Ca^{2+}	0.96	
Zn^{2+}	0.04	
Fe^{3+}	0.075	
Na^+	0.091	
Cu^{2+}	0.006	
Cl^-		1.28
Phytate		0.82
Phospholipids[b]	1.50	1.93
Gliadin[c]	2.10	1.62
Glutenin[c]	2.97	2.36
Albumin + globulin[c]	1.80	2.00

[a] At 14% moisture basis, 70% extraction, 15% protein (Wehrli and Pomeranz, 1969a).
[b] Mainly lecithin, phosphatidylethanolamine, and phosphatidylserine.
[c] Calculated from data in Holme (1966) and Wall (1964).

deamidated gluten were studied by Matsumodi et al (1981, 1982). Since acetylation greatly reduces the ability of the gluten proteins to enter into hydrogen bonding, these bonds seem to play a significant role in the cohesive properties of gluten. Deuterium bonds in most cases have somewhat higher bond energies than hydrogen bonds. When deuterium oxide is used instead of water in baking, the gluten is strengthened and the elasticity increased (Vakar et al, 1965). The preparation and properties of acid-solubilized gluten were described by Wu et al (1976). Lasztity (1969, 1980) reported on the correlation between the chemical structure and rheological properties of gluten and viscoelasticity of chemically modified gluten.

The modification of wheat flour proteins with succinic anhydride was described by Grant (1973). Some functional properties of acylated wheat gluten were described by Barber and Warthesen (1982). Modification and derivatization of gluten and their influence on the manufacture of baked goods was reported by Gebhardt et al (1982). Ma et al (1986) modified vital wheat gluten by deamidation and succinylation. Deamidation caused a progressive degradation of gliadin with an increase in low-molecular-weight (LMW) components; glutenin was not affected. Deamidation also increased the net negative charge and surface hydrophobicity of gluten, and the bread loaf volume and dough extensibility were decreased. The most significant change in physicochemical properties of gluten caused by succinylation was an increase in net negative charge. Succinylation decreased dough extensibility but had no significant effect on loaf volume. The data indicated the importance of hydrogen bonding offered by the amide groups of gluten in the breadmaking process. The authors suggested that changes in molecular weight distribution and hydrophobic interaction may also affect the baking performance of gluten. Ionic interaction may be involved in dough development but is less critical in controlling the overall baking performance of gluten.

Adding 3M urea destroys dough structure (Jankiewicz and Pomeranz, 1965). Urea forms weaker hydrogen bonds than water and cannot lower the hydrogen bond energy of water. Whitney and Tanford (1962) therefore suggested that urea breaks hydrophobic bonds.

The most effective hydrogen-bond-forming components in wheat flour, in addition to water, are pentosans. The total (soluble and insoluble) pentosan content of flour is only about 1.5%, but pentosans can absorb 15 times their own weight of water (Bushuk, 1966); therefore, up to 13% of the water in dough could be bound to pentosans. Although most hydroxyl groups of the flour are in the starch, the starch granules are so tightly packed that only the groups at the surface are available for hydrogen binding. The specific surface of starch is 0.1249 m^2/g (Gracza and Greenberg, 1963). In 100 g of dough there is about 44 g of starch, which absorbs 8.7 g of water, corresponding to about 3.5 monomolecular layers of water around the starch granules. During baking, the starch granules swell and gelatinize and absorb most of the water. The role of glutamine, the major amino acid and a strong hydrogen-bond-forming residue of wheat proteins, is well established. Naturally present or added sugars further increase the content of polar substances that may participate in hydrogen bonding. Finally, naturally occurring or added polar lipids may form hydrogen bonds with any of the previously mentioned components.

Hydrogen bonds between small molecules, such as water, significantly increase the viscosity of a liquid. The mobility and much of the plasticity of doughs is affected by such bonds. In macromolecules, hydrogen bonds may lead to elastic structures (e.g., DNA, certain proteins) or even rigid structures (e.g., cellulose, starch). The bonds are weak enough to be temporarily extended, exchanged, or broken. As soon as the stress is released, the molecule returns to its original stable form.

E. Van der Waals Bonds

Van der Waals bonds and dipole-induced-dipole interactions provide very weak bonds (up to 0.5 kcal per mole). Although they occur between all atoms separated by at least four covalent bonds, they are not significant in the presence of stronger bonds and at distances longer than 5 Å. They may play a role in attraction between nonpolar amino acid residues or fatty acid side chains in systems in which hydrophobic bonds are impossible because of limited water. The starch-glyceride complex is the only type of complex that is stabilized by dipole-induced-dipole interaction and has been postulated to affect baking and bread properties, Such complexes occur naturally in starch granules (Acker and Schmitz, 1967) or are formed between starch and artificial surfactants or between starch and sucroesters.

F. Hydrophobic Bonds

Many of the amino acids in proteins have nonpolar side chains. The contribution of these groups to secondary and tertiary structures was postulated by Kauzmann (1959). He proposed the term *hydrophobic bond* to account for the forces responsible for the tendency of the nonpolar residues to adhere to one another and to avoid contact with the aqueous surroundings. Klotz (1960), on the other hand, visualized the nonpolar side chains forming crystalline hydrates, with water and hydration "icebergs" (Frank and Evans, 1945) coalescing to produce stable ice lattices over the protein surface.

Without free water, no hydrophobic bonds can be produced. According to Toledo et al (1968), only about 50% of the water in dough is bound. Consequently, hydrophobic bonds are possible. Experimental evidence for the importance of hydrophobic bonds was obtained by modifying dough properties by organic solvents (Ponte et al, 1967b) and by hydrocarbons (Pomeranz et al, 1966b, Elton and Fisher, 1968). The existence of hydrophobic bonds is also indicated by nuclear magnetic resonance (NMR) (Wehrli and Pomeranz, 1970a). Lipids in dough can form hydrophobic bonds. The feasibility of interaction between lipids and nonpolar groups of proteins can be calculated by the method of Bigelow (1967), provided the amino acid composition of a wheat protein is known. The hydrophobicities of proteins range from 440 to 2,000 cal per average amino acid residue. The hydrophobicities of glutenin and gliadin, from data on amino acid composition reported by Woychik et al (1961), are 1,016 and 1,109 cal, respectively. Thus, theoretically, the hydrophobicity of both gluten proteins is high enough to make intra- and intermolecular hydrophobic bonds.

Hydrophobic bonds may contribute to both plasticity (as in forming bimolecular lipid layers) and elasticity (as in the α-helix of proteins). The bond energies are low enough to allow rapid interchange at room temperature and to contribute to dough plasticity. On the other hand, they might stabilize conformations with small surfaces. Protein that has been stretched and has had its surface thus increased would tend to return to the original globular conformation as soon as the stress is released. Hydrophobic bonds could thus contribute to elasticity.

Hydrophobic bonds could also be important in early stages of baking, especially in oven-spring formation. All chemical bonds are weakened as temperature increases. However, hydrophobic bond formation is an endothermic process favored by increasing temperature up to about 60° C.

G. Conclusions

Studies with model systems or individual flour components frequently tempted researchers to attribute the central role to a single type of bond and to discount the role of other bonds. Despite the appeal of such hypotheses, it seems that practically all types of bonds contribute to the structure of dough. According to Vakar and Kolpakova (1976), the contribution of hydrogen bonds, ionic bonds, and hydrophobic interactions to the aggregation of soluble and insoluble glutenin proteins were 56.3, 17.3, and 26.4%, respectively, for a tough-gluten variety and 80.1, 12.8, and 7.1%, respectively, for a weak-gluten variety.

Lutsishina et al (1980) and Lutsishina and Matyash (1984) conducted wide-band proton NMR spectroscopic studies of gluten proteins. Glutens from hard and soft wheats differed in their wide-band proton NMR spectrum parameters (δH, M2, and S were higher in hard varieties), which was explained by differences in the conformational state of protein complexes. Macromolecules of gluten proteins were more compact, with stronger hydrogen bonds. There were positive correlations between the NMR parameters and rheological and technological properties of wheat. These parameters may serve as indexes for the determination of wheat quality. Whereas covalent and ionic bonds primarily increase cohesiveness of doughs, dipole, hydrogen, and hydrophobic bonds contribute to elasticity and plasticity. Van der Waals interactions are apparently of limited significance. Such a diversity of forces complicates the interpretation of factors that govern dough structure. At the same time, it opens several avenues for modifying the functionality of wheat flours.

III. COMPOSITION AND FUNCTIONALITY: GENERAL CONCEPTS

Much of our knowledge of the functionality of wheat flour components comes from investigations of the suitability of flours for breadmaking. Consequently, a large part of this chapter is devoted to these studies. Basically, relating the chemical composition and structure of wheat flour components to functional properties requires knowledge of the components present; methods by which to extract, fractionate, and characterize flour components; techniques by which to reconstitute the isolated moieties; and tools with which to ascertain that neither the isolation nor the reconstitution procedure impairs the functional properties of the components.

Historically, the last requirement was met first. Investigations of Finney and Barmore (1945a, 1945b, 1948) led to the development of an optimized baking test. In this test, five factors—water absorption, mixing time, oxidation level, yeast concentration, and fermentation time—are optimized and balanced. In addition, none of the added ingredients (shortening, sugar, and malt) essential to produce an optimum loaf of bread is permitted in a limiting capacity. Evaluating breadmaking potentialities in an optimized system (rather than determining the performance of a wheat flour under a set of fixed and arbitrarily selected conditions) provided a meaningful and reproducible analytical tool for determining quality characteristics inherent in wheat and flour components. The original test involved baking 100-g flour samples; the baking test was scaled down to 10 g of flour for studies with chromatographically separated fractions (Shogren et al, 1969). The optimized baking test can be performed with the same analytical precision as most generally accepted biological assays. Several parameters are determined in the test. The most important is loaf volume, because it cannot be determined by any other method and because over a relatively wide range it is highly and positively correlated with desirable rheological properties, consumer acceptance, and inherent breadmaking potentialities.

In spite of the variety of pastry products baked from soft wheat flours, evaluations of soft wheat breeding lines in laboratories of the U.S. Department of Agriculture are directed toward two end products—cookies and cakes (Yamazaki, 1969). To measure usefulness in cookie making, straight-grade flours from soft wheats are tested for moisture, ash, protein, no-time acid viscosity, mixogram area, and alkaline water retention capacity. Of all the tests used, however, the most meaningful probably is the cookie-baking test. The cookies are baked from 40 g of flour. The criterion used in evaluating the quality of the sugar-snap cookie is the diameter of the product. Cookie-spread potential appears to be a wheat varietal characteristic. Within a variety, spread is somewhat affected by protein content, and some workers adjust cookie spread to a uniform protein content in making intervarietal comparisons.

The large number of available cake formulas, special ingredients, and operational variations complicate the process of standardizing tests to evaluate cake flours, whether in a commercial laboratory or as part of a breeding program. At the Soft Wheat Quality Laboratory of the U.S. Department of Agriculture, finely milled 50% extraction flours are impact-milled to reduce average particle size and are treated with chlorine gas to attain a pH in the range of 4.6–4.8. Chemical tests are similar to those used to evaluate cookie flours. In addition, the original pH value and the change in the pH value after treatment with chlorine are determined. Cakes are baked by a formula that was developed to subject flours to maximum strain through high sugar content and lack of milk solids and egg whites (Kissell, 1959). Data for optimum liquid level, cake volume, and internal score are then recorded. The no-egg, no-milk formula gives responses of wider range from cake-making flours than does a formula of optional ingredients. None of the available chemical or physicochemical tests seems to be highly correlated with cake quality. Similarly, it is questionable whether a single cake-baking test can be used to predict the performance of a flour with other types of cakes.

Tests for durum products have also been developed. Dexter et al (1987) studied the relationship of durum wheat test weight (TW) to milling performance and spaghetti quality. A 1-kg decrease in TW was associated with a linear 0.7% decrease in semolina yield. A negative relationship between semolina ash and TW resulted in a further 1% loss in semolina yield potential per 1-kg decrease in TW at constant semolina ash. The semolina became duller as the TW decreased, resulting in a positive relationship between TW and spaghetti brightness. The only beneficial effect of lower TW was a higher protein percentage in the wheat and semolina. This, in turn, led to improved firmness and resilience of the cooked spaghetti. The stickiness of cooked spaghetti was related to TW at the 5% level of significance. Spaghetti stickiness is influenced little by protein content. Stickiness and cooking loss are related to starch gelatinization and lipid-starch interactions.

Taha and Sagi (1986) studied the relationship between total, soluble, and insoluble proteins of semolina and quality of macaroni in six Hungarian durum wheats. Total and insoluble protein contents of semolina were associated positively with strength, whereas soluble protein content was correlated positively with cooking time and swelling properties of macaroni.

IV. WATER-SOLUBLE COMPONENTS

A. Proteins

When flour is slurried with water, part of it becomes soluble. The water-soluble material contains soluble carbohydrates, nitrogenous compounds, and glycoproteins. The water-solubles were necessary to normal baking characteristics in two of three reconstituted flours (Finney, 1943). Pence et al (1951) also found that soluble components were required for maximum performance of all glutens studied except that from a durum wheat. A crude albumin fraction isolated from the water-solubles was responsible for the largest volume increase. Pence (1962) stated that albumins are implicated in the baking performance of flours and may account for a significant part of the differences in baking characteristics.

Pence et al (1954a) found that both the albumin and the globulin contents, as well as the ratio of albumins to globulins, varied significantly among 32 flours of widely varying types and baking qualities. The amount of soluble protein increased with increasing flour protein, but it decreased with increasing flour protein when expressed as a percentage of the total flour protein. Electrophoretic patterns of albumins prepared from durum wheat flours differed from electrophoretic patterns of albumins from club and common wheat flour (Pence et al, 1954b).

Although the amount of total protein was higher in the 13 hard wheat flours than in the 11 soft wheat flours studied by Cluskey et al (1961), the amount of water-soluble proteins recovered from gluten washing was essentially the same. Maes (1966) found the baking quality of flours to be correlated negatively with the percentage of protein soluble in water.

Studies on the water-soluble and salt-soluble proteins of wheat flour are limited (Simmonds, 1963; Strobel and Holme, 1963a, 1963b). Bell and Simmonds (1963) found in 26 flour samples a negative correlation between total nitrogen and the percentage of that total extracted into sodium pyrophosphate. Yet they concluded that total flour nitrogen was the best single chemical index of baking quality. Koenig et al (1964) found that better-quality flours (which also required longer mixing) contained more salt-soluble proteins. However, in studies reported by Mullen and Smith (1965), similar amounts of $0.1M$ NaCl-soluble nitrogen were obtained from short- and long-mixing flours, although electrophoresis indicated quantitative differences in protein composition.

Bread wheat contains 21 chromosome pairs arranged in three groups (A, B, and D genomes) of seven (i.e., 1A, 2A...7A; 1B, 2B...7B; 1D, 2D...7D). Each chromosome consists of two arms, which for some chromosomes are readily distinguishable and differ in length. Cereal rye contains only one group (R genome) of seven chromosomes (i.e., 1R, 2R...7R). In 1B/1R substitution lines, chromosome 1B of wheat has been replaced by 1R of cereal rye, whereas in the 1B/1R translocation stocks, the short arm of 1B has been replaced by the short arm of 1R (Dhaliwal et al, 1987).

In contrast to its substantial and beneficial agronomic effects, the 1B/1R translocation has detrimental effects on wheat quality. Doughs derived from such wheats develop stickiness with high-speed mixing, reduced dough strength, and intolerance to overmixing.

Some wheat breeding programs use material derived from rye to improve yield and disease resistance. Many of these cultivars exhibit intense dough stickiness. This defect becomes evident during mixing in the test bakery and causes serious problems when dough handling equipment is used in pilot bakery trials. Studies of Martin and Stewart (1987) have shown that laboratory mixing equipment developed by the Grain Research Laboratory (GRL) of the Canadian Grain Commission can be used to prepare the doughs before the evaluation of dough surface properties. This equipment consists of a GRL 200 mixer and a GRL direct reading energy input meter. Intense dough stickiness has been found in crossbreds from Australian wheat breeding programs and in wheat varieties or lines from other countries such as the Veery lines (Glennson, Ures, Genaro, and Seri), Amigo (a breeding line from Oklahoma), and the Nebraska variety, Siouxland.

Backcross derivatives of three commercial Australian bread wheat varieties carrying the 1B/1R translocation were compared by Dhaliwal et al (1987) with their recurrent parents for milling and quality characteristics. The 1B/1R translocation generally had no major deleterious effects on test weight, 1,000-kernel weight, protein hardness, flour yield and color, or farinograph water absorption. In the hard wheats, the 1B/1R translocation substantially and consistently reduced the volume of sedimentation in sodium dodecyl sulfate (SDS) and dough development time. There was also a tendency toward reduced extensigraph resistance and extensibility. For the latter characteristics, however, the best 1B/1R derivatives were similar to the recurrent parents. The soft wheat derivatives showed no evidence that the 1B/1R translocation significantly reduced SDS-sedimentation volume, dough development time, or extensigraph resistance. Dough extensibility, however, was consistently reduced in the 1B/1R derivatives.

In a subsequent study, flour proteins were extracted sequentially with a range of solvents from two hard and one soft wheat cultivar and their 1B/1R derivatives and were examined by acidic buffer- and SDS-polyacrylamide gel electrophoresis (PAGE) (Dhaliwal et al, 1988). Wheat cultivars and their derivatives showed a similar quantitative protein distribution pattern except for an increase in the proportion of water-soluble protein in the 1B/1R lines. On the basis of electrophoretic and immunological properties, this difference was attributed to the presence of rye secalin proteins, which are more water-soluble than their wheat counterparts. Total and water-soluble pentosans also were determined. Water-soluble pentosan content was slightly higher in some 1B/1R lines but did not exceed the range normally found in bread wheats.

B. Glycoproteins and Pentosans

Interest in the role of water-soluble pentosans in breadmaking originated with Baker et al (1943), who found that wheat flours contain about 1% water-soluble gums (Pomeranz 1961; Kulp and Bechtel, 1963a, 1963b) that form an irreversible gel when reacting with certain oxidizing agents used as dough improvers. Pence et al (1950) described the preparation of wheat flour pentosans for use in reconstituted doughs. Soluble pentosans had little effect on the baking performance of doughs reconstituted from gluten and starch, but they modified the handling properties of the doughs. Improvements in volume, grain, and texture were small. In a later study, solubles were isolated from 15 flours of widely varying characteristics (Pence et al, 1951). The nondialyzable fraction of the solubles, designated crude albumin, was rich in pentosans. It produced in all flours—except durum—a positive volume response and shortened mixing time when added to gluten-starch doughs. Mattern and Sandstedt (1957) suggested that the principal factor responsible for determining the mixing requirements of wheat flours was water-soluble materials. According to their findings, the most effective water-soluble component for shortening mixing requirements was gliadin rather than pentosans.

Holme (1962) studied the effects of various factors on the composition of soluble proteins and of pentosans in cake flour. Kuendig et al (1961a, 1961b) showed that a glycoprotein was responsible for the gelation of aqueous extracts of wheat flour in the presence of small amounts of oxidizing agents. The UV spectra of the glycoproteins showed an absorption maximum at 320 nm in addition to the protein maximum at 275 nm. The maximum at 320 nm was increased and shifted to longer wavelengths when borate and germanate were added. This finding indicated the presence of an *o*-dihydroxy or an aliphatic ene-diol structure in the glycoproteins. Furthermore, the maximum at 320 nm disappeared on oxidation. It was suggested that the gelation was caused by oxidation of the dihydroxy grouping, followed by the formation of cross-links between the glycoprotein molecules (Neukom et al, 1967).

The principal glycoprotein fraction of the water-soluble wheat flour pentosans was subjected to various degradation reactions (Fausch et al, 1963). The fraction contained about 13–17% protein as well as three phenolic compounds, one of which was ferulic acid (1.5% on xylan basis). The ferulic acid could not be detected after oxidative treatment. In a subsequent study, Painter and Neukom (1968) reported that the capacity of a wheat flour glycoprotein fraction, containing small amounts of esterified ferulic acid, to gel on the addition of hydrogen peroxide is lost when traces of bound Fe^{3+} or Ca^{2+} are removed by electrodialysis. The gelation capacity can be restored by subsequent addition of both ions. On the basis of studies of model systems, the authors suggested that the gelation of wheat flour glycoprotein resembles the oxidative gelation of caffeoyl guaran, which involves the conversion of caffeic acid ester groups into hydroxy-quinoid structures, followed by the formation of intermolecular cross-linkages of the chelate-bridge type.

Tracey (1964) studied the role of wheat flour pentosans in baking. The carbohydrate, rather than the protein component, of pentosan preparations affected breadmaking quality. Adding enzyme preparations containing pentosanase activity to a bread dough decreased loaf volume. Pentosanase was responsible for the deleterious effect. Adding flour solubles to gluten-starch mixtures markedly increased loaf volume. Flour solubles treated with a proteolytic enzyme of wide specificity effectively increased loaf volume, but an endogenous flour enzyme and snail digestive enzyme both rendered flour solubles ineffective, although the protein pattern was unchanged by this treatment. This showed that protein in the solubles was not responsible for the volume increase and suggested that pentosans were responsible. Treatments that destroyed the viscosity of flour solubles also destroyed their effect on loaf volume. Viscosity alone, however, was not responsible for the effect on loaf volume; adding certain viscous gums did not increase volume (Cawley, 1964). Wrench (1965) used enzymes from snail digestive juice to degrade two of five fractions obtained when flour pentosans were chromatographed on diethyl-aminoethyl cellulose. The fractions were both glycoproteins containing arabinose and galactose, with one also containing xylose. Solutions of the last fractions gelled when oxidizing agents were added, but gelling was lost after incubation with snail juice enzymes. Loaves baked from doughs in which the pentosans had been degraded by snail juice enzymes were lower in loaf volume, but the effect was not observed in the presence of oxidizing agents. The effect of oxidizing agents, however, did not appear to be related directly to pentosans. Tao and Pomeranz (1967) extracted water-soluble pentosans from flours milled from HRW, HRS, soft red winter (SRW), durum, and club wheats. Pentosan preparations from the durum flour differed from preparations from other flours in carbohydrate composition, electrophoretic mobility of associated proteins, and infrared spectra. The amino acids in proteins associated with the pentosans resembled amino acids in salt-soluble proteins. The pentosans from HRW, HRS, SRW, and club—but not durum—increased water absorption. The pentosans decreased dough development time and dough stability, as measured by a microfarinograph technique, but little change was detected in alveograms. Durum pentosans lowered amylograph peak viscosity; pentosans from other flours increased or had no effect on peak viscosity. Adding pentosans slightly increased oxidation requirements. Loaf volumes were increased slightly by adding pentosans from HRW or club flours and were decreased by adding pentosans from SRW and durum flours.

Hoseney et al (1969a) found that the amount and composition of material soluble in aqueous solutions during the washing of gluten from flour depended on the salt concentration in the wash water and therefore on the ratio of flour to water. When a flour-water ratio was selected to exclude gliadin proteins, the water-soluble fraction of the flour was not responsible for varietal differences in loaf volume. Nevertheless, the water-soluble fraction was required, to produce a normal loaf of bread. Increasing the amount of water-solubles above the amount normally present in flour did not increase loaf volume. The water-solubles were fractionated according to the scheme given in Fig. 3. The albumin or globulin proteins were not involved in breadmaking performance. The water-solubles were found to have the dual role of contributing to gassing power and modifying the physical properties of the gluten. The dialysate (containing mainly soluble carbohydrates, amino acids, and peptides) contributed to gas production. This contribution could be replaced by synthetic yeast food or even ammonium chloride. One of the nondialyzable fractions contained the water-soluble pentosans and glycoproteins and was involved in the modification of gluten.

In a study on wheat flour fractions involved in the oxidation (bromate) reaction, Hoseney et al (1972b) found that defatted flour (extracted with petroleum ether) gave essentially no bromate response. Fractionating and reconstituting techniques implied that potassium bromate improves gas retention by blocking a normally occurring, deleterious reaction and that a nondialyzable entity from the water-soluble fraction is required for the bromate reaction. In addition, both the dialyzable and the nondialyzable fractions from the most soluble gluten fraction are required for the bromate reaction.

Suckow et al (1983) conducted studies to determine the anomalous technological properties of wheats of inferior breadmaking quality (feed wheats). The feed wheats were characterized by high amounts of soluble starch and soluble proteins, low swelling power of gluten proteins, somewhat reduced molecular weight of pentosans from starch tailings, and a significantly reduced gelling power of arabinoxylan-protein complexes. The glycoproteins of feed

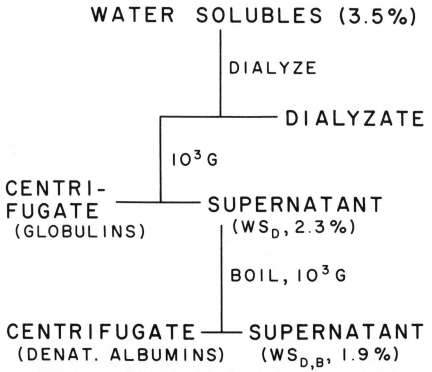

Fig. 3. Fractionation scheme employed to obtain certain water-soluble fractions of flour, i.e., dialyzed water-solubles (WS_D) and dialyzed and boiled water-solubles ($WS_{D,B}$). Percentages are based on total flour weight. (Reprinted, with permission, from Hoseney et al, 1969a)

wheats were low in polyphenols and had a poor oxidation response. Related studies were conducted by Booth and Melvin (1979). Functional properties of some European wheats grown in Europe and in the United States (in Kansas) were studied by Finney et al (1985).

Sollars (1959) found that LMW components, which constitute 68–78% of the water-solubles, had a negligible effect on cookie diameter. The high-molecular-weight (HMW) components of the water-solubles had a small effect; the water-soluble polysaccharides from the HMW fraction reduced cookie diameter.

Pentosans associated with gluten are similar to pentosans extracted from flour by water (D'Appolonia and Gilles, 1971a). Pentosans extracted from durum glutens had a high degree of branching and a high intrinsic viscosity. The authors suggested that the pentosans associated with gluten may affect its functional properties. Water-soluble and water-insoluble wheat flour pentosans exhibit an extremely hydrophilic nature and increase water absorption. The water-soluble pentosans may reduce mixing time and increase loaf volume (D'Appolonia, 1971); the effects are small. The water-insoluble pentosans may reduce loaf volume. According to D'Appolonia and Kim (1976), the water-soluble pentosans slow down the rate of retrogradation in bread by affecting the amylopectin fraction of starch; the water-insoluble pentosans retard the extent of retrogradation by affecting both amylose and amylopectin. Jelaca and Hlynka (1972) reported that pentosans, especially the water-soluble fraction, affect rheological properties of doughs and of gluten in a manner similar to that of the oxidant iodate. The mechanism of action of the pentosans likely involves an interaction between gluten proteins and pentosan polysaccharides.

Microbaking tests conducted by Patil et al (1976) showed that water-soluble pentosans were required for normal loaf volume from reconstituted gluten and starch doughs. Pentosans and bromate, in the absence of other water-soluble components, had an additive effect of overoxidation, which caused dough rigidity and lowered loaf volume. A HMW glycoprotein greatly increased the loaf volume of gluten-starch loaves in the absence of bromate. Reconstituting the water-soluble pentosans, in place of the water-solubles, produced (in the absence of bromate) a loaf-volume-improving effect equal to that of the water-solubles plus bromate. It was postulated that the rigidity of reconstituted doughs containing pentosans and bromate (generally accompanied by reduced loaf volume) results from a combination of two factors: first, the removal of water-soluble components responsible for gluten-protein extensibility, oxidation requirement, or both (to suppress the detrimental effect of overoxidation) and, second, oxidation of the pentosan-glycoprotein interaction product.

The significance of oxidative gelation of wheat flour pentosans as a way of cross-linking polymers and its potential involvement in dough formation or bread structure was reviewed by Neukom and Markwalder (1978). Hoseney and Faubion (1981) studied the oxidative gelation of wheat flour water-soluble pentosans with viscometry. Several common oxidants (potassium bromate, potassium iodate, and ascorbic acid) did not increase the viscosity of wheat flour solubles; viscosity increased after the addition of hydrogen peroxide, ammonium persulfate, and formamidine disulfide. Hydrogen peroxide was effective only in the presence of wheat flour peroxidase or tyrosinase. Ferulic acid, fumaric acid, and cystine stopped the formation of the gel when hydrogen peroxide was added. The authors suggested a cross-linking mechanism responsible for the oxidative gelation of wheat flour water-solubles. The proposed cross-linking system involves the addition of a protein radical to the activated double bond of ferulic acid and esterification of ferulic acid to the arabinoxylan fraction of the pentosan. The gelation appears to govern only part of the oxidative changes in bread doughs.

Ferulic acid was associated with five water-soluble pentosan fractions isolated from five wheat flours (Ciacco and D'Appolonia, 1982). The most abundant polysaccharide was an arabinoxylan isolated as fractions I and II. Fraction II contained the highest amount of ferulic acid. Durum arabinoxylan had the highest degree of branching; SRW arabinoxylan had the lowest. The highest intrinsic viscosity values were obtained for the arabinoxylan isolated from western white, and the lowest values were for that from SRW. Viscosity measurements after the addition of hydrogen peroxide-peroxidase to the arabinoxylan showed a higher gelling capacity in fraction II than in fraction I. For both fractions, the maximum increase in viscosity was directly related to the intrinsic viscosity. A sharp decrease in viscosity in some of the arabinoxylans after the maximum was reached indicated that an oxidative degradation of the carbohydrate chain was taking place competitively with the gelling reaction.

C. Monosaccharides, Disaccharides, and Oligosaccharides

Nonenzymatic browning of the Maillard type, which involves complex reactions between free amino groups and reducing sugars, is also generally considered responsible for bread crust color (Bertram, 1953).

Wheat starch pastes (10%) with and without single and multiple additions of wheat gluten, glucose, lactose, shortening, and nonfat dry milk, prepared at pH values of 4.4, 5.4, and 6.4, were stored at 75° C for 25 days (Stenberg and Geddes, 1960a, 1960b). The sugars were the principal ingredients that caused browning, as measured by reflectance of the dried pastes at 600 nm.

Ponte et al (1963) reported that L-arabinose and D-xylose inhibited the gassing power of simple flour-water doughs. In doughs also containing 10% D-glucose, 1% L-arabinose had little effect on gas production. Baking tests with 2.5 and 5.0% of the pentoses in a formula containing 5.5 and 8.0% total sugar showed the pentoses to have no effect on loaf volume but to prolong proof time, to dull crumb color, and to produce a markedly thicker and darker crust.

Rubenthaler et al (1963) investigated the effects of glycine, lysine, and glutamic acid, alone or in combination with each of 17 sugars, on the breadmaking potentialities and crust coloration of bread. Addition of 0.2–0.8 g of free amino acids or up to 4 g of sugars (per 100 g of flour) had no measurable effects on the bromate requirements, mixing time, or water absorption of wheat flour doughs. Glycine had the most adverse effect on loaf volume and caused pronounced browning of bread crust. The effect of lysine on loaf volume was insignificant, despite increased crust browning. Glutamic acid generally improved loaf volume and increased browning slightly. None of the amino acids, and only ribose among the sugars, affected crumb color. Added on an equimolar basis (0.011 g·mol/100 g of flour), raffinose and the pentoses imparted the deepest color to crust; melibiose, sorbose, lactose, and galactose followed in order. The hexoses glucose, levulose, and mannose had little effect and were followed by the disaccharides cellobiose, sucrose, and maltose. The smallest effect was exerted by melezitose and trehalose. The effect of saccharides on loaf volume, except those contributing to the level of fermentable sugars, was small. Adding certain amino acids (0.0026 g·mol/100 g of flour) along with sugars augmented the effects of each separately on loaf volume or crust color.

The results summarized in Table III show that under conditions of panary fermentation, the effects of free amino acids and sugars on loaf volume or bread crust might not be causatively related but might be two separate, independent reactions. Although adding sugars, at the levels employed and in the absence of added free amino acids, resulted in pronounced darkening of crust color, it did not adversely affect loaf volume. The increase in loaf volume from adding sucrose results from the contribution of additional fermentable sugar. Adding glycine increased crust coloration, whereas lysine and glutamic acid had no effect when added at the level of 0.2 g without sugar supplementation. Adding lysine had a small effect on loaf volume (which decreased); adding glycine gave the smallest loaf; and glutamic acid, again, had a beneficial effect. These results show that, although amino acids added with sugars might increase bread crust coloration, the extent of crust browning results primarily from sugars added or originally present in the dough. The findings were confirmed in a subsequent study in which sugars and amino acids were added to a semisynthetic model system, in which a mixture of native and pregelatinized wheat starch was substituted for wheat flour (Rubenthaler et al, 1964).

Brownness in macaroni has been attributed to a Maillard-type reaction, to an enzymatic reaction, and to bran contamination. Menger (1964) suggested that nonenzymatic browning in wheat pastes could result from condensation processes involving soluble carbohydrates. Matsuo and Irvine (1967) found that brownness arising from a varietal characteristic of durum wheat was due to a basic, water-soluble protein. The component was apparently associated with copper.

D. Enzymes

The role of enzymes in baking was reviewed by Anderson (1946). The role of enzymes in cereals is reviewed in Chapter 8 of Volume I of this monograph. Consequently, only studies pertaining to wheat varietal differences in enzymes

TABLE III
Effects of Sugars or Amino Acids on Loaf Volume and Bread Crust Color[a]

Sugar	Sugar Level (g)	Amino Acid	Amino Acid Level (g)	Loaf Volume (cm^3)[b]	Crust Color[c] Bottom	Crust Color[c] Top[b]
Basic formula	0	None	0	711	62.5	41.0
Arabinose	2	None	0	700	31.5	13.5
Sucrose	2	None	0	782	47.5	23.0
Galactose	2	None	0	725	36.0	15.0
None	0	Glycine	0.2	650	53.0	31.0
None	0	Lysine	0.2	685	60.5	44.0
None	0	Glutamic acid	0.2	740	61.5	42.0

[a] Source: Rubenthaler et al (1963).
[b] Per 100 g of flour.
[c] Reflectance measurements; the higher the figure, the less the color.

as related to functional properties are reviewed briefly in this section.

Flour from ungerminated grain contains a substantial amount of β-amylase, but α-amylase activity is the limiting factor in maltose formation. β-Amylase, the thermolabile saccharifying enzyme, is unable to attack native, mechanically undamaged starch, and it hydrolyzes damaged starch only slowly. α-Amylase, a relatively thermostable starch-liquefying or dextrinizing enzyme in germinated wheat, can attack starch that is not sufficiently damaged to be susceptible to β-amylase.

Prolonged storage generally reduces proteolytic activity. Upon germination, proteolytic activity increases up to 20 times, and the increase is detrimental to breadmaking. The amount of germinated or malted flour with high proteolytic activity that can be used as a supplement in breadmaking is generally limited to about 0.5%. The proteinases in flour milled from sound wheat appear, however, of little significance in breadmaking. Proteolytic activity is reduced by dough mixing and the addition of salt.

Protease systems are of great interest to cereal chemists from both practical and theoretical standpoints. The use of proteases in sponge doughs reduces mixing requirements at the dough stage. Protease-treated doughs are more extensible and can be sheeted more thinly and molded with greater ease, which is particularly beneficial with "bucky" doughs. Most proteases isolated in pure form split specific peptide linkages. Studies with pure enzymes, therefore, can provide useful information on wheat protein composition and structure and on enzymatic effects on dough rheology and the breadmaking potentialities of wheat flour doughs.

Oxygen uptake in doughs is enzyme-catalyzed and mediated by lipoxygenase. Although lipoxygenase is involved in bleaching flour pigments, dough maturation and improvement is not necessarily involved. Endogenous lipoxygenase activity of wheat and especially of patent flour is low. It is highest in the scutellum and embryo and lowest in the endosperm (Todd et al, 1954; Hawthorn and Todd, 1955).

Red wheats (HRS, HRW, and SRW), as a class, have higher lipoxygenase activities than white wheats or newer varieties of amber durum wheat. Lipoxygenase activity is particularly important in alimentary pastes, as excessive enzymatic activity impairs the color of the manufactured products. Catalase bleaches carotene in the presence of slightly peroxidized linoleic acid. The germ contains about a 100-fold higher concentration of catalase than the endosperm.

Relationships between semolina color indexes (yellow: SYI, brown: SBI) and some chemical components as well as oxidative enzyme activities of semolina were studied in seven durum wheats (Taha and Sagi, 1987). Semolina carotene content was highly correlated with SYI and SBI. High semolina peroxidase and polyphenoloxidase activities decreased SYI. High SBI was associated with higher values of semolina ash content, peroxidase activity, and protein content. Carotene processing loss was strongly affected by semolina ash content. Milling and processing caused considerable losses of oxidative enzyme activities in grain and semolina, respectively. Selection of varieties with high endosperm carotene content and low oxidative enzyme activities, small reduction of semolina yield during milling, and drying of macaroni at high temperatures improved macaroni color.

Spring wheats contain higher levels of catalase than winter wheats grown under comparable or variable conditions. HRS wheat flours generally have higher catalase activity than flours from HRW, SRW, white, or amber durum wheats. Variation among varieties is as high as threefold, and among locations as high as fourfold. The catalase activity of ground whole wheat was about six times that of semolina. Wheat germ contains a potent peroxidase. Relative peroxidase activity is substantially higher in wheat than in barley, corn, and rice.

Strelnikova et al (1964) determined flour strength, gluten viscosity and hydration, vitamin C, and catalase, peroxidase, and polyphenol oxidase (PPO) activities in strong and weak wheats. Strong wheats were characterized by higher activities of the oxidative enzymes and by more vitamin C at different vegetative periods than weak wheats. Oxygen consumption during the respiration of immature grains was higher in strong than in weak wheats.

Ascorbic acid that is of significance in metabolism is of interest among flour improvers, as it is a reducing agent. The usual improving agents are oxidants. Ascorbic acid is readily oxidized chemically by air in the presence of metallic catalysts or enzymatically by certain oxidases. The first step in the enzymatic system involves a reversible oxidation to dehydroascorbic acid.

Sandstedt and Hites (1945) found that extracts of a second-clear flour oxidized ascorbic acid fairly slowly, although more rapidly than a boiled extract, suggesting enzymatic oxidation. However, oxidation of ascorbic acid by extracts of a high-grade flour was slight.

Kuninori and Matsumoto (1963, 1964a, 1964b) showed that ascorbic acid is oxidized rapidly to dehydroascorbic acid during dough mixing and that flour contains a dehydroascorbic acid reductase, which is active at about pH 6.0. Dehydroascorbic acid reductase is active in dough in the presence of reduced glutathione; in the absence of reduced glutathione, activity increases slightly when protein thiol groups of dough are increased.

Carter and Pace (1965) showed that the activity of ascorbic acid oxidase in wheat endosperm is low. However, ascorbic acid is rapidly oxidized to dehydroascorbic acid in dough mixed in air. Dehydroascorbic acid was reduced rapidly in doughs to which reduced glutathione was added, indicating an active dehydroascorbic acid reductase in wheat. The enzymatic activity was about five times as high in the embryo and scutellum as in the endosperm and aleurone and was negligible in the pericarp and testa. Since the endosperm constitutes about 80% of the grain, most of the grain's total activity is in the endosperm. This accounts for the relatively high activity in white flour. The improving action of ascorbic acid is due to the oxidation of SH groups in dough by dehydroascorbic acid (Tsen, 1965).

Several dehydrogenases were detected in wheat germ and endosperm. Honold et al (1966) determined dehydrogenases in whole wheat and five milling fractions of two HRW wheats (Triumph and Bison) and two spring wheats (Lee and Selkirk). The number of isoenzyme bands and the level of dehydrogenase activity were generally higher in the two spring wheats than in the two winter wheats.

The researchers suggested that enzymes may affect enzymatic and nonenzymatic reactions involved in polymerization of proteins in the dough and in flour dough improvement and that enzymes in wheat flour that catalyze oxidation or reduction reactions may function in the maturing of flour and influence its baking properties. Quantitative determinations indicated that, generally, spring wheats contained significantly higher dehydrogenase activity than winter wheats (Honold et al, 1967). The HRS and HRW wheats contained a major lipoxygenase; the HRS wheat also contained a minor isoenzyme absent in HRW wheat (Guss et al, 1967). The activities of catalases (two isoenzymes) and peroxidases (eight isoenzymes) were much higher in HRS than in HRW wheats (Honold and Stahmann, 1968). Wheats from both classes contained ascorbate and PPOs, and indolacetic acid oxidase activity was detected in HRW only.

Differences in PPO activities and substrate specificities have been noted between wheat varieties. High levels detected in certain dwarf varieties may be responsible for the darkening of whole wheat dough and chapati.

Lamkin et al (1981) studied PPO activity as a possible method for the identification of wheat varieties or classes. Thirty wheat cultivars, representing HRW, SRW, HRS, white common, club, and durum wheats were studied. Different samples of the same cultivar from different growing locations and crop years gave comparable activities. In many cases, differences in PPO activities could be used to distinguish between cultivars. The durum wheats, which had lower activities, could be differentiated from varieties belonging to other classes by PPO activity alone. Lesser differences in activities were noted between the HRW and SRW wheats, between the HRW and HRS wheats, and between the white and club wheats.

Marsh and Galliard (1986) determined PPO activity in wheat and milling fractions. Bran contained higher levels of PPO than wheat; white flour contained relatively low levels, and no PPO activity was detected in a germ-rich fraction.

Godon and Herard (1986) attempted to correlate the composition of wheat protein extracts—prepared as described by Godon and Herard (1984), using the solvent sequence of water, sodium sulfite, and acetic acid—with a number of functional properties (solubility [nitrogen solubility index], water absorption, oil absorption, foaming capacity, and foam stability). The water extracts, containing mainly gliadins and some albumins and globulins, displayed a high water solubility, an average water absorption, a low oil absorption, a low foaming capacity, and a high foam stability. The sulfite extracts, containing LMW subunits of glutenins, presented a low water solubility, a low water absorption, a low foaming capacity, and a low foam stability. The acetic acid extracts, containing HMW subunits of glutenins, displayed a highly variable solubility as a function of pH (with a slight minimum at pH 8), an average water absorption capacity, a high oil absorption, a high foaming capacity, and a high foam stability. The functional properties of the extracts differed among themselves and also differed from those of gluten obtained from the same dough. Interactions between the gluten proteins decreased water hydration, oil absorption, foaming capacity, and foam stability. Thus, functional properties of total gluten proteins are not additive, because the net effects depend on interactions among the protein components.

It is possible to modify the functional properties by modifying the composition of the dough. Adding salt and, especially, oil to the dough decreases the quantities of glutenins, albumins, and globulins in the water extracts and the higher-molecular-weight subunits of glutenins in acetic acid extracts. Adding oil increases the quantities of lower-molecular-weight subunits of glutenins in sulfite extracts. Those modifications increase the solubility of components, sulfite extracts, and acetic acid extracts when their solubility is minimal, at basic pH. As a consequence, there is a decrease in the oil-absorption capacity of the acetic extracts and an increase in that of the water and sulfite extracts. Adding salt decreases the oil-absorption capacity of the water extracts and increases the capacity of the sulfite extracts; the presence of oil has an opposite effect. Adding salt or oil decreases foaming capacity of the water and acetic extracts, which have a low foam stability; they increase foaming capacity of sulfite extracts without improving the foam stability.

V. PROTEINS

A. Proteins and Loaf Volume

Reviews of functional (breadmaking) and biochemical properties of wheat flour components were published by Hoseney and Finney (1971), Wall (1971), Tipples et al (1982), Zimmermann (1976), Finney (1978), Pomeranz (1980b, 1980c, 1980d, 1983), and MacRitchie (1984). The properties and role of proteins in breadmaking were reviewed by Kasarda (1970), Kasarda et al (1976), Bushuk and Wrigley (1974), Lasztity (1975), Vakar et al (1975); Wall (1979a, 1979b), Huebner (1977), MacRitchie (1980a), Miflin et al (1983), Pyler (1983), Bushuk (1984a, 1984b, 1985), Hoseney (1984a), Lasztity et al (1984), Lasztity and Golenkov (1985); and Shewry and Miflin (1985). See also Chapter 5 in Volume I of this monograph.

A weak, over-extensible flour was modified by additions of gluten fractions that had been obtained from the same flour (MacRitchie, 1973). The physical dough-testing characteristics of a strong, well-balanced flour could be closely matched by additions of a suitably proportioned mixture of the whole gluten and a HMW protein fraction of the gluten, prepared by fractional extraction.

Bread was baked by an optimized formula from flour samples milled from wheat varieties grown under a wide range of climatic and soil conditions and from commercial wheat samples (Finney and Barmore, 1948). The major factor accounting for variation in loaf volume within a variety was protein content; the relation between loaf volume and protein content was essentially linear between 8 and 18% protein, the range encountered. The regression of loaf volume on protein content was different with different varieties. The regression lines for loaf volume versus protein content for the varieties represent differences in protein quality (Fig. 4). A similar relation between loaf volume and flour protein was reported by Fifield et al (1950) for 589 samples representing 10 varieties of HRS wheat grown in four crop years under a wide range of climatic and soil conditions. The relation between loaf volume and flour protein for each variety was linear within the limits of protein encountered, approximately 8.5–18%. Regression lines for loaf volume versus protein content for any variety were similar for four crop years, indicating that the bread-baking quality of each variety was essentially the same in different years. Again, the level and slope of the regression lines for loaf volume on protein content for the varieties differed significantly, indicating differences between varieties in protein quality. The linearity of the relation between protein content and loaf volume within a variety greatly simplifies the determination of the bread-baking quality of new wheat varieties. In practice, even two or three samples of a new variety at different protein levels establish with remarkable accuracy the regression of protein content versus loaf volume for the new variety in relation to that of standard

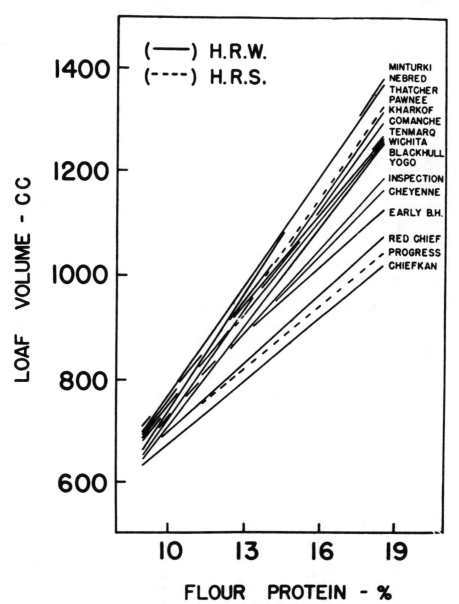

Fig. 4. Loaf volume-protein content regression lines for hard red winter (HRW) and hard red spring (HRS) wheat varieties. Each variety regression line represents many samples harvested throughout the Great Plains during several crop years. (Reprinted, with permission, from Finney and Barmore, 1948)

varieties. An increase in flour protein is accompanied by an increase in oxidation requirements. The correlation is low, however, as a result of varietal differences, variation in the mixing requirement within a variety, and the interrelation of mixing time and the oxidation requirement. For samples with a narrow range of protein contents, oxidation requirements decrease materially with longer mixing time. Loaf volume at a protein level of 13% increases sharply with mixing time from about ⅞ min to about 3 min (in the mixograph). With mixing times longer than 3 min, loaf volume at 13% protein is approximately constant.

B. Proteins of Maturing Wheat

The desirable qualities of good bread wheat are largely associated with the quantity and quality of its protein. The quantity is influenced mainly by environmental factors, but the quality of protein is mainly a heritable characteristic. A high-quality variety produces good bread over a fairly wide range of protein percentages, whereas a low-quality variety produces relatively poor-quality bread even when its protein content is high.

The protein content of wheat depends on the relative amounts of carbohydrates and nitrogenous compounds made available to the maturing wheat (Simmonds and O'Brien, 1981). The limiting factor in protein production appears to be the amount of available nitrogen in the soil at different stages of crop development in relation to soil moisture, mineral nutrients in the soil, and environmental factors that determine yield. Total rainfall, seasonal distribution of rainfall, and temperature have a profound effect on the amount of protein and, in some cases, on the characteristics of the protein produced. Differences in protein content between varieties of wheat grown under comparable conditions are small compared with differences due to environment. In general, higher yield per acre has been associated with lower protein content for different varieties grown under different environments. Agronomists have developed varieties that average 2–3% more functional protein than the varieties replaced. The new varieties also yield more per acre.

Finney and co-workers (Scott et al, 1957; Finney, 1965) showed that the synthesis of proteins involved in gluten formation, as far as loaf volume potentialities are concerned, begins generally three weeks before HRW wheat is ripe. The wheat reaches optimum potentialities of volume, crumb grain, mixing requirements, and mixing tolerance as early as two weeks before it ripens. Although the protein content of wheat changes relatively little during the last two weeks of ripening, the gluten-forming capacity of the proteins improves dramatically. The transformation is assumed to result from increases in the size and complexity of the proteins (Hoseney et al, 1966).

Finney (1965) showed that the synthesis of gluten proteins, as measured by loaf volume potentialities of 109–118% of those for ripe grain, was reached as early as 10–14 days before the wheat was ripe. Excellent crumb grain scores accompanied maximum loaf volume. Loaf volume values for Pawnee, Red Chief, and Triumph wheat samples harvested at various stages of maturity at Manhattan, Kansas, in 1954 indicated that the synthesis of gluten proteins was complete 9–14 days before the grain was ripe (Scott et al, 1957). Mixograms and pictures of loaves baked from wheat harvested at various stages of maturity are shown in Figs. 5 and 6. The quantity of proteins involved in gluten formation in wheat harvested at various stages of development was inversely related to absorbance at 280 nm of $3M$ urea extracts. The amount of protein, as determined by the biuret method, in the $3M$ urea extracts did not change during maturation. Thus the decreasing absorbance (at 280 nm) with increasing maturity between 27 and 17 days before ripening seems to indicate a gradual increase in the molecular weight and complexity of the synthesized proteins. Changes in the formation of gluten proteins paralleled improvement in the breadmaking properties of the flours (Fig. 7).

C. Early Fractionation Studies

According to Finney et al (1982), breadmaking, involving biochemical and physicochemical systems, is an essential analytical test in the scheme of relating physical and biochemical properties of important wheat flour fractions and components to functional properties. The best apparent way to bridge the gap between functional baking research and biochemical research on wheat and flour is by fractionation and reconstitution techniques. Normal functional properties of wheat fractions and components can be demonstrated when each

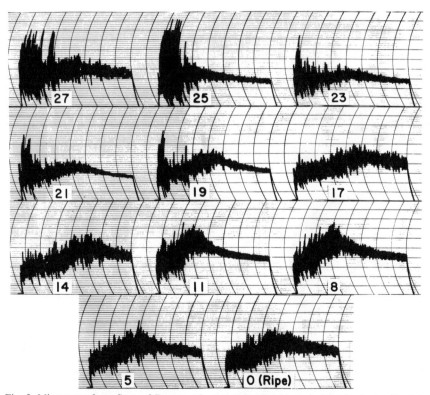

Fig. 5. Mixograms from flour of Pawnee wheat samples that were air-dried in shade after being harvested at various stages of maturity (numbers show days before ripe) in 1961. (Reprinted, with permission, from Hoseney et al, 1966)

fraction is unaltered and allowed to perform and interact singly and in various combinations with the other essential ingredients of a fermenting and optimum dough. Fractions and components are considered to be unaltered when the functional properties of the reconstituted flour or dough are the same as those of the original flour.

A study of the different steps in the separation, fractionation, and reconstitution of wheat flours to determine where detrimental changes to functional properties might arise was described by MacRitchie (1985).

Wheat flours representing a wide range in quality characteristics were fractionated into starch, gluten, and water-soluble fractions; fractions from one variety were recombined in the original proportion and different proportions; and various fractions of some varieties were interchanged (Finney, 1943). In each case, flours recombined in their original proportions gave bread equal to that made with the original, unfractionated flour. Gluten fractions accounted for the recognized differences in bread quality of the varieties studied.

Results of a study (Hoseney, 1968; Hoseney et al, 1969a) along similar lines are summarized in Table IV. The three flours included one composite flour, one

Fig. 6. Loaves of bread baked from a skeletal flour series of Kaw wheat samples that were air-dried in shade after being harvested at various stages of maturity (numbers show days before ripe) in 1961. (Reprinted, with permission, from Hoseney et al, 1966)

TABLE IV
**Baking Data for Three Original (Unfractionated) Flours and for Reconstituted Flours
in Which Fractions of Good and Poor Varieties Were Interchanged[a]**

Flours	Source of Gluten	Source of Starch	Source of Water-Solubles	Mixing Time (min)	Absorp-tion (%)	Potassium Bromate Requirement (ppm)	Loaf Volume (cm³)[b]
Unfractionated	RBS[c]			3¾	65.0	30	83
	C.I. 12995			6½	65.0	20	79
	K501099			1⅞	65.0	40	65
Reconstituted	RBS	RBS	RBS	3	65.0	30	82
	C.I. 12995	C.I. 12995	C.I. 12995	5½	65.0	20	78
	K501099	K501099	K501099	1⅜	65.0	40	64
	C.I. 12995	C.I. 12995	K501099	4⅝	65.0	30	78
	C.I. 12995	K501099	C.I. 12995	5⅜	69.0	20	79
	K501099	C.I. 12995	K501099	1⅛	62.0	40	66
	K501099	K501099	C.I. 12995	1	65.0	30	66

[a] Source: Hoseney (1969a).
[b] Per 10 g of flour.
[c] Regional baking standard.

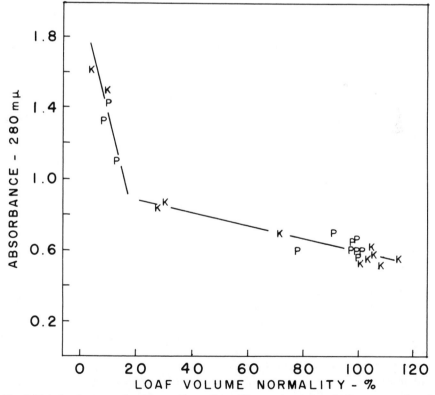

Fig. 7. Relation between absorbance of urea-dispersible proteins and loaf-volume normality of flours milled from Pawnee (P) and Kaw (K) wheats harvested at various stages of maturity in 1961. (Reprinted, with permission, from Hoseney et al, 1966)

high-quality flour (C.I. 12995), and one low-quality flour (KS501099). Bread was baked under optimized conditions on a 10-g flour scale. The results clearly confirm early fractionation studies. Although the above and many other investigations document that wheat gluten governs differences in the breadmaking potential of wheat varieties, an acceptable loaf of bread cannot be produced unless the other components of wheat flour are also present in approximately the same amounts as in a normal, unfractionated flour. Bread cannot be baked without starch, though starches from several cereals (rye and barley) can satisfactorily replace wheat starch (Hoseney et al, 1971). In reconstitution studies, an optimum bread is produced only if the water-soluble components or lipids are restored.

D. Fractionation of Gluten Proteins

Interrelationships between protein, wet gluten, damaged starch, pentosans, amylograph viscosity, and six water-absorption values (farinograph and baking absorption in two baking methods with variable malt level) were examined statistically by Tipples et al (1978) for flours and flour streams milled from Canadian HRS wheat. The strongest single predictor variable for farinograph absorption was starch damage. The addition of protein content as a second variable gave an excellent prediction of absorption; the addition of further variables to the prediction equation was not advantageous. By contrast, baking absorption in both long and short processes was largely a function of gluten and protein content, and the best explanatory combinations were any two of protein, wet gluten, and wet gluten plus protein. Damaged starch not only had little influence on baking absorption but also appeared to have a negative effect, particularly when malt was used in the formula in a method involving bulk fermentation. Differences in pentosan content and amylograph viscosity between flour streams did not appear to have a major influence on fluctuations in either farinograph or baking absorption.

Factors contributing to the baking quality differences in HRS wheat were studied by Marais and D'Appolonia (1981a). When baked at a fixed bromate level (10 ppm), the flours from two HRS wheat varieties, Coteau and Butte, had the same loaf volumes. Interchanges of the starch, gluten, tailings, and water-soluble fractions showed little difference for these fractions. The superior loaf volume potential of Coteau resides in its ability to respond to bromate. Both the albumin and globulin protein classes were implicated in an effect. The small variations in pentosan contents in the two varieties did not affect the bromate requirements. Most variability in the bromate requirement could not be explained by the variables that were measured. At optimum bromate levels, loaf volume correlated positively with total protein content but negatively with the percentage of glutenin and residue protein.

In a subsequent study (Marais and D'Appolonia, 1981b), the bases for different mixing properties of two HRS wheats (Olaf and F01277) were investigated by fractionation and reconstitution techniques. The stronger mixograph curve characteristics of Olaf resided primarily in the gluten fraction. The higher residue protein content of this variety appeared to explain most of the gluten effect. Total protein content related negatively to peak time and positively to peak height (at a constant water absorption). Total pentosan content had a small positive effect on peak height and height after 8 min. The albumin protein content affected peak time and curve width positively. However, the albumin proteins appear to sometimes have a negative effect on peak time, depending on the manner in which they interact with entities in the gluten and tailings fractions. Gliadin content showed a positive association with mixograph peak area in 21 progeny lines.

Wrigley (1972) suggested that molecular size distribution is an important property governing rheological and baking properties. This size distribution is a measure of the proportions of gliadin and glutenin. The HMW glutenin confers certain rheological properties (i.e., resistance to extension), which are sometimes identified as strength. The gliadin (or gliadins), probably in combination with lipids, modify those properties and contribute to the breadmaking potential of wheat flours.

According to Wall (1979b), the unique, cohesive properties of wheat dough are due to its water-insoluble proteins. This conclusion is evidenced by the fact that one can extract lipids from a flour and wash away its starch, water-soluble carbohydrates and proteins, and other components and still retain a hydrated rubbery mass, the gluten, which is 80% protein. The addition of reducing agents, such as sulfites, to glutenin protein destroys its viscoelastic properties and decreases its viscosity in solution. Sulfite reduction markedly lowers the molecular weight of the glutenin protein, as evidenced by ultracentrifugation and migration into gels during electrophoresis. There is evidence for both limited intramolecular and intermolecular bonds in glutenin, resulting in an extended molecule with minimal branching. The solubility and viscoelastic properties of the reduced protein can be restored by reoxidation. Both transmission and scanning electron microscopy have been employed to examine the transformation of flour into dough. The protein is integrated to yield a continuous fiberlike matrix that covers and binds the starch granules into the dough (Wall, 1979a). Glutenin from bread wheat has a fibrous structure, and that of durum wheat is ribbonlike and has a film structure, when viewed by scanning electron microscopy (Orth et al, 1973a, 1973b). A confirmatory bio-chemical and submicroscopic study of glutenins was conducted by Lefebvre et al (1979).

According to Bietz et al (1973) and Bietz and Huebner (1980), glutenin governs the strength and elasticity of dough. HMW glutenin molecules result in long mixing times and high dough stability. In the dough, the glutenin interacts with gliadin and other proteins to produce optimum flour performance. Glutenin constitutes about 35–45% of the flour proteins (Bushuk, 1974). It comprises about 17 subunits, ranging in molecular weight from 12,000 to 134,000. The subunits are held together, in a functional particle, by noncovalent interactions and by SS cross-links (Khan and Bushuk, 1978; Bushuk et al, 1980). The glutenin can be fractionated into glutenin that is soluble in acetic acid (Glu I) and glutenin that is insoluble in acetic acid (Glu II). The loaf volume potential of a bread flour depends on the relative quantities of the two types of glutenin. Khan and Bushuk (1978, 1979a, 1979b) and Khan et al (1980) proposed a model for functional glutenin, in which Glu I interacts with other molecules of Glu I and with Glu II molecules through noncovalent linkages. Hamada et al (1982) fractionated proteins in HRS wheat flours into glutenin, gliadin, albumin, and nonprotein nitrogen by gel filtration on Sephadex G-100. A positive relationship appeared to exist between glutenin in flour protein and mixing time. A HMW fraction separated by gel filtration on Sepharose CL-2B and a fraction insoluble in $0.05M$ acetic acid seem to be responsible for the mixing and breadmaking characteristics.

Lee and MacRitchie (1971) found that proteins extractable with $0.1M$ sodium hydroxide improved mixing stability and elasticity. The fractionation and properties of gluten proteins were described by MacRitchie (1972). According to Orth and Bushuk (1972), both glutenin and residue protein had a direct effect on baking performance. The proportion of glutenin in total flour protein was negatively correlated with loaf volume per unit protein, whereas residue protein and loaf volume per unit protein were positively correlated. The ratios of gliadin to glutenin and albumin to globulin were also significantly positively correlated with loaf volume per unit protein.

The aggregation behavior of wheat gluten was investigated by Arakawa and co-workers (Arakawa and Yonezawa, 1975; Arakawa et al, 1976, 1977). Aggregation behavior (as measured by changes in the turbidity of gluten suspensions) is related to rheological dough properties. Aggregation is determined mainly by the nature of glutenins. Replacement of gliadin by a HMW fraction (about mol wt 100,000) improved the strength of mixing curves for a synthetic dough system, consisting of gliadin, glutenin, and starch (Preston et al, 1975). In contrast, replacement with a lower-molecular-weight gliadin fraction (mol wt 27,000 or 44,000) weakened the curves. According to Huebner and Wall (1976), strong flours exhibiting long mixing times have high ratios of higher-molecular-weight to lower-molecular-weight glutenins. Flours of weak-dough wheats generally have lesser amounts of HMW glutenins and unextracted protein. Reversible aggregation of α-gliadin to fibrils that may be involved in dough structure was reported by Kasarda et al (1967). A correlation between the amount of aggregated proteins and baking quality was postulated by Field et al (1983b).

A suggested molecular unit of glutenin is the concatenation (Ewart, 1968, 1972; Greenwood and Ewart, 1975; Ewart, 1977). This consists of a variable number of polypeptide chains, joined together by SS bonds to form a supermolecule possessing a linear configuration. Concatenations can adopt complex conformations in water and can be entangled with one another. Regions of strong interaction and entanglement points form cross-links essential for rubberlike elasticity. Individual polypeptide chains may be unfolded by stress. Such chains tend to return to a compact state of lower free energy, and this accounts for elasticity. Entanglement points depend on secondary forces only and can be unraveled by stress. This is a special case of molecular slip and accounts for viscous flow in gluten and dough. The probability of entanglement decreases rapidly as the average length of a concatenation falls.

Assuming a random distribution of cysteine residues within a glutenin polypeptide chain, Ewart (1978) calculated that about one sixth of an average chain is available for unfolding under stress. As the unfolding of chains under stress has a magnifying effect on the effective molecular length of glutenin, a small range in the average number of chains per molecule can encompass the levels of dough tenacities encountered in practice.

The fall in the viscosity of dispersions of gluten treated with excess mercaptoethanol does not exhibit an initial induction period (Ewart, 1979). This finding appears to rule out the idea that glutenin is highly cross-linked in a branching mode. It is compatible with a linear model in which one SS bond joins adjacent chains, or a model with low levels of branching. The insoluble residue fraction of wheat flour protein is mainly linear, HMW glutenin. A linear hypothesis can account for an effect in which baking quality is positively and negatively correlated with insoluble and soluble protein, respectively. It was suggested that gliadin may act as a plasticizer and aid the dispersion of glutenin (Ewart, 1979).

The polypeptide chains of glutenin are believed to be joined by SS bonds. Because each chain has more than one SS bond, an unsolved problem is how many SS bonds are between the chains. If there is only one per chain, glutenin molecules must be linear concatenations. If there are more than one, the molecules are probably branched, but a linear molecule with two SS bonds between neighboring chains is conceivable (Ewart, 1988).

Two attempts to measure the average number of interchain SS bonds per chain in glutenin were made by Ewart (1988). The first was to let reduced glutenin slowly reoxidize in dilute solution and to look for oligomers in samples taken at intervals. It was hoped that intrachain SS bonds would form first because they are stabler and because dilution minimized interchain contacts. The appearance of dimer bands would show when interchain SS bonds were starting to form: the SH groups still unoxidized would then be a measure of interchain SS bonds. Oligomers did not appear as discrete bands but as marks on the origin and possibly tailing. Oxidation to glutenin was rapid. The tailing that occurs in the electrophoretic patterns of low-molecular-weight glutenin suggests that there is little preference for specific pairings of subunits and that they are randomly distributed in glutenin molecules. In the other attempt, glutenin was reduced to varying extents, and SH groups were blocked. Analysis measured the fraction (τ) of broken SS bonds. The intrinsic viscosity $[\eta]_\mu$ was determined and plotted against the fraction of SS bonds broken. These two quantities, τ and $[\eta]_\mu$, were calculated from theory, assuming a range of values for the fraction of junctions with two SS bonds in linear glutenin (δ). The theoretical curves were compared with the experimental points by computer, in the hope that δ would give the best fit; it was found that δ was unlikely to be greater than 0.3 and could be zero. Thus, if glutenin has linear molecules, at least two thirds, and possibly all, of the junctions are likely to consist of a single SS bond.

Graveland et al (1984) reported on the structure of glutenins and their breakdown during dough mixing by a complex oxidation-reduction system. A rather complex model for the molecular structure of wheat flour glutenins was proposed by Graveland et al (1985). The intrinsic viscosities of glutens and glutenins from 36 English-grown wheat samples from 25 varieties were highly correlated with their loaf volumes per gram of protein (Ewart, 1980). If intrinsic viscosity of wheat proteins is correlated with molecular weight, molecular weight of the glutenin is an important contributor to the baking quality of wheat. The intrinsic viscosity of glutenin from a Canadian western red spring wheat was higher than that of English wheats. Heat damage may be due to the cross-linking of glutenin molecules, so that the molecular weight remains high but the ability to swell and form a continuous matrix is impaired. Genetic factors, rather than environmental, seem the more important in controlling the intrinsic viscosities of gluten and glutenin, since no significant variations were detected when varieties were grown at more than one site or in different years.

Studies conducted by Legouar et al (1979) indicated that mixing properties (strength) of wheat flours were due to the glutenin soluble in acetic acid and that superior loaf volume depended on the glutenins insoluble in acetic acid. Opposite results were reported by Preston and Tipples (1980). The dough-strengthening effects obtained when gluten proteins were added to base flours were mainly due to proteins in the acid-soluble gluten fraction; the acid-insoluble gluten proteins had a dough-weakening effect. Increases in loaf volume were obtained by the addition of the acid-soluble gluten proteins. The addition of acid-insoluble gluten proteins significantly reduced loaf volumes.

Subda and Biskupski (1982) found that good baking values were associated with a lower content of acid-soluble glutenins and a correspondingly higher content of glutenins soluble in $HgCl_2$ and in mercaptoethanol. Rheological dough properties were mainly determined by protein quantity and proteolytic activity. Water absorption of flour and development, stability, softening, and valorimetric value of dough were positively correlated with the quantity of glutenins soluble in $HgCl_2$ and mercaptoethanol but insoluble in acid. Softening of the dough was positively correlated with proteolytic activity. Good baking value of flour from spring and winter varieties was associated with a high content of acid-insoluble glutenins.

Bietz and Wall (1980) proposed that at least six major types of polypeptides contribute to wheat gluten and that each has distinct structures, properties, and origins. Differences in molecular weight and in properties of each type of protein may change dough properties.

Despite progress in the fractionation of wheat proteins, for many years little progress has been made in establishing by actual performance tests the contribution of isolated protein fractions. Several attempts in this general area have been rather unsuccessful. The failure apparently resulted from using harsh solvent systems that irreversibly modify proteins during extraction and fractionation. Old methods, in which gluten proteins were indiscriminately fractionated with organic solvents and concentrated inorganic solutions, are unable to reveal the nature of factors responsible for the physical properties of the dough and changes in those properties during breadmaking. Protein moieties from grain, flour, and dough should be isolated and studied under conditions that prevent the denaturation and alteration of physicochemical properties.

In addition to establishing the conditions for separating flour components without impairing their functional properties, a baking method for testing 10 g of flour, either original or reconstituted from separated fractions, was developed (Shogren et al, 1969). Most of the gluten separated from starch and water-solubles can be suspended in dilute lactic acid (0.005N). Centrifugation of the suspension at about 1,000 × g for 20 min yields two fractions: an insoluble fraction (centrifugate) that constitutes 5–8% of the total flour protein, together with starch not removed during the gluten-washing process, and a supernatant from which the undamaged gluten proteins can be collected by adjusting the pH to 6.1 and by centrifugation. Loaf volume potentials of the insoluble fractions of a good-breadmaking and a poor-breadmaking variety did not differ. When the insoluble fractions were omitted in reconstituted flours, loaf volumes were at least equal to those of the corresponding unfractionated flours. Thus, the insoluble fraction has no specific role in breadmaking, and loaf volume potential is governed by the soluble gluten fraction (Hoseney et al, 1969b). The soluble glutens from a strong flour (C.I. 12995) and from a weak flour (KS501099) were further fractionated by ultracentrifugation at 100,000 × *g* for 5 hr. Approximately 15% of the protein was recovered as centrifugate (100-5C), and 85% remained in the supernatant (100-5S). Starch gel electrophoresis characterized the fractions as proteins retained at the origin (centrifugate) and proteins migrating into the starch gel (supernatant). The two fractions, when reconstituted and baked into bread, produced breads with loaf volume and crumb grain equal to those of the original flour. The 100-5S and the 100-5C fractions from both varieties were also reconstituted (to the protein contents of the original flours) at various ratios with their respective starch and water-solubles and were baked into bread. The 100-5S and 100-5C fractions in ratios of 80:20 to 90:10 for the strong flour gave loaf volumes comparable to that of the control. When gluten proteins of both varieties were replaced by their respective 100-5C fractions only, the reconstituted flours had practically indefinite mixing times and gave strikingly similar baking results, irrespective of variety. Replacing the gluten proteins with the respective 100-5S fractions indicated varietal differences.

The 100-5S and 100-5C fractions from C.I. 12995 and KS501099 were also interchanged. The samples were reconstituted at an 85:15 ratio of 100-5S to 100-5C, at 12.5% protein, and with starch and water-solubles from C.I. 12995. The baking results (Table V and Fig. 8) show that the 100-5S fraction controlled the volume of the loaf. The 100-5C fraction from the low-strength KS501099 was

TABLE V
Baking Data for Reconstituted Flours of Two Wheat Varieties and Reconstituted Flours
in Which the Fractions of the Two Varieties Were Interchanged[a]

Original or Reconstituted Flour	Flour Fraction	Mixing Time (min)	Baking Absorption (%)	Potassium Bromate Requirement (ppm)	Loaf Volume (cm³)
Original unfractionated					
C.I. 12995		6½	65.0	20	79
K501099		1⅞	65.0	40	61[b]
Reconstituted[c]					
C.I. 12995	100-5S 100-5C	2⅝	61.0	20	81
K501099	100-5S 100-5C	⅞	64.0	40	53
Reconstituted[d] and interchanged					
C.I. 12995 K501099	100-5S 100-5C	2	61.0	30	79
K501099 C.I. 12995	100-5S 100-5C	1¼	59.0	30	52

[a] Source: Hoseney et al (1969b).
[b] Loaf volume adjusted to the protein content (12.5%) of C.I. 12995; per 10 g of flour.
[c] With an 85:15 ratio of 100-5S to 100-5C.
[d] To 12.5% protein with C.I. 12995 starch and water-solubles.

just as good as that from C.I. 12995 when each was baked with the 100-5S fraction from C.I. 12995. The 100-5C fraction from C.I. 12995 was no better than that from KS501099 when each was baked with the 100-5S fraction from KS501099. Thus, the factor or factors responsible for differences in loaf volume potential were in the 100-5S fraction.

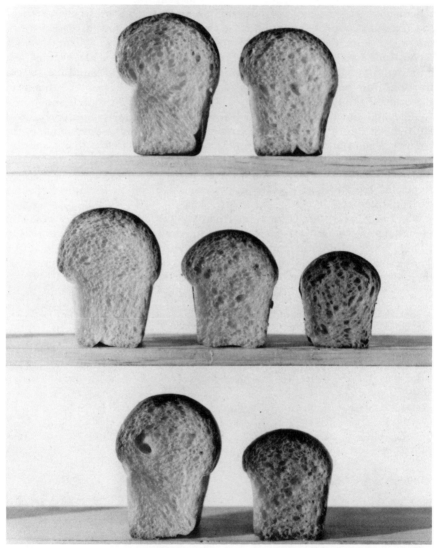

Fig. 8. Cut loaves of bread baked from reconstituted flours containing the 100-5S and 100-5C fractions of C.I. 12995, and reconstituted flours in which those two fractions of C.I. 12995 and K501099 were interchanged. All reconstituted flours contained 12.5% protein and C.I. 12995 starch and water-solubles. Loaves represent, from left to right: first row, original flours of C.I. 12995 and K501099; second row, 100-5S plus fractions 100-5C, 100-5S, and 100-5C of C.I. 12995; third row, fractions 100-5S of C.I. 12995 plus 100-5C of K501099 and fractions 100-5S of K501099 plus 100-5C of C.I. 12995. (Reprinted, with permission, from Hoseney et al, 1969b)

It was postulated that since all the proteins in the 100-5S fraction migrate electrophoretically into the starch gel, they apparently also represent LMW proteins of the gluten. The 100-5S fraction contains all the gliadin proteins, along with a significant part of the glutenin proteins. If the glutenin proteins perform a role in baking similar to that of the glutenin proteins in the 100-5C fraction, then the gliadin proteins would be responsible for differences in loaf volume potential between good- and poor-quality flours.

Gluten was also fractionated into gliadin and glutenin with 70% ethanol (Hoseney et al, 1969c). The ratio of gliadin to glutenin was essentially constant at 53% gliadin and 47% glutenin for four wheat varieties that varied widely in breadmaking quality. The gluten proteins retained their functional properties, provided that certain rheological properties were restored by remixing.

Reconstitution of the glutenin fractions of good- and poor-quality varieties with a fixed gliadin-rich fraction showed that the gliadin proteins control the loaf volume potential of a wheat flour. Similar reconstitution studies showed that the glutenin fraction governed the mixing requirement of a wheat flour. Additional evidence supporting those conclusions was obtained by reconstituting and baking protein fractions that were obtained by partially solubilizing gluten in dilute lactic acid. The fractions were characterized by starch gel electrophoresis as gliadin-rich (soluble) and glutenin-rich (insoluble).

Studies reviewed by Hoseney and Finney (1971) yielded evidence that the glutenin proteins were responsible for the mixing requirement and the gliadin proteins for the loaf volume potential of bread wheat flours. Goforth and Finney (1976) separated glutenin from gliadin by centrifuging the acid-soluble gluten at $435,000 \times g$. Properties of bread made from the reconstitutions of the protein fractions indicated that the gliadin and glutenin fractions were functional. In a subsequent study, effects of the strength and concentration of acid on the functional properties of solubilized glutens of good- and poor-quality bread flours were reported by Goforth et al (1977). On the basis of those studies, Jones et al (1983) found that the fractions of acid-solubilized glutens of good- and poor-quality wheat flours sedimented at different rates. They determined the time at $100,000 \times g$ required to sediment the 11–15% HMW protein fraction (pellet) and the time at $435,000 \times g$ to sediment an amount of gel protein (31–35%), so that about 54% of the acid-solubilized gluten protein would remain in the combined viscous layer and supernatant protein fractions of both good- and poor-quality bread wheats.

The gluten proteins sedimented at $435,000 \times g$ in the gel, viscous layer and supernatant are essentially the same proteins that were unseparated in the 100-5S fraction described in the review of Hoseney and Finney (1971). Finney et al (1982) interchanged those corresponding protein fractions of good- and poor-quality wheat flours and related them and their physical and biochemical properties to functional properties.

Good-quality (RBS-76) and poor-quality (76-412) hard winter wheat flours were fractionated into crude gluten and starch plus water-soluble fractions (Fig. 9). Gluten proteins, solubilized in 0.0045–0.005N lactic acid, were sedimented by ultracentrifugation into a pellet (relatively insoluble HMW glutenins), a gel (relatively soluble LMW glutenins), a viscous layer (soluble HMW gliadins), and a supernatant (soluble LMW gliadins). The corresponding gel and viscous layer plus supernatant fractions of the good- and poor-quality flours were interchanged singly in reconstituted flours containing the starch plus water-soluble fraction and baked into bread (10 g of flour). The gel glutenin proteins of the acid-soluble gluten proteins controlled the mixing requirement and baking absorption, and the viscous layer and supernatant gliadin proteins controlled loaf volume and crumb grain. Sedimentation rates and physical properties of the different protein fractions varied greatly within a quality level, and the sedimentation rate of a given protein fraction varied materially between the two quality levels. In comparison to the PAGE patterns of the supernatant, those of the gel protein fractions showed decreased densities of the rapidly moving bands (small proteins) and increased number and densities of the slowly moving bands (large proteins) (Finney et al, 1982).

The relative ease with which the HMW pellet glutenins sedimented suggests that they are relatively free compared to the LMW gel glutenins. The gel

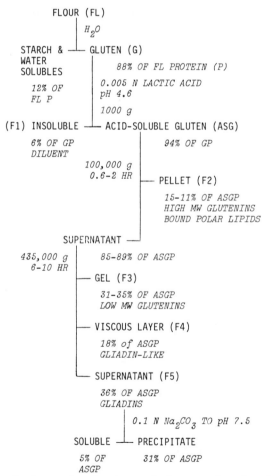

Fig. 9. Scheme to fractionate wheat flour into crude gluten protein and starch plus water-solubles and to fractionate the acid-soluble gluten into two glutenin and two gliadin fractions. (Reprinted, with permission, from Finney et al, 1982)

glutenins appear to be relatively tenaciously interacted or bound to the gliadin proteins. The extent to which the bound polar lipids interacted with the pellet proteins to produce HMW aggregates may render them relatively nonreactive with the gel glutenin and gliadin proteins, so that the pellet glutenin proteins are relatively free to sediment at relatively low centrifugal forces. For a discussion of the properties and role in breadmaking of gel proteins see also Jeanjean and Feillet (1980).

When flour is wetted and mixed into a dough or when gluten is forming, all free polar and about half of the free nonpolar lipids are no longer extractable in petroleum ether and are presumed to form lipoprotein complexes with the glutenin fraction of gluten protein. The ease of sedimenting the pellet glutenins indicates that they are not involved in the formation of additional lipoprotein complexes and that the free lipids probably interact with reactive gel glutenin and reactive gliadin proteins. Similarly, when the dough is formed, reactive gel glutenins probably interact with reactive gliadins. When the centrifugation forces are greater than the protein interaction forces but less than the molecular forces that keep the relatively small gliadin proteins in solution, the gel glutenins sediment.

A variety with a very short mixing requirement almost invariably has a low loaf volume potential. Similarly, a variety with a medium to medium-long mixing requirement almost invariably has a good to very good loaf volume potential (Finney and Yamazaki, 1967). Because loaf volume is a function of gliadin quality and the mixing requirement is a function of glutenin quality, good- and poor-quality glutenins are related to or associated with good- and poor-quality gliadins, respectively. Those relationships are diagrammed in Fig. 10.

MacRitchie (1987) fractionated the proteins from glutens of six wheat varieties, three of high and three of low baking performance, into either nine or 10 fractions by successive extraction with dilute HCl. Fractions were added to base flours to increase protein level by one percentage point; the fortified flours were assessed in terms of their mixograph peak development times and loaf

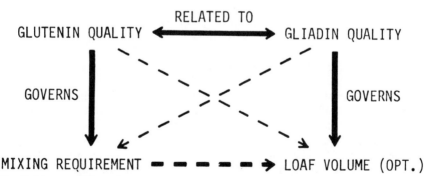

Fig. 10. Diagram of likely direct and indirect relationships between glutenin quality, gliadin quality, and the functional properties, mixing requirement and loaf volume (optimum). Broken lines indicate indirect relationships. The wider the line, the stronger the relationship. Also (not shown), glutenin quality governs mixing tolerance and dough absorption (attributable to flour proteins). Oxidation requirement and dough stability are strongly related indirectly to mixing requirement. (Reprinted, with permission, from Finney et al, 1982)

volumes in an optimized bread-baking test. Early-extracted fractions (gliadins) decreased mixing requirements and slightly depressed loaf volumes. Intermediate fractions (glutenins) increased dough development times and loaf volumes. This trend was reversed by the latest-extracted fractions, including the final residue. Larger amounts of protein (believed to be the main protein present in the latest fractions) were extractable with 0.75 M NaCl from glutens of the poorer flours than from those of the better-performing flours. The relative proportion of these globulin-type proteins to the glutenins appears to be important in determining baking quality.

E. Electrophoretic and Chromatographic Bands and Wheat Quality

Many attempts have been made to correlate pasta and baking quality with types of gluten proteins separated by chromatography or electrophoresis.

DURUM WHEATS

Gliadin and glutenin proteins form a gluten complex important for making pasta from durum wheat. Most attention has been focused on the gliadin proteins. This follows the discovery by Damidaux et al (1978) of a relationship between gliadin bands 42 and 45 and gluten strength; gliadin 42 is associated with poor elastic recovery and gliadin 45 with high elastic recovery.

Wasik and Bushuk (1975a) reported that the amount of certain glutenin polypeptides was related to spaghetti-making quality. These polypeptides are, however, of medium molecular weight and appear to be gliadin types.

Damidaux et al (1978, 1980) fractionated gliadin proteins extracted from 117 durum wheat varieties of different genetic origins by PAGE. Component mobilities were established by reference to the standard 51 band, in agreement with the common wheat gliadin nomenclature. Wheats were classified into two main groups according to the γ-gliadin region; one was characterized by the presence of a strong band 45 and the absence of a band in the 38–42 region, and the other by the absence of a band 45 and the presence of a strong band 42. Sixty-six varieties belonged to the 45 type, 47 to the 42 type, and four to neither. Excellent agreement was found between the electrophoretic patterns of the durum varieties and their viscoelastic properties (Fig. 11). In 59 of the 66 varieties (89%) of the gliadin 45 type, the elastic recovery of gluten was above 1.2 mm. In 46 of the 47 varieties (98%) of the gliadin 42 type, the elastic recovery was below 1.2 mm. The results were confirmed by Cottenet et al (1983) and Autran and Berrier (1984).

The gliadin proteins of seven durum wheat varieties of diverse origins were analyzed by two two-dimensional electrophoresis techniques by Payne et al (1984). Three of the varieties and biotype A of the Australian cultivar Duramba, with good pasta-cooking qualities, contained a γ-gliadin band 45 and an ω-gliadin band 35. These wheats also contained the same group of LMW subunits of glutenin, termed LMW-2. Genes coding for γ-gliadin 45 and ω-gliadin 35, linked tightly to each other on the short arm of chromosome 1B, are also linked tightly to genes for LMW-2. In similar studies, three other varieties and biotype D of Duramba, all of poor pasta-cooking quality, contained γ-gliadin 42; ω-gliadins 33, 35, and 38; and the group of LMW glutenin subunits LMW-1.

The inheritance of gliadin protein differences in durum wheat was investigated by du Cros and Hare (1985). These inheritance studies were conducted on crosses between the two main biotypes of Duramba. Results from F_1, F_2, and BC_1F_1 populations were consistent with the hypothesis that the syntheses of gliadin bands 45 and 42 are controlled by alleles; however, band 42 displayed a greater degree of dominance than did band 45.

An electrophoretic study by du Cros (1987) of the glutenin proteins of durum wheat, using 102 breeding lines, investigated associations between proteins and gluten strength. The LMW glutenin polypeptides were highly correlated with gluten strength. In contrast, the HMW glutenin polypeptides were poor indicators of viscoelastic properties. However, the presence of the lowest-mobility HMW band had a positive effect on gluten strength. Breeding lines that contained gliadin band 42 and had intermediate dough strength, instead of the dough weakness normally associated with this gliadin, contained this HMW glutenin polypeptide, which was not present in other lines examined.

Burnouf and Bietz (1984) analyzed durum wheat gliadins by reversed-phase high-performance liquid chromatography (RP-HPLC). Gliadins extracted from each durum variety were resolved to give a complex chromatographic

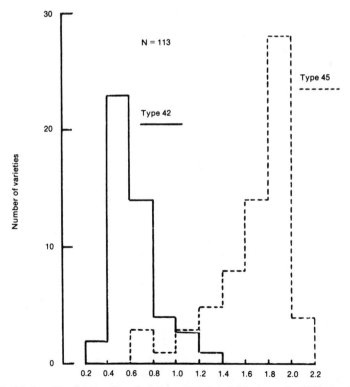

Fig. 11. Relationship of electrophoretic patterns and viscoelastic properties of durum wheat samples with gliadin bands 42 and 45. (Reprinted, with permission, from Damidaux et al, 1980)

pattern that was unique to that variety. Four chromatographic fractions, which were eluted under hydrophobic conditions, could be used to classify durum varieties into two groups. Proteins in these fractions were purified by preparative RP-HPLC and analyzed by aluminum lactate PAGE, which indicated that two of these fractions contained gliadin bands 45 or 42, shown previously to be correlated closely with gluten strength and weakness, respectively.

HMW glutenin subunit composition of 80 Italian durum cultivars and of seven commercial brands of spaghetti was investigated by PAGE in the presence of SDS (Vallega, 1986). The wheats carried 12 *Glu-A1* and *Glu-B1* alleles, accounting for nine different electrophoretic patterns. SDS-PAGE of commercial brands of spaghetti showed the presence of the same HMW glutenin subunits found in unprocessed whole meals. No obvious relation between HMW glutenin subunit composition and spaghetti-cooking quality was observed in 19 cultivars.

Principal component analysis was carried out by Autran et al (1986a, 1986b) on results of nine quality tests applied to 112 durum wheat cultivars and breeding lines. Gluten firmness and elasticity and SDS sedimentation were strongly associated, typically variety-dependent, and independent of protein content. Surface characteristics (stickiness, clumping) of cooked spaghetti were independent of rheological characteristics and were influenced by protein content, growing location, and genotype. In the relation between gluten rheological characteristics and the specific allele (γ-gliadin plus LMW glutenin as evidenced by PAGE or SDS-PAGE) present at the *Gli-B1* chromosome locus of durum wheat genotypes, allele 45 had a marked positive effect, whereas allele 42 was deleterious as far as gluten characteristics were concerned. Surface characteristics of cooked pasta could not be predicted by rheological measurements, making essential the separate evaluation of the two major components of durum pasta-cooking quality.

BREAD WHEATS

Since the report of Damidaux et al (1978) on the relationship between certain gliadin components and processing characteristics of pasta products, numerous investigations have studied similar relationships between various protein components and breadmaking potential (Branlard et al, 1984). Most of the studies involved the determination of rheological properties and indirect measures of breadmaking potential (such as SDS sedimentation). Both gliadin proteins and glutenin subunits were implicated. In addition, determination of breadmaking wheats on the basis of reduced albumin globulin fractions was described by Ohms (1980).

In one of the earlier studies in this field, Cluskey et al (1961) compared by electrophoresis the protein compositions of gluten and water-soluble constituents fractionated from selected samples of hard and soft wheat flours. The gluten proteins of the hard wheat flours contained more α, less β, and identical amounts of γ and ω components. The differences reported were small. Electrophoretic comparisons of the glutens of short- and long-mixing flours indicated similar amounts of α-, γ, and ω-gluten components (Mullen and Smith, 1965). The short-mixing flour had more of the β-gluten component and less acid-insoluble protein.

Gluten proteins from eight commercial varieties of bread wheat (*T. vulgare*) and four tetraploid *Triticum* species were separated by chromatography on carboxymethyl cellulose and by PAGE (Lee and Wrigley, 1963). The wheat proteins differed in protein profiles. The tetraploid species differed in many respects from *T. vulgare*. It was not possible, however, to relate any type of chromatographic pattern to the baking quality of the eight commercial varieties of *T. vulgare* studied.

Coulson and Sim (1964, 1965) studied endosperm proteins of several varieties of *T. vulgare* and found distinct differences in the composition of fractions of low electrophoretic mobility. No satisfactory correlation was found between those components and the rheological characteristics of the flour, and there appeared to be no obvious relation between closely related varieties. There was evidence, however, that some fractions reflected varietal morphological characteristics and also were influenced by environmental factors. In contrast, fractions of intermediate electrophoretic mobility were similar for all varieties. Lee and Ronalds (1967) demonstrated that the combined effects of locality, soil type, and season, resulting in up to a twofold variation in total protein content, have little effect on the distribution of gliadin components. This suggested that the composition of gliadin could be used in wheat genetics or in detecting differences in breadmaking potential.

The HMW glutenins account for only about 1% of the dry weight of the mature endosperm. Based on previous work (Payne et al, 1984) that related individual HMW glutenin subunits to breadmaking quality by genetic analysis, quality scores were assigned to each of the commonly occurring subunits (Payne et al, 1987). The grain proteins of 84 homegrown wheat varieties were fractionated by SDS-PAGE to determine their HMW glutenin subunit composition. The quality scores of each of the subunits were summed to create a *Glu-1* quality score for each variety. Between 47 and 60% of the variation in breadmaking qualities could be accounted for by variation in HMW subunits of glutenin. The presence of a translocated chromosome, which consisted of the long arm of 1B and the short arm of 1R from rye, was confirmed by its association with poor breadmaking quality. A correction factor was applied to the *Glu-1* quality score of varieties that contained the 1BL/1RS chromosome. The variations in the rye-adjusted *Glu-1* quality scores were compared with those of the breadmaking qualities of the varieties, and the proportion of the variation in quality accounted for increased to 55–67%. The *Glu-1* quality score and the biscuit-making qualities of the same set of varieties were negatively related.

The evidence of the relationship between the presence of specific HMW glutenin polypeptides and dough strength as measured by rheological properties was based primarily on correlation analyses. The finding that a number of Australian wheat cultivars consist of two or three biotypes that differ in HMW glutenin subunit composition made it possible to examine these relationships more directly (Lawrence et al, 1987). Biotypes from each cultivar were identical (with one exception) with respect to gliadin composition as determined by gel electrophoresis. Dough properties were determined with the extensigraph. The study, using genetically similar stocks for comparison, demonstrated that subunits 5 + 10 at the *Glu-D1* locus confer significantly greater dough resistance than subunits 2 + 12. The results indicate a key role for specific HMW glutenin polypeptides in relation to dough properties.

Baking quality results of 15 winter wheat varieties were correlated with gliadin and glutenin genes on the 1A, 1B, 1D, 6A, 6B, and 6D chromosomes (Sasek et al, 1986). Gliadin marker 1B3 and glutenin markers 1B 6 + 8 and 1D 2 + 12 were associated with low baking quality. The combined use of gliadin and glutenin markers gave the best results in predicting baking quality.

GLUTENINS

Using 26 hexaploid bread wheats of diverse baking quality, Orth and Bushuk (1973a) found no relationship between their quality differences and the SDS-PAGE patterns of their reduced glutenins. The glutenins from all of these wheats contained HMW subunits regardless of quality.

In a subsequent study (Orth and Bushuk, 1973b), glutenins from four extracted AABB tetraploid wheats, their hexaploid (AABBDD) common wheat counterparts, a synthetic (AABBDD) hexaploid, its parents, and seven accessions of *Aegilops squarrosa* were isolated, reduced, and analyzed by SDS-PAGE. The glutenin subunits of the synthetic hexaploid were simply inherited from its tetraploid (AABB) and diploid (DD) parents. Each extracted tetraploid lacked three glutenin subunits and showed a decrease in the amount of one electrophoretic band present in its hexaploid parent. Three of the affected subunits were of the same molecular weight for each hexaploid-tetraploid pair, the fourth being different for one of the four pairs. In each pair, the four deleted (or diluted) subunits were of the same molecular weight as subunits present in the glutenin of the *A. squarrosa* samples studied. Common (bread) wheats contained HMW glutenin subunits that were absent in durum wheats; their presence appears to be a necessary condition for breadmaking quality.

In the classical study of Payne and co-workers (1979), the subunit composition of glutenin in two varieties of contrasting pedigrees was analyzed by SDS-PAGE. Maris Widgeon, a variety of good breadmaking quality, contained two glutenin subunits not present in Maris Ranger, a much higher-yielding variety that is unsuitable for breadmaking. A third subunit was found only in Maris Ranger glutenin. To determine if any of these subunits are directly related to breadmaking quality, 60 randomly derived F_2 progeny from a cross of Maris Widgeon and Maris Ranger were analyzed for breadmaking quality and for glutenin subunit composition. A strong correlation was demonstrated between quality and the presence of one of the two subunits inherited from Maris Widgeon. This subunit (termed subunit 1 glutenin) had an approximate molecular weight of 145,000. The subunit was traced back through Holdfast to White Fife, a Canadian hard spring wheat of excellent breadmaking quality. Some 67 varieties were screened for the presence of subunit 1, which was found in 31% of them. Several unrelated varieties of good breadmaking quality did not contain subunit 1.

The homoeologous group 1 chromosomes of wheat contain in total five major gene groupings that code for the HMW subunits of glutenin (Payne et al, 1981). Each gene group displays allelic variation that is detectable by SDS-PAGE. Some of the allelic subunits are related to breadmaking quality, as determined by the SDS sedimentation test. Two strong correlations with quality were detected in the progeny of several crosses: the presence of subunit 1, a polypeptide coded by genes of chromosome 1A, and the presence of subunits 5 and 10, coded by chromosome 1D.

According to Payne (1983) and Payne et al (1980, 1983), glutenin, which gives elasticity to a dough, is built up of at least 12 different subunits. When fractionated by SDS-PAGE, the subunits fall into three groups. One group, the HMW subunits, is important in determining breadmaking quality. Hexaploid wheat varieties have different combinations of HMW subunits, the variation being due to allelic genes. Each variety contains three to five major subunits, two coded by genes on chromosome 1D, one or two by chromosome 1B, and none or one by chromosome 1A.

The endosperm proteins of wheat are determined by nine complex loci (Law and Payne, 1983). *Glu-A1*, *Glu-B1*, and *Glu-D1*, responsible for HMW glutenin subunits, are located proximally on the long arms of chromosomes 1A, 1B, and 1D, respectively. *Gli-A1*, *Gli-B1*, and *Gli-D1*, coding for the ω- and γ-gliadins, are located distally on the short arms of each of these chromosomes. The remaining three loci, predominantly coding for the α- and β-gliadins, are on the short arms of 6A, 6B, and 6D and are designated *Gli-A2*, *Gli-B2*, and *Gli-D2*. Variation at each of these loci gives an enormous range of proteins. Some of this variation among the HMW glutenin subunits is correlated with breadmaking quality.

Protein bodies were isolated by Field et al (1983a) from developing grain of wheat, barley, and rye. The protein was solubilized in a solvent consisting of $0.01M$ acetic acid, $6M$ urea, and $55mM$ cetyltrimethylammonium bromide (commonly known as the AUC mixture) and separated by column chromatography. A proportion of the protein eluted in the excluded volume of the column indicated the presence of multimeric aggregates in the protein bodies. The aggregates contained an increased proportion of HMW polypeptides and certain sulfur-rich prolamins. The proportion of aggregated material in wheat protein bodies was similar to that in gluten preparations. Two-dimensional electrophoretic analyses showed that the protein bodies and gluten consisted of almost identical polypeptides. The authors suggested that the aggregates of gluten proteins, which have been implicated as important in determining breadmaking quality, are laid down in protein bodies shortly after synthesis in the developing seed.

Burnouf and Bouriquet (1980) studied by SDS-PAGE the subunit composition of glutenins from 47 genetically related European wheat cultivars. Sixteen electrophoretic types were distinguished. Two subunits with molecular weights of 108,000–122,000 were found in cultivars of good quality and were absent in cultivars unsuitable for making French bread.

Moonen et al (1982) postulated that differences in breadmaking quality of wheat cultivars, as measured by the SDS sedimentation test, are governed by the quantity of SDS-insoluble glutenins. The SDS-insoluble glutenin was separated by SDS-PAGE into 15 subunits. The positive effects of the HMW subunits 3 + 10 and 2* on breadmaking quality were reported by Moonen et al (1983) and Moonen and Zeven (1985). Biochemical properties of some HMW subunits of wheat glutenin were studied by Moonen et al (1984, 1985). Results of studies conducted on eastern European wheats were reported by Pallagi-Bankfalvi (1984), Pallagi-Bankfalvi and Orsi (1985), and Orsi et al (1985).

The relationships between the protein content and the quality characteristics of 70 wheat cultivars were studied by Branlard and Dardevet (1985b) by analyzing the HMW subunits of glutenin by SDS-PAGE. Simple correlations were calculated between the proportion of each of the HMW glutenin subunits and results of tests with the Chopin Alveograph, which measures strength (W index), tenacity (P index), swelling (G index), and extensibility (L index); results of the Zeleny sedimentation test; and results of the Pelshenke swelling test. All characteristics were correlated with at least four subunits. Subunits 6 and 8 were the only subunits not correlated significantly with at least one of the characteristics. W, P, Zeleny sedimentation, and Pelshenke swelling were correlated positively with bands 5 and 10 and negatively with bands 1 and 12. Subunit 9 was correlated positively with W, P, and Zeleny sedimentation. Subunits 1 and 2* were correlated positively with G and W, respectively. Holding grain protein constant did not change the sign or significance of the correlations observed. Moreover, when the influence of the heterogeneity of the gliadin bands was eliminated, most of the correlations with P, G, and L disappeared. Only subunits 17 and 18 remained favorably correlated with swelling (G). This suggests that gliadins and HMW glutenin subunits interact in governing the rheological characteristics of the dough.

A comprehensive review of the relation between glutenin subunits and baking quality was published by Kubanek and Cerny (1985).

GLIADINS

Popineau et al (1986) separated ω-gliadin components by ion exchange chromatography from an ω-gliadin fraction. Residual glutenin, strongly adsorbed on the gel, was removed from gliadin fractions. All the isolated components were characterized by their electrophoretic mobilities at acid pH, their amino acid components, and especially their tryptophan contents and molecular weights and were detected by SDS-PAGE and gel filtration in the presence of $6M$ guanidine-HCl. These were compared with ω-gliadin components previously purified in other laboratories. The existence of a minimum of four types of ω-gliadins with different molecular weights or amino acid components was demonstrated.

The surface hydrophobicities of wheat α-, β-, γ-, and ω-gliadins were investigated by hydrophobic interaction chromatography on octyl- and phenyl-Sepharose CL-4B, by RP-HPLC on octadecylsilane, and by apolar ligand (2-p-toluidinylnaphthalene-6-sulfonate, TNS) binding (Popineau and Pineau, 1987). The results obtained with the three methods were in general agreement. The experiments suggested that the surface hydrophobicities of α-, β-, and γ-gliadins depend on both aromatic and aliphatic amino acid side chains, whereas that of ω-gliadins depends mainly on aromatic side chains. The binding of TNS by gliadin molecules was cooperative and resulted in protein aggregation and masking of hydrophobic sites at high ligand-protein ratios. Gliadins differed in the number of binding sites, with values ranging from 11 (β-gliadins) to 30–40 (ω_{28}); the binding constants also varied, being about 10 times lower for ω-gliadins than for α-, β-, and γ-gliadins.

Fractions containing LMW subunits of glutenin and aggregated gliadins were prepared from four cultivars of winter wheat that varied in breadmaking quality (Tatham et al, 1987). The fractions were analyzed by SDS-PAGE, and their amino acid compositions were determined. Their secondary structures were studied by circular dichroism spectroscopy, and the proportions of α-helix and β-sheet were calculated from the spectra. These values were compared with the secondary structure predicted from the published amino acid sequence of an aggregated gluten protein. The proportions of α-helix varied from 34 to 37% and of β-sheet from 18 to 24%, compared with calculated values of 37% for α-helix and 15% for β-sheet. Heating to 80°C decreased the secondary-structure contents of the reduced LMW subunits of glutenin more than those of the unreduced aggregated gliadins, indicating that the latter were partially stabilized by the intact SS bonds. There were no major differences in the secondary-structure contents or thermal stabilities of the fractions from the four cultivars. The circular dichroism spectra of the aggregated gliadin and LMW subunit fractions were similar to those of the α-, β-, and γ-gliadins rather than to those of the ω-gliadins or HMW subunits of glutenin. This is consistent with their known amino acid sequence relationships.

Analysis of the relationship between the spectrum of the gliadins and baking performance showed that the type of gliadin influences the water imbibition of wheat flour proteins (Bebyakin and Balabolina, 1980; Bebyakin and Goshitskaya, 1985), dough properties (Zeleny value, Pelshenke time, and alveograph parameters); and loaf volume (Sozinov et al, 1974; Peruanskii and Nadirov, 1978; Branlard and Rousset, 1980; Sozinov and Poperelya, 1980; Branlard and Bellot, 1983).

Seventy wheat cultivars of different genetic origins and quality characteristics were examined by Branlard and Dardevet (1985a) with respect to the relationship between the protein content of the grains and the breadmaking quality of the wheat. Gliadins were examined by electrophoresis. For each of the cultivars, dough strength, tenacity, and extensibility were assessed using the Chopin Alveograph, and the samples were also tested by the Zeleny and Pelshenke tests. Correlations were calculated between functionality and the protein contents and/or the patterns of component polypeptide types in the α-, β-, γ-, and ω-gliadin fractions. Most of the values from the functionality tests were correlated significantly with grain protein content (except the Pelshenke test). None of the gliadin fractions were correlated with tenacity. Correlations were calculated between each functionality test and the protein in the 55 bands observed for the 70 cultivars. Five α-, three β-, four γ-, and seven ω-gliadins were correlated significantly with at least one of the functionality tests. Several correlations were the results of the technological tests and certain bands being influenced by the protein content of the wheat. When the effect of protein content was held constant, 10 bands were correlated positively and eight negatively with the technological criteria of quality. Multiple regressions were calculated between the functionality tests and the gliadin bands. Prediction of 37–54% of the variation in tests from one cultivar to another required consideration of only seven to 12 gliadin bands.

γ-Gliadins 40 and 43.5, the presence of which correlates with breadmaking quality, were purified and partially characterized by Dal Belin Peruffo et al (1985). These proteins had the same molecular weight (42,000), as determined by SDS-PAGE, and similar amino acid compositions. Peptide mapping, after enzymic hydrolysis, suggested that these proteins resemble each other closely with respect to their primary structures but are not identical.

Bebyakin et al (1984) conducted a hybridological analysis of the electrophoretic components of gliadins and their significance in defining flour quality.

In soft wheats, the gliadins were inherited as a group, and this affected flour quality. A block of component gliadins identified in the cultivar Albidum 1616 affected the protein and gluten positively, and blocks identified in gliadin of the cultivar Red River 68 resulted in a high flour sedimentation value.

Studies of wheat flour glutenins by a modified Landry-Moureaux method and of gliadins by isoelectric focusing were conducted by Ryadchikov et al (1980, 1981). Wrigley (1980) and Wrigley et al (1981, 1982) described results of studies on the genetic, chemical, and functional significance of varietal differences in the gliadin and glutenin proteins.

Campbell et al (1987) selected 71 wheat lines for a study of statistical correlations between quality attributes and the protein composition of grain. No protein components were related to dough extensibility or development time, but highly significant relationships were obtained between resistance to dough extension and specific gliadins and between this resistance and HMW glutenin subunits. The protein components most consistently implicated were the HMW glutenin subunits 5 and 10 and gliadin 59 (correlated with high resistance) and glutenin subunits 2 and 12 and gliadin 58 (with dough weakness).

In discussing the basis of associations between protein and quality, the authors emphasized that there may be no cause-and-effect relationship between the protein component or components and the observed quality attribute (e.g., resistance of dough to extension). A dough-modifying compound may be the product of another gene, closely linked to the protein gene. The main accent has been on statistical studies, which can be misleading because of unconsidered factors. However, evidence for the value of such bands as markers of quality is starting to be provided by actual breeding experiments. For example, evidence that glutenin subunits 5 and 10 produce "stronger" wheats than glutenin subunits 2 and 12 has come from producing reconstituted hexaploids and from quality evaluation of contrasting biotypes of Australian wheats (Lawrence et al, 1987).

Although attention has focused primarily on extreme effects, the studies also point to components that modify dough properties to a lesser extent. They could be valuable in regular breeding programs where extremes of strength or weakness are not desired.

Since grain quality is also modified by genetic factors other than protein composition, it is precarious to suggest that quality differences can be explained merely in terms of quality-associated protein components. Evidence is accumulating, however, to indicate the existence of major genes that can simplify breeding for grain quality by conventional means. Such evidence also opens new possibilities for genetic engineering. Some gluten components might raise or lower resistance to extension. The effects might vary with the degree of correlation of the specific component to dough properties and with its interaction with a new genetic background.

The study of Cressey et al (1987) provided general validation for the possibility described by Campbell et al (1987) of selecting for high or low resistance to extension in a breeding program according to a few specific proteins.

Milled grain of advanced lines from New Zealand breeding programs was analyzed for dough quality (extensigraph, farinograph, mechanical development), baking quality, grain hardness, and protein composition (gel electrophoresis at pH 3 and in SDS) and by small-scale tests (sedimentation, residue). Highly significant correlations ($P = 0.001$) were found between the resistance of dough to extension and specific gliadins and HMW glutenin subunits (particularly subunits 5 and 10 or subunits 2 and 12). Segregation of the lines according to the presence of these glutenins and of gliadins 58 and 59 successfully identified a group of lines with mean dough resistance to extension.

Glutenins 5 and 10 and glutenins 2 and 12 appeared to be more effective than gliadins 58 and 59 if only one combination of protein components were to be used. Classification based on the presence of all these proteins was most effective for selecting lines with greatest resistance. Presumably these proteins indicate major genes that can be used to select for larger quality differences; other proteins associated with more minor effects could serve for fine tuning of quality in breeding.

According to Cressey et al (1987), segregation for dough resistance can be performed with protein markers or specific probes, such as monoclonal antibodies or complementary DNA.

F. Sulfur, SH Groups, and SS Groups

PROTEIN SULFUR CONTENT

Several studies in Australia and England have indicated the desirability of providing supplemental sulfur to increase yield and improve the quality of wheat (Archer, 1974; Spencer and Freney, 1980; Wrigley et al, 1980, 1984; Randall et al, 1981; Randall and Wrigley, 1986). Grain containing below 0.12% sulfur and having nitrogen-sulfur ratios above 17 was considered to be sulfur-deficient.

Pot experiments (in glasshouse and controlled environment studies) of Byers and Bolton (1979) produced grain with a wide range of nitrogen and sulfur contents; the latter ranged from 0.06 to 0.21%. Grain from sulfur-deficient plants had nitrogen-sulfur ratios above 15 and proteins that were deficient in cystine, methionine, and several essential amino acids and high in asparagine and aspartic acid. Wrigley et al (1980) found that restriction of sulfur in five wheat varieties grown in sand cultures increased the proportion of low-mobility gliadins at the expense of high-mobility gliadins, depleted all essential amino acids, and increased arginine and aspartic acid. In field experiments in Australia conducted by Moss et al (1981), the yield of Olympic wheat responded to both nitrogen and sulfur fertilization. Concentrations of nitrogen in the grain were 1.38–2.56%, and concentrations of sulfur were 0.08–0.18%. Low-sulfur grain was hard, and the dough had a great resistance to extension; sulfur fertilization remedied these deficiencies. Similar responses to sulfur deficiency were confirmed for a hard bread wheat, a soft biscuit wheat, and a multipurpose wheat (Moss et al, 1983).

In the United States, sulfur deficiencies have been reported in parts of Idaho and Washington. Sulfur deficiency has been attributed to a decline in soil organic matter, the use of concentrated pure nitrogen fertilizers, and the use of high nitrogen rates and high-yielding varieties that deplete reserve soil sulfur (Harder and Thiessen, 1971). Rates of up to 25 kg of sulfur per hectare increased wheat yields from 2.65 to 3.36 t/ha. Concentrations of grain nitrogen and of sulfur and nitrogen-sulfur ratios were 1.96, 0.086, and 22.7, respectively, in wheat grown in sulfur-deficient soil and 1.74, 0.108, and 16.2, respectively, in wheat grown in sulfur-fertilized soils. SRW wheat from sulfur-fertilized soils produced high yields of flour with low protein and ash contents; the flour yielded extensible doughs, which produced cookies with a good spread and top grain.

In the study conducted by Timms et al (1981), three samples of field-grown winter wheat (*T. aestivum* cv. Atou) were produced by late application of urea as a nitrogenous fertilizer. The breadmaking quality of the flours increased as the protein content was raised from the lowest to an intermediate level; flours of intermediate protein content were equivalent in breadmaking quality to those with the highest protein content. Loaves were also baked from "flours" reconstituted to equivalent protein levels using the isolated glutens. The flours of low and intermediate protein content yielded glutens of similar baking quality. The gluten derived from the flour with the highest protein content gave a lower loaf volume and a lower texture score. This was due to an effect of the relative levels of nitrogen and sulfur available to the plants. The ratio of sulfur to nitrogen fell as the grain protein content increased, and this correlated with a lower proportion of the sulfur amino acids cystine (cysteine) and methionine. Gel electrophoresis studies revealed an increase in the proportion of the sulfur-deficient ω-gliadin species as the grain protein content increased. Agarose gel filtration chromatography of the flour and gluten proteins suggested a correlation between the aggregation of their glutenin components (mediated by SS bonds involving cystine residues) and their functional properties. The results suggested that late application of high levels of a nitrogenous fertilizer in the absence of sulfur fertilization may lead to a change in the balance between available nitrogen and available sulfur, such that the available sulfur levels become insufficient for "normal" grain development. Considerable alteration in the biochemical characteristics of the flour proteins occurred before gluten baking quality was noticeably affected.

Moss et al (1983) studied alteration in grain, flour, and dough quality in three wheat types with variation in soil sulfur supply. Shortim (a hard bread wheat), Egret (a soft biscuit wheat), and Olympic (a multipurpose wheat) were grown at a field site where yield responses to the application of both sulfur and nitrogen fertilizer had been obtained previously. Grain ranging in sulfur content from 0.10 to 0.21% (w/w, dry basis) was produced. From a grain sulfur content of 0.1%, the grain became softer as the sulfur content increased. The flour sulfur content of the three cultivars showed a close relationship with dough extensibility and an inverse relationship with resistance to stretching. At flour sulfur contents from 0.08 to 0.17%, these relationships remained linear. Flour color improved with increasing sulfur content. Mixing responses were variable and showed no uniform trend either in water absorption or dough development time.

Wrigley et al (1984) examined qualitative and quantitative differences in protein composition in 12 samples of flour milled from wheat (cultivar Olympic) grown under a wide range of sulfur and nitrogen inputs. Changes in gliadin composition associated with sulfur deficiency involved the proportions of individual gliadins but not their charge, size, or isoelectric point characteristics. Sulfur deficiency increased the proportions of polypeptides in the molecular weight ranges of 51,000–80,000 and above but decreased the proportions of lower-molecular-weight polypeptides (mol wt 8,000–28,000, mainly albumins). The proportions of polypeptides in five size ranges were highly correlated with the ratio of sulfur to nitrogen in the flour samples and with dough quality characteristics, particularly as measured with the extensigraph. Doughs produced from the variety Olympic usually are extensible, but the increased proportion of HMW polypeptides in the sulfur-deficient samples was associated with an increase in the toughness and a decrease in the extensibility of the dough.

Gliadin protein patterns were determined by PAGE and HPLC for wheats grown with various fertilizer treatments in sulfur-deficient soils in the United States (Washington state) and in Australia (Lookhart and Pomeranz, 1985). The patterns of samples grown in sulfur-deficient soil to which nitrogen was added exhibited relative intensities of their relative mobility band 73 that were higher than those of the corresponding samples from unfertilized soil. The samples from Australia, which differed in their response to a semiquantitative sulfur deficiency test, also differed in PAGE and HPLC patterns. Wheat grown in sulfur-deficient soil that had been fertilized with sulfur showed more-intense high-mobility PAGE bands and less-intense low-mobility bands than wheat grown in unfertilized sulfur-deficient soil. Similarly, for Australian wheat grown in sulfur-fertilized soil, the peaks eluting in the 20- to 25-min range from the HPLC column were smaller and those eluting in the 43- to 55-min range were larger than those from wheat grown in unfertilized soil. The researchers concluded that alterations in gliadin PAGE and HPLC patterns of wheat resulted from sulfur fertilization of soils severely deficient in sulfur. Some Australian wheat soils are potentially sulfur-deficient; sulfur deficiency, however, is not common in wheat crops. Therefore, wheat must have a minimal requirement for sulfur, which is relatively easily furnished in most soil conditions.

As stated before, evidence for the importance of sulfur-containing amino acids in gluten function has been provided by the observation that sulfur deficiency during grain development alters dough properties (Randall and Wrigley, 1986). Sulfur deficiency increases the synthesis of low-sulfur proteins, particularly the ω-gliadins and the HMW glutenin subunits. Those changes in protein composition result in a loss of dough extensibility and an increase in resistance to extension. It was postulated (Fullington et al, 1987) that sulfur deficiency reduces the amounts of SH and SS groups, which are required for an orderly extensibility of molecules or complexes of molecules within the dough as it is stretched to the point of rupture.

With a SDS-Tris buffer either with or without reducing agent, Fullington et al (1987) identified specific groups of proteins associated with the baking properties of flour. The proteins were extracted from flour from normal and sulfur-deficient wheats. Total proteins (with reducing agent), extract proteins (without reducing agent), and residue proteins were fractionated by SDS-PAGE. The proportion of HMW components with relative molecular weight (M_r) of 80,000 (corresponding to HMW glutenin subunits) in the residue was positively correlated with resistance to extension and negatively correlated with dough extensibility and breakdown. Of the proteins extracted mainly by SDS without reduction, ω-gliadins in the range of M_r 51,000–80,000 showed the same correlations as the HMW glutenin subunits. Proteins in the ranges of M_r 38,000–50,000 and 28,000–39,000 were positively correlated with dough extensibility; these proteins correspond mainly to α-, β-, and γ-gliadins and LMW glutenin subunits. The proteins with M_r <28,000 (albumins and some globulins) appeared mainly in the unreduced extract; they correlated negatively with resistance to extension and positively with dough breakdown and extensibility. Electrophoretic patterns of the residue proteins were similar to those from residues in the SDS and Zeleny tests, suggesting that the sedimentation test residues provide a measure of the proportion of HMW glutenin subunits in the flour. The SDS sedimentation test depends on determining the settled volume of insoluble material after suspension in an SDS solvent similar to the solvent used in the study of Fullington et al (1987). The SDS sedimentation test is used as a rapid test to predict dough properties.

PROTEIN SH GROUPS AND SS GROUPS

The role of protein SH and SS groups in breadmaking is well established (see also Chapter 4 in this volume). It is generally assumed (Sokol and Mecham, 1960; Bloksma, 1963, 1964a, 1964b; Tsen and Bushuk, 1963; Mecham, 1968) that the improver action of oxidizing agents on bread results from their reaction with SH groups. For optimum improver effect, only a fraction of the total SH groups is oxidized. The SS bonds are important because they form cross-links between polypeptide chains and because, by reacting with SH groups, they can interchange and provide mobility for the relatively semisolid structure of the dough (Nielsen et al, 1962).

As part of an investigation of the factors that determine the baking quality of flour, Wostmann (1950) determined by a polarographic method the cystine contents of protein preparations from 23 wheat flours and correlated his results with the physical properties of salt-water doughs, as determined by means of the extensigraph. The cystine contents of the wheat flour proteins varied from 2.2 to 4.5%. The physical properties of salt-water doughs, as reflected by the surface area of the extensigrams for 23 flours taken after a rest time of 135 min, were correlated positively with protein content, the cystine content of the flour proteins, and the total cystine content of the flour. Correlation coefficients between the extensigram areas and those three variables were 0.82, 0.77, and 0.42, respectively. Wostmann concluded that the quality of flour protein for breadmaking increases with an increase in the number of possible SS linkages, as determined by its cystine content. He suggested that the total cystine (that is, the total number of possible SS linkages) largely determines the properties of flour that are measured by the extensigram.

The SH contents, expressed as cystine, of gluten samples prepared from flours representing 13 Dutch varieties varied from 0.28 to 0.61%; the corresponding values for the soluble proteins varied from 0.46 to 2.50% (DeLange and Hintzer, 1955a, 1955b). Estimations of apparent SH content were made polarographically. The cystine values of the gluten and soluble proteins were 1.84–2.78% and 4.01–5.96%, respectively. Baking tests and redox-potential measurements suggested that the ratio of SH to SS groups in flour proteins may be related to the improving action of oxidants. A low but significant correlation ($r = -0.52$) was found between the increase in loaf volume from adding an oxidant and the ratio, $\log(RSSR)/(RSH)^2$.

Miller et al (1950) observed differences in cystine with respect to environment for several varieties of HRW wheat grown during one crop year. Percent cystine differed significantly between samples grown in 1946 and those grown in 1947. The wheat grown in 1947 contained the most cystine and also required longer mixing for optimum dough development. The data suggested a relation between percent cystine and dough-mixing time as influenced by environment. In contrast to short-mixing varieties, the longer-mixing varieties tended to reflect a greater change in mixing requirement for a small change in cystine content.

In a specific histochemical test for SH groups of proteins, the major sites of SH groups were the aleurone layer and the germ (Pomeranz and Shellenberger, 1961). The amount of SH groups in the endosperm increased from the center to the outer layers.

Axford et al (1962) found an inverse relation between the SS value and the protein content of flour from single wheat varieties or a mixed grist. The values (per 1,000 nitrogen atoms) were constant in various air-classified flour fractions, regardless of protein content. Work reporting an inverse relation between the SS content of flour protein and the protein content of flour was followed by measurements of the SS content of the soluble protein in flour (Axford et al, 1964). These varied with the protein content of the flour in a way similar to that of the SS contents of the total proteins, but over a smaller range.

Tsen and Anderson (1963) reported that hard wheat flours contained the most SS bonds, and soft wheat flours the fewest SS bonds and SH groups. On a protein basis, soft wheat flours contained the most SS bonds, and hard wheat flours the fewest. Durum wheat flours contained more SH groups than other wheat flours. Matsuo and McCalla (1964) found the lowest number of gluten SH groups for hard wheat, and the highest for durum.

According to Fabriani et al (1970), the cooking quality of pasta was better when the ratio of reactive SS to total SS content was higher in a cultivar. Fabriani et al (1975) found that the thiol group content of water-soluble durum wheat proteins decreased as a consequence of milling and pasta manufacturing. Alary and Kobrehel (1987) reported that the surface state of cooked pasta is positively correlated with the amount of SH plus SS groups in glutenins.

Durum wheat glutenin fractions with SH plus SS groups content of 185 and 140 μmol/g of protein for cultivars Mondur and Kidur, respectively, were extracted from gluten with Na-tetradecanoate (2.5 mg/g of gluten) after extraction of gliadins (Kobrehel et al, 1988). The fractions were high in cysteine. SDS-PAGE, without reduction, showed these proteins to consist of two major low-molecular-weight proteins, DSG-1 and DSG-2 (durum-wheat sulfur-rich glutenin fractions) with molecular mass of 14.1 and 17.7 kDa, respectively. After reduction, SDS-PAGE showed DSG-1 to be composed of two subunits of lower mobility than DSG-1, having apparent molecular masses of 14.6 and 14.7 kDa. The SDS-PAGE mobility of DSG-2 increased after reduction to approximately 15.8 kDa.

The DSG fractions appear to represent a unique protein type. Their high cysteine content and possibly strong interaction with lipids may contribute to their functional properties in cooking quality of pasta.

Doughs made from Japanese domestic wheat generally have a much lower ratio between resistance to stretching and extensibility than doughs from Canadian red spring or U.S. western white wheats. Much of this trend had been explained by the difference in bound SH compounds in flour (Yasunaga and Uemura, 1964). Flours from Japanese wheat contained more SH compounds soluble in a sodium chloride solution and also were higher in proteolytic activity. The decrease of SH compounds during storage was much less rapid in Japanese than in Canadian or U.S. wheats, which contain fewer SH compounds. The significance of accessible SH groups in dough was stressed by Bushuk (1961). Some aspects of the SH-SS system in flour and bread were studied by Matsumoto and Hlynka (1959).

Rohrlich and Essner (1965) determined the SH and SS contents and the S quotient (SS/SH ratio) in six flours, shorts, and bran from low-protein German wheat and a high-protein Canadian wheat. The sum of SH and SS groups was consistently higher in German than in Canadian wheat. Consequently, the S quotient was substantially higher in Canadian than in German wheat. Protein SH increased with an increase in the ash contents of German wheat, but not of Canadian wheat. Significant positive correlations (respectively, 0.65 and 0.70) were established between the S quotient and loaf volume for flours milled from 50 German summer and winter wheats.

According to Mecham (1968), the SH content of patent and straight-grade flours is near 1 μeq per gram, with values over a range of 0.6–1.8 μeq per gram of flour, or 4–19 μeq per gram of protein. The SH content of freshly ground wheat decreases by 20–30% within 3 hr in air but is maintained in wheat ground and stored under nitrogen. The SS content ranges from 7.4 to 16.9 μeq per gram of flour, or from 83 to 130 μeq per gram of protein. The SS content per unit protein is inversely related to flour protein content. In addition, the SS content per unit protein decreases as flour extraction decreases. The SH content of durum samples is consistently higher on a protein basis than that of common wheats (Tsen and Anderson, 1963). The concentration of both SH and SS is higher in water or saline extracts of flour than in gluten or the water- or saline-extracted residues. When doughs are mixed in air, the SH content decreases. The rate of decrease differs among flours; the course of the changes may be correlated with the stability of doughs to mixing (Sokol et al, 1960). The oxidation with air or oxidizing agents (bromate and iodate) is much faster during mixing than in doughs at rest. Loss of SH groups in air appears to proceed by direct oxidation by occluded atmospheric oxygen and also by oxidation through a lipoxygenase fatty acid-fatty acid peroxide system, which competes for oxygen and for SH groups.

Bloksma (1972a, 1972b) found no unequivocal relation between the total SH content and various rheological properties of doughs. Similarly, no correlations were found between the total, reactive, or nonreactive SS bonds and the elastic properties of doughs.

The stiffening of dough is apparently due, not primarily to the formation of additional SS bonds, but rather to the removal of thiol groups (Bloksma, 1975). Experiments with doughs that differed in their thiol and SS contents demonstrated no unequivocal relation either between the SS content and the elastic deformation or between the thiol content and the viscous deformation. Only small fractions of the analytically determined thiol and SS groups are rheologically effective. The rheologically effective thiol groups are probably located in small peptides (see also Chapter 4 in this volume).

According to Gorshkova et al (1979), the number of SS groups in a protein varies linearly with the elasticity of the protein. Similarly, according to Kaczkowski and Mieleszko (1980) and Mieleszko and Kaczkowski (1980), the quality of wheat flour is proportional to the amount of easily reduced SS bonds in glutenin.

Seeger and Belitz (1981) observed that a reduction of only 3–4% of the total SS bonds in succinylated gluten led to an 80% decrease in the content of HMW proteins. The ease of reduction of SS bonds of gluten varies widely.

LMW COMPOUNDS

Reduced glutathione was isolated from wheat germ by Sullivan (1940); its presence in flour was postulated by Sullivan et al as early as 1936. The dough-improving effect of ascorbic acid and the SH-SS interchange reaction may involve glutathione in flour. Kuninori and Matsumoto (1964b) separated glutathione from a flour extract by means of Sephadex G-50 and characterized it by several methods. The estimated apparent reduced glutathione in flour was 0.90 and 0.06 mg% of flour 1 day after milling and 14 days after milling, respectively. Reduced plus oxidized glutathione totaled 8.0–9.6 mg% in two flours.

The specificity of the alloxan method used by Kuninori and Matsumoto (1964b) has not been established completely. The method, using ethylenediamine tetraacetic acid and urea, indicated 0.7–0.99 μeq of LMW SH and SS compounds per gram of flour (Proskuryakov and Zueva, 1964). A specific enzymatic method utilizing glutathione reductase (Kuninori et al, 1968) showed that the glutathione contents in the middlings of HRW and HRS wheats were 2.4 and 1.4 μeq, respectively, per 100 g of flour. These levels include both oxidized and reduced glutathione and are extremely small compared with the results from other investigators. The involvement of glutathione in SS-SH interchanges in wheat flour was reported by Kuninori and Sullivan (1968).

Thioctic acid in wheat was reported by Dahle and Sullivan (1960). Its presence in wheat flour was confirmed by Morrison and Coussin (1962). Flour contained thioctic acid at approximately 1–10 ppm; wheat germ contained about 200–300 ppm of the coenzyme (Sullivan et al, 1961).

Reduced thioctic acid weakened the mixing character of doughs—an effect not obtained with cyclic thioctic acid (Dahle and Hinz, 1966). Adding cyclic thioctic acid to yeasted systems, however, reduced thioctic acid through the action of the thioctic-reducing enzyme system known to be present in yeast. Baking data revealed flour to be sensitive to exogenous thioctic acid levels as low as 0.025 μmol/g of flour. Incubation of flour slurries containing reduced thioctic acid resulted in increased amounts of water-soluble and acid-soluble proteins. The very low levels of endogenous thioctic acid in the strong flour used suggested a possible connection between endogenous thioctic levels and flour strength characteristics. However, as the endogenous thioctic acid is present in a bound form, it is not easily accessible to yeast enzymes and has little effect on rheological properties (Dahle and Pinke, 1968).

Mecham et al (1966) reported that hydrogen sulfide is given off by wheat flour doughs as they are mixed at 30° C under nitrogen. These workers suggested that hydrogen sulfide in dough may be related to some features of dough behavior important in baking technology (such as SH-SS interchange among proteins in doughs), to modification of HMW proteins, and to reaction with oxidants in maturing of doughs.

Adding SH-blocking agents prevented or considerably lowered hydrogen sulfide formation, indicating that SH groups were a primary source of hydrogen sulfide in dough mixing. Similarly, such oxidants as potassium bromate or iodate diminished the amounts of hydrogen sulfide released. Doughs from 50 g of flour were mixed under nitrogen in a farinograph at 189 rpm (Mecham and Bean, 1968). In 30 min at 30°C, hydrogen sulfide was released in amounts of 10–65 μg when 16 flours of widely varied source and baking properties were compared. No consistent correlation between dough or baking properties and hydrogen sulfide release was established.

Jones and Carnegie reported (1969a) that wheat flour contains a family of SS peptides. Glutathione is only a minor member of that family; the rest are basic and have a molecular weight of about 2,000. The basic peptides could not be fractionated into individual components, but they contained some SS groups that were very reactive toward glutathione at pH 5.8, which is approximately the pH of dough. The SS peptides were further separated into acidic glutathione and basic peptides (Jones and Carnegie, 1969b). The only peptide normally present in flour and having a marked activity in dough was glutathione. The other SS peptides were only slightly active, and their sole effect was to decrease the tolerance of dough to prolonged mixing. When the peptides were in the SH form, they were slightly more reactive and decreased dough development time and maximum resistance.

Tkachuk (1969) presented some indirect evidence of the involvement of LMW SH peptides in oxidative dough improvement. The amount of such peptides (including cysteinylglycine and glutathione) is about 0.1 μmol per gram of protein.

OXIDATION REQUIREMENTS AND MATURING

Maturing of flour is one of the most challenging and intricate problems of cereal chemistry. The action of the optimum amount of a maturing agent on flour produces a dough with better machining properties and an improved baked product. It is generally believed that SH groups are involved in improving flour by aging and by maturing agents. But despite much work on the subject, the entire mechanism remains obscure.

Storage of flour in air causes slow oxidation of thiol groups. Mixing dough in air or oxygen oxidizes thiol groups more rapidly. The disappearance of thiol groups after the addition of bromate or iodate has also been demonstrated. Iodate causes thiol groups to disappear more rapidly than does an equivalent amount of bromate, which confirms the faster reaction rate of iodate.

Tsen (1968) studied the oxidation of SH groups of flour by bromate under various conditions, including the baking stage. The effects of bromate during fermentation were small; the major bromate effect occurred at elevated temperatures—mainly during the early stage of baking.

Reagents that reduce the SS group to SH groups increase the rate of bromate reaction; the reagents include bisulfite, thioglycolic acid, cysteine, glutathione, thiolated gelatin, and sodium borohydride (Anderson, 1961). However, improvement by the formation of SS cross-links between protein molecules from oxidation of two thiol groups is not a complete explanation.

Jørgensen (1936) hypothesized that oxidants inhibit proteinases, which attack gluten proteins, and that, as a result of inhibitory action on proteolysis, the gas-retaining capacity of the dough proteins and the baking strength both increase. Shen and Geddes (1942) summarized evidence for and against the mechanism of protease inhibition. The very rapid effect of glutathione, wheat germ, and other reducing substances on dough properties and the reduction of their harmful effects on the extension of fermentation time, coupled with the "excess bromate effect" and its amelioration by purely physical means (remixing after fermentation), strongly indicate that proteinase activity is not of prime importance to the baking effects of oxidizing agents. Olcott et al (1943) showed that reducing agents produce their characteristic effects in the complete absence of proteinase. They therefore concluded that the primary effect of reducing agents on gluten was a chemical one on the proteins and that its effect was only secondarily that of enzyme activation. Baker et al (1946) reached similar conclusions in studies of supercentrifugates from dough. Studies by Howe (1946) strongly suggested that the proteolytic theory of oxidants in dough was not valid and that some other mechanism was responsible. Her studies were supported by the fact that flour proteases are not inhibited by bromate concentrations used in commercial breadmaking. Similarly, Freilich and Frey (1947) found that the protease content of patent and baker's-grade flours is so small that the protease theory must assume a minor role in explaining oxidation effects.

Hites et al (1953), studying the kinetics of the action of native proteases of flour, found that reducing agents did not activate these proteases; nor did oxidizing agents inactivate them.

A concept that has received a great deal of consideration is that of SH-SS interchange. Low levels and availability of SH and SS groups make it unlikely that the formation of new SS bonds is the only cause of changes that result from adding oxidizing agents. It has been postulated that viscous deformation of dough involves the breaking and reformation of SS bonds and that those reactions proceed as a result of thiol-SS interchange.

Goldstein (1957) showed that small amounts of ascorbic acid and potassium bromate improved dough properties considerably and that when all available SH groups were blocked with *p*-chloromercuribenzoate, the addition of ascorbic acid and potassium bromate had no effect. Goldstein pointed out that because of the very small amounts of free SH groups contained in gluten, it is unlikely that two SH groups would be in close enough proximity to be oxidized to an SS bond. He postulated that improvement could be more adequately described by the action of maturing agents hindering exchange between SH and SS groups.

A group of Australian workers suggested that qualitative differences between wheats of the same protein content are related to the number of SS bonds and the rate at which they interchange with free SH groups—according to the scheme proposed by Goldstein. Frater et al (1960, 1961) designed experiments to test the hypothesis that relates the strength and elasticity of dough to thiol groups, intermolecular SS bonds, and the rate at which they interchange. From experiments with an Australian flour, they concluded that at any given protein content of flour, the rheological properties of dough appear related to the number of intermolecular SS bonds and the rate at which they can interchange with thiol groups. Quality differences in wheats seem to be related to either or both of these properties.

They suggested that the action of improvers can be related to the strengthening of dough through the inhibition of SS exchange reactions. With oxidizing improvers, there is the additional possibility that the dough increases in strength through the formation of new intermolecular SS bonds. The action of reducing agents in destroying the structure of dough can be interpreted as lowering the number of intermolecular SS bonds and/or increasing their rate of exchange. Strains introduced in dough by the usual procedure of balling and shaping were shown to relax much more slowly in the presence of iodate and N-ethylmaleimide. In the presence of cysteine, the strains relax more quickly. It was concluded that the relaxation of strains occurs through SS exchange reactions and that the formation of a strained system during mixing is also accompanied by the same type of exchange reaction.

Some experimental evidence indicating an exchange of SS and SH groups was presented by McDermott and Pace (1961) and by Villegas et al (1963), who demonstrated that thiolated gelatin becomes bound to gluten proteins, and by Mauritzen and Stewart (1963), who examined the system by using fractions containing labeled cysteine and separated by ultracentrifugation of dough.

Formamidine disulfide was found by Sullivan and Dahle (1966) to mature flour—probably by blocking its SH groups. The compound also was used for measuring the SS-SH interchange in flour extracts and doughs. These studies indicated that the interchange may not, in fact, take place to any significant degree in the normal pH range of doughs. Evidence of interchange in glutathione was obtained only in an alkaline medium. Subsequent studies with labeled glutathione (Lee and Lai, 1968) pointed to interchange reactions. Some experimental evidence for SS-SH interchange was provided by Redman and Ewart (1967), who studied the interaction of LMW thiols and SS groups with dough proteins. The incorporation of labeled cystine, cysteine, and N-ethyl-maleimide into wheat flour doughs was studied by Mauritzen (1967). Lee and Lai (1968) reported interchange reactions between inactive SH compounds and flour into which ^{35}S had been incorporated. The results suggested that LMW thiols such as reduced glutathione can easily come into reactive contact with the SS sites on a protein. Protein thiols, possibly because of their larger steric requirements, would approach and interchange with protein SS groups with much greater difficulty. The time-dependent interaction of oxidizing and reducing agents in breadmaking was reported by Moss et al (1979).

McDermott et al (1969) studied SS interchange between gliadin and glutenin and three LMW SS groups. They examined mixed SS reaction products by gel filtration. The results indicated SS interchange that under suitable conditions approached theoretical values for complete reaction. The reaction products exhibited modified molecular weight distributions, which were interpreted in terms of inter- and intrachain SS bondings of the original proteins.

The mechanism of oxidant action is still a challenging problem for speculation and experiment. Proponents of the proteolytic enzyme theory are fewer than previously, although many are reluctant to discard it entirely. Majority opinion seems to assign it a minor role in explaining improver effects.

At the same time, some oppose the theory that oxidizing agents used by the baking industry improve rheological properties entirely through modifications involving thiol groups and SS bonds of wheat flour proteins. Kovats and Lasztity (1965) summarized the limitations of the theory regarding the oxidation of SH groups to SS groups by commonly employed oxidants. The amount of oxidants used is extremely small. The formation of a SS linkage would require close proximity of two thiol groups in a three-dimensional complex network.

Only small amounts of thiol groups disappear, without consistent and significant formation of SS groups. In many cases, changes in rheological properties as a result of added oxidants are slow, suggesting a mechanism different from simple chemical oxidation.

It is difficult to answer some of the questions raised by Kovats and Lasztity. Present analytical methods are not sensitive enough to detect the small increase in SS groups that would result from oxidation of all available SH groups. Under normal conditions, only a small portion of flour SH groups would be oxidized for optimum improvement. Generally, changes in rheological properties follow the rate of oxidant reaction and seem to be part of the improver effect.

OXIDANT RESPONSE

Early work suggested that the relation between the protein content and the loaf volume of bread was curvilinear, or that 1% of protein at the 11% level had more influence on loaf characteristics than 1% at the 16% level. Many early studies on the relation of protein content to loaf volume of bread were made with lean formulas, fixed mixing times, and other techniques that did not provide conditions to express the potential strength of flour. Extensive studies showed that when rich formulas, optimum oxidation levels, and proper breadmaking procedures were used, the expression of protein strength was not limited, and the relation between protein and loaf volume was linear within the limits of protein encountered.

Thus, unless potassium bromate was included in the formula, no significant correlation could be established between loaf volume and protein content (Larmour, 1930). Similarly, Aitken and Geddes (1939) found that loaf volume bore a linear relation to flour protein, but the range in loaf volume was much smaller when diastatic malt and bromate were omitted from the baking formula. They concluded that when protein content is the only variable, it is a reliable index of flour strength if an oxidizing agent is used to assist gluten development and if gas production during proofing is adequate. Small amounts of oxidizing agents, e.g., potassium bromate, improve handling properties, loaf volume, and crumb grain.

Finney and Barmore (1945a, 1945b) advocated using a rich, highly bromated formula to increase the correlations of loaf volume with protein quality and quantity. In loaves from such a formula, differences in grain and texture tended to disappear, whereas loaf volume differences increased. Their formula specified 6% sugar, 1.5% salt, 3% shortening, 2% yeast, 4% nonfat dry milk, and 0.25–0.50% malt syrup. Optimum potassium bromate levels (or mixtures containing ascorbic acid at 50–100 ppm and potassium bromate at 10–20 ppm), absorptions, and mixing times were used (Finney, 1984). Recent studies showed that the regression lines are slightly curvilinear if the protein content is below 12% (Finney, 1985).

The question of whether large amounts of ascorbic acid are detrimental has no clear-cut answer. Limited observations of the author indicate that such a detrimental effect (of ascorbic acid at about 50 or 100 ppm) is more likely to take place in a low-protein, lean formula than in a high-protein, rich formula.

G. Dough Mixing

In breadmaking by conventional methods, the mixing of a dough is generally considered a critical step that influences overall bread quality. Less protein is extracted with dilute acetic acid from a strong flour than from a dough (Mecham et al, 1962). The insoluble material in the flour forms a gelatinous mass, the volume of which decreases with increases in mixing time (Mecham et al, 1963). The rate of solubilization of flour proteins during dough formation varies with flours of differing mixing characteristics. Both the rate and the extent of solubilization increase if SH-blocking reagents are added to doughs. These observations led to the suggestion that protein agglomerates or aggregates are broken down during dough mixing.

Tsen (1967) extracted proteins from flours and doughs with $0.05 N$ acetic acid. Increasing the mixing time and mixing speed increased the amount of extracted protein. The extracted proteins were fractionated by gel filtration. A fraction containing the highest-molecular-weight proteins increased during dough development. The change was also observed for wet gluten mixed for various periods. When the changes were examined in relation to the mixing properties of various flours, the HMW component was found to increase faster in weak flours than in strong flours. Extracts of flours from soft white winter wheat contained proportionally more of the HMW component than extracts of HRS wheat flours. The results supported findings of Smith and Mullen (1965), that α-glutenins of short- and long-mixing flours differ in their molecular size, and of Pomeranz (1965), that poor- and good-breadmaking flours differ in their protein dispersibilities in $3 M$ urea. When equiprotein extracts were fractionated, the relative concentrations of LMW proteins decreased (Tsen, 1967). Proteins dispersible in $3 M$ urea increased during the mixing of doughs in a farinograph (Mamaril and Pomeranz, 1966). The increase was positively correlated with mixing time. Equiprotein amounts of the extracts were fractionated on Sephadex G-100 columns. Mixing in air, but not in a nitrogen atmosphere, substantially increased HMW proteins and decreased LMW proteins. The effects of mixing bromated doughs were similar to those of mixing plain doughs in air, but the changes were more pronounced in the bromated doughs. No such changes were observed in iodate-treated doughs.

Meredith and Wren (1969) used gel filtration to fractionate proteins extracted with a dissociating aqueous solution containing $0.1 M$ acetic acid, $3 M$ urea, and $0.01 M$ cetyltrimethylammonium bromide (AUC). The results indicated no significant intermolecular SS reactions between glutenins and other proteins of doughs mixed in air or with potassium iodate. The authors speculated that the solubilization of proteins in doughs resulted from the action of endogenous enzymes. They emphasized that their findings were incompatible with the hypothesis of several investigators (Beckwith et al, 1965; Elton and Ewart, 1967) concerning modifications in wheat proteins during the mixing of dough. It may be noted that electrophoresis patterns of proteins in doughs (Pomeranz et al, 1968a) and glutens (Hoseney, 1968) were shown to differ from patterns of flour proteins. The results indicated intermolecular reactions that were not reversed under conditions relatively free from association effects.

Changes in the solubility of flour proteins during mixing were studied with three flours of widely different mixing properties (Tanaka and Bushuk, 1973a). Extended mixing in the presence of relatively high concentrations of N-ethyl-maleimide and potassium iodate accentuated mixing breakdown. Extraction of the proteins from freeze-dried doughs with $0.05 N$ acetic acid indicated that under mixing conditions where doughs showed a marked decrease in consistency, the amount of protein soluble in this solvent increased markedly. This was attributed to the depolymerization of the HMW glutenin. Results of Osborne-type solubility fractionation were consistent with the depolymerization hypothesis. Concomitant with the decrease in the amount of the insoluble protein, increases were observed in the amount of the alcohol-soluble and the acetic acid-soluble protein fractions.

Gel filtration of AUC extracts of doughs on Sephadex G-150 showed that during accentuated dough breakdown the amount of HMW glutenin decreased (Tanaka and Bushuk, 1973b). The amount of LMW glutenin and gliadin increased. There were only minor changes in the SDS-PAGE patterns of the protein fractions obtained by Osborne-type fractionation. The alcohol-soluble fraction of doughs that suffered extensive mixing breakdown had several new bands in the HMW region. Components of similar molecular weight were observed in the patterns of the reduced acetic acid-soluble proteins and the reduced residue proteins. The SDS-PAGE patterns of reduced acetic acid-soluble and residue proteins were essentially identical and were not affected by mixing.

The number of SH groups decreased during mixing in all doughs except in the controls mixed in nitrogen (Tanaka and Bushuk, 1973c). The rate of loss was slightly different for three varieties of flour. There was no change in the number of SS groups with mixing. The proteolytic activity and the content of amino groups of the three flours increased with decreasing strength. A possible mechanism for the breakdown of doughs by excessive mixing (with or without SH reagents) was presented. Whereas previous workers emphasized the disaggregation mechanism of dough breakdown during mixing, Tanaka and Bushuk (1973c) placed greater emphasis on the depolymerization mechanism according to the following equation:

$$P_1-SH + P_2-SS-P_3 \rightarrow P_1-SS-P_3 + P_2-SH$$

If P_1-SH is considerably lower in molecular weight than P_2-SH, then the contribution of the two components on the left of the equation to dough consistency would be much greater than that of the two components on the right. The SS interchange would be facilitated by the presence of LMW SH compounds in the original flour or added to the dough.

H. Proteins in Cookies and Cakes

Water-retention properties of wheat flours as they relate to cookie making were reported by Sollars (1973a, 1973b) and Sollars and Rubenthaler (1975).

Yamazaki et al (1977) separated three pure-variety straight-grade untreated flours into five fractions: free lipids, starch, gluten, tailings, and water-solubles. For each variety, blends with restored lipids were baked into cookies. The functions of the fractions were similar for the cultivars Thorne and Blackhawk and different for Shawnee. Within the "valid area" of compositional variation, and at high starch levels, high water-soluble content was associated with small cookie diameter for Shawnee but with large diameter for Thorne and Blackhawk. High tailings content was associated with poorer cookies in

Shawnee blends. The starch fraction did not show varietal differences. For all three varieties, the fraction effect on diameter was associated with the alkaline-water-retention capacities of the gluten, starch, and tailings fractions, and the effect was additive. In a study of the effect of fat and sugar on sugar-snap cookie spread, the protein content appeared to have a much smaller effect than unidentified genetic factors (Abboud et al, 1985a).

A soft wheat flour milled for high-ratio layer cakes usually consists of a blend of mill streams with low ash and low protein contents. The flour is fine in average particle size and before being baked is further reduced in average particle size and treated with chlorine gas. The reaction between chlorine and flour is surface-dependent and is effective with flours of small average particle size. Yamazaki and Donelson (1972) found a high negative correlation for white layer cake volume versus mass-median diameter of laboratory-milled cake flours obtained from pure-variety wheats. Cake volume was associated inversely with the mass-median diameters of straight-grade and coarsely milled flours and directly with the quantity of sifted meal from wheats milled to obtain patent flours for cake baking. Varietal differences in cake potential for these wheats appeared to be associated largely with inherent differences in endosperm friability.

Gaines and Donelson (1982) evaluated cakes from two flours for stickiness and cake volume by interchanging fractions (i.e., prime starch, tailings, gluten, water solubles, and hexane-extracted lipids) from chlorinated and unchlorinated flours. The stickiness of cakes made from chlorinated flour resulted primarily from chlorine alteration of the prime starch fraction. Cake volume improvement was caused primarily by chlorine alteration of flour lipids.

The effect of artificially varying cake flour protein content on the size and tenderness of white layer cake and angel food cake was evaluated by Gaines and Donelson (1985) by varying the cake flour protein content from approximately 7 to 16% by air-classification techniques and by adding gluten. The volume and tenderness of white layer cakes were not significantly affected by flour protein content. The height and tenderness of angel food cake decreased as flour protein content increased. However, relatively large (approximately 2%) increases in protein content were required to effect the change.

In a comprehensive study reported by Gaines (1985), 83 soft red and white wheat test lines or cultivars were evaluated for kernel hardness, kernel and flour protein, flour ash, straight-grade and cake patent flour particle size, and cake patent flour chlorine response. Each characteristic was evaluated for associations with sugar-snap cookie diameter and high-ratio white layer cake volume. An additional 136 soft wheats were included to evaluate associations among cookie diameter, flour particle size, and protein content. Across cultivars, both cookie diameter and cake volume were positively associated with soft textured wheats having lower protein contents, which produced more break flour and flour having smaller particle size. Wheats producing straight-grade flour with small particle size also produced patent flours with small particles (both before and after pin milling). Wheats having better milling quality were more coarsely granulating during milling and produced smaller cakes and cookies. Wheats producing more break flour were finer-granulating. Kernel hardness (particle size index) was not correlated with milling quality.

Time-lapse photographs were taken to observe changes that occur in cookie diameter during baking (Abboud et al, 1985b). Cookie diameter is a function of the rate of spreading and the setting point of the cookie dough. The rate of spreading was greater and the expansion time longer for good-quality (soft wheat) cookie doughs compared to poor-quality (hard wheat) cookie doughs. Compression tests showed that both doughs underwent a decrease in viscosity as temperature was raised. When compression was measured on doughs heated to a relatively low temperature (40°C), there was a large difference in force between doughs made from hard wheat flours and those made from soft wheat flours, with soft wheat dough exhibiting less resistance to compression.

Baking tests for hard sweet and short sweet cookies were used to examine samples from wheats grown in 1980–1983 in England (Wainright et al, 1985). According to the authors, such tests are of limited commercial value for cookie flour specification. Flours from soft-milling wheats required less water to give cookie doughs of standard consistency than did flours from hard-milling wheats. Doughs made with soft-milling wheat flours gave greater oven spring, i.e., cookies of lower bulk density, than doughs made from hard-milling wheat flours. The difference in the mean values of both parameters for flours milled from soft-milling and hard-milling wheats was highly significant. Within both soft-milling and hard-milling varieties, significant differences were observed between seasons for the mean values of both measured cookie parameters. The effect of flour particle size on the cookies was studied by regrinding a number of flours. With flours from both soft-milling and hard-milling wheats, reduction in particle size resulted in hard, sweet cookies of higher density and short, sweet cookies of lower density.

I. Proteins in Pasta Products

Sheu et al (1967) studied the effect of interchanging various components of HRS and durum wheats on macaroni quality. Differences in cooking quality in terms of cooked weight, residue, and firmness were attributed primarily to the gluten fraction. Macaroni color was affected by gluten and water-soluble fractions.

The effect of different types of gluten on spaghetti cooking quality was studied by Matsuo and Irvine (1970). Gluten characteristics were measured using the farinograph, alveograph, and gluten stretching test. Gluten of medium strength as assessed by the physical tests appeared to produce spaghetti of optimum cooking quality. Reconstitution studies indicated that gluten quality is the major factor determining cooking quality.

Walsh and Gilles (1971) conducted research to establish the influence of protein composition on quality by investigating the proteins of eight wheat varieties representing a wide range of spaghetti properties. Proteins from each variety were separated by extraction and Sephadex G-200 gel filtration chromatography into albumins, globulins, gliadins, glutenins, and base-solubles. Protein composition was related to several spaghetti quality factors. Poor spaghetti color was shown for varieties that had high albumin and glutenin contents. High spaghetti firmness was associated with high glutenin and low gliadin contents.

For a number of durum wheat varieties of varying qualities, protein quantity had a marked effect on the cooking quality of spaghetti (Matsuo et al, 1972;

Dexter and Matsuo, 1977c). In all cases, the cooking quality improved with higher protein content. Proteins from rapeseed meal, fish protein concentrate, soy flour, and egg albumin, as well as protein components from durum wheat, were added to semolina and processed into spaghetti. The most pronounced improvement in cooking quality was found with egg albumin and with glutenin. Improvement was also found with gliadin, whereas proteins from other sources had little effect.

The influence of protein content on quality characteristics of durum wheat was investigated in two Canadian durum wheats of differing spaghetti-making quality (Dexter and Matsuo, 1977a). A moderate increase in protein content was accompanied by a marked decrease in farinogram mixing time, concomitant with an increase in maximum consistency and tolerance index. Cooking quality and tolerance to overcooking improved as protein increased. The proportion of nongluten protein (albumins and globulins) decreased significantly with increasing protein content. Gluten characteristics, as measured by the Berliner turbidity test, improved as protein increased. However, this improvement could not be related to the Osborne solubility distribution of the gluten proteins.

Three durum wheats of widely differing spaghetti cooking quality and a HRS wheat were processed into spaghetti in a laboratory extruder (Dexter and Matsuo, 1977b). There was a substantial loss of SH groups by the extrusion stage; the greatest loss occurred during the first 6 hr of the 29-hr drying cycle. No relation could be established between cooking quality and the loss of SH groups. SS bond levels did not change significantly during processing. Protein extractability in dilute acetic acid had decreased substantially by the time the dough entered the extruding auger. No further change in solubility occurred during the remainder of processing. An increase in insoluble residue protein was concomitant with a decrease in salt-soluble protein. Gel filtration elution profiles on Sephadex G-150 in AUC showed no significant changes in the molecular weight distribution of the proteins during spaghetti processing. This, in conjunction with electrophoretic results, which revealed no qualitative changes in protein electrophoretic patterns during spaghetti processing, suggested that the loss of salt-soluble proteins may be due to binding by the insoluble components of the semolina rather than by polymerization.

Subsequently, gluten proteins from a durum wheat and a HRS wheat were fractionated by precipitation from $0.005N$ lactic acid at various pH levels (Dexter and Matsuo, 1978). As pH increased, the proportion of glutenins within each fraction decreased concomitantly with an increase in gliadins. Reconstitution experiments demonstrated that a decrease in farinograph mixing time at pasta-processing absorptions accompanied this decrease in glutenins and increase in gliadins. The addition of gluten fractions, with the exception of the fraction insoluble in lactic acid, had no detrimental effect on spaghetti color. The fraction insoluble in lactic acid and the most gliadin-rich fraction improved spaghetti cooking quality somewhat. Differences in cooking quality were consistent with differences in the distribution of gluten proteins. Qualitative differences in gluten proteins, however, apparently outweighed quantitative differences in spaghetti-cooking quality.

Three durum wheats and a HRS wheat of differing cooking quality were processed into spaghetti in a laboratory extruder and cooked for 3–28 min (Dexter and Matsuo, 1979b). Protein extractability in dilute acetic acid rapidly decreased during cooking up to about 12 min. The poorer-quality wheats had significantly greater protein extractability than the two better-quality wheats. Osborne protein solubility fractionation showed this to be mainly due to a greater proportion of extractable gluten protein for the poor-quality wheats. Gel filtration elution profiles of acetic acid extracts revealed no significant quantitative differences in their pattern of protein denaturation. The relation between the protein extractability of cooked spaghetti and the spaghetti-cooking quality was confirmed for 18 durum wheat lines of differing spaghetti-cooking quality.

Thirty durum wheat lines were evaluated for farinograph properties of pasta dough, protein characteristics, and spaghetti-cooking quality (Dexter and Matsuo, 1980a). Farinograph properties were strongly influenced by gluten strength. Farinograph band width was a better indicator of spaghetti-cooking quality than mixing time and tolerance index. Insoluble residue was the protein fraction most responsible for variations in gluten strength, farinograph properties, and spaghetti-cooking quality. The intrinsic viscosity of whole gluten dissolved in a dissociating medium was significantly correlated with the proportion of residue protein, gluten strength, farinograph properties, and cooking quality. Intrinsic viscosity measurements and gel filtration profiles of lactic acid-soluble gluten proteins in a dissociating medium suggested that differences in the soluble gluten proteins may also be a quality factor for durum wheat gluten.

Conversion of flour into spaghetti was accompanied by a decrease in SH groups, proteins extractable in $0.05N$ acetic acid, and salt-soluble proteins (Dexter and Matsuo, 1980b). The SS group did not change. Gel chromatography and electrophoresis patterns on polyacrylamide gel showed little changes during processing. Differences in surface hydrophobicity of durum gliadins may be responsible for their cooking quality (Godon and Popineau, 1981). Hydrophobic properties of gliadins, in general, were studied by Popineau and Godon (1982) and Popineau et al (1980).

Dexter et al (1983) processed spaghetti in a semicommercial-scale laboratory press from a range of raw materials, dried by a low-temperature and a high-temperature drying cycle. Cooked spaghetti processed with the high-temperature drying cycle was generally less sticky, more resilient, and firmer and exhibited lower cooking loss than spaghetti processed with the low-temperature drying cycle. As the hardness of the cooking water increased, the spaghetti became stickier and cooking loss increased. Stickiness was influenced by cultivar, wheat class, the granulation of raw materials, and protein content, but it was not related to sprout damage. Stickiness was significantly correlated with cooking loss, cooked weight, degree of swelling, compressibility, recovery, and firmness. However, even when all these factors were included in a step-up regression, less than 50% of the variance in stickiness could be predicted.

Gluten strength, mixing properties, baking quality, and spaghetti quality of some Canadian durum and common wheats were compared by Dexter et al (1981).

The relation between glutenin and spaghetti-making quality was investigated by Wasik and Bushuk (1975a), using 14 durum wheat varieties of different quality. Gel filtration fractionation of the AUC-soluble endosperm proteins revealed that varieties with high ratios of glutenin to gliadin were generally superior in cooking quality to those with low ratios. Significant correlations were obtained between glutenin peak area and farinograph mixing tolerance index (−0.661), gluten strength (0.845), and tenderness index (−0.681). The ratio of glutenin to gliadin correlated linearly with the farinograph mixing tolerance index (−0.666). These results suggested differences in protein composition among durum wheat varieties of different quality.

SDS-PAGE of reduced durum wheat glutenins, purified by fractional salt precipitation of the AUC-soluble flour proteins, revealed varietal differences in the molecular weight distribution and relative concentration of the six largest subunits (Wasik and Bushuk, 1975b). The differences appeared to be related to the spaghetti-making quality of the variety from which the glutenin had been isolated. The glutenin of varieties with excellent spaghetti-making quality had more of subunit 6 (mol wt 53,000) than subunit 5 (mol wt 60,000). Varieties with low or mediocre spaghetti-making quality appeared to have less of subunit 6 than subunit 5.

In a subsequent study on the role of glutenin proteins in spaghetti-making quality of durum proteins, Wasik (1978) used solubility fractionation to investigate the protein components of semolinas from 12 durum wheat varieties differing widely in pasta quality. The proportion of residue protein appeared to have a direct relationship with specific dough and gluten properties. The effects of wet extrusion, drying, and cooking on the proteins from five varieties of durum wheat were investigated by protein solubility distribution, gel filtration chromatography on Sephadex G-200, and electrophoresis in SDS. Processing changed both the protein solubility distribution and gel filtration properties; no significant changes occurred in the electrophoresis patterns of the fractions. Cooking insolubilized the proteins in pasta. The electrophoretic patterns of the protein fractions obtained from the cooked sample only remotely resembled those from corresponding semolina and uncooked pasta fractions. Wasik (1978) concluded that the proportion of insoluble protein in cooked pasta appears to be related to the tenderness index of cooked pasta, suggesting that the greater the proportion of insoluble protein, the firmer the product and the better the cooked quality.

According to Grzybowski and Donnelly (1979), protein quantity (up to a certain limit) and especially protein quality (strength, as determined by rheological characteristics) affect the cooking quality of spaghetti with respect to the maintenance of firmness and cooking stability. According to Dexter et al (1985a), high concentrations of amylose on the surface of cooked spaghetti may contribute to stickiness.

The principal factors governing the eating quality of Chinese noodles were protein content, dough strength, and starch paste viscosity (Miskelly and Moss, 1985). There was an optimum range outside which quality declined. Noodle brightness was related inversely to protein content and to flour color grade.

Miskelly (1984) investigated the influence of components contributing to the color and brightness of flour, flour paste, and Chinese- and Japanese- style noodles. Differences in brightness and yellowness were attributable to wheat cultivar, milling extraction rate, protein content, starch damage, and brown and yellow pigments.

Oh et al (1985) found that the optimum cooking time of Asian dry noodles increased linearly with protein content. High-protein noodles were darker and stronger and were firmer internally when cooked than low-protein noodles. However, protein content was not correlated with surface firmness. At high rates of flour extraction, noodle color darkened, but the internal firmness of cooked noodles did not change. The optimum absorption of noodle dough increased with starch damage and the fineness of granulation. Increasing starch damage reduced both internal and surface firmness of cooked noodles; decreasing particle size improved the strength of uncooked noodles but did not affect noodle color or the firmness of cooked noodles.

The effects of alkaline conditions on the properties of wheat flour dough and Cantonese-style noodles were reported by Moss et al (1986). Starch granules were swollen and the protein network was disrupted by the addition of 1% NaOH. The 1% NaOH gave the brightest and yellowest noodles, but they were slightly sticky.

The suitability of various Canadian wheats for noodle processing for the People's Republic of China was described by Preston et al (1986). With the exception of Canada western red winter wheat, all the wheats produced noodles of good quality.

VI. LIPIDS

A. Lipids of Wheat and Flour

Lipids may affect baking in many ways. During progressive stages in the baking process, the lipids may: 1) modify gluten structure at the mixing stage; 2) catalyze the oxidation of SH groups; 3) catalyze the polymerization of proteins through a process that involves lipid peroxidation; 4) act as lubricants; 5) improve gas retention by sealing gas cells; 6) prevent interaction between starch granules during gelatinization; 7) give some structural support to the gluten; 8) retard water transport from protein to starch; 9) retard starch gelatinization; and 10) act as an antistaling agent. It is possible that some of these effects (e.g., retardation of starch gelatinization and the antistaling effect) are due to the same mechanism (Wehrli, 1969).

In wheat, lipids form 1–2% of the endosperm, 8–15% of the germ, and about 6% of the bran, averaging 2–4% of the whole kernel (Morrison, 1976, 1978a, 1978b). (See also Chapter 7 in Volume I of this monograph.) The thickness of the aleurone layer is unaffected, and that of the outer bran layers slightly affected, by the size of unshriveled kernels. Consequently, larger grains contain a higher proportion of endosperm, by weight, than smaller grains. As a result of the higher lipid content of the germ and of the aleurone than of the endosperm, large kernels contain less lipid than small kernels (Chiu and Pomeranz, 1967).

The composition and distribution of lipids in wheat and milled wheat products were studied by several investigators (Morrison, 1978a, 1978b, 1979). Grain tissues differ not only in the amounts of lipid they contain but also in the composition of lipids in these tissues. Consequently, small differences in the highly empirical milling flow sheet and in the extraction rate can affect considerably the amounts and kinds of lipids present in the flour. Such differences are especially pronounced when flours are milled from various wheat classes and varieties.

Patent wheat flour from a mixed grist of bread wheats contains about 0.8%

free lipids extractable with petroleum ether (PE). The free lipids can be separated into two fractions according to their elution from a silicic acid column. One lipid fraction, about 0.6% of the flour, can be eluted with chloroform and is arbitrarily called the nonpolar fraction. This fraction contains mainly triglycerides and smaller amounts of hydrocarbons, sterols, steryl esters, monoglycerides, diglycerides, and free fatty acids. The second fraction (0.2% of the flour) can be eluted from the column with a more polar solvent, such as methanol, and is a mixture of free polar lipids. Free polar lipids are rich in glycolipids and contain relatively small amounts of phospholipids (Fig. 12).

In addition to 0.8% free lipids, patent wheat flour also contains 1.0% total bound lipids. About 0.6% bound lipids (mainly polar) can be extracted from flour with water-saturated butanol, after the PE extraction. However, during gluten washing or dough mixing, part of the nonpolar components becomes bound (unable to be extracted with PE). The extractability of lipids from whole wheat flour or milled wheat products depends also on particle size of the material. Thus, more lipids will remain unextracted with PE from wheat ground to pass a 20-mesh sieve, than from wheat ground to pass a 60-mesh sieve. In flour milled to pass a 100-mesh sieve, particle size distribution is less important.

B. Lipids in Wheats from Various Classes and Varieties

Shollenberger et al (1949) reported that in 481 samples of wheat from five crop years and 63 locations in the United States and Canada, the PE-extractable lipids were a varietal characteristic.

Lipids from two HRW wheat flours, Ponca and Red Chief, that differ in breadmaking quality, were extracted with water-saturated butanol and

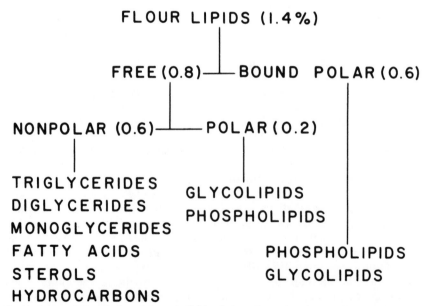

Fig. 12. Schematic description of the main lipids of wheat flour. (Reprinted, with permission, from Hoseney et al, 1969d)

fractionated by countercurrent distribution (Mason and Johnston, 1958). Red Chief flour lipids contained more choline phospholipids and less phosphatidylethanolamine than Ponca flour lipids. It was suggested that the differences might be related to the poor and good baking quality of flour from Red Chief and Ponca wheat, respectively.

The total lipid content of wheats from Yugoslavia and Italy ranged from 1.66 to 2.48%; the range of nonpolar lipids was 0.53–1.55% and that of polar lipids 0.49–0.61% (Garton et al, 1963a). Contents of free fatty acids, nonpolar lipids, ether-soluble nonsaponifiable lipids, and polar lipids were not significantly correlated with the baking quality of high-yielding wheat varieties. By expressing amounts of fatty acids in grain in percentage of total acids present, or in molar ratios, practically constant values were obtained regardless of varietal characteristics.

In an attempt to modify lipid contents and composition (Garton et al, 1963b), a high-yielding poor-quality wheat was grown in fields fertilized by various levels and ratios of nitrogen and phosphorus fertilizer. The total, nonpolar, nonsaponifiable, and polar lipid contents were unaffected by fertilization. The nitrogen added as fertilizer was utilized by grain to increase protein content. Similarly, fertilizer had no effect on fatty acid content or composition.

Free lipids isolated from gluten of eight European wheat varieties contained only a small amount of phosphorus (Mihailovic et al, 1963). The results indicated that practically all the phospholipids in gluten were in the bound form, i.e., as a lipoprotein complex. In four wheat varieties of good baking quality, about 50% of bound lipids were phospholipids. The amount of phospholipids in whole-meal flour was similar to the amount in gluten, indicating that the available phospholipids in the grain of wheat varieties of good baking quality are mainly involved in the formation of lipoprotein complexes. Separations by thin-layer chromatography (TLC) indicated no differences in the composition of nonpolar, polar, or ether-soluble nonsaponifiable lipids of wheat varieties varying in breadmaking strength. Of the total fatty acids present in each fraction, about 74–81% consisted of unsaturated acids, with linoleic acid predominant. Palmitic acid was the main saturated fatty acid. Small amounts of two fatty acids with an odd number of carbon atoms (pentadecanoic and heptadecanoic) were detected in nonpolar lipids, but not in phospholipids. The study did not confirm a previously proposed (Garton et al, 1963a) correlation between the contents of the metastable form of linoleic acid and oxidation requirement and baking quality. Small differences were found in the composition of polar lipids. The strong flour contained somewhat more lipoproteins and less glycolipids than the poor flour. There was, however, no difference in the kinds and relative concentrations of amino acids associated with the complex lipids.

Flours of 70% extraction milled from five wheat varieties grown in the United States and two varieties grown in Britain were examined for lipid content and composition (Fisher et al, 1964). Varietal, seasonal, and environmental differences were demonstrated. TLC showed no qualitative differences in the lipid compositions of the seven varieties. Diglycerides were separated and identified among the less-polar lipids and lysolecithin among the polar lipids. Two-dimensional TLC of flour lipids revealed 23 components, including phosphatidylcholine, phosphatidylethanolamine, glycerophosphorylinositol, and glycerophosphoryl glycerol. The lipid content of HRS wheats was higher than that of HRW wheats. U.S. wheats showed significant varietal differences in lipid nitrogen and lipid phosphorus, but the differences were not consistently correlated with breadmaking strength. No significant and consistent correlations could be established between loaf volume and choline, phosphatidylinositol, total glycolipid, or fatty acid distribution.

A later study (Fisher et al, 1966) demonstrated substantial varietal and environmental differences in lipid content and composition. Evidence suggested that only in very special circumstances would the performance of a flour be limited by the level of unsaturated fat available for the lipoxygenase-thiol oxidation mechanism.

Acker et al (1968) presented data indicating differences in lipid composition of wedge and adhering proteins. The wedge-protein fraction contained 3.63% lipids, 11.7% of which were phospholipids; the adhering protein fraction contained only 1.22% lipids, but the lipid fraction was rich in phospholipids (27.3%). Calculated on a protein basis, the wedge fraction (with 84.0% protein) contained 4.3% lipids, and the adhering fraction (7.4% protein), 7.9% lipids. Lipids in air-fractionated flours were studied by Daftary et al (1966). A shift in protein and ash contents was accompanied by a shift in lipid contents.

C. Effects of Lipids on Rheological Properties

Johnson (1928) found that extraction of flours with ether did not affect the absorption and lowered only slightly the extensibility of the doughs made from the flours. Sullivan et al (1936a) reported that fresh wheat-germ fat and most of its constituents, such as the unsaponifiable material and triglycerides, had little or no effect on rheological properties of a patent flour. More-highly saturated fatty acids had little effect on farinograph curves, and unsaturated fatty acids increased dough development times.

According to Moore and Herman (1942), adding up to 8% hydrogenated cottonseed oil, dairy butter, or lard significantly lowered water-absorption as measured by the farinograph. The effects on dough development time and stability were generally small and varied with the lipids tested. Mecham and Pence (1957) found that doughs from butanol-extracted flours required a fourfold increase in mixing time to reach optimum development, as compared to doughs from the untreated flour. Similarly, Bloksma (1966) reported that flours extracted with an acetone-water mixture required an extended mixing time for dough formation and development as compared to the original flour.

Merritt and Bailey (1945) found that fats differed in their effects on doughs made from flours of different strength. The fats tended to decrease the extensibility and extensigraph area of doughs from strong flours. Extensibility, resistance to extension, and extensigraph areas of doughs from weak flours were increased by adding fats. Tests with the Simon Extensometer indicated that the effect of defatting varied in flours of various extractions and produced little or no changes in low-extraction flours (Cookson et al, 1957).

The results obtained with the amylograph indicated that peak viscosity was reached in defatted flours at a slightly higher temperature than in unextracted flours (Cookson and Coppock, 1956). Medcalf et al (1968) reported that pasting characteristics of defatted starches depended on the polarity of the solvents used to extract the fat.

Tao and Pomeranz (1968) studied the effects of shortening, refined corn oil, and wheat-flour lipids (total and nonpolar) on rheological properties of dough. The lipids were added to seven HRW wheat flours comparable in milling extraction and protein contents but varying widely in protein quality. Total (unfractionated) and nonpolar wheat-flour lipids substantially increased the length of time needed to reach the point of minimum mobility in the farinograph and in the mixograph. Water-absorption was unaffected, and mixing tolerance was improved by adding nonpolar or total free lipids. The baking strength of the flour from which lipids were extracted had no effect on the contribution of the lipids to mixing characteristics. The effects of lipids on mixing increased with an increase in levels of added lipids. Nonpolar and total flour lipids exerted similar effects on untreated flour and PE-extracted flour. The temperature of peak hot-paste viscosity (as assessed by the amylograph) was lowered about 4° C by the addition of 2% flour lipids. Nonpolar lipids substantially increased peak viscosity; polar lipids had little effect.

The effects of lipids on rheological properties differed from the effects of the lipids on loaf volume and crumb-freshness retention. Whereas vegetable shortening and polar wheat-flour lipids substantially improved the size and quality of bread, as compared to that baked from untreated flour, they had little or no significant effects on rheological properties of the doughs. At the same time, nonpolar lipids of wheat flour, which had no effect on loaf volume or crumb grain of bread baked from untreated flour and adversely affected bread baked from PE-extracted flours, substantially increased dough development time. The effects of nonpolar lipids on mixing times depended on the protein contents and protein quality of the flours to which the lipids were added. Equal amounts of nonpolar lipids increased mixing times much more for strong flours than for poor flours.

In a later study, Pomeranz et al (1968c) compared effects of commercially available natural and modified soybean lecithins on rheological properties and breadmaking potentialities of untreated flour and PE-extracted flour. Hydrogenated lecithin added to doughs containing shortening reduced loaf volume to, or materially below, that of bread baked from doughs containing no shortening. Hydrogenated lecithin did not substitute for free flour lipids, as did eight other soybean phospholipids when added to PE-extracted flours baked with shortening. Alcohol-soluble phosphatides containing a 2:1 mixture of phosphatidylcholine and phosphatidylethanolamine were the most effective in improving loaf volume and crumb grain of bread baked from untreated and PE-extracted flour without shortening. It is noteworthy that 0.5% alcohol-soluble phospholipids substituted for both 0.8% free flour lipids and 3% shortening when added to PE-extracted flours. Softness retention of bread was generally correlated with loaf volume. Adding alcohol-soluble phosphatides lowered the amylograph temperature and the hot-paste viscosity peak. Hydroxylated phosphatides increased farinograph dough development times, valorimeter values, and mixogram times. No correlation could be established between bread quality and the effects of phospholipids on rheological properties.

D. Functionality of Defatted Flours

Information on the bread quality of defatted flours is contradictory, or at best conjectural. Johnson (1928) found that bread of better color, texture, and volume was produced from ether-extracted flour than from natural flour. Low-grade flours were improved more than were patent flours. Extraction of flours with ether did not affect absorption, yields of wet or dry gluten, or viscosity of acidulated suspensions. The extensibility of doughs from ether-extracted flours was slightly less than that of corresponding natural flours. Doughs prepared from ether-extracted flours were markedly superior to natural flour doughs in ability to retain the carbon dioxide produced during fermentation (Johnson and Whitcomb, 1931). Adding fat to doughs prepared from ether-extracted flours reduced their gas-retaining powers. Martin and Whitcomb (1932) observed that ether extraction improved the baking quality of flour from Marquis wheat but harmed that from Kubanka and Federation wheats. Sullivan et al (1936b) reported that ether extraction damaged the baking quality of both strong and weak flours baked with lard in the bread formula. Flour oil, but not germ oil, restored the quality of ether-extracted flour. The results indicated that flour lipids contained components essential for good baking quality. Several lipids from vegetable and animal sources were studied, but none could fully restore the baking quality of ether-extracted flour.

Cookson and Coppock (1956) found that flours that were defatted (with carbon tetrachloride) gave doughs of greater resistance and lower extensibility and required slightly more time to reach peak viscosity in the amylograph than controls. Flours extracted with ether, acetone, carbon tetrachloride, or PE were similar in behavior and showed only slight reduction in dough extensibility compared with unextracted flour. Alcohol-extracted flours had low extensibility-to-resistance ratios and gave a tough, inextensible gluten with a low proportion of wet and dry gluten; their doughs had low gas production, unlike those of the other defatted flours. Only alcohol extraction damaged bread quality considerably. In a later study (Cookson et al, 1957), some defatted flours gave poorer volume and texture than undefatted flours, although previously the reverse was more usual.

Free lipids (about 0.8%) were extracted with PE from two HRW and one HRS, one SRW, one durum, and one club wheat flour (Pomeranz et al, 1968b). The original and defatted flours were baked by a straight-dough procedure in a rich formula with and without 3% commercial vegetable shortening. Adding shortening to the original flours improved loaf volume and crumb grain substantially and consistently. Adding shortening to the defatted flours impaired the crumb grain. Loaf volumes of bread baked from defatted strong flours were decreased, but those from from defatted poor flours increased, by addition of shortening to the dough formula (Table VI). The shortening response of strong flours was completely restored by reconstitution with free lipids from any of the six flours tested. Reconstitution techniques established that the amount of free lipids required to give the original loaf volume was at least half the amount in the original flour.

The reconstitution study emphasized the importance of free lipids in breadmaking. It also explained the reasons for previous conflicting results on the role of free lipids in breadmaking, for the effect of shortening varied with

TABLE VI

Effects of Free Flour Lipids and Shortening on Loaf Volume and Crumb Grain of Bread Baked from Various Flours[a]

| Class | Variety | Number[b] | Original Flour | | | | PE[d]-Extracted Flour | | | | Reconstituted[e] Flour with 3 g of Shortening | |
| | | | Without Shortening | | With 3 g of Shortening | | Without Shortening | | With 3 g of Shortening | | | |
			Loaf Volume (cm³)	Crumb Grain[c]	Loaf Volume (cm³)	Crumb Grain	Loaf Volume (cm³)	Crumb Grain	Loaf Volume (cm³)	Crumb Grain	Loaf Volume (cm³)	Crumb Grain
HRW	Qv-Tm × Mql-Oro[f]	12995	755	S	922	S	813	S	810	Q	918	S
HRW	Chiefkan × Tenmarq	K501099	620	U	764	U	635	Q-S	765	U	760	Q-U
HRS	Marquis	3641	750	S	877	S	775	S	720	S	885	S
SRW	Seneca	12529	615	Q	740	Q-S	563	Q-S	675	Q	750	Q-S
Durum	Wells	13333	435	U	443	U	345	U	393	U	385	U
Club	Omar	13072	465	U	588	Q-U	473	Q	598	Q-U	617	U

[a] Source: Pomeranz et al (1968b).
[b] Cereal investigation (C.I.) or Kansas Experiment Station number.
[c] S = satisfactory, Q = questionable, U = unsatisfactory.
[d] Petroleum ether.
[e] By blending 100 g of flour with 0.8 g of free wheat-flour lipids.
[f] Quivira-Tenmarq × Marquillo-Oro.

wheat-flour strength and depended on the shortening being added to untreated or PE-extracted flours.

In addition to 0.8% free flour lipids extractable with PE, the flour also contained 0.6% bound lipids (primarily polar lipids) extractable with water-saturated *n*-butanol. Several early attempts to extract bound lipids (Mecham and Mohammad, 1955; Bloksma, 1959, 1966) without irreversibly damaging rheological properties, fermentation, and bread quality were essentially unsuccessful. Hoseney et al (1969d) studied the flour fraction(s) damaged by water-saturated butanol, investigated techniques of extracting practically all bound lipids without damaging the flour, and compared the role of the free polar, free nonpolar, and bound polar lipids in baking when reconstituted with an almost completely defatted flour. They found that water-saturated butanol, used to extract bound lipids, formed a complex with starch and inhibited gas production. A reconstituted flour almost completely free of lipid (except for 0.08%) was prepared by first extracting free lipids from the flour with PE, followed by washing out the gluten, solubilizing the gluten in 0.005*N* lactic acid, and centrifuging at $100,000 \times g$ for 5 hr; the lyophilized centrifugate was then extracted with water-saturated butanol. Dough from the reconstituted flour, after premixing to restore certain rheological properties, was comparable in its breadmaking potentialities to dough from the original flour.

E. Role of Wheat-Flour Lipid Fractions

Working (1924) reported that prolonged washing of soft gluten from low-grade flour removed phosphatides and gradually increased tenacity until the gluten was nearly equal to that from patent flour. Phosphatides, added to flour in small quantities, lowered gluten quality as measured by the feel of hand-washed gluten, by viscosity as determined by the MacMichael viscometer, and by baking tests. Working concluded (1928) that ordinary flour contains much more phosphatide than necessary to develop the dough properly. The phosphatides increase the ductility of dough, by lubricating gluten strands so that they slip readily.

Coppock et al (1954) added to dough 0.003% of an acetone-insoluble phosphatide fraction separated from the precipitate formed on the aging of flour oil. The phosphatide substantially improved the loaf volume of bread baked from defatted flour. Lipids were separated into fractions by countercurrent distribution; the fractions were added to extracted flour, and bread was baked from doughs with or without lard. The fractions differed markedly in their effects on loaf volume and crumb texture.

Pomeranz et al (1965) studied the effects of polar and nonpolar wheat flour lipids on the quality of bread baked from untreated wheat flour. Adding up to 3 g of vegetable shortening per 100 g of a hard winter wheat flour composite improved crumb grain and increased loaf volume from 802 to 948 ml. Higher levels of shortening had no additional improving effect. When the bread formula included 3 g of vegetable shortening, adding 0.5 g of nonpolar or polar lipids or the unfractionated lipids from wheat flour had no significant effect on crumb grain or loaf volume. However, the loaf volume of bread baked without vegetable shortening was increased strikingly, from 802 to 958 ml, by adding 0.5 g of polar flour lipids. Adding 0.5 g of nonpolar lipids did not increase loaf volume significantly. Adding 0.5 g of unfractionated flour lipids gave an intermediate volume of 893 ml, which was expected from the contribution of the lipid components.

That investigation was followed by a study of the effects on loaf volume and bread characteristics of commercial vegetable shortening and polar and nonpolar lipids isolated from wheat flours varying widely in breadmaking potentialities (Pomeranz et al, 1966a). Bread was baked from flour milled from HRW, HRS, SRW, durum, and club (white) wheat varieties, each from 1963 and 1964 crops. Loaf volumes were increased 87–195 ml, and crumb grains were improved, by adding 3 g of vegetable shortening per 100 g of flour. The improving effect increased steeply from additions of up to 1.5 g of shortening and thereafter increased only slightly with additions up to 4.5 g of shortening. Adding 0.5 g of polar lipids isolated from six flours to a composite HRW flour almost equaled the improving effect of 3 g of shortening; adding 0.5 g of nonpolar flour lipids had very little effect; and adding 0.5 g of unfractionated, original flour lipids had an intermediate effect. Neither shortening nor any of the tested wheat flour lipids affected gassing power. Loaf volume increase and crumb grain improvement were accompanied by parallel retardation of crumb firming during storage. The effects on bread quality of polar lipids were independent of the source of flour lipids.

Daftary et al (1968) extracted free (PE-soluble) and bound (soluble in water-saturated butanol, following PE) lipids from a composite HRW flour; from two other HRW flours varying widely in breadmaking potentialities; and from a HRS flour. The lipids were fractionated into polar and nonpolar fractions by silicic-acid-column chromatography and subfractionated by diethylaminoethyl-cellulose-column chromatography. In bread baked without shortening, free polar lipids substantially increased loaf volume; the increase was smaller when bound polar lipids were added. Lipid fractions isolated from various flours indicated no varietal differences. Total free lipids containing a mixture of nonpolar and polar components (in a ratio of 3:1) improved bread quality less than polar lipids alone. Nonpolar lipids decreased the loaf volume and impaired the crumb grain of bread baked from PE-extracted flours; the deleterious effects were counteracted by polar lipids. Effects on bread depended on the levels and ratios of nonpolar to polar lipids (Fig. 13). Fractions rich in galactosylglycerides increased loaf volume of bread baked from PE-extracted flours substantially more than did those rich in phospholipids. Ponte and DeStefanis have shown (1969) similar effects in commercial bread baked by the sponge method.

The synthesis of glycosylglycerides (Wehrli and Pomeranz, 1969b) made it possible to determine the effects in breadmaking of glycolipids varying in composition (Pomeranz and Wehrli, 1969). The effects of synthetic glycosyl-glycerides were compared with the effects of wheat-flour polar lipids, soybean polar lipids (natural and fractionated), phospholipids, and glycolipids of the sucroester type. The natural and synthetic polar lipids restored breadmaking potentialities of PE-defatted wheat flour. Among the synthetic glycosyl-glycerides, cellobiosyl derivatives were more effective improvers than monogalactosyl derivatives. The optimum chain length for monogalactosyl-glycerides was that of octanoic acid; for cellobiosylglycerides it was that of decanoic acid. Linoleoyl derivatives were more effective than stearoyl derivatives, although double bonds are not essential for the effect of glycolipids.

Studies with flours defatted with water-saturated butanol showed that small amounts of free polar or bound polar lipids were detrimental to loaf volume unless accompanied by nonpolar lipids or bound lipids in their native state. High amounts of bound polar or free polar lipids restored loaf volume. However, larger quantities of bound polar than of free polar lipids were required. The results indicated that bound polar lipids performed a different role in baking when in their native state than when extracted and reconstituted.

Cole et al (1960) found that when lipid-free flours were supplemented with a phosphorus-free fraction from flour oil, no improvement occurred in the cookie-baking quality of those flours; the phosphorus-containing fraction, however, produced complete recovery of cookie characteristics. Soybean or corn phosphatide preparations and a purified soybean lecithin also brought about a similar improvement.

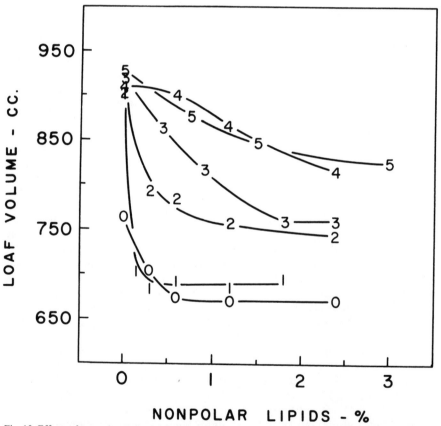

Fig. 13. Effects of nonpolar and polar lipids on loaf volume of bread from petroleum ether-extracted flour baked without shortening and with various combinations of nonpolar and polar lipids. Numbers 1–5 denote 0.1–0.5 g of polar lipids per 100 g of flour. (Reprinted, with permission, from Daftary et al, 1968)

F. Some Recent Studies of Functionality
of Lipids in Wheat Products

Lipids in wheat and their importance in wheat products were reviewed by Ponte and Baldwin (1972), Chung and Pomeranz (1977, 1981), Chung et al (1978a, 1978b), and Pomeranz (1981). Daniels et al (1971) discussed the role of flour lipids in dough development, and Frazier et al (1979) authored an excellent review on lipoxygenase action and lipid binding in breadmaking. A study on phase equilibria and structures in the aqueous system of wheat lipids and their relation to functionality was reported by Carlson et al (1978). Physical structure and phase properties of aqueous systems of lipids from rye and triticale in relation to wheat lipids were investigated by Carlson et al (1980). Carlson (1981) is the author of a dissertation on the colloidal aspects of the wheat flour dough and its lipid and protein constituents in aqueous media.

Hoseney et al (1972a) investigated the effectiveness of using synthetic surfactants in replacing total free lipids in breadmaking. Adding sucrose monotallowate to a standard baking formula increased loaf volume 7–18%, depending on the level of shortening in the formula. When two sucroglycerides (sucrose monotallowate and sucrose monopalmitate) were added to PE-defatted flour, each replaced the total free flour lipids, and they increased loaf volumes 18 and 22%, respectively, above that of the original flour (in a formula with shortening). Native free flour lipids could also be replaced with sodium or calcium stearoyl-2-lactylates. However, adding either lactylate, unlike adding the sucroglycerides, did not increase loaf volumes above that of the control flour. When two nonionic surfactants (pluronic polyols F-108 and F-68) were baked with PE-defatted flour, loaf volumes were generally comparable to that of the control flour. However, all loaves baked with pluronic polyols had impaired crumb grains.

Lin et al (1974b) extracted both total lipids and individual classes of lipids from HRS wheat flour and from durum wheat semolina. The lipids were also separated into five fractions (Lin et al, 1974a). Extraction of wheat flour with PE resulted in bread with reduced loaf volume and poorer crumb and crust characteristics. The baking quality was restored with the addition of any of the five lipid fractions at the 0.5% level, if shortening was present in the bread formula. Without shortening, the baking properties were restored and improved only by the fraction rich in digalactosyldiglyceride (DGDG). The effect of the nonpolar lipids and phospholipids on bread properties baked from the untreated flour containing shortening was small. However, DGDG at 0.4–0.6% consistently improved the loaf volume; this improvement was more pronounced without shortening in the bread formula.

Total lipids of a commercial wheat flour were separated by DeStefanis and Ponte (1976) into polar and nonpolar fractions. When added at the dough stage, the nonpolar lipids were detrimental in baking. Nonpolar lipids were fractionated into steryl esters, triglycerides, free fatty acids, and diglycerides, each of which was then tested in three dough systems: 1) defatted flour, 2) intact flour, and 3) intact flour with 3% lard. The deleterious effects were found to be caused by free fatty acids added to either the defatted or intact flour. In the presence of 3% lard, the ill effects were not as evident as in the other two systems. Within the fatty acids class, detrimental effects in bread were directly related to linoleic acid. Complementary studies showed that the free fatty acids both increased the peak viscosity and delayed the peak time of starch during the gelatinization cycle.

Wheat-flour lipids were extracted with each of nine solvents, PE, n-hexane, n-heptane, benzene, chloroform, acetone, water-saturated 1-butanol, methanol, and 95% ethanol (Finney et al, 1976). Nonpolar solvents extracted substantially less lipids than the more polar solvents. Lipids extracted with nonpolar solvents contained less polar lipids than those extracted with polar solvents. Generally, as extracted bound or total lipids increased, mixing time increased and loaf volume decreased for reconstituted flours. Extracting lipids with each of the three alcohols reduced to zero or greatly impaired the gas-retaining capacity of gluten protein.

Chung et al (1977a, 1977b, 1979, 1980b) conducted a series of studies on defatted and reconstituted flours. The first article in the series (Chung et al, 1977a) concerned the effects of lipid-extracting conditions on functionality of reconstituted flours in breadmaking. Lipid extractability increased with solubility in the series Skellysolve B, benzene, acetone, and 2-propanol, and it was higher in a regular than in a vacuum Soxhlet. Extraction temperature was lowered 12–18° C by pressure reduction with a vacuum pump connected to the condenser of a modified Soxhlet. An increase in total extractable lipids was primarily from an increase in extracted polar lipids. Rheological properties and baking characteristics of dough were little affected, whether the flour was extracted in a regular or a vacuum Soxhlet with Skelly B. Deleterious effects were small for flours treated with benzene and acetone and substantial for flour treated with 2-propanol. The adverse effects were lowered by extraction of the flour in a vacuum Soxhlet. Extraction of lipids by 2-propanol was effective in a vacuum Soxhlet, and damage to functional breadmaking properties of the flour was much less than to those of a flour extracted in a regular Soxhlet.

In a subsequent study (Chung et al, 1977b), reconstituting techniques were used to study the effects of extraction methods and conditions on extractability of lipids and on functional properties of good and poor breadmaking flours. Extractability of lipids in three widely differing bread flours increased with solubility of the solvent (Skellysolve B < benzene < acetone < 2-propanol) and extraction temperature (30–75° C). For any solvent-temperature combination, more lipids usually were extracted by Soxhlet than by the shaker technique used. Temperature-controlled extraction (for 2 hr) by shaker had little effect on rheological properties of doughs made with the reconstituted flours. Without shortening, breads baked with the reconstituted strong flours that had been extracted with Skellysolve B, benzene, or acetone had larger loaf volumes than those of the controls. The improving effects were masked, however, when shortening was used. Soxhlet extraction with 2-propanol appeared to infinitely increase mixing time and irreversibly impair the functionality of reconstituted, good breadmaking flours; however, it substantially improved the functionality of a poor quality flour, especially when the flour was made into bread with 3% shortening.

Shortening, or fat, increases loaf volume and improves dough handling properties, crumb grain, and retention of freshness. Native flour lipids, especially polar lipids, are essential for obtaining the beneficial effects of shortening on loaf volume and crumb grain. In the absence of flour nonpolar lipids, loaf volume is affected by the quantity of flour polar lipids, presence of shortening, and interaction of shortening and polar lipids.

Studies have demonstrated the significant contributions that both the quantity and the quality of wheat flour proteins make to the loaf volume of bread. The quantity of protein is influenced mainly by environmental factors, but the quality is mainly a heritable characteristic. Chung et al (1980a) conducted studies 1) to determine whether the shortening effects depend on inherent differences, presumably in protein quality, among wheat flours; 2) to verify the interaction of shortening and polar lipids in wheat flours that vary in breadmaking quality; and 3) to determine how the binding of lipids to other flour constituents affects the shortening-lipid interaction. The study was designed to explain what makes good-quality bread-wheat flours good and poor ones poor, insofar as interaction with lipids is concerned.

To this end, 11 wheat flours that varied in breadmaking quality were defatted at 75° C with Skellysolve B or 2-propanol. All defatted flours contained only small amounts of residual nonpolar lipids: flours defatted by Skellysolve B contained more residual bound polar lipids than the flours defatted by 2-propanol. Lipid removal increased mixing time; the increase was greater for 2-propanol extraction than for Skellysolve B extraction. Removal of lipids affected mixing time substantially more than addition of 3% shortening. Small differences in amounts of free polar lipids in flours that vary in breadmaking quality accentuated differences in "loaf volume potential" of flours through the interaction of the free polar lipids and shortening: the better the inherent quality of a flour, the greater the benefits derived from adding shortening. Removal of most of the bound polar lipids could result in improvements of loaf volume in bread baked from propanol-defatted flour without shortening (Figs. 14 and 15). The amount of improvement was related to the inherent quality of the flours, probably because of differences in their protein quality and protein-protein interactions. The authors concluded that good loaf volume and crumb grain can be obtained from flours with good protein quality in which adequate protein aggregation is enhanced by free polar lipids that can interact with proteins and shortening. As a mechanism to bring out the maximum loaf volume potential of flours, the multiple interaction of protein-lipid (free polar)-protein in the presence of shortening seems to be superior to the mere formation of protein-protein aggregates.

According to MacRitchie and Gras (1973), curves of loaf volume as a function of lipid content for several flours with a range of properties all showed minima at lipid contents intermediate between those of the defatted and whole flours (Fig. 16). Polar and nonpolar lipid fractions have different effects on loaf volume. Variations in loaf volume as a result of different lipid treatments become apparent only during the baking stage in the oven. Lipid content and bromate level both affect loaf volume but appear to act independently.

MacRitchie (1977) separated wheat flour lipid into five fractions by selective elution from silica gel. Two fractions, which contained mainly nonpolar components, depressed loaf volume when added to defatted flour. The three other fractions contained polar components and maintained or increased loaf volume. The effects of the fractions as foam stabilizers paralleled their effects on loaf volume, supporting the theory that flour lipids exert their action in baking through their role as surfactants in stabilizing or destabilizing the gas bubble structure during expansion of the loaf. Flour lipids can be classified into three groups according to their effects in baking. The polar galactolipids and phospholipids increase loaf volume; compounds of intermediate polarity depress loaf volume; and the more nonpolar compounds such as triglycerides have little effect in the absence of other lipids but beneficial effects when other lipids are present.

MacRitchie (1977) postulated that effects of lipids become apparent during the baking stage in the oven although the causes of these effects likely are established during the mixing and molding stage. The requirements of a flour for good loaf volume and texture in terms of lipid content and composition may be reduced to two generalizations: 1) it is desirable that the lipid content be as high as possible so as not to fall near the minimum in the loaf volume-lipid content curve, and 2) the higher the proportion of polar lipid, the better the flour.

MacRitchie (1981) determined the effects on loaf volume when comparable amounts of lipid and gluten protein are removed from a flour (Fig. 17). Loaf volume decreases as both lipid and protein contents decrease from their natural values. The decrease for protein is linear; however, the lipid curve reaches a

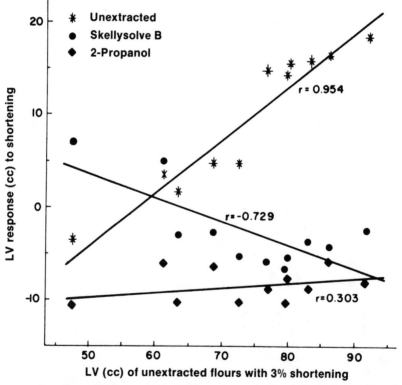

Fig. 14. Loaf volume (LV) responses to shortening in extracted flours or flours defatted with Skellysolve B or 2-propanol, plotted against LV values of breads baked with 3% shortening from untreated control flours that vary in breadmaking quality. The LV response was calculated by subtracting the LV of bread baked without shortening from the LV with shortening. (Reprinted, with permission, from Chung et al, 1980b)

minimum, thereafter increasing on further removal of lipid. For the flour represented in Fig. 17, the volume increment was 6.5 cm^3 per percent of gluten protein, whereas the volume increment for lipid in the steep parts of the curve on either side of the minimum was about 50 cm^3 per percent of lipid. On the other hand, complete removal of protein from a flour destroys its dough-forming properties and bread-baking capacity, whereas removal of lipid does not. Dough properties and bread-baking capacity are retained by a defatted flour.

Chung and Pomeranz (1981) calculated that polar lipids are about 50 times more effective as a loaf volume improver than good quality gluten proteins. However, they had to recognize that flour proteins are structurally the backbone of dough, whereas flour lipids primarily strengthen that structural backbone and bring out the best in its performance.

Flours that differ in baking performance give rise to different loaf volume-lipid content curves. This effect can be related to the gluten protein component (MacRitchie, 1978). When the gluten from a good baking flour was replaced by that from a poor flour, the loaf volume-lipid curve was modified (Fig. 18). The minimum was displaced to a higher lipid content. As a result, the loaf volume was relatively low at the natural lipid content and corresponded to a point on the

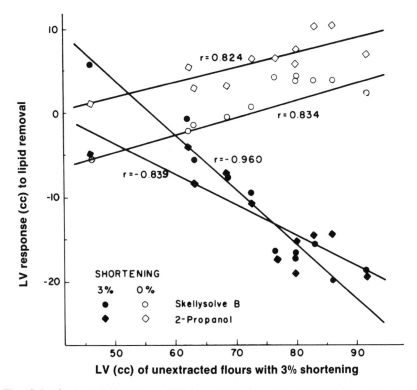

Fig. 15. Loaf volume (LV) responses to lipid removal of breads baked from flours extracted with Skellysolve B or 2-propanol plotted against LV values of breads baked with 3% shortening from unextracted control flours that vary in breadmaking quality. The LV response was calculated by subtracting the LV value of defatted flour from the LV value of the untreated flour. (Reprinted, with permission, from Chung et al, 1980b)

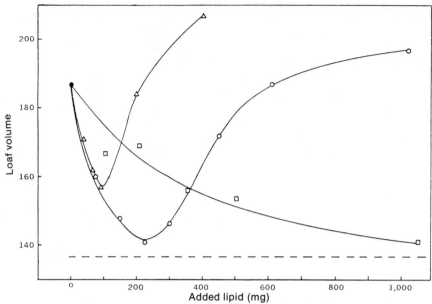

Fig. 16. Effects on loaf volume of additions of polar and nonpolar lipid fractions to defatted flour. Circles = whole lipid, triangles = polar lipid, squares = nonpolar lipid, dashed line = volume at end of proofing stage. (Reprinted, with permission, from MacRitchie and Gras, 1973)

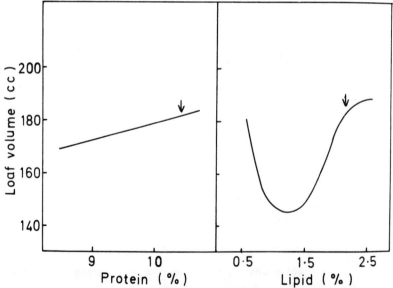

Fig. 17. Variation of loaf volume with changing protein (left) and lipid (right) contents of the same flour, redrawn from results of MacRitchie (1978). Arrows indicate natural values for the flour. Lipid is expressed as hydrolysate lipid. Starch contains approximately 0.5% hydrolysate lipid; curve shows effects of additions of chloroform-extracted lipid. (Reprinted, with permission, from MacRitchie, 1981)

steeply rising part of the curve, whereas the loaf volume of the good flour had approached its maximum at the natural lipid content.

This effect is often observed when flours of varying baking performance are compared. A poor flour may approach a good one in loaf volume at sufficiently high lipid contents. This explains the shortening response of certain flours. Once a flour contains natural lipid, shortening can be as effective in increasing loaf volume as the natural lipid. In some cases, the poorer flour does not attain the loaf volume of the better one even at high lipid additions. Interchange experiments have shown this to be due to a deficiency in gluten protein quality.

The contribution of lipids in breadmaking is complicated by many factors and interactions. Flour free lipids play an important role in interacting mainly with proteins in gluten development during the mixing stage and further interacting with nonprotein components at the subsequent stages of breadmaking. Although the baking quality of flour does not seem to be governed by lipids, studies on interchanging lipids from good and poor quality flours have made the role of native flour lipids in baking well established, documented, and irrefutable (Pomeranz, 1967).

G. Lipids and Breadmaking Potential

Significant contributions of wheat flour proteins to loaf volume of breads have been well demonstrated. Two major factors account for variations in loaf volume of wheat varieties. One is protein content, which is influenced mainly by

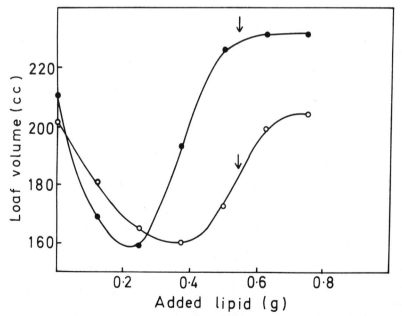

Fig. 18. Loaf volume-lipid content curves for a good baking flour (●) and for the same flour with its gluten protein replaced, at the same protein level, by gluten protein from a poorer baking flour (o). Arrows indicate the natural lipid level of flour. Lipid additions refer to the weight of whole lipid added to a defatted flour of 30.2 g dry weight. (Reprinted, with permission, from MacRitchie, 1981)

environmental factors; the other is protein quality, which is primarily genetically controlled. Lipids, minor components of wheat flour, function importantly in breadmaking. Shollenberger et al (1949) reported that the PE-extractable lipids were a varietal characteristic. Many scientists tried to correlate lipid content or composition with genetic differences in breadmaking quality of wheats, but no significant relationship was established. Fisher et al (1964, 1966), however, demonstrated varietal and environmental effects on the quantity and quality of lipids that could be extracted with water-saturated 1-butanol. Those and other studies implied that sound wheats of the same class and unexposed to extremes in environment might best differentiate wheats according to breadmaking quality.

The composition and amounts of lipids that can be extracted from a flour depend on the genetic makeup of the wheat from which the flour was produced, milling yield of the flour, particle size and moisture content of the sample, and conditions of lipid extraction, including time and temperature of extraction, type of extractor, and type of solvent. Solubility values or polarities of extractants can be varied by the use of different solvents alone or in combination with water. A preliminary study (Chung et al, 1980a) showed the conditions of lipid extraction that differentiate HRW wheat flours that vary in breadmaking potential. Six solvents (PE, Skellysolve B, benzene, acetone, 2-propanol, and water-saturated 1-butanol) were compared. The ratio of nonpolar lipids to polar lipids extracted with PE or Skellysolve B best differentiated the five flours according to loaf volume potential.

Chung et al (1982) extended that preliminary study to HRW wheats grown in the Great Plains of the United States and to their straight-grade flours. Lipids were extracted (with PE) from 21 samples of HRW wheats and 23 samples of experimentally milled straight-grade flours that varied in breadmaking potential. Wheat protein content varied from 11.5 to 15.7%, flour mixing time from $\frac{7}{8}$ to 9 min, and loaf volume per 100 g of flour from 523 to 1,053 cm^3. The total lipids from 10 g of flour (db) were fractionated into polar and nonpolar lipids; total lipids were analyzed colorimetrically for carbohydrates, mainly galactose. Polar lipid content varied from 14.8 to 28.1 mg per 10 g of wheat and from 10.6 to 27.3 mg per 10 g of flour; ratios of nonpolar lipids to polar lipids were in the ranges of 6.31–11.32 for wheat and 2.47–6.91 for flour; lipid galactose ranges were 1.61–5.49 mg and 2.64–5.61 mg in 10 g of wheat and flour, respectively. Significant linear correlations were found between loaf volume and the following variables: polar lipid content ($r = 0.877$ for wheat and 0.888 for flour), nonpolar-polar lipid ratio ($r = -0.902$ for wheat and -0.907 for flour), and lipid galactose ($r = 0.745$ for wheat and 0.905 for flour) (Table VII). Polar lipids, the nonpolar-polar lipid ratio, and lipid galactose were curvilinearly related to mixing time requirement. The correlation coefficients of loaf volume with polar lipids, nonpolar-polar lipid ratio, and lipid galactose generally were somewhat improved when loaf volume and lipid contents were corrected to a constant protein basis (Figs. 19 and 20). The data indicated that the quantity of polar lipids or galactolipids occurring naturally in wheat is related to breadmaking (functional) properties and may govern or be closely related to other factors that govern functional properties of good and poor varieties of wheat. The highly significant correlations pointed to the potential usefulness of polar lipids, nonpolar-polar lipid ratio, and lipid galactose for estimating loaf volume potential of HRW wheat flours.

Subsequently, Zawistowska et al (1984) described lipid content and composition of flours and the distribution of dry matter, protein, and lipids in Osborne protein fractions of five HRS wheat cultivars with diverse baking quality. The hexane-soluble lipid (free-lipid) contents of the flours were markedly different among the cultivars. Prediction equations showed excellent agreement between measured and predicted loaf volumes, using the ratio of nonpolar lipids to polar lipids in the free-lipid fraction. For the five cultivars, the albumin + globulin Osborne fraction contained 21.8–30.3% of the flour lipids. Analogous data for the gliadin, acetic-acid-soluble glutenin, and residue fractions were 27.5–39.8, 13.3–21.7, and 20.5–26.3%, respectively. The gliadin fraction contained relatively more polar lipids than the glutenin fraction. The lipid content of both fractions was significantly correlated with loaf volume, which indicated that the lipid binding capacity of the gluten proteins, gliadin and glutenin, is an important factor in the breadmaking quality of flour. Glutenin fractions soluble in AUC ($0.1M$ acetic acid, $3M$ urea, and $0.01M$ cetyltrimethylammonium bromide) or acetic acid had the same lipid binding capacity. Determination of the lipid content and composition of three fractions (I, > 200,000 mol wt; IIa, 27,500–75,000; and IIIa, 15,500–21,000) of Osborne gliadin fractionated by gel filtration on Sephadex G-200 showed that fraction I had an extremely high binding capacity for polar lipids. Only trace amounts of lipids were associated with the other fractions. The protein composition of fraction I showed that it contained mostly the high-molecular-weight (HMW)

TABLE VII
Linear Correlation Coefficients of Wheat or Flour
Components and Loaf Volume[a]

Wheat or Flour Component(s)[b]	Loaf Volume	
	As Received	Corrected[c]
Wheat		
TL	0.343	⋯
NL	0.089	⋯
PL	0.877[d]	0.907[d]
Ratio of NL to PL	−0.902[d]	−0.925[d]
Lipid galactose	0.745[d]	0.854[d]
Protein	0.225	⋯
Flour		
TL	0.453[e]	⋯
NL	−0.186	⋯
PL	0.888[d]	0.905[d]
Ratio of NL to PL	−0.907[d]	−0.904[d]
Lipid galactose	0.905[d]	0.928[d]
Protein	−0.157	⋯

[a] Source: Chung et al (1982).
[b] TL = petroleum ether-extractable total lipids, NL = nonpolar lipids, PL = polar lipids.
[c] Loaf volume and flour lipid content corrected to a 12% and wheat lipid content to a 13% protein basis. Corrected ratio of NL to PL equals NL content (as received) divided by PL content (corrected to 12% protein for flour or 13% for wheat).
[d] Significant at 0.01 level.
[e] Significant at 0.05 level.

gliadin components (mol wt > 68,000), and small amounts of low-molecular-weight (LMW) components. For the five cultivars, the amount of aggregated fraction I was negatively correlated with the loaf volume attainable by the remix baking procedure.

The lipid contents of Australian bakers' flours were examined and their relationships to breadmaking qualities were studied by Marston and

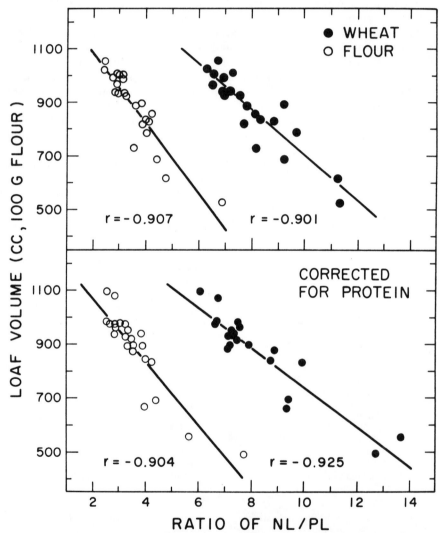

Fig. 19. Relation between loaf volume (LV) of bread baked from 100 g of flour and the ratio of nonpolar lipids (NL) to polar lipids (PL) of wheat or flour. Top, LV and NL/PL ratio on the as-received basis of protein content; bottom, LV and NL/PL ratio corrected to a constant protein basis. The corrected ratio of NL to PL was obtained from NL content (as received) divided by PL content corrected to 12% protein for flour or 13% for wheat. (Reprinted, with permission, from Chung et al, 1982)

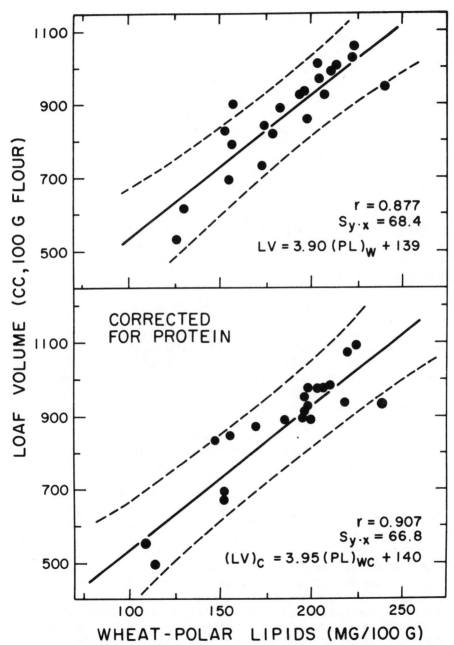

Fig. 20. Relation between loaf volume (LV) of bread baked from 100 g of flour and petroleum ether-extractable polar lipids of wheat $(PL)_w$. Top, LV and $(PL)_w$ on the as-received protein basis; bottom, LV corrected to 12% and $(PL)_w$ to 13% protein content. c = corrected, $S_{Y \cdot X}$ = standard deviation. (Reprinted, with permission, from Chung et al, 1982)

MacRitchie (1985). Differences in lipid content and composition were observed, but no useful relationship was found between these characteristics and breadmaking quality. The addition of fat to dough as in normal breadmaking practice blanketed the effects of differences in indigenous lipid content and composition. Differences in flour protein quality also appear to mask differences among indigenous lipids and their effects in breadmaking.

The breadmaking performance of flours milled from wheat varieties grown in the United Kingdom was studied by Bell et al (1987) in relation to the composition of the free lipids extractable by light petroleum (b.p. 60–80° C) from wheats. The results with six varieties grown at three sites in 1983 confirmed reports of significant correlations between loaf volume and free polar lipid contents in wheat samples from the United States and Canada (Chung et al, 1982; Békés et al, 1986). However, with 15 winter varieties and six spring varieties grown at three sites in 1984, no significant correlations were found between any lipid parameter and loaf volume or quality score of breads prepared by either the Chorleywood bread process or long fermentation methods.

A study of European wheats conducted by O. K. Chung and Y. Pomeranz (unpublished data, 1980–1982) indicated no correlation between lipid content and composition and breadmaking potential. The reasons for differences between North American wheats, on one hand, and Australian and European wheats, on the other hand, are unknown.

H. Effects on Cookie Quality

Cole et al (1960) found that cookies baked from flours that had been extracted with water-saturated butanol were generally darker in color and had diameters appreciably smaller than those baked from untreated flours. Restoration of lipids to the extracted flours resulted in an increase in cookie diameter and an apparently complete recovery of the characteristics of cookies baked from the original flours. Approximately three fourths of the total lipid must be removed from a flour before definite changes in cookie-baking behavior (top grain and color) are noted. When the lipids of two flours (that differed in cookie-spread properties) were interchanged, the diameter of the cookies was the same as that of cookies from untreated flours; that is, the diameter appeared to depend on constituents other than flour lipids.

Flours of four wheat varieties, defatted with PE, produced cookies that were smaller and had reduced top-grain definition compared to cookies from parent whole flours (Kissell et al, 1971). Return of unfractionated free lipids to defatted flours at normal concentration restored original spread and top-grain quality. Polar and nonpolar lipid fractions alone were only partially effective in improving defatted flour; both were required for full restoration of quality. Thin-layer chromatography of lipid extracts revealed no detectable varietal differences. Flour-lipid interchange by variety produced no cookie quality differences owing to free lipid source. Cookie characteristics at normal lipid concentration were determined by varietal properties of defatted flour residues. Both whole (parent) and defatted flours increased progressively in cookie spread and top-grain score when treated with free lipids at one and a half, two, and three times the normal levels.

In a subsequent study (Yamazaki and Donelson, 1976), soft wheat flour was fractionated into five components—free lipids, gluten, tailings, starch, and water-solubles. Cookies baked from a blend of these fractions were identical to those baked from the parent flour. The good quality of cookies baked from blends of fractions indicated that a composite structure obtained by doughing the fractions is not necessary for the production of normal cookies. The interaction of free lipids with tailings and water had an adverse effect on the internal structure of cookies, and this effect was intensified when gluten was added as an interactant. Inclusion of starch as a further interactant moderated the effect and also controlled the spread of cookies.

Cookies baked from hexane-extracted flour exhibited limited spread and top grain as a result of breakdown of internal structure (Clements and Donelson, 1981). Functionality was restored when total free lipids were added back to the extracted flour. To determine the source of functionality, free flour lipids were separated by preparative thin-layer chromatography and added back to cookie mixes containing hexane-extracted flours. Two fractions (about 13% of the free lipids), corresponding to digalactosyldiglyceride (DGDG) (plus phosphatidylcholine) and monogalactosyldiglyceride, exhibited high degrees of restoration. Pure commercial DGDG added alone at 0.1% of flour weight (dry basis) or pure phosphatidylcholine from egg yolk added at 0.5% resulted in essentially complete restoration. Monogalactosyldiglyceride added at levels up to 0.15% gave little response. The data suggests that DGDG and/or phosphatidylcholine are the primary contributors to functionality. An unidentified glycolipid with chromatographic mobility similar to that of monogalactosyldiglyceride also appears to contribute to functionality.

I. Effects on Cake Quality

When pan-baking tests of cakes made with flour extracted with ethyl ether were conducted, cake volume was smaller and cake cells were finer than those obtained with the normal flour (Seguchi and Matsuki, 1977). Normal cake characteristics were restored when 0.2% flour lipids extracted with ethyl ether or water-saturated 1-butanol were added to the extracted flour. Flour lipids extracted with water-saturated 1-butanol were fractionated into polar (62%) and nonpolar (38%) lipids, and each fraction was added in turn to the extracted flour. Pan-baking tests showed that 0.2% polar lipid fraction restored cake volume and cell structure, whereas the nonpolar lipid fraction did not. Addition of a fatty acid sucrose ester to the extracted flour also restored cake volume and cell structure, but the addition of acetylated sucrose fatty acid ester or lard shortening had no effect on this restoration. These facts indicated that hydrophilic groups in lipids are necessary for good quality in pan-baked cake.

The results of a study on the role of free flour lipids in batter expansion in layer cakes were reported by Clements and Donelson (1982a, 1982b). White layer cakes baked from unbleached patent flours that had been aged (exposed to moving air) for 9 or 14 weeks exhibited greater oven expansion than did cakes baked from bleached control flours. A measurable increase in expansion occurred in cakes baked from flours aged two weeks. Expansion in cakes baked from defatted, unbleached and bleached flours reconstituted with free lipids from aged flours suggests that expansion induced by aging is a result of changes in free lipids. Cake volumes were retained when the lipids were added back to defatted, bleached flours, but some degree of collapse usually resulted when the lipids were added back to defatted, unbleached flours. Free lipids were extracted from unbleached flours and heated at 100°C for different periods, either unsupported or supported on diatomaceous earth. Cakes were then baked from defatted, bleached and unbleached flours reconstituted with the heat-treated lipids. Heating unsupported lipids for up to 60 min usually resulted in oven expansion greater than that of unbleached controls. Heating supported lipids for shorter periods resulted in oven expansion equal to or exceeding expansion from bleached control flours. Differences in sensitivity to heating were noted among flours. Expansion was retained as volume when heat-treated lipids were added to defatted, bleached flours, but cakes usually collapsed when lipids were added to defatted, unbleached flours.

Chlorinated and unchlorinated flours from four wheat varieties were extracted with hexane, and the free lipids were removed (Spies and Kirleis, 1978). The free lipids were reconstituted with extracted-chlorinated flour and baked using lean and rich cake formulas. Flours without free lipids produced smaller cakes with poorer textures than did flours with free lipids. Most of the original quality was restored when the free lipids were returned. Flours reconstituted with unchlorinated lipids produced larger cakes with poorer textures than did those reconstituted with chlorinated lipids.

The authors concluded that the free flour lipids are definitely needed to produce a satisfactory cake when using either a lean or rich formula. The ability of the free lipids to perform their functional role appears to be altered somewhat when the flour is chlorinated. Differences in cake-baking potential of flours apparently are determined by factors in the parent flour, rather than by differences in the free lipid composition. A relationship may exist, however, between cake baking potential and the amount of free lipid available as determined by the particle size of the flour.

Commercial unchlorinated soft red winter patent flour (pH 5.8) was treated to pH 5.2 (low), 4.8 (intermediate), and 4.0 (high) levels using 560, 1,120, and 2,240 ppm of chlorine gas, respectively (Kissell et al, 1979). Hexane-extractable (free) lipids were removed by exhaustive refluxing with the solvent. Baking performances of the hexane-extracted (defatted) flours were poor and about equal, regardless of chlorine treatment. When lipids were returned to the respective extracted flours, the original baking quality of the chlorinated flours was restored. Cake volume increased with increasing chlorination to a maximum at pH 4.8, then decreased at pH 4.0. When lipids from pH 4.0 (highly chlorinated) flour were added to the defatted flours of low and intermediate chlorine treatment, baking performance was inferior to the responses with their own lipid extracts. Addition of lipids from untreated flours of low and intermediate chlorination improved baking function of the highly chlorinated flour residue. In a parallel test in which the same chlorine treatments were applied to hexane-extracted unchlorinated flour, a similar set of responses was obtained, but the combination of variables yielding acceptable performance was restricted to low and intermediate chlorine rates. Interpolated factor space of cake volume responses for different combinations of free-lipid extracts and flour residues of differentially chlorinated flours are shown in Fig. 21. Figure 22 summarizes interpolated factor space showing cake volume responses for combinations of free-lipid extracts from serially chlorinated flours and residues of flour that were defatted and then serially chlorinated. The results of the study confirmed the importance of free lipids in situ at the time of flour chlorination.

In subsequent study (Donelson et al, 1984), chlorinated and untreated cake flours were hexane-extracted to remove free lipids and then fractionated by an aqueous procedure into prime starch, tailings, gluten, and water-soluble fractions. The fractions were reconstituted into dry blends such that one component from the respective flours was interchanged at a time. Bake results with a high-ratio layer-cake formulation showed that chlorinated lipids were the primary component contributing to the cake-quality potential. Lipid-interchange studies between varieties and wheat classes indicated that this functionality was independent of the lipid source. Differences in cake volume between flours were shown to reside in the hexane-extracted flour residues.

Objective measurements were used by Donelson and Clements (1986) to study the components of batter expansion in white layer cakes. They concluded that free flour lipids are necessary for expansion, but the hexane-extracted flour residue controls the degree of batter expansion in chlorinated patent flours.

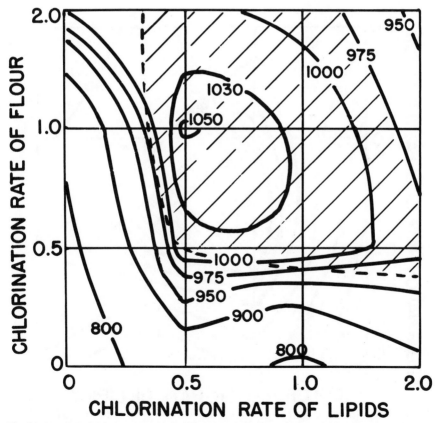

Fig. 21. Interpolated factor space showing cake volume responses for different combinations of free-lipid extracts and flour residues of differentially chlorinated flours. Shaded area represents limits of acceptable performance. (Reprinted, with permission, from Kissell et al, 1979)

Cake volume and batter expansion data indicated that shortening emulsifiers probably influence baking performance by contributing to batter aeration, whereas free flour lipids probably contribute to foam stability.

J. Effects on Pasta Products

The role of lipids in pasta products has been studied little. A review of the literature on lipids in pasta and pasta processing was published by Laignelet (1983).

Dahle and Muenchow (1968) removed various amounts of lipids and protein by extracting spaghetti with water-saturated butanol, 70% ethanol, and three concentrations of acetic acid. The extracted samples were cooked for different periods of time under reflux, and cooking characteristics were compared. Removal of lipid or protein increased the amylose concentration in the cooking water. The cooking quality of spaghetti was impaired more by protein removal than by lipid removal. The deleterious effects of lipid extraction were reduced in lipoprotein-enriched spaghetti. The authors concluded that protein is an

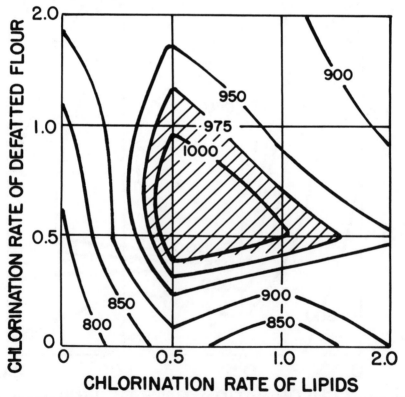

Fig. 22. Interpolated factor space showing cake volume responses for combinations of free-lipid extracts from serially chlorinated flours and residues of flours that were defatted and then serially chlorinated. Shaded area represents limits of acceptable performance. (Reprinted, with permission, from Kissell et al, 1979)

essential structural component of spaghetti and other pasta products. Without it, the strands tend to disintegrate and lose their form on cooking. The lipids supplement the function of protein and minimize other consequences of cooking such as stickiness.

Lin et al (1974a) separated total lipids of wheat flour and semolina into five fractions. Nonpolar lipids added at the 0.5% level to either the untreated or the PE-defatted flour caused an increase in farinogram mixing time and mixing tolerance and a slight decrease in water absorption. DGDG and phospholipids generally showed no pronounced effect on absorption, mixing time, or mixing tolerance of the dough from the untreated flour; however, a reduction in water absorption and mixing time was noted with the addition of DGDG to the defatted flour. Extraction of durum semolina with PE resulted in a higher water absorption in pasta and a loss of yellow color in the spaghetti. Addition of 0.6% nonpolar lipids to the defatted semolina restored and slightly improved the spaghetti color. Nonpolar lipids and monogalactosyldiglyceride slightly increased water absorption of the untreated semolina and the firmness of spaghetti, and DGDG and phospholipids decreased these factors. In general, neither nonpolar nor polar lipids affected the cooking quality of the spaghetti to any extent.

Fabriani et al (1968) found less extractable lipid in pasta than in semolina, suggesting that under the mechanical action of the extrusion screw, lipids undergo chemical changes, or are complexed, or both.

Barnes et al (1981) found that about 90% of the free lipids in semolina became bound during commercial pasta processing, especially during the drying stage.

Niihara et al (1982) studied the effect of lipids on Japanese "hand-stretched" noodles. They reported that fatty acids produced during storage contribute to the texture of cooked noodles by inhibiting the swelling of starch granules and affecting the viscosity of gelatinized starch after cooking. A study of the interaction between starch and fatty acid esters in frozen starch noodles showed that fatty acid esters, especially those of higher molecular weight, were effective in decreasing the adhesive force of starch (Mohri, 1980).

According to Matsuo et al (1986), durum semolina nonpolar lipids influence surface stickiness of cooked spaghetti. Removal of nonpolar lipids with PE increases stickiness, whereas enrichment with nonpolar lipids decreases stickiness. Commercial monoglycerides decreased stickiness and improved tolerance to overcooking. Semolina amylograph characteristics were affected by removal and addition of lipids. Reconstitution of defatted semolina with extracted lipids demonstrated that the extraction procedure did not influence amylograph characteristics of spaghetti quality. Effects of lipids on farinograph characteristics were not related to cooking quality factors. Processing spaghetti by a laboratory-scale continuous press and drying it at high and low temperatures verified the strong improving effect of monoglycerides on spaghetti stickiness and tolerance to overcooking.

K. Role of Lipids in Oxidation

The effects of mixing doughs in a vacuum and in the presence of various gases were studied by Baker and Mize (1937). The reaction of oxygen with flour components influenced mixing and breadmaking properties of doughs. The effects of oxygen on dough properties probably involve oxidation of flour lipids mediated by a lipoxygenase system (Miller and Kummerow, 1948).

Smith and Andrews (1957) found that mixing of flour-water doughs is accompanied by oxygen uptake. Removing lipids with pentane reduced oxygen uptake. Oxygen uptake could be restored by adding small amounts of linoleic acid. The oxygen-uptake capacity of flours increased during storage, paralleling the increase in free fatty acids. Oxygen uptake involved both lipids and water-soluble components. Heating the soluble flour components destroyed oxygen-uptake ability, indicating that enzymes are involved in the uptake mechanism. Gluten also may be involved; when it was absent, oxygen uptake by the remaining flour components was reduced significantly. Adding a commercial lipoxygenase alone, or especially with flour fat, increased oxygen uptake substantially and modified farinogram characteristics slightly.

In a second report, Smith et al (1957) reported that the oxidation of poly-unsaturated fatty acids in flour doughs, apparently by the action of lipoxygenase, alters the extensigram patterns of doughs in relation to mixing and increases the oxidation of sulfhydryl (SH) groups during mixing. The amount of oxygen that is used by doughs in the absence of fatty acids is small, but the amount markedly affects dough SH content and rheological properties. However, changes in dough SH content due to oxidation were not correlated with changes in dough physical properties as measured by the extensigraph.

Cunningham and Hlynka (1958) studied the relation between flour lipids and the effects of potassium bromate on dough properties. Prolonged mixing in the presence of lipids accelerates subsequent bromate reaction. In the absence of lipids, the bromate reaction is depressed. The authors concluded that bromate reaction in doughs from defatted flours mixed in air or oxygen is depressed through concurrent removal of natural antioxidants in the lipid fraction. Lee and Tkachuk (1959) found that the bromate reaction was more intensive in the original flour than in PE-extracted flour.

Treatment of flour with benzoyl peroxide reduced bromate consumption. Persulfate in doughs and in water slurries was reduced more rapidly than bromate (Lee and Small, 1960); extraction of lipids had, however, no effect on the rate of persulfate decomposition.

Bushuk and Hlynka (1961) suggested that the consumption of bromate in doughs occurs, at least to a major extent, by the reaction of SH groups of flour proteins with bromate. At the same time, lipid hydroperoxides, produced by lipoxygenase action, compete with bromate for reaction with SH groups. Results obtained with the thiobarbituric acid method on lipids extracted from flour indicated that lipid peroxides were formed in dough during mixing in air or oxygen (Tsen and Hlynka, 1962, 1963). Adding lipoxygenase increased the rate of peroxidation of flour lipids. On the other hand, adding antioxidants inhibited peroxidation. Several synthetic organic peroxides increased the structural relaxation constant of dough, indicating a similar improving role for flour lipid peroxides. Incorporating SH-blocking reagents and improving oxidants increased lipid peroxidation. This suggested that flour lipids competed with SH groups for available oxygen in dough. Free lipids were responsible for the peroxidation.

Narayanan and Hlynka (1962) conducted rheological studies on the role of lipid peroxidation in dough. Doughs from untreated flour showed only modest response to mixing in atmospheric oxygen, as indicated by structural relaxation data. Doughs from defatted flour showed a very marked effect of air, indicating that lipids protect against the improver effect of oxygen. After reconstitution, PE-extracted flour regained the properties of normal flour. Adding oleic and linoleic, but not palmitic, acid to extracted flour produced an initial improver effect of atmospheric oxygen and a subsequent protective effect similar to that of the lipid fraction. The unsaponifiable fraction of flour lipids, tocopherol, and several antioxidants had an improving effect when doughs were mixed in air. The researchers suggested that the improving agents, bromate and iodate, gave greater response to doughs from defatted flour on mixing in air, and that the response was due to the marked effect of oxygen on defatted flour. Bread from defatted, strong flours baked with shortening appeared to be overoxidized (as assessed by crust and crumb-grain character) at an oxidation level optimum for unextracted flours (Pomeranz et al, 1968b). Samples from PE-extracted flours gave an optimum loaf volume at a shorter mixing time than samples from untreated flours (Narayanan and Hlynka, 1962; Tao and Pomeranz, 1968).

Tsen and Hlynka (1962) showed that flour dough lost its SH groups faster than did defatted flour dough when mixed in oxygen or air. The difference between the normal flour dough and the defatted flour dough was attributed to the oxidation of lipids. Oxidized flour lipids or oxidized methyl linoleate increased the SH oxidation and exerted an improving effect (greater extensigram height), as did such simple peroxides as *tert*-butyl hydroperoxide, methyl ethyl ketone peroxides, and acetone peroxides. Apparently, when enough oxygen is available in dough, it reacts with both SH groups and lipids, and oxidized lipids also oxidize SH groups and thus exert an improving effect.

Hawthorne (1961) suggested that, on the basis of available information, no conclusive evidence can be presented to correlate oxygen incorporation by the lipoxygenase-fatty acid system with improvement of baking performance by oxygen uptake.

Similarly, Dahle and Sullivan (1963) compared the improver action of peroxidic linoleate and of a lipoxygenase-linoleate system with the improver action of hydrogen peroxide and found the latter much greater. The reactivity of a lipoxygenase-linoleate of glutathione was low. Assay of oxidized lipid by the 2-thiobarbituric acid method confirmed the oxidation of lipids in dough during mixing, and established, qualitatively, a slight interaction of SH and oxidizing lipid. The evidence indicated that peroxidic lipid, as a SH-oxidizing agent, has a minor role in the maturing action during mixing. On the other hand, lipoxygenase from soybeans can be involved in lipid binding and oxidation, modification of rheological properties of a dough, and oxidative improvement (Frazier, 1979).

L. Lipids in Protein-Enriched Bread

Commercially available, synthetic glycolipids (of the sucroester type) can be used to counteract the highly deleterious effects on loaf volume of high concentrations of soy flour in nutritionally improved bread (Pomeranz et al, 1969a, 1969b). The effects of synthetic glycolipids increase with an increase in the hydrophilic-lipophilic balance (i.e., with decrease in numbers and chain lengths of fatty acids attached to the carbohydrate molecule). Natural and synthetic glycolipids also improve loaf volume and crumb texture of bread baked from wheat flour enriched with defatted cottonseed flour, fish protein concentrate, sesame-seed flour, and food-grade yeast.

Adding polar lipids of wheat flour increased the loaf volume and crumb grain of bread nutritionally enriched with either heat-treated germ or germ proteins extracted with salt solutions (Pomeranz et al, 1970a, 1970b). In bread enriched with germ, the phospholipids rather than glycolipids in the polar wheat-flour lipids effected the improvement. Loaf volume of bread baked with up to 30% heat-treated germ was significantly increased by adding partly purified lecithin from soy flour. The amount of lecithin required for maximum loaf volume increased with increase in germ level. In bread containing 10–30% wheat germ, loaf volume was highest if the lecithin-to-germ ratio was in the range of 1:10 to 1.5:10.

The partly purified soy lecithin used was a mixture of phospholipids. Microloaves were, therefore, baked with 10% of heat-treated germ and various purified phospholipids. Some of the phospholipids were better improvers than others. Best results were obtained with pure synthetic phosphatidylcholine, phosphatidylethanolamine, and inositolphosphatide. The latter two, and especially phosphatidylethanolamine, imparted a yellow color to the bread.

In view of the high price of pure phospholipids, these results were mainly of academic interest. In practice, fractions of soy phospholipids containing mixtures of selected components, or modified phospholipids, can be used. Best results were obtained with alcohol-insoluble soy phospholipids (a mixture of two parts of inositolphosphatides, one part phosphatidylethanolamine, and a small amount of phosphatidylcholine), alcohol-soluble soy phospholipids (a mixture of two parts of phosphatidylcholine, one part phosphatidyl-ethanolamine, and small amounts of inositolphosphatides), or hydroxylated lecithin (phospholipids made by treatment with hydrogen peroxide and lactic acid to increase the hydrophilic properties of natural soy lecithin). Bread baked with 15% heat-inactivated wheat germ and any of the above commercial lecithins was comparable (except for color) to bread baked by a regular rich formula.

M. Changes in Lipids During Storage

Storing surplus grain to keep an "ever-normal granary" and to maintain strategic supplies makes it important to recognize early stages of deterioration, to prevent economic losses. Deterioration of grain and cereal products in storage is accompanied by increased acidity, including free fatty acids, acid phosphates, and amino acids (Pomeranz, 1971a). At early stages of deterioration, fat acidity increases at a much greater rate than either of the other two types, or all types of acidity combined. Consequently, it has been suggested that fat acidity be used as one of the best measures of grain damage (Zeleny and Coleman, 1938; Baker et al, 1959).

Oxidative rancidity does not take place in viable grain seeds; therefore, in general, only hydrolytic rancidity becomes of consequence in stored whole grains. Upon crushing or milling, the biological equilibrium characterizing viable seeds is altered. Hydrolytic, enzymatic, and oxidative rancidities may develop at a rate that is governed by the temperature and moisture content of the product.

Studies on the relation between the formation of fatty acids during storage of cereals and breadmaking performance produced somewhat conflicting results. Fenton and Swanson (1930) observed that poor baking quality of wheat flour was associated with high fat rancidity. Storage deterioration was shown to harm the milling properties of the wheat. Sorger-Domenigg et al (1955) and Pomeranz et al (1956) confirmed that flour produced from wheat having high fat acidity had high ash content, poor color, and poor baking strength.

The hydrolysis of lipids during storage of flour has been reported to damage gluten structure. For example, Barton-Wright (1938a, 1938b) found that saturated acids of high molecular weight produced only a slight shortening effect, whereas unsaturated acids damaged gluten seriously. Sinclair and McCalla (1937) reported that adding lipids from fresh flour (but not from germ) improved the loaf volume of both a fresh and a severely deteriorated flour. Sullivan et al (1936b) also observed that fat from fresh germ was not deleterious to the baking quality of a flour. The only constituents injurious to germ fat were the unsaturated fatty acids. Individual unsaturated fatty acids influenced flour in direct relation to the number of double bonds. Unsaturated fatty acids exposed to oxygen had a marked action on the baking quality of flour. The dough felt dead, lacked elasticity, and tore easily. Under those conditions, loaf volume was decreased considerably. Thus, the damaging effect of unsaturated fatty acids on breadmaking characteristics occurred only when fatty acids became oxidized. Later studies (Sullivan, 1940) showed that the unsaturated fatty acids influenced the colloidal behavior of the gluten by thickening the gel. Saturated acids showed no such effect.

The development of fat acidity in wheat or in wheat flour has been considered an index of breadmaking quality. Decrease of loaf volume was reported when more than 30% of flour lipids was hydrolyzed (Greer et al, 1954). Other workers, however, found no such deterioration, even after complete hydrolysis. An examination of many white flours stored in British lighthouses is particularly interesting (Greer et al, 1954). Although the free fatty acid content of the fat increased considerably (up to about 70%, compared with 5–10% in freshly milled flours), the flours retained good baking quality long after extensive hydrolysis of fat had occurred. Neither oxidative rancidity nor fungal mustiness was related to free fatty acid content. Fifield and Robertson (1959) found that wheat stored up to 33 years failed to germinate; the baking quality of flour milled from the wheat so stored was not affected materially, despite a fairly regular increase in fat acidity. Shellenberger et al (1958) found no correlation between bread quality and fat acidity of wheat flours stored at 4° C. The increase of fat acidity was related to the baking performance of flours stored at about 40° C.

According to Hutchinson (1961), there are two main objections to using fat acidity as the sole criterion of damage. First, there is a large scatter of experimental points about the regression line between seed viability (probably the best direct index of grain soundness) and fat acidity. Second, storage for short periods under unsatisfactory conditions changes fat acidity little but has a marked effect on the storage potentiality of the grain. Linko and Sogn (1960) emphasized that fat acidity indicates lowered viability, although fat acidity is not necessarily associated with any of the known deteriorative processes.

Hutchinson (1961) concluded that fat acidity did not provide valid information for evaluation of the baking quality of stored wheat or flour. Wheat flour may maintain its baking quality for many years, despite a progressive increase in fat acidity. Actually, a certain improvement in baking quality may occur in the early stages of storage, as reported by Shellenberger (1939).

In damp wheat stored at ambient temperature and with ample air access, fat acidity is likely to increase. In spite of the value of fat acidity as an index of changes occurring in stored grain, one cannot specify a meaningful limit for fat acidity without considering the previous history of the grain. Neither is there any proof that an increase in fat acidity is identical to damage to grain. Rate of increase in fat acidity is generally an index of changes occurring in grain. Yet, two samples may have comparable fat-acidity levels and differ substantially in breadmaking quality. In one, fat acidity may increase as a result of mold growth and concomitant damage to lipids, proteins, lipoproteins, and other flour components. In the other sample, fat acidity may increase slowly over a long period of storage at relatively low moisture. The second sample may have a higher fat-acidity value than the first one and (provided no damage was done to functional moieties governing breadmaking potentialities) still be well suited for breadmaking.

Daftary and Pomeranz (1965) followed changes in lipid composition during grain deterioration by qualitative and quantitative TLC and fractionation on silicic acid columns. Deterioration of wheat was accompanied by the formation of at least four unidentified compounds that showed autofluorescence under ultraviolet light. Grain deterioration was accompanied by a lowering of polar lipids and a rapid disappearance of at least five ninhydrin-positive or Dragendorff reagent-positive polar lipids. The breakdown of polar lipids was more rapid and more intensive than the formation of free fatty acids or the disappearance of triglycerides.

When flour containing 14.7% moisture was stored for six months at room temperature, it was lumpy and had an objectionable color and a musty odor (Pomeranz et al, 1968a). Gassing power was unaffected, but mixing time was doubled in mold-damaged flours. Bread baked from the damaged flour was substantially lower in loaf volume and poorer in crumb grain than bread baked from sound flour. Damage to breadmaking quality was accompanied by a substantial decrease of bound lipids and almost complete breakdown of free lipids. Baking of fractionated and reconstituted doughs indicated that damage to breadmaking quality was mainly due to breakdown of lipids.

In flours containing 18% moisture, free lipids and other flour components were damaged (Daftary et al, 1970a). The breakdown of bound lipids was accompanied by transformation of polar components to nonpolarlike components. The ratio of nonpolar to polar components increased as the storage temperature of the flours increased.

The breakdown of free lipids was accompanied by disappearance of lipoprotein, which was present in PE extracts of the sound flours. Starch-gel electrophoresis and baking of reconstituted doughs showed damage to gluten proteins and water-solubles (Daftary et al, 1970b).

Bell et al (1979a) stored flours milled from weak, medium, and strong wheats at three temperatures (12, 25, and −20° C) for up to 66 months. Lipolysis was rapid in the medium flour stored at 25° C and slow in control flours stored at

−20° C in an inert atmosphere. Changes in breadmaking quality during storage were determined by 1) the Chorleywood bread process, 2) a long fermentation process, and 3) the activated dough development process. Changes in loaf volume varied with the flour type and test method and largely depended on the presence of fat in the breadmaking recipe. For the medium and strong flours, loaf volumes were practically constant for up to 24 months and decreased rapidly thereafter. Gradual deterioration became apparent when the second and third tests were made with fat. When fat was omitted, short-term improvement (during the first 12 months) occurred with all three baking methods and was followed by a consistent decrease in loaf volume. The study confirmed that, whereas an increase in free fatty acids is correlated with a decrease in breadmaking potential, fatty acids may not be causatively related to that decrease. Consequently, placing a fixed upper limit on the level of fatty acids above which deterioration is evident does not seem possible, and fat acidity alone cannot give a reliable indication of deterioration of stored flours.

In a subsequent study (Bell et al, 1979b), a fresh flour dough containing 0.7% of a fat prepared from saturated and unsaturated triglycerides was supplemented with 0.7% pure oleic or linoleic acids, an amount sufficient to raise the level of unesterified fatty acids to that obtained with a flour that had badly deteriorated as a result of storage. Both fatty acids were detrimental to the volume of bread baked by the Chorleywood bread process. Margaric acid was not deleterious. Both unsaturated fatty acids behaved similarly during dough mixing; they became largely "bound" and increased binding of glycerol trioleate and flour lipids. Margaric acid, however, remained largely "free" and did not affect the binding of other lipid components in the dough. The similarity of the effects of oleic and linoleic acids on loaf volume and lipid binding may not apply to all flours and all fatty acid levels. However, their detrimental effect was not proportional to the number of double bonds and did not depend on oxidation by lipoxygenase, for which oleic acid is not a substrate. The adverse effects of unsaturated fatty acids are likely to take place mainly through their influence on the associations of other lipids with gluten or starch.

Warwick et al (1979) monitored changes in lipids in three flours stored for five years. A small but significant loss of polyenoic acid was related to the rate of lipolysis in the flours. A marked increase in free fatty acids and a corresponding decrease in triglycerides were found. Triglycerides and phospholipids were hydrolyzed to free fatty acids and water-soluble components without accumulation of partial hydrolysis products. The glycerides showed accumulation of monoacyl derivatives and some evidence of further hydrolysis. The most rapid increase in free fatty acids took place in the initial storage period, during which no significant changes in baking quality occurred. The most likely cause of the loss of polyenoic acids was postulated to be oxidation—probably involving lipoxygenase or, less probably, autooxidation. The implications of oxidative changes are important because free radical products of lipid oxidation can denature proteins and decrease baking quality. During early storage, some oxidative processes, although not necessarily those affecting lipids, may improve breadmaking quality. Thus, the oxidative changes seem to have a dual effect that may explain the complex changes in stored flours (Warwick et al, 1979).

N. Lipid Binding and Dough Structure—Interactions

The evidence for and role and significance of interactions between lipids (mainly glycolipids) and proteins and starch were reviewed by Pomeranz (1971b, 1973b, 1980a, 1984), Pomeranz and Chung (1978), MacRitchie (1980a, 1980b), Nierle and ElBaya (1981), Schuster (1984), and Chung (1986).

Rohrlich and Muller (1968) and Rohrlich and Niederauer (1967a, 1967b) used the Folch method to prepare a proteolipid from wheat, rye, and oats. They postulated that the wheat-flour complex is responsible for the improving effect of adding strong HRS flours to soft, low-protein European flours.

Zentner (1958, 1960) found lipid-protein complexes in moist-acetone extracts of gluten and flour. Zentner (1964) also suggested that cysteic acid groups of oxidized wheat proteins may react with electropositive groups carried by the lipids that form part of the gluten complex, and that phosphoric acid residues that occur in the flour lipids may be similarly bound to proteins.

Gluten washed from flour contains substantially more lipids than the original flour. Although much lipid in flour is in free form, only a small fraction can be extracted from gluten with ether or petroleum ether. McCaig and McCalla (1941) suggested that a protein-lipid complex is formed during gluten preparation or dough formation. Evidence of the binding of added lipid to flour gluten was also provided by experiments in which only part of the added lipid was released during washing-out of gluten from a developed dough (Baker and Mixe, 1942).

Olcott and Mecham (1947) studied changes in lipid binding during dough mixing. A high-protein patent flour contained 1.5% total lipids, 70% of which was extractable with ether. The flour was mixed with water with minimum doughing and was dried by lyophilization. Only 40% of the dough lipids was extractable with ether. After the flour was kneaded into a dough and dried, less than 10% of the lipids could be extracted. The capacity of the flour to "bind" lipids during wetting and doughing was ascertained by determining the extractability of added flour lipids. At least three times the amount of lipid normally present could be bound by the doughing procedure. Phospholipids were bound preferentially. Most of the bound lipid was associated with the gluten, rather than with the nonprotein constituents of flour, and, when gluten was fractionated, the lipid was bound to the glutenin rather than to gliadin fractions (Ponte et al, 1967a).

Salt decreased lipid binding in doughs, both of total lipid and of phospholipid, to 20–40% less than that occurring in the absence of salt (Mecham and Weinstein, 1952). Shortening appeared to decrease phospholipid binding slightly but did not affect total lipid binding appreciably. Other bread ingredients had no detectable effects. The lipid contents of glutens washed out in salt solutions were lower than those of glutens washed out in water, paralleling observations on lipid binding in doughs.

Baldwin et al (1963, 1965) studied lipid binding in doughs and bread made by conventional and continuous processes. Daniels et al (1966) investigated the distribution of lipids in bread produced by three mixing methods, using bread formulas that included low levels of added shortening and full-fat soy flour.

Chiu and Pomeranz (1966) extracted free and total lipids from a flour and from dough, fermented dough, bread crumb, and bread crust. Dough formulations used in breadmaking included—in addition to a basic formula of flour, water, yeast, and sodium chloride—either sugar, commercial vegetable shortening, or nonfat dry milk, or their combinations. PE-soluble flour lipids were reduced to one third during dough mixing or fermentation; subsequent baking lowered the residual free lipids to half. PE-soluble free lipids were affected little by dough composition. Only small amounts of hydrogenated vegetable shortening were bound during dough mixing, but about one third to one half of the added shortening lipids became bound during baking. Processing flour into bread had no effect on the amounts of total lipids extractable by a chloroform-methanol mixture. Fractionation of extracted lipids by TLC showed that much more polar wheat flour lipids than nonpolar components were bound during dough mixing.

Daniels et al (1968) made the interesting observation that binding of shortening fat in premixed doughs decreased during prolonged mixing at relatively high speeds (up to 378 rpm). Daniels and co-workers (1969) studied the binding of specific lipids during high-speed mixing of doughs and in the resulting bread as related to mixer atmosphere. The presence of air reduced the linoleic acid content of triglyceride lipids, an effect apparently related to the reduction in lipid binding. Replacing free flour lipids (the main source of triglyceride linoleate) by extra shortening fat increased bound lipid in doughs mixed either in air or in nitrogen, and also decreased bread quality and loaf volume. The authors concluded that the role in breadmaking of natural free lipid of flour is to respond to the atmosphere in the dough mixer and to affect the binding of shortening triglyceride during dough development.

Under ideal conditions, the quality of bread produced by high-energy dough development is improved by mixing under reduced pressure. The mixing at reduced pressure also increased lipid binding (Daniels et al, 1969). These authors suggested that the vacuum acts on lipid binding through the resultant reduction in the oxygen content of the dough atmosphere, rather than through the physical effect of reduced pressure alone.

Extractability of lipids, by various solvents, from gluten and dough was reported by Wootton (1966). Youngs et al (1967) studied lipid binding in various wheat-flour fractions.

Chiu et al (1968) fractionated a composite HRW wheat flour, containing 12.9% protein, into gluten and a mixture of starch and water-solubles. The original flour, the mixture of starch and water-solubles, and reconstituted flours containing respectively 10, 13, and 16% protein were baked by a formula that included 3% shortening. Similarly, PE-defatted flour was fractionated, reconstituted, and baked (Fig. 23). In addition, 16 flours milled from eight HRW, five HRS, and one each SRW, club, and durum wheat varieties were baked. Free lipids were extracted with PE and bound lipids with water-saturated butanol following PE. The gluten contained up to five times as much protein and lipids as the original flour (Table VIII). The mixture of starch and water-solubles contained little lipids, and distribution of its nonpolar and polar lipid components varied from the distribution of lipids in gluten. No measurable amounts of shortening lipids were bound during mixing of doughs from untreated flours and about 0.2% were bound in doughs from PE-extracted flours. During the baking stage, about one third to one half of the shortening lipids were bound; the extent of binding increased with increasing protein contents. Lipid binding during dough-mixing of flours that varied in bread-making potential increased with protein content and mixing time and decreased in flours of poor protein quality.

Pomeranz et al (1968d) studied the effects of lipid fractions, mixing time, and dough ingredients on lipid binding. Up to 1.5% unsaturated corn oils were bound in dough mixed from PE-extracted flour; less of saturated corn oils and fats were bound. Much more unsaturated oils than saturated fats were bound in bread crumb. Increasing the length of dough mixing increased the binding of free flour lipids; binding decreased during prolonged overmixing. Adding 2–4% NaCl to the dough had little effect on the binding of polar flour lipids (added to a PE-extracted flour), but it significantly reduced binding of nonpolar lipids. The

TABLE VIII
Protein and Lipid Contents[a] of a Composite Hard Red Winter
Wheat Flour and Its Fractions[b]

Material	Protein[c] (%)	Free (%)	Bound (%)
		Lipids	
Untreated flour	12.9	0.85	0.39
Gluten	64.5	0.68	5.31
Starch and water-solubles	2.5	0.12	0.13
Starch	0.5	0.05	0.08
Petroleum ether-extracted flour	13.0	0.10	0.35
Gluten	57.1	0.07	1.45
Starch and water-solubles	2.4	0.05	0.05

[a] Source: Chiu et al (1968).
[b] 14% moisture basis.
[c] $N \times 5.7$.

Fig. 23. Loaves baked from 10 g of flour. Top row, from untreated flours; bottom row, from petroleum ether-extracted flours. From left to right: original flour containing 13% protein, mixture of starch and water-solubles, flours reconstituted from a mixture of starch, gluten, and water-solubles to give 10, 13, and 16% protein, respectively. (Reprinted, with permission, from Chiu et al, 1968)

results suggested that the action of salt was on the gluten proteins and not on saltlike linkages. The presence of nonpolar lipids reduced the binding of polar lipids. More bound flour lipids were released by remixing a dough with a NaCl solution than by remixing with water. Practically no added nonpolar lipids were bound by dry-mixing with flour, but substantial amounts of free polar lipids were bound during dry-mixing with a flour containing 4.4% moisture. The binding increased with increases in moisture contents and was highest in dough.

Fabriani et al (1968) studied changes in lipids in the processing of pasta products from durum wheat semolina. Infrared analysis of acetone-extracted lipids indicated quantitative changes in triglyceride, phospholipid, and fatty acid contents in extruded and sheet-formed pasta with respect to the corresponding semolina; it also showed a decrease of absorbance at the wave length of β-sitosterol. Gas chromatography confirmed the above quantitative changes in sterol and fatty acid contents.

Hess (1954) proposed, on the basis of X-ray, electron-microscope, and optical measurements, a structural relationship of protein, lipid, and starch in wheat flour, in which wedge-protein deposits are surrounded by a lipid (and lipoprotein) layer, beyond which lie adhesive protein layers and corresponding starch granules.

Traub et al (1957) investigated lipid-protein complexes by studying diffraction patterns provided by X-ray photographs of the endosperm of cereal grains and of the flours, doughs, glutens, starches, and fats derived therefrom. Endosperm sections of wheat grains showed spacing at 47 Å not found in barley, corn, rice, or oats, although present to a lesser extent in rye. The 47-Å spacing appeared very early in the development of the wheat grain and persisted in flour after milling. It was due to fatty materials associated with the gluten proteins of the endosperm, and the attachment was strengthened during doughmaking or gluten-washing. The orientation of spacing in gluten, gliadin, and certain lipid fractions suggested that the involved fat molecules occur within or near the protein as layers of bimolecular leaflets, with the fat molecules roughly perpendicular to the protein fibers. Most of the fatty material could be extracted from fresh flour by cold acetone, leaving a flour with no spacing at 47 Å, although still with much of its breadmaking and gluten-forming capacities. The workers therefore concluded that the association between lipids and proteins in flour, dough, and gluten may have little relevance to the formation of gluten or to major differences between weak and strong wheats.

When the lipid extract involved in the strong diffraction spacing at 47 Å was fractionated by chromatography on silicic acid columns, it gave at least six bands (Daniels, 1958). All but band V were rich in carbohydrates, indicating glycolipids. Band V gave the 47-Å X-ray spacing. Analyses showed choline, ester groups, and phosphorus to be present in molar ratio 1:2:1, indicating that a main constituent of band V was a lecithin. There was no evidence that lecithin had any relation to the breadmaking capacity of wheat flour.

Grosskreutz (1960) showed that proteins in moist gluten consist of folded polypeptide chains in the alpha-helix configuration, arranged into flat platelets of the order of 70 Å thick. However, no significant differences in physical microstructure were found between the glutens of two flours that varied widely in dough-development time.

Extraction of the phospholipids did not affect the basic platelet but seriously impaired its ability to bond into sheets capable of sustaining large plastic deformations (Grosskreutz, 1961). X-ray evidence of the phospholipid structure in gluten favored the assumption that there exist well-oriented bimolecular leaflets of the type found in myelin figures, and a lipoprotein model was postulated that occupies 2–5% of the elastic gluten structure. Consideration of the nature and strength of the bonds present in a hypothetical gluten sheet showed that the proposed lipoprotein model is capable of providing gluten with the plasticity necessary for optimum baking characteristics. Grosskreutz (1961) postulated that, as part of the lipoprotein structure, protein chains are bound to the outer edge of a phospholipid bimolecular leaflet array, probably by saltlike linkages between acidic groups of the phospholipid and the basic protein groups.

Seckinger and Wolf (1967) used electron microscopy to study lipid distribution in the protein matrix of the wheat endosperm. Free lipids were distributed throughout the protein matrix; bound lipids appeared in small inclusions in the matrix and were assumed to be remnants of cytoplasmic structures occurring in endosperm cells at maturity. No difference was observed in the protein matrix, which would suggest that it could be separated into wedge- and adhering-protein fractions.

Phospholipoprotein complexes may arise through combinations of protein with phospholipid and with metallic ions, to form three-way complexes (Fullington, 1969). These complexes are stable and may persist in doughs in preference to other arrangements. The lipids seem to exist and function in the form of micelles, or self-contained small aggregates, rather than as the sheets proposed by Grosskreutz (1961).

Studies based on differences in solubility of gluten proteins have shown that wheat-flour polar lipids (principally glycolipids) are bound to the glutenin protein by hydrophobic bonds and to gliadin protein by hydrogen bonds (Hoseney et al, 1970). Results indicated that, in unfractionated gluten, the lipid apparently is bound to both protein groups at the same time and forms a complex. The complex was pictured as units of gliadin—and glutenin—protein bound together by polar lipids.

To study the interaction of glycolipids with wheat flour macromolecules, complexes between galactosylglycerides and starch, gliadin, and glutenin were investigated by infrared and nuclear magnetic resonance (NMR) spectroscopy (Wehrli and Pomeranz, 1970a). Infrared spectroscopy indicated that hydrogen bonds exist between glycolipids and gelatinized starch and gluten components, and Van der Waals bonds between glycolipids and gluten components. By definition, hydrophobic interactions cannot be ascertained by infrared spectroscopy in the absence of water. NMR spectra showed an inhibition of the methylene signal (at 8.7τ) in glutenin mixed with glycolipid, indicating hydrophobic bonding.

The two previous studies concerned interaction of glycolipids with isolated starch and gluten. To investigate which interactions take place in dough and bread containing both starch and proteins, additional studies were made (Wehrli and Pomeranz, 1970b). Tritium-labeled galactosyldidecanoylglycerol was synthesized. Sections 5 μm thick prepared from dough and bread containing the labeled galactolipid were studied by autoradiography. In the dough, the galactolipid was distributed in the gluten and, to some extent, in the starch; in the bread, most of the galactolipid was in oven heat-gelatinized starch granules.

The findings on the interaction of glycolipid and wheat flour macromolecules are summarized in Table IX. The results indicate that, in dough, glycolipid interacts with gluten protein according to the scheme based on differences in solubility. A limited interaction with starch granules is also indicated. Basically, increase in loaf volume during oven-spring can be attributed to the formation of a complex between gluten proteins and glycolipids. In the baked bread, a complex between glycolipids and starch seems of primary importance and could be responsible for the improved freshness retention of bread baked with glycolipids.

Simmonds and Wrigley (1972) used extraction with $6M$ urea to separate the more readily soluble gliadin and glutenin fractions of freeze-dried gluten, and of wheat storage protein, from the insoluble residue. Less protein was extractable from gluten than from storage protein, which had been prepared with organic solvents and was thus depleted in lipid. Reconstitution experiments, which involved the wetting of storage protein in the presence of readded flour lipid, suggested that the difference in protein solubility in the case of gluten is due to lipid-protein association during dough formation.

Chung and Pomeranz (1978) examined extractability and composition of proteins from flours of good and poor baking quality, and from these flours after delipidation. Proteins were extracted with $0.05 N$ acetic acid from nondefatted and defatted flours. Extracts were dialyzed, lyophilized, and fractionated by gel filtration on a Sephadex G-100 column. The extracts and the fractions were characterized by starch gel electrophoresis and sodium dodecyl sulfate (SDS)-polyacrylamide gel electrophoresis (PAGE). Both the nondefatted and the defatted poor (Chiefkan/Tenmarq) flours showed higher protein extractability than did the good flours (Shawnee). More proteins were extracted from good and poor control flours than from the corresponding defatted flours. In elution profiles from the gel column, all protein extracts showed similar retention times for the glutenin, gliadin, and water-soluble fractions. Quantitatively, the glutenins were the most variable among the gel-separated fractions in the acetic acid extracts from the flours. Whereas concentration of gliadins were similar in all extracts, concentrations of glutenins depended on the flour samples and the

TABLE IX
Bonds in Glycolipid and Wheat-Flour Macromolecule Complexes[a]

Method of Study	Type of Bond Between Glycolipid and		
	Starch	Gliadin	Glutenin
Solvent extraction of gluten proteins	...	Hydrogen	Hydrophobic
Lipid binding in starch dough	Hydrogen	...	...
Infrared	Hydrogen	Van der Waals, hydrogen	Van der Waals, hydrogen
Nuclear magnetic resonance	Hydrogen, some induced dipole interaction	...	Hydrophobic, hydrogen
Autoradiography	Strong interaction in bread	...	Interaction in dough
Baking test	(Hydrophobic and hydrogen bonds found to be essential for improvement in breadmaking.)		

[a] Source: Wehrli and Pomeranz (1969a).

delipidation solvent. Delipidation decreased glutenin-gliadin ratios. Glutenin-gliadin fractions showed differences between strong and weak flours and between nondefatted and defatted flours in number and intensity of electrophoretic bands.

In a subsequent study (Chung and Pomeranz, 1979), binding of acid-soluble proteins from untreated and defatted flours of good (Shawnee C.I. 14157) and poor (Chiefkan/Tenmarq, KS501097) baking quality to phenyl-Sepharose CL-4B was examined by batch-elution and column-elution techniques. When a mixture of hydrophobic-gel and acid-soluble flour proteins was eluted batchwise with five solvents, total absorbance at 280 nm of the eluates was higher for the poor than for the good baking quality flour. Acid-soluble proteins of defatted good and poor baking flours differed little from those of nondefatted flours in hydrophobic binding capacity. Glutenins from isopropanol-defatted flour of poor baking quality (KS501097) were less hydrophobic, and gliadins were more hydrophobic, than glutenins and gliadins from the flour with good baking quality. The difference in apparent hydrophobic interaction was more pronounced for glutenin than for gliadin. The relation between the amount of protein eluted and the amount of protein adsorbed by the hydrophobic gel varied with protein sources and their concentrations. Among protein eluting agents, 1% SDS in $0.02N$ acetic acid was most effective. The results indicate that acid-soluble proteins from good and poor breadmaking flours differ in apparent hydrophobic properties.

Pomeranz and Chung (1982) compared interactions between lipids and proteins in flour, dough, and gluten. Two flours were selected that contained comparable amounts of protein (about 13%) but had widely different mixing times, stability, and loaf volume potential. The good breadmaking flour contained much more free polar lipids than the poor breadmaking flour.

Extracting lipids reduced dispersibility of wheat flour proteins in $0.05N$ acetic acid or $3M$ urea. In each protein extractant and each level of extracted lipids, less protein was extracted from the good than from the poor quality flour. That difference was largest in the control flour, from which no lipids were removed, and decreased as amount of extracted lipids increased; this was another indirect confirmation for another role of lipids in breadmaking: protein extractability varies most widely when the total lipid content is left intact.

The types of protein that are present in extracts of flours defatted to various degrees were measured by exclusion and elution from a Sephadex G-100 column. Lipid extraction seems to have little effect on extractability of gliadin and water-soluble proteins. The reduction in dispersible glutenin proteins accounts for the large drop in dispersibility as the amount of extracted lipids exceeds about 1%. This large effect on extractability of glutenin proteins and lack of effect on gliadin proteins explains a steep drop in glutenin-gliadin ratios as percent of total extracted lipids increases.

The drop in protein dispersibility effected by extraction of lipids can be reversed, to some extent, by recombining the lipids with the defatted flour. Restoring polar lipids had little effect on restoring protein dispersibility and, in the case of the good flour, actually decreased it somewhat; nonpolar lipids restored dispersibility. Restoration was best in flours from which only free lipids were removed and restored. The effects of acetone and benzene were intermediate, and that of 2-propanol was least reversible. Those differences reflect, no doubt, a combination of the irreversible effects of polar solvents on proteins and the effects of removing bound lipids on binding sites of proteins.

Studies were also conducted on lipid effects and interactions in flours and doughs treated with various levels of cysteine (Pomeranz and Chung, 1982). The two flours selected for those studies were C.I. 12995, which has a long mixing time, and RBS-75, a composite of good breadmaking varieties grown throughout the Great Plains. No poor breadmaking flour was included because its short mixing time would preclude determining the effects of a reducing agent. Mellowing of the somewhat bucky and excessively long-mixing flour by cysteine reduced mixing time and increased loaf volume. Although a minimum mixing time, and the mixing stability associated with it, is desirable, a long mixing time is not desirable and cannot be equated with good breadmaking potential.

Defatting with 2-propanol increased mixing time more than threefold and reduced loaf volume in flours baked with or without 3% shortening. Reconstitution with the extracted lipids did not restore loaf volume or normalize mixing time. However, when cysteine was added, both loaf volume and mixing time were largely restored. The results seem to indicate that the improving effect of cysteine is related to enhanced lipid binding of somewhat degraded high-molecular-weight gluten proteins. Removing lipids increases protein aggregation and reduces protein extractability. If the aggregation is excessive, mixing time increases unduly and loaf volume is impaired. Cysteine reduces excessive aggregation and increases lipid binding. This mellowing effect, involving lipid-protein interaction, improves overall breadmaking quality. Both nonpolar and polar lipids are involved in this interaction. The involvement of polar lipids seems to be of greater significance (despite their low total levels) because of their well-established functional properties and because their binding increases as cysteine levels increase.

Good breadmaking flours can tolerate lower pH levels than poor flours in gluten solubilization without adversely affecting breadmaking properties. Gluten solubilization changes are accompanied by changes in the ratio of bound lipids (BL) to free lipids (FL). The changes depend on the pH of solubilization, but not on whether lactic or acetic acids are used. The BL-FL ratio in the neutralized gluten increases as pH of solubilization increases. Slopes of the lines for two flours (C.I. 129950 and Cfk/Tm, of good and poor breadmaking quality, respectively) differed by a factor of 2.22 (9.06/4.08). The breaking points (with regard to loaf volume) coincide with BL-FL ratios of about 13 for C.I. 12995 and about 9 for Cfk/Tm.

Those results point to the facts that lipid binding in gluten proteins differs in flours that differ in breadmaking quality, that binding depends on the pH that was used to solubilize the gluten, and that binding reflects lipid-protein interaction and must reach a certain critical level to assure functionality in breadmaking. The extent of lipid-protein binding can bring out the best in a flour; it cannot, however (at this time, at least) raise the optimum above a level that is genetically controlled.

The level of lipid binding, reflecting protein interaction, differs in flour, dough, and gluten. The overall BL-FL ratio is largest in crude gluten, and it is largest in acid-soluble proteins of the optimally mixed dough. The BL-FL ratios of gluten proteins from both good and poor quality flours increase when the pH is raised to 6.1. The BL-FL ratios of the solubilized gluten from good and poor flours differ by a factor of about 2.2 (3.3/1.5). The factor decreases to about 1.7 (15/9) for the functional gluten that was precipitated by neutralization to pH 6.1 and reflects inherent differences in breadmaking quality of wheat flour proteins.

The interaction of protein with lipid in wheat gluten was studied by electron spin resonance (Nishiyama et al, 1981). The gluten in the flour suspension was spin-labeled with a fatty acid spin label and washed from the flour. The electron spin resonance spectra of the spin label incorporated into gluten exhibited clearly separated parallel and perpendicular hyperfine splittings. The orientation of the gluten lipid and its fluidity showed temperature dependence. A phase transition was observed at 25°C. Compared with gluten, vesicles of lipids from the flour were in a less oriented, highly fluid state, and with much lower activation energy for rotational viscosity, whereas the reconstituted gluten, which was prepared by mixing purified gluten protein and the extracted lipids, had a lipid environment similar to that of gluten. The results indicate that the lipid was immobilized in the gluten matrix by strong interaction with protein.

Békés et al (1983a) obtained gliadin preparations from glutens with various lipid contents and compositions in studies on protein-lipid complexes. Lipid contents of the glutens and the gliadins varied with the defatting solvent (*n*-hexane, *n*-butanol, water-saturated *n*-butanol). Gliadin preparations were fractionated into five fractions by gel filtration chromatography on Sephadex G-200. Fraction I (eluted in the void volume) contained mainly polar lipids, and fraction III contained mainly nonpolar lipids. Gliadin preparations from completely defatted flour or gluten contained no carbohydrate, whereas carbohydrate and lipid contents of other gliadin preparations were related. Protein subunit composition of fraction I, determined by SDS-PAGE, depended on whether or not lipids were present in the gliadin preparation. Fraction I of the preparation from completely defatted flour or gluten contained only subunits of high (>100,000) molecular weight, whereas the analogous fraction from undefatted or partially defatted flour also contained subunits of low (<40,000) molecular weight. These results indicate that in the gliadin preparations from undefatted flour or gluten, lipids (likely galactolipids) cause aggregation of low-molecular-weight subunits with the high-molecular-weight subunits.

In a subsequent study (Békés et al, 1983b), gliadins from undefatted and defatted flour were fractionated by gel filtration chromatography on a Sephadex G-200 column. Fraction I (excluded) from undefatted flour contained almost all of the SDS-PAGE subunits of the original unfractionated gliadin. After removal of lipids from this fraction, its protein was fractionated into five peaks on rechromatography on Sephadex G-200. The five protein fractions of the gliadin from defatted flour were reconstituted with the original lipids and chromatographed on Sephadex G-200. SDS-PAGE showed that the subunit composition of fraction I was similar to that of the original gliadin from undefatted flour. The researchers postulated that the lipid-mediated aggregation of high- and low-molecular-weight subunits of gliadin is reversible. They also concluded that gluten lipids are approximately equally divided between the gliadin and glutenin fractions.

Frazier et al (1981) reported the isolation of a lipoprotein, ligolin, from a 0.05*M* acetic acid extract of lyophilized dough. The lyophilized acid-soluble proteins were dissolved in AUC. Glutenins and gliadins were precipitated with ammonium sulfate, and the supernatant contained ligolin. Ligolin contained about 10% of the total gluten proteins; it had a molecular weight of 9,000; and the molar ratio of radioactive-labeled triglyceride to protein was about 1:1.

Gluten proteins were fractionated by pH precipitation (method A) and ammonium sulfate precipitation (method B) to determine whether the association of lipids with the gliadin fraction prepared by method A results from mutual solubility of the lipids and the proteins in 70% aqueous ethanol (Zawistowska et al, 1985). The distribution of lipids and proteins among fractions obtained by the two methods were different. For method A, 56% of the gluten lipid was associated with the gliadin fraction and 44% with the glutenin fraction. Almost all of the polar lipid in gliadin A, which contained 75% of the glycolipid of the original gluten, coeluted with fraction I when gliadin A was fractionated by gel filtration chromatography on Sephadex G-200. For method B, acetic acid-soluble gluten lipid was distributed almost entirely between two fractions, the first precipitate, P_1, and the supernatant, S. Most of the polar lipid of P_1 was associated with its ethanol-soluble subfraction, P_1S. The two fractions of method B, P_1S and S, seemed to have a particularly strong affinity for polar lipid; they contained 69% of the polar lipid present in the acetic-acid-soluble gluten that was fractionated. SDS-PAGE and solubility characteristics showed that fraction P_1S of method B and fraction I of gliadin A contained common protein components that appear to be responsible for binding polar lipid. The results suggested an intrinsic affinity of specific gluten proteins for specific lipids. The presence of the lipids in gliadin was influenced entirely by use of the ethanol as solvent, as both gliadin subfractions P_1S and 1 contained similar lipids.

The study also showed that various methods of gluten fractionation can distribute proteins differently among fractions that may be considered the same or similar. This is in line with the findings of Olcott and Mecham (1947), Ponte et al (1967a), and Hoseney et al (1969c). The results of all those studies demonstrated that lipids associated with gluten fractions can be compared only if the fractions were prepared by the same procedures.

Lipoprotein S isolated by method B by Zawistowska et al (1985) is the same as ligolin isolated by Frazier et al (1981). Zawistowska et al (1985) suggested that protein S may have an important functional role in breadmaking because of its unique affinity for polar lipids. As the amounts of protein S in two flours that varied widely in breadmaking were similar and the lipids associated with them (amounts and compositions) were also similar, the precise contribution of this unique protein is yet to be established (Chung, 1986).

The involvement of both carbohydrates and lipids in the aggregation of glutenin proteins was postulated by Zawistowska et al (1985). Glutenins were prepared from untreated flour, from flour partially defatted by *n*-butanol and then by *n*-hexane, and from flour defatted by water-saturated *n*-butanol. The glutenins differed in lipid contents, but all contained similar amounts of carbohydrate that upon hydrolysis yielded almost entirely glucose. Glutenins were then reduced, alkylated, and fractionated on Sephadex G-200. Of the resulting three fractions, fraction I, which eluted at the void volume, had unusually high (for gluten) contents of lysine and aspartic acid, lower contents of glutamic acid and proline, and all the carbohydrate and lipid originally present in the glutenin. SDS electrophoresis showed that fraction I contained subunits of lower molecular weight than expected from their elution volumes, indicating aggregation. Fractions II and III were characteristic of gluten proteins rich in glutamic acid and proline. They contained no carbohydrate or lipid. After partial or total defatting, the proportion of fraction I changed, indicating that lipid contributed to the aggregation of the proteins in this fraction. Removal of carbohydrate from fraction I by digestion with amyloglucosidase produced further disaggregation. SDS electrophoresis of subfractions of fraction I from which the carbohydrate had been removed showed that they contained the same glutenin subunits as in the original fraction I but in different proportions, The authors concluded that the strong aggregative tendency of fraction I proteins of glutenin is determined mainly by protein structure but that specific lipids and carbohydrates contribute significantly to this behavior. Because such protein aggregation may contribute substantially to dough functionality, further research in that area is warranted (Zawistowska et al, 1985).

VII. CARBOHYDRATES

A. Starch in Baked Products

Starch is the major component of wheat kernels; it comprises about 54–72% of the kernel's dry weight, depending on both variety and growing conditions (see also Chapter 6 in Volume I of this monograph).

Physical properties of starch in concentrated systems such as dough and bread were the subject of a dissertation by Eliasson (1983). Gelatinization and retrogradation of starch in concentrated aqueous systems were studied using differential scanning calorimetry and a stress-relaxation technique. An aqueous suspension of wheat starch exhibits three endothermic transitions. The high-temperature endotherm is due to an amylose-lipid complex and the other two endotherms are directly related to the gelatinization process. According to the thermal behavior, we can describe gelatinization as a diluent-dependent melting of the crystalline regions in the starch granules. The low-temperature endotherm emerges from a combined melting in excess water, and the second endotherm represents a melting phenomenon when no excess water is present. When the rheological properties of concentrated wheat starch gels were measured in stress-relaxation experiments, the starch gels were described as composite materials with swollen granules dispersed in an aqueous amylose solution (Lindahl and Eliasson, 1986).

Starches isolated from wheats from the Australian gene pool have higher viscosities than starches from wheats developed elsewhere. Hot-pasting abilities of flours of 68 wheats of U.S. origin were compared with a Brabender Amylograph in the absence of amylase activity (Meredith and Pomeranz, 1982). Starches prepared from eight of the wheats with minimal granule damage were examined similarly, and their susceptibility to amylase attack was determined. The inherent hot-pasting abilities of flours of U.S. wheat cultivars in the absence of amylase were similar in range to those previously determined for New Zealand wheat cultivars. Flours from soft and club wheats and one sample of durum wheat had poor pasting ability. A marked seasonal effect on pasting was observed. Part of the variation in pasting can be explained by differences in protein contents and starch damage.

Greenwood (1976) showed schematically the state of organization of the starch granules in various types of baked goods as shown by optical and scanning electron microscopy (Table X). In Scottish short bread (a mixture of flour, butter, and sugar only), the granules fully retain their natural characteristics and are merely swollen or partly gelatinized. Depending on the type of biscuit (cookie), the granules can vary from being in the swollen to being in the disrupted state. In cakes, the granules are gelatinized and often disrupted. In bread, the whole spectrum, from gelatinized to enzymatically degraded granules, are apparent. Wafers are unique in that they are a matrix of dispersed starch material with no visible organized structure.

B. Starch in Breadmaking

The importance of starch in breadmaking was reviewed by Schoch (1965), Torsinskaja et al (1967), Dennet and Sterling (1979), Medcalf (1968), and Medcalf and Gilles (1968). Bushuk (1966) described the essentiality of the granular wheat starch for bringing about a proper balance and relocation of water in bread. Water, originally associated with gluten and pentosans, becomes associated with gelatinized starch by the end of baking.

Alsberg (1927) postulated that starch contributes to breadmaking because it affects water absorption, dough consistency, and viscosity. The starches of various flours differ in susceptibility to enzymatic digestion and in the extent to which starch granules are freed during the milling process. The role of starch as a factor in dough formation was studied by Stamberg (1939).

According to Sandstedt (1961), starch in breadmaking has the following functions: it dilutes the gluten to an optimum level, furnishes fermentable sugars, provides a surface suitable for a strong union with the gluten adhesive, becomes flexible during gelatinization in conditions of limited water to permit the gas-cell film to stretch, and absorbs water from the gluten by gelatinization, thus causing the film to set and become rigid and providing a bread structure permeable to gas so that the baked bread does not collapse on cooling.

According to Hoseney et al (1978) and Hoseney (1984b), starch acts as a temperature-triggered water sink in baked cereal products. As temperature increases and starch gelatinizes, the starch competes with other components for

TABLE X
Stages of Granular Dispersion in Baked Goods[a]

Stages	Baked Goods
Swollen	Scottish short bread
↓	
Gelatinized	Biscuits
↓	Cakes
Disrupted	
↓	
Dispersed	Bread Wafers
↓	
Enzymatically degraded	

[a] Source: Greenwood (1976).

the available water in the system. Thus, starch "sets" the structure of the baked product system. The extent of starch pasting is a function of the availability of water to the starch granules. The final state of the starch contributes to the textural attributes of the baked product.

Starches isolated by Soulaka and Morrison (1985a) from 23 bread wheats and 26 durum wheats contained 26.3–30.6% (mean 29.1%) total amylose, 19.3–25.1% (mean 22.9%) apparent amylose, and 783–1,144 mg/100 g (mean 977 mg 100/g) lysophospholipids. Gelatinization temperatures were 57.3–64.9° C (mean 61.8° C) and enthalpies 6.4–11.8 J/g (mean 9.7 J/g) in excess water, measured by differential scanning calorimetry. There were no correlations between any of these parameters. A-granule mean volumes were 1,235–2,585 μm^3 (av. 1,778), modal volumes 863–1,804 μm^3 (av. 1,264), mean diameters 13.9–16.0 μm (av. 13.99), and specific surface areas 0.236–0.302 m^2/g. B-granule mean volumes were 35.4–100.4 μm^3 (av. 55.9), modal volumes 16.5–54.5 μm^3 (av. 27.7), mean diameters 3.66–5.07 μm (av. 4.09), and specific surface areas 0.684–0.920 m^2/g. The B-granule contents of the starches were 12.8–34.6% (av. 27.3) by weight and 13.0–37.3% (av. 24.0) by volume.

Harris and Sibbitt (1941) prepared starches from a series of wheat flours, milled from HRS, HRW, SRW, durum, and white wheat classes. A constant gluten substrate prepared from HRS wheat was used in all the bakes. The gluten was mixed with starch to give blends of 10.0, 13.2, and 16.0% protein (on a 13.5% moisture basis). The level of protein in the blend had a consistent effect on loaf volume and absorption. Starches varied widely in their effects on loaf volume and crumb grain of bread baked from gluten-starch blends. Starch differences could not be correlated, however, with the breadmaking properties of the flours from which the starches were obtained. In a later study, 48 samples of starch and gluten were prepared from 13 varieties of HRS wheat flours (Harris and Sibbitt, 1942). The glutens were blended with a HRS wheat starch, and the starches were blended individually with a common gluten substrate. The blends were made into doughs and baked. Loaf volumes of the blends were shown to differ significantly among wheat varieties as well as among locations of growth.

Pence et al (1959) determined relationships between protein content and loaf volume for several flours of widely varying type and quality by 1) diluting each flour with its own starch or enriching it with its own protein components, 2) enriching a low-protein base flour with protein components of the different flours, and 3) interchanging starch and protein components of three of the flours in completely reconstituted doughs. Results with the reconstituted doughs indicated that protein, starch, and protein-starch interaction effects are all significant factors governing flour quality.

Medcalf and Gilles (1965) isolated starches from 17 varieties representing HRS, HRW, durum, and soft wheat classes. Amylose and amylopectin were isolated from both HRS and durum wheat starches and characterized by intrinsic viscosity and amount of iodine absorbed. No significant differences were observed between the polymers from the two wheat types. Amylose ranged from 23.4 to 27.6%; durum starches tended to be on the high end of the range. Intrinsic viscosities varied for starches within the same wheat class and even for different samples of the same variety; however, no significant differences between classes were observed. Starches from durum wheat generally had larger water-binding capacities, greater rates of iodine absorption, and slightly lower temperatures of initial pasting than starches from other wheat classes. The researchers could not explain the observed differences in gelatinization characteristics among the various wheat starches. Gunzel (1966) reported that starches isolated from Canadian and Australian wheat flours had higher hot-paste peaks and cooling-curve viscosities than starches from German wheat flours. Differences were not correlated with particle size. Similarly, no correlation could be established between pasting characteristics and bread-making potentialities.

D'Appolonia and Gilles (1971b) isolated starch from 12 varieties of HRS wheat flour; three HRS composite flours that had yielded loaves of good, intermediate, and poor baking quality; three durum semolinas; two club wheat flours; and a soft white wheat flour. In addition, starch was fractionated into a large- and small-granule fraction. One commercial sample each of corn starch and wheat starch also were included in the study. The temperature of initial pasting and the peak height ranged from 55 to 59°C and from 540 to 865 BU, respectively, for the different wheat starches. Absolute density, water-binding capacity, and starch-damage values were 1.466–1.496 g/cm^3, 81.5–100.0%, and 2.3–12.2 Farrand units equivalent, respectively, for the wheat starches. Farinograph mixing time and stability ranged from 1.5 to 4 min and 3 to 6 min, respectively, for the gluten-starch blends containing the different wheat starches. The same gluten-starch blends when baked into bread showed a wide range in loaf volume. The addition of common water-solubles to the gluten-starch blends showed great variations in loaf volume. The results suggested an interaction between water-soluble components and the starches.

Fractionation and reconstitution studies by Hoseney et al (1971) showed that, contrary to previous claims, wheat starch is not unique in its contribution to breadmaking. Barley and rye starches were as functional as wheat starch; the similarity in the contribution of those starches seemed related—in part at least—to their physical characteristics.

The properties of mature wheat starch granules and the effects of milling were investigated by Kulp (1972). Starches were separated from three hard winter, three hard spring, and two soft wheat flours and from their parent wheats. The granules from different wheats varied slightly in physicochemical properties except that those of the soft-wheat group were more soluble and yielded hot pastes of lower stabilities than those from hard wheats. The breadmaking potential of winter-wheat starches was highest, followed by spring- and soft-wheat starches. As compared with the parent-wheat starches, the flour starches differed little in swelling power and gelatinization-temperature ranges, but they had lower Brabender consistencies, slightly higher solubilities, greater susceptibility to enzymolysis, and a higher water-hydration capacity. The baking potential of flour starch was less than that of the parent-wheat starch, indicating that native granules are desirable for optimum starch performance. Of the flour starches, best results were obtained with those from spring wheat, followed in order by those from soft wheat and then hard winter wheat.

In a subsequent study, Kulp (1973a, 1973b) evaluated breadmaking properties of small wheat-starch granules and compared them with those of regular wheat starches. The starches were isolated from commercially milled flours. From the same flours, a fraction (fines) containing principally small granules was separated by air classification and used as a source of small granules. These granules were lower in iodine affinity, indicating differences in amylose levels or some fundamental structural differences; the hot paste consistencies of both starches were similar, but the small-granule starch was lower in hot paste stability and produced a cold paste consistency below that of the regular starch. Defatting of starch before testing did not remove these differences. Swelling powers were comparable from 65 to 95° C, except that at 95° C the swelling power of small granules was higher. Solubilities were generally below those of regular starches. Gelatinization temperature ranges, water-binding capacities, and enzymatic susceptibilities of small granules were higher than those of regular ones. Doughs made from starch-gluten systems had lower stabilities when small-granule starch was used than when they contained the corresponding regular starch. Baking tests using blends of different starches with a single gluten preparation showed that the small granules have a lower baking potential than the regular ones. All differences attributable to granule size were consistent, regardless of the source or method of starch preparation.

Kulp (1973b, 1973c) concluded that starch performs two functions in bread production: it serves as a structure-forming component, and it may supply fermentable carbohydrates. Since U.S. bread formulas contain adequate levels of sugars, the latter function is of minor importance, whereas the structure-forming characteristics are the main quality-determining index of the baking potential of starch. In this respect, the intact native starch granules afford the best structure-forming material. Bread systems in which the supply of fermentable sugars is limited (such as French bread) benefit by the presence of some damaged starch as a source of fermentable material, but its amount should be restricted to preserve the baking potential of the starch.

Lelievre et al (1987) prepared starch granules of three size classes from a New Zealand wheat by sedimentation fractionation. Each was recombined with gluten from the same wheat at protein concentrations of 8, 12, and 16% and test baked. The pup loaves were judged for volume, crust, and crumb characteristics. Crumb was tested for compression and shear. A different optimum starch size fraction was found for each protein concentration, i.e., protein and starch interacted in forming loaf texture and volume. The most notable result was that the starch fraction greatly and directly affected the toughness or bite of the bread crumb, but it affected the softness of chew only in a complex manner dependent on the protein level. The percentage of protein had large effects on both bite and chew properties.

The crumb of bread baked from wheat flour, rye flour, and rye meal was examined by light microscopy and scanning electron microscopy (Pomeranz and Meyer, 1984). Whereas the wheat bread crumb is held together by a matrix of denatured protein, in the rye bread crumb, highly expanded starch granules fulfill that role. Examination by scanning electron microscopy of residues of bread crumb macerated to wash out soluble starch demonstrated the presence of a residual coherent structure of apparently denatured gluten proteins in wheat bread. Rye bread contained only a few similar, less coherent, structures.

C. The Role of Mechanically Damaged Starch

When wheat is milled into flour, a portion of the wheat starch becomes mechanically modified as a consequence of the grinding action of the mill rolls.

The granules thus physically altered have been referred to as *damaged, available*, or *susceptible* starch (Jones, 1940). Dadswell and Wragge (1940) found that injury to starch granules in milling correlated with variety but not with place of growth. Jones (1940) found that the amounts of damaged granules varied according to conditions of milling. Damage to starch granules increases susceptibility to enzymatic action. The harder the wheat, the higher the maltose value of a flour. Differences in diastatic activity of flours from different types of wheat are attributable at least partly to differences in the physical hardness of the endosperm, for hardness affects the extent of starch damage during milling.

The damage that occurs to starch during wheat flour milling has long been known to influence the baking properties of flour (Miller et al, 1964). Starch damage is largely responsible for differences in water absorption, handling properties of the dough, sugar production, and slackening during fermentation. Starch damage also affects loaf volume and crumb tenderness.

Flour contains about 1-2% readily fermentable carbohydrates, sufficient to sustain yeast in dough for about 2 hr. Then, in the absence of added sugars, additional gas must be derived from the fermentation of maltose produced as a result of the action of starch-degrading enzymes on "available" starch, which is about 5-10% of the flour. However, excessive starch damage may be detrimental. Excessive starch damage can harm the machining properties of dough and impair loaf volume and crumb resilience.

Native starch granules placed in cold water swell to only a limited extent. Hydration is a process of simple adsorption that causes the granule to increase about 10% in diameter or about 33% in volume. The sorption is not confined to the surface of the granule. However, starch heated in water to its gelatinization temperature or above will take up 10 times or more of its weight of water, and it becomes readily susceptible to enzymatic action (Sandstedt, 1961). Like gelatinized granules, damaged granules and the damaged portions of granules swell extensively in cold water and are readily digested by amylases.

The effect of starch damage on water absorption was first observed by Alsberg and Griffing (1925). Greer and Stewart (1959) measured the water absorptions of flours from 23 samples of two wheat varieties. Comparison of the results with estimates of flour protein and damaged-starch content indicated that differences in absorption were largely accounted for by variations in these factors. Water uptakes were 0.44 and 2.00 g per gram of native and damaged starch, respectively.

Sandstedt et al (1939) stated that the undesirable characteristics of very hard wheat flours for breadmaking (e.g., Chiefkan) can be attributed to the damage done to the starch during milling. Johnson and Miller (1959) also found that the integrity of the starch-granule structure was important to the production of high-quality bread. An excessive amount of damaged starch was detrimental, and pregelatinized starch did not perform in bread as well as native starch. Introduction of new milling processes (mainly air classification) and continuous breadmaking reemphasized the significance of starch damage in wheat flour processing.

Meredith (1966) examined for three seasons both commercial and experimentally milled New Zealand wheat flours. Their water-absorbing capacity depended on the content of both protein and damaged-starch granules. Although protein was more important, it was possible to vary the water absorption of flour by varying the amount of starch damage produced in milling.

A U.K. bread-flour grist used for conventional and mechanically developed doughs was commercially milled at three levels of starch damage (Farrand, 1972). The flours were studied at three levels of water absorption and three levels of yeast for each level of starch damage. Dough consistencies were measured with a Brabender Farinograph and also a Do-Corder at variable mixing speeds; bread quality was characterized in terms of a standard laboratory baking test. Farrand (1972) concluded that the physical state of the starch component of a flour has important effects on the rheological parameters, sedimentation values, and yeast utilization in relation to loaf quality obtained by conventional and mechanical development techniques.

According to Kozlov (1980), the volume of baked bread is inversely related to the water-binding capacity of starch. Water absorbed by the starch is unavailable to gluten and its expansion during baking. The findings, actually, may reflect the fact that adding insufficient amounts of water to flour containing large amounts of damaged starch may limit the amount of water for gluten development.

Dexter et al (1985b) reported on the relationship between flour starch damage and flour protein as they affect the quality of Brazilian-type hearth bread.

Moss (1961) proposed to relate water absorption (W) to protein content (P) and starch damage (S) with the equation:

$$W = 41.6 + 1.32P + 0.34S .$$

According to Farrand (1972), the optimum level of starch damage is

$$(\text{Protein})^2 / 6 .$$

Optimum starch damage is 20% at 11% protein and 24% at 12% protein in the flour. The recommended numerical value is affected to a large extent by the method used to determine damaged starch.

Breadmaking characteristics can be controlled over a wide range of potential quality for wheats of several types (for reviews, see Pomeranz, 1985, 1987; Pomeranz et al, 1984). In many cases, differences in flour characteristics may be related to effects of milling on particle size and starch damage rather than to intrinsic differences in the chemical, biochemical, and physical structures of flour components. A similar conclusion was reached by Meuser et al (1977).

Baking tests were made by the rapid-mix test on flours from two soft and three hard German wheat cultivars from a single location and on 12 composites, six each, of soft and hard wheat selections from many locations (Pomeranz et al, 1984). Maltose contents, amylose values, and starch damage were substantially and consistently lower in the soft than in the hard wheat flours. The difference was accompanied by higher water absorption in the hard wheat flours. Tested by the standard rapid-mix test, the soft wheat flours ranked lower in volume of baked rolls than the hard wheat flours. The soft wheat flours showed a larger volume response to increased sugar levels in the formula, added to compensate for limited starch damage, than the hard wheat flours. When the sugar in the formula was optimized, the volume potential of the soft wheat flours was realized and was equal to that of hard wheat flours on a constant protein basis.

Two deductions can be made from that study. First, the relatively extensive starch damage in hard wheats provides adequate levels of fermentable sugars to bring out the loaf volume potential of the flour proteins. That potential cannot be realized, however, in flours from soft wheats that have limited starch damage. Second, if the increases in loaf volume per 1% protein are viewed as an index of the functional breadmaking quality of protein, there is no basic difference between the proteins in soft and hard wheats, although there were clearly varietal differences within both the soft and the hard wheats that were examined.

The loaf volume potential in soft wheats can be brought out by the addition of sugar. In a more general sense, this may not compensate completely, however, for the effects of increased starch damage in hard wheats (e.g., for the increase in water absorption). To compensate for the difference in water absorption, more extensive starch damage in soft wheats would be required. In addition, the inherent milling properties of hard wheats (response to conditioning, ease of sieving, and flour yield) are definitely desirable features of wheats with intermediate hardness. Other factors that must be considered are the increased or even excessive crust browning and impaired dough properties at high sugar levels.

Yet, it is clear that soft wheats contain proteins that are equal on a constant protein basis, in terms of functional properties, to proteins of hard wheats and that, if there is a definite economic advantage in growing certain soft wheat varieties (e.g., because of increased yield), those wheats can be made to perform more satisfactorily by increasing the sugar in the formula to compensate for the low starch damage.

Yamazaki (1959) concluded that both flour granulation and starch damage must be considered in assessing differences among cookie flours. Mertz and Nordstrom (1960) and Miller et al (1967a) have shown that while a fine granulation is desirable in cake flour, high starch damage (above 5%) is detrimental.

D. Starch and Bread Staling

Bread staling was reviewed by, among others, Bechtel (1961), Bice and Geddes (1953), Pelshenke and Hampel (1962), Schoch (1965), and D'Appolonia and Morad (1981). A series of studies on the role of wheat flour constituents in bread staling was reported by Kim and D'Appolonia (1971, 1977a, 1977b, 1977c). The effects of loaf volume, moisture content, and protein quality on the softness and the the staling rate of bread were reported by Maleki et al (1980). Kulp and Ponte (1981) are the authors of an excellent and comprehensive review of the fundamental causes of staling of white pan bread.

It is generally agreed that bread from strong flours keeps better than bread from weak flours. The results indicate that both the quantity and quality of flour proteins are involved in bread staling. However, extensive research has shown that changes in the starch fraction are responsible for some of the major changes in bread during staling. Katz (1934) was the first to show that the X-ray diffraction pattern of fresh bread was similar to that of freshly gelatinized wheat starch; the pattern for stale bread was similar to that of retrograded starch. Hellman et al (1954) measured the rate of crystallization of starch gels at moisture levels comparable to those found in bread. The curve relating extent of crystallinity to age of gel was similar to that relating crumb firmness to age of bread, and it supported the assumption that the crystallization of starch is important in bread staling.

Schoch and French (1947) found that water-soluble starch leached from the crumb of fresh bread at 30° C was predominantly the branched B-fraction. The linear A-fraction must be insolubilized by retrogradation during baking and hence cannot contribute to the staling process. The progressive decline in solubles during the aging of bread indicates that the cause of staling may be spontaneous aggregation of the branched B-fraction. As additional evidence, the percentage of the soluble B-fraction from stale bread was restored to a constant high level when it was heated at 50° C. This was in accord with the refreshening of bread by moderate heating. The behavior of starch in bread and the nature of the staling action was influenced by the presence of malt and free fatty acids.

Prentice et al (1954) studied how flour fractions from HRS and SRW wheat flours influence firming of bread crumb. Gluten, starch, starch tailings, and water-soluble fractions separated from the flours were combined in various proportions to yield synthetic flours. Increasing the protein content of the synthetic flours, but maintaining a constant ratio of gluten to water-solubles, increased absorption and loaf volume but decreased average crumb firmness and crumb-firming rate. Those findings were confirmed with bread baked from soft wheat flour that had been enriched with gluten to 13.5 and 16.5% protein. Substituting soft wheat flour starch or gluten for hard wheat flour starch or gluten increased average crumb firmness but did not affect firming rate. Starch tailings had no effect on loaf volume or crumb-firming rate, but the tailings from both hard and soft wheat flour decreased average crumb firmness. Hard and soft wheat flour water-solubles increased water absorption and loaf volume and decreased both average crumb firmness and crumb-firming rate.

Bechtel (1959) baked bread made of gluten and wheat starch or a mixture of wheat starch and a cross-linked starch derivative that contained up to 80% of the derivative. Part of the bread was supplemented with bacterial α-amylase at 4 and 24 SKB units per pound. Freshly baked bread was increasingly stale and firm as the proportion of cross-linked starch was increased. During six days of storage, bread staled and firmed less with increasing levels of bacterial α-amylase. Changes in starch were shown to be of major importance in the staling and firming of bread. The results indicated, however, that retrogradation or crystallization cannot be regarded as the only property of starch involved in the staling of bread.

Zobel and Senti (1959) investigated the relation between crumb firmness and starch crystallization in breads made with reconstituted flours containing 40% of a starch cross-linked with epichlorohydrin and supplemented with a heat-stable bacterial amylase. X-ray patterns showed that up to three or four days after baking, starches in the experimental breads were more crystalline than in conventional bread. Extensive starch crystallization in amylase-supplemented bread was not paralleled, however, by a corresponding increase in crumb firmness. X-ray patterns of isolated starch gels showed that the proportion of retrograded structure was greater in gels of the cross-linked starch than in comparable gels of unmodified wheat starch. Upon aging, starch in either gel developed increased crystallinity irrespective of the presence of a heat-stable bacterial amylase.

According to Yasunaga et al (1968), various degrees of starch gelatinization can be obtained during baking. This, in turn, can effect progressive changes in crumb grain as the bread ages. The main factors that control the degree of gelatinization are baking absorption, temperature, and time. Higher absorption, higher baking temperature, and longer baking time each produce more extensive gelatinization. Other factors and ingredients that affect the starch in the dough (such as physical damage to starch granules and the presence of malt, shortening, and monoglycerides in the dough formula) also affect the degree of gelatinization during baking.

Schoch (1965) summarized studies on changes in starch during bread-baking. The available evidence indicates that, during baking, wheat-starch granules undergo restricted swelling, limited by the relatively small amount of water present. During that stage, a portion of amylose dissolves and diffuses out of the granules into the surrounding aqueous phase. In a cooled loaf, the solution containing the amylose molecules sets to a gel structure. Hence, normal fresh bread represents swollen starch granules embedded in a firm gel of the linear fraction. No changes seem to occur in the gel network, and staling during storage seems to result from crumb hardening due to retrogradation of the amylopectin molecules in the swollen starch granules.

Starch was isolated by Soulaka and Morrison (1985b) from six types of wheat and separated into large A-granule and small B-granule fractions. The A- and B-granule fractions were remixed in various proportions and reconstituted with freeze-dried gluten and freeze-dried water-solubles from flour. The specific volumes of the loaves were not affected by starch amylose or lipid content, but were significantly correlated with the gelatinization temperature of either the total starches or their A-granules. The optimum proportion of B-granules was 25–35% by weight, but their gelatinization temperature had no effect on loaf specific volume. Staling changes were quantified from crumb compressibility measurements and the enthalpy of the endotherm for gelatinization of retrograded amylopectin in stored bread crumb, measured by differential scanning colorimetry. Initial and limiting moduli of crumb firmness were correlated with loaf specific volume and starch gelatinization temperature, but rate and time constants calculated from the Avrami equation were independent. Rate and time constants and limiting enthalpy values calculated from the differential scanning colorimetry results were independent of all other measured parameters.

While it is generally accepted that bread staling is governed or even controlled by changes in the starch component(s), according to Banecki (1984) gluten has a much greater influence on bread staling than starch does.

E. Composite (Gluten-Free) Bread Flours

Rotsch (1949, 1953, 1954) made bread from dough in which wheat protein was replaced by other gel-forming substances but failed to find a suitable substitute for starch. He concluded that the bread crumb owes its coherence to partly gelatinized starch and attributed differences in crumb properties to varying degrees of gelatinization of starch. He assumed, however, that carbon dioxide is retained in the dough by the gluten gel. Therefore, he suggested that doughs without gluten can retain gas only if another gel replaces gluten.

Jongh (1961) prepared dough from starch instead of flour. The dough behaved like a concentrated stable suspension; it showed, among other properties, dilatancy. Such a dough acquired plastic properties when small quantities of glyceryl monostearate were added. Starch bread without the additive had a stiff crumb with an irregular, coarse structure. Adding 0.1% glyceryl monostearate gave a loose crumb with a fine and regular structure. Although these findings show that presumably acceptable bread can be baked from starch doughs without gluten, the character of the starch doughs and of the resulting bread are basically different from those obtained in conventional breadmaking.

The use of starches from various tubers was described by Ciacco and D'Appolonia (1977). The results suggested the importance of pasting properties of the tuber starches when used in composite flours. Preparation of composite (gluten-free) flours and their use in breadmaking were reviewed by Kulp et al (1974), de Ruiter (1978), and Pomeranz (1969).

The investigations of Jongh (1961) led to interesting developments in which nonwheat baked products were produced from mixtures of tuber flours and defatted oil seeds (Jongh et al, 1968; Kim and DeRuiter, 1968). On the basis of studies by Rotsch, Jongh, and Sandstedt, it has been suggested (Medcalf and Gilles, 1965, Medcalf et al, 1968) that varietal differences in breadmaking might be due to differences in the quality of starch (rather than in gluten). The central theme in most of these claims is that wheat starch is indispensable in breadmaking (Pomeranz, 1969). Various researchers have suggested that starch affects or even governs rheological properties of dough. Doguchi and Hlynka (1967) showed that gluten from a lower grade of HRS wheat was slightly weaker, and gluten from durum wheat was considerably weaker, than gluten from top-grade HRS wheat. The authors interpreted the results as support of the view that physical properties of dough are in large measure attributable to its gluten, which provides a matrix of other constituents. This interpretation is strengthened by investigations that demonstrated (Heaps et al, 1968) that gluten controls basic rheological parameters and that starch acts as a diluent. This, however, is by no means the only function of starch in breadmaking, certainly not in the oven-baking and postbaking stages and not even in the dough mixing and fermentation stages.

F. Starch in Cookies and Cakes

Sollars and Rubenthaler (1971) substituted starches from rye, barley, corn, rice, and potatoes for wheat starch in reconstituted flours and subjected these flours to cookie-baking, cake-baking, and MacMichael viscosity tests. For cake-baking, the starches were used untreated and treated with chlorine at three levels. Reconstituted flours with rye and barley starches proved very good for cakes and cookies and had viscosities close to those of flours with wheat starch. Corn and potato starches generally gave fair-quality cakes and cookies and intermediate viscosity values. Rice starch gave very poor cakes and cookies and low viscosities. The results indicate that starch must have certain physical and chemical properties for satisfactory performance. Wheat starch was not unique, however, since rye starch in particular was virtually the equal of wheat starch.

Miller and Trimbo (1965) described a method to evaluate cake flour and the effects of batter formulation and composition on the gelation of starch in white layer cakes. Conditions that changed the gelatinization temperature of starch granules could be used to overcome certain deficiencies in cake-flour performance (Miller et al, 1967b).

Howard et al (1968) studied the role of starch in cake-baking by substituting granular starches and two starch fractions for cake flour. In aeration of the batter during mixing, granular starch increased viscosity, but aeration depended primarily on water-soluble proteins. During the early stages of heating a fluid batter in the oven, granular starch is important, as the rate of granule swelling, accompanied by some absorption of water, affects the viscosity and increases the emulsion stability of the batter. At the thermal-setting stage, at which the batter changes from a fluid, aerated emulsion to a solid and stable porous structure, the water absorption of the starch granule controls the quality of the baked cake.

Greenwood (1976) replaced wheat starch by various maize starches in high-ratio yellow cakes. Starch from normal maize (but not from waxy or amylomaize) had no adverse effect on cake structure or texture.

Studies on the role of wheat starch in production of chemically leavened baked products are limited. Zaehringer et al (1956) fractionated a soft wheat pastry flour and a hard wheat bread flour into gluten, starch, water-solubles, and tailings. Biscuits (cookies) containing hard wheat gluten were larger and had paler and less tender crusts than biscuits containing soft wheat gluten. They were smaller in volume and had less tender crusts and crumb when made from flour containing hard wheat solubles. Biscuits had larger volumes and browner and more tender crusts when made with reconstituted flour containing hard wheat starch. Flours containing hard wheat tailings gave less compressible doughs, and the biscuits were smaller in volume with browner crusts.

G. Starch in Pasta Products

In a study conducted by Dexter and Matsuo (1979a), seven wheats, four barleys, three corns, rye, triticale, oats, and buckwheat were used to investigate the effect of starch properties on pasta dough rheology and spaghetti cooking quality. The diverse nature of the starches was reflected by significant differences in water absorption, amylose contents, microscopic structure, and amylograph pasting properties. A wide range in mixing properties was obtained for starch and gluten blends when absorption was held constant at 36% (14% moisture basis). These differences appeared to be mainly attributable to variations in water absorptions of the starches. Waxy maize and waxy barley starches were detrimental to spaghetti cooking quality, but high-amylose corn starch appeared to impart a slight improvement in cooked spaghetti firmness. Assessment of the cooking quality of starches with more normal amylose content suggested that other starch properties may supersede amylose content as a criterion of spaghetti cooking quality once a threshold level of amylose is attained.

The optimum cooking time of Asian dry noodles increased linearly with protein content (Oh et al, 1985). High-protein noodles were darker and stronger and were internally firmer when cooked than low-protein noodles. However, protein content was not correlated with surface firmness. At high rates of flour extraction, noodle color darkened, but the internal firmness of cooked noodles did not change. The optimum absorption of noodle dough increased with starch damage and fineness of granulation. Increasing starch damage reduced both internal and surface firmness of cooked noodles; decreasing particle size improved the strength of uncooked noodles but did not affect noodle color or firmness of cooked noodles.

H. Starch-Adhering Components

The nature of the starch-protein interface in wheat endosperm was studied by Barlow et al (1973). Several studies point to the diverse roles of starch-adhering substances and surface characteristics of granules. The adhering components include water-soluble and water-insoluble proteins, starch polymer fragments, pentosans, and enzymes (Kulp and Lorenz, 1981). The researchers postulated that the adhering substances may serve as a linkage between starch and other flour components (proteins and lipids).

A salt-extractable protein was isolated and purified from large (A-type) wheat flour starch granules (*T. aestivum* cv. Maris Huntsman) by Lowy et al (1981). This protein was the major species extractable with NaCl. It was tentatively thought to be associated with the surface of the granules. It had a molecular weight of approximately 30,000 and an isoelectric point in excess of 10 as judged by analytical isoelectric focusing in polyacrylamide gel. The amino acid composition of the protein was different from those of wheat gluten proteins. A protein with similar properties was present on small (B-type) Maris Huntsman starch granules and on A-type granules from another variety of hexaploid bread wheat (Flinor) as well as on A-type granules from tetraploid durum wheat. This protein was not present on starch granules prepared from germinated grains. The protein had α-amylase activity or inhibitory activity against wheat or hog pancreatic α-amylases. Other proteins could be extracted from A-type starch granules only after gelatinization of the starch granule in the presence of SDS. The electrophoretic behavior of these proteins was different from that of the NaCl-extracted protein. It is of interest that some of the minor components at the surface of the starch granule also may be affected by chlorination of flour for high-ratio cakes.

Gough et al (1985) reported that lipid-containing starch granules from wheat, maize, and rice can sequester SDS from an aqueous solution at room temperature. On heating in water, SDS destabilizes the granules and causes gelatinization and pasting at temperatures lower than normal. Proteins, lipids and some amylose are lost in the process. After gelatinization, the SDS is complexed with amylose in place of lipid. In solution, SDS-amylose complexation is thermodynamically favored at 25 and 45° C and may provide energy to overcome intermolecular bonding within the granule. Waxy maize starch (with no amylose) and potato starch (with no lipid) are destabilized by SDS to a lesser extent. The authors suggested that an interaction between SDS and amylopectin, together with solubilization of the proteins, may play a role in the destabilization of all starches by SDS, in addition to the amylose-SDS complex formation in lipid-containing starches.

Greenwell et al (1985) used scanning electron microscopy to follow the progressive digestion of wheat starch granules to soluble carbohydrate, by incubation with amyloglucosidase from *Aspergillus niger*. The surface of native granules was eroded into numerous small, closely spaced pits. Pretreatments of the starch with chlorine gas, with pronase, or with aqueous alkali, did not alter the extent of solubilization by amyloglucosidase, but the granule surfaces were eroded in a different way; they remained almost as smooth as before incubation. All these treatments degraded and/or removed traces of protein associated with the surfaces of the water-washed granules. In contrast, extraction of starch with toluene, which changes some of the surface properties of the granules and removes a fraction of the surface protein, did not prevent the formation of pits by amyloglucosidase. Greenwell et al (1985) concluded that degradation of wheat starch catalyzed by amyloglucosidase is localized, but not retarded, by structures containing a protein fraction on or near the granule surface. The results suggest that chlorine reacts with proteins located at or near the surface of starch granules and that this reaction may play a role in the technologically important process of flour chlorination.

Greenwell and Schofield (1986) reported a positive association between the presence of a starch protein with relative molecular mass of 15,000 and endosperm softness, for wheats of widely different genetic backgrounds. This protein may play a role in conferring endosperm softness on wheats. The mechanism by which the protein causes this effect is not known. Since the protein is associated with the surface of the starch granule, they speculated that it has some "non-stick" property that reduces the adhesion between the granule and the protein matrix of the endosperm. (See also Chapter 5 in Volume I and Chapter 7 in this volume.)

I. Starch Tailings

Wheat flour fractions rich in water-soluble pentosans have been called amylodextrin (Sandstedt et al, 1939), tailings (MacMasters and Hilbert, 1944), or squeegee starch (Clendenning and Wright, 1950). The fraction was reported to contribute to the handling properties of dough. Prentice et al (1954) showed that starch tailings had no effect on loaf volume or crumb-firming rate but that the tailings from both hard and soft wheat flours decreased average crumb softness. The tailings fraction increased water absorption and produced bread of higher moisture content and a tendency toward decreased staling. Kulp and Bechtel (1963a) isolated from flour tailings a fraction containing 55% pentosans. One percent of the fraction increased flour absorption 5.0–5.6% but did not significantly affect gas evolution and retention, amylogram characteristics, or extensigraph parameters. Increasing the levels of added insoluble pentosans to 2% lowered the loaf volume by impairing oven-spring and crumb grain. Similar effects of pentosans were demonstrated in the baking of reconstituted doughs.

When 5, 10, and 15% of flour was replaced with flour tailings isolated from untreated patent flour, a progressive decrease in loaf volume and grain quality was observed (Kulp and Bechtel, 1963b). Prime starch used in the controls did not show a similar adverse effect. Components of the tailings fraction were studied with the objective of localizing the factor responsible for its adverse baking performance. Damaged starch, small-granule starch, lipids, enzymes, and pentosans were evaluated. Analytical data and the results of baking tests with model systems showed that the pentosan-rich fraction accounted for the poor baking characteristics of tailings.

Tryptic digestion lowered the protein content of tailings by 50% but had no appreciable effect on functional properties in breadmaking. Thus it appeared that the adverse effects of water-insoluble pentosans more likely were due to the pentosans than to the pentosan-protein complex.

A gradual enzymolysis of water-insoluble pentosans of tailings decreased water absorption of this fraction (Kulp, 1968a, 1968b). The modified tailings were superior in baking quality to those isolated from flour. At the optimum level of treatment, the modified pentosans were comparable in breadmaking quality to prime starch. Casier and Soenen reported (1967) that insoluble pentosans can be extracted with dilute alkaline solutions and rendered water soluble. Adding 1% of solubilized pentosans increased water absorption, mixing time, and dough stability of soft wheat flours. If the pentosans were added to a lean dough formula, loaf volume increased and crumb grain, color, texture, and freshness retention were improved.

Yamazaki (1955) studied the effects of insoluble pentosans on cookie quality. Variation in pentosan content in soft wheat flours appeared to be a varietal characteristic. The tailings fraction has a deleterious effect on the cookie spread of reconstituted flours. A purified fraction was highly hydrophilic; its deleterious effects were proportional to the amount added to a control flour. Additions of similar hydrophilic materials to flour also decreased spread to various degrees, indicating that the effect of tailings was perhaps due in great part to their physical properties. Straight-grade flours from hard wheats contained more starch tailings than those from soft wheats, suggesting that varietal differences in cookie-baking may be attributable to quantitative differences in tailings. Low-grade flours that are rich in starch tailings gave poor cookies. The purified tailings were rich in pentosans and low in starch and nitrogenous compounds. However, no definite correlation could be established between chemical composition of tailings and cookie spread.

Sollars (1956) separated wheat flours into water-solubles, gluten, tailings, and prime starch by an acetic acid fractionation procedure. The fractions were reconstituted into a flour suitable for baking cookies. Fractions from flours varying in pastry quality were also interchanged. At the same absorption level, blends of soft wheat fractions had longer mixing time than blends of hard wheat fractions. In flours with interchanged fractions, tailings affected dough development most; prime starch had little effect. The tailings fraction had a far greater effect on cookie diameter than any other and appeared to be the most important fraction in cookie quality. The water-solubles had a small but consistent decreasing effect; starch had a slight effect, and the effect of gluten was erratic.

Starch tailings isolated from Pacific Northwest wheat flours and substituted at 5 and 10% levels in a standard cookie flour significantly decreased cookie diameter (Sollars and Bowie, 1966). When the tailings were fractionated, the pentosan-rich subfractions and the damaged starch had much larger diameter-decreasing effects than the original tailings. Proteins and enzymes appeared to have negligible effects, and lipids and small-granule starch had small, diameter-reducing effects.

Adding water-insoluble pentosans to a marginal-quality soft wheat flour improved grain, texture, and volume of yellow layer cake (Gilles, 1960). Glutens from durum semolinas leave on the washing sieve variable amounts of a gelatinous residue that is rich in pentosans (Bains and Irvine, 1965). Semolinas milled from the lower grades of durum wheat contain relatively higher amounts of the crude residue, and addition of the isolated material to the higher grades produced rheological effects on the doughs similar to those normally observed with the lower grades: increased water absorption, dough-development time, and stability to mixing.

VIII. MINERALS

Numerous studies concerned the effects in breadmaking of mineral components in water and in yeast foods used by the baker. The deleterious effects on panary fermentation of adventitious trace amounts of heavy metals are well established. Little is known, however, about the effects of flour mineral contents and composition. Mineral content increases with increased protein content and with flour extraction rate. In neither case has it been shown that the increased amount of minerals per se affects breadmaking (Miller and Johnson, 1954).

Sullivan and Near (1927) analyzed the ash of 20 wheat samples of widely varying character and reported that magnesium content was related directly to the strength of wheats as determined by protein percentage and gluten quality.

Gericke (1934a, 1934b) grew wheat in hydroponic solutions and studied the effects on bread scores of nitrate and chloride salts supplied during the later period of plant growth. The quality of flour from wheat grown in cultures where nitrogen was supplied in the form of ammonia was inferior to that milled from wheat grown in cultures supplied with nitrogen in the nitrate form. The use of calcium nitrate or chloride resulted in higher bread scores than did similar salts of potassium, sodium, and magnesium. The use of ammonium nitrate resulted in the poorest loaf produced from wheat grown in solutions containing different nitrate salts. McCalla and Woodford (1935) employed the liquid culture technique of Gericke (1934a) and reported that limiting the supply of potassium to wheat plants impaired the quality of grain, as determined by gluten and baking tests.

Bequette et al (1963) studied the influence of environment and variety on the total ash and elemental composition of gluten, starch, and water-soluble fractions separated from 40 HRW wheat flours and the relationships of these data to flour quality. Location and variety were both important in determining the significance of the differences in the elemental composition of flour fractions. The total amounts of ash and of phosphorus, iron, potassium, manganese, magnesium, sodium, and calcium were distributed differently among the three flour fractions.

Ash and concentration of phosphorus in the gluten were negatively correlated with most measures of protein quality. The concentration of manganese in the gluten also was negatively correlated with most measures of protein quality. The concentrations of calcium in the water-solubles and starch fractions, however, were positively correlated with most measures of protein quality. No consistent pattern was found in the relation between concentration of any of the other elements and the measures of protein quality. The conclusions were based on partial correlation coefficients, with protein held constant.

Studies conducted by Watson et al (1963) indicated that total phosphorus in gluten was correlated with flour quality. Bourdet and Feillet (1967) determined phosphorus compounds (total, lipid, phytic, and nucleic) in albumins, globulins, gliadins, and glutenins of four soft, four hard, and two durum wheats. The varietal differences were not correlated with breadmaking quality. Two wheat samples were obtained from fertilizer test plots located on sulfur-deficient soils, one from a plot on which sulfur had been used consistently as a fertilizer for about 25 years, and the other from an unfertilized plot. The levels of sulfur in the samples were 0.18 and 0.10%, respectively. The difference decreased in the flours, and decreased further in the glutens, but was still highly significant. The proportion of nongluten nitrogen in the sulfur-deficient sample was much larger; the gluten from this sample contained smaller amounts of cystine and methionine, and a lower content of sulfhydryl and disulfide groups. No difference was found in the content of sulfhydryl groups of the flour, nor in sedimentation behavior of the dispersed gluten. (See also the section on sulfur in this chapter.)

A preliminary report from Douglas and Tyson (1985) indicated a relationship between baking quality and mineral composition of 35 wheat samples grown in New Zealand. Positive relationships were found between bake score and the nitrogen and sulfur concentrations in the grain and negative correlations with potassium and molybdenum.

Copper is required for many essential functions of plants, in particular the production of viable pollen and lignin. It is a necessary component of many enzymes and affects ascorbic acid oxidase, cytochrome oxidase, and phenolase. Copper is relatively immobile in the growing plant, being retranslocated from older tissue only with the onset of senescence and the mobilization of nitrogen. Consequently, it must be continually taken up by the plant if a deficiency is to be avoided. Copper deficiency in cereals is widespread throughout the world and has been responsible for severe reductions in grain yields over large areas. Very little is known about copper's effect on grain quality.

Wheat was sown at a site deficient in soil copper to investigate the effect of copper deficiency on the physical dough and baking properties of flour (Flynn et al, 1987). Copper was applied as a foliar spray at late tillering and/or booting stages of plant growth. Plants on plots that were not treated had classic signs of copper deficiency. The application of copper increased grain yield, with the grain being well filled and with no visual evidence of copper deficiency. There was also a slight improvement in the dough and baking quality. When a second application of copper was made at the booting stage, which occurs after pollen production, a marked improvement in both dough rheology and loaf volume occurred.

The low charge density on the surface of the proteins combined with hydrophobic areas and an abundance of residues capable of forming hydrogen bonds result in a high sensitivity to salt concentration (Bernardin, 1978). Even a low salt concentration effectively masks the repulsion of one charged storage protein molecule for another of like charge. This allows hydrophobic and hydrophilic interactions to form.

The effect of salt concentration on protein aggregation follows ionic strength. Particularly interesting are the organic acids and lipid molecules. These compounds strongly bind to the hydrophobic regions on the molecule. Since the binding constants for this type of interaction are frequently large, an effect is seen at very low ligand concentration. Extraction of wheat proteins with salts of fatty acids and their electrophoretic characterization were described by Kobrehel (1980).

Kobrehel and Bushuk (1977) have demonstrated that a glutenin fraction can be solubilized by the addition of stearic acid. This is an example of a decrease in protein-protein interaction brought about by small ion binding and increasing the charge on the protein. This is also evidence for the hypothesis that wheat storage proteins interact primarily through secondary chemical bonds (in contrast to covalent disulfide bonds). The quantity and nature of proteins solubilized with different amounts of soaps may be related to the breadmaking quality of wheat varieties (Kobrehel and Matignon, 1980).

LITERATURE CITED

ABBOUD, A. M., RUBENTHALER, G. L., and HOSENEY, R. C. 1985a. Effect of fat and sugar in sugar-snap cookies and evaluation of tests to measure cookie flour quality. Cereal Chem. 62:124-129.

ABBOUD, A. M., HOSENEY, R. C., and RUBENTHALER, G. L. 1985b. Factors affecting cookie flour quality. Cereal Chem. 62:130-133.

ACKER, L., and SCHMITZ, H. J. 1967. Uber die Lipide der Weizenstarke. Staerke 19:275-280.

ACKER, L., SCHMITZ, H. J., and HAMZA, Y. 1968. Uber die Lipide des Weizens. Getreide Mehl 18:45-50.

AITKEN, T. R., and GEDDES, W. F. 1939. The relation between protein content and strength of gluten-enriched flours. Cereal Chem. 16:223-231.

ALARY, R., and KOBREHEL, K. 1987. The sulfhydryl plus disulfide content in the proteins of durum wheat and its relationship with the cooking quality of pasta. J. Sci. Food Agric. 39:123-136.

ALSBERG, C. L. 1927. Starch in flour. Cereal Chem. 4:485-492.

ALSBERG, C. L., and GRIFFING, E. P. 1925. Effect of fine grinding upon flour. Cereal Chem. 2:325-344.

ANDERSON, J. A., ed. 1946. Enzymes and Their Role in Wheat Technology. Interscience Publishers, Inc., New York.

ANDERSON, J. A. 1961. Bromate reaction in dough. Proc. R. Aust. Chem. Inst. 28:283.

ARAKAWA, T., and YONEZAWA, D. 1975. Compositional difference of wheat flour glutens in relation to their aggregation behaviors. Agric. Biol. Chem. 39:2123-2128.

ARAKAWA, T., MORISHITA, H., and YONEZAWA, D. 1976. Aggregation behaviors of glutens, glutenins and gliadins from various wheats. Agric. Biol. Chem. 40:1217-1220.

ARAKAWA, T., YOSHIDA, M., MORISHITA, H., HONDA, J., and YONEZAWA, D. 1977. Relation between aggregation behaviour of glutenin and its polypeptide composition. Agric. Biol. Chem. 41:995-1001.

ARCHER, M. J. 1974. A sand culture experiment to compare the effects of sulfur on five wheat cultivars (*T. aestivum* L.). Aust. J. Agric. Res. 25:369-380.

AUTRAN, J.-C., and BERRIER, R. 1984. Durum wheat functional protein subunits revealed through heat treatments. Biochemical and genetical implications. Pages 175-183 in: Gluten Proteins. Proc. Int. Workshop Gluten Proteins, 2nd. A. Graveland and J. H. E. Moonen, eds. Inst. Cereals, Flour and Bread, TNO, Wageningen, The Netherlands.

AUTRAN, J. C., ABECASSIS, J., and FEILLET, P. 1986a. Statistical evaluation of different technological and biochemical tests for quality assessment in durum wheats. Cereal Chem. 63:390-394.

AUTRAN, J. C., GODON, B., KOBREHEL, K., LAIGNELET, B., and POPINEAU, Y. 1986b. Studies on tests for the prediction and estimation of quantitative and qualitative characteristics of wheat gluten. Sci. Aliments 6:447-469.

AXFORD, D. W. E., CAMPBELL, J. D., and ELTON, G. A. 1962. Disulphide groups in flour proteins. J. Sci. Food Agric. 13:73-78.

AXFORD, D. W. E., ELTON, G. A. H., and TICE, B. B. P. 1964. The disulphide group content of flour and dough. J. Sci. Food Agric. 15:269-273.

BAINS, G. S., and IRVINE, G. N. 1965. The quality of Canadian amber durum wheat grades and the role of a pentosan-rich fraction in macaroni dough quality. J. Sci. Food Agric. 16:233-240.

BAKER, D., NEUSTADT, M. H., and

ZELENY, L. 1959. Relationships between fat acidity values and types of damage in grain. Cereal Chem. 36:308-311.

BAKER, J. C., and MIZE, M. D. 1937. Mixing doughs in vacuum and in the presence of various gases. Cereal Chem. 14:721-734.

BAKER, J. C., and MIZE, M. D. 1942. The relation of fats to texture, crumb, and volume of bread. Cereal Chem. 19:84-94.

BAKER, J. C., PARKER, H. K., and MIZE, M. D. 1943. The pentosans of wheat flour. Cereal Chem. 20:267-280.

BAKER, J. C., PARKER, H. K., and MIZE, M. D. 1946. Supercentrifugates from dough. Cereal Chem. 23:16-30.

BALDI, V., LITTLE, L., and HESTER, E. E. 1965. Effect of the kind and proportion of flour components and of sucrose level on cake structure. Cereal Chem. 42:462-475.

BALDWIN, R. R., JOHANSEN, R. G., KEOGH, W. J., TITCOMB, S. T., and COTTON, R. H. 1963. Continuous bread-making: The role that fat plays. Cereal Sci. Today 8:273, 274, 276, 284, 296.

BALDWIN, R. R., TITCOMB, S. T., JOHANSEN, R. G., KEOGH, W. J., and KOEDDING, D. 1965. Fat systems for continuous mix bread. Cereal Sci. Today 10:452-457.

BANECKI, H. 1984. Effect of glutens on the shelf life of bread products. Mlyn. Pek. Prum. Tech. Skladovani Obili 30:239-241.

BARBER, K. J., and WARTHESEN, J. J. 1982. Some functional properties of acylated wheat gluten. J. Agric. Food Chem. 30:930-934.

BARLOW, K. K., BUTTROSE, M. S., SIMMONDS, D. H., and VESK, M. 1973. The nature of the starch-protein interface in wheat endosperm. Cereal Chem. 50:443-454.

BARNES, P. J., DAY, K. W., and SCHOFIELD, J. D. 1981. Commercial pasta manufacture: Changes in lipid binding during processing of durum wheat semolina. Z. Lebensm. Unters. Forsch. 172:373-376.

BARNEY, J. E., II, POLLOCK, H. B., and BOLZE, C. C. 1965. A study of the relationship between viscoelastic properties and the chemical nature of wheat gluten and glutenin. Cereal Chem. 42:215-236.

BARTON-WRIGHT, E. C. 1938a. Studies on the storage of wheaten flour: III. Changes in the flora and the fats and the influence of these changes on gluten character. Cereal Chem. 15:521-541.

BARTON-WRIGHT, E. C. 1938b. Observations of the nature of the lipids of wheat flour, germ and bran. Cereal Chem. 15:723-738.

BAXTER, E. J., and HESTER, E. E. 1958. The effect of sucrose on gluten development and the solubility of the proteins of a soft wheat flour. Cereal Chem. 35:366-374.

BEBYAKIN, V. M., and BALABOLINA, T. G. 1980. Composition of gliadin, its biochemical polymorphism and the quality of wheat flour. Skh. Biol. 15:549-555.

BEBYAKIN, V. M., and GOSHITSKAYA, N. A. 1985. Blocks of gliadin components and the quality of wheat grain. Biol. Nauki (Moscow) 7:75-80.

BEBYAKIN, V. M., DUSHAEVA, N. A., and CHERVAKOVA, T. G. 1984. Hybridological analysis of the electrophoretic gliadin components of *Triticum aestivum* varieties and its significance in defining flour quality. Genetika 20:1528-1535.

BECCARI, A. 1745. De frumento. Instituto atque Academia Commentarii 2, Part I, 122.

BECHTEL, W. G. 1959. Staling studies of bread made with flour fractions. V. Effect of a heat-stable amylase and a cross-linked starch. Cereal Chem. 36:368-377.

BECHTEL, W. G. 1961. Progress in the study of the staling phenomenon. Bakers Dig. 35(5):48-50, 172, 174.

BECKWITH, A. C., WALL, J. S., and JORDAN, R. W. 1965. Reversible reduction and reoxidation of the disulfide bonds in wheat gliadin. Arch. Biochem. Biophys. 112:16-24.

BÉKÉS, F., ZAWISTOWSKA, U., and BUSHUK, W. 1983a. Protein-lipid complexes in the gliadin fraction. Cereal Chem. 60:371-378.

BÉKÉS, F., ZAWISTOWSKA, U., and BUSHUK, W. 1983b. Lipid-mediated aggregation of gliadin. Cereal Chem. 60:379-380.

BÉKÉS, F., ZAWISTOWSKA, U., ZILLMAN, R. R., and BUSHUK, W. 1986. Relationship between lipid content and composition and loaf volume of twenty-six common spring wheats. Cereal Chem. 63:327-331.

BELDEROK, B. 1967. Bedeutung der Thiol und Disulfidgruppen fur die Zuchtung des Weizens auf Backqualitat. Getreide Mehl 17:20-26.

BELL, P. M., and SIMMONDS, D. H. 1963. The protein composition of different flours and its relationship to nitrogen content and baking performance. Cereal Chem. 40:121-128.

BELL, B. M., CHAMBERLAIN, N., COLLINS, T. H., DANIELS, D. G. H., and FISHER, N. 1979a. The composition, rheological properties, and breadmaking behavior of stored flours. J. Sci. Food Agric. 30:1111-1122.

BELL, B. M., DANIELS, D. G. H., and FISHER, N. 1979b. The effects of pure saturated and unsaturated fatty acids on breadmaking and on lipid binding, using

Chorleywood Bread Process doughs containing a model fat. J. Sci. Food Agric. 30:1123-1130.

BELL, B. M., DANIELS, D. G. H., FEARN, T., and STEWART, B. A. 1987. Lipid compositions, baking qualities and other characteristics of wheat varieties grown in the U.K. J. Cereal Sci. 5:277-286.

BENNET, R., and EWART, J. A. D. 1965. The effect of certain salts on doughs. J. Sci. Food Agric. 16:199-205.

BEQUETTE, R. K., WATSON, C. A., MILLER, B. S., JOHNSON, J. A., and SCHRENK, W. G. 1963. Mineral composition of gluten, starch, and water-soluble fractions of wheat flour and its relation to flour quality. Agron. J. 55:537-542.

BERNARDIN, J. E. 1978. Gluten protein interaction with small molecules and ions—The control of flour properties. Bakers Dig. 52(4):20-23.

BERTRAM, G. L. 1953. Studies on crust color. I. The importance of the browning reaction in determining the crust color of bread. Cereal Chem. 30:127-139.

BICE, C. W., and GEDDES, W. F. 1953. The role of starch in bread staling. In: Starch and Its Derivatives. J. A. Radley, ed. Chapman and Hall, London.

BIETZ, J. A., and HUEBNER, F. R. 1980. Structure of glutenin: Achievements at the Northern Regional Research Center. Ann. Technol. Agric. 29:249-277.

BIETZ, J. A., and WALL, J. S. 1980. Identity of high molecular weight gliadin and ethanol-soluble glutenin subunits of wheat: Relation to gluten structure. Cereal Chem. 57:415-421.

BIETZ, J. A., HUEBNER, F. R., and WALL, J. S. 1973. Glutenin—The strength protein of wheat flour. Bakers Dig. 47(1):26-31, 34, 35, 67.

BIGELOW, C. C. 1967. The average hydrophobicity of proteins and the relation between it and protein structure. J. Theor. Biol. 16:187-211.

BLOKSMA, A. H. 1959. The influence of the extraction of lipids from flour on gluten development and breakdown. Chem. Ind. (London) 1959:253-254.

BLOKSMA, A. H. 1963. Oxidation by molecular oxygen of thiol groups in unleavened doughs from normal and defatted wheat flours. J. Sci. Food Agric. 14:529-535.

BLOKSMA, A. H. 1964a. Oxidation by potassium iodate of thiol groups in unleavened wheat flour doughs. J. Sci. Food Agric. 15:83-94.

BLOKSMA, A. H. 1964b. The role of thiol groups and flour lipids in oxidation-reduction reactions in dough. Bakers Dig. 38(2):53-60.

BLOKSMA, A. H. 1966. Extraction of flour by mixtures of butanol-1 and water. Cereal Chem. 43:602-622.

BLOKSMA, A. H. 1972a. Flour composition, dough rheology, and baking quality. Cereal Sci. Today 17:380-386.

BLOKSMA, A. H. 1972b. The relation between the thiol and disulfide contents of dough and its rheological properties. Cereal Chem. 49:104-118.

BLOKSMA, A. H. 1975. Thiol and disulfide groups in dough rheology. Cereal Chem. 52:170r-183r.

BOOTH, M. R., and MELVIN, M. A. 1979. Factors responsible for the poor breadmaking quality of high yielding European wheat. J. Sci. Food Agric. 30:1057-1064.

BOURDET, A., and FEILLET, P. 1967. Distribution of phosphorus compounds in the protein fractions of various types of wheat flours. Cereal Chem. 44:457-482.

BRANLARD, G. and BELLOT, P. 1983. Improvement of bread wheat gluten quality. Qual. Plant. Plant Foods Hum. Nutr. 33:121-126.

BRANLARD, G., and DARDEVET, M. 1985a. Diversity of grain proteins and bread wheat quality. I. Correlation between gliadin bands and flour quality characteristics. J. Cereal Sci. 3:329-343.

BRANLARD, G., and DARDEVET, M. 1985b. Diversity of grain proteins and bread wheat quality. II. Correlation between high molecular weight subunits of glutenin and flour quality characteristics. J. Cereal Sci. 3:345-354.

BRANLARD, G., and ROUSSET, M. 1980. Les caractéristiques électrophoretiques des gliadines et la valeur en panification du blé tendre. Ann. Amelior. Plant. 30:133-149.

BRANLARD, G., ROUSSET, M., VILLEMONT, P., and MOUSSET, C. 1984. Prediction of the technological quality of bread wheat from the gliadin and glutenin polymorphism. Pages 195-205 in: Gluten Proteins. Proc. Int. Workshop Gluten Proteins, 2nd. A. Graveland and J. H. E. Moonen, eds. Inst. Cereals, Flour and Bread, TNO, Wageningen, The Netherlands.

BURNOUF, T., and BIETZ, J. A. 1984. Reversed-phase high-performance liquid chromatography of durum wheat gliadins: Relationships to durum wheat quality. J. Cereal Sci. 2:3-14.

BURNOUF, T., and BOURIQUET, R. 1980. Glutenin subunits of genetically related European hexaploid wheat cultivars: Their relation to bread-making quality. Theor. Appl. Genet. 58:107-111.

BUSHUK, W. 1961. Accessible sulfhydryl

groups in dough. Cereal Chem. 38:438-448.

BUSHUK, W. 1966. Distribution of water in dough and bread. Bakers Dig. 40(5):38-40.

BUSHUK, W. 1974. Glutenin—Functions, properties and genetics. Bakers Dig. 48(4):14-16, 18, 19, 21, 22.

BUSHUK, W. 1984a. Functionality of wheat proteins in dough. Cereal Foods World 29:162-164.

BUSHUK, W. 1984b. Plant proteins. Pages 210-225 in: Food Science and Technology: Present Status and Future Directions. J. V. McLoughlin and B. M. McKenna, eds. Boole Press, Dublin.

BUSHUK, W. 1985. Wheat flour proteins: Structure and role in breadmaking. Pages 187-198 in: Analyses as Practical Tools in the Cereal Field. K. M. Fjell, ed. Norwegian Grain Corp., Oslo.

BUSHUK, W., and HLYNKA, I. 1961. The bromate reaction in dough. V. Effect of flour components and some related compounds. Cereal Chem. 38:316-325.

BUSHUK, W., and WRIGLEY, C. W. 1974. Proteins, composition, structure and function. Pages 119-145 in: Wheat, Production and Utilization. G. E. Inglett, ed. Avi Publ. Co., Inc., Westport, CT.

BUSHUK, W., KHAN, K., and MacMASTER, G. 1980. Functional glutenin: A complex of covalently and noncovalently linked components. Ann. Technol. Agric. 29:279-294.

BYERS, M., and BOLTON, J. 1979. Effects of nitrogen and sulphur fertilizers on the yield, N and S content, and amino acid composition of the grain of spring wheat. J. Sci. Food Agric. 30:251-263.

CAMPBELL, W. P., WRIGLEY, C. W., CRESSEY, P. J., and SLACK, C. R. 1987. Statistical correlations between quality attributes and grain-protein composition for 71 hexaploid wheats used as breeding parents. Cereal Chem. 64:293-299.

CARLSON, T. L.-G. 1981. Law and order in wheat flour dough. Colloidal aspects of the wheat flour dough and its lipid and protein constituents in aqueous media. Ph.D. thesis, Univ. of Lund, Lund, Sweden.

CARLSON, T., LARSSON, K., and MIEZIS, Y. 1978. Phase equilibria and structures in the aqueous system of wheat lipids. Cereal Chem. 55:168-179.

CARLSON, T. L.-G., LARSSON, K., and MIEZIS, Y. 1980. Physical structure and phase properties of aqueous systems of lipids from rye and triticale in relation to wheat lipids. J. Dispersion Sci. Technol. 1:197-208.

CARTER, J. E., and PACE, J. 1965. Some interrelationships of ascorbic acid and dehydroascorbic acid in the presence of flour suspensions and in dough. Cereal Chem. 42:201-208.

CASIER, J. P. J., and SOENEN, M. 1967. Die wasserunloslichen Pentosane aus Roggen und Weizen und ihr Einfluss auf die Backwerte. Getreide Mehl 17:46-49.

CAWLEY, R. W. 1964. The role of wheat flour pentosans in baking. II. Effect of added flour pentosans and other gums on gluten starch loaves. J. Sci. Food Agric. 15:834-838.

CHIU, C.-M., and POMERANZ, Y. 1966. Changes in extractability of lipids during bread-making. J. Food Sci. 31:753-756.

CHIU, C.-M., and POMERANZ, Y. 1967. Lipids in wheat kernels of varying size. J. Food Sci. 32:422-425.

CHIU, C.-M., POMERANZ, Y., SHOGREN, M. D., and FINNEY, K. F. 1968. Lipid binding in wheat flours varying in bread-making potential. Food Technol. (Chicago) 22:1157-1162.

CHUNG, K. H., and POMERANZ, Y. 1978. Acid-soluble proteins of wheat flours. I. Effect of delipidation on protein extraction. Cereal Chem. 55:230-243.

CHUNG, K. H., and POMERANZ, Y. 1979. Acid-soluble proteins of wheat flours. II. Binding to hydrophobic gels. Cereal Chem. 56:196-201.

CHUNG, O. K. 1986. Lipid-protein interactions in wheat flour, dough, gluten, and protein fractions. Cereal Foods World 31:242-244, 246-247, 249-252, 254-256.

CHUNG, O. K., and POMERANZ, Y. 1977. Wheat flour lipids, shortening and surfactants; a three way contribution to breadmaking. Bakers Dig. 51(5):32-34, 36-38, 40, 42-44, 153.

CHUNG, O. K., and POMERANZ, Y. 1981. Recent research on wheat lipids. Bakers Dig. 55(5):38-50, 55.

CHUNG, O. K., POMERANZ, Y., FINNEY, K. F., HUBBARD, J. D., and SHOGREN, M. D. 1977a. Defatted and reconstituted wheat flours. I. Effects of solvent and Soxhlet types on functional (breadmaking) properties. Cereal Chem. 54:454-465.

CHUNG, O. K., POMERANZ, Y., FINNEY, K. F., and SHOGREN, M. D. 1977b. Defatted and reconstituted wheat flours. II. Effects of solvent type and extracting conditions on flours varying in breadmaking quality. Cereal Chem. 54:484-495.

CHUNG, O. K., POMERANZ, Y., and FINNEY, K. F. 1978a. Wheat flour lipids in breadmaking. Cereal Chem. 55:598-618.

CHUNG, O. K., POMERANZ, Y., FINNEY, K. F., and SHOGREN, M. D. 1978b. Surfactants as replacements for natural lipids in bread baked from defatted wheat flour. J. Am. Oil Chem. Soc. 55:635-641.

CHUNG, O. K., POMERANZ, Y., HWANG, E. C., and DIKEMAN, E. 1979. Defatted and reconstituted wheat flours. IV. Effects of flour lipids on protein extractability from flours that vary in bread-making quality. Cereal Chem. 56:220-226.

CHUNG, O. K., POMERANZ, Y., JACOBS, R. M., and HOWARD, B. G. 1980a. Lipid extraction conditions to differentiate among hard red winter wheats that vary in bread-making. J. Food Sci. 45:1168-1174.

CHUNG, O. K., POMERANZ, Y., SHOGREN, M. D., FINNEY, K. F., and HOWARD, B. G. 1980b. Defatted and reconstituted wheat flours. VI. Response to shortening addition and lipid removal in flours that vary in bread-making quality. Cereal Chem. 57:111-117.

CHUNG, O. K., POMERANZ, Y., and FINNEY, K. F. 1982. Relation of polar lipid content to mixing requirement and loaf volume potential of hard red winter wheat flour. Cereal Chem. 59:14-20.

CIACCO, C. F., and D'APPOLONIA, B. L. 1977. Characterization of starches from various tubers and their use in bread-baking. Cereal Chem. 54:1096-1107.

CIACCO, C. F., and D'APPOLONIA, B. L. 1982. Characterization of pentosans from different wheat flour classes and of their gelling capacity. Cereal Chem. 59:96-99.

CLEMENTS, R. L., and DONELSON, J. R. 1981. Functionality of specific flour lipids in cookies. Cereal Chem. 58:204-206.

CLEMENTS, R. L., and DONELSON, J. R. 1982a. Role of free flour lipids in batter expansion in layer cakes. I. Effects of "aging." Cereal Chem. 59:121-124.

CLEMENTS, R. L., and DONELSON, J. R. 1982b. Role of free flour lipids in batter expansion in layer cakes. II. Effects of heating. Cereal Chem. 59:125-128.

CLENDENNING, K. A., and WRIGHT, D. E. 1950. Separation of starch and gluten. V. Problems in wheat starch manufacture arising from flour pentosans. Can. J. Res. Sect. F 28:390-400.

CLUSKEY, J. E., TAYLOR, N. W., CHARLEY, H., and SENTI, F. R. 1961. Electrophoretic composition and intrinsic viscosity of glutens from different varieties of wheat. Cereal Chem. 38:325-335.

COLE, E. W., MECHAM, D. K., and PENCE, J. W. 1960. Effect of flour lipids and some lipid derivatives on cookie-baking characteristics of lipid-free flours. Cereal Chem. 37:109-121.

COOKSON, M. A., and COPPOCK, J. B. M. 1956. The role of lipids in baking. III. Some further breadmaking and other properties of

defatted flours and of flour lipids. J. Sci. Food Agric. 7:72-87.

COOKSON, M. A., RITCHIE, M. L., and COPPOCK, J. B. M. 1957. The role of lipids in baking. IV. Some further properties of flour lipids and defatted flours. J. Sci. Food Agric. 8:105-116.

COPPOCK, J. B. M., COOKSON, M. A., LANEY, D. H., and AXFORD, D. W. E. 1954. The role of lipids in baking. II. The influence of flour oils on the behavior of glycerinated fats in baking and the effect of natural monoglycerides present in flour oils and baking fats on the pharmacological desirability of using glycerinated fats in baked products. J. Sci. Food Agric. 5:19-26.

COTTENET, M., AUTRAN, J.-C., and JOUDRIER, P. 1983. Plant physiology — Isolation and characterization of γ-gliadins 45 and 42 compounds associated with viscoelastic characteristics of the gluten of durum wheat. C. R. Hebd. Seances Acad. Sci. Ser. C 297:149-154.

COULSON, C. B., and SIM, A. K. 1964. Proteins of various species of wheat and closely related genera and their relationship to genetical characteristics. Nature 202:1305-1308.

COULSON, C. B., and SIM, A. K. 1965. Proteins of wheat flour. Nature 208:583.

CRESSEY, P. J., CAMPBELL, W. P., WRIGLEY, C. W., and GRIFFIN, W. B. 1987. Statistical correlations between quality attributes and grain-protein composition for 60 advanced lines of crossbred wheat. Cereal Chem. 64:299-301.

CUNNINGHAM, D. K., and HLYNKA, I. 1958. Flour lipids and the bromate reaction. Cereal Chem. 35:401-410.

DADSWELL, I. W., and WRAGGE, W. B. 1940. The autolytic digestion of flour in relation to variety and environment. Cereal Chem. 17:584-601.

DAFTARY, R. D., and POMERANZ, Y. 1965. Changes in lipid composition of wheat during storage deterioration. J. Agric. Food Chem. 13:442-446.

DAFTARY, R. D., WARD, A. B., and POMERANZ, Y. 1966. Distribution of lipids in air-fractionated flours. J. Food Sci. 31:897-901.

DAFTARY, R. D., POMERANZ, Y., SHOGREN, M., and FINNEY, K. F. 1968. Functional breadmaking properties of lipids. II. The role of flour lipid fractions in breadmaking. Food Technol. (Chicago) 22:79-82.

DAFTARY, R. D., POMERANZ, Y., HOSENEY, R. C., SHOGREN, M. D., and FINNEY, K. F. 1970a. Changes in wheat flour damaged by mold during storage. Effect in breadmaking. J. Agric. Food Chem. 18:617-619.

DAFTARY, R. D., POMERANZ, Y., and SAUER, D. B. 1970b. Changes in wheat flour damaged by mold during storage. Effects on lipid, lipoprotein, and protein. J. Agric. Food Chem. 18:613-616.

DAHLE, L. K., and HINZ, R. S. 1966. The weakening action of thioctic acid in unyeasted and yeasted doughs. Cereal Chem. 43:682-688.

DAHLE, L. K., and MUENCHOW, H. L. 1968. Some effects of solvent extraction on cooking characteristics of spaghetti. Cereal Chem. 45:464-468.

DAHLE, L. K., and PINKE, P. 1968. A note comparing effects on dough mixing of thioctic acid (lipoic acid) and yeast-fermented flour extracts. Cereal Chem. 45:287-289.

DAHLE, L., and SULLIVAN, B. 1960. Presence and probable role of thioctic acid in wheat flour. Cereal Chem. 37:679-682.

DAHLE, L. K., and SULLIVAN, B. 1963. The oxidation of wheat flour. V. Effect of lipid peroxides and antioxidants. Cereal Chem. 40:372-384.

DAL BELIN PERUFFO, A., POGNA, N. E., TEALDO, E., TUTTA, C., and ALBUZIO, A. 1985. Isolation and partial characterisation of γ-gliadins 40 and 43.5 associated with quality in common wheat. J. Cereal Sci. 3:355-361.

DAMIDAUX, R., AUTRAN, J. C., GRIGNAC, P., and FEILLET, P. 1978. Mise en évidence de relation applicable en selection entre l'électrophoregramme des gliadines et les propriétés viscoélastiques du gluten de *Triticum durum* Desf. C. R. Hebd. Seances Acad. Sci. Ser. D 287:701-704.

DAMIDAUX, R., AUTRAN, J.-C., and FEILLET, P. 1980. Gliadin electrophoregrams and measurements of gluten viscoelasticity in durum wheats. Cereal Foods World 25:754-756.

DANIELS, D. G. H. 1958. Polar lipids in wheat flour. Chem. Ind. (London) 1958:653-654.

DANIELS, N. W. R., RICHMOND, J. W., RUSSELL EGGITT, P. W., and COPPOCK, J. B. M. 1966. Studies on the lipids of flour. III. Lipid binding in breadmaking. J. Sci. Food Agric. 17:20-29.

DANIELS, N. W. R., RICHMOND, J. W., RUSSELL EGGITT, P. W., and COPPOCK, J. B. M. 1968. Studies on the lipids of flour. IV. Factors affecting lipid binding in bread-making. J. Sci. Food Agric. 20:129-136.

DANIELS, N. W. R., WOOD, P. S., RUSSELL EGGITT, P. W., and COPPOCK, J. B. M. 1969. Effect of vacuum on lipid binding during high energy dough development. Chem. Ind. (London) 1969:167-168.

DANIELS, N. W., FRAZIER, P. J., and WOOD, P. S. 1971. Flour lipids and dough development. Bakers Dig. 45(4):20-25, 28.

D'APPOLONIA, B. L. 1971. Role of pentosans in bread and dough. A review. Bakers Dig. 45(6):20-23, 63.

D'APPOLONIA, B. L., and GILLES, K. A. 1971a. Pentosans associated with gluten. Cereal Chem. 48:427-436.

D'APPOLONIA, B. L., and GILLES, K. A. 1971b. Effect of various starches in baking. Cereal Chem. 48:625-636.

D'APPOLONIA, B. L., and KIM, S. K. 1976. Recent developments on wheat flour pentosans. Bakers Dig. 50(3):45-49, 53.

D'APPOLONIA, B. L., and MORAD, M. M. 1981. Bread staling. Cereal Chem. 58:186-190.

DE LANGE, P., and HINTZER, H. M. R. 1955a. Studies on wheat proteins. I. Polarographic determination of the apparent sulfhydryl content of wheat proteins. Cereal Chem. 32:307-313.

DE LANGE, P., and HINTZER, H. M. R. 1955b. Studies on wheat proteins. II. Significance of sulfhydryl groups and disulfide bonds for baking strength. Cereal Chem. 32:314-324.

DENNET, K., and STERLING, C. 1979. Role of starch in bread formation. Staerke 31:209-213.

DE RUITER, D. 1978. Composite flours. Pages 349-385 in: Advances in Cereal Science and Technology, Vol. 2. Y. Pomeranz, ed. Am. Assoc. Cereal Chem., St. Paul, MN.

DE STEFANIS, V. A., and PONTE, J. G., Jr. 1976. Studies on the breadmaking properties of wheat-flour nonpolar lipids. Cereal Chem. 53:636-642.

DEXTER, J. E., and MATSUO, R. R. 1977a. Changes in semolina proteins during spaghetti processing. Cereal Chem. 54:882-894.

DEXTER, J. E., and MATSUO, R. R. 1977b. The spaghetti-making quality of developing durum wheats. Can. J. Plant Sci. 57:7-16.

DEXTER, J. E., and MATSUO, R. R. 1977c. Influence of protein content on some durum wheat quality parameters. Can. J. Plant Sci. 57:717-727.

DEXTER, J. E., and MATSUO, R. R. 1978. The effect of gluten protein fractions on pasta dough rheology and spaghetti-making quality. Cereal Chem. 55:44-57.

DEXTER, J. E., and MATSUO, R. R. 1979a. Effect of starch on pasta dough rheology and spaghetti cooking quality. Cereal Chem. 56:190-195.

DEXTER, J. E., and MATSUO, R. R. 1979b. Changes in spaghetti protein solubility during cooking. Cereal Chem. 56:394-398.

DEXTER, J. E., and MATSUO, R. R. 1980a.

Relationship between durum wheat protein properties and pasta dough rheology and spaghetti cooking quality. J. Agric. Food Chem. 28:899-902.

DEXTER, J. E., and MATSUO, R. R. 1980b. Changes in semolina proteins during spaghetti production. Tec. Molitoria 31:857-867.

DEXTER, J. E., MATSUO, R. R., PRESTON, K. R., and KILBORN, R. H. 1981. Comparison of gluten strength, mixing properties, baking quality and spaghetti quality of some Canadian durum and common wheats. Can. Inst. Food Sci. Technol. J. 14:108-111.

DEXTER, J. E., MATSUO, R. R., and MORGAN, B. C. 1983. Spaghetti stickiness: Some factors influencing stickiness and relationship to other cooking quality characteristics. J. Food Sci. 48:1545-1551, 1559.

DEXTER, J. E., MATSUO, R. R., and MacGREGOR, A. W. 1985a. Relationship of instrumental assessment of spaghetti cooking quality to the type and the amount of material rinsed from cooked spaghetti. J. Cereal Sci. 3:39-53.

DEXTER, J. E., PRESTON, K. R., TWEED, A. R., KILBORN, R. H., and TIPPLES, K. H. 1985b. Relationship of flour starch damage and flour protein to the quality of Brazilian-style hearth bread and remix pan bread produced from hard red spring wheat. Cereal Foods World 30:511-514.

DEXTER, J. E., MATSUO, R. R., and MARTIN, D. G. 1987. The relationship of durum wheat test weight to milling performance and spaghetti quality. Cereal Foods World 32:772-777.

DHALIWAL, A. S., MARES, D. J., and MARSHALL, D. R. 1987. Effect of 1B/1R chromosome translocation on milling and quality characteristics of bread wheats. Cereal Chem. 64:72-76.

DHALIWAL, A. S., MARES, D. J., MARSHALL, D. R., and SKERRITT, J. H. 1988. Protein composition and pentosan content in relation to dough stickiness of 1B/1R translocation wheats. Cereal Chem. 65:143-149.

DOGUCHI, M., and HLYNKA, I. 1967. Some rheological properties of crude gluten mixed in the farinograph. Cereal Chem. 44:561-575.

DONELSON, D. H., and WILSON, J. T. 1960a. Effect of the relative quantity of flour fractions on cake quality. Cereal Chem. 37:241-262.

DONELSON, D. H., and WILSON, J. T. 1960b. Studies on the effect of flour-fraction interchange upon cake quality. Cereal Chem.

37:683-710.

DONELSON, J. R., and CLEMENTS, R. L. 1986. Components of cake batter expansion in white layer cakes. Cereal Chem. 63:109-110.

DONELSON, J. R., YAMAZAKI, W. T., and KISSELL, L. T. 1984. Functionality in white layer cake of lipids from untreated and chlorinated patent flours. II. Flour fraction interchange studies. Cereal Chem. 61:88-91.

DOUGLAS, J. A., and TYSON, C. B. 1985. Preliminary investigations of concentrations of minerals and nitrogen in wheat grain, and their relationship with baking quality and grain weight. N.Z. J. Agric. Res. 28:81-85.

DuCROS, D. L. 1987. Glutenin proteins and gluten strength in durum wheat. J. Cereal Sci. 5:3-12.

DuCROS, D. I.., and HARE, R. A. 1985. Inheritance of gliadin proteins associated with quality in durum wheat. Crop Sci. 25:674-677.

ELIASSON, A. C. 1983. Physical properties of starch in concentrated systems such as dough and bread. Ph.D. thesis, Univ. of Lund, Lund, Sweden.

ELTON, G. A. H., and EWART, J. A. D. 1967. Some properties of wheat proteins. Bakers Dig. 41(1):36-39, 42-44.

ELTON, G. A. H., and FISHER, N. 1968. Effect of solid hydrocarbons as additives in breadmaking. J. Sci. Food Agric. 19:178-181.

EWART, J. A. D. 1968. A hypothesis for the structure and rheology of gluten. J. Sci. Food Agric. 19:617-623.

EWART, J. A. D. 1972. A modified hypothesis for the structure and rheology of glutelins. J. Sci. Food Agric. 23:687-699.

EWART, J. A. D. 1977. Re-examination of the linear glutenin hypothesis. J. Sci. Food Agric. 28:191-199.

EWART, J. A. D. 1978. Glutenin and dough tenacity. J. Sci. Food Agric. 29:551-556.

EWART, J. A. D. 1979. Glutenin structure. J. Sci. Food Agric. 30:482-492.

EWART, J. A. D. 1980. Loaf volume and the intrinsic viscosity of glutenin. J. Sci. Food Agric. 31:1323-1336.

EWART, J. A. D. 1982. Baking quality and gluten amide level. J. Food Technol. 17:365-372.

EWART, J. A. D. 1988. Studies on disulfide bonds in glutenin. Cereal Chem. 65:95-100.

FABRIANI. G., LINTAS, C., and QUAGLIA, G. B. 1968. Chemistry of lipids in processing and technology of pasta products. Cereal Chem. 45:454-463.

FABRIANI, G., LELLI, M. E., LINTAS, C., and QUAGLIA, G. B. 1970. Titrage amperométrique des groupes -SH plus S-S dans les semoules des différentes variétés de blé dur. Proc. Int. Congr. Bread Flour Wheat, 6th, Dresden, pp. 61-65.

FABRIANI, G., QUAGLIA, G. B., and MAFFEI, A. 1975. Effetto della molitura e della pastificazione sul contento in gruppi tiolici reattivi e totali di varieta pure di frumento duro. Tec. Molitoria 26:81-85.

FARRAND, E. A. 1972. Controlled levels of starch damage in a commercial United Kingdom bread flour and effects on absorption, sedimentation value, and loaf quality. Cereal Chem. 49:479-488.

FAUSCH, H., KUENDIG, W., and NEUKOM, H. 1963. Ferulic acid as a component of a glycoprotein from wheat flour. Nature 199:287.

FENTON, F. C., and SWANSON, C. O. 1930. Studies on the qualities of combined wheats as affected by type of bin, moisture and temperature conditions. I. Cereal Chem. 7:428-448.

FIELD, J. M., SHEWRY, P. R., BURGESS, S. R., FORDE, J., PARMAR, S., and MIFLIN, B. J. 1983a. The presence of high molecular weight aggregates in the protein bodies of developing endosperms of wheat and other cereals. J. Cereal Sci. 1:33-41.

FIELD, J. M., SHEWRY, P. R., and MIFLIN, B. J. 1983b. Solubilisation and characterisation of wheat gluten proteins: Correlations between the amount of aggregated proteins and baking quality. J. Sci. Food Agric. 34:370-377.

FIFIELD, C. C., and ROBERTSON, D. W. 1959. Milling, baking, and chemical properties of Marquis and Kanred wheat grown in Colorado and stored 25 to 33 years. Cereal Sci. Today 4:179-183.

FIFIELD, C. C., WEAVER, R., and HAYES, J. F. 1950. Bread loaf volume and protein content of hard red spring wheats. Cereal Chem. 27:383-390.

FINNEY, K. F. 1943. Fractionating and reconstituting techniques as tools in wheat flour research. Cereal Chem. 20:381-396.

FINNEY, K. F. 1965. Evaluation of wheat quality. Page 73 in: Food Quality Effects of Production Practices and Processing. G. W. Irving and S. R. Hoover, eds. Publ. 77. Am. Assoc. Adv. Sci., Washington, DC.

FINNEY, K. F. 1978. Contribution of individual chemical constituents to the functional (breadmaking) properties of wheat. Pages 139-158 in: Cereals '78: Better Nutrition for the World's Millions. Y. Pomeranz, ed. Am. Assoc. Cereal Chem., St. Paul, MN.

FINNEY, K. F. 1984. An optimized, straight-dough, bread-making method after 44 years. Cereal Chem. 61:20-27.

FINNEY, K. F. 1985. Experimental bread-making studies, functional (breadmaking) properties, and related gluten protein fractions. Cereal Foods World 30:794-796, 798-799, 801.

FINNEY, K. F., and BARMORE, M. A. 1945a. Varietal responses to certain baking ingredients essential in evaluating the protein quality of hard winter wheats. Cereal Chem. 22:225-243.

FINNEY, K. F., and BARMORE, M. A. 1945b. Optimum vs. fixed mixing time at various potassium bromate levels in experimental bread baking. Cereal Chem. 22:244-254.

FINNEY, K. F., and BARMORE, M. A. 1948. Loaf volume and protein content of hard winter and spring wheats. Cereal Chem. 25:291-312.

FINNEY, K. F., and YAMAZAKI, W. T. 1967. Quality of hard, soft, and durum wheats. Pages 471-503 in: Wheat and Wheat Improvement. K. S. Quisenberry and L. P. Reitz, eds. Am. Soc. Agron., Madison, WI.

FINNEY, K. F., POMERANZ, Y., and HOSENEY, R. C. 1976. Effects of solvent extraction on lipid composition, mixing time, and bread loaf volume. Cereal Chem. 53:383-388.

FINNEY, K. F., JONES, B. L., and SHOGREN, M. D. 1982. Functional (bread-making) properties of wheat protein fractions obtained by ultracentrifugation. Cereal Chem. 59:449-453.

FINNEY, K. F., HEYNE, E. G., SHOGREN, M. D., BOLTE, L. C., and POMERANZ, Y. 1985. Functional properties of some European wheats grown in Europe and Kansas. Cereal Chem. 62:83-88.

FINNEY, K. F., YAMAZAKI, W. T., YOUNGS, V. L., and RUBENTHALER, G. L. 1987. Quality of hard, soft, and durum wheats. Pages 677-748 in: Wheat and Wheat Improvement. E. G. Heyne, ed. Am. Soc. Agron., Crop Sci. Soc. Am., Soil Sci. Soc. Am., Madison, WI.

FISHER, N., BROUGHTON, M. E., PEEL, D. J., and BENNETT, R. 1964. The lipids of wheat. II. Lipids of flours from single wheat varieties of widely varying baking quality. J. Sci. Food Agric. 15:325-341.

FISHER, N., BELL, B. M., RAWLINGS, C. E. B., and BENNETT, R. 1966. The lipids of wheat. III. Further studies of the lipids of flours from single wheat varieties of widely varying baking quality. J. Sci. Food Agric. 17:370-382.

FLYNN, A. G., PANOZZO, J. F., and GARDNER, W. K. 1987. The effect of copper deficiency on the baking quality and dough properties of wheat flour. J. Cereal Sci.

6:91-98.

FRANK, H. S., and EVANS, M. W. 1945. Free volume and entropy in condensed systems. III. Entropy in binary liquid mixtures; partial molal entropy in dilute solutions; structure and thermodynamics in aqueous electrolytes. J. Chem. Phys. 13:507.

FRATER, R., HIRD, F. J. R., MOSS, H. J., and YATES, J. R. 1960. A role for thiol and disulphide groups in determining the rheological properties of dough made from wheaten flour. Nature 186:451-454.

FRATER, R., HIRD, F. J. R., and MOSS, H. J. 1961. Role of disulphide exchange reactions in relaxation of strains introduced in dough. J. Sci. Food Agric. 12:269-273.

FRAZIER, P. J. 1979. Lipoxygenase action and lipid binding in breadmaking. Bakers Dig. 53(6):8-10, 12, 13, 16, 18, 20, 29.

FRAZIER, P. J., BRIMBLECOMBE, F. A., DANIELS, N. W. R., and EGGITT, P. W. R. 1979. Better bread from weaker wheats—A rheological attack. Getreide Mehl Brot 33:268-271.

FRAZIER, P. J., DANIELS, N. W. R., and RUSSELL-EGGITT, P. W. 1981. Lipid-protein interactions during dough development. J. Sci. Food Agric. 32:877-897.

FREILICH, J., and FREY, C. N. 1947. Dough oxidation and mixing studies. VII. The role of oxygen in dough mixing. Cereal Chem. 24:436-448.

FULLINGTON, J. G. 1969. Lipid-protein interaction. Bakers Dig. 43(6):34-36, 38, 61.

FULLINGTON, J. G., MISKELLY, D. M., WRIGLEY, C. W., and KASARDA, D. D. 1987. Quality-related endosperm proteins in sulfur-deficient and normal wheat grain. J. Cereal Sci. 5:233-245.

GAINES, C. S. 1985. Associations among soft wheat flour particle size, protein content, chlorine response, kernel hardness, milling quality, white layer cake volume, and sugar-snap cookie spread. Cereal Chem. 62:290-292.

GAINES, C. S., and DONELSON, J. R. 1982. Contribution of chlorinated flour fractions to cake crumb stickiness. Cereal Chem. 59:378-380.

GAINES, C. S., and DONELSON, J. R. 1985. Effect of varying flour protein content on angel food and high-ratio white layer cake size and tenderness. Cereal Chem. 62:63-66.

GARTON, G. A., MIHAILOVIC, M. L., ANTIC, M., and HADZIJEV, D. 1963a. Chemical investigation of wheat. 6. Grain lipids of some high-yielding wheat varieties. Doc. Chem. Yugoslavia 28:555-571.

GARTON, G. A., MIHAILOVIC, M. L., ANTIC, M., and HADZIJEV, D. 1963b. Chemical investigation of wheat. 7. Grain lipids of wheat in relation to fertilizer treatment. Bull. Chem. Soc. (Belgrade) 28:303-325.

GEBHARDT, E., NEUMANN, C., and ECKERT, I. 1982. Modification and derivatization of gluten and influence on the manufacture of baked goods. Backer Konditor 30(2):37.

GERICKE, W. F. 1934a. Effect of nitrate salts supplied to wheat grown in liquid media on bread scores. II. Cereal Chem. 11:141-152.

GERICKE, W. F. 1934b. Effect of chloride salts supplied to wheat grown in liquid media on bread scores. III. Cereal Chem. 11:335-343.

GILLES, K. A. 1960. The present status of the role of pentosans in wheat flour quality. Bakers Dig. 34(5):47-52.

GODON, B., and HERARD, J. 1984. Extractibilité des protéines du blé. 1: Incidences sur les associations protéiques du pétrissage des pâtes en présence de divers composés actifs. Sci. Aliments 4:287-303.

GODON, B., and HERARD, J. 1986. Extractability of wheat proteins. Sci. Aliments 6:601-621.

GODON, B., and POPINEAU, Y. 1981. Differences in surface hydrophobicity of gliadins in two *Triticum durum* wheat varieties according to cooking quality. Agronomie 1:77-82.

GOFORTH, D. R., and FINNEY, K. F. 1976. Separation of glutenin from gliadin by ultracentrifugation. Cereal Chem. 53:608-612.

GOFORTH, D. R., FINNEY, K. F., HOSENEY, R. C., and SHOGREN, M. D. 1977. Effect of strength and concentration of acid on the functional properties of solubilized glutens of good- and poor-quality bread flours. Cereal Chem. 54:1249-1258.

GOLDSTEIN, S. 1957. Sulfhydryl- und Disulfidgruppen der Klebereiweisse und ihre Beziehung zur Backfähigkeit der Brotmehle. Mitt. Geb. Lebensmittelunters. Hyg. 48:87-93.

GORSHKOVA, N. S., BUTKO, V. P., and SILINA, O. I. 1979. Level of sulfhydryl groups and disulfide bonds in wheat proteins. Izv. Vyssh. Uchebn. Zaved. Pishch. Tekhnol. 3:22-24.

GOUGH, B. M., GREENWELL, P., and RUSSELL, P. L. 1985. On the interaction of sodium dodecyl sulphate with starch granules. Pages 99-108 in: New Approaches to Research on Cereal Carbohydrates. R. D. Hill and L. Munck, eds. Elsevier Sci. Publ., Amsterdam.

GRACZA, R., and GREENBERG, S. I. 1963. The specific surface of flour and starch granules in a hard winter wheat flour and in its five subsieve-size fractions. Cereal Chem. 40:51-61.

GRANT, D. R. 1973. The modification of wheat flour proteins with succinic anhydride. Cereal Chem. 50:417-428.

GRAVELAND, A., BOSVELD, P., LICHTENDONK, W. J., and MOONEN, J. H. E. 1984. Structure of glutenins and their breakdown during dough mixing by a complex oxidation-reduction system. Pages 59-68 in: Gluten Proteins. Proc. Int. Workshop Gluten Proteins, 2nd. A. Graveland and J. H. E. Moonen, eds. Inst. Cereals, Flour and Bread, TNO, Wageningen, The Netherlands.

GRAVELAND, A., BOSVELD, P., LICHTENDONK, W. J., MARSEILLE, J. P., MOONEN, J. H. E., and SCHEEPSTRA, A. 1985. A model for the molecular structure of the glutenins from wheat flour. J. Cereal Sci. 3:1-16.

GREENWELL, P., and SCHOFIELD, J. D. 1986. A starch granule protein associated with endosperm softness in wheat. Cereal Chem. 63:379-380.

GREENWELL, P., EVERS, A. D., GOUGH, B. M., and RUSSELL, P. L. 1985. Amyloglucosidase-catalysed erosion of native, surface-modified and chlorine-treated wheat starch granules. The influence of surface protein. J. Cereal Sci. 3:279-293.

GREENWOOD, C. T. 1976. Starch. Pages 119-157 in: Advances in Cereal Science and Technology, Vol. I. Y. Pomeranz, ed. Am. Assoc. Cereal Chem., St. Paul, MN.

GREENWOOD, C. T., and EWART, J. A. D. 1975. Hypothesis for the structure of glutenin in relation to rheological properties of gluten and dough. Cereal Chem. 52:146r-153r.

GREER, E. N., and STEWART, B. A. 1959. The water absorption of wheat flour; relative effects of protein and starch. J. Sci. Food Agric. 10:248-252.

GREER, E. N., JONES, C. R., and MORAN, T. 1954. The quality of flour stored for periods up to 27 years. Cereal Chem. 31:439-450.

GROSSKREUTZ, J. C. 1960. The physical structure of wheat protein. Biochim. Biophys. Acta 38:400-409.

GROSSKREUTZ, J. C. 1961. A lipoprotein model of wheat gluten structure. Cereal Chem. 38:336-349.

GRZYBOWSKI, R. A., and DONNELLY, B. J. 1979. Cooking properties of spaghetti: Factors affecting cooking quality. J. Agric. Food Chem. 27:380-384.

GUNZEL, G. 1966. Verkleisterungseigenschaften sortenreiner Weizenstaerken und Weizenmehle. Getreide Mehl 16:75-77.

GUSS, P. L., RICHARDSON, T., and STAHMANN, M. A. 1967. The oxidation-reduction enzymes of wheat. III. Isoenzymes

of lipoxidase in wheat fractions and soybean. Cereal Chem. 44:607-610.

HAMADA, A. S., McDONALD, C. E., and SIBBITT, L. D. 1982. Relationship of protein fractions of spring wheat flour to baking quality. Cereal Chem. 59:296-301.

HANDLEMAN, A. R., CONN, J. F., and LYONS, J. W. 1961. Bubble mechanics in thick foams and their effects on cake quality. Cereal Chem. 38:294-305.

HARDER, R. W., and THIESSEN, W. L. 1971. Sulfur fertilizer improves yield and quality of soft white winter wheat. Curr. Inf. Ser. No. 165. Agric. Exp. Stn., Univ. of Idaho, Moscow.

HARRIS, R. H., and SIBBITT, L. D. 1941. The comparative baking qualities of starches prepared from different wheat varieties. Cereal Chem. 18:585-604.

HARRIS, R. H., and SIBBITT, L. D. 1942. The comparative baking qualities of hard red spring wheat starches and glutens as prepared by the gluten-starch blend baking method. Cereal Chem. 19:763-772.

HAWTHORN, J. 1961. Oxygen in the mixing of bread doughs. Bakers Dig. 35(4):34-35, 38, 40, 42, 43.

HAWTHORN, J., and TODD, J. P. 1955. Catalase in relation to the unsaturated-fat oxidase activity of wheat flour. Chem. Ind. (London) 1955:446-447.

HEAPS, P. W., WEBB, T., RUSSELL EGGITT, P. W., and COPPOCK, J. B. 1968. Rheological testing of wheat glutens and doughs. Chem. Ind. (London) 1968:1095-1096.

HEHN, E. R., and BARMORE, M. A. 1965. Breeding wheat for quality. Adv. Agron. 17:85-114.

HEID, W. G. 1979. U.S. Wheat Industry. Agric. Econ. Rep. 432. U.S. Dep. Agric., Washington, DC.

HELLMAN, N. N., FAIRCHILD, B., and SENTI, F. R. 1954. The bread staling problem. Molecular organization of starch upon aging of concentrated starch gels at various moisture levels. Cereal Chem. 31:495-505.

HESS, K. 1954. Protein, Kleber, und Lipoid im Weizenkorn und Mehl. Kolloid Z. 136:84-99.

HITES, B. D., SANDSTEDT, R. M., and SCHAUMBURG, L. 1953. Study of proteolytic activity in wheat flour doughs and suspensions. III. The misclassification of the proteases of flour as papainases. Cereal Chem. 30:404-412.

HOLME, J. 1962. Characterization studies on the soluble proteins and pentosans of cake flour. Cereal Chem. 39:132-146.

HOLME, J. A. 1966. A review of wheat flour proteins, and their functional properties.

Bakers Dig. 40(6):38-42, 78.

HONOLD, G. R., and STAHMANN, M. A. 1968. The oxidation-reduction enzymes of wheat. IV. Qualitative and quantitative investigations of the oxidases. Cereal Chem. 45:99-108.

HONOLD, G. R., FARKAS, G. L., and STAHMANN, M. A. 1966. The oxidation-reduction enzymes of wheat. I. A qualitative investigation of the dehydrogenases. Cereal Chem. 43:517-529.

HONOLD, G. R., FARKAS, G. L., and STAHMANN, M. A. 1967. The oxidation-reduction enzymes of wheat. II. A quantitative investigation of the dehydrogenases. Cereal Chem. 44:373-382.

HOSENEY, R. C. 1968. Functional (breadmaking) and biochemical properties of wheat flour components. Ph.D. thesis, Kansas State Univ., Manhattan.

HOSENEY, R. C. 1984a. Gas retention in bread doughs. Cereal Foods World 29:305-308.

HOSENEY, R. C. 1984b. Starch and other polysaccharides; basic structure and function in food. Pages 27-45 in: Carbohydrates, Proteins, Lipids: Basic Views and New Approaches in Food Technology. F. Meuser, ed. Inst. Lebensmitteltechnologie, Getreidetechnologie, Berlin, Federal Republic of Germany.

HOSENEY, R. C., and FAUBION, J. M. 1981. A mechanism for the oxidative gelation of wheat flour water-soluble pentosans. Cereal Chem. 58:421-424.

HOSENEY, R. C., and FINNEY, K. F. 1971. Functional (breadmaking) and biochemical properties of wheat flour components. XI. A review. Bakers Dig. 45(4):30-36, 39, 40, 64.

HOSENEY, R. C., and FINNEY, K. F., and POMERANZ, Y. 1966. Changes in urea-dispersibility of proteins during wheat maturation. J. Sci. Food Agric. 17:273-276.

HOSENEY, R. C., FINNEY, K. F., SHOGREN, M. D., and POMERANZ, Y. 1969a. Functional (breadmaking) and biochemical properties of wheat flour components. II. Role of water-solubles. Cereal Chem. 46:117-125.

HOSENEY, R. C., FINNEY, K. F., SHOGREN, M. D., and POMERANZ, Y. 1969b. Functional (breadmaking) and biochemical properties of wheat flour components. III. Characterization of gluten protein fractions obtained by ultracentrifugation. Cereal Chem. 46:126-135.

HOSENEY, R. C., FINNEY, K. F., POMERANZ, Y., and SHOGREN, M. D. 1969c. Functional (breadmaking) and biochemical properties of wheat flour components. IV. Gluten protein fractionation

by solubilizing in 70% ethyl alcohol and in dilute lactic acid. Cereal Chem. 46:495-502.

HOSENEY, R. C., FINNEY, K. F., POMERANZ, Y., and SHOGREN, M. D. 1969d. Functional (breadmaking) and biochemical properties of wheat flour components. V. Role of total extractable lipids. Cereal Chem. 46:606-613.

HOSENEY, R. C., FINNEY, K. F., and POMERANZ, Y. 1970. Functional (breadmaking) and biochemical properties of wheat flour components. VI. Gliadin-lipid-glutenin interaction in wheat gluten. Cereal Chem. 47:135-140.

HOSENEY, R. C., FINNEY, K. F., POMERANZ, Y., and SHOGREN, M. D. 1971. Functional (breadmaking) and biochemical properties of wheat flour components. VIII. Starch. Cereal Chem. 48:191-201.

HOSENEY, R. C., FINNEY, K. F., and SHOGREN, M. D. 1972a. Functional (breadmaking) and biochemical properties of wheat flour components. IX. Replacing total free lipid with synthetic lipid. Cereal Chem. 49:366-371.

HOSENEY, R. C., FINNEY, K. F., and SHOGREN, M. D. 1972b. Functional (breadmaking) and biochemical properties of wheat flour components. X. Fractions involved in the bromate reaction. Cereal Chem. 49:372-378.

HOSENEY, R. C., LINEBACK, D. R., and SEIB, P. A. 1978. Role of starch in baked goods. Bakers Dig. 57(4):11-14, 16, 18, 40.

HOWARD, N. B., HUGHES, D. H., and STROBEL, R. G. K. 1968. Function of the starch granule in the formation of layer cake structure. Cereal Chem. 45:329-338.

HOWE, M. 1946. Further studies on the mechanism of the action of oxidation and reduction on flour. Cereal Chem. 23:84-88.

HUEBNER, F. R. 1977. Wheat flour proteins and their functionality in baking. Bakers Dig. 51(5):25-28, 30, 31, 154.

HUEBNER, F. R., and WALL, J. S. 1976. Fractionation and quantitative differences of glutenin from wheat varieties varying in baking quality. Cereal Chem. 53:258-269.

HUTCHINSON, J. B. 1961. Hydrolysis of lipids in cereals and cereal products. SCI Monogr. 11:137-148.

JANKIEWICZ, M., and POMERANZ, Y. 1965. Comparison of the effects of N-ethyl-maleimide and urea on rheological properties of dough. Cereal Chem. 42:37-43.

JEANJEAN, M. F., and FEILLET, P. 1980. Properties of wheat gel proteins. Ann. Technol. Agric. 29:295-308.

JELACA, S. L., and HLYNKA, I. 1972. Effect of wheat-flour pentosans in dough, gluten,

356 / *Wheat: Chemistry and Technology*

and bread. Cereal Chem. 49:489-495.

JOHNSON, A. H. 1928. Studies of the effect on their bread-making properties of extracting flours with ether. Cereal Chem. 5:169-180.

JOHNSON, A. H., and WHITCOMB, W. O. 1931. Wheat and flour studies. XIX. Studies of the effect on their bread-making properties of extracting flours with ether, with special reference to the gas retaining powers of doughs prepared from ether-extracted flours. Cereal Chem. 8:392-403.

JOHNSON, J. A., and MILLER, D. 1959. Determination of the feasibility of producing non-staling bread-like products. U.S. Army, Quartermaster Contract No. DA 19-129-QM-1119.

JONES, B. L., FINNEY, K. F., and LOOKHART, G. L. 1983. Physical and biochemical properties of wheat protein fractions obtained by ultracentrifugation. Cereal Chem. 60:276-280.

JONES, C. R. 1940. The production of mechanically damaged starch in milling as a governing factor in the diastatic activity of flour. Cereal Chem. 17:133-169.

JONES, I. K., and CARNEGIE, P. R. 1969a. Isolation and characterization of disulphide peptides from wheat flour. J. Sci. Food Agric. 20:54-60.

JONES, I. K., and CARNEGIE, P. R. 1969b. Rheological activity of peptides, simple disulphides and simple thiols in wheaten dough. J. Sci. Food Agric. 20:60-64.

JONGH, G. 1961. The formation of dough and bread structures. I. The ability of starch to form structures, and the improving effect of glyceryl monostearate. Cereal Chem. 38:140-152.

JONGH, G., SLIM, T., and GREVE, H. 1968. Bread without gluten. Bakers Dig. 42(3):24-29.

JØRGENSEN, H. 1936. On the existence of powerful but latent proteolytic enzymes in wheat flour. Cereal Chem. 13:346-355.

KACZKOWSKI, J., and MIELESZKO, T. 1980. The role of disulfide bonds and their localisation in wheat protein molecules. Ann. Technol. Agric. 29:377-384.

KASARDA, D. D. 1970. The conformational structure of wheat proteins. Bakers Dig. 44(6):20-26.

KASARDA, D. D., BERNARDIN, J. E., and THOMAS, R. S. 1967. Reversible aggregation of α-gliadin to fibrils. Science 155:203-205.

KASARDA, D. D., BERNARDIN, J. E., and NIMMO, C. C. 1976. Wheat proteins. Pages 158-236 in: Advances in Cereal Science and Technology, Vol. I. Y. Pomeranz, ed. Am. Assoc. Cereal Chem., St. Paul, MN.

KATZ, J. R. 1934. The physical chemistry of starch and breadmaking. XX. The changes in

starch when bread becomes stale and when starch paste retrogrades. Z. Phys. Chem. A. 169:321-338.

KAUZMANN, W. 1959. Some factors in the interpretation of protein denaturation. Adv. Protein Chem. 14:1-63.

KHAN, K., and BUSHUK, W. 1978. Glutenin: Structure and functionality in breadmaking. Bakers Dig. 52(2):14-16, 18-20.

KHAN, K., and BUSHUK, W. 1979a. Structure of wheat gluten in relation to functionality in breadmaking. Pages 191-206 in: Functionality and Protein Structure. A. Pour-El, ed. Am. Chem. Soc., Washington, DC.

KHAN, K., and BUSHUK, W. 1979b. Studies of glutenin. XII. Comparison by sodium dodecyl sulfate-polyacrylamide gel electrophoresis of unreduced and reduced glutenin from various isolation and purification procedures. Cereal Chem. 56:63-68.

KHAN, K., BUSHUK, W., and McMASTER, G. 1980. Functional glutenin: A complex of covalently and non-covalently-linked components. Ann. Technol. 29:279-294.

KIM, J. C., and DE RUITER, D. 1968. Bread from nonwheat flours. Food Technol. (Chicago) 22:867-874, 876, 878.

KIM, S. K., and D'APPOLONIA, B. L. 1971. The role of wheat flour constituents in bread staling. Bakers Dig. 51(1):38-44, 57.

KIM, S. K., and D'APPOLONIA, B. L. 1977a. Bread staling studies. 1. Effect of protein content on staling rate and bread crumb pasting properties. Cereal Chem. 54:207-215.

KIM, S. K., and D'APPOLONIA, B. L. 1977b. Bread staling studies. II. Effect of protein content and storage temperature on the role of starch. Cereal Chem. 54:216-224.

KIM, S. K., and D'APPOLONIA, B. L. 1977c. Bread staling studies. III. Effect of pentosans on dough, bread, and bread staling rate. Cereal Chem. 54:225-229.

KISSELL, L. T. 1959. A lean-formula cake method for varietal evaluation and research. Cereal Chem. 36:168-175.

KISSELL, L. T., POMERANZ, Y., and YAMAZAKI, W. T. 1971. Effects of flour lipids on cookie quality. Cereal Chem. 48:655-662.

KISSELL, L. T., DONELSON, J. R., and CLEMENTS, R. L. 1979. Functionality in white layer cake of lipids from untreated and chlorinated patent flours. I. Effects of free lipids. Cereal Chem. 56:11-14.

KLOTZ, I. M. 1960. Non-covalent bonds in protein structure. Brookhaven Symp. Biol. 13:25-48.

KOBREHEL, K. 1980. Extraction of wheat proteins with salts of fatty acids and their electrophoretic characterization. Ann.

Technol. Agric. 29:125-132.

KOBREHEL, K., and BUSHUK, W. 1977. Studies of glutenin. X. Effect of fatty acids and their sodium salts on solubility in water. Cereal Chem. 54:833-839.

KOBREHEL, K., and MATIGNON, B. 1980. Solubilization of proteins with soaps in relation to the bread-making properties of wheat flour. Cereal Chem. 57:73-74.

KOBREHEL, K., REYMOND, C., and ALARY, R. 1988. Low molecular weight durum wheat glutenin fractions rich in sulfhydryl plus disulfide groups. Cereal Chem. 65:65-69.

KOENIG, V. L., OGRINS, A., TRIMBO, H. B., and MILLER, B. S. 1964. The electrophoretic analysis of flour from several varieties of hard red winter wheat grown at several locations. J. Sci. Food Agric. 15:492-497.

KOVATS, L. T., and LASZTITY, R. 1965. Die Bedeutung und Rolle der Sulfhydrylgruppen in der Weizenchemie und Weizenverarbeitung. Period. Polytech. Chem. Eng. 9:253-262.

KOZLOV, G. F. 1980. Effect of starch on the baking properties of wheat flour. Izv. Vyssh. Uchebn. Zaved. Pishch. Tekhnol. 1980:132-133.

KRULL, L. H., and WALL, J. S. 1966. Synthetic polypeptides containing side-chain amide groups. Water-soluble polymers. Biochemistry 5:1521-1524.

KRULL, L. H., WALL, J. S., ZOBEL, H., and DIMLER, R. J. 1965. Synthetic polypeptides containing side-chain amide groups. Water-insoluble polymers. Biochemistry 4:626-633.

KUBANEK, J., and CERNY, J. 1985. Chemistry and genetics of glutenin proteins of wheat. Genet. Slechteni 21:1-22.

KUCHUMOVA, L. P., and STRELNIKOVA, M. M. 1968. Sulfhydryl groups and disulfide bonds of the gluten of hard and soft wheats. Prikl. Biokhim. Mikrobiol. 3:592. (Chem. Abstr. 68:38259h)

KUENDIG, W., NEUKOM, H., and DEUEL, H. 1961a. Investigations of the cereal pentosans. I. Chromatographic fractionation of water-soluble wheat flour pentosans on diethylaminoethyl cellulose. Helv. Chim. Acta 44:823-829.

KUENDIG, W., NEUKOM, H., and DEUEL, H. 1961b. Investigations of the cereal pentosans. II. The gelling properties of water-soluble solutions of wheat flour pentosans caused by oxidative materials. Helv. Chim. Acta 44:969-976.

KULP, K. 1968a. Enzymolysis of pentosans of wheat flour. Cereal Chem. 45:339-350.

KULP, K. 1968b. Pentosans of wheat endosperm. Cereal Sci. Today 13:414-417, 426.

KULP, K. 1972. Physicochemical properties of starches of wheats and flours. Cereal Chem. 49:697-706.

KULP, K. 1973a. Characteristics of small-granule starch of flour and wheat. Cereal Chem. 50:666-679.

KULP, K. 1973b. Properties of starch granules derived from flours and parent wheats. Bakers Dig. 47(1):55-59.

KULP, K. 1973c. The physicochemical properties of wheat starch as related to bread. Bakers Dig. 47(5):34-38.

KULP, K., and BECHTEL, W. G. 1963a. The effect of tailings of wheat flour and its subfractions on the quality of bread. Cereal Chem. 40:665-675.

KULP, K., and BECHTEL, W. G. 1963b. Effect of water-insoluble pentosan fraction of wheat endosperm on the quality of white bread. Cereal Chem. 40:493-504.

KULP, K., and LORENZ, K. 1981. Starch functionality in white pan breads. Bakers Dig. 55(5):24, 25, 27, 28, 36.

KULP, K., and PONTE, J. G. 1981. Staling of white pan bread: Fundamental causes. CRC Crit. Rev. Food Sci. Nutr. 15:1-48.

KULP, K., HEPBURN, F. N., and LEHMANN, T. A. 1974. Preparation of bread without gluten. Bakers Dig. 48(3):34-37, 58.

KUNINORI, T., and MATSUMOTO, H. 1963. L-ascorbic acid oxidizing system in dough and dough improvement. Cereal Chem. 40:647-657.

KUNINORI, T., and MATSUMOTO, H. 1964a. Dehydro-L-ascorbic acid reducing system in flour. Cereal Chem. 41:39-46.

KUNINORI, T., and MATSUMOTO, H. 1964b. Glutathione in wheat and wheat flour. Cereal Chem. 41:252-259.

KUNINORI, T., and SULLIVAN, B. 1968. Disulfide-sulfhydryl interchange studies of wheat flour. II. Reaction of glutathione. Cereal Chem. 45:486-495.

KUNINORI, T., YAGI, M., YOSHINO, D., and MATSUMOTO, H. 1968. Glutathione in wheat flour. II. An enzymatic determination. Cereal Chem. 45:480-485.

LAIGNELET, B. 1983. Lipids in pasta and pasta processing. Pages 269-286 in: Lipids in Cereal Technology. P. J. Barnes, ed. Academic Press, London.

LAMKIN, W. M., MILLER, B. S., NELSON, S. W., TRAYLOR, D. D., and LEE, M. S. 1981. Polyphenol oxidase activities of hard red winter, soft red winter, hard red spring, white common, club, and durum wheat cultivars. Cereal Chem. 58:27-31.

LARMOUR, R. K. 1930. Relation between protein content and quality of wheat as shown by different baking methods. Cereal Chem. 7:35-48.

LASZTITY, R. 1969. Investigation of the rheological properties of gluten. II. Viscoelastic properties of chemically modified gluten. Acta Chem. Acad. Sci. Hung. 62:75-85.

LASZTITY, R. 1975. The technological properties of wheat gluten and their relations to molecular parameters. Nahrung 19:749-757.

LASZTITY, R. 1980. Correlation between chemical structure and rheological properties of gluten. Ann. Technol. Agric. 29:339-361.

LASZTITY, R., and GOLENKOV, V. F. 1985. Research on the wheat proteins in eastern Europe. Pages 175-186 in: Analyses as Practical Tools in the Cereal Field. K. M. Fjell, ed. Norwegian Grain Corp., Oslo.

LASZTITY, R., VARGA, J., ORSI, F., and RAGASITS, I. 1984. The effect of the fertilizers on the protein distribution and technological quality of wheat. Pages 69-80 in: Gluten Proteins. Proc. Int. Workshop Gluten Proteins, 2nd. A. Graveland and J. H. E. Moonen, eds. Inst. Cereals, Flour and Bread, TNO, Wageningen, The Netherlands.

LAW, C. N., and PAYNE, P. I. 1983. Review— Genetical aspects of breeding for improved grain protein content and type in wheat. J. Cereal Sci. 1:79-93.

LAWRENCE, G. J., MOSS, H. J., SHEPHERD, K. N., and WRIGLEY, C. W. 1987. Dough quality of biotypes of eleven Australian wheat cultivars that differ in high-molecular-weight glutenin subunit composition. J. Cereal Sci. 6:99-101.

LEE, C. C., and LAI, T.-S. 1968. Studies with radioactive tracers. XIV. A note on the disulfide-sulfhydryl interchange in doughs made with ^{35}S-labeled flour. Cereal Chem. 45:627-630.

LEE, C. C., and SMALL, D. G. 1960. Studies with radioactive tracers. V. The reduction of S^{35}-labeled persulfate to sulfate in flour and dough systems. Cereal Chem. 37:280-288.

LEE, C. C., and TKACHUK, R. 1959. Studies with radioactive tracers. III. The effects of defatting and of benzoyl peroxide on the decomposition of Br^{82}-labeled bromate by water-flour doughs. Cereal Chem. 36:412-420.

LEE, J. W., and MacRITCHIE, F. 1971. The effect of gluten protein fractions on dough properties. Cereal Chem. 48:620-625.

LEE, J. W., and RONALDS, J. A. 1967. Effect of environment on wheat gliadin. Nature 213:844-846.

LEE, J. W., and WRIGLEY, C. W. 1963. The protein composition of gluten extracted from different wheats. Aust. J. Exp. Agric. Anim. Husb. 3:85-88.

LEFEBVRE, J., CROZET, N., GUEGUEN, J., SOUVANAVONG, S. V., and PETIT, L.

1979. Biochemical and submicroscopic study on glutenins. Ann. Technol. Agric. 28:153-179.

LEGOUAR, M., COUPRIE, J., and GODON, B. 1979. Behavior of protein constituents in wheat flour mixtures or some of their fractions. Lebensm. Wiss. Technol. 12:266-272.

LELIEVRE, J., LORENZ, K., MEREDITH, P., and BARUCH, D. W. 1987. Effects of starch particle size and protein concentration on breadmaking performance. Staerke 39:347-352.

LIN, M. J. Y., D'APPOLONIA, B. L., and YOUNGS, V. L. 1974a. Hard red spring and durum wheat polar lipids. II. Effect on quality of bread and pasta products. Cereal Chem. 51:34-45.

LIN, M. J. Y., YOUNGS, V. L., and D'APPOLONIA, B. L. 1974b. Hard red spring and durum wheat polar lipids. I. Isolation and quantitative determinations. Cereal Chem. 51:17-33.

LINDAHL, L. and ELIASSON, A. C. 1986. Effects of wheat proteins on the viscoelastic properties of starch gels. J. Sci. Food Agric. 37:1125-1132.

LINKO, P., and SOGN, L. 1960. Relation of viability and storage deterioration to glutamic acid decarboxylase in wheat. Cereal Chem. 37:489-499.

LOOKHART, G. L., and POMERANZ, Y. 1985. Gliadin high-performance liquid chromatography and polyacrylamide gel electrophoresis patterns of wheats grown with fertilizer treatments in the United States and Australia on sulfur-deficient soils. Cereal Chem. 62:227-229.

LOWY, G. D. A., SARGEANT, J. G., and SCHOFIELD, J. D. 1981. Wheat starch granule protein: The isolation and characterisation of a salt-extractable protein from starch granules. J. Sci. Food Agric. 32:371-377.

LUTSISHINA, E. G., and MATYASH, A. I. 1984. Wide-band proton NMR spectroscopic study of gluten proteins of wheat. Deposited Doc. 1984. VINITI 3547-3584. (Chem. Abstr. 103:36378p, 1985)

LUTSISHINA, E. G., KOSTROMINA, N. A., VOROB'EVA, L. A., and TIKHONOVA, R. V. 1980. Study of gluten by proton NMR. Fiziol. Biokhim. Kult. Rast. 12:42-48.

MA, C.-Y., OOMAH, B. D., and HOLME, J. 1986. Effect of deamidation and succinylation on some physicochemical and baking properties of gluten. J. Food Sci. 51:99-103.

MacMASTERS, M. M., and HILBERT, G. E. 1944. The composition of the "amylodextrin" fraction of wheat flour. Cereal Chem. 21:548-555.

MacRITCHIE, F. 1972. The fractionation and

properties of gluten proteins. J. Macromol. Sci. Chem. 6:823-829.

MacRITCHIE, F. 1973. Conversion of a weak flour to a strong one by increasing the proportion of its high molecular weight gluten protein. J. Sci. Food Agric. 24:1325-1329.

MacRITCHIE, F. 1977. Flour lipids and their effects in baking. J. Sci. Food Agric. 28:53-58.

MacRITCHIE, F. 1978. Differences in baking quality of wheat flours. J. Food Technol. 13:187-194.

MacRITCHIE, F. 1979. A relation between gluten protein amide content and baking performance of wheat flours. J. Food Technol. 14:595-601.

MacRITCHIE, F. 1980a. Physicochemical aspects of some problems in wheat research. Pages 271-326 in: Advances in Cereal Science and Technology, Vol. III. Y. Pomeranz, ed. Am. Assoc. Cereal Chem., St. Paul, MN.

MacRITCHIE, F. 1980b. Protein-lipid interaction in baking quality of wheat flours. Bakers Dig. 54(5):10-13.

MacRITCHIE, F. 1981. Flour lipids: Theoretical aspects and functional properties. Cereal Chem. 58:156-158.

MacRITCHIE, F. 1984. Baking quality of wheat flours. Adv. Food Res. 29:201-277.

MacRITCHIE, F. 1985. Studies of the methodology for fractionation and reconstitution of wheat flours. J. Cereal Sci. 3:221-230.

MacRITCHIE, F. 1987. Evaluation of contributions from wheat protein fractions to dough mixing and breadmaking. J. Cereal Sci. 6:259-268.

MacRITCHIE, F., and GRAS, P. W. 1973. The role of flour lipids in baking. Cereal Chem. 50:292-302.

MAES, E. E. A. 1966. Protein solubility and baking quality. Cereal Sci. Today 11:200-202.

MALEKI, M., HOSENEY, R. C., and MATTERN, P. J. 1980. Effects of loaf volume, moisture content, and protein quality on the softness and staling rate of bread. Cereal Chem. 57:138-140.

MAMARIL, F. P., and POMERANZ, Y. 1966. Isolation and characterization of wheat flour proteins. IV. Effects on wheat flour proteins of dough mixing and oxidizing agents. J. Sci. Food Agric. 17:339-343.

MARAIS, G. F., and D'APPOLONIA, B. L. 1981a. Factors contributing to baking quality differences in hard red spring wheat. I. Bases for different loaf volume potentials. Cereal Chem. 58:444-447.

MARAIS, G. F., and D'APPOLONIA, B. L. 1981b. Factors contributing to baking quality differences in hard red spring wheat. II. Bases for different mixing properties. Cereal Chem.

58:448-453.

MARSH, D. R., and GALLIARD, T. 1986. Measurement of polyphenol oxidase activity in wheat-milling fractions. J. Cereal Sci. 4:241-248.

MARSTON, P., and MacRITCHIE, F. 1985. Lipids: Effects on the breadmaking qualities of Australian flours. Food Technol. Aust. 37:362-365.

MARTIN, D. J., and STEWART, B. G. 1987. Dough stickiness in rye-derived wheats. Cereal Foods World 32:672-673.

MARTIN, W. M., and WHITCOMB, W. O. 1932. Physical and chemical properties of ether-soluble constituents of wheat flour in relation to baking quality. Cereal Chem. 9:275-288.

MASON, L. H., and JOHNSTON, A. E. 1958. Comparative study of wheat flour phosphatides. Cereal Chem. 35:435-448.

MATSUDOMI, N., KANEKO, S., KATO, A., and KOBAYASHI, K. 1981. Functional properties of deamidated gluten. Nippon Nogei Kagaku Kaishi 55:983-989.

MATSUDOMI, N., KATO, A., and KOBAYASHI, K. 1982. Conformation and surface properties of deamidated gluten. Agric. Biol. Chem. 46:1583-1586.

MATSUMOTO, H., and HLYNKA, I. 1959. Some aspects of the sulfhydryl-disulfide system in flour and dough. Cereal Chem. 36:513-521.

MATSUO, R. R., and IRVINE, G. N. 1967. Macaroni brownness. Cereal Chem. 44:78-85.

MATSUO, R. R., and IRVINE, G. N. 1970. Effect of gluten on the cooking quality of spaghetti. Cereal Chem. 47:173-180.

MATSUO, R. R., and McCALLA, A. G. 1964. Physicochemical properties of gluten from three types of wheat. Can. J. Biochem. 42:1487-1498.

MATSUO, R. R., BRADLEY, J. W., and IRVINE, G. N. 1972. Effect of protein content on the cooking quality of spaghetti. Cereal Chem. 49:707-711.

MATSUO, R. R., DEXTER, J. E., BOUDREAU, A., and DAUN, J. K. 1986. The role of lipids in determining spaghetti cooking quality. Cereal Chem. 63:484-489.

MATTERN, P. J., and SANDSTEDT, R. M. 1957. The influence of the water-soluble constituents of wheat flour on its mixing and baking characteristics. Cereal Chem. 34:252-267.

MAURITZEN, C. M. 1967. The incorporation of cysteine-^{35}S, cystine-^{35}S, and N-ethyl-maleimide-^{14}C into doughs made from wheat flour. Cereal Chem. 44:170-182.

MAURITZEN, C. A. M., and STEWART, P. 1963. Disulphide-sulphydryl exchange in dough. Nature 197:48-49.

McCAIG, J. D., and McCALLA, A. G. 1941. Changes in the physical properties of gluten with aging of flour. Can. J. Res. Sect. C 19:163-176.

McCALLA, A. G., and WOODFORD, E. K. 1935. The effect of potassium supply on the composition and quality of wheat. II. Can. J. Res. Sect. C 13:339-354.

McDERMOTT, E. E., and PACE, J. 1961. Modification of the properties of flour protein by thiolated gelatin. Nature 192:657.

McDERMOTT, E. E., STEVENS, D. J., and PACE, J. 1969. Modification of flour proteins by disulphide interchange reactions. J. Sci. Food Agric. 20:213-217.

MECHAM, D. K. 1968. The sulfhydryl and disulfide contents of wheat flours, doughs, and proteins. Bakers Dig. 42(1):26-28, 30, 59.

MECHAM, D. K., and BEAN, M. M. 1968. The release of hydrogen sulfide during dough mixing. Cereal Chem. 45:445-453.

MECHAM, D. K., and MOHAMMAD, A. 1955. Extraction of lipids from wheat products. Cereal Chem. 32:405-415.

MECHAM, D. K., and PENCE, J. W. 1957. Flour lipids, an unappreciated factor governing flour quality. Bakers Dig. 31(1):40-44.

MECHAM, D. K., and WEINSTEIN, N. E. 1952. Lipid binding in doughs. Effects of dough ingredients. Cereal Chem. 29:448-455.

MECHAM, D. K., SOKOL, H. A., and PENCE, J. W. 1962. Extractable protein and hydration characteristics of flours and doughs in dilute acid. Cereal Chem. 39:81-93.

MECHAM, D. K., COLE, E. G., and SOKOL, H. A. 1963. Modification of flour proteins by dough-mixing: Effects of sulfhydryl-blocking and oxidizing agents. Cereal Chem. 40:1-9.

MECHAM, D. K., BEAN, M. M., and KNAPP, C. 1966. Hydrogen sulphide from wheat flour doughs. Chem. Ind. (London) 1966:989-990.

MEDCALF, D. G. 1968. Wheat starch properties and their effect on bread baking quality. Bakers Dig. 42(4):48-50, 52, 65.

MEDCALF, D. G., and GILLES, K. A. 1965. Wheat starches. I. Comparison of physicochemical properties. Cereal Chem. 42:558-568.

MEDCALF, D. G., and GILLES, K. A. 1968. The function of starch in dough. Cereal Sci. Today 13:382-385, 388, 392-393.

MEDCALF, D. G., YOUNGS, V. L., and GILLES, K. A. 1968. Wheat starches. II. Effect of polar and nonpolar lipid fractions on pasting characteristics. Cereal Chem. 45:88-95.

MENGER, A. 1964. Uber die Beinflussung der Teigwarenfarbe durch Verlust und Neubildung von Pigmenten wahrend der Herstellung.

Getreide Mehl 14:85-94.

MEREDITH, O. B., and WREN, J. J. 1969. Stability of the molecular weight distribution in wheat flour proteins during dough mixing. J. Sci. Food Agric. 20:234-237.

MEREDITH, P. 1966. Dependence of water absorption of wheat flour on protein content and degree of starch granule damage. N. Z. J. Sci. 9:324-328.

MEREDITH, P., and POMERANZ, Y. 1982. Inherent amylograph pasting ability of U.S. wheat flours and starches. Cereal Chem. 59:355-360.

MERRITT, P. P., and BAILEY, C. H. 1945. Preliminary studies with the extensograph. Cereal Chem. 22:372-391.

MERTZ, L. C., and NORDSTROM, K. L. 1960. Flour milling process. U.S. patent 2,941,730 (June 21).

MEUSER, F., REIMERS, H., and FISCHER, G. 1977. Physical and chemical studies of flours from wheats milled in different ways, with special reference to their baking quality. Muehle Mischfuttertech. 114:180-188.

MIELESZKO, T., and KACZKOWSKI, J. 1980. The distribution of disulfide bonds in some protein fractions of gluten complex. Acta Aliment. Pol. 6:207-214.

MIFLIN, B. J., FIELD, J. M., and SHEWRY, P. R. 1983. Cereal-storage proteins and their effects on rheological properties. Pages 255-319 in: Seed Proteins. J. Daussant, J. Mosse, and J. Vaughan, eds. Academic Press, London.

MIHAILOVIC, M. L., GARTON, G. A., ANTIC, M., and HADZIJEV, D. 1963. Chemical investigation of wheat. 5. Some chromatographic investigations of wheat lipids. Bull. Chem. Soc. (Belgrade) 28:179-199.

MILLER, B. S., and JOHNSON, J. A. 1954. A review of methods for determining the quality of wheat and flour for breadmaking. Kans. Agric. Exp. Stn. Tech. Bull. 7.

MILLER, B. S., and KUMMEROW, F. A. 1948. The disposition of lipase and lipoxidase in baking and the effect of their reaction products on consumer acceptability. Cereal Chem. 25:391-398.

MILLER, B. S., and TRIMBO, H. B. 1965. Gelatinization of starch and white layer cake quality. Food Technol. (Chicago) 19:640-648.

MILLER, B. S., SEIFFE, J. Y., SHELLENBERGER, J. A., and MILLER, G. D. 1950. Amino acid content of various wheat varieties. I. Cystine, lysine, methionine, and glutamic acid. Cereal Chem. 27:96-106.

MILLER, B. S., KOENIG, V. L., TRIMBO, H. B., and OGRINS, A. 1964. Effects of cobalt-60 γ-irradiation and mechanical damage on soft wheat flour. J. Sci. Food

Agric. 15:701-711.

MILLER, B. S., TRIMBO, H. B., and POWELL, K. R. 1967a. Effects of flour granulation and starch damage on the cake making quality of soft wheat flour. Cereal Sci. Today 12:245-247, 250-252.

MILLER, B. S., TRIMBO, H. B., and SANDSTEDT, R. M. 1967b. The development of gummy layers in cakes. Food Technol. (Chicago) 21(3A):59A-68A.

MISKELLY, D. M. 1984. Flour components affecting paste and noodle colour. J. Sci. Food Agric. 35:463-471.

MISKELLY, D. M., and MOSS, H. J. 1985. Flour quality requirements for Chinese noodle manufacture. J. Cereal Sci. 3:379-387.

MOHRI, Z. 1980. Interaction between starch and fatty acid esters in frozen starch noodles. Agric. Biol. Chem. 44:1455-1458.

MOONEN, J. H. E., and ZEVEN, A. C. 1985. Association between high molecular weight subunits of glutenin and bread-making quality in wheat lines derived from backcrosses between *Triticum aestivum* and *Triticum speltoides*. J. Cereal Sci. 3:97-101.

MOONEN, J. H. E., SCHEEPSTRA, A., and GRAVELAND, A. 1982. Use of the SDS-sedimentation test and SDS-polyacrylamide gel electrophoresis for screening breeder's samples of wheat for bread-making quality. Euphytica 31:677-690.

MOONEN, J. H. E., SCHEEPSTRA, A., and GRAVELAND, A. 1983. The positive effect of the high molecular weight subunits 3 + 10 and 2* of glutenin on the bread-making quality of wheat cultivars. Euphytica 32:735-742.

MOONEN, J. H. E., MARSEILLE, J. P., SCHEEPSTRA, A., and GRAVELAND, A. 1984. Genetic aspects and biochemical properties of the high-molecular-weight subunits of glutenin in relation to bread-making quality. Pages 185-193 in: Gluten Proteins. Proc. 2nd Int. Workshop Gluten Proteins. A. Graveland and J. H. E. Moonen, eds. Inst. Cereals, Flour and Bread, TNO, Wageningen, The Netherlands.

MOONEN, J. H. E., SCHEEPSTRA, A., and GRAVELAND, A. 1985. Biochemical properties of some high molecular weight subunits of wheat glutenin. J. Cereal Sci. 3:17-27.

MOORE, C. L., and HERMAN, R. S. 1942. The effect of certain ingredients and variations in manipulations on the farinograph curve. Cereal Chem. 19:568-587.

MORRISON, W. R. 1976. Lipids in flour, dough and bread. Bakers Dig. 50(4):29-34, 36, 47.

MORRISON, W. R. 1978a. Cereal lipids. Pages

221-348 in: Advances in Cereal Science and Technology, Vol. II. Y. Pomeranz, ed. Am. Assoc. Cereal Chem., St. Paul, MN.

MORRISON, W. R. 1978b. Wheat lipid composition. Cereal Chem. 55:548-558.

MORRISON, W. R. 1979. Lipids in wheat and their importance in wheat products. Pages 313-335 in: Recent Advances in the Biochemistry of Cereals. D. L. Laidman and R. G. Wyn Jones, eds. Academic Press, London.

MORRISON, W. R., and COUSSIN, B. R. 1962. Thioctic acid in wheat flour. J. Sci. Food Agric. 13:257-260.

MOSS, H. J. 1961. Milling damage and quality evaluation of wheat. Aust. J. Exp. Agric. Anim. Husb. 1:133-139.

MOSS, H. J., WRIGLEY, C. W., MacRITCHIE, F., and RANDALL, P. J. 1981. Sulfur and nitrogen fertilizer effects on wheat. II. Influence on grain quality. Aust. J. Agric. Res. 32:213-226.

MOSS, H. J., RANDALL, P. J., and WRIGLEY, C. W. 1983. Alteration to grain, flour and dough quality in three wheat types with variation in soil sulfur supply. J. Cereal Sci. 1:255-264.

MOSS, H. J., MISKELLY, D. M., and MOSS, R. 1986. The effect of alkaline conditions on the properties of wheat flour dough and Cantonese-style noodles. J. Cereal Sci. 4:261-268.

MOSS, R., STENVERT, N. L., POINTING, G., WORTHINGTON, G., and BOND, E. E. 1979. The time-dependent interaction of oxidizing and reducing agents in breadmaking. Bakers Dig. 53(2):10-17.

MULLEN, J. D., and SMITH, D. E. 1965. Studies on short- and long-mixing flours. I. Solubility and electrophoretic composition of proteins. Cereal Chem. 42:263-274.

NAGAO, S., IMAI, S., SATO, T., KANEKO, Y., and OTSUBO, H. 1976. Quality characteristics of soft wheats and their use in Japan. I. Methods of assessing wheat suitability for Japanese products. Cereal Chem. 53:988-997.

NAGAO, S., ISHIBASHI, S., IMAI, S., SATO, T., KANBE, T., KANEKO, Y., and OTSUBO, H. 1977a. Quality characteristics of soft wheats and their utilization in Japan. II. Evaluation of wheats from the United States, Australia, France, and Japan. Cereal Chem. 54:198-204.

NAGAO, S., ISHIBASHI, S., IMAI, S., SATO, T., KANBE, T., KANEKO, Y., and OTSUBO, H. 1977b. Quality characteristics of soft wheats and their utilization in Japan. III. Effects of crop year and protein content on product quality. Cereal Chem. 54:300-306.

NARAYANAN, K. M., and HLYNKA, I. 1962.

Rheological studies of the role of lipids in dough. Cereal Chem. 39:351-363.

NEUKOM, H., and MARKWALDER, H. U. 1978. Oxidative gelation of wheat flour pentosans: A new way of cross-linking polymers. Cereal Foods World 23:374-376.

NEUKOM, H., GEISMANN, T., and PAINTER, T. J. 1967. New aspects of the function and properties of the soluble wheat-flour pentosans. Bakers Dig. 41(5):52-55.

NIELSEN, H. C., BABCOCK, G. E., and SENTI, F. R. 1962. Molecular weight studies on glutenin before and after disulfide-bond splitting. Arch. Biochem. Biophys. 96:252-258.

NIERLE, W., and ELBAYA, A. W. 1981. Zur Wechselwirkung zwischen Eiweiss und Lipiden im Teig. Getreide Mehl Brot 35:124-127.

NIIHARA, R., NISHIDA, Y., and YONEZAWA, D. 1982. Role of fatty acids produced in the storage process of tenobesomen (a kind of noodle) called "Yaku." Pages 973-978 in: Progress in Cereal Chemistry and Technology. Proc. World Cereal Bread Congr. 7th, Prague. J. Holas and J. Kratochvil, eds. Elsevier, Amsterdam.

NISHIYAMA, J., KUNINORI, T., MATSUMOTO, H., and ATSUSHI, H. 1981. ESR studies on lipid-protein interaction in gluten. Agric. Biol. Chem. 45:1953-1958.

OH, N. H., SEIB, P. A., WARD, A. B., and DEYOE, C. W. 1985. Noodles. IV. Influence of flour protein, extraction rate, particle size, and starch damage on the quality characteristics of dry noodles. Cereal Chem. 62:441-446.

OHMS, J. P. 1980. Electrophoretic differentiation of grain (cultivars). Characterization of wheat cultivars of A-baking quality by the electrophoretic patterns of the reduced albumin/globulin-fraction. Z. Lebensm. Unters. Forsch. 170:27-30.

OLCOTT, H. S., and MECHAM, D. K. 1947. Characterization of wheat gluten. I. Protein-lipid complex formation during doughing of flours. Lipoprotein nature of the glutenin fraction. Cereal Chem. 24:407-414.

OLCOTT, H. S., SAPIRSTEIN, L. A., and BLISH, M. J. 1943. Stability of wheat gluten dispersions toward reducing agents in the presence and absence of a gluten proteinase. Cereal Chem. 20:87-97.

ORSI, F., PALLAGI-BANKFALVI, E., and LASZTITY, R. 1985. Analysis of dependence between flour quality and electrophoretic protein spectrum. Acta Aliment. Acad. Sci. Hung. 14:49-57.

ORTH, R. A., and BUSHUK, W. 1972. A comparative study of the proteins of wheats of diverse baking qualities. Cereal Chem. 49:268-275.

ORTH, R. A., and BUSHUK, W. 1973a. Studies of glutenin. II. Relation of variety, location of growth, and baking quality to molecular weight distribution of subunits. Cereal Chem. 50:191-197.

ORTH, R. A., and BUSHUK, W. 1973b. Studies of glutenin. III. Identification of subunits coded by the D-genome and their relation to breadmaking quality. Cereal Chem. 50:680-687.

ORTH, R. A., DRONZEK, B. L., and BUSHUK, W. 1973a. Studies of glutenin. IV. Microscopic structure and its relations to breadmaking quality. Cereal Chem. 50:688-696.

ORTH, R. A., DRONZEK, B. L., and BUSHUK, W. 1973b. Scanning electron microscopy of bread wheat proteins fractionated by gel filtration. Cereal Chem. 50:696-701.

PAINTER, T. J., and NEUKOM, H. 1968. The mechanism of oxidative gelation of a glycoprotein from wheat flour. Evidence from a model system based upon caffeic acid. Biochim. Biophys. Acta 158:363-381.

PALLAGI-BANKFALVI, E. 1984. Protein subunit distribution in wheat flours and their relation to baking quality. Acta Aliment. Acad. Sci. Hung. 13:3-11.

PALLAGI-BANKFALVI, E., and ORSI, F. 1985. Assay into the correlation between protein composition and baking quality of wheat flours. Acta Aliment. Acad. Sci. Hung. 14:29-48.

PATIL, S. K., FINNEY, K. F., SHOGREN, M. D., and TSEN, C. C. 1976. Water-soluble pentosans of wheat flour. III. Effect of water-soluble pentosans on loaf volume of reconstituted gluten and starch doughs. Cereal Chem. 53:347-354.

PAYNE, P. I. 1983. Breeding for protein quantity and protein quality in seed crops. Pages 223-253 in: Seed Proteins. J. Daussant, J. Mosse, and J. Vaughan, eds. Academic Press, London.

PAYNE, P. I., CORFIELD, K. G., and BLACKMAN, J. A. 1979. Identification of a high-molecular-weight subunit of glutenin whose presence correlates with bread-making quality in wheats of related pedigree. Theor. Appl. Genet. 55:153-159.

PAYNE, P. I., HARRIS, P. A., LAW, C. N., HOLT, L. M., and BLACKMAN, J. A. 1980. The high-molecular-weight subunits of glutenin: Structure, genetics and relationship to bread-making quality. Ann. Technol. Agric. 29:309-320.

PAYNE, P. I., CORFIELD, K. G., HOLT, L. M., and BLACKMAN, J. A. 1981. Correlations between the inheritance of certain high-molecular weight subunits of glutenin and bread-making quality in progenies of six crosses of bread wheat. J. Sci. Food Agric. 32:51-60.

PAYNE, P. I., HOLT, L. M., and LAWRENCE, G. J. 1983. Detection of a novel high molecular weight subunit of glutenin in some Japanese hexaploid wheats. J. Cereal Sci. 1:3-8.

PAYNE, P. I., JACKSON, E. A., and HOLT, L. M. 1984. The association between γ-gliadin 45 and gluten strength in durum wheat varieties: A direct causal effect or the result of genetic linkage? J. Cereal Sci. 2:73-81.

PAYNE, P. I., NIGHTINGALE, M. A., KRATTIGER, A. F., and HOLT, L. M. 1987. The relationship between HMW glutenin subunit composition and the bread-making quality of British-grown wheat varieties. J. Sci. Food Agric. 40:51-65.

PELSHENKE, P. F., and HAMPEL, G. 1962. Starch retrogradation in various bread products. Bakers Dig. 36(3):48-50, 51-54, 56, 57, 85.

PENCE, J. W. 1962. The flour proteins. Cereal Sci. Today 7:178-180, 208.

PENCE, J. W., ELDER, A. H., and MECHAM, D. K. 1950. Preparation of wheat flour pentosans for use in reconstituted doughs. Cereal Chem. 27:60-66.

PENCE, J. W., ELDER, A. H., and MECHAM, D. K. 1951. Some effects of soluble flour components on baking behavior. Cereal Chem. 28:94-104.

PENCE, J. W., WEINSTEIN, N. E., and MECHAM, D. K. 1954a. The albumin and globulin contents of wheat flour and their relationship to protein quality. Cereal Chem. 31:303-311.

PENCE, J. W., WEINSTEIN, N. E., and MECHAM, D. K. 1954b. Differences in the distribution of components in albumin preparations from durum and common wheat flours. Cereal Chem. 31:396-406.

PENCE, J. W., EREMIA, K. M., WEINSTEIN, N. E., and MECHAM, D. K. 1959. Studies on the importance of starch and protein systems of individual flours in loaf volume production. Cereal Chem. 36:199-214.

PERUANSKII, Y. V., and NADIROV, B. T. 1978. Gliadin subfractions and the quality of wheat grain. Vestn. Skh. Nauki Kaz. 12:130-131.

POMERANZ, Y. 1961. Cereal gums. Qual. Plant. Mater. Veg. 8:157-200.

POMERANZ, Y. 1965. Dispersibility of wheat proteins in aqueous urea solutions—A new parameter to evaluate breadmaking potentialities of wheat flour. J. Sci. Food Agric. 16:586-593.

POMERANZ, Y. 1967. Wheat flour lipids, a minor component of major importance in breadmaking. Bakers Dig. 41(5):48-50, 170.

POMERANZ, Y. 1968. Relation between chemical composition and breadmaking potentialities of wheat flour. Adv. Food Res. 16:335-455.

POMERANZ, Y. 1969. Bread without bread. Bakers Dig. 43(1):22, 26-28.

POMERANZ, Y. 1971a. Biochemical and functional changes in stored cereal grains. Crit. Rev. Food Technol. 2:45-60.

POMERANZ, Y. 1971b. Glycolipid-protein interaction in breadmaking. Bakers Dig. 45(1):26-35, 58.

POMERANZ, Y. 1973a. From wheat to bread: A biochemical study. Am. Sci. 61:683-691.

POMERANZ, Y. 1973b. Interaction between glycolipids and wheat flour macromolecules in breadmaking. Adv. Food Res. 20:153-188.

POMERANZ, Y. 1980a. Interaction between lipids and gluten proteins. Ann. Technol. Agric. 29:385-397.

POMERANZ, Y. 1980b. Molecular approach to breadmaking—An update and new perspectives. Bakers Dig. 54(1):20-27; 54(2):12, 14, 16, 18, 20, 24.

POMERANZ, Y. 1980c. What? How much? Where? What function? in bread making. Cereal Foods World 25:656-662.

POMERANZ, Y. 1980d. Wheat flour components in breadmaking. Pages 201-232 in: Cereals for Food and Beverages: Recent Progress in Cereal Chemistry and Technology. G. E. Inglett and L. Munck, eds. Academic Press, New York.

POMERANZ, Y. 1981. Wrap-up of symposium on theory and application of lipid-related materials in breadmaking: Today and tomorrow (not yesterday). Cereal Chem. 58:190-193.

POMERANZ, Y. 1983. Molecular approach to breadmaking: An update and new perspectives. Bakers Dig. 57(4):72-86.

POMERANZ, Y. 1984. Carbohydrate-protein-lipid interaction: The total perspective. Pages 115-144 in: Carbohydrates, Proteins, Lipids: Basic Views and New Approaches in Food Technology. F. Meuser, ed. Inst. Lebensmittel-technologie, Getreidetechnologie, Berlin, Federal Republic of Germany.

POMERANZ, Y. 1985. Functional Properties of Food Components. Academic Press, Orlando, FL.

POMERANZ, Y. 1987. Modern Cereal Science and Technology. VCH Publishers, Weinheim, W. Germany.

POMERANZ, Y., and CHUNG, O. K. 1978. Interaction of lipids with proteins and carbo-hydrates in breadmaking. J. Am. Oil Chem.

Soc. 55:285-289.

POMERANZ, Y., and CHUNG, O. K. 1982. Lipid-protein interactions: New findings and interpretations. Pages 825-830 in: Progress in Cereal Chemistry and Technology. Proc. World Cereal and Bread Congress, 7th, Prague. J. Holas and J. Kratochvil, eds. Elsevier, Amsterdam.

POMERANZ, Y., and MEYER, D. 1984. Light and scanning electron microscopy of wheat- and rye-bread crumb. Interpretation of specimens prepared by various methods. Food Microstruct. 3:159-164.

POMERANZ, Y., and SHELLENBERGER, J. A. 1961. Histochemical characterization of wheat and wheat products. V. Sulfhydryl groups: Their localization in the wheat kernel. Cereal Chem. 38:133-140.

POMERANZ, Y., and WEHRLI, H. P. 1969. Synthetic glycosylglycerides in breadmaking. Food Technol. (Chicago) 23(9):109-111.

POMERANZ, Y., HALTON, P., and PEERS, F. G. 1956. The effects on flour dough and bread quality of molds grown in wheat and those added to flour in the form of specific cultures. Cereal Chem. 33:157-169.

POMERANZ, Y., RUBENTHALER, G., and FINNEY, K. F. 1965. Polar vs. nonpolar wheat flour lipids in breadmaking. Food Technol. (Chicago) 19:1724-1725.

POMERANZ, Y., RUBENTHALER, G. L., DAFTARY, R. D., and FINNEY, K. F. 1966a. Effects of lipids on bread baked from flours varying widely in breadmaking potentialities. Food Technol. (Chicago) 20(9):131-134.

POMERANZ, Y., RUBENTHALER, G. L., and FINNEY, K. F. 1966b. Studies on the mechanism of the bread-improving effect of lipids. Food Technol. (Chicago) 20(11):105-108.

POMERANZ, Y., DAFTARY, R. D., SHOGREN, M. D., HOSENEY, R. C., and FINNEY, K. F. 1968a. Changes in biochemical and breadmaking properties of storage-damaged flour. Agric. Food Chem. 16:92-96.

POMERANZ, Y., SHOGREN, M. D., and FINNEY, K. F. 1968b. Functional bread-making properties of lipids. I. Reconstitution studies and properties of defatted flours milled from wheats varying widely in bread making potentialities. Food Technol. (Chicago) 22:76-79.

POMERANZ, Y., SHOGREN, M. D., and FINNEY, K. F. 1968c. Natural and modified phospholipids: Effects on bread quality. Food Technol. (Chicago) 22:897-900.

POMERANZ, Y., TAO, R. P.-C., HOSENEY, R. C., SHOGREN, M. D., and FINNEY, K. F. 1968d. Evaluation of factors affecting lipid binding in wheat flours. Agric. Food Chem. 16:974-978.

POMERANZ, Y., SHOGREN, M. D., and FINNEY, K. F. 1969a. Improving bread-making properties with glycolipids. I. Improving soy products with sucroesters. Cereal Chem. 46:503-511.

POMERANZ, Y., SHOGREN, M. D., and FINNEY, K. F. 1969b. Improving bread-making properties with glycolipids. II. Improving various protein-enriched products. Cereal Chem. 46:512-518.

POMERANZ, Y., CARVAJAL, M. J., HOSENEY, R. C., and WARD, A. B. 1970a. Wheat germ in breadmaking. I. Composition of germ lipids and germ protein fractions. Cereal Chem. 47:373-380.

POMERANZ, Y., CARVAJAL, M. J., SHOGREN, M. D., HOSENEY, R. C., and FINNEY, K. F. 1970b. Wheat germ in breadmaking. II. Improving breadmaking properties by physical and chemical methods. Cereal Chem. 47:429-437.

POMERANZ, Y., FINNEY, K. F., and HOSENEY, R. C. 1970c. Molecular approach to breadmaking. Science 167:944-949.

POMERANZ, Y., BOLLING, H., and ZWINGELBERG, H. 1984. Wheat hardness and baking properties of wheat flours. J. Cereal Sci. 2:137-143.

PONTE, J. G., Jr., and BALDWIN, R. R. 1972. Studies on the lipid system of flour and dough. Bakers Dig. 46(1):28-31, 34, 35.

PONTE, J. G., Jr., and DE STEFANIS, V. A. 1969. Note on the separation and baking properties of polar and nonpolar wheat flour lipids. Cereal Chem. 46:325-329.

PONTE, J. G., Jr., TITCOMB, S. T., and COTTON, R. H. 1963. Effect of L-arabinose and D-xylose on dough fermentation and crust browning. Cereal Chem. 40:78-82.

PONTE, J. G., Jr., DE STEFANIS, V. A., and COTTON, R. H. 1967a. Studies of gluten lipids. I. Distribution of lipids in gluten fractions separated by solubility in 70% ethanol. Cereal Chem. 44:427-435.

PONTE, J. G., Jr., DE STEFANIS, V. A., TITCOMB, S. T., and COTTON, R. H. 1967b. Study of gluten properties as influenced by certain organic solvents. Cereal Chem. 44:211-220.

POPINEAU, Y., and GODON, B. 1982. Surface hydrophobicity of gliadin components. Cereal Chem. 59:55-62.

POPINEAU, Y., and PINEAU, F. 1987. Investigation of surface hydrophobicities of purified gliadins by hydrophobic interaction chromatography, reversed-phase high performance liquid chromatography and apolar ligand binding. J. Cereal Sci. 5:215-231.

POPINEAU, Y., LEFEBVRE, J., and GODON, B. 1980. Hydrophobic properties of the gliadins of some varieties of *Triticum vulgare* wheats. Ann. Technol. Agric. 29:191-204.

POPINEAU, Y., LE GUERRONE, J. L., and PINEAU, F. 1986. Purification and characterization of ω-gliadin components from common wheat. Lebensm. Wiss. Technol. 19:266-271.

PRENTICE, N., CUENDET, L. S., and GEDDES, W. F. 1954. Studies on bread staling. V. Effect of flour fractions and various starches on the firming of bread crumb. Cereal Chem. 31:188-206.

PRESTON, K. R., and TIPPLES, K. H. 1980. Effects of acid-soluble and acid-insoluble gluten proteins on the rheological and baking properties of wheat flours. Cereal Chem. 57:314-320.

PRESTON, K., WOODBURY, W., and BENDELOW, V. 1975. The effects of gliadin fractions of varying molecular weight on the mixing properties of a synthetic-dough system. Cereal Chem. 52:427-430.

PRESTON, K. R., MATSUO, R. R., DEXTER, J. E., TWEED, A. R., KILBORN, R. H., and TULLY, D. 1986. The suitability of various Canadian wheats for steamed bread and noodle processing for the People's Republic of China. Can. Inst. Food Sci. Technol. J. 19:114-120.

PROSKURYAKOV, N. I., and ZUEVA, E. S. 1964. Enzymic reduction of disulfide bonds in proteins and low molecular weight substances in wheat meal. Dokl. Akad. Nauk SSSR 158:232-234. (Chem. Abstr. 61:15034e)

PYLER, E. J. 1983. Flour proteins role in baking performance. Bakers Dig. 55(3):24-28, 33.

RANDALL, P. J., and WRIGLEY, C. W. 1986. Effects of sulfur supply on the yield, composition, and quality of grain from cereals, oilseeds, and legumes. Pages 171-206 in: Advances in Cereal Science and Technology, Vol. VIII. Y. Pomeranz, ed. Am. Assoc. Cereal Chem., St. Paul, MN.

RANDALL, P. J., SPENCER, K., and FRENEY, J. R. 1981. Sulfur and nitrogen fertilizer effects on wheat. I. Sulfur and nitrogen concentrations in the grain in relation to yield response. Aust. J. Agric. Res. 32:203-219.

REDMAN, D. G., and EWART, J. A. D. 1967. Disulphide interchange in dough proteins. J. Sci. Food Agric. 18:15-18.

ROHRLICH, M., and ESSNER, W. 1965. Menge und Verteilung der SH-Gruppen und SS-Verbindungen in Weizenmehlprodukten. Page 9 in: Arbeitsgemeienschaft Getreideforsch. Detmold, Federal Republic of Germany.

ROHRLICH, M., and MULLER, V. 1968. Untersuchungen uber Fett-Eiweiss-Komplexe in Cerealien. IV. Beziehung zwischen Fett-Eiweiss-Komplex und dem Backverhalten eines Mehles. Muehle 105:638-641.

ROHRLICH, M., and NIEDERAUER, T. 1967a. Untersuchungen uber Fett-Eiweiss-Komplexe in Cerealien. I. Isolierung eines Proteolipids aus Weizen, Roggen und Hafer. Part I. Fette Seifen Anstrichm. 69:63-67.

ROHRLICH, M., and NIEDERAUER, T. 1967b. Untersuchungen uber Fett-Eiweiss-Komplexe in Cerealien. I. Isolierung eines Proteolipids aus Weizen, Roggen und Hafer. Part II. Fette Seifen Anstrichm. 69:226.

RONALDS, J. A. 1974. Determination of protein content of wheat and barley by direct alkaline distillation. J. Sci. Food Agric. 25:179-185.

ROTSCH, A. 1949. Untersuchungen uber die Ursachen der Backfahigkeit. Getreide Mehl 3:153.

ROTSCH, A. 1953. Uber die Bedeutung der Starke fur die Krumenbildung. Brot Gebaeck 7:121.

ROTSCH, A. 1954. Chemische und technische Untersuchungen an kunstlichen Teigen. Brot Gebaeck 8:129.

RUBENTHALER, G., POMERANZ, Y., and FINNEY, K. F. 1963. Effects of sugars and certain free amino acids on bread characteristics. Cereal Chem. 40:658-665.

RUBENTHALER, G., POMERANZ, Y., and FINNEY, K. F. 1964. Semisynthetic model system for study of browning during baking. Cereal Chem. 41:424-427.

RYADCHIKOV, V. G., ZIMA, V. G., ZHMAKINA, O. A., LEBEDEV, A. V., and PHILIPAS, T. B. 1980. Studies of wheat flour glutenins by modified Landry-Moureaux method and gliadins by isoelectric focusing. Ann. Technol. Agric. 29:321-335.

RYADCHIKOV, V. G., ZIMA, V. G., ZHMAKINA, O. A., LEBEDEV, A. V., and PHILIPAS, T. B. 1981. Study of glutenins and gliadins of wheat flour. Prikl. Biokhim. Mikrobiol. 17(1):25-35.

SANDSTEDT, R. M. 1961. The function of starch in the baking of bread. Bakers Dig. 35(3):36-44.

SANDSTEDT, R. M., and HITES, B. D. 1945. Ascorbic acid and some related compounds as oxidizing agents in doughs. Cereal Chem. 22:161-187.

SANDSTEDT, R. M., JOLITZ, C. E., and BLISH, M. J. 1939. Starch in relation to some baking properties of flour. Cereal Chem. 16:780-792.

SASEK, A., KUBANEK, J., CERNY, J.,

MALY, J., and PLOCEK, J. 1986. Parallel effects of gliadin and glutenin markers during evaluation of baking quality in common wheat (*Triticum aestivum* L.). Sci. Agric. Bohemoslov. 18(2):87-95.

SCHOCH, T. J. 1965. Starch in bakery products. Bakers Dig. 39(2):48-52.

SCHOCH, T. J., and FRENCH, D. 1947. Studies on bread staling. I. The role of starch. Cereal Chem. 24:231-249.

SCHUSTER, G. 1984. Lipide, Struktur und Eigenschaften. Pages 79-113 in: Carbohydrates, Proteins, Lipids: Basic Views and New Approaches in Food Technology. F. Meuser, ed. Inst. Lebensmitteltechnologie, Getreidetechnologie, Berlin, Federal Republic of Germany.

SCOTT, G. E., HEYNE, E. G., and FINNEY, K. F. 1957. Development of the hard red winter wheat kernel in relation to yield, test weight, kernel weight, moisture content, and milling and baking quality. Agron. J. 49:509-512.

SECKINGER, H. L., and WOLF, M. J. 1967. Lipid distribution in the protein matrix of wheat endosperm as observed by electron microscopy. Cereal Chem. 44:669-674.

SEEGER, R., and BELITZ, H. D. 1981. Relation between the reduction of disulfide bonds and the molecular weight distribution of wheat gluten. Z. Lebensm. Unters. Forsch. 172:182-184.

SEGUCHI, M., and MATSUKI, J. 1977. Studies on pan-cake baking. II. Effect of lipids on pan-cake qualities. Cereal Chem. 54:918-926.

SHELLENBERGER, J. A. 1939. Variation in the baking quality of wheat during storage. Cereal Chem. 16:676-682.

SHELLENBERGER, J. A., MILLER, D., FARRELL, E. P., and MILNER, M. 1958. Effect of wheat age on storage properties of flour. Food Technol. (Chicago) 12:213-216.

SHEN, T., and GEDDES, W. F. 1942. Protease activity and reducing matter content of wheat flour doughs in relation to baking behavior. Cereal Chem. 19:609-631.

SHEU, R.-Y., MEDCALF, D. G., GILLES, K. A., and SIBBITT, L. D. 1967. Effect of biochemical constituents on macaroni quality. I. Differences between hard red spring and durum wheats. J. Sci. Food Agric. 18:237-239.

SHEWRY, P. R., and MIFLIN, B. J. 1985. Seed storage proteins of economically important cereals. Pages 1-83 in: Advances in Cereal Science and Technology, Vol. VII. Y. Pomeranz, ed. Am. Assoc. Cereal Chem., St. Paul, MN.

SHOGREN, M. D., FINNEY, K. F., and

HOSENEY, R. C. 1969. Functional (breadmaking) and biochemical properties of wheat flour components. I. Solubilizing gluten and flour protein. Cereal Chem. 46:93-102.

SHOLLENBERGER, J. H., CURTIS, J. J., JAEGER, C. M., EARLE, F. R., and BAYLES, B. B. 1949. The chemical composition of various wheats and factors influencing their composition. U.S. Dep. Agric. Tech. Bull. 995.

SIMMONDS, D. H. 1963. Proteins of wheat and flour. The separation and purification of the pyrophosphate-soluble proteins of wheat flour by chromatography on DEAE-cellulose. Cereal Chem. 40:110-120.

SIMMONDS, D. H., and O'BRIEN, T. P. 1981. Morphological and biochemical development of the wheat endosperm. Pages 5-70 in: Advances in Cereal Science and Technology, Vol. IV. Y. Pomeranz, ed. Am. Assoc. Cereal Chem., St. Paul, MN.

SIMMONDS, D. H., and WRIGLEY, C. W. 1972. The effect of lipid on the solubility and molecular-weight range of wheat gluten and storage proteins. Cereal Chem. 49:317-323.

SINCLAIR, A. T., and McCALLA, A. G. 1937. The influence of lipoids on the quality and keeping properties of flour. Can. J. Res. Sect. C 15:187-203.

SMITH, D. E., and ANDREWS, J. S. 1957. The uptake of oxygen by flour dough. Cereal Chem. 34:323-336.

SMITH, D. E., and MULLEN, J. D. 1965. Studies on short- and long-mixing flours. II. Relationship of solubility and electrophoretic composition of flour proteins to mixing properties. Cereal Chem. 42:275-287.

SMITH, D. E., VAN BUREN, J. P., and ANDREWS, J. S. 1957. Some effects of oxygen and fat upon the physical and chemical properties of flour doughs. Cereal Chem. 34:337-349.

SOKOL, H. A., and MECHAM, D. K. 1960. Review of the functional role and significance of the sulfhydryl group in flour. Bakers Dig. 34(6):24-26, 28, 30, 74.

SOKOL, H. A., MECHAM, D. K., and PENCE, J. W. 1960. Sulfhydryl losses during mixing of doughs: Comparison of flours having various mixing characteristics. Cereal Chem. 37:739-748.

SOLLARS, W. F. 1956. Evaluation of flour fractions for their importance to cookie quality. Cereal Chem. 33:121-128.

SOLLARS, W. F. 1958a. Cake and cookie flour fractions affected by chlorine bleaching. Cereal Chem. 35:100-110.

SOLLARS, W. F. 1958b. Fractionation and reconstitution procedures for cake flours. Cereal Chem. 35:85-99.

SOLLARS, W. F. 1959. Effects of the water-soluble constituents of wheat flour on cookie diameter. Cereal Chem. 36:498-513.

SOLLARS, W. F. 1973a. Fractionation and reconstitution techniques for studying water retention properties of wheat flours. Cereal Chem. 50:708-716.

SOLLARS, W. F. 1973b. Water-retention properties of wheat flour fractions. Cereal Chem. 50:717-722.

SOLLARS, W. F., and BOWIE, S. M. 1966. Effect of the subfractions of starch tailings on cookie diameter. Cereal Chem. 43:244-260.

SOLLARS, W. F., and RUBENTHALER, G. L. 1971. Performance of wheat and other starches in reconstituted flours. Cereal Chem. 48:397-410.

SOLLARS, W. F., and RUBENTHALER, G. L. 1975. Flour fractions affecting farinograph absorption. Cereal Chem. 52:420-427.

SORGER-DOMENIGG, H., CUENDET, L. S., CHRISTENSEN, C. M., and GEDDES, W. F. 1955. Grain storage studies. XVII. Effect of mold growth during temporary exposure of wheat to high moisture contents upon the development of germ damage and other indices of deterioration during subsequent storage. Cereal Chem. 32:270-285.

SOULAKA, A. B., and MORRISON, W. R. 1985a. The amylose and lipid contents, dimensions, and gelatinisation characteristics of some wheat starches and their A- and B-granule fractions. J. Sci. Food Agric. 36:709-718.

SOULAKA, A. B., and MORRISON, W. R. 1985b. The bread baking quality of six wheat starches differing in composition and physical properties. J. Sci. Food Agric. 36:719-727.

SOZINOV, A. A., and POPERELYA, F. A. 1980. Genetic classification of prolamines and its use for plant breeding. Ann. Technol. Agric. 29:229-245.

SOZINOV, A. A., POPERELYA, F. A., and STAKANOVA, A. J. 1974. Use of electrophoresis of gliadin for selection of wheat by quality. Vestn. Skh. Nauki (Moscow) 7:99-108 (Chem. Abstr. 81:166298a).

SPENCER, K., and FRENEY, J. R. 1980. Assessing the sulfur status of field-grown wheat by plant analysis. Agron. J. 72:469-472.

SPIES, R. D., and KIRLEIS, A. W. 1978. Effect of free flour lipids on cake-baking potential. Cereal Chem. 55:699-704.

STAMBERG, O. E. 1939. Starch as a factor in dough formation. Cereal Chem. 16:769-780.

STENBERG, R. J., and GEDDES, W. F. 1960a. Some chemical changes which accompany the browning of canned bread during storage. Cereal Chem. 37:614-622.

STENBERG, R. J., and GEDDES, W. F. 1960b. Studies on accelerated browning in starch pastes containing various bread ingredients. Cereal Chem. 37:623-637.

STRELNIKOVA, M. M., VERTIL, S. A., and MEERSON, E. E. 1964. Biochemical peculiarities of strong wheats. Biokhim. Zerna Khlebopech. 7:167-179. (Chem. Abstr. 61:16435b)

STROBEL, R. G., and HOLME, J. 1963a. Curtain electrophoretic separation of the water-soluble constituents of bleached cake flour. Cereal Chem. 40:351-360.

STROBEL, R. G., and HOLME, J. 1963b. Chemical composition of the water-soluble constituents of bleached cake flour. Cereal Chem. 40:361-372.

SUBDA, H., and BISKUPSKI, A. 1982. Evaluation of protein quality and quantity in wheat and their effect on the baking quality. Hodowla Rosl. Aklim. Nasienn. 26:479-491.

SUCKOW, P., ABDEL-GAWAD, A., and MEUSER, F. 1983. Versuche zur Aufklarung des Anomalen Technologischen Verhaltens nicht Backfahiger Weizen. Schriftenreihe aus dem Fachgebiet Getreidetechnologie. Vol. 6. F. Meuser, ed. Inst. Getreidetechnologie, Technische Univ., Berlin, Federal Republic of Germany. 160 pp.

SULLIVAN, B. 1940. The function of the lipids in milling and baking. Cereal Chem. 17:661-668.

SULLIVAN, B., and DAHLE, L. K. 1966. Disulfide-sulfhydryl interchange studies of wheat flour. I. The improving action of formamidine disulfide. Cereal Chem. 43:373-383.

SULLIVAN, B., and NEAR, C. 1927. Relation of magnesium in the ash and the lipoid-protein ratio to the quality of wheats. J. Am. Chem. Soc. 49:467-472.

SULLIVAN, B., NEAR, C., and FOLEY, G. H. 1936a. The role of the lipids in relation to flour quality. Cereal Chem. 13:318-331.

SULLIVAN, B., NEAR, C., and FOLEY, G. H. 1936b. The harmful action of wheat germ on the baking quality of flour and the constituents responsible for this effect. Cereal Chem. 13:453-462.

SULLIVAN, B., DAHLE, L. K., and PETERSON, D. A. 1961. The oxidation of wheat flour. III. The isolation of thioctic acid. Cereal Chem. 38:463-465.

TAHA, S. A., and SAGI, F. 1986. Relationships between chemical composition of durum wheat semolina and macaroni quality. I. Total, soluble and insoluble protein. Cereal Res. Commun. 14:259-266.

TAHA, S. A., and SAGI, F. 1987. Relationships between chemical composition of durum wheat semolina and macaroni quality. II. Ash, carotenoid pigments and oxidative enzymes. Cereal Res. Commun. 15:123-129.

TANAKA, K., and BUSHUK, W. 1973a. Changes in flour proteins during dough-mixing. I. Solubility results. Cereal Chem. 50:590-596.

TANAKA, K., and BUSHUK, W. 1973b. Changes in flour proteins during dough-mixing. II. Gel filtration and electrophoresis results. Cereal Chem. 50:597-605.

TANAKA, K., and BUSHUK, W. 1973c. Changes in flour proteins during dough-mixing. III. Analytical results and mechanisms. Cereal Chem. 50:605-612.

TAO, R. P.-C., and POMERANZ, Y. 1967. Water-soluble pentosans in flours varying widely in bread-making potentialities. J. Food Sci. 32:162-168.

TAO, R. P.-C., and POMERANZ, Y. 1968. Functional breadmaking properties of wheat flour lipids. III. Effects of lipids on rheological properties of wheat flour doughs. Food Technol. (Chicago) 22:1145-1149.

TATHAM, A. S., FIELD, J. M., SMITH, S. J., and SHEWRY, P. R. 1987. The conformations of wheat gluten proteins. II. Aggregated gliadins and low molecular weight subunits of glutenin. J. Cereal Sci. 5:203-214.

TIMMS, M. F., BOTTOMLEY, R. C., ELLIS, R. S., and SCHOFIELD, J. D. 1981. The baking quality and protein characteristics of a winter wheat grown at different levels of nitrogen fertilisation. J. Sci. Food Agric. 32:684-698.

TIPPLES, K. H., MEREDITH, J. O., and HOLAS, J. 1978. Factors affecting farino-graph and baking absorption. II. Relative influence of flour components. Cereal Chem. 55:652-660.

TIPPLES, K. H., PRESTON, K. R., and KILBORN, R. H. 1982. Implications of the term "strength" as related to wheat and flour quality. Bakers Dig. 56(6):16-18, 20.

TKACHUK, R. 1969. Involvement of sulfhydryl peptides in the improver reaction. Cereal Chem. 46:203-205.

TKACHUK, R., and IRVINE, G. N. 1969. Amino acid compositions of cereals and oilseed meals. Cereal Chem. 46:206-218.

TODD, J. P., HAWTHORN, J., and BLAIN, J. A. 1954. The bleaching and improvement of bread doughs. Chem. Ind. (London) 1954:50.

TOLEDO, R., STEINBERG, M. P., and NELSON, A. I. 1968. Quantitative deter-mination of bound water by NMR. J. Food Sci. 33:315-317.

TORSINSKAJA, L. R., et al. 1967. The role of starch in the baking quality of flours. Tr. Vses. Nauchno Issled. Inst. Zhirov 58-59:400. (In Russian)

TRACEY, M. V. 1964. The role of wheat flour pentosans in baking. I. Enzymic destruction of pentosans in situ. J. Sci. Food Agric. 15:607-611.

TRAUB, N., HUTCHINSON, J. B., and DANIELS, D. G. H. 1957. X-ray studies of the wheat protein complex. Nature 179:769-770.

TSEN, C. C. 1965. The improving mechanism of ascorbic acid. Cereal Chem. 42:86-97.

TSEN, C. C. 1967. Changes in flour proteins during dough mixing. Cereal Chem. 44:308-317.

TSEN, C. C. 1968. Oxidation of sulfhydryl groups of flour by bromate under various conditions and during the breadmaking process. Cereal Chem. 45:531-538.

TSEN, C. C., and ANDERSON, J. A. 1963. Determination of sulfhydryl and disulfide groups in flour and their relation to wheat quality. Cereal Chem. 40:314-323.

TSEN, C. C., and BUSHUK, W. 1963. Changes in sulfhydryl and disulfide contents of doughs during mixing under various conditions. Cereal Chem. 40:399-408.

TSEN, C. C., and BUSHUK, W. 1968. Reactive and total sulfhydryl and disulfide contents of flours of different mixing properties. Cereal Chem. 45:58-62.

TSEN, C. C., and HLYNKA, I. 1962. The role of lipids in oxidation of doughs. Cereal Chem. 39:209-219.

TSEN, C. C., and HLYNKA, I. 1963. Flour lipids and oxidation of sulfhydryl groups in dough. Cereal Chem. 40:145-153.

VAKAR, A. B., and KOLPAKOVA, V. V. 1976. Solubility of the glutenin fraction of gluten. Vestn. Skh. Nauki (Moscow) 1976:45-50.

VAKAR, A. B., PUMPYANSKII, A. Y., and SEMENOVA, L. V. 1965. The influence of D_2O on the physical properties of gluten and wheat dough. Prikl. Biokhim. Mikrobiol. 1:5-24. (Chem. Abstr. 63:3539d)

VAKAR, A. B., PRISHCHEP, E. G., and KOLPAKOVA, V. V. 1975. Structural differences between glutens of varying quality and between their protein components. Nahrung 19:759-768.

VALLEGA, V. 1986. High-molecular-weight glutenin subunit composition of Italian *Triticum durum* cultivars and spaghetti cooking quality. Cereal Res. Commun. 14:251-257.

VILLEGAS, E., POMERANZ, Y., and SHELLENBERGER, J. A. 1963. Effects of thiolated gelatins and glutathione on rheological properties of wheat doughs. Cereal Chem. 40:694-703.

WAINRIGHT, A. R., COWLEY, K. M., and WADE, P. 1985. Biscuit making properties of flours from hard and soft milling single variety wheats. J. Sci. Food Agric. 36:661-668.

WALL, J. S. 1964. Cereal proteins. Pages 315-334 in: Proteins and Their Reactions. H. W. Schultz and A. F. Anglemier, eds. AVI Publ. Co., Westport, CT.

WALL, J. S. 1971. A review of the composition, properties, and distribution of some important wheat flour constituents. Cereal Sci. Today 16:412, 414-417, 429.

WALL, J. S. 1979a. Properties of proteins contributing to functionality of cereal foods. Cereal Foods World 24:288-292, 313.

WALL, J. S. 1979b. The role of wheat proteins in determining baking quality. Pages 275-308 in: Recent Advances in the Biochemistry of Cereals. D. L. Laidman and R. G. Wyn Jones, eds. Academic Press, London.

WALSH, D. E., and GILLES, K. A. 1971. The influence of protein composition on spaghetti quality. Cereal Chem. 48:544-554.

WARWICK, M. J., FARRINGTON, W. H. H., and SHEARER, G. 1979. Changes in total free acids and individual lipid classes on prolonged storage of wheat flour. J. Sci. Food Agric. 30:1131-1138.

WASIK, R. J. 1978. Relationship of protein composition of durum wheat with pasta quality and the effect of processing and cooking on these proteins. Can. Inst. Food Sci. Technol. J. 11:129-133.

WASIK, R. J., and BUSHUK, W. 1975a. Relation between molecular-weight distribution of endosperm proteins and spaghetti-making quality of wheats. Cereal Chem. 52:322-328.

WASIK, R. J., and BUSHUK, W. 1975b. Sodium dodecyl sulfate-polyacrylamide gel electrophoresis of reduced glutenin of durum wheats of different spaghetti-making quality. Cereal Chem. 52:328-334.

WATSON, C. A., MILLER, B. S., JOHNSON, J. A., and BEQUETTE, R. K. 1963. Relationship of types of phosphorus in wheat flour and gluten to flour quality. Agron. J. 55:526-528.

WEHRLI, H. P. 1969. The synthesis of glycolipids and their role in breadmaking. Ph.D. thesis, Kansas State Univ., Manhattan.

WEHRLI, H. P., and POMERANZ, Y. 1969a. Chemical bonds in dough. Bakers Dig. 43(6):22-31.

WEHRLI, H. P., and POMERANZ, Y. 1969b. Synthesis of galactosylglycerides and related lipids. Chem. Phys. Lipids 3:357-370.

WEHRLI, H. P., and POMERANZ, Y. 1970a. A note on the interaction between glycolipids and wheat flour macromolecules. Cereal Chem. 47:160-166.

WEHRLI, H. P., and POMERANZ, Y. 1970b. A note on autoradiography of tritium-labeled galactolipids in dough and bread. Cereal Chem. 47:221-224.

WHITNEY, P. L., and TANFORD, C. 1962. Solubility of amino acids in aqueous urea solutions and its implication for the denaturation of proteins by urea. J. Biol. Chem. 237:PC1735.

WILSON, J. T., and DONELSON, D. H. 1963. Studies on the dynamics of cake-baking. I. The role of water in formation of layer cake structure. Cereal Chem. 40:466-481.

WILSON, J. T., and DONELSON, D. H. 1965. Studies on the dynamics of cake-baking. II. The interaction of chlorine and liquid in the formation of layer-cake structure. Cereal Chem. 42:25-37.

WOOTTON, M. 1966. Binding and extractability of wheat flour lipid after dough formation. J. Sci. Food Agric. 17:297-301.

WORKING, E. B. 1924. Lipoids, a factor influencing gluten quality. Cereal Chem. 1:153-158.

WORKING, E. B. 1928. The action of phosphatides in bread dough. Cereal Chem. 5:223-234.

WÖSTMANN, B. 1950. The cystine content of wheat flour in relation to dough properties. Cereal Chem. 27:391-397.

WOYCHIK, J. H., BOUNDY, J. A., and DIMLER, R. J. 1961. Amino acid composition of proteins in wheat gluten. J. Agric. Food Chem. 9:307-310.

WRENCH, P. M. 1965. The role of wheat flour pentosans in baking. III. Enzymatic degradation of pentosan fractions. J. Sci. Food Agric. 16:51-54.

WRIGLEY, C. W. 1972. The biochemistry of the wheat protein complex and its genetic control. Cereal Sci. Today 17:370-375.

WRIGLEY, C. W. 1980. The genetic and chemical significance of varietal differences in gluten composition. Ann. Technol. Agric. 29:213-227.

WRIGLEY, C. W., DuCROS, D. L., ARCHER, M. J., DOWNIE, P. G., and ROXBURGH, C. M. 1980. The sulfur content of wheat endosperm proteins and its relevance to grain quality. Aust. J. Plant Physiol. 7:755-766.

WRIGLEY, C. W., ROBINSON, P. I., and WILLIAMS, W. T. 1981. Association between electrophoretic patterns of gliadin proteins and quality characteristics of wheat cultivars. J. Sci. Food Agric. 32:433-442.

WRIGLEY, C. W., LAWRENCE, G. J., and SHEPHERD, K. W. 1982. Association of glutenin subunits with gliadin composition and grain quality in wheat. Aust. J. Plant

Physiol. 9:15-30.

WRIGLEY, C. W., DuCROS, D. L., FULLINGTON, J. G., and KASARDA, D. D. 1984. Changes in polypeptide composition and grain quality due to sulfur deficiency in wheat. J. Cereal Sci. 2:15-24.

WU, C. H., NAKAI, S., and POWRIE, W. D. 1976. Preparation and properties of acid-solubilized gluten. J. Agric. Food Chem. 24:504-510.

YAMAZAKI, W. T. 1953. An alkaline water retention capacity test for the evaluation of cookie baking potentialities of soft winter wheat flours. Cereal Chem. 30:242-246.

YAMAZAKI, W. T. 1955. The concentration of a factor in soft wheat flours affecting cookie quality. Cereal Chem. 32:26-37.

YAMAZAKI, W. T. 1959. Flour granularity and cookie quality. II. Effects of changes in granularity on cookie characteristics. Cereal Chem. 36:52-59.

YAMAZAKI, W. T. 1969. Soft wheat flour evaluation. Bakers Dig. 43(1):30-32, 63.

YAMAZAKI, W. T., and DONELSON, D. H. 1972. The relationship between flour particle size and cake-volume potential among Eastern soft wheats. Cereal Chem. 49:649-653.

YAMAZAKI, W. T., and DONELSON, J. R. 1976. Effects of interactions among flour lipids, other flour fractions, and water on cookie quality. Cereal Chem. 53:998-1005.

YAMAZAKI, W. T., DONELSON, J. R., and KWOLEK, W. F. 1977. Effects of flour fraction composition on cookie diameter. Cereal Chem. 54:352-360.

YASUNAGA, T., and UEMURA, M. 1964. Factors affecting the difference in qualitative characteristics of dough obtained from various wheat flours. I. The significance of sulfhydryl groups in Japanese domestic flour on their dough strength. Shokuryo Kenkyujo Kenkyu Hokoku 18:88-93. (Chem. Abstr. 61:13803)

YASUNAGA, T., BUSHUK, W., and IRVINE, G. N. 1968. Gelatinization of starch during bread-baking. Cereal Chem. 45:269-279.

YOUNGS, V. L., MEDCALF, D. G., and GILLES, K. A. 1967. The distribution of lipids in the four major fractions of hard red spring and durum wheat flour. (Abstr.) Cereal Sci. Today 12:111.

ZAEHRINGER, M. V., BRIANT, A. U., and PERSONIUS, C. J. 1956. Effects on baking powder biscuits of four flour components used in two proportions. Cereal Chem. 33:170-180.

ZAWISTOWSKA, U., BEKES, F., and BUSHUK, W. 1984. Intercultivar variations in lipid content, composition, and distribution and their relation to baking quality. Cereal Chem. 61:527-531.

ZAWISTOWSKA, U., BEKES, F., and BUSHUK, W. 1985. Gluten proteins with high affinity to flour lipids. Cereal Chem. 62:284-289.

ZELENY, L., and COLEMAN, D. A. 1938. Acidity in cereals and cereal products, its determination and significance. Cereal Chem. 15:580-595.

ZENTNER, H. 1958. Flour lipids, a note on the acetone soluble fraction. Chem. Ind. (London) 1958:129-130.

ZENTNER, H. 1960. The continuous electrophoresis of wheat gluten. Chem. Ind. (London) 1960:317-318.

ZENTNER, H. 1964. The oxidation of mechanically developed doughs. J. Sci. Food Agric. 15:629-634.

ZIMMERMANN, R. 1976. Functional significance of flour components in baking potential. Backer Konditor 30:40-44.

ZOBEL, H. F., and SENTI, F. R. 1959. The bread staling problem. X-ray diffraction studies on breads containing a cross-linked starch and a heat-stable amylase. Cereal Chem. 36:441-451.

CHAPTER 6

BREAD INDUSTRY AND PROCESSES

KAREL KULP
Cereal Science Research
American Institute of Baking
Manhattan, Kansas

I. INTRODUCTION

A major portion of wheat flour in the United States is utilized in commercial bakeries for the manufacture of breads and similar bakery foods. This sector of the baking industry consumed 13.81 billion pounds of flour in 1982, which was only slightly more than in 1972, when the use of 13.56 billion pounds of flour by bread bakers was reported. The bread industry is not a homogeneous producer of white pan bread any more. Its product mix includes (in addition to white pan bread) hearth breads, hot dog buns, hamburger buns, and variety breads, e.g., whole wheat, wheat, rye, cracked wheat, and multigrain breads. The changes in the distribution of manufactured breads are evident in Tables I and II, which give the production volume of breads and related products reported in the Census of Manufactures (Anonymous, 1977, 1985a) for the years 1972 and 1982 and estimated by the U.S. Department of Commerce (Anonymous, 1986) for 1985 and 1986.

Traditional white pan bread remains the predominant type in the United States, but other bread types have gained a major portion of the market. The increase in that market was 55.8% from 1972 to 1982 and 63.3% from 1972 to 1986. It is generally accepted that the trend is leveling off, and more market mobility is expected within the varieties, rather than a further loss in the market position of white pan bread. This trend is attributed to changes in taste, nutritional considerations, economic factors, living style, and demographic changes.

II. STRUCTURE OF THE INDUSTRY

The centralization of the industry into large corporations continues (Anonymous, 1986). On the other hand, the tendency to concentrate production is limited to certain production operations, with the final bake-off conducted in in-store bakeries. Retail stores are also on an increase, due to the emergence of specialty bakery shops, which are often operated under a franchise. Figure 1 shows the market size (in billions of dollars) for the period 1981–1986 for various types of bakeries. In-store bakeries produce bakery foods from frozen doughs shipped from a center plant for the bake-off operation or from scratch, starting with individual ingredients or mixes. Some use combination methods (e.g., scratch or mix and bake-off). The distribution of these methods is evident from Fig. 2.

In view of the various procedures employed by today's baking industry and the growing spectrum of products manufactured, both the requirements for flour performance and the technological methodology are changing and becoming more diverse and complex. This chapter summarizes the current technological practices and indicates factors affecting flour performance. Supplemental information is available in other publications (Matz, 1972a; Pyler, 1973; Reed and Peppler, 1973; Ponte and Reed, 1982; Kent, 1983; Anonymous, 1985b; Blanshard et al, 1986).

III. BAKING METHODS

Breads in the United States are produced by five major processes: the straight

TABLE I
U.S. Production of White Pan and Variety Breads in 1972, 1982, and 1986[a]

Product Category	1972 Millions of Pounds	%[b]	1982 Millions of Pounds	%	1986 Millions of Pounds	%
All breads	11,418.7	100.00	10,709.6	100.00	10,922.2	100.00
White pan bread	8,657.3	76.82	6,410.2	59.85	6,414.1	58.73
Variety breads	2,761.4	24.18	4,299.4	40.15	4,508.1	41.27

[a] Data from Anonymous (1977, 1985a, 1986).
[b] Percentages are expressed on the basis of all breads.

TABLE II
U.S. Production of Bread Varieties in 1972, 1982, and 1986[a]

Product Category	1972 Millions of Pounds	%[b]	1982 Millions of Pounds	%	1986 Millions of Pounds	%
All bread varieties[c]	2,761.4	100.00	4,299.4	100.00	4,508.1	100.00
White, hearth, including French, Italian, etc.	549.7	19.90	918.8	21.37	976.1	21.65
Whole wheat, cracked wheat, and other dark wheat breads	777.7	28.16	1,956.1	45.50	2,196.7	48.73
Rye, including pumpernickel	426.0	15.43	383.0	8.91	398.8	8.85
Other specialty (raisin, potato, salt-free, salt-rising, etc.)	339.0	12.28	455.0	10.58	445.9	9.89
Bread, white, wheat, rye, and not specified kind	669.0	24.23	586.5	13.64	490.6	10.88

[a] Data from Anonymous (1977, 1985a, 1986).
[b] Percentages are expressed on the basis of all bread varieties.
[c] White in this group refers to special white breads.

dough, sponge-dough, liquid ferment, continuous-mix, and no-time methods. Overseas, the Chorleywood bread process, which is a special example of the no-time procedure, is widely used.

Of these, the sponge-dough process still predominates for the production of white pan bread (accounting for over 50% of total production), followed closely by the liquid ferment method. The straight dough procedure is generally used in retail operations and for the production of bread varieties. The use of the continuous-mix process has decreased dramatically, because breads manufactured by this method have met with poor acceptance by consumers. It is still employed to a limited extent in the production of hamburger rolls, hot dog buns, and certain types of breadings (Dubois and Vetter, 1987).

The straight dough and sponge-dough processes are described in this section. Other processes are described in sections V–VIII.

A. Formulation

Typical commercial formulas for white pan bread produced by various processes are given in Table III. Some variations in ingredient levels are

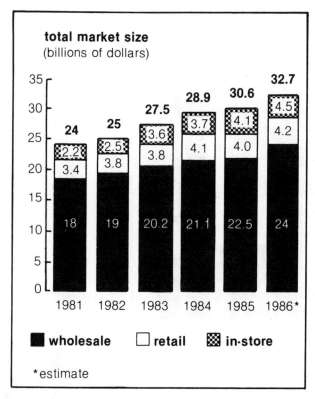

Fig. 1. Distribution of sales volumes of wholesale, retail, and in-store bakeries during 1981–1986. (Data from Anonymous, 1985a, 1986)

expected from plant to plant. Certain other minor additives might also be used for special purposes; these are discussed below.

B. Straight Dough Process

As indicated above, the straight dough process finds minor applications in wholesale bread production. It involves the combination of all ingredients by single-stage mixing into a dough of optimum physical properties. The procedural conditions are indicated in Table III.

Compared to the sponge-dough method, the straight dough method requires less processing time, labor, power, and equipment. Straight dough breads have a blander flavor than sponge-dough breads. The straight dough system is basically

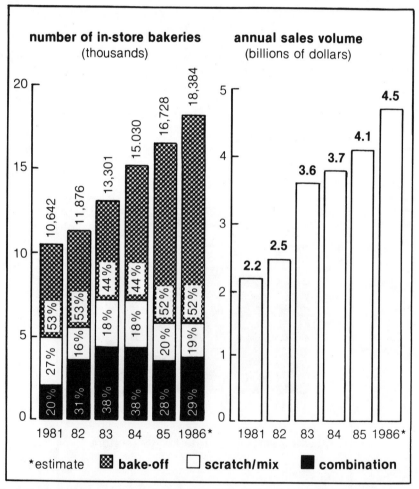

Fig. 2. Number of in-store bakeries operating during 1981–1986 using bake-off, scratch or mix, and combination procedures (left) and their annual sales volumes (right). (Reprinted, with permission, from Anonymous, 1986)

less flexible than the sponge-dough with respect to schedule adherence; relatively small variations in processing may lead to noticeable variations in the final quality of breads. A modification of the straight dough method is the straight dough remix procedure, in which the fermented straight dough is subjected to a second mixing. Such doughs are hybrids between straight dough and sponge-dough and yield breads with characteristics intermediate to those of breads produced by those two systems.

C. Sponge-Dough Process

The sponge-dough formula currently used by the U.S. industry is presented in Table III, along with the normal operational conditions (Kulp and Dubois,

TABLE III
White Pan Bread Formulations[a]

Ingredient	Sponge/Dough[b,c] Sponge	Sponge/Dough[b,c] Dough	Straight Dough[d]	Straight No-Time Dough[e]	Brew Bread[f]
Flour[g]	70.0	30.0	100.0	100.0	100.0
Brew	...	...	...	...	35.0
Water	40.0	24.0	64.0	65.0	32.0
Yeast, compressed	3.0	...	2.5	3.5	...
Salt	...	2.0	2.0	2.0	1.4
Sugar or sweetener solids	...	8.5	8.0	6.0	7.0
Shortening	...	3.0	2.75	2.75	2.75
Yeast food	0.5	...	0.5	0.6	0.5
Nonfat dry milk or milk replacer	...	2.0	2.0	2.0	2.0
Fungal protease	0.5	...	0.25	0.5	...
L-Cysteine, ppm	...	...	...	40.0	...
Potassium bromate, ppm	...	...	15.0	30.0	30.0
Ascorbic acid, ppm	...	...	...	60.0	...
Vinegar (100 grain)	...	...	...	0.5	...
Monocalcium phosphate[h]	...	...	...	0.25	...
Mono- and diglycerides, hydrate	...	0.5	0.5	0.75	0.5
Dough strengtheners	...	...	...	...	0.25
Calcium propionate	...	0.2	0.2	0.2	0.2

[a] Source: Kulp and Dubois (1982); used by permission.
[b] Values are percentages, based on flour, except as otherwise noted.
[c] The sponge-dough ratio is 70:30. The sponge is mixed for 1 min at low speed and 3 min at high speed at 77°F (25°C) and fermented for 3–5 hr at 86°F (30°C). The dough is mixed for 1 min at low speed and 10–12 min at high speed at 80°F (27°C); rested for 15–20 min; divided into 18-oz (515-g) pieces for 1-lb (454-g) loaves; rounded and given a 7-min rest period; sheeted, shaped, and panned; proofed for 55 min at 107°F (42°C) and 85% relative humidity; and baked for 18 min at 445°F (230°C).
[d] One-step process. The dough is mixed for 1 min at low speed and 15–20 min at high speed at 86°F (30°C) and fermented for 2 hr at 86–95°F (30–35°C) and 85% relative humidity. This procedure is used by retailers or for specialty breads.
[e] The dough is mixed for 1 min at low speed and 15–18 min at high speed at 82°F (28°C) and proofed for 55 min at 107°F (42°C) and 85% relative humidity.
[f] Brew ingredients are dispersed by high-speed agitation for 5 min at 82°F (28°C). The brew is fermented (with low agitation) for 1.5 hr at 88–92°F (31–33°C) and added to the dough ingredients. The process then continues like the sponge-dough process. The brew consists of 80.95% water, 7.75% sweetener solids, 8.0% compressed yeast, 3.8% salt, and 0.2–0.5% buffer (calcium carbonate plus ammonium sulfate); 35% of this brew is added to the dough.
[g] About 11.2–11.7% protein and 0.44–0.46% ash, enriched according to U.S. standards (14% mb).
[h] Used instead of vinegar.

1982). Figure 3 describes a production line for the sponge-dough process (Seiling, 1969).

SPONGE MIXING AND FERMENTATION

Water, yeast, and yeast food are blended in an ingrediator (slurry tank) and then pumped into the mixer. Flour is pneumatically conveyed from storage bins into automatic overhead scales, which automatically disperse appropriate quantities of flour into the mixer. The sponge ingredients are then mixed; the objective is to combine the components into a uniform mass without the full development of a gluten matrix. The mixed sponge is transferred into troughs and kept in a fermentation room at 86° F (30° C) with 86% relative humidity. A sponge temperature in the mixer of 77° F (25° C) is usually sought; a rise of about 10° F (5.6° C) is caused by the fermentation reactions. The sponge time may range from 3 to 5 hr, the typical length for wholesale white bread being 3.5 hr at 86° F (30° C). The sponge fermentation time is influenced by several factors, e.g., the yeast level, the percentage of flour in the sponge (typical for today's practice is 70%), and quality expectations. The volume of the sponge increases four- to fivefold during fermentation, and the mixture develops a characteristic spongy or web structure, which may be observed when its surface is pulled apart. After achieving the maximum volume, the sponge collapses. This "drop," or "break," is used by some baking technologists as a reference point, at which 66–70% of the sponge fermentation has been completed. Depending on whether the process employed requires old (overfermented) or young (underfermented) sponge, the duration of the sponge fermentation is adjusted.

The panary fermentation, of which the sponge preparation is a part, is a complex biochemical process. During the first hour, bakers' yeast ferments simple sugars (glucose, fructose) present in the flour, and when these have been metabolized, the fermentation of maltose (originally present or arising by

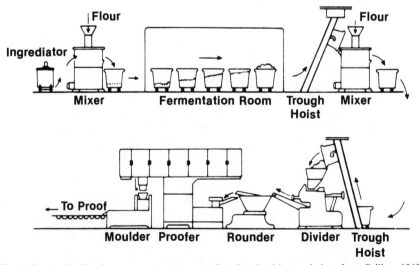

Fig. 3. Production line for sponge-dough bread. (Reprinted, with permission, from Seiling, 1969)

amylolysis of susceptible damaged starch) sets in (Lee and Geddes, 1959; Lee et al, 1959). Ethanol and carbon dioxide are the major products, and organic acids, carbonyl compounds, and higher alcohols are minor metabolites of the process. The physical changes in the dough properties, attributable to the development of acidity by fermentation and enzymatic and oxidation-reduction reactions, are referred to by baking technologists, in a trivial generic term, as *mellowing* of the dough (Pyler, 1973). Additional factors involved in these changes are ions; e.g., calcium toughens the gluten. Oxidizing agents have similar dough-strengthening effects, the result of regulating the sulfhydryl-disulfide interchange during dough preparation. The opposite effects are produced by reducing agents. Oxidizing agents and reducing agents are both used as additives to modify dough properties, but they are also indigenous components of flours, causing functional differences in performance. There is little information on the enzyme reactions in the dough process. Our knowledge is generally limited to proteolytic and amylolytic enzymes, which cause important changes in dough rheology. Other enzymes, oxidoreductases and hemicellulases, have received only limited attention. Although they may be less important in white pan breads, they may play significant roles in variety breads.

As carbon dioxide diffuses from the yeast cells into the aqueous phase of the sponge or dissolves in liquid ferments, it forms carbonic acid, which, although weakly ionizable, alters the acidity of the medium. The pH changes are affected not only by the amounts of generated carbon dioxide but also by the buffering capacities of the sponge-dough or liquid ferment systems. Excess carbon dioxide then collects as gas bubbles around loci in the aqueous phase. These loci originate from small air bubbles incorporated during mixing (Baker and Mize, 1941) or perhaps from air absorbed by the flour; about one fifth of the volume of flour is air entrapped among the flour particles (Bushuk et al, 1968). The significance of air bubbles was also demonstrated by mixing dough in oxygen. The oxygen formed bubbles in loci; these bubbles were metabolized by yeast during fermentation, and the resulting structure of the bread crumb lacked the normal porosity (Chamberlain et al, 1962).

There is substantial evidence indicating that yeast does not multiply in panary fermentation.

DOUGH MIXING

At the completion of fermentation, the sponges have either a plastic or a liquid consistency. Plastic sponges are hoisted in troughs to a position above the mixers; liquid ferments are pumped into a keeping tank, kept at about 40° F (4.44° C), and when needed transferred into the mixer. The dough ingredients (Table III) are added to the sponge in the mixer, and the mixing process is performed. The addition of salt may be delayed, since this ingredient extends the dough-mixing time and increases the energy requirement when added initially. Many bakers add salt during the last 2 min or so of mixing. Coated salt would have a similar effect; however, this is not a commonly used ingredient.

Proper dough mixing produces a dough of an optimal degree of development. The main factors affecting this reaction are the mixing time and optimal water absorption. These conditions are influenced by the type of flour, formula, temperature, dough consistency, and speed and type of mixer. The mixing behavior may be predicted to a certain degree by farinograph, mixograph, or similar dough-testing instruments. During the initial stages of mixing, the flour components hydrate; some ingredients dissolve or are dispersed in the aqueous phase of the dough. The dough at this point has a wet, lumpy appearance and lacks coherence. As mixing progresses, complex interactions take place, the net result being the formation of a thin gluten film that surrounds the starch granules. This film originates from protein and lipid components of the flour. The interaction of proteins through sulfhydryl-disulfide interchange, association by secondary bonds (hydrogen, hydrophobic), and the formation of lipid-protein complexes are involved. The proper rheological properties are essential for good machinability of the dough and concomitant speedy and uninterrupted production. It should have proper strength, should not be sticky, and should have adequate stability. Lack of these properties would result in doughs of poor handling properties, production breaks, and finally nonuniform breads of poor quality. The literature of the mixing process was recently evaluated by MacRitchie (1986).

The formation of the optimal dough in the mixer is evident from its appearance. The mass, which was initially lumpy during hydration, becomes coherent, exhibits elasticity, and has a lessening tendency to stick to the side of the mixer. At some later point, the entire dough gathers about the mixing bars and begins to slap against the mixer bowl. This is the "cleanup" stage, regarded by many bakers as one of the criteria for determining the length of the dough-mixing operation. At the peak of development, the dough has a silky, glossy appearance and elastic properties; when a piece is pulled between the fingers, it can be stretched into a thin membrane in which small bubbles and gluten strands can be seen. The completion of mixing is judged by "dough feel" and by practical baking experiments relating variations of mixing time to end-product quality. Peak dough development is related to power input to the mixer; wattmeters have been used to gauge dough-mixing times.

IV. BAKERY OPERATIONS AND EQUIPMENT

A. Mixers

The most common commercial mixer, the horizontal conventional mixer, is illustrated in Fig. 4. A heavy-duty frame, housing the electric motor, supports a U-shaped mixer bowl. The agitator drive shaft, which rotates at 60–85 rpm, is horizontally positioned and mounts two "spiders," which support two to four mixing arms. One or more of the arms may be free-rolling, and the arms may be straight or curved. A fluted mixing arm has also been used in the design of mixers (Sternberg, 1968). Agitators are so designed that as the dough rotates around the shaft, one bar carries the dough up while the other rolls it forward; during this movement the dough is subjected to throwing, stretching, and shearing actions, which result in gluten development. Heat of friction is generated during the process and is dissipated by refrigerated jackets on the mixer; the dough is usually taken out at 80°F (26.7°C). These mixers accommodate up to 2,000 lb of dough. The mixed dough is discharged, either by a bowl-tilting mechanism or by dropping the front door of the mixer, into a second trough, where it is fermented for a short time, usually about 20–30 min. This "dough time," or "floor time," allows the dough a recovery period.

Other mixers in use are high-speed mixers, including the Tweedy mixer (French and Fish, 1981), which operates at higher speed than the horizontal mixer, and the Stephan mixer (Niamtu et al, 1981), which operates at still higher speed (Figs. 5 and 6). As an illustration, a horizontal mixer develops bread doughs from a typical U.S. bread flour (11% protein) within 12–14 min, a Tweedy mixer in 6–8 min, and a Stephan mixer in 2–3 min. The doughs from high-speed operations are subjected to a high shear action, resulting in soft doughs, which require high levels of oxidants for recovery to good consistency. The mixing operations are sometimes followed by means of wattmeters attached to those units. Tweedy mixers equipped with this type of instrument may be set for a predetermined energy requirement, and the mixer is automatically shut off when the required energy input has been reached.

B. Dividing

The first step in dough makeup is dividing the dough into pieces of appropriate size. Figure 7 shows a commercial volumetric dough divider. All dough dividers perform their function volumetrically. The dough is fed into a hopper of the machine, flows downward, and by gravity and suction enters an adjustable chamber. A piston operates in the chamber on a cycle during which the dough enters, is cut off, and is deported on a moving belt. Dividers may have up to eight pockets (i.e., chambers) and have velocity ratings of 20–26 strokes per minute. They are generally run at speeds of 15–16 strokes per minute. At the maximum speed of 26 strokes, an eight-pocket divider produces 12,480 dough pieces per hour. Dividers are operated mechanically or hydraulically.

Because of continuing fermentation during dividing, resulting in additional formation of gas, the dough decreases in density. This change requires

Fig. 4. Horizontal commercial bread mixer. (Courtesy AMF, Inc., Richmond, VA)

Fig. 5. Tweedy batch mixer. (Courtesy Lantham Machinery Co. Inc., Atlanta, GA)

Fig. 6. Stephan bread mixer. (Courtesy Stephan Machinery Corporation, Columbus, OH)

continuous monitoring of the weight of dough pieces during the operation, which is generally performed manually. Automatic adjustment devices using weight cells with microprocessor controls are built into some divider units, but they have not been perfected yet (Kulp et al, 1983a).

A new divider is of the rotary type (Fig. 8). It includes an extruder that transports the dough for dividing by means of a rotating knife. The difficulty of using this approach is that the extrusion acts as an additional dough mixing, and proper adjustment of the mixing time at the mixer is required in order to compensate for the extrusion. This adjustment is not always successful.

C. Rounding

When discharged from the divider, the dough piece is rough and has a sticky, cut surface through which the continuously generated carbon dioxide can escape. The objective of the rounding operation is to form a skin around the dough piece to minimize gas diffusion and also to cause some redistribution of gas cells; the retained gas may then accumulate in the gas vesicles, which upon subsequent division form the cell structure of bread crumb. Skin formation also promotes the smooth, dry, and nonsticky character of doughs, which facilitates their further handling in the subsequent processing steps. During this operation, the importance of the dough stability, which is governed by its rheological properties, becomes evident. Excessive extensibility causes the dough to spread and lose its shape; too much elasticity makes rounding difficult. In rheological

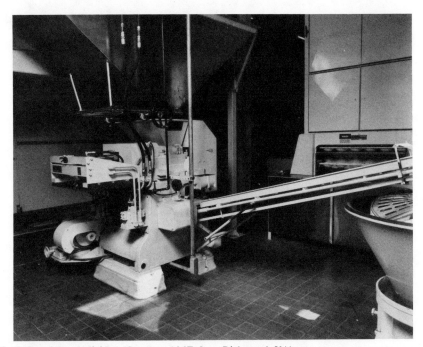

Fig. 7. Bread dough divider. (Courtesy AMF, Inc., Richmond, VA)

terms, rounding structurally activates or work-hardens the dough and thus promotes the properties of stability and elasticity (Hlynka, 1970).

Figure 9 shows three types of rounders: bowl, umbrella, and cone. In each case, the dough piece is forced to travel against a moving surface, continually rotating as it travels. Because of differences in shape, doughs travel through rounders at different speeds: in conical rounders the rate of travel of the dough

Fig. 8. Rotary divider for bread dough. (Courtesy AMF, Inc., Richmond, VA)

piece is initially slow and progressively increases as the dough reaches the end of the spiral; in umbrella rounders the reverse is true; and in drum rounders the dough piece travels at a relatively constant rate from beginning to end of the operation.

D. Intermediate Proof

The dividing and rounding operations subject the dough piece to considerable mechanical abuse, which alters its rheological properties and removes much of the carbon dioxide that has been formed. The dough at that stage has lost a considerable part of its pliability and extensibility and is not well suited for molding. To restore the necessary rheological characteristics, the dough is given a rest period of about 6–8 min, called the *intermediate proof*. The rounded dough pieces are transferred directly from the rounder into closed units called overhead proofers, an example of which is shown in Fig. 10. Endless belt or

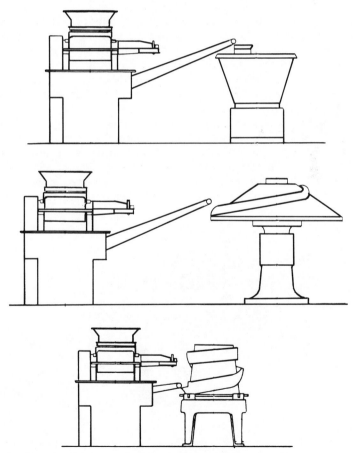

Fig. 9. Three types of rounders: bowl, umbrella, and cone (top to bottom), and their position in relation to the dough divider. (Courtesy Dutchess Bakers' Machinery Corp., Superior, WI)

tray-type conveyors are employed. As the dough pieces leave the intermediate proofer, they have expanded in volume as a result of gas formation and have become more pliable and extensible.

E. Molding

The molding stage is critical; it is the last dough-manipulating stage, at which the gas cells built in during processing are finally subdivided and uniformly distributed through the dough piece.

A proper balance of viscoelastic dough properties is essential for optimum molding. For proper sheeting, the dough should exhibit good extensibility, and its elastic response should not be excessive; too much elasticity causes quick retraction of the sheeted piece.

Molding proceeds in three steps: sheeting, curling, and rolling and scaling. The principle of this operation, using a drum-type molder, is evident from Fig. 11. As the dough piece emerges from the intermediate proofer, it is a flattened spheroid. This spheroid first passes through a set of two or more head rolls of the molder and is reduced to a thickness of about 1/4 in.; sheeting expels the gas, particularly that in large cells. The reduction must be done carefully to avoid tearing the dough. The sheet is then loosely curled by various means, depending on the type of molder. The curled dough is sealed and extended in length along the axial dimension by being forced between a compression plate and a revolving drum; the clearance between the plate and the drum is gradually decreased.

Drum molding has largely given way to reverse sheeting and cross-grain molding. These developments stem from the observation that dough sheeting causes a movement of moisture in the dough toward the trailing end. Since normally, in straight molding, the trailing end of the dough sheet forms the core of the molded dough loaf, moisture is unevenly distributed in the dough piece during the initial stages of the final proof. In reverse sheeting, the distribution of

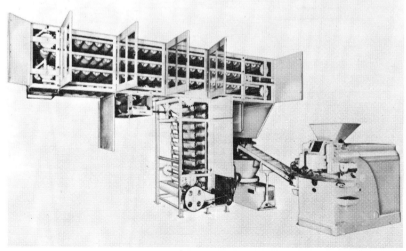

Fig. 10. Overhead proofer. (Courtesy AMF, Inc., Richmond, VA)

moisture is improved by sheeting the dough in two directions, thus reversing the position of the trailing end in the molded dough loaf. In cross-grain molding, the sheeted dough piece is turned 90°. This manipulation improves the moisture distribution and produces an elongated cell structure in the grain of the bread crumb. A cross-grain molder is shown in Fig. 12.

Other types of molders used for special applications are curl and bread-twisting molders.

Modern molders are generally equipped with an automatic panning device,

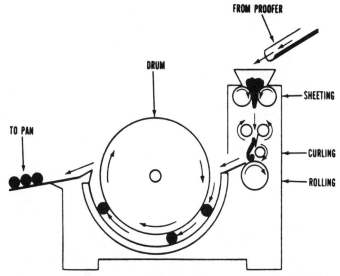

Fig. 11. Schematic diagram of drum molder. (Courtesy U.S. Department of Defense, Washington, DC)

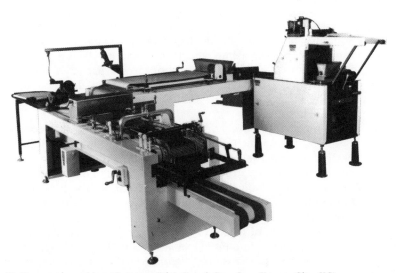

Fig. 12. Cross-grain molder. (Courtesy Stickelber & Son, Inc., Kansas City, KS)

which consists of a pan conveyor mounted integrally into the frame of the molder. This conveyor carries the empty pans under the molder head to a position just beyond the compression board, where an operated mechanism deposits a dough loaf into each pan.

F. Final Proof

The panned loaves are transported to a proof box for a final fermentation, or proofing. Proof boxes are units into which pan-laden racks are transported by hand or by means of monorail, or the system may be completely automated, sometimes as part of a proofer-oven. The proofer part of the system is either a conveyor or rack-type proofer. The latter uses either a center-mounted chain or side-mounted chains. Wells (1983) gave details of this type of operation.

For most production, proofing requires 45–55 min and is conducted at 110° F (43.3° C) at a relative humidity of 80–85%. As in most of the processing steps, an optimum level of proofing depends on many factors, which include flour strength, enzymatic activity of the flour, bread formulations, yeast activity, and the type of product desired.

G. Baking

The baking stage is essential to the conclusion of the breadmaking process. The extent to which fundamental qualities determined by ingredients, biochemical reactions, and processing are built into the dough is finally revealed in the oven. Proofed loaves going into the oven have a skin, formed during proofing; the skin soon thickens and becomes elastic in the oven, depending to some extent on the moisture content of the oven atmosphere—high humidity improves crust formation (Matz, 1972b). At the same time, as the dough temperature increases in the oven, carbon dioxide in the gaseous phase expands according to the general Charles gas law, $V = kt$, by which the volume (V) is proportional (by a constant, k) to the temperature (t) of the gas at a constant weight and pressure of the gas. In addition to carbon dioxide present in the gas phase, the solubility of carbon dioxide dissolved in the aqueous portion of the dough is reduced by the increasing temperature, and it acts as a gas. Another source of carbon dioxide is yeast fermentation, which proceeds at high rates until the yeast activity is arrested in the oven, at the thermal death point, 140° F (60° C). Generally, in U.S. formulas, sufficient formula sugar is available to support the fermentation.

Other important changes occurring in the oven include physicochemical reactions of flour proteins, which undergo denaturation, and thermal changes of starch, generally designated gelatinization and swelling. Both reactions, conformational reorientation in the structure of starch granules and proteins, cause important moisture interchange among dough components. Starch, present originally in the form of semicrystalline granules, changes its order by undergoing gelatinization and swelling. There is no general agreement among starch chemists on the definition of gelatinization and swelling. The degree of both changes is dependent on the amount of available water and levels of sugars and salts in the system. In this discussion, *gelatinization* of wheat starch refers to changes of order that occur within the temperature range of 55–63° C. Their initiation is detectable by microscopy as a loss of birefringence of starch granules in polarized light. Swelling, which can also be observed, starts at the same time as gelatinization and proceeds further throughout the baking process. The water in bread doughs limits the degree of swelling, but it is not sufficiently low to alter the gelatinization temperature range (Banks and Greenwood, 1975; Greenwood, 1976; Kulp and Lorenz, 1981). During both processes, amylose is leached from the granules. This linear starch component undergoes a rapid retrogradation upon the cooling of baked bread and contributes to the initial crumb firmness of freshly baked bread. Starch granules in their native state are resistant to the action of amylases. When heated, they become susceptible to amylases and start absorbing water after reaching the gelatinization temperature. Since the cereal amylases are still active at that point, an excessive degradation of starch may occur, resulting in adverse properties of baked bread: sticky crumb, low loaf volume, and open crumb grain. This condition may occur when the flour has been supplemented with excessive amounts of barley malt at the mill or when it has been milled from sprouted wheats. This damage is attributed to the excessive level of α-amylase, but other enzymes (e.g., proteases) may also contribute to the deleterious effects (Kulp et al, 1983a, 1983b; Lorenz et al, 1983).

Figure 13 shows the time-temperature relationship for crumb of pan bread baked at two oven temperatures. This variable affects the enzymatic reactions, since the optimum temperature for α-amylase is 60–70° C and that of β-amylase is above 50° C, and thermal inactivation of α-amylase occurs at 70–85° C and that of β-amylase occurs between 55 and 75° C (Audidier, 1968). The time between starch swelling and amylase activity before denaturation is obviously a critical period for starch modification. The fungal amylases, which are also used as a bakery ingredient, are less heat-stable than the cereal ones. They probably act more on damaged starch during the dough stage than on the gelatinizing starch in the early stages.

Figure 13 also shows that at the higher baking temperature the amylase

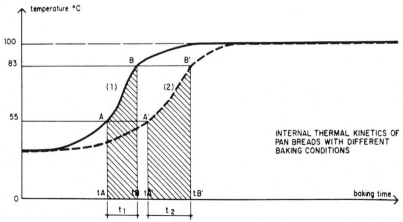

Fig. 13. Time-temperature relationship of baking bread at two oven temperatures. Amylase activity, measured from 55 to 83° C, occurs in the time span from tA to tB at the higher temperature (solid line) and from tA' to tB' at the lower temperature (dashed line). (Reprinted, with permission, from Audidier, 1968)

activity proceeds over the time span from tA to tB; at the lower baking temperature the duration is the obviously longer period from tA' to tB'. Also, the progressing swelling exposes the susceptible starch to enzymolytic action over a longer period when a lower baking temperature is applied. In a study encompassing three different baking temperatures, the temperature was shown to exert effects on crumb firmness, loaf weight, and grain score (Ponte et al, 1963).

Wheat protein denaturation starts at about 70° C (Audidier, 1968) and is of major importance in establishing bread structure. Another functional feature of wheat protein is its hydration during dough formation and the transfer of water from gluten to the starch component during baking, to support the swelling of starch granules. The denaturation of gluten is accompanied by decreased solubility and proceeds to a point where the gas vesicle walls are fixed and expansion is terminated. Denaturation is more extensive in the crust regions than in the crumb; the temperature rise is much faster in the crust, and much higher temperatures are reached there. Crumb temperatures do not exceed 100° C, whereas crust temperatures reach 195° C (Walden, 1955).

Another important set of reactions deals with color formation in the crust. Two main reactions take place: browning and caramelization. The first one, also called the Maillard reaction, occurs between reducing sugars (mainly glucose and fructose and, to a lesser extent, lactose) and the amino group of amino acids and proteins. Caramelization is a complex series of reactions of sugars effected by heat. The work of Rubenthaler et al (1963) suggests that the extent of crust color formation depends primarily on sugars added or originally present in the dough. Various proteins, peptides, and amino acids from dairy ingredients and milk replacers (blend of soy and whey) play an important role. Both reactions proceed faster at high pH than at low pH.

H. Bakery Ovens

Three types of ovens are in use in baking plants: tunnel, single-lap, and double-lap ovens. The last two are forms of traveling-tray ovens.

TUNNEL OVENS

Tunnel ovens were the first approach to high-capacity production units for large-scale baking operations. The first was built in 1913, and after more flexible heating systems were designed, tunnel ovens became the standard of the industry (Matz, 1972b). The early ovens were massively constructed of brick, but later ones are built of corrosion-resistant metal (Fig. 14).

Bread pans filled with dough enter at one end of the tunnel oven and make a single pass directly through the baking chamber on a traveling hearth; the baked loaves are discharged at the opposite end. These ovens usually consist of several heating zones, set at temperatures found to provide good bread characteristics; for example, a four-zone oven could be set at 345, 425, 440, and 380° F (174, 218, 227, and 193° C, respectively) in sequence for zones 1 through 4 (Anderson, 1966). A typical 1-lb loaf of bread is baked in 15–18 min. The high accuracy and consistency of temperature control in the various heating zones provide for uniformity in baking results. Although the tunnel oven is simple in design and construction, it is difficult to adapt to automatic unloaders or to use in a compact bakery layout. Its requirement for large floor space and its high initial cost hindered adoption of the tunnel oven and led to the development of traveling-tray ovens.

TRAVELING-TRAY OVENS

Traveling-tray ovens are a modification of the reel oven, consisting of trays moving over two parallel endless chains that connect two reels. Each tray, holding several pans, is permanently fixed to the conveying chain. The chain pulls the trays from the front of the oven through the hot-air chambers to the back, then to a lower track, where the product continues to bake on its return to the front of the oven, where it is unloaded.

Traveling-tray ovens are designed as either single-lap or double-lap ovens. Principles of these ovens are shown in Figs. 15 and 16. Single-lap ovens require a low ceiling and provide good steam characteristics. They have very good control of top and bottom heat and are compact, requiring only half the floor space occupied by a comparable tunnel oven. The trays of double-lap ovens travel through the oven several times, producing the product more economically, in large volumes, and this type of oven requires less floor space than others. A double-lap oven requires a minimum of 30 trays for good economy. Zoning of this type of oven is difficult, and added height lowers the desirable steam characteristics. Most innovations in ovens are confined to heating methods (Varilek and Walker, 1983a, 1983b, 1984a, 1984b) and improvements of efficiencies (Koch, 1983).

Fig. 14. Tunnel oven. (Courtesy D. K. Dubois)

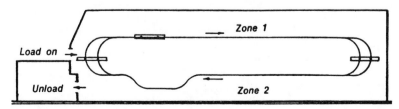

Fig. 15. Single-lap bread oven. (Reprinted, with permission, from Varilek and Walker, 1983a)

V. LIQUID FERMENT PROCESSES

Techniques have been developed wherein the use of fermented brews or barns is a part of the breadmaking process (Pyler, 1973). Some interest in liquid ferments was generated in 1954, when the Stable process of the American Dry Milk Institute was announced (McLaren, 1954). Certain aspects of liquid ferments, called also pre-ferments or broth, were developed as part of the continuous process at the time.

The present liquid ferment process has found wide industrial application as an independent process. It is essentially a modification of the conventional sponge-dough process, which is a long-established method of commercial bread production. The basic difference between conventional sponges (plastic sponges) and liquid ferments (semiliquid sponges) is the lower ratio of flour to water in the latter. This change produces sponges that can be transferred by pumping. Modifications introduced in recent years utilize advances in the engineering design of pumping systems for handling liquid ferments. General advantages of the process are savings in plant space and labor, increased production flexibility, and easier cleaning of equipment (sanitation).

Almost 50% of plants in the United States today are using the liquid ferment process (Dubois, 1984). Despite its wide acceptance for manufacturing white pan bread and hamburger rolls and its use to a certain extent for variety breads, its application varies from plant to plant in formulation and technology. The technological aspects and their effects on bread quality were investigated by Kulp and co-workers at the American Institute of Baking (Martinez-Anaya and Kulp, 1984; Martinez-Anaya et al, 1984; Kulp et al, 1985).

A. Fermentation Line

A typical production line using liquid flour or water ferments for bread production is illustrated in Fig. 17. It consists of a dry-ingredient mixer, to blend all dry ingredients except flour; a flour-blending tank; fermentation units; a heat exchanger, to cool liquid ferments, after the completion of fermentation, to 40–50° F (5–10° C); a holding tank, to keep the liquid at a lower temperature until it is used; and a dough mixer. When water brews are used, the ingredient mixer and blending tank are omitted, and water and dry ingredients are fed directly into the fermentation unit (Kulp, 1983; Thompson, 1983).

Flour liquid ferments are transferred from the holding tank into the mixer through a batch weight tank; water ferments are fed into the mixer through a

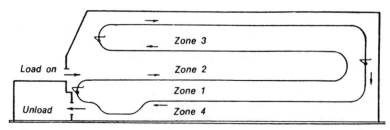

Fig. 16. Double-lap bread oven. (Reprinted, with permission, from Varilek and Walker, 1984a)

flowmeter. This production line, which represents a batch operation, may be replaced by a continuous system (e.g., that developed by APV Crepaco, Inc., Chicago).

B. Changes During Fermentation

During the fermentation of liquid ferments, changes in acidity, as indicated by pH and total titratable acidity (TTA), depend on the buffering capacity of the medium. Water ferments lack buffering capacity; therefore, calcium carbonate is used at suitable levels for adjustment and control of acidity. In flour ferments, the flour provides an adequate adjustment.

The effect of the buffer concentration (0, 0.2, and 0.5%, based on flour) on pH and TTA of water ferments is evident from Figs. 18 and 19, respectively. In commercial bread production, the 0.2% level would be considered an adequate buffer.

Similar studies (Kulp et al, 1985) conducted with 0, 20, and 50% flour ferments demonstrated the buffering capacity of flours. Trends in pH and TTA values during fermentation of the flour ferments were comparable to those of buffered water ferments. In all cases, TTA values reached maxima after approximately 1.5 hr, when an inflection point was observed, attributable to the exhaustion of

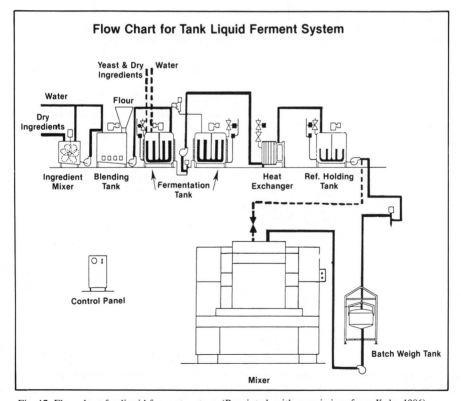

Fig. 17. Flow chart for liquid ferment system. (Reprinted, with permission, from Kulp, 1986)

rapidly fermentable glucose, while fructose continued to ferment at a somewhat slower rate.

The subsequent decrease in TTA values was explained by a loss of carbon dioxide; a secondary increase in TTA of flour ferments was due to maltose fermentation. The continued increases in TTA of unbuffered water ferments showed a different pattern than that of the flour and buffered ferments, because of the greater capacity of unbuffered water ferments to dissolve carbon dioxide. The addition of calcium carbonate buffer to water ferments and flour constituents to liquid flour ferments reduced the amount of dissolved carbon dioxide.

C. Sugar Fermentation

Sucrose was the sugar used in liquid ferment experiments. As in the straight dough and sponge-dough processes, as reported by other investigators (Lee et al, 1959), sucrose was rapidly inverted into glucose and fructose in liquid ferments, and the fermentation of the monosaccharides followed. From studies of water ferments containing sucrose, it was concluded that the fermentation rate of fructose was slower than that of glucose and was sensitive to the pH of the medium. In liquid flour ferments (Fig. 20), the glucose was also utilized at faster rates than fructose, regardless of the flour level in the ferment (20 and 50% flour). When fructose syrup was used instead of sucrose, similar results were obtained.

Maltose fermented more slowly than the monosaccharides. During the early stages of fermentation, more maltose was generated by the amylolysis of starch than was utilized by fermentation, but fermentation predominated over this reaction once the yeast enzymes became adapted to maltose fermentation.

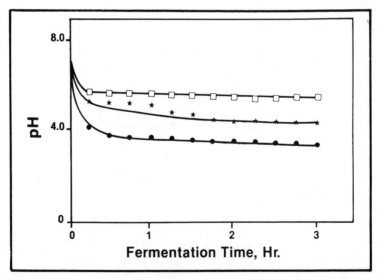

Fig. 18. Changes of pH during fermentation of water ferments. • = 0% buffer; * = 0.2% buffer (flour basis); □ = 0.5% buffer. (Reprinted, with permission, from Kulp et al, 1985)

Liquid flour ferments containing 50% flour generated more maltose than those containing 20% flour.

D. Ethanol Formation

Ethanol formation, determined by means of immobilized alcohol dehydrogenase, showed that ethanol yield was affected by the pH of liquid ferments, being slightly lower in unbuffered than in buffered brews. The flour ferments (20 and 40% flour) yielded similar levels of ethanol, indicating that the alcohol was derived mainly from added sucrose rather than from maltose generated by starch enzymolysis (Fig. 21) during fermentation.

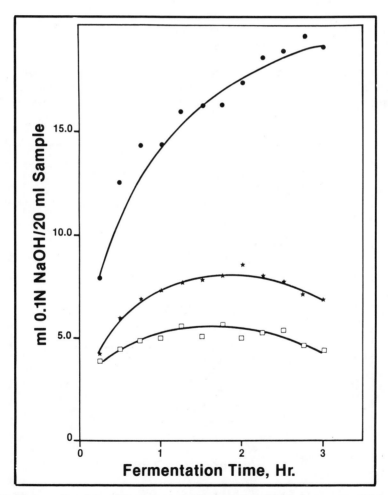

Fig. 19. Changes of total titratable acidity during fermentation of water ferments. ● = 0% buffer; * = 0.2% buffer (flour basis); □ = 0.5% buffer. (Reprinted, with permission, from Kulp et al, 1985)

E. Dough Properties

Properties of dough produced with liquid ferments were evaluated by means of farinograph and mixograph. Liquid ferments (after the completion of fermentation) were added with the balance of the flour to the farinograph, and the consistency curves were recorded (Fig. 22); the same procedure was used in the determination of mixograms. These data indicated that fully fermented water ferments did not alter the consistency and stability of doughs; i.e., optimally buffered water ferments (0.2%) had no appreciable effect on the dough absorption and improved the dough stability slightly. Flour in liquid ferments lost some of its water-absorptive capacity and produced softer doughs. These effects increased with the level of flour in the ferments, but in all cases, they were less pronounced than those observed in the sponge-dough system. Similar conclusions were reached on the basis of mixograms.

F. Quality of Breads Using Liquid Ferments with Varied Flour Contents

In judging bread quality, the industry uses several criteria. These include external and internal characteristics and the keeping quality of breads. The reference breads considered of optimal quality in the United States are those

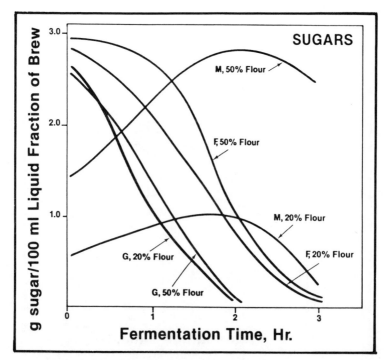

Fig. 20. Fermentation of liquid flour ferments. G = glucose; F = fructose; M = maltose. (Reprinted, with permission, from Kulp, 1986)

manufactured by a sponge-dough process in which generally 70% of the total flour is used in the sponge, with a 4-hr fermentation period. The quality of breads from liquid ferments is compared with that of sponge-dough breads in Table IV, and their residual sugar compositions are compared in Table V. The scores of breads from liquid ferments were similar to those produced by the sponge-dough process. Breads produced from flour ferments with less than 50% flour were given only slightly lower ratings than breads produced from the water brews. The increased flour level reduced proof times and slightly improved the specific loaf volume.

Adjustment of the pH of the water ferment mainly influenced the length of proof time; it had no appreciable effect on the specific loaf volume. Also, other bread quality characteristics were similar, including the pH and TTA values.

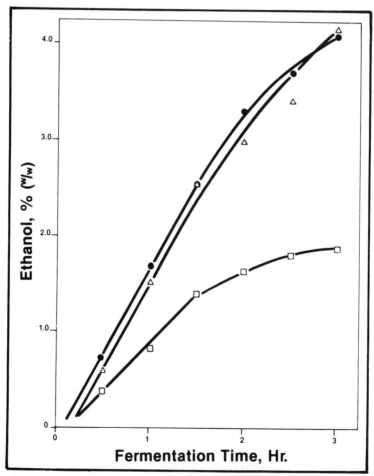

Fig. 21. Ethanol production in water and liquid flour ferments (ethanol in liquid portion). □ = 0% flour and 0.2% buffer; ● = 20% flour; Δ = 50% flour. (Reprinted, with permission, from Martinez-Anaya and Kulp, 1984)

Breads from the sponge-dough process were lower in maltose than breads from liquid ferments. In organoleptic testing with a 20-member panel, these variations were of no consequence (Kulp, 1986).

The differences between sponge-dough breads and those from flour liquid ferments were not appreciable except for the lower flavor intensity of the latter. This was especially noticeable in breads produced with water ferments.

G. Significance of Fermentation Indexes

The industry generally depends on pH and TTA values in estimating the

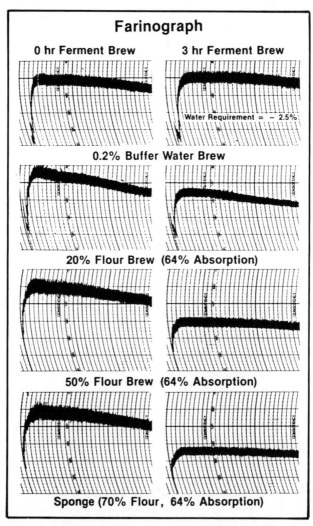

Fig. 22. Effects of fermentation and type of sponge or liquid ferment on dough properties determined by farinograph. (Reprinted, with permission, from Kulp, 1986)

progress of fermentation and adjusts these values where they deviate from the optima. The experimental results suggest that good-quality breads may be produced without meeting rigid pH and TTA requirements. This was illustrated by a series of breads prepared using water ferments with and without adjustments of acidity. In all cases, good-quality bread was obtained (Kulp, 1986).

The evaluation of dough properties also showed little effect of buffers on these properties. This indicates that buffered and unbuffered doughs are expected to be comparable in machinability. The pH and TTA values merely reflect the yeast activity and are an index of its quality without being as critical to the process as believed by the industry. It appears that increases in ethanol or decreases in sugars during fermentation would offer a more reliable index to follow the fermentation process of liquid ferments than monitoring pH and TTA.

H. Shelf Life (Softness of Breads Using Ferments with Varied Flour Contents)

The shelf life of breads produced from water and flour ferments was estimated by the determination of firmness with the Instron Universal Testing Machine and organoleptic evaluation of freshness (Kulp, 1986). Firmness values (Fig. 23) obtained one, four, and seven days after baking demonstrate that breads produced from 0% flour ferments were firmer throughout the entire test period than those produced from 40% flour ferments or by the sponge-dough process (70% flour). The 40% flour ferment bread and the sponge-dough bread did not differ in firming rates.

The perceived freshness of breads, estimated by organoleptic testing, led to the

TABLE IV
Characteristics of Breads with Liquid Flour Ferments and Sponge-Dough Processes[a]

Process	Proof Time (min)	Specific Volume (ml/g)	pH	TTA[b]	Total Score[c]
20% Flour brew	52.5	5.21	5.20	2.7	84.5
50% Flour brew	51.0	5.34	4.20	2.8	86.5
Sponge/dough	61.0	4.56	5.36	3.1	87.5

[a] Source: Kulp (1986); used by permission.
[b] Total titratable activity.
[c] Perfect score = 100.

TABLE V
Sugar Composition of Breads (at 38% Moisture Basis) from Liquid Flour Ferments[a]

Process	Sugar (%)			
	Sucrose	Glucose	Fructose	Maltose
20% Flour brew	ND[b]	1.17	1.87	1.14
50% Flour brew	ND	1.04	1.73	1.03
Sponge/dough	ND	1.22	1.74	0.77

[a] Source: Kulp (1986); used by permission.
[b] Nondetectable.

same conclusion. The water ferment breads were rated less fresh than the 40 and 70% flour breads throughout seven days of testing. The differences between the 40% flour ferment bread and the sponge-dough bread were not significant.

VI. CONTINUOUS PROCESSES

An important goal in bread baking, as indeed in most major manufacturing processes, is to achieve production methods as rapid and as free of human manipulation (i.e., automated) as feasible. Considerable efforts were expended, as Parker (1965) pointed out, in the years from 1927 to 1950, on the development of methods for continuous mixing. At the same time, work was conducted to develop mechanical feeding and mixing devices for the milling industry, the principles of which were found applicable to continuous breadmaking processes. The pioneering work of Swanson and Working (1926) showed that an input of intense mechanical energy in the dough can replace fermentation to a large extent. This concept was applied in the 1950s and became widely used commercially during the period from the late 1950s to the early 1970s in the United States. Basically similar processes are commonly used in Europe, Asia, and Africa. Continuous processes lost their position in the United States in the 1970s because of poor consumer acceptance of breads manufactured by these methods. The objections raised by the U.S. public were that the breads were of poor quality, especially in lacking flavor and resilience of the crumb. In some instances, continuous operations were converted for the production of other bakery products, e.g., hamburger rolls and specialty bread products.

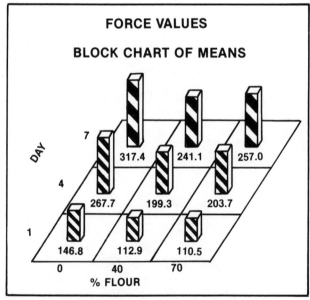

Fig. 23. Changes in firmness of breads during storage, determined by Instron. Force values are given in grams per 1.0179×10^{-3} m^2. (Reprinted, with permission, from Kulp, 1986)

A. Continuous Methods in the United States

Two continuous processes found wide applications in the United States, the Do-Maker procedure (Fig. 24) and the Amflow procedure (Fig. 25). Both

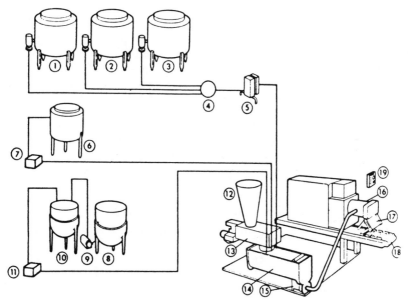

Fig. 24. Flow diagram for Do-Maker process. 1–3 = broth tanks, 4 = broth selector valve, 5 = broth heat exchanger, 6 = oxidation solution tank, 7 = oxidation solution feeder, 8 = shortening blending kettle, 9 = shortening transfer pump, 10 = shortening holding kettle, 11 = shortening feeder, 12 = flour hopper, 13 = flour feeder, 14 = premixer, 15 = dough pump, 16 = developer, 17 = divider, 18 = panner, 19 = control panel. (Reprinted, with permission, from Snyder, 1963)

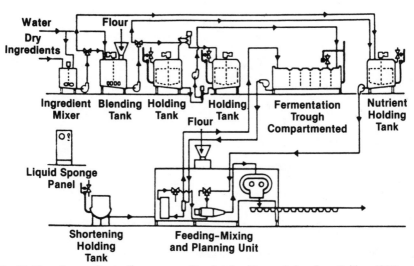

Fig. 25. Flow diagram for Amflow process. (Reprinted, with permission, from Seiling, 1969)

processes share several essential features. A ferment is first made (with or without flour, in the case of the Do-Maker); the ferment is blended with the remaining ingredients into a homogeneous mass, which is sent to a pump that regulates the dough flow. From the pump, the dough goes to the developer for intense mixing, is extruded into baking pans, and then is transferred to a final proofer.

The Amflow process was introduced by the American Machine and Foundry Company in 1954. It is based on the same principle of continuous mixing and mechanical dough development as the Do-Maker but features a horizontal instead of a vertical development chamber.

Continuous-mix formulas for both the Do-Maker and the Amflow methods are given in Tables VI and VII. They may be considered examples of typical formulas at the time when the continuous-mix process was in favor. The remaining users have undoubtedly altered their formulas in response to current trends, using fructose syrups, milk replacers, and oils in place of sucrose, nonfat dry milk, and shortening, respectively. The oxidation is provided by bromate, azodicarbonamide (ADA), and possibly ascorbic acid. When iodates

TABLE VI
Continuous Breadmaking Formula—Nonflour Brew[a]

Brew Ingredient	%[b]	Dough Composition	
Water	60.32	Flour, %	100
Sugar	7.61	Brew, %	74
Yeast	2.67	Shortening blend, %	3
Yeast food	0.50	Water, %	2
Salt	2.10	$KBrO_3$, ppm	60
Milk powder	0.33	KIO_3, ppm	12
Mold inhibitor[c]	0.10		

[a] Source: Redfern et al (1964); used by permission.
[b] Total flour basis.
[c] Calcium propionate.

TABLE VII
Continuous Breadmaking Formula—Flour Brew[a]

Ingredient	Formula (%)	Stage 1 (%)	Stage 2 (%)	Dough Stage (%)
Flour	100	30		70
Water	67	56	4.0	7.0
Yeast	3.0	3.0		
Yeast food	0.5	0.5		
Salt	2.0		2.0	
Sugar	6.0		1.0	5.0
Milk	3.0			3.0
Monocalcium phosphate	0.1	0.1		
Inhibitor[b]	0.1			0.1
Shortening	3.0			3.0
Oxidation	variable[c]			

[a] Adapted from Trum (1964).
[b] Calcium propionate.
[c] Total of 75 ppm ($KBrO_3$ and KIO_3 in a ratio of 3:1 to 8:1; KIO_3 generally replaced by azodicarbonamide).

are added, their level is kept at a minimum (up to 200 ppm). The flour level in the ferment varies from 0 to 50% of the total flour. From a recent survey, Dubois and Vetter (1987) estimated production variables (Table VIII) used by the few plants still applying the continuous process. The dough temperature is considerably higher than in the other processes, and there is no floor time, since the dough is deposited directly from the mixer into the bread pans. The baking time is shorter and the baking temperature generally higher than in the other production methods. Readers interested in details of this technology are referred to Ponte (1971) and Collins (1985).

In the R'Real Bread process, described by Freed (1963), undermixed but fully fermented straight dough is subjected to high-speed continuous development in a 65-ft-long segmented cylinder equipped with a screw dough conveyor in its first section and an agitator, provided with four blades set at right angles, in its developer section. From the developer chamber, the dough passes into a scaling head, is extruded to cutting knives for dividing, and is directly deposited in pans. Continuous flow is obtained by appropriate scheduling of the individual batch doughs so that fermented dough reaches the continuous dough developer in an uninterrupted flow.

The "S" Full Flavor process (Stickelber, 1968) combines conventional dough-handling equipment with a special machine consisting of two chambers with an agitator for dough preparation.

B. Systems Outside the United States

CHORLEYWOOD BREAD PROCESS

The most widely accepted continuous system in bakeries in more than 30 countries (including some European countries) is the Chorleywood bread process (CBP) (Chamberlain, 1984). Bakers have long been familiar with no-time doughs, and as early as 1927 a patent was filed for a process involving the overmixing of dough with fast-acting oxidizing agents to replace bulk fermentation (Elton, 1965). The process was introduced confidentially to

TABLE VIII
Typical Continuous-Mix Process for White Bread[a]

	Range
Brew	0–50% of total flour
Brew set temperature	76–92° F
Fermentation time	
0% Flour	2–2.5 hr
50% Flour	2–3 hr
Temperature rise	to 88–98° F
Temperature after cooling	46–58° F
Dough temperature	102–104° F
Proof time	44–60 min
Proof temperature	
Dry bulb	108–120° F
Wet bulb	104–112° F
Typical degrees spread	6–8° F
Bake time	14–18 min
Bake temperature	400–490° F

[a] Adapted from Dubois and Vetter (1987).

subscribing members of the Flour Milling and Research Association in 1961 and described publicly in 1962 (Chamberlain et al, 1962). The main features of the process are the following:

1. Work input during mixing is about 11 W·hr/kg (5 W·hr/lb) in a time of less than 5 min, preferably between 2 and 4 min. The quantity of mixing energy is about five to eight times that consumed by a conventional dough mixer for a bulk fermentation process.

2. The process requires higher levels of oxidants than bulk fermentation processes do. The original CBP required ascorbic acid at 75 ppm for U.K. flours treated at the mill with potassium bromate (about 20–30 ppm). In countries where potassium bromate is not permitted, ascorbic acid is the sole oxidant; in some countries, fast-acting oxidants such as ADA are used in complex mixtures.

3. The inclusion of fat or a surfactant or a blend of both is considered essential. The level of fat with a high slip point (50° C) is generally 0.7% (flour basis) or less. This type of fat, prepared exclusively for the CBP, may also contain the desired level of oxidants.

4. To maintain acceptable dough consistency, water is added to about 3.5% (flour basis) above the absorption used in the sponge-dough process (Chamberlain et al, 1962).

5. To maintain normal proof times, the process requires yeast in amounts 50–100% above the conventional level. This increase in the amount of yeast compensates for the failure of yeast to attain full activity in the CBP.

6. Both laboratory and commercial experience indicate that for a given flour, better bread can be made by the CBP than by conventional methods. The reason for this effect is not clear, but it has been suggested that possible "antifactors" (e.g., proteases, reactive thiol compounds) present in flour have less opportunity for deleterious action during the CBP than in conventional methods, because of the shorter reaction time in the CBP.

In evaluation of the CBP and the quality of the final bread, one should consider that the bread quality criteria applied by the inventors of the process were based on the properties of the standard English bread (Chamberlain, 1984) and may not be valid elsewhere. For instance, Kilborn and Tipples (1979) found that, in Canada, bread volumes of 1,000 ml per 100 g of dough from high-protein Canadian flour are common and desirable but almost double what is regarded as normal in the United Kingdom.

Other variations to be considered are oxidation and rheological properties arising from mechanical dough development. Much of the discussion has centered on the desirability of mixing the dough to peak consistency or minimum mobility, as represented by a peak in the graph of mixing torque versus time. It was concluded that peak torque is not directly related to (or correlated with) mixing in an English commercial bakery. The basis for optimum dough development was investigated by Frazier (1979). Kilborn and Tipples (1979) demonstrated that the optimum mixing for short-process breadmaking, e.g., the CBP, depends to a large extent on the level of chemical oxidants added. To produce satisfactory bread in the absence of added oxidants, doughs must be mixed considerably past peak consistency, to a stage normally considered well into the dough breakdown region. Under these conditions, a period of intermediate proof is essential for minimizing or eliminating "green" external loaf characteristics and for producing a thin-walled cell structure resembling the sponge-dough crumb cell pattern. The loaf volume of breads is smaller when formulation without oxidants is used. When amounts of oxidants are optimized, the greatest tolerance to the intermediate proof is obtained with doughs mixed to peak consistency. Work requirements for mixing appear to be inversely proportional to the amount of added oxidants. An insufficient intermediate proof cannot be adequately compensated for by extension of the final proof.

The CBP may be practiced as either a batch or a continuous process. A number of batch mixers were designed specifically for the CBP, the first being the Tweedy machine. The newest model developed by this company is the Tweedy 7700 high-speed dough mixer (Fig. 5), which produces 12 mixes per hour with an energy input of 11 W·hr/kg (5 W·hr/lb) when used to develop 3,280 kg of dough per hour. It is often operated along with other production steps from a central panel.

A renewed interest is being shown in applying the CBP to the production of American white pan bread and buns (French and Fish, 1981; Chaney and Wilson, 1983a, 1983b; Chamberlain, 1984). Early attempts in the mid-1960s failed, because fine crumb structure could not be achieved by the process at the high loaf volume customary in the United States. Inflating the loaf volume by higher work input (as high as 19 W·hr/kg), with a simultaneous increase of oxidants, and application of a partial vacuum was a partial solution. Modification of the mixing action of the Tweedy unit, with normal energy input, created more and smaller air bubbles and seemed to be a more satisfactory solution.

BRIMEC PROCESS

A mechanical dough development, the Brimec process, which resulted from independent studies (Marston, 1967) by the Bread Research Institute of Australia, is similar to the CBP.

BLANCHARD BATTER PROCESS

A somewhat different approach to short-time breadmaking is that of the Blanchard batter process (Blanchard, 1966). Gluten development is achieved through a two-stage mixing process, the first of which involves mixing a batter.

The process originally was a batch method but has also been used on a continuous basis (Anonymous, 1969). The first stage involves mixing 75% of the flour, most of the water, 2% yeast, and 0.7% soy flour; this batter is beaten to the stage at which complete hydration and development of gluten occur. The remainder of the flour and water, 1.8% salt, 1.7% fat, and ascorbic acid at 35 ppm are added to the batter, and the dough is mixed. The process attains physical development of gluten with a relatively lower work input than that of the CBP. Work input by the continuous batter method is 0.2 hp·lb^{-1}·min^{-1} Dough temperatures are also low, and oxidant requirements are about half that of other mechanical development processes.

VII. CHEMICAL DOUGH DEVELOPMENT

Another means of rapid conditioning of dough to the stage at which it is mature involves chemical development. Reducing agents and oxidants are employed for this purpose. L-Cysteine is the specific reducing compound, and the oxidants are potassium bromate, ascorbic acid, and ADA. These agents do not react simultaneously. L-Cysteine reduces the mixing time and reacts with wheat proteins. This reaction is stopped by the rapid oxidants (ascorbic acid and ADA) and later by the slow oxidant (bromate). The process is used in Canada and the United Kingdom; it is less common in the United States, except for buns, French bread, and other specialty breads (e.g., Henika and Zenner, 1960; Tipples, 1967; Tsen, 1969; Smerak, 1973).

VIII. MISCELLANEOUS METHODS

Finally, several other breadmaking systems are reported in the literature. Marston and Bond (Marston, 1966; Marston and Bond, 1966; Marston, 1967) described a no-time dough system specifically designed for the slow-speed mixers used in many smaller bakeries in Australia. The method uses bromate at 30 ppm and ascorbic acid at 100 ppm, extra mixing, and slightly higher dough temperatures, and fat is included in the formula. Bread produced by this method is reportedly of better quality than that produced by traditional no-time systems but not quite as good as bread made by conventional or mechanical dough development. Elias and Wragg (1963) and Russell Eggitt and Coppock (1965), in England, utilized the Strahmann continuous mixer (developed in Germany) in systems including rounding and molding to obtain conventional bread grain characteristics. The use of carbon dioxide to provide aeration is reported in two patented systems: the Oakes system (Baker, 1960) and the Suntheimer method (Suntheimer, 1962). Stenvert et al (1979) described bread production by dough rollers.

LITERATURE CITED

ANDERSON, R. C. 1966. Oven temperature and humidity control. Bakers Dig. 40(6):60-63.

ANONYMOUS. 1969. A new development—Blanchard continuous batter process. Br. Baker 159(10):90-92.

ANONYMOUS. 1977. Bakery products. Census of Manufactures, SIC 2051 and 2052. Bur. Census, U.S. Dep. Commer., Washington, DC.

ANONYMOUS. 1985a. Bakery products. Census of Manufactures, SIC 2501 and 2052. Bur. Census, U.S. Dep. Commer., Washington, DC.

ANONYMOUS. 1985b. The Master Bakers' Book of Breadmaking, 2nd ed. Natl. Assoc. Master Bakers, Confectioners, and Caterers, Ware, Herts., England.

ANONYMOUS. 1986. Trends '86. Bakery Prod. Mark. 21(5):80-84, 86-87, 90, 92, 94-96, 98-99.

AUDIDIER, Y. 1968. Effects of thermal kinetics and weight loss kinetics on biochemical reactions in dough. Bakers Dig. 42(5):36-38, 40, 42.

BAKER, J. C. 1960. U.S. patent 2,953,460. Sept. 20.

BAKER, J. C., and MIZE, M. D. 1941. The origin of the gas cell in bread dough. Cereal Chem. 18:19-34.

BANKS, W., and GREENWOOD, C. T. 1975. Starch and Its Components. Halsted Press, New York.

BLANCHARD, G. 1966. Blanchard batter process for bread. (Abstr.) Milling 147:519.

BLANSHARD, J. M. V., FRAZIER, P. J., and GALLIARD, T., eds. 1986. Chemistry and Physics of Baking. R. Soc. Chem., London.

BUSHUK, W., TSEN, C. C., and HLYNKA, I. 1968. The function of mixing in breadmaking. Bakers Dig. 42(4):36-38, 40.

CHAMBERLAIN, N. 1984. The Chorleywood bread process: International prospects. Cereal Foods World 29:656-658.

CHAMBERLAIN, N., COLLINS, T. H., and ELTON, G. A. H. 1962. The Chorleywood bread process. Bakers Dig. 36(5):52-53.

CHANEY, D., and WILSON, D. 1983a. Fuchs battle for tight south Florida market. Bakery

Prod. Mark. 18(3):98-102, 104, 106-107.

CHANEY, D., and WILSON, D. 1983b. High tech powers bun baker's drive. Bakery Prod. Mark. 18(1):72-75, 78, 80, 82.

COLLINS, B. 1985. Breadmaking processes. Pages 1-47 in: The Master Bakers' Book of Breadmaking, 2nd ed. Natl. Assoc. Master Bakers, Confectioners, and Caterers, Ware, Herts., England.

DUBOIS, D. K. 1984. Processing and ingredient trends in U.S. breadmaking. Page B-1-14 in: Int. Symp. Adv. Bakery Sci. Technol. Dep. of Grain Science, Kansas State Univ., Manhattan.

DUBOIS, D. K., and VETTER, J. L. 1987. White, whole wheat, wheat, and multigrain breads—A survey of formulas and processes. Am. Inst. Baking, Res. Dep. Tech. Bull. 9(2).

ELIAS, D. G., and WRAGG, B. H. 1963. New British techniques. Cereal Sci. Today 8:271-272, 295-296.

ELTON, G. A. H. 1965. Mechanical dough development. Bakers Dig. 39(4):38, 43-36, 78.

FRAZIER, P. 1979. A basis for optimum dough development. Baking Ind. J. 12(2):20-21, 25, 27.

FREED, R. 1963. A new system of rapid dough development. Bakers Dig. 37(3):55-59.

FRENCH, F. F., and FISH, A. R. 1981. High speed mechanical dough development. Bakers Dig. 55(5):80-82.

GREENWOOD, C. T. 1976. Starch. Pages 119-157 in: Advances in Cereal Science and Technology, Vol. 1. Y. Pomeranz, ed. Am. Assoc. Cereal Chem., St. Paul, MN.

HENIKA, R. G., and ZENNER, S. F. 1960. Baking with the new instant dough develop-ment process. Bakers Dig. 34(3):36-37, 40-42.

HLYNKA, I. 1970. Rheological properties of dough and their significance in the bread-making process. Bakers Dig. 44(2):40-41, 44-46, 57.

KENT, N. L. 1983. Technology of Cereals. Pergamon Press, Oxford.

KILBORN, R. H., and TIPPLES, K. H. 1979. The effect of oxidation and intermediate proof on work requirements for optimum short-process bread. Cereal Chem. 56:407-412.

KOCH, A. 1983. Oven energy efficiency. Pages 90-97 in: Proc. ASBE Annu. Meet., 59th, Chicago. Am. Soc. Bakery Eng., Chicago.

KULP, K. 1983. Technology of brew systems in bread production. Bakers Dig. 57(6):20, 22-23.

KULP, K. 1986. Influence of liquid ferments on quality characteristics of white pan bread. Res. Dep. Tech. Bull. 8(9). Am. Inst. Baking, Manhattan, KS. 6 pp.

KULP, K., and DUBOIS, D. K. 1982. Breads and sweet goods in the United States. Am.

Inst. Baking, Res. Dep. Tech. Bull. 4(6).

KULP, K., and LORENZ, K. 1981. Starch functionality in white pan bread—New developments. Bakers Dig. 55(5):24-28, 36.

KULP, K., PONTE, J. G., Jr., and JONES, G. Y. 1983a. Trends in using microprocessors in baking technology. Pages 979-984 in: Progress in Cereal Chemistry and Technology. J. Holas and J. Kratochvil, eds. Elsevier, Amsterdam.

KULP, K., ROEWE-SMITH, P., and LORENZ, K. 1983b. Preharvest sprouting of winter wheat. I. Rheological properties of flours and physicochemical characteristics of starches. Cereal Chem. 60:355-359.

KULP, K., CHUNG, H., MARTINEZ-ANAYA, M. A., and DOERRY, W. 1985. Fermentation of water ferments and bread quality. Cereal Chem. 62:55-59.

LEE, J. W., and GEDDES, W. F. 1959. Studies on the brew process of bread manufacture: The effect of sugar and other nutrients on baking quality and yeast properties. Cereal Chem. 36:1-18.

LEE, J. W., CUENDET, L. S., and GEDDES, W. F. 1959. The fate of various sugars in fermenting sponges and doughs. Cereal Chem. 36:522-533.

LORENZ, K., ROEWE-SMITH, P., KULP, K., and BATES, L. 1983. Preharvest sprouting of winter wheat. II. Amino acid composition and functionality of flour and flour fractions. Cereal Chem. 60:360-366.

MacRITCHIE, F. 1986. Physicochemical processes in mixing. Pages 132-146 in: Chemistry and Physics of Baking. J. M. V. Blanshard, P. J. Frazier, and T. Galliard, eds. R. Soc. Chem., London.

MARSTON, P. E. 1966. Fresh look at no-time doughs for bread production with normal bakery equipment. Cereal Sci. Today 11:530-532, 542.

MARSTON, P. E. 1967. The use of ascorbic acid in bread production. Bakers Dig. 41(6):30-33, 70.

MARSTON, P. E., and BOND, E. E. 1966. Rapid dough development for no-time breadmaking processes. Milling 147:67.

MARTINEZ-ANAYA, M. A., and KULP, K. 1984. Fermentation of liquid ferments and bread quality. Page D-1-12 in: Int. Symp. Adv. Baking Sci. Technol. Dep. of Grain Science, Kansas State Univ., Manhattan.

MATZ, S. A. 1972a. Bakery Technology and Engineering, 2nd ed. Avi Publ. Co., Inc., Westport, CT.

MATZ, S. A. 1972b. Ovens and associated equipment. Pages 410-437 in: Bakery Technology and Engineering, 2nd ed. Avi Publ. Co., Inc., Westport, CT.

McLAREN, L. H. 1954. Practical aspects of the stable ferment baking process. Bakers Dig. 28(3):23-24, 30.

NIAMTU, J., Jr., LABRIOLA, T., and GORTON, L. A. 1981. Mixer automatically loads and empties to feed 4-high volume pretzel lines. Baking Ind. 148(1811):13-15.

PARKER, H. K. 1965. Continuous mixing and baking to date. Cereal Sci. Today 10:272, 274-276.

PONTE, J. G., Jr. 1971. Bread. Pages 675-742 in: Wheat: Chemistry and Technology, 2nd ed. Y. Pomeranz, ed. Am. Assoc. Cereal Chem., St. Paul.

PONTE, J. G., Jr., and REED, G. 1982. Bakery foods. Pages 246-292 in: Prescott and Dunn's Industrial Microbiology. Avi Publ. Co., Inc., Westport, CT.

PONTE, J. G., Jr., TITCOMB, S. T., and COTTON, R. H. 1963. Some effects of oven temperature and malted barley level on breadmaking. Bakers Dig. 37(3):44-48.

PYLER, E. J. 1973. Baking Science and Technology, Vols. 1 and 2. Siebel Publ. Co., Chicago.

REDFERN, S., BRACHFELD, B. A., and MASELLI, J. A. 1964. Laboratory studies of processing temperatures in continuous breadmaking. Cereal Sci. Today 9:190-191.

REED, G., and PEPPLER, H. J. 1973. Yeast Technology. Avi Publ. Co., Inc., Westport, CT.

RUBENTHALER, G., POMERANZ, Y., and FINNEY, K. F. 1963. Effects of sugars and certain free amino acids on bread characteristics. Cereal Chem. 40:658-665.

RUSSELL EGGITT, P. W., and COPPOCK, J. B. M. 1965. An approach to continuous mixing. Cereal Sci. Today. 10:406-408, 410, 474-475.

SEILING, S. 1969. Equipment demands of changing production requirements. Bakers Dig. 43(5):54-56, 58-59.

SMERAK, L. 1973. Effective commercial no-time dough processing for bread and rolls. Bakers Dig. 47(4):12-15, 18, 20.

SNYDER, E. 1963. Continuous baking process: Its success and future. Bakers Dig. 37(4):50-52, 54-60.

STENVERT, N. L., MOSS, R., and BOND, E. E. 1979. Bread production by dough rollers. Bakers Dig. 53(2):22-27.

STERNBERG, G. 1968. A new concept in conventional dough mixing. Bakers Dig. 42(1):60, 66.

STICKELBER, D. 1968. A new approach to continuous dough processing. Bakers Dig. 42(2):30-32, 37.

SUNTHEIMER, F. J. 1962. U.S. patent 3,015,565. Jan. 2.

SWANSON, C. O., and WORKING, E. B. 1926. Mechanical modification of dough to make it possible to bake bread with only the fermentation in the pan. Cereal Chem. 3:65-83.

THOMPSON, D. R. 1983. Liquid sponge technology applied to high-speed dough mixing. Bakers Dig. 57(6):11-12, 14, 16-17.

TIPPLES, K. H. 1967. Recent advances in baking technology. Bakers Dig. 41(3):18-20, 24, 26-27.

TRUM, G. W. 1964. The AMF pilot plant in continuous bread experimentation. Cereal Sci. Today 9:248, 250, 252, 254.

TSEN, C. C. 1969. Effects of oxidizing and reducing agents on changes of flour proteins during dough mixing. Cereal Chem. 46:435-442.

VARILEK, P., and WALKER, C. E. 1983a. Baking and ovens; history of heat technology. Bakers Dig. 57(5):52-54, 56-57, 59.

VARILEK, P., and WALKER, C. E. 1983b. Baking and ovens; history of heat technology. II. Bakers Dig. 57(6):24-27.

VARILEK, P., and WALKER, C. E. 1984a. Baking and ovens; history of heat technology. III. Bakers Dig. 58(1):24-26, 29.

VARILEK, P., and WALKER, C. E. 1984b. Baking and ovens; history of heat technology. IV. Bakers Dig. 58(2):12-15

WALDEN, C. C. 1955. The action of wheat amylases on starch under conditions of time and temperature as they exist during baking. Cereal Chem. 32:421-431.

WELLS, R. A. 1983. Proofing and baking systems. Pages 119-125 in: Proc. ASBE Annu. Meet., 59th, Chicago. Am. Soc. Bakery Eng., Chicago.

CHAPTER 7

SOFT WHEAT PRODUCTS

R. CARL HOSENEY
Department of Grain Science and Industry
Kansas State University
Manhattan, Kansas

PETER WADE *(retired)*
United Biscuits (U.K.) Ltd.
Stoke Poges, Bucks., England

JOHN W. FINLEY
Nabisco Brands, Inc.
East Hanover, New Jersey

I. INTRODUCTION

A. Types of Wheat

The terms *hard* and *soft* as applied to wheats are descriptions of the texture of the kernel. A hard wheat kernel requires greater force to cause it to disintegrate than does a soft wheat kernel. The flour obtained from a hard wheat kernel has a coarser particle size than does flour from a soft wheat kernel. *Hard* and *soft* are used with slightly different connotations in different parts of the world. In Western Europe, *hard* is applied primarily to durum wheats, of the species *Triticum turgidum* var. *durum*, the term *soft* being applied to all wheats of the species *T. aestivum*. Elsewhere the terms are applied to different cultivars within the species *T. aestivum*. Wheats of the species *T. durum* are used primarily in the manufacture of pasta-type products. Flour from hard wheats of the species *T. aestivum*, if of suitable protein content (11–13%, N × 5.7), is primarily used for breadmaking. Low-protein flour (7–9% protein) milled from soft cultivars of *T. aestivum* is most suitable for making cakes and biscuits. It is in this sense that the term *soft wheat* is used in this chapter.

The hardness of a given cultivar of wheat is genetically controlled and is not directly correlated with the protein content of the kernel (Simmonds, 1974; Yamazaki and Donelson, 1983; Miller et al, 1984). This fact is of particular importance in the United Kingdom, as many of the new varieties of wheat introduced during the last 20 years have hard-texture kernels and, although they have relatively low protein content, yield flours unsuitable for the manufacture of soft wheat products. This caused problems for the manufacturer because no reliable tests other than a baking test existed for identifying flours milled from mixtures of hard and soft wheats (Wade, 1971a).

B. Products Made from Soft Wheat Flours

Many products of many different types are made from soft wheat, and it would be difficult to compile a comprehensive list (Yamazaki and Lord, 1971; Yamazaki and Greenwood, 1981). The most important groups of products include biscuits, cookies, crackers, wafers, and pretzels; cakes of all types and sizes, from sponges and high-ratio cakes to heavily fruited cakes and from cupcakes to wedding cakes, including the wide range of prepared cake mixes; pastry products, from piecrusts to sweet Danish pastries; waffles, pancakes, and doughnuts; and oriental noodles. Soft wheat flour is also used as a thickener for soups and soup mixes, in the manufacture of crumb for coating fish and meat products, and as the basis for some breakfast cereals. All of these products have a better appearance and better eating quality when made from soft rather than hard wheat flour. Another major use of soft wheat flour is in the production of flat breads. In North Africa, the Middle East, and India, these breads are a major source of calories and protein. Soft wheat is equal to or superior to hard wheat and generally preferred for flat breads.

C. Product Terminology

Some confusion in product terminology exists between the biscuit industry in the United States and that in the United Kingdom.

In the United States, the term *crackers* is applied to products made from a viscoelastic dough in which the flour protein has been converted to gluten. These doughs are formed into sheets by rolling, and the crackers are cut from these sheets, hence the alternative description *cutting-machine products*. In the United Kingdom, such doughs are described generically as *hard doughs* and those containing sugar as *hard sweet doughs*. In the United States, *crackers* are made from flour and fat with amounts of sugar varying from 0 to about 40% of the flour weight; they include at one extreme saltines and at the other graham crackers sweetened with honey. In the United Kingdom, the term is traditionally reserved for products made from flour and fat with little, if any, added sugar. They include cream crackers (made from fermented dough) and water biscuits. Another group of products, known as semisweet biscuits, have a sugar content of about 20–30% of the flour weight. Although the formula for flour, fat, and sugar in semisweet biscuits is not dissimilar to that of graham crackers, in appearance and eating qualities these British products—rich tea, morning coffee, thin arrowroot, etc.—bear no resemblance to American crackers.

In recent years the situation in the United Kingdom has begun to change, as new cracker products containing sugar are being introduced into the market. It is, however, too early to be certain if they will become an established part of the product range.

Cookies (rotary, wire-cut, and bar press) are made from doughs containing high levels of fat or sugar (or both) in which a gluten network has not developed. These doughs are not cohesive under tension and are referred to in the United Kingdom as *short doughs* or *short sweet doughs*. The products described as *rotary cookies* in the United States are known as *sweet biscuits* in the United Kingdom. Sweet biscuits are frequently used as shells for cream sandwich biscuits (e.g., custard creams), and many are intended to be eaten without any secondary processing (e.g., digestive, Lincoln, and shortcake). Wire-cut and bar press cookies are made and sold in the United Kingdom but do not at present form a large proportion of the market. The high-sugar products, some flavored with ginger, are known as *snaps* on both sides of the Atlantic.

D. Types of Soft Wheats

Soft wheats may be classified according to various characteristics. One common method of classification is related to the time of sowing: spring wheats are sown in the spring, and winter wheats are sown in the late summer or fall. The majority of soft wheats now grown are winter wheats. Wheats may also be classified by the color of their seed coat, conventionally known as red or white. In the United States, wheats of both colors are grown and produce flours of similar properties. Only red wheats are now grown in the United Kingdom. This is because of the link between the white seed coat and the tendency to premature sprouting of the grain if it is harvested under damp or wet conditions—a perpetual hazard in the United Kingdom. Any tendency to premature sprouting leads to the presence of elevated levels of α-amylase in the resultant flour. High levels of amylase most seriously affect products made from fermented doughs, or doughs given a prolonged standing time between mixing and baking. It has been claimed that with an excessively damp harvest and consequent high levels of amylase activity in the flour, biscuits may color excessively during baking (J. B. S. Wilson, personal communication; Lorenz and Valvano, 1981).

II. SOFT AND HARD WHEATS

As discussed briefly above, the cultivars of common wheats (*T. aestivum*) separate into two distinct classes of hardness. These are the hard and soft wheats of commerce in the United States, United Kingdom, and Australia. The hardness of wheat is under genetic control. Symes (1965, 1969) showed that hardness appeared to be controlled by one gene and was simply inherited. To demonstrate this, he prepared a set of nearly isogenic lines of hard and soft wheats.

An excellent discussion of the mechanics of fracture and its application to wheat and milling was given by MacRitchie (1980). The mechanics of fracture are relatively well known, but the biochemical basis for the differences in hardness is not completely clear.

A. Theories of Grain Hardness

Barlow et al (1973a) used a micropenetrometer to test the hardness of starch and matrix protein from a number of hard and soft cultivars. They reported that the matrix protein was about 15% harder than the starch and that no significant difference existed in the hardness of either the protein or the starch from different varieties. On the basis of those results, the authors suggested that the difference between hard and soft wheats was in the strength of the bond between the protein and the starch.

Further evidence that hardness is related to the adhesion of starch and protein is the fact that starch samples prepared by flotation in a nonaqueous solvent were quite different depending on whether the starting material was hard or soft flour (Simmonds, 1971). Starches from hard wheat flour have a large quantity of protein adhering to them; soft wheat starch is relatively free of the adhering protein. Simmonds et al (1973) suggested that a water-soluble material was responsible for the adhesion. They also reported no differences in the electrophoretic patterns of gliadin proteins from near-isogenic lines of hard and soft wheats. Hard wheat starches prepared by the nonaqueous procedure gave much larger amounts of soluble material than did similar starches from soft wheats. The soluble material was found to contain carbohydrate and protein in a ratio of 2:1. Barlow et al (1973a, 1973b), using a fluorescent antibody technique, showed that the water-soluble protein is concentrated at the starch-protein interface. Recent reports (Greenwell and Schofield, 1986) implicated a low-molecular-weight protein associated with soft wheat starch but present at very low levels or absent from hard wheat starch (Fig. 1).

Similar conclusions concerning the strength of the starch-protein interface were reached by Hoseney and Seib (1973). Using scanning electron microscopy (SEM), they showed that the wetting of protein on starch was much better with

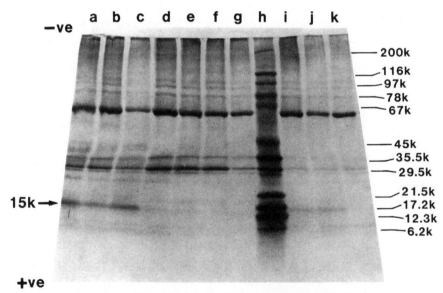

Fig. 1. Fractionation of proteins extracted at 50°C with 1% sodium dodecyl sulfate from starches of soft wheat (tracks a–c, i, and j) and hard wheat (tracks d–g, and k) by sodium dodecyl sulfate polyacrylamide gel electrophoresis, in a gel containing a linear gradient of acrylamide from 7.5 to 25%. Molecular weight standards are in track h. The arrow at the left-hand side of the gel indicates the mol wt 15,000 protein present in soft wheat starches. (Reprinted, with permission, from Greenwell and Schofield, 1986)

hard wheat than with soft wheat. The better wetting gives rise to a stronger bond, as evidenced by the number of starch granules that are broken when the kernel is fractured. Soft wheats have essentially no broken granules, hard wheats a moderate level, and durum wheats a high level. This was interpreted as evidence of increasing strength of the protein-starch bond.

An alternative theory was suggested by Stenvert and Kingswood (1977). They suggested that hard wheat has a continuous protein matrix that physically traps starch, and this filled matrix results in the hardness. On the other hand, a discontinuous structure with many air spaces or an unfilled matrix results in a weak matrix and, thereby, a soft wheat. The theory is attractive at first glance, but it does appear to have certain deficiencies. For example, the degree of discontinuity appears to be affected mainly by environment, although hardness is mainly affected by genetics. In addition, many relatively high-protein soft wheats are vitreous but still soft, and, conversely, many relatively low-protein hard wheats are opaque (contain many air spaces) but still hard.

B. Significance of Hard and Soft Wheats

As might be expected, hard and soft wheats mill differently. Hard wheats tend

Fig. 2. Scanning electron micrograph of a cross section of a hard winter wheat, showing breakage at the cell walls. Bar is 100 μm.

to fracture at cell walls, a point of weakness in the kernel (Fig. 2). Soft wheats fracture through the cell with no apparent pattern (Fig. 3). They are reduced to a finer particle size than are hard wheats. The fine particle size and the rough, irregular surface of the particles make soft wheat flour agglomerate easily and, thus, sieve poorly (Neel and Hoseney, 1984a, 1984b).

Reducing hard wheat to flour requires more work than reducing soft wheat. Therefore, more of the hard wheat starch is damaged as a result. Damaged starch is generally viewed as a negative characteristic for soft wheat flours.

C. Methods of Measuring Hardness

The absolute hardness of wheat is difficult to measure. The grain has a nonuniform shape, and thus it is difficult to know the geometry of what is being broken if the grain is stressed until it fractures. Therefore, how the grain breaks is usually measured rather than the absolute hardness.

It has been known for many years (Biffin, 1908) that soft wheat breaks into a fine powder and hard wheat breaks into angular fragments that produce a gritty flour. Given the above information, Cutler and Brinson (1935) developed the particle size index to measure wheat hardness. Wheat is ground with a standard grinder and then sifted with a standard set of sieves. Hard wheat gives a coarser

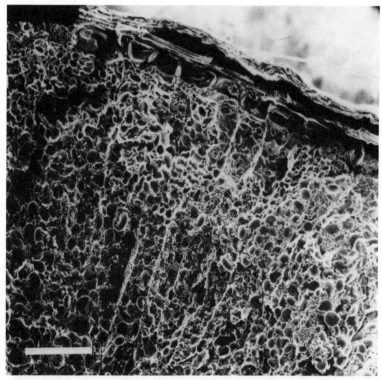

Fig. 3. Scanning electron micrograph of a cross section of a soft winter wheat, showing breakage through the cells. Bar is 100 μm.

product than soft wheat. The various methods used to measure wheat hardness were reviewed by Kosmolak (1978) and by Simmonds (1974).

Yamazaki and Donelson (1972, 1973) and Yamazaki (1972) described the procedure used to determine the particle size index at the Soft Wheat Quality Laboratory of the U.S. Department of Agriculture. Wheat is ground on a burr mill and then sifted. They cautioned that the correct grinder must be used and suggested that a hard and a soft wheat and a wheat of intermediate hardness be used to calibrate the system. Once the procedure is in place, it is relatively fast and reproducible.

Another characteristic used to differentiate between hard and soft wheats is grinding time (Kosmolak, 1978). Hard wheats grind much faster than do soft wheats (Miller et al, 1981).

A somewhat similar test is the pearling index (McCluggage, 1943). In this case, the whole grain is abraded from the outside rather than being fractured to small pieces. Because it is ground from the outside, the bran layers have more of an influence on the pearling index. In general, soft wheats are abraded more during the same time of pearling than are hard wheats.

Near infrared reflectance has also been suggested as a means of measuring the hardness of wheat (Bruinsma and Rubenthaler, 1978; Williams, 1979). It is generally believed that the measurement corresponds to particle size and, thus, that the method provides another means of measuring how the wheat grinds.

In studying the factors affecting hardness, Yamazaki and Donelson (1983) showed that protein content is not related to hardness. However, they reported that grain moisture does affect the measurement of kernel hardness. Grosh and Milner (1959), using a penetrometer measurement, showed that wheat hardness decreased as moisture was increased. On the other hand, McCluggage (1943) and Chesterfield (1971), using a barley pearler, reported that moisture had little influence on the results, and Kramer and Albrecht (1948) reported increased hardness with increased moisture.

Those apparently conflicting results can be explained if one considers that moisture is known to toughen the bran and, thus, results in an apparent increase in hardness as measured by pearling. Moisture is also known to soften the endosperm and thereby decreases hardness as measured with a penetrometer. It might also be expected that the time during which the water was in contact with the wheat would be important. Yamazaki and Donelson (1983) also showed that different wheats responded differently to the application of water (Fig. 4).

Although a number of procedures have been shown to do a reasonable job of determining the hardness of pure types of wheat, the problem is much harder with mixed types. Pomeranz et al (1985) reported measurements of the hardness of hard and soft red winter wheats in bulk samples. Lai et al (1985) reported similar studies in which the hardness of individual kernels was determined.

III. SOFT WHEAT FLOUR

Soft wheat products are often less dense than typical products containing hard wheat; however, soft wheat products have a more uniform internal structure, softer bite, greater tenderness, and more desirable height or spread characteristics. The physicochemical explanation for these differences is not clear. They appear to be related to the relatively low water absorption, fine granulation, and low protein content of soft wheat flour. Recently, there has been a growing concern about the increasingly hard characteristics of genetically soft wheat varieties. The newer cultivars of soft wheats in the United States generally are harder-milling and tend to produce harder products. In the United Kingdom, hard wheats have replaced soft wheat because of their agronomic advantages (Wainwright et al, 1985). Wade (1985) proposed that cookie manufacturers may wish to set specifications for the amount of damaged starch and the particle size that are acceptable for flour from these harder wheats.

A. Milling

To accommodate the wide variety of applications of soft wheat flour in the food industry, a wide variety of flours are produced through careful recombination of mill streams. Soft wheat is also frequently blended with hard wheat

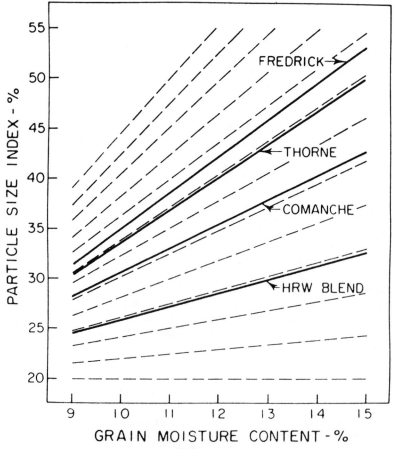

Fig. 4. Idealized fan-shaped family of regression lines of particle size index versus moisture content, and the regression lines for four wheats. (Reprinted, with permission, from Yamazaki and Donelson, 1983)

before milling, or, alternatively, the flours are blended after milling to yield flours of specific characteristics. Typically, all-purpose flour for home use in the United States varies from strong blends (containing relatively high proportions of hard wheat), in the Midwest, to blends containing more soft wheat, in the Southeast. Soft wheat is also frequently fine-ground and air-classified to produce high-protein or high-starch flours for special purposes or for blending.

Seeborg (1953) established a milling score to describe the milling quality of a wheat cultivar with a single number. Data on yield, ash, milling time, long patent extraction, and temper moisture were combined and weighted to obtain the flour quality score. Elder et al (1953) demonstrated that those scores correlated highly with the content of pentosans extractable with dilute acid.

B. Quality Tests

The applications of soft wheat are very different from those of hard and durum wheats; therefore, the testing criteria are also significantly different. Attempts have been made to understand the physical chemistry of the interactions of soft wheat flour with water, lipids, sweeteners, and other ingredients in soft flour products. Quality testing of soft wheat flours has been performed by baking tests for sugar cookies and cakes. Since most rheological testing of these flours has been based on techniques developed to measure bread flours, the results have been of only limited value in predicting cookie- or cake-making performance. Typical quality tests for soft wheat flours include tests for moisture, protein, ash; MacMichael viscosity and alkaline-water-retention capacity (AWRC) tests; mixograph and alveograph tests; and baking tests for sugar cookies and cakes. Cake-baking tests involve milling the flour to produce a 50% patent flour, which is then impact-milled, chlorinated, and baked into a cake. Scores are based on crumb quality and cake volume.

The utilization of soft wheat for cakes, cookies, crackers, and various snack foods has stimulated interest in soft wheat flour quality and a growing emphasis on determining the wheat and flour factors responsible for the quality of soft wheat. Although a number of chemical and rheological methods are available for assessments of flour quality, baking tests (whether performed in the laboratory or pilot plant or production environment) remain the ultimate quality test. The evaluation of soft wheat quality is complicated by the many different products made and by the large number of flour blends available. The American Association of Cereal Chemists (AACC) currently recognizes two baking tests for the evaluation of soft wheat quality: one for sugar-snap cookies (Method 10-50D) and one for high-ratio white cakes (Method 10-90) (AACC, 1983).

C. Wheat Color and Class

In addition to results of certain tests, flour specifications may call for flour to be milled exclusively from soft red winter or soft white winter wheat. This specification is based on real or perceived differences between the classes of wheat. Differences between soft white and soft red wheats tend to be seen more in baking tests than in tests on the flour. Both white and red soft wheats, along with club wheats, are soft when compared by kernel hardness testing or milling characteristics. As mentioned earlier, a trend toward increased hardness in soft white wheats has been noted in recent years. The new soft white winter wheats, for example, tend to produce cookies and crackers that are slightly flinty, as one would expect from a hard wheat. Certain producers prefer white wheat flours for certain types of cookies and piecrusts. Soft red wheat flour is preferred for the manufacture of crackers, because it makes them slightly more tender and gives them a better color.

Soft red wheat flour is preferred for soup and gravy thickeners, because white wheat is known to be more susceptible to sprout damage. The resistance to sprouting appears to be only genetically linked to the color of the wheat. Several cultivars of white wheat are known to be resistant to sprouting (McCrate et al, 1981; Upadhyay et al, 1984). When thickening is the critical functional characteristic, however, most manufacturers play it safe and use red wheat, because the level of α-amylase is high in sprouted wheat, and this affects starch-gelling characteristics. In the United Kingdom, flours intended for use as soup thickeners, even those from red wheats, are usually heat-treated to ensure complete inactivation of any cereal amylase.

D. Cookie Flour Quality

Depending on its product application, various components of a flour are critical to its overall functional characteristics. For example, Yamazaki (1954) found a strong correlation between the hydration characteristics of flour and cookie-baking quality. On the basis of that finding, he developed the AWRC test (Yamazaki, 1953). Results of this test generally correlate well ($r = -0.70$ to $r = -0.85$) with the spread of cookies made with straight-grade untreated flours. A purified starch tailings fraction isolated from soft wheat flour was shown to have a great capacity for holding water (Yamazaki, 1955). Small amounts of the starch tailings, added to flour, significantly decreased cookie spread in the standard test. This further illustrates the importance of water-holding capacity to soft wheat flour quality. Sollars (1956) confirmed the observation and demonstrated that varietal differences in cookie-baking quality could be accounted for by differences in the starch tailings fraction. Later Sollars (1959) reported that a water-soluble fraction from wheat also influenced cookie-baking quality, but to a lesser extent than the starch tailings fraction. The starch tailings were further subdivided into three fractions, all of which were rich in pentosans (Upton and Hester, 1966). Gilles (1960) also demonstrated that water-insoluble pentosans reduced the spread of cookies. Cookie flours are generally low in protein (7–8%), and protein quality does not appear to be strongly correlated with cookie-baking quality. In the preparation of dough for sugar cookies, such as those made according to the AACC cookie test procedure, the gluten does not develop to any appreciable extent, because of the competition for water among sugars, salts, and pentosans. The proteins appear to function during baking and help form the basis for the cookie structure. This subject appears to need further study.

Yamazaki (1955) demonstrated the importance of particle size in cookie flour quality. Flours sifted through a 325-mesh screen had lower protein and ash contents and produced cookies with greater spread than did coarser fractions. It is important to note that this procedure minimized starch damage. Brenneis

(1965a) progressively milled flour to obtain smaller particle sizes. This technique increased starch damage and reduced cookie spread. Pratt (1963) air-classified flour and found that low-protein flour with fine particle size produced cookies with the greatest spread. The effects of particle size, hardness, protein content, and chlorine treatment on cake volume and cookie spread, for products made with flours from several cultivars, were studied by Gaines (1985). Both cookie diameter and cake volume were positively correlated with flour softness and lower protein content. Gaines and Donelson (1985b) found that flours from softer wheat cultivars produced cookies with greater spread. Whole wheat flours were also studied and shown to correlate well with straight-grade flours in cookie spread tests.

E. Cracker Flour Quality

Flour for cracker products is higher in protein (9–10%) and generally stronger than cookie flour. The quantity and quality of protein are important, because cracker dough contains relatively low levels of soluble ingredients such as sugar and salts, and thus the protein becomes hydrated and forms gluten during sheeting. Proper development of the gluten during sheeting is essential for gas retention, volume, and final product texture. If too much protein is present or if it is too strong, excessive product shrinkage and cracker toughness result. The ideal cracker flour generally has poor spread in the cookie test. The flour must withstand long fermentation times and retain enough strength to yield the open texture and oven spring of saltines (Matz, 1968; Thomson, 1976). Cracker flours can be straight soft wheat flours or blends of soft wheat and hard wheat flours. The increased strength of the hard wheat flour must be traded off against the poorer product characteristics that result if the hard wheat content of the blend is too high.

Criteria for measuring flour quality continue to present problems for the commercial baker. Tests preferred by various workers include those for moisture, ash, protein, particle size, starch damage, AWRC, and the spread factor from cookie baking. The rheology of a cracker dough is important to the processor; therefore, a battery of rheological tests have been applied. These include farinograph, alveograph, and mixograph tests and the MacMichael viscosity test. Uniformity is the most critical attribute of flour in the cracker process.

F. Cake Flour Quality

Cake flours have been studied more extensively than other soft wheat flours. One of the most important properties of a cake flour is its ability to carry the high amounts of sugar in high-ratio cake formulas (1.3–1.4 times the weight of flour). Cake flour must be able to develop a strong protein structure without making the product tough. The protein and starch components of the flour must also hydrate rapidly. With the knowledge that the proteins in a cake formula had to support a high sugar content, Kissell (1959) proposed a lean cake formula to evaluate flour for cake production. Because milk and eggs were omitted from the formula, the flour proteins had to bear the entire burden of structure in the cake. The test, therefore, is a very sensitive means of differentiating flours.

G. Air-Classification

Air-classification can be used to produce flours for specific purposes. Several workers have described the types of flours that can be produced by air-classification (Wichser, 1958; Jones et al, 1959; Gracza, 1962; Pratt, 1963; Bode et al, 1964). A low-protein, high-starch fraction with particle size in the range of 15–35 μm appears to be optimum for cookie production. Gaines and Donelson (1985a) varied flour protein through air-classification and gluten addition to obtain cake flours containing from 7 to 16% protein. The volume and tenderness of white cakes were not significantly influenced by the protein content. However, the height and tenderness of angel food cakes decreased with increasing protein content. Changes of 2% or more in the protein content were needed to effect a significant change. Generally, the variation in flour protein of soft wheat flour is more on the order of 0.5%.

Air-classification has been the classical means of demonstrating the influence of particle size on cake quality. Wichser (1958) demonstrated that high-starch, low-protein flour with intermediate particle size was most suitable for cake baking. Microscopic examination of this fraction showed that it contained large amounts of large starch granules that had been completely separated from the protein. Miller et al (1967a) observed that pin milling of flour improved cake-baking quality but excessive grinding caused quality to decline below the optimum. Yamazaki and Donelson (1972, 1973) showed with several flours that cake-baking quality after chlorination correlated with smaller particle size or smaller mean diameter of the flour particles. This suggests that the greater surface area of the finer flour may allow more efficient chlorination. Generally, bleaching agents other than chlorine are not added to soft wheat flours.

H. Flour Chlorination

Chlorination of flour for cake baking was first reported by Montzheimer (1931), and its effects were confirmed by Smith (1932). Both workers observed that the crumb of cakes made with chlorinated flour had a smoother, more even texture than that of cakes made with unchlorinated flour or flour treated with other bleaching agents. Smith also observed greater volume and better symmetry in cakes made with chlorinated flour. The normal range of chlorination is 1,100–2,300 ppm. This is usually accomplished on a continuous basis by injecting gas into a stream of freshly milled flour. Since hydrochloric acid is produced in the process, the reduction in the pH of the flour provides a convenient analytical tool to monitor the extent of chlorination. Flour pH is usually defined as the pH of a slurry of 10% flour in water (Sollars, 1958b; AACC, 1983, Method 02-52). Chlorinated flour generally has a pH in the range of 4.5–5.2. Specialty products may require slightly higher or lower levels of chlorination. Some suppliers specify that all soft wheat flour be treated with a very light chlorination to help improve flour uniformity.

Holme (1962) investigated the influence of pentosans and soluble proteins on cake-baking quality and found that cake flours contained unusually high levels of soluble gluten. Five separate proteins were identified electrophoretically in the albumin fraction, but it was not demonstrated whether they were results of chlorination or whether they became more extractable as a result of chlorination. Kissell (1971) treated flour with serial levels of chlorine and found that maximum protein solubility corresponded with optimum cake-baking performance.

The mechanism by which chlorination improves cake baking has been investigated rather extensively. Kissell et al (1974) compared the effects of low, optimal, and high levels of chlorination. At low levels, the batters expanded well initially but later shrank significantly. At high levels, expansion was severely inhibited, with the result that cake volume was limited. Optimum chlorination therefore is a compromise between the amount of expansion and the amount of shrinkage.

Several workers have investigated the mode of action of chlorine in flour. When chlorinated flour is fractionated and analyzed for chlorine distribution, most of the chlorine is found in the protein fraction. Sollars (1961) found most of the chlorine distributed in the gluten and the water-soluble protein fractions. In an attempt to determine the fate of chlorine while minimizing the effects of sample preparation, Chamberlain (1962) air-classified chlorinated and unchlorinated flour into high-protein and low-protein fractions. Analysis of the fractions showed that one third of the chlorine was taken up by flour lipids, one half by protein, and one fifth to one seventh by carbohydrates. Wilson et al (1964) concluded that the chlorine distribution was related to the surface area of the particles; therefore the finer high-protein fractions bound more of the chlorine. Sollars (1961) utilized fractionation of chlorinated flour by water and acetic acid and determined the chlorine distribution. He found half of the chlorine in the water-soluble portion, which suggested that the chloride had been washed out of the insoluble fractions during the separation step. Most of the gluten containing chloride was extracted with butanol. Some portions of this butanol-extractable lipid may have been hexane-extractable before extraction with water and acetic acid (Olcott and Mecham, 1947; Davies et al, 1969). When one considers the differences between bound and free lipids, Sollars and Chamberlain are in good agreement. Gilles et al (1964) reported that 90% of the chlorine was recovered in the lipid and water-soluble fractions and that essentially none was found in the starch fraction. Ewart (1968) observed that the reaction of chlorine with cysteine and methionine yielded cysteic acid and methionine sulfoxide, respectively. The change from the hydrophobic cystine disulfide region and the methionine to the hydrophilic oxidized forms could help explain the increased solubility and conformational changes in the proteins of chlorinated flour. Ewart also observed destruction of tyrosine and histidine.

From that work it was clear that the major reaction site of chlorine was with protein and lipid fractions. However, Lamb and Bode (1963) wet-fractionated flour, chlorinated the fractions, and recombined them with unchlorinated fractions. When the recombined flours were baked into cakes, the results demonstrated that chlorination of the starch was primarily responsible for the quality improvement. Sollars (1958a) had reported earlier that both the starch and the gluten fractions were involved in the improvement. The changes that are critical to cake baking, therefore, may not be reflected in the quantitative distribution of the chlorine after reaction.

Several groups have clearly shown that the starch is the primary site affected by chlorine and that this reaction results in an improvement in cake-baking quality (Sollars, 1958a, 1958b; Lamb and Bode, 1963; Sollars and Rubenthaler,

1971; Johnson and Hoseney, 1979). The question remains, How is starch affected? A physicochemical understanding of the function of starch in a cake batter helps to elucidate the effects of chlorination. Allen et al (1982) demonstrated by differential scanning calorimetry (DSC) that chlorination of flour did not affect the transition temperature or the enthalpies of either flour or starch isolated from it. The results suggested that any direct effect of chlorine on starch molecules is too small to result in measurable changes in the starch. Chamberlain (1962) proposed that the degree of gelatinization in relation to the expansion of the batter was critical in preventing the collapse of cakes. If starch granules do not expand and gelatinize sufficiently before the batter sets, the proper three-dimensional matrix is not established to support the foam and prevent collapse. Frazier et al (1974) demonstrated the greater strength of crumb from chlorinated flour. Their results support the concept that the greater crumb strength of chlorinated flour produces better cakes; this conclusion was confirmed by Ngo et al (1985). If chlorination were to increase the permeability of the starch granule surface, one would anticipate greater hydration, increased swelling and gelatinization of the starch, and consequently greater crumb strength.

Little experimental evidence of the chemical attack on starch during the chlorination of cake flour has been reported. Whistler et al (1966) showed considerable chemical damage to starch at very high levels of chlorination. They demonstrated substantial oxidation of the glucose residues at C-2 and C-3. Since then Johnson et al (1980) observed oxidation of prime starch at normal chlorination levels. They observed baking improvements from oxidative treatments (such as chlorination, bromination, or hydration and redrying of starch in air), which resulted in oxidative depolymerization of glucose residues of the starch chain. Huang et al (1982), in a detailed study, demonstrated that the oxidative damage at normal levels of chlorination was qualitatively similar to the damage reportedly caused by high levels of chlorination (Uchino and Whistler, 1962; Ingle and Whistler, 1964).

Chlorine does not appear to affect the crystallinity of the starch granule (Cauvain et al, 1977; Huang et al, 1982). Since amylose and amylopectin do not show any significant changes, one is tempted to speculate that the main effect of chlorination involves the lipids or the protein-lipid complex on the starch granule. Gough and Pybus (1971) proposed that the chlorination reaction disrupts the lipid-protein complex on the surface of the starch granule, allowing greater permeability by water. Varriano-Marston (1985) presented evidence of changes on the surface of the starch granule suggesting that it should be more hydrophobic after chlorination. Seguchi and Matsuki (1977) had shown earlier that starch from chlorinated flour appeared to be more hydrophobic than starch from untreated flour. Seguchi (1984) also found greater oil-binding capacity, suggesting greater hydrophobicity. The increased hydration of the starch allows for more even total hydration, improved moisture retention during baking, and a reduced tendency to collapse after baking.

Varriano-Marston (1985) proposed that for all practical purposes the consequences of oxidative damage to flour are related to the water-binding characteristics of the high-starch fraction of chlorinated flour. Chlorination was shown to alter the distribution pattern of water in the flour at moisture levels up to 16%. Kulp et al (1972) found increased hydration capacity at increased dosages of chlorine. Further research and more sensitive methods are needed to distinguish between the tightly bound and the loosely bound water in the starch system.

Alternatives to chlorination, reviewed in detail by Hodge (1975), suggest that the proposed mechanism may be valid. Doe and Russo (1968, 1969, 1970) described a heat treatment for flour that, with accompanying formula modifications, resulted in good-quality cakes. The flour was adjusted to 7% moisture, air-classified, and milled to assure the maximum number of free starch granules. It was then heated for 30 min at 120–140° C. To obtain the highest-quality high-ratio cakes, egg solids were replaced with egg albumin. Russo and Doe (1970) reported that in this treatment protein solubility was reduced to varying degrees, depending on the flour. Frazier et al (1974) found that the increase in the strength of gels of cooked flour and water and compressed cake crumb prepared from heat-treated flour was of the same order of magnitude as the changes caused by chlorination. Their conclusion agreed with that of Russo and Doe (1970), that the effect on starch was the most critical aspect of the change. Hodge (1975) reported that starch granules do not have to be freed from the protein for the thermal treatment to effect the desired changes. SEM did not reveal any changes in the granule surface of heat-treated starch (Russo and Doe, 1970; Cauvain and Gough, 1975). Although not observed by SEM, changes at the molecular level could occur at the surface of the granule (Gough et al, 1978).

Gaines and Donelson (1982) evaluated fractions from chlorinated and unchlorinated flours for their influence on cake crumb stickiness. The stickiness of cakes appeared to be related to the influence of protein on the prime starch fraction, and the improvement in cake volume was associated with the action of chlorine on the lipid portion of the flour. Kissell and Yamazaki (1979) observed that when flours were serially chlorinated, the crumb changed from gummy to dry with increasing levels of chlorine treatment.

I. Effect of Chlorination on Flour Lipids

Spies and Kirleis (1978) reported that the cake-baking potential of a flour was not governed by the free lipid fraction. However, cakes containing lipids extracted from chlorine-treated flours were larger and had better texture than those containing lipids from flours not treated with chlorine. Later, Donelson et al (1984) concluded that the chlorination of lipids was the primary effect of chlorination in baking tests for high-ratio cakes.

From the distribution of chlorine in chlorinated flour one can expect significant effects of chlorination on the lipid fraction. This speculation is reinforced by knowledge of the reaction between unsaturated fatty acid side chains and halogens. Because of the low concentration and complex nature of flour lipids, this has been difficult to prove. Xanthophylls make up substantial portions of flour pigments. The immediate color loss upon chlorination of flour has been associated with the formation of colorless addition compounds of these carotenoids (Sollars, 1961). Tsen et al (1971) confirmed this reaction by demonstrating a rapid loss in carotenoids after chlorination.

Daniels et al (1963) reported that chlorination reduced the essential fatty acids of flour by 60%. They also reported the development of several new fatty acid peaks in a gas chromatogram, including a peak thought to represent dichloro-stearic acid. The results suggested that a preferential reaction occurred with monounsaturated oleic acid rather than with the polyunsaturated fatty acids. One could also speculate that the hypochlorite ion was competing to oxidize the polyunsaturated lipid while chlorine was chlorinating the fatty acids. There appears to be no direct evidence in the literature to support this idea. The chlorination of lipids may influence their functionality or their interactions with nonlipid components in the flour. Changes in the extractability of the lipids from gluten suggest that changes in the lipid or in the protein alter the physicochemical nature of the protein-lipid complex. Changes in the lipid could also significantly change the way in which starch and lipids interact (Gracza, 1960; Youngquist et al, 1969; Rees, 1971). Seguchi (1984) reported that when prime starch is recovered from serially chlorinated flour, the oil-binding capacity of the starch increased with the level of chlorination. Treatment of the flour with protease, amylolytic enzymes, or dilute acid decreased this trend. This oil-binding appeared to be associated with the protein on the surface of the starch granule.

IV. SOFT WHEAT FLOUR EVALUATION

Most of our physical methods of evaluating wheat flours were developed to measure the properties of bread doughs. With most of the products made from soft wheat flours, the doughs are not mixed to development. The exceptions to this rule are certain cracker doughs, which are developed by sheeting, and sheeted cookie doughs. Therefore, it would appear unusual to use techniques designed for hard wheat flours to evaluate soft wheat flours.

The use of physical dough-testing machines such as the mixograph, farinograph, and alveograph to evaluate soft wheat flours appears to be based on the assumption that the rheological properties of soft wheat flour are the opposite of those of hard wheat flours. Stated in other terms, if a good hard wheat gives a strong curve, then a good soft wheat flour should give a weak curve. There appears to be little evidence to support such an assumption. The physicochemical properties of a good-quality soft wheat flour are a unique blend of properties. They are not just the properties of a poor-quality hard wheat flour.

From the preceding discussion it would appear on theoretical grounds that physical dough instruments would be of little value in characterizing soft wheat flours. This has been found to be generally true.

The MacMichael viscometer is another instrument that was originally developed for use with bread flours, to measure the apparent viscosity of a flour-water suspension. The standard method (AACC, 1983, Method 56-80) calls for the addition of increments of lactic acid until the viscosity no longer increases. We are apparently measuring the swelling and, thereby, the water-binding capacity of the protein at acidic pH. The relation of the water-binding properties of wheat protein under acidic conditions to the performance of a flour in cookies made under basic conditions is not clear. Correlations of MacMichael values with cookie diameters were low (Abboud et al, 1985a). The test appears to be used more widely to measure the properties of cracker flours. Because cracker sponges are acidic, this appears to be more reasonable. The apparent (MacMichael) viscosity is strongly affected by the ions in the flour-water suspension and therefore is affected strongly by the ash content of the flour.

It is generally believed that the protein content of soft wheat flours should be relatively low. Protein contents of 8.5–9.0% are often specified for cookie flours. Evidence available in the scientific literature, however, fails to show a relationship between performance and protein content (Yamazaki, 1954; Abboud et al, 1985a). The protein content seems to be not very important if the flour is milled from a good-quality soft wheat. The misconception of the effect of the protein on product quality probably stems from the fact that in the United States hard wheat flours are generally of higher protein content than soft wheat flours. In the industry, they may occasionally be mixed, particularly when it is an economic advantage to do so. When the percentage of hard wheat flour in the mill mix increased, two factors were noted. First, the protein content of the flour increased; second, the product quality declined. Therefore, it seemed obvious that higher protein content was related to poorer product quality. This scenario is to some extent supported by the observation that the hardness of both semisweet biscuits and cream crackers, as measured by a texture meter (Fig. 5), increased with increasing protein content of the flour (Wade, 1972a, 1972c).

Starch damage is often mentioned as an important quality factor for soft wheat flours, and thus its measurement is considered important. A high level of starch damage is undoubtedly detrimental in cookies. However, such damage is invariably linked with wheat hardness. Thus, it is difficult to tell which characteristic (hardness or starch damage) affects the product quality. Reported

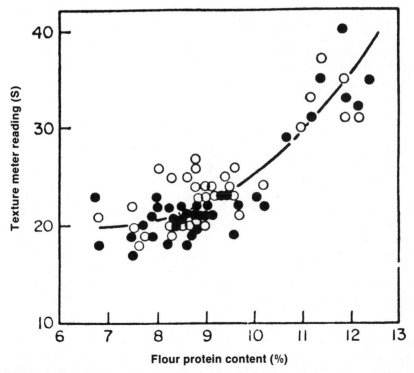

Fig. 5. Relations between the eating properties of semisweet biscuits, as measured by a texture meter, and the protein content of the flour. o = Unsulfited doughs; ● = sulfited doughs.

correlations of starch damage and cookie diameter are low (Abboud et al, 1985a). This may indicate that the hardness was much more important than the absolute amount of starch damage. Wainwright et al (1985) and Wade (1985) reached essentially the same conclusion. Damaged starch is well known to absorb much higher levels of water than does undamaged starch. The amount of water absorbed by the flour is an important quality of cookie flours: one that absorbs small amounts of water is desired. Therefore, damaged starch must be detrimental. However, the level of damaged starch in soft wheat flours is low. In fact, the measured level may not represent damaged starch at all but instead may be a background level given by the test. This would explain the absence of a high correlation between damaged starch and cookie diameter.

Another important quality of soft wheat flours is particle size (Yamazaki, 1959a, 1959b). This appears to be not very important in cookie flours, perhaps because the production of fine particles also results in a high level of starch damage. The net result is no improvement in cookie quality as particle size is reduced. In cake flours, the relationship between particle size and quality is much clearer (Yamazaki and Donelson, 1972). Reduction of particle size improves the apparent quality of the cake flours, which are not sensitive to small changes in the level of damaged starch.

Probably the most useful test to evaluate soft wheat flours is the AWRC test. This relatively simple test (Finney and Yamazaki, 1953; Yamazaki, 1953) does a reasonable job of evaluating soft wheat flours and roughly predicts flour performance in cookie baking. It was designed to simulate the role of soft wheat flours in cookies. For example, the alkaline conditions of the test provide the same environment for the flour as in a cookie. In addition, the ability of a flour to hold water against centrifugal force measures its affinity for water. Good-quality soft wheat cookie flours bind water poorly.

A. Milling Quality of Soft Wheats

Soft wheat, because of its soft character, is easy to grind to the fineness of flour, much easier than is hard wheat. The major problems with milling soft wheat are cleanup of the bran and sifting of the stocks. Both problems appear to be caused by the same factors. During grinding, the endosperm is reduced to fine particles. The small particle size allows the endosperm particles to get close together and to make good contact with the bran and with each other. That contact, together with the fat and water on the surface of the particles, causes the particles to adhere to each other (Neel and Hoseney, 1984b). This causes the bran to clean up poorly and the resultant soft wheat flours to sift poorly. The flours are sometimes referred to as "woolly flours." Examination of a soft wheat mill shows that it has a much larger sifter area than does a hard wheat mill of the same capacity.

A common practice in soft wheat milling is to add little or even no temper water before the wheat is milled. The endosperm of the soft wheat is weak to begin with and thus does not need additional weakening. Because moisture is involved in the adhesion of the flour particles, lower moisture in the flour leads to better sifting (bolting) properties (Neel and Hoseney, 1984a).

B. Baking Properties of Soft Wheat Flour

As is clear from the foregoing discussion, none of the measurements applied to flour tells us its baking quality. To learn its baking quality we must bake the flour. Baking the flour, however, does not necessarily tell us how it performs in any particular product or on any particular baking line. Certainly we cannot bake cookies and expect to predict how the flour performs in crackers or cakes. In fact, we should not expect a cookie-baking test to tell us how the flour performs in all cookie formulas or lines, as there are large differences in what is expected from flours in different systems.

The most widely used baking procedure to evaluate soft wheat flours is the microbaking procedure developed by Finney et al (1950b). The major advantages of the method are that it is quite reproducible and very sensitive to changes in both flours and formulations (Finney et al, 1950a). The major disadvantages are that the results may not correlate with the use of the flour in the bakery. The procedure is quite good but must be used with care. Oven settings and many other production-line variables or changes in formula can alter the performance of the flour.

C. Function of Flour Components and Formula Ingredients During Cookie Baking

Most of the study of what happens during cookie baking and of the function of ingredients has been conducted with a sugar-snap cookie formula. This is both an advantage and a disadvantage. It is easy to compare work and to build an understanding of what happens in this system. The disadvantage is that we know little about other systems. The work described here was conducted with sugar-snap cookies, unless specified otherwise.

The formula for simple sugar-snap cookies (Table I) is high in both sugar and shortening and relatively low in water. As a result of those ratios and the creaming step, only part of the sugar is dissolved during mixing (Yamazaki and Lord, 1971; Kissell et al 1973). This can be shown (Abboud and Hoseney, 1984) by DSC of the mixed cookie dough (Fig. 6). At a relatively low temperature the shortening melts. The next broad endotherm corresponds to the dissolving of the crystalline sucrose. Thus, only about half of the sucrose is dissolved during mixing, and the remainder dissolves during baking. When sucrose dissolves in

TABLE I
Formula for Simple Sugar-Snap Cookies

Ingredient	Percent of Flour Weight	Percent of Total
Flour	100	45.7
Sugar	60	27.5
Shortening	30	13.7
Water (optimum)	22.75	10.4
Sodium bicarbonate	1.0	0.5
Ammonium bicarbonate	0.75	0.3
Nonfat milk solids	3.00	1.4
Sodium chloride	1.00	0.5

water, each gram creates about $0.6\,cm^3$ of additional solvent (Ghiasi et al, 1983). In dilute systems, this may be an insignificant amount. However, in a cookie system, with extremely limited water, that fact is of great importance. Completely dissolving the sugar (heating to dissolve, cooling, and using the supersaturated system rapidly) gives a very sticky dough (Curley and Hoseney, 1984). Therefore, the use of crystalline sugar has obvious advantages for dough handling. It is also obvious that problems arise if noncrystalline sugar or syrups are used.

Because of the limited amount of water and the creaming step, a question might arise as to whether the flour is hydrated during mixing. If the flour was not hydrated, we would expect a hydration peak (exothermic) after the shortening melts in DSC scans. Because no such peak is found, we must assume that the flour is hydrated during mixing.

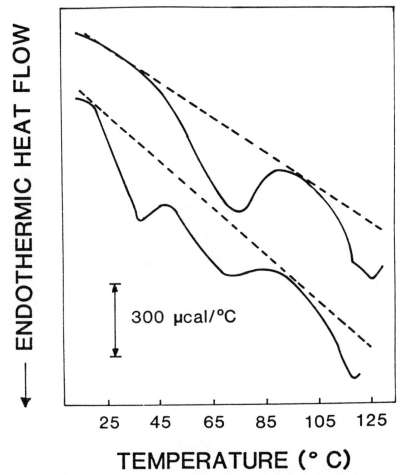

Fig. 6. Thermogram of a standard cookie dough (lower curve). The dashed line is the baseline. The upper curve is the thermogram of a dough containing no fat.

In time-lapse photographs of cookies made from good- and poor-quality flours (Fig. 7), we can see that the dough both expands and flows as it is heated. The expansion is the result of leavening. The flowing is due to the force of gravity. Doughs made with good-quality flour flow much faster than those made with poor-quality flour. This is shown in Fig. 8, in which the expansion of cookie dough is plotted against baking time (Abboud et al, 1985b). From those plots it

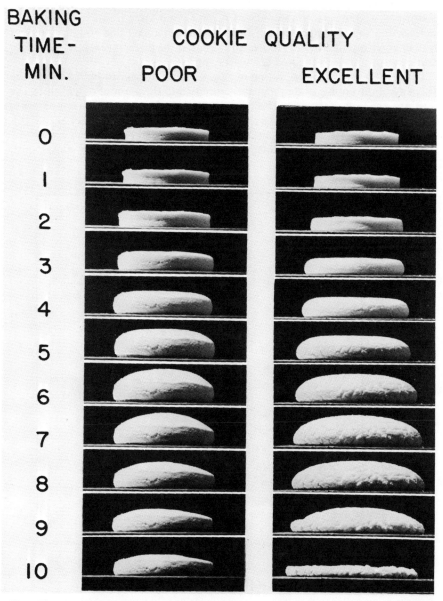

Fig. 7. Time-lapse photographs of cookies made from an excellent and a poor-quality cookie flour. (Reprinted from Yamazaki and Lord, 1971)

is obvious that the final size of a cookie is determined by two factors: the rate of flow (represented by the slope of the expansion line) and the set time (the point at which expansion stops).

Because all cookie doughs are cut to the same size, the difference between cookies produced from good-quality flours and those from poor-quality flours does not express itself until the dough is heated. Such thought prompted Yamazaki (1959c) to apply heat to various tests for cookie quality. He performed heated mixograph, heated AWRC, and viscograph tests. Viscosity increased at lower temperatures in flours that produced poor-quality cookies than in good-quality flours. He also showed that the differences were not due to heat penetration or to the loss of volatile ingredients.

The slope of the expansion line in Fig. 8 would appear to be related to the viscosity of the cookie dough. Because gravity and the amount of leavening are constant, the flow of the dough is controlled by the viscosity. As the temperature of the dough increases, the apparent viscosity decreases (Abboud et al, 1985b), because of the effect of temperature on viscosity and also because of the dissolving of the crystalline sucrose. The cookie apparently expands (flows) until

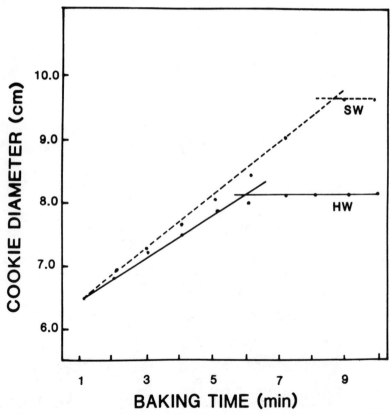

Fig. 8. Changes in cookie diameter during baking, for cookies made with hard wheat flour (HW) and soft wheat flour (SW).

the viscosity suddenly increases, as shown by Yamazaki (1959c). Clearly this rapid increase in viscosity is what determines the set point of the dough. The viscosity must be determined by the amount of water in the dough that is free to act as a solvent. The more free water, the greater the dissolved portion of the total sugar, and the lower the apparent viscosity. Thus, a good-quality cookie flour does not bind as much water as a poor-quality flour. This is the reason that the AWRC test does a reasonable job of determining cookie flour quality. It essentially measures how much water is bound or held by the flour. Of course, it does not measure differences in the set points of the doughs.

The set point of cookie dough is undoubtedly related to the rapid increase in viscosity noted by Yamazaki (1959c). This, of course, raises the question of what is responsible for that rapid increase. Our first guess might be that it is caused by starch gelatinization. However, a high sugar concentration is known to increase the gelatinization temperature of starch (Spies and Hoseney, 1982). Abboud and Hoseney (1984), performing DSC on raw cookie dough and baked cookies, showed that none of the starch in sugar-snap cookies is gelatinized. Our next guess might be that the apparent viscosity increases as water is lost from the dough. However, measurements of the rate of water loss from doughs made from good- and poor-quality flours did not reveal any difference between the doughs (Doescher et al, 1987b).

The elimination of starch gelatinization and water loss as possible factors controlling the rapid increase in viscosity brings us to the protein fraction. When cookie dough was heated in an electrical resistance oven, the dough expanded rapidly at higher temperatures, even though leavening was removed from the formula (Doescher et al, 1987a). The temperature at which the rapid expansion started correlated well with the set time in baking and appeared to be related to a change in the protein.

During cookie dough mixing, the flour is hydrated but remains as individual particles (Flint et al, 1970; Doescher et al, 1987a). As the dough is heated, the flour goes through an apparent glass transition and as a result becomes mobile. The mobility allows it to interact with neighboring particles, to form a continuous matrix. The viscosity of this matrix is too high for the cookie dough to continue to expand under the force of gravity (Doescher et al, 1987a).

V. CHEMICAL LEAVENING

Most, if not all, soft wheat products are chemically leavened. Yeast is widely used as the leavening agent in breads and other hard wheat products but is only rarely used in soft wheat products. Most soft wheat products have a tender "bite" when chewed. Yeast presumably would make the bite tougher and thus would be detrimental. Whether that is the reason that yeast is not used with soft wheat products is open to conjecture, but the facts are that it is not used, with the important exceptions of saltine crackers and Arabic flat breads.

Four major leavening gases are used in the leavening of baked products. These are air (actually a mixture of gases), carbon dioxide, ammonia, and water vapor. Air is used inadvertently by every baker with every product (Dunn and White, 1939; Kichline and Conn, 1970). The nitrogen in air is extremely important in most baked products. Nitrogen is not very soluble in water and therefore remains as a gas during dough mixing. It forms small gas cells, which serve as nucleation sites for the expansion of leavening gases produced later in baking. Air, of course, also expands as it is heated and therefore has a leavening action by itself.

Water, also a part of every baked product, acts as a leavening gas when converted to vapor. Because of its relatively high boiling point, water vapor does not appear to be a major leavening gas in most baked products (Moore and Hoseney, 1985). It appears to have a significant effect in semisweet biscuits (Wade, 1971b) and to be a major leavening gas in products that are heated rapidly, such as saltine crackers and most flat breads (Hoseney 1986).

Ammonia gas is produced, along with carbon dioxide and water, by the decomposition of ammonium bicarbonate heated in water (Van Wazer and Arvan, 1954a, 1954b). The major advantage of ammonium bicarbonate as a leavening agent is that it decomposes completely to gases and therefore leaves no residual salts to affect either the taste or the rheological properties of the product. The major disadvantage is that the products must be baked to near dryness. Even relatively small amounts of water left in the product retain ammonia, and even small amounts of ammonia make the product inedible. Therefore, the use of ammonium bicarbonate is limited to certain dry cookies and crackers.

Carbon dioxide is widely used as a leavening gas. With chemical leavening, the source of the carbon dioxide is one of the bicarbonates. The above-mentioned ammonium salt and the potassium and sodium salts are all potential sources. The potassium salt is not used widely, because it is hygroscopic and reported to give a slightly bitter taste. Sodium carbonate could be used as a source of CO_2 but generally is not. It gives a more strongly alkaline reaction, which could be detrimental. Sodium bicarbonate (commonly known as baking soda) is by far the most widely used. The advantages of sodium bicarbonate are many (LaBaw, 1982). For example, it is readily obtainable at relatively low cost and at high purity. It is relatively tasteless and nontoxic. It is relatively stable and stores well. Because the salt is alkaline, it produces no gas at room temperature unless the pH of the dough is lowered by an acid. At higher temperatures, the bicarbonate disproportionates to form Na_2CO_3, H_2O, and CO_2 (Van Wazer and Arvan, 1954a, 1954b).

Some doughs contain acid as a part of their formulation, those containing acidic fruits or buttermilk, for example. If the dough does not contain an acid, then an acidic salt is needed. The type of acidic salt determines when the leavening gas is produced. The most common leavening acids are given in Table II.

The combination of sodium bicarbonate and a leavening acid produces a baking powder (Conn, 1965; Reiman, 1977). Baking powders consist of a mixture of baking soda, one or more leavening acids, and a diluent. By U.S. law, a baking powder must produce at least 12% carbon dioxide upon complete reaction. That regulation establishes the level of soda in the baking powder. The amount of acid required in the formulation depends upon the neutralization value of the acid. The stoichiometry of the reaction of bicarbonate and acidic salts is not well known. Therefore, the concept of neutralization values was developed. The neutralization value is defined as the number of grams of soda neutralized by 100 g of acidic salt (Conn, 1981; LaBaw, 1982). This value is supplied by the manufacturer. The diluent in a baking powder is usually dried starch and is used to separate the two active components in the baking powder. A double-acting baking powder is one that contains two leavening acids, usually one that reacts at room temperature and one that reacts in the oven at higher temperatures.

The choice of leavening acid is based on a number of factors (Parks et al, 1960), the most important of which is probably the rate of reaction (Barackman, 1931; Anonymous, 1983). Kichline and Conn (1970) published the relative reaction rates of the common leavening acids (Fig. 9). Interestingly, the pyrophosphates have different reaction rates, depending on how they are made. Another important factor concerning the choice of leavening acid is the taste of the product. The choice of acid can affect the taste of the product. For example, the "pyro" taste of products using pyrophosphates is well known. Actually, the pyro taste does not come from pyrophosphate at all. The pyrophosphates are broken down to sodium phosphate by enzymes in dough. The taste appears to come from an interchange of calcium for sodium on the surface of the teeth. The problem has been partially removed by adding small amounts of calcium lactate to the baking powder (Barch, 1959). Equally well known is the bitter taste associated with the use of glucono-δ-lactone. Another important factor is the residual salts that are produced as a result of the reaction. These are known to affect the dough rheology to a great extent (Guy et al, 1967; Kinsella and Hale, 1984; Holm, 1986). It appears that this phenomenon needs to be studied in much more detail.

VI. COOKIES, OR SWEET BISCUITS

A. Types of Manufacture

As mentioned above, the doughs used in the manufacture of cookies and sweet biscuits contain relatively large amounts of fat and sugar. They do not possess a developed gluten network and are therefore not coherent under tension but have a short fracture, i.e., break easily. Although possessing this feature in common, short doughs nevertheless exhibit considerable variation in consistency, which is related to the methods used to form the pieces of dough before baking.

TABLE II
Leavening Acids[a]

Acid	Neutralization Value	Reaction Rate[b]
Monopotassium tartrate	45	1
Monocalcium phosphate monohydrate	80	1
Anhydrous monocalcium phosphate	83.5	2
Sodium acid pyrophosphate	72	3
Sodium aluminum phosphate	100	4
Sodium aluminum sulfate	100	4
Dicalcium phosphate	33	5[c]
Glucono-δ-lactone	50	...

[a] Adapted from Kichline and Conn (1970).
[b] 1 = Reactive at room temperature; 5 = requires oven temperature.
[c] Reacts too slowly to be a leavening acid; adjusts final pH.

Drier, more crumbly doughs cohere under pressure and are processed on a rotary machine (in British terminology, a rotary moulder). The principles of a rotary machine are shown diagrammatically in Fig. 10A. The dough is compressed by a grooved roller into biscuit-shaped dies engraved on a forming roller. Surplus dough is trimmed from the surface of the latter roll by means of an oscillating knife and is returned to the feed hopper via the grooved roller. The biscuit-shaped dough pieces are extracted from the engravings in the forming roll and placed on a canvas web, which is pressed firmly against the forming roll by means of a third, rubber-covered roller. The dough pieces thus formed are transferred from the canvas web to the oven band for baking. This method of forming permits complex and sharply defined patterns to be produced on the surface of the dough pieces. Because rotary pieces undergo relatively small

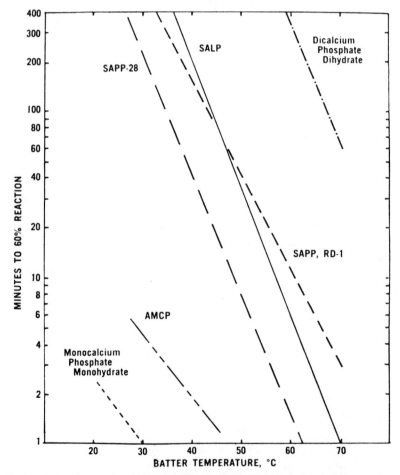

Fig. 9. Reaction rate versus batter temperature for various leavening acids. SAPP = sodium acid pyrophosphate (grades SAPP-28 and RD-1); SALP = sodium aluminum phosphate; AMCP = anhydrous monocalcium phosphate. (Reprinted, with permission, from Kichline and Conn, 1970)

changes in dimension during baking, these complex patterns are carried through to the finished product.

Softer doughs are formed into pieces ready for baking by an extrusion process. The dough is forced through dies of an appropriate size and shape by two contrarotating grooved rollers. For wire-cut cookies, the extruded dough is cut at the die face by a reciprocating wire; the disk-shaped dough pieces then usually fall directly to the oven band. For bar cookies, the dough is deposited in a series of parallel ribbons, either directly on the oven band or on a transfer web. The forming machine is then known as a rout press. These ribbons may be cut into suitable short lengths by a guillotine either before or after baking. Some filled biscuits, e.g., Fig Newtons (in the United Kingdom, Fig Rolls) are made by coextruding the filling inside a hollow tube formed from a short dough. Again, the extruded rope may be cut into short pieces before or after baking. A more detailed description of the machinery outlined above was given by Manley (1983).

B. Role of Flour Components

The principal components of flour are starch, protein, pentosans, moisture, and lipids. Moisture content apart, each of these components is made up of a complex mixture of chemically related molecules, but little information is available concerning the function of the component molecules in the making of cookies or sweet biscuits.

STARCH

Starch is present in wheat flour both as intact and as damaged granules. The damage occurs during the milling process. Damaged granules in soft wheats usually form a relatively small proportion of the total starch (10–20%, as measured by the Farrand enzymatic method, and a much lower proportion, by the AACC procedure [AACC, 1983, Method 76-30A]). The proportion of damaged granules is important, as their power to absorb water is about three times that of undamaged granules (Farrand, 1964).

PROTEIN

In the manufacture of short-dough biscuits, formulation and processing

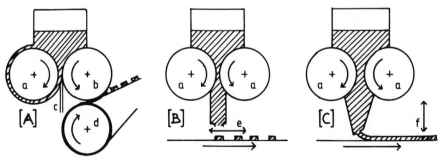

Fig. 10. Principles of operation of rotary (A), wire-cut (B), and bar or rout (C) presses. a = grooved roller, b = forming roller, c = oscillating knife, d = rubber-covered roller, e = wire, f = guillotine.

conditions are chosen to prevent the conversion of flour protein to gluten. When examined under the microscope, short biscuit doughs reveal the presence of many large fragments of intact endosperm with the starch granules still embedded in their protein matrix (Flint et al, 1970). Other observations have indicated the significance of flour particle size on the properties of flour in the baking of several types of biscuits (Brenneis, 1965b; Seibel et al, 1976; Gaines, 1985; Wainwright et al, 1985). These observations suggest that one important function of the flour protein in relation to the making of short-dough biscuits is to control the breakdown of wheat endosperm during milling.

PENTOSANS

Pentosans, although present as only 2–3% of the flour, are important for their ability to absorb up to 10 times their weight of water (Kulp, 1968). In association with other hemicellulosic substances, they form the basic structure of the endosperm cell walls in wheat grain and are found in the starch tailings fraction of the flour. Early work using wet fractionation of cookie flour yielded a starch tailings fraction that, when added in small amounts to a control flour, decreased the spread of cookies baked from the flour (Yamazaki, 1955; Sollars, 1956). Further work (Gilles, 1960; Sollars and Bowie, 1966) confirmed these results. However, a recent study (Abboud et al, 1985a) found only a poor correlation between the total pentosan content of 56 samples of flour and the spread of cookies baked from these flours ($r = -0.21$).

MOISTURE CONTENT

The moisture content of flour in equilibrium with atmospheric humidity is 14.0%. This moisture is distributed between the hydrophilic components of the flour (protein, starch, and pentosans) according to the isotherm appropriate to each. In the standard cookie spread test, normal practice is to adjust the weights of flour and added water to compensate for deviations in flour moisture from the nominal 14.0%. However, Doescher and Hoseney (1985b) demonstrated that such compensation does not necessarily produce cookies having the same degree of surface cracking when the moisture content of the flour was deliberately reduced by artificial drying. The mechanism underlying this observation may be associated with a pronounced hysteresis of the desorption-adsorption isotherms of the flour. This suggestion is supported by measurements made by Bushuk and Winkler (1957).

LIPIDS

Lipids make up only about 1.2% of wheat flour. This is a small amount in comparison with the amount of recipe fat added in cookie manufacture (approximately 30% of flour weight in the standard cookie spread test and nearly twice this amount in some other cookies). However, this small amount plays a significant role in the making of cookies (Kissell et al, 1971). Cole et al (1960) showed that cookies made with flour from which the natural lipid had been extracted by means of an appropriate solvent were of smaller diameter and different appearance than those made from the original flour. Restoration of the extracted lipid to the flour resulted in the complete recovery of the normal cookie-making properties. These observations were confirmed and extended by later workers. Research on wheat lipids up to 1981 was reviewed by Chung and Pomeranz (1981).

C. Role of Formula Ingredients

After flour, the two most important structural components of a cookie recipe are the fats and sugars. The types and amounts of these ingredients have a great effect on the appearance and eating qualities of the finished product. However, the relative proportions of fat to flour and of sugar to flour in commercial short-dough recipes can each vary independently over a wide range, and therefore generalizations concerning the role of these ingredients are somewhat speculative. Water, although added in significant amounts in most recipes, should be regarded as a processing aid rather than an ingredient, since all but a small amount of the moisture in the dough (the added water plus the moisture already present in the flour and other ingredients) is removed during the baking process. The ingredients added in smaller amounts—aerating agents, sodium bicarbonate, salt, dairy products, etc.—serve primarily to develop texture, color, and flavor. The use of eggs and egg products, once common in biscuits and cookies, has largely been abandoned on economic grounds.

FATS AND EMULSIFIERS

The fats most commonly used in the manufacture of cookies and sweet biscuits are plastic fats of vegetable origin. The presence of some solid fat in the dough during mixing and forming is essential; the use of liquid oil or too high a dough temperature, resulting in the complete liquefaction of the plastic fat, produces adverse changes in the handling characteristics of the dough (e.g., Abboud et al, 1985a). On the other hand the fat must be sufficiently plastic to enable it to become well distributed during mixing, and little if any solid fat can be present at temperatures above body temperature ($37°$ C), or else the product leaves a greasy or waxy feel in the mouth. In some, if not all, short biscuit doughs, the fat acts as the principal agent binding the other ingredients (Flint et al, 1970).

The addition of emulsifiers to short doughs has been shown to make the products more tender for eating. A large number of emulsifiers have been tested, and their use has been suggested as a means of reducing the amount of fat in the recipe (Burt and Thacker, 1981) or of permitting the use of hard wheat flours in cookies (Tsen et al, 1975). However, the use of these materials has not been widely taken up by the industry.

SUGARS

The principal sugar in short doughs is sucrose. Smaller amounts of other sugars, such as invert sugar, honey, or one of the glucose syrups, are also added to promote browning and aid in the development of flavor during baking. An additional role of such sugars is to influence the crystallization of the sucrose in the finished product (Curley and Hoseney, 1984; Doescher and Hoseney, 1985b). Most commercial short-dough formulations contain insufficient added water to dissolve all the sugar during mixing. Abboud and Hoseney (1984) used DSC to study the effect of heating sugar-snap dough and demonstrated the presence of an endothermic peak corresponding to the solution of crystal sugar in the dough water. However, as they pointed out in another context, no water was lost from their doughs under the conditions of measurement, whereas during baking water is lost continuously from the cookie dough. Thus, under commercial baking conditions some crystal sugar may remain unchanged in the finished product. The particle size of sugar crystals may affect the flow of short doughs during baking, although Abboud et al (1985a) found an appreciable effect only with coarse crystals (>35-mesh).

WATER

Under commercial conditions, for a given formulation, with all other factors being constant (e.g., ingredient temperatures and condition of the fat), the amount of water added is adjusted so as to produce a dough of acceptable consistency for processing. The method of assessing dough consistency is usually subjective, based on the mixer operator's experience. Some instrumental methods have been developed (Wade and Watkin, 1967; Steele, 1977; Gaines and Tsen, 1980) and used in bench-scale baking tests (Gaines and Tsen, 1980; Wainwright et al, 1985). From a commercial point of view, since all the added water must be removed during baking, it is important that the amount of water added be kept to a minimum. From a research point of view, it is important to ensure that the conditions of model dough systems are as closely related as possible to those of commercial operations, since the amount of water added affects not only the initial consistency but also the changes in consistency occurring between mixing and processing (Gaines, 1982).

VII. BATTERS

The principal products made from simple batters containing soft wheat flour (other than cake products) include wafers, waffles, and pancakes. Before these products are described, it is necessary to clarify a point of terminology concerning wafers. In the United States, the term *ice-cream wafers* is applied to thin biscuitlike products cut and baked from a sheeted dough in a manner similar to that used in the manufacture of crackers. In the United Kingdom, the term describes thin sheets of material manufactured by the process used to make sugar wafers.

A. Wafers

Wafers are baked from a fluid batter containing 35–40% dry solids (by weight). The bulk of these solids is supplied by soft wheat flour. The remaining solids are made up of small amounts of oil or lecithin (2–3% of the weight of the flour) together with salt (0.25%) and sodium bicarbonate (0.3%). Small amounts of other ingredients are added by some manufacturers to influence the flavor, color, or texture of the finished product. The batter is baked between pairs of heated metal plates. A typical plate is 470 × 290 mm. Each plate is engraved with crisscross lines or some other suitable pattern, which is transferred to the wafer during baking and facilitates the adhesion of fillings, which may be applied during secondary processing. The pairs of wafer plates, each pair hinged together down one edge to form a "book," are mounted on an endless carrier and are heated by passage through a tunnel oven. The baked wafer consists of dry, porous starch gel. A detailed study of factors affecting the properties of the finished product was described by Pritchard and co-workers (Pritchard and Wade, 1972; Pritchard and Stevens, 1973, 1974). The wafer sheets are used to make sandwiches consisting of three to five layers of wafer separated by layers of fat-sugar filling (sugar wafers) or by layers of caramel or toffee (caramel wafers).

A batter of similar composition, baked in an appropriately shaped mold, is used to make ice-cream cones. The addition of sucrose to a wafer batter results in the sticking of the wafer to the baking surfaces when baked in normal equipment. However, with specially designed baking equipment, wafers containing high levels of sucrose (up to 60% of flour weight) can be produced. Such products, while still hot from the baking process, are quite flexible and may be rolled round a mandrel to form cones or folded to form a decorative fan or other shapes. Such high-sugar products are served and eaten with ice cream.

B. Waffles and Pancakes

The batter used in the manufacture of waffles and pancakes is much richer than that used for wafers. Appreciable amounts of fat (25–30% of flour weight) and sugar (10–25% of flour weight) are added, together with eggs, milk solids, and baking powder. Waffle batter is baked between deeply indented plates, to produce a thick product with a crisp outer surface and a soft, chewy interior. It is normally eaten immediately while still hot or deep-frozen for subsequent reheating. Pancakes are baked on a flat surface and are turned over partway through the baking process. Dried fruit or other ingredients may be added to the batter.

VIII. CAKES

Cakes, like cookies, are quite high in both sugar and shortening. The major difference between the two is that cake formulas tend to contain much more water. The much higher water level has a profound effect on the final product. When baked, cakes set to give a light, tender product, whereas cookies collapse during or immediately after baking. The difference appears to be controlled by the concentration of sugar in the available water. The molar concentration of sugar controls the gelatinization temperature of starch (Bean and Osman, 1959; Bean and Yamazaki, 1978; Bean et al, 1978; Savage and Osman, 1978; Spies and Hoseney, 1982).

If the concentration of sugar is too high, the starch does not gelatinize during baking. Because of the water in the baked products, the temperature of the crumb only slightly exceeds $100°$ C. If the starch is not gelatinized, the cookie or cake collapses. This happens in such cookies as sugar cookies to give a relatively flat product. Perhaps the best example of this phenomenon is brownies. Cake brownies set and have a cakelike structure; however, if the sugar content is increased, the starch no longer gelatinizes, and the brownie collapses, resulting in a fudge brownie. The level of sugar, then, is used to control the gelatinization temperature of starch and, thereby, the setting point of cakes. In general, the setting point should be just below $100°$ C.

How does sugar control the starch gelatinization temperature? The mechanism is not completely understood. Spies and Hoseney (1982) showed that water activity affected the gelatinization temperature to some extent but that other factors were also important. Plotting gelatinization temperature as a function of water activity, they found straight lines for sugars, according to whether they were monosaccharides, disaccharides, or trisaccharides (Fig. 11). From this data they suggested that the sugars penetrate the starch granule and interact (by hydrogen bonding) with the amorphous parts of the starch molecules. That binding tends to lower the kinetic energy of the amorphous part of the granule, and, therefore, higher temperature is required to melt the crystals. Evidence to support this concept was given by Chinachoti and Steinberg (1984). The interaction of starch and sugar has been studied in detail (Chinachoti and Steinberg, 1986a, 1986b, 1986c).

What makes a good-quality cake flour is not known. In general, soft wheat flour is clearly preferred. For high-ratio cakes, the flour must be treated with chlorine, and the smaller the particles, the better the flour quality (Miller et al, 1967a, 1967b; Yamazaki and Donelson, 1972).

A. Mixing of Cake Batters

The mixing of cake batters is different from the mixing of a cookie dough or a bread dough, although the baker is trying to accomplish many of the same things. The object is to produce a uniform mixture of the ingredients; this, of

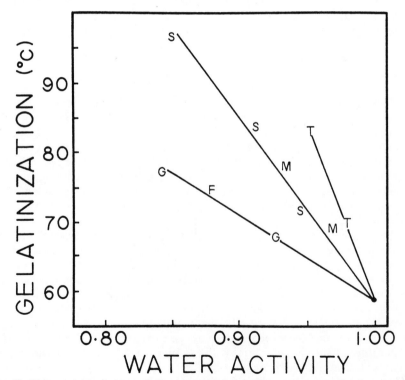

Fig. 11. Water activity of sugar solutions versus gelatinization temperature. F = fructose; G = glucose; M = maltose; S = sucrose; T = maltotriose.

course, results in hydration of the flour particles. The gluten is not developed into a cohesive mass, such as that of a bread dough, primarily because of three factors. First, flour protein is traditionally low in cake flours; second, the sugar competes for the water and slows the gluten development; and, third, the basic pH also slows gluten development.

One of the important things that must be accomplished during mixing is to incorporate air as small bubbles into the batter. These small bubbles act as nucleation sites for the gas produced during leavening. Without these nucleation sites, the cake does not develop a uniform grain.

Bubble mechanics (Handleman et al, 1961) states that $p = 2\gamma/r$, where p is the pressure inside a gas cell, r is the radius of the cell, and γ is the interfacial tension. For any given sample, γ is a constant. Therefore, as r becomes small, p increases, and vice versa. The formation of a bubble with a very small radius requires a very high pressure. This leads to the conclusion that no new bubbles can be formed as a result of leavening. With no nuclei present, the leavening gas (CO_2) saturates the system and diffuses to a lower concentration. The lowest concentration (or pressure) is the extremely large bubble that is the atmosphere above the batter. If the nuclei are of uneven size, then the gas diffuses to the large bubbles instead of the small bubbles. Also, as a result of the relationship expressed in the equation, if a cake batter is allowed to stand after mixing, the gas from the small bubbles diffuses into the aqueous phase as it becomes unsaturated by the diffusion of gas to the surface of the batter. Because the larger bubbles have a lower pressure, a "flow" of CO_2 is set up from the small bubbles, which eventually disappear, to the large bubbles. As a result, the large bubbles become larger and eventually attain such buoyancy that they rise to the top of the batter and are lost. Thus, if the cake batter is allowed to stand on the counter overnight, it loses the nuclei that required a lot of energy to put in. Methods of producing the nuclei vary from one type of cake to another and are discussed later.

Another important attribute of cake batter is its viscosity. A viscous batter can easily be distinguished from one that is not viscous, but it is nevertheless difficult to measure the difference in viscosity (Lee and Hoseney, 1982). The major problem is the complex nature of the batter, an aqueous system in which discontinuous air, fat, and starch granules are suspended.

A high viscosity is required for a number of reasons. First, gas diffusion is slower in a more viscous system. Second, the migration of buoyant bubbles is slower in a more viscous medium. Third, starch granules do not settle out and form a separate layer. Thus, a viscous batter produces a cake of better volume and a more uniform grain, both of which are positive qualities.

The number of air cells can be decreased in a cake batter, either by the coalescing of two cells into one, by a decrease in the size of the bubbles until they disappear, or by an increase in their size that allows them to rise to the surface and escape. On the other hand, no new cells can be formed. Thus, it is very important to incorporate as many air cells as possible during mixing.

Double-stage leavening systems are often used in cakes. The first stage reacts during mixing, and as a result the air cells entrained during mixing are enlarged. The larger cells can be more easily subdivided into smaller cells. Therefore, the total number of cells is increased. Because the second leavening acid is not soluble at room temperature, an excess of soda stays solubilized in the batter until the cake is placed in the oven. This avoids loss of gas during any bench time.

The timing of the gas release by the high-temperature leavening acid is important. If the carbon dioxide is released too early, then part of it is lost before the cake is set. If, on the other hand, the cake is set before the gas is released, the increased gas does not increase the cake volume and may lead to destruction of the grain as the pressure builds up. Also important but poorly understood is the effect of various ions on the cake structure.

B. Types of Cakes

Much of the confusion in the literature on cake making results from a lack of a clear definition of the type of cake being studied. There are three distinct types of cakes, defined by how they are mixed.

The oldest of the three types is made by the multistage mixing procedure, in which the sugar and shortening are mixed together first to produce a cream. In later stages, the water, other ingredients, and finally the flour are added. This may involve a total of only two mixing steps or may involve three or more. The advantage of the creaming step is that it is very effective in incorporating air into the shortening phase. A second advantage is that the air bubbles are quite stable as long as they remain in the shortening (until the shortening melts). This procedure produces excellent cake with a fine grain.

The second type is the so-called box cake. It has this name because essentially all the ingredients come in a box. The stability of the leavening system in these mixes was reported by Conn and Kichline (1960). The cook need only add water and usually eggs. With a relatively small amount of mixing the batter is ready to bake. In a box cake, air is incorporated directly into the aqueous phase. Two steps appear to be necessary to obtain a large number of small gas cells. The first is the use of certain surfactants (α-tending surfactants) that are very effective in incorporating air into the aqueous phase. The second step is to "finish" the cake flour. This involves grinding the flour-shortening mixture during the production of the mix (Lee et al, 1982). The grinding may be done with roller mills or with a bran duster. The net result appears to be that the shortening is "hidden" on the flour. Because of the grinding, the flour and shortening interact. Evidence for this is that shortening, even if it is dyed, cannot be found in the mix with a microscope. Free shortening destroys an air-in-water emulsion, and presumably the finishing of the cake flour is a way of avoiding this.

Box cakes are relatively easy to make and are extremely tolerant of variations in formula and processing, but they are not outstanding cakes. The grain is not as uniform as that of the other types, they are not stable during bench time, and they are very tender. The tenderness makes them undesirable for commercial shipping. For their intended purpose—to be made at home and consumed there—they are quite satisfactory.

The poor shipping quality of box cakes is apparently related to the surfactant used. As any good physical chemist knows, anything that can be done with a surfactant can probably be done by mechanical means. To produce an air-in-water emulsion one needs only to subdivide the air bubbles finely enough. The Oakes mixer works on this principle and makes the third type of cake. The cakes produced are quite good and are stable for shipment.

Most of the cakes sold in the United States are high-ratio cakes. The name means that they contain more sugar than flour. This ratio results in the very sweet cakes preferred by U.S. consumers. Chlorine-treated flour is necessary for producing high-ratio cake. The reason for chlorine treatment of flour is discussed earlier in this chapter. Typical formulas for various high-ratio cakes are given in Table III.

Another popular cake in the United States is angel food (Table III). For an angel food cake, an aqueous egg-air foam is produced and the flour added carefully, so as not to disturb the foam. Angel food foams are very sensitive to even small amounts of fat. The best-selling commercial cakes are the sponge snack cakes (Table III).

C. Dynamics of Cake Baking

The dynamics of cake baking are quite interesting (Trimbo et al, 1966; Trimbo and Miller, 1973). For the most part, the heat enters the cake by way of the pan. Therefore, the batter that is in contact with the pan expands first, as illustrated in Fig. 12. As the batter is heated, its viscosity decreases, as would be expected. However, the higher temperature also triggers the leavening system and

TABLE III
Typical Formulas for White Layer, Angel Food, and Sponge Cakes
(percent of flour weight)

Ingredient	White Layer Cake	Angel Food	Sponge Cake
Flour	100	100	100
Sugar	140	500	110
Shortening	55	0	0
Eggs, whites	76	500	11[a]
Milk, fresh	95	0	9[b]
Baking powder	1.3	0	5
Cream of tartar	0	20	0
Salt	0.7	0	2.5
Water	0	0	100

[a] Whole eggs, dry.
[b] Nonfat dried milk.

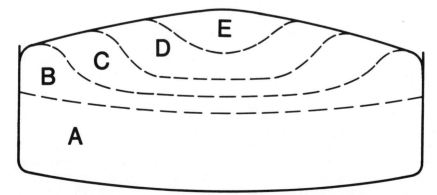

Fig. 12. Diagram of the expansion of cake batter in a pan in successive stages of baking.

produces gas. As the bubbles increase in size, the batter becomes a foam. The foam results in a higher apparent viscosity of the system. As the temperature of the system increases, the starch gelatinizes, and much of the free water becomes bound. This sets the cake into a rigid system that does not collapse as it cools. As can be seen in Fig. 12, the top center of the cake is the last part to bake. Interestingly, if the cake has too little or too much water, a dip forms at the center. With the correct amount of water, a smooth-contoured crown is produced.

IX. OTHER SOFT WHEAT PRODUCTS

A wide range of products is made from or contains soft wheat flour. A few of these are discussed here.

A. Piecrusts

To produce a tender, flaky piecrust the flour must be shortened. The term *shorten* means that the flour, when wetted, does not produce a long or cohesive dough. This is accomplished by cutting fat into the flour. This, of course, is why fat became known as *shortening*. The type of flour is also important (Miller and Trimbo, 1970). A good piecrust flour does not absorb water readily and does not form a cohesive dough. This is, of course, very similar to the definition of a good-quality cookie flour.

B. Biscuits

To keep the terminology straight, biscuits should be referred to as *chemically leavened bread*. It is doubtful, however, that this terminology will catch on in fast-food establishments in the southern United States. Many of these products are made from self-rising flour, which is flour to which soda, a leavening acid (usually monocalcium phosphate), and salt have been added at the mill (Reiman, 1977). Thus, to produce the biscuits one needs only to cut shortening into the flour, add water, mix and sheet the dough, cut the biscuits, and bake. A soft wheat flour is usually used. Little or no work on the type of flour required has been reported in the scientific literature.

C. Refrigerated Biscuits and Rolls

Refrigerated biscuits and rolls include many variations of the products described above. They vary widely, from plain biscuits to those containing considerable sugar and various toppings. These products owe their existence to the discovery that biscuit dough wrapped in aluminum foil is stable. The dough is mixed and sheeted, cut, and placed in a can without completely filling it. The can is generally made of cardboard with a layer of aluminum foil covering its inner surface. It is designed to withstand the pressure that develops but also to open easily. The can is sealed and placed in a proof box. The temperature is raised to trigger the leavening system. As the biscuits expand, they drive out the air trapped in the can. The can is sealed by the dough, and sufficient gas is produced to generate about 15 psi. The dough is then anaerobic and is stored refrigerated under pressure. Its shelf life is on the order of 90 days. All of the leavening reaction occurs during the proof period (LaBaw, 1982). The leavening acid used is quite important. It must be essentially inactive during dough makeup but also must perform 100% of its action in the relatively short proofing time. Only certain sodium pyrophosphates meet those requirements (Conn, 1981). The shelf life of the product is not long, as it is only refrigerated, and enzymes and microorganisms are still active in the system. Certain microorganisms can lower the pH and also produce gas if the product is mishandled. This can lead to dangerously high pressures.

Soft wheat flour is usually used in these products, although there is little in the scientific literature to suggest a reason for its use or, for that matter, what is desired in a good biscuit flour.

D. Soft Wheat Products in Asia and Africa

Much of the soft wheat grown in the Pacific Northwest is exported to Asia and Africa (Finney et al, 1987). The most popular products made from these wheats are flat breads, noodles, and various steamed products.

The flours used for flat breads range from whole meal to rather pure white flour. Consumer acceptance is associated with a light crust and crumb color. Desirable flours generally have a short mixing time and a weak gluten (Faridi and Rubenthaler, 1983).

The major unbaked product from soft wheat in Asia is noodles (Nagao et al, 1976). The flour required for noodles generally is white and finely granulated and has rapid water uptake and low enzymatic activity.

X. CRACKERS

Crackers can generally be subdivided into three basic categories: soda crackers, or saltines (in the United Kingdom, cream crackers); sprayed crackers; and savory crackers. Leavening can be accomplished either by yeast fermentation or by chemical leavening. A fourth group, with higher levels of added sugar, e.g., graham crackers, is invariably chemically leavened. Most yeast-leavened crackers are based on a sponge-and-dough fermentation process, as outlined for saltines in Fig. 13; some cream crackers are made with a single fermentation of the dough (Wade, 1972b). Cracker doughs differ from bread doughs by being much stiffer at both the sponge and the dough stages. In addition to the ingredients shown in Table IV, small amounts of proteases are commonly added to make the dough more mellow and improve extensibility. Depending on the product, sponge fermentation can vary from 3 to 18 hr. As shown in Table IV, a wide variety of crackers can be produced with minor variations in ingredients or processes. Figure 14 outlines the process for chemically leavened crackers. Typically, the carbon dioxide in chemical leavening is formed by the reaction of sodium bicarbonate with an acidic salt.

The fermentation process changes the rheology of cracker doughs, as well as influencing the flavor, color, and texture of the final product. To produce a light cracker with a flaky but not brittle texture, a moderately strong wheat flour is required. After the dough is machined and fermented, it should be extensible to allow sheeting without tearing. If the dough is too tough, the resulting crackers are small and deformed. For saltine crackers, the sponge flour is usually an unbleached soft red wheat flour (with a protein content of 8.5–10%). For cream crackers, Wade (1972c) demonstrated a strong relationship between flour properties, particularly protein content, and optimum processing conditions. Separation of the laminations was used as the criterion for cracker quality. The best crackers were produced from flours containing 10.5–11.5% protein. For both saltines and cream crackers, the flour added at dough-up (the dough-mixing stage) can be somewhat weaker, with a protein content of 8–9%.

In the past, bisulfite was used to weaken doughs and improve the extensibility of strong soft wheat or even hard wheat flours for cracker production. Recently, because of concerns about bisulfite in the diet, the amount of bisulfite used in most cracker applications has been reduced, even though it has been convincingly shown that essentially no bisulfite remains in the finished product (Thewlis and Wade, 1974).

The relatively long fermentation is responsible for the unique flavor and texture of soda crackers produced by the sponge-and-dough process (Faridi and Johnson, 1978). Micka (1955) found that efforts to shorten the fermentation time resulted in products with atypical flavor and texture. The chemistry of crackers has not been studied extensively (Heppner, 1959; Matz, 1968; Pieper, 1971). Micka (1955) and Sugihara (1978a, 1978b) performed pioneering work on

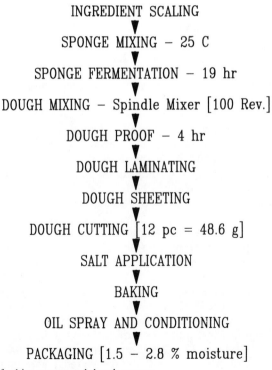

INGREDIENT SCALING
▼
SPONGE MIXING – 25 C
▼
SPONGE FERMENTATION – 19 hr
▼
DOUGH MIXING – Spindle Mixer [100 Rev.]
▼
DOUGH PROOF – 4 hr
▼
DOUGH LAMINATING
▼
DOUGH SHEETING
▼
DOUGH CUTTING [12 pc = 48.6 g]
▼
SALT APPLICATION
▼
BAKING
▼
OIL SPRAY AND CONDITIONING
▼
PACKAGING [1.5 – 2.8 % moisture]

Fig. 13. Outline of saltine sponge-and-dough process.

the microflora of saltine fermentation. The microbiology of the fermentation process was further investigated by Fields et al (1982), who found that the limited water and dense physical structure of the dough inhibited the growth of the microflora. Lactic acid bacteria dominate the dough, because they are more successful than the yeast, particularly as the pH drops, in competing for the limited amount of simple carbohydrates needed for fermentation. Pizzinatto and Hoseney (1980) reported that the bacteria in the added yeast were responsible for the drop in pH of doughs during fermentation. They also demonstrated that dough strength decreased with fermentation time, presumably as a result of a naturally occurring proteolytic enzyme in flour with an optimum activity at pH 4.1. Salt increased resistance to extension; soda and mixing time increased extensibility.

During a study of the role of fermentation in the manufacture of cream crackers (Wade, 1972b), evidence was obtained to support the suggestion that the amount of separation of the laminations during baking was related to the rate of CO_2 production at the end of the fermentation period. By increasing the amount of yeast in the formula and controlling the temperature of a mechanically developed dough, the fermentation time could be reduced to 30 min. Under these conditions, the dough was processed satisfactorily on

TABLE IV
Typical Cracker Formulas

		Sprayed Cracker		Savory Cracker	
	Saltines	**Sponge and Dough**	**Chemically Leavened**	**Cheddar Type**	**Blue-Cheese Type**
Sponge ingredients (lb)					
Flour	70	50	⋯	75	80
Yeast	0.23	0.25	⋯	0.3	0.2
Water	30	24	⋯	23.5	21
Shortening	4	⋯	⋯	7.5	⋯
Cheese (cheddar or blue)	⋯	⋯	⋯	10	17.5
Diastatic malt	0.02	⋯	⋯	⋯	⋯
Salt	⋯	⋯	⋯	⋯	1
Sponge time (hr)	18	18	⋯	18	18
Dough ingredients (lb)					
Flour	30	50	100	25	20
Shortening	5.8	8	7.5	⋯	20
Sugar	⋯	2.5	5	⋯	⋯
Salt	1.4	1.5	1	1	⋯
Cheese	⋯	⋯	⋯	4	⋯
Sodium bicarbonate	0.63	0.52	0.9	0.45	0.75
Malt	⋯	2.5	1.5	⋯	⋯
Malt syrup	0.92	⋯	⋯	1	⋯
Nonfat dry milk/ dry buttermilk	⋯	⋯	2.5	⋯	⋯
Invert syrup	⋯	⋯	2.5	⋯	⋯
Monocalcium phosphate	⋯	⋯	0.75	⋯	⋯
Water	0.8	4	28	0.8	⋯
Fermentation					
Time (hr)	4	5.5	⋯	4–6	4–6
Temperature, °F	82	82	⋯	88	88

commercial-scale laminating and cutting equipment and produced finished crackers of commercially acceptable structure and appearance. During baking, the mechanically developed dough produced little aroma. However, after the crackers had been cooled, packed, and stored for 24 hr, taste panels were unable to distinguish between those made from mechanically developed doughs and control crackers made by a process that included a long fermentation. These observations suggest that texture plays an important role in the perceived flavor of this type of product.

The rheological changes occurring in a cracker sponge were investigated by Doescher and Hoseney (1985a). They demonstrated that the 18-hr sponge process was critical to the development of good-quality crackers. Changes in rheology were monitored over the entire fermentation time by means of an extensigraph. The greatest drop in resistance occurred in the early hours of fermentation. In later hours of fermentation, after the pH had dropped, a flour protease became active.

Cream and soda crackers differ in several attributes. Primarily, soda crackers have a slightly higher pH ($\sim$7.2), because of the addition of sodium bicarbonate, and cream crackers are slightly more acidic. This difference results in flavor differences. Cream crackers are sold with slightly higher moisture levels (3–4%) than soda crackers (1–2%). Some soda crackers are sprayed with a light coat of oil as they emerge from the oven, to develop a gloss on the surface. Excellent

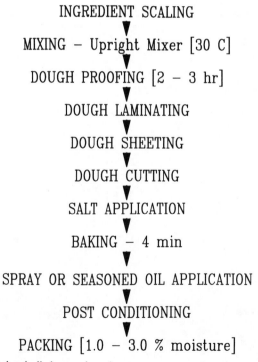

Fig. 14. Outline of chemically leavened cracker process.

discussions of the practical aspects of cracker manufacturing were provided by Matz and Matz (1978) and Manley (1983).

Sprayed crackers are generally chemically leavened and have a final pH of about 6.5. After dough-up and a relatively short rest time, the dough is sheeted to form a continuous ribbon, which is then laminated, with a light application of dusting flour between the layers. The use of dusting flour or occasionally an application of shortening between the layers results in better separation of the layers during baking and improved product tenderness. The dough is then rolled and stamped to the appropriate shape. Docker holes are punched in the dough to inhibit the formation of large blisters or puffing of the crackers. The dough is then transferred to a mesh band and baked for 3–5 min. After baking, the final product is sprayed with a bland shortening at 65–70° C, to yield a cracker with 20–25% fat.

Flavored or savory crackers are well accepted in the U.S. market and account for much of the recent growth in cracker sales. The intense savory flavors are produced by adding the appropriate flavoring agents directly to the dough or to the surface of the crackers after baking. Savory or cheese crackers are generally produced from fermented doughs. The yeast products and the lower pH improve the cheese flavor. The formulation is basically similar to that of soda crackers, with adjustments to compensate for the fat and moisture content of the cheese.

Physicochemical definition and understanding of the cracker-making process is essentially missing. The process involves a complex fermentation and the blending of a number of variable ingredients. Some of the basic studies suggest that a careful study of this process from the microbiological and physico-chemical point of view could be extremely interesting and potentially valuable in improving product consistency and reducing the tendency for high breakage in crackers.

XI. PRETZELS

Pretzels are made from a straight or sponge-dough system (Table V) that is quite similar to a cracker formula. Pretzel dough, however, is somewhat drier and, therefore, stiffer. As with cracker doughs, dough development during mixing is minimized. This prevents overdevelopment during extrusion.

Typically, the sponge fermentation time is on the order of 6–10 hr. After mixing, the dough receives either a very short or no fermentation time. It is then

TABLE V
Typical Pretzel Formula

Ingredient	Quantity (lb)
Sponge	
Flour	20
Water	10
Yeast	0.63–1.25
Dough	
Flour	80
Water	25
Salt	1.2
Shortening	3

extruded at relatively low pressure through dies, cut into strips, and rolled to the desired thickness between two belts. The strands are clipped, twisted, and lightly rolled to set the knot in the pretzel. Alternatively, the dough is extruded in the traditional pretzel shape and cut at the desired thickness, to produce a dough of uniform thickness with no knots. The pretzels are then proofed for a short time and passed quickly ($\sim$25 sec) through a caustic bath. The lye bath consists of 1.25% sodium hydroxide, which is held at 85–90°C. The pretzels are salted to about 2% with large-crystal salt and baked for 5 min at about 246°C and somewhat lower temperatures in the later zones of the oven. Baking in air converts the NaOH to innocuous Na_2CO_3. The pretzels are further dried for 20–90 min at temperatures between 105 and 135°C. This final drying step takes the moisture from as high as 15% down to 2–4% (Matz and Matz, 1978). The stepwise baking is necessary to prevent checking. The knot causes variations in the moisture in a pretzel and causes stress to develop. If dried too quickly, pretzels check and break into pieces.

The uniqueness of the pretzel process is in the alkaline dipping step. Several changes would be expected during this stage of the process. Proteins and starch become solubilized on the surface. Upon drying at high temperatures, these

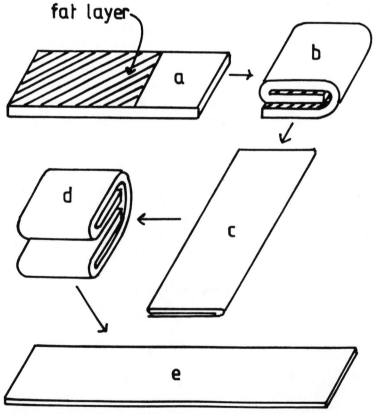

Fig. 15. Sequence of dough folding and rolling in the lamination of Danish and puff pastry.

materials harden and caramelize, forming the typical surface glaze of dry pretzels. The alkaline conditions also cause more color formation in the phenolic components of the flour, which along with browning contribute to the color and flavor of the product.

Soft pretzels are made by a similar process but with Na_2CO_3 replacing $NaOH$ in the bath. In addition, they are not dried to the lower moisture level during baking.

Pretzels are steeped in tradition and have a rich history. For that history the reader is referred to Reiman (1969) and Anonymous (1971).

XII. SWEET GOODS

The term *sweet goods* is used in the United States to cover a wide range of baked products. The term, and some of the products, are unknown in the United Kingdom, where the nearest equivalent (although not with exactly the same meaning) might be *morning goods*, i.e., flour confectionery products that originally were baked and sold fresh daily (Daniel, 1971). The common characteristic of the two groups of products is that they are derived from a fermented bread-type dough enriched with a selection from the following ingredients, in various proportions: fat, sugar, eggs, milk products, dried fruit, candied peel, chopped nuts, etc. The flour used to make these products frequently contains a proportion of hard wheat flour and has a protein content in the region of 11%. American coffee cakes, cinnamon rolls, and similar products are characterized by their light, fluffy texture. British products have a denser texture, more akin to that of British breads; typical examples are fruited tea cakes, Chelsea buns, and hot cross buns (a spiced bun sold especially at Easter time).

A. Doughnuts

Doughnuts differ from other sweet goods in being heat-treated by deep-fat frying rather than baking. They are made in various shapes—in the traditional ring or in balls or fingers. In the United Kingdom, the ball- and finger-shaped kinds are usually filled by injection with jam (fruit preserve). After frying and draining, doughnuts are sugar-dusted or glazed to improve their appearance and eating quality.

B. Danish and Puff Pastry

Although classed with sweet goods, Danish pastry is like puff pastry in some aspects of its manufacture. In both cases, the dough, after mixing (and fermentation in the case of Danish pastry), is formed into a moderately thick sheet (about ½ in. thick), which is then coated over two thirds its surface with a suitable fat (Fig. 15). The dough is then folded to produce a sandwich of three dough layers separated by two fat layers. This is achieved manually in small bakeries or by mechanical means in larger plants. In a novel piece of equipment, the dough and fat are coextruded in the form of a hollow tube of dough lined with fat on its inner surface. This tube is then flattened and folded into a multilayered sandwich (Cleven and Fluckiger, 1977). In either case, the resultant sandwich is reduced in thickness by careful rolling, and the dough sheet is subjected to further folding and rolling, to build up the required number of layers of thin dough sheets separated by layers of fat (Fig. 15). The laminated dough is then cut and formed into the required final shape, and additional ingredients, such as dried fruit and chopped nuts, for Danish pastries, are added.

The fat used in the lamination process is selected so as to have the property that a large proportion of it is solid at the dough temperature, to prevent absorption by the dough, but little of it is solid at body temperature (37° C), to prevent a waxy or greasy feel on the palate when the pastry is eaten. In addition, the fat must be plasticized so as to have a consistency similar to that of the dough with which it is to be combined.

XIII. CONCLUSIONS

The products made from soft wheat flour are many and varied. Each has its own processing problems along with a unique chemistry and physics. Most of the products have been developed over countless years and the formulas refined by unlimited trials. Although, as noted in this chapter, good progress is being made, we are only beginning to understand the basic differences between hard and soft wheats and the chemistry that makes them different. Investigations of the chemistry and physics of what happens in a simple cookie system will keep us busy for many years to come.

LITERATURE CITED

AACC. 1983. Approved Methods of the AACC, 8th ed. Method 02-52, approved 1961, reviewed 1982; Method 10-50D, approved 1985, revised 1981 and 1984; Method 10-52, approved 1985; Method 10-90, approved 1976, revised 1982; Method 56-80, approved 1961, revised 1982; Method 76-30A, approved 1969, revised 1984. The Association, St. Paul, MN.

ABBOUD, A. M., and HOSENEY, R. C. 1984. Differential scanning calorimetry of sugar cookies and cookie doughs. Cereal Chem. 61:34-37.

ABBOUD, A. M., RUBENTHALER, G. L., and HOSENEY, R. C. 1985a. Effect of fat and sugar in sugar-snap cookies and evaluation of tests to measure cookie flour quality. Cereal Chem. 62:124-129.

ABBOUD, A. M., HOSENEY, R. C., and RUBENTHALER, G. L. 1985b. Factors affecting cookie flour quality. Cereal Chem. 62:130-133.

ALLEN, J. E., SHERBON, J. W., LEWIS, B. A., and HOOD, L. F. 1982. Effect of chlorine treatment on wheat flour and starch: Measurement of thermal properties by differential scanning calorimetry. J. Food Sci. 47:1508-1511.

ANONYMOUS. 1971. Pretzels: Steeped in legend and caustic soda. Snack Food (July):39-41.

ANONYMOUS. 1983. Sodium Bicarbonate Applications Leavening Agents. Church and Dwight Co., Inc., Princeton, NJ.

BARACKMAN, R. A. 1931. Chemical leavening agents and their characteristic action in doughs. Cereal Chem. 8:423-433.

BARCH, W. E. 1959. Phosphate baking powders. U.S. patent 2,870,017.

BARLOW, K. K., BUTTROSE, M. S., SIMMONDS, D. H., and VESK, M. 1973a. The nature of the starch-protein interface in wheat endosperm. Cereal Chem. 50:443-454.

BARLOW, K. K., SIMMONDS, D. H., and KENRICK, K. G. 1973b. The localization of water-soluble proteins in the wheat endosperm as revealed by fluorescent antibody techniques. Experientia 29:229-331.

BEAN, M. M., and OSMAN, E. M. 1959. Behavior of starch during food preparation. II. Effects of different sugars on the viscosity and gel strength of starch pastes. Food Res. 24:665-671.

BEAN, M. M., and YAMAZAKI, W. T. 1978. Wheat starch gelatinization in sugar solutions. I. Sucrose: Microscopy and viscosity effects. Cereal Chem. 55:936-944.

BEAN, M. M., YAMAZAKI, W. T., and

DONELSON, D. H. 1978. Wheat starch gelatinization in sugar solutions. II. Fructose, glucose, and sucrose: Cake performance. Cereal Chem. 55:945-952.

BIFFIN, R. H. 1908. On the inheritance of strength in wheat. J. Agric. Sci. 3:86-101.

BODE, C. E., KISSELL, L. T., HEIZER, H. K., and MARSHALL, B. D. 1964. Air-classification of a soft and a hard wheat flour. Cereal Sci. Today 9:432-435, 442.

BRENNEIS, L. S. 1965a. Flour granulation vs. cookie spread. Biscuit Bakers Inst., Annu. Training Conf., 40th, Montreal. Biscuit and Cracker Manuf. Assoc., Washington, DC.

BRENNEIS, L. S. 1965b. Qualitative factors in the evaluation of cookie flours. Bakers Dig. 39(1):66-69.

BRUINSMA, B. L., and RUBENTHALER, G. L. 1978. Estimation of lysine and texture in cereals by NIR. Page X-BR-1 in: Proc. Technicon Intl. Congr., 8th. Technicon Instrum. Co., Ltd., Basingstroke, U.K.

BURT, D. J., and THACKER, D. 1981. Use of emulsifiers in short dough biscuits. Food Trade Rev. 51:344, 346-347, 349-350.

BUSHUK, W., and WINKLER, C. A. 1957. Sorption of water vapor on wheat flour, starch, and gluten. Cereal Chem. 34:73-86.

CAUVAIN, S. P., and GOUGH, B. M. 1975. High ratio yellow cake. The starch cake as a model system for response to chlorine. J. Sci. Food Agric. 26:1861-1868.

CAUVAIN, S. P., GOUGH, B. M., and WHITEHOUSE, M. E. 1977. The role of starch in baked goods. II. The influence of purification procedure on the surface properties of the granule. Starch 29:91-93.

CHAMBERLAIN, N. 1962. What goes on at Chorleywood: How confectionery fits into research programme. Baker's Rev. 79:2014, 2016, 2033.

CHESTERFIELD, R. S. 1971. A modified barley pearler for measuring hardness of Australian wheat. J. Aust. Inst. Agric. Sci. 37:148-151.

CHINACHOTI, P., and STEINBERG, M. P. 1984. Interaction of sucrose with starch during dehydration as shown by water sorption. J. Food Sci. 49:1604-1608.

CHINACHOTI, P., and STEINBERG, M. P. 1986a. Interaction of solutes with raw starch during desorption as shown by water retention. J. Food Sci. 51:450-452.

CHINACHOTI, P., and STEINBERG, M. P. 1986b. Moisture hysteresis is due to amorphous sugar. J. Food Sci. 51:453-455.

CHINACHOTI, P., and STEINBERG, M. P. 1986c. Crystallinity of sucrose by X-ray diffraction as influenced by absorption versus desorption, waxy maize starch content, and

water activity. J. Food Sci. 51:456-463.

CHUNG, O. K., and POMERANZ, Y. 1981. Recent research on wheat lipids. Bakers Dig. 55(5):38-50, 55, 96-97.

CLEVEN, F., and FLUCKIGER, W. 1977. A new method of continuous puff pastry and flaky pastry production. Getreide Mehl Brot 31:73-74.

COLE, E. W., MECHAM, D. K., and PENCE, J. W. 1960. Effect of flour lipids and some lipid derivatives on cookie-baking characteristics of lipid-free flours. Cereal Chem. 37:109-121.

CONN, J. F. 1965. Baking powders. Bakers Dig. 39(2):66-70.

CONN, J. F. 1981. Chemical leavening systems in flour products. Cereal Foods World 26:119-123.

CONN, J. F., and KICHLINE, T. P. 1960. Leavening acids: Their effect on the shelf life of cake mixes and on cake grain. Cereal Sci. Today 5:143-147.

CURLEY, L. P., and HOSENEY, R. C. 1984. Effects of corn sweeteners on cookie quality. Cereal Chem. 61:274-278.

CUTLER, G. H., and BRINSON, G. A. 1935. The granulation of whole meal and a method of expressing it numerically. Cereal Chem. 12:120-129.

DANIEL, A. R. 1971. The Bakers Dictionary, 2nd ed. Elsevier Appl. Sci. Publ., London.

DANIELS, N. W. R., FRAPE, D. L., RUSSELL-EGGITT, P. W., and COPPOCK, J. B. M. 1963. Studies on the lipids of flour. II. Chemical and toxicological studies on lipid of chlorine treated flour. J. Sci. Food Agric. 14:883-893.

DAVIES, R. J., DANIELS, N. W. R., and GREENSHIELDS, R. N. 1969. An improved method of adjusting flour moisture in studies on lipid binding. J. Food Technol. 4:117-123.

DOE, C. A. F., and RUSSO, J. V. B. 1968. Flour treatment process. British patent 1,110,711.

DOE, C. A. F., and RUSSO, J. V. B. 1969. Flour treatment process. Canadian patent 808,591.

DOE, C. A. F., and RUSSO, J. V. B. 1970. Flour treatment process. U.S. patent 3,490,917.

DOESCHER, L. C., and HOSENEY, R. C. 1985a. Saltine crackers: Changes in cracker sponge rheology and modification of a cracker-baking procedure. Cereal Chem. 62:158-162.

DOESCHER, L. C., and HOSENEY, R. C. 1985b. Effect of sugar type and flour moisture on surface cracking of sugar-snap cookies. Cereal Chem. 62:263-266.

DOESCHER, L. C., HOSENEY, R. C., and

MILLIKEN, G. A. 1987a. A mechanism for cookie dough setting. Cereal Chem. 64:158-163.

DOESCHER, L. C., HOSENEY, R. C., MILLIKEN, G. A., and RUBENTHALER, G. L. 1987b. Effect of sugars and flours on cookie spread evaluated by time-lapse photography. Cereal Chem. 64:163-167.

DONELSON, J. R., YAMAZAKI, W. T., and KISSELL, L. T. 1984. Functionality in white layer cake of lipids from untreated and chlorinated patent flours. II. Flour fraction interchange studies. Cereal Chem. 61:88-91.

DUNN, J. A., and WHITE, J. R. 1939. The leavening action of air included in cake batter. Cereal Chem. 16:93-100.

ELDER, A. H., LUBISICH, T. M., and MECHAM, D. K. 1953. Studies on the relation of the pentosans extracted by mild acid treatments to milling properties of Pacific Northwest wheat varieties. Cereal Chem. 30:103-114.

EWART, J. A. D. 1968. Action of glutaraldehyde, nitrous acid or chlorine on wheat proteins. J. Sci. Food Agric. 19:370-373.

FARIDI, H. A., and JOHNSON, J. A. 1978. Saltine cracker flavor. I. Changes in organic acids and soluble nitrogen constituents of cracker sponge and dough. Cereal Chem. 55:7-15.

FARIDI, H. A., and RUBENTHALER, G. L. 1983. Experimental baking techniques for evaluating Pacific Northwest wheats in North African breads. Cereal Chem. 60:74-79.

FARRAND, E. A. 1964. Flour properties in relation to the modern bread processes in the United Kingdom, with special reference to alpha-amylase and starch damage. Cereal Chem. 41:98-111.

FIELDS, M. L., HOSENEY, R. C., and VARRIANO-MARSTON, E. 1982. Microbiology of cracker sponge fermentation. Cereal Chem. 59:23-26.

FINNEY, K. F., and YAMAZAKI, W. T. 1953. An alkaline viscosity test for soft wheat flours. Cereal Chem. 30:153-159.

FINNEY, K. F., YAMAZAKI, W. T., and MORRIS, V. H. 1950a. Effects of varying quantities of sugar, shortening, and ammonium bicarbonate on the spreading and top grain of sugar-snap cookies. Cereal Chem. 27:30-41.

FINNEY, K. F., MORRIS, V. H., and YAMAZAKI, W. T. 1950b. Micro versus macro cookie baking procedures for evaluating the cookie quality of wheat varieties. Cereal Chem. 27:42-49.

FINNEY, K. F., YAMAZAKI, W. T., YOUNGS, V. L., and RUBENTHALER, G. L. 1987. Quality. Pages 677-748 in: Wheat and Wheat Improvement. Am. Soc. Agron., Madison, WI.

FLINT, O., MOSS, R., and WADE, P. 1970. A comparative study of the microstructure of different types of biscuits and their doughs. Food Trade Rev. 40(4):32-39.

FRAZIER, P. J., BRIMBLECOMBE, F. A., and DANIELS, N. W. R. 1974. Rheological testing of high ratio cake flours. Chem. Ind. p. 1008.

GAINES, C. S. 1982. Influence of dough absorption level and time on stickiness and consistency in sugar-snap cookie doughs. Cereal Chem. 59:404-407.

GAINES, C. S. 1985. Associations among soft wheat flour particle size, protein content, chlorine response, kernel hardness, milling quality, white layer cake volume, and sugar-snap cookie spread. Cereal Chem. 62:290-292.

GAINES, C. S., and DONELSON, J. R. 1982. Contribution of chlorinated flour fractions to cake crumb stickiness. Cereal Chem. 59:378-380.

GAINES, C. S., and DONELSON, J. R. 1985a. Effect of varying flour protein content on angel food and high-ratio white layer cake size and tenderness. Cereal Chem. 62:63-66.

GAINES, C. S., and DONELSON, J. R. 1985b. Evaluating cookie spread potential of whole wheat flours from soft wheat cultivars. Cereal Chem. 62:134-136.

GAINES, C. S., and TSEN, C. C. 1980. A baking method to evaluate flour quality for rotary-molded cookies. Cereal Chem. 57:429-433.

GHIASI, K., HOSENEY, R. C., and VARRIANO-MARSTON, E. 1983. Effects of flour components and dough ingredients on starch gelatinization. Cereal Chem. 60:58-61.

GILLES, K. A. 1960. The present status of the role of pentosans in wheat flour quality. Bakers Dig. 34(5):47-52.

GILLES, K. A., KAELBLE, E. F., and YOUNGS, V. L. 1964. X-ray spectrographic analysis of chlorine in bleached flour and its fractions. Cereal Chem. 41:412-424.

GOUGH, B. M., and PYBUS, J. N. 1971. Effect of gelatinization temperature on starch granules of prolonged treatment at 50°C. Staerke 23:210-212.

GOUGH, B. M., WHITEHOUSE, M. E., and GREENWOOD, C. T. 1978. The role and function of chlorine in the preparation of high-ratio cake flour. 4. Crit. Rev. Food Sci. Nutr. 11:91-113.

GRACZA, R. 1960. Flour research problems. Cereal Sci. Today 5:166-173.

GRACZA, R. 1962. Average particle size and specific surface of flours and air-classified flour fractions. Cereal Sci. Today 7:272-274,

276, 284.

GREENWELL, P., and SCHOFIELD, J. D. 1986. A starch granule protein associated with endosperm softness in wheat. Cereal Chem. 63:379-380.

GROSH, G. M., and MILNER, M. 1959. Water penetration and internal cracking in tempered wheat grain. Cereal Chem. 36:260-273.

GUY, E. J., VETTEL, H. E., and PALLANSCH, M. J. 1967. Effect of the salts of the lyotropic series on the farinograph characteristics of milk-flour dough. Cereal Sci. Today 12:200-203.

HANDLEMAN, A. R., CONN, J. F., and LYONS, J. W. 1961. Bubble mechanics in thick foams and their effects on cake quality. Cereal Chem. 38:294-305.

HEPPNER, W. A. 1959. The fundamentals of cracker production. Bakers Dig. 33(2):68-70, 85-86.

HODGE, D. G. 1975. Alternatives to chlorination for high ratio cake flours. Baking Ind. J. 8(1):12, 14, 16-17.

HOLM, J. T. 1986. The use of chemical leavening in frozen doughs. M.S. thesis, Kansas State Univ., Manhattan.

HOLME, J. 1962. Characterization studies on the soluble proteins and pentosans of cake flour. Cereal Chem. 39:132-144.

HOSENEY, R. C. 1986. Principles of Cereal Science and Technology. Am. Assoc. Cereal Chem., St. Paul, MN.

HOSENEY, R. C., and SEIB, P. A. 1973. Structural differences in hard and soft wheats. Bakers Dig. 47(6):26-28, 56.

HUANG, G., FINN, J. W., and VARRIANO-MARSTON, E. 1982. Flour chlorination. II. Effects on water binding. Cereal Chem. 59:500-506.

INGLE, T. R., and WHISTLER, R. L. 1964. Action of chlorine on semidry starch. Cereal Chem. 41:474-483.

JOHNSON, A. C., and HOSENEY, R. C. 1979. Chlorine treatment of cake flours. III. Fractionation and reconstitution techniques for Cl₂-treated and untreated flours. Cereal Chem. 56:443-445.

JOHNSON, A. C., HOSENEY, R. C., and GHIASI, K. 1980. Chlorine treatment of cake flours. V. Oxidation of starch. Cereal Chem. 57:94-96.

JONES, C. R., HALTON, P., and STEVENS, D. J. 1959. The separation of flour into fractions of different protein contents by means of air-classification. J. Biochem. Microbiol. Technol. Eng. 1:77-98.

KICHLINE, T. P., and CONN, T. F. 1970. Some fundamental aspects of leavening agents. Bakers Dig. 44(4):36-40.

KINSELLA, J. E., and HALE, M. L. 1984. Hydrophobic associations and gluten consistency: Effects of specific anions. J. Agric. Food Chem. 32:1054-1056.

KISSELL, L. T. 1959. A lean-formula cake method for varietal evaluation and research. Cereal Chem. 36:168-175.

KISSELL, L. T. 1971. Chlorination and water-solubles content in flours of soft wheat varieties. Cereal Chem. 48:102-108.

KISSELL, L. T., and YAMAZAKI, W. T. 1979. Cake-baking dynamics: Relation of flour-chlorination rate to batter expansion and layer volume. Cereal Chem. 56:324-327.

KISSELL, L. T., POMERANZ, Y., and YAMAZAKI, W. T. 1971. Effects of flour lipids on cookie quality. Cereal Chem. 48:655-662.

KISSELL, L. T., MARSHALL, B. D., and YAMAZAKI, W. T. 1973. Effect of variability in sugar granulation on the evaluation of flour cookie quality. Cereal Chem. 50:255-264.

KISSELL, L. T., YAMAZAKI, W. T., and DONELSON, D. H. 1974. Relation of batter expansion and contraction to cake volume. (Abstr.) Cereal Sci. Today 19:400-401.

KOSMOLAK, F. G. 1978. Grinding time—Screening test for kernel hardness in wheat. Can. J. Plant Sci. 58:415-420.

KRAMER, H. H., and ALBRECHT, H. R. 1948. The adaption to small samples of the pearling test for kernel hardness in wheat. J. Am. Soc. Agron. 40:422-431.

KULP, K. 1968. Pentosans of wheat endosperm. Cereal Sci. Today 13:414-417, 426.

KULP, K., TSEN, C. C., and DALY, C. J. 1972. Effect of chlorine on the starch component of soft wheat flour. Cereal Chem. 49:194-200.

LaBAW, G. D. 1982. Chemical leavening agents and their use in bakery products. Bakers Dig. 56(1):16-21.

LAI, F. S., ROUSSER, R., BRABEC, D., and POMERANZ, Y. 1985. Determination of hardness in wheat mixtures. II. Apparatus for automated measurement of hardness of single kernels. Cereal Chem. 62:178-184.

LAMB, C. A., and BODE, C. E. 1963. Quality evaluation of soft winter wheat. Ohio Agric. Exp. Stn. Res. Bull. 926.

LEE, C. C., and HOSENEY, R. C. 1982. Optimization of the fat-emulsifier system and the gum-egg white-water system for a laboratory-scale single-stage cake mix. Cereal Chem. 59:392-395.

LEE, C. C., HOSENEY, R. C., and VARRIANO-MARSTON, E. 1982. Development of a laboratory-scale single-stage cake mix. Cereal Chem. 59:389-392.

LORENZ, K., and VALVANO, R. 1981. Functional characteristics of sprout-damaged soft wheat flours. J. Food Sci. 46:1018-1020.

MacRITCHIE, F. 1980. Physicochemical aspects of some problems in wheat research. Pages 271-326 in: Advances in Cereal Science and Technology, Vol. III. Y. Pomeranz, ed. Am. Assoc. Cereal Chem., St. Paul, MN.

MANLEY, D. J. R. 1983. Technology of Biscuits, Crackers and Cookies. Ellis Horwood Ltd., Chichester, England.

MATZ, S. H. 1968. Cookie and Cracker Technology. Avi Publ. Co., Inc., Westport, CT.

MATZ, S. H., and MATZ, T. D. 1978. Cookie and Cracker Technology, 2nd ed. Avi Publ. Co., Inc., Westport, CT.

McCLUGGAGE, M. E. 1943. Factors influencing the pearling test for kernel hardness in wheat. Cereal Chem. 20:686-700.

McCRATE, A. J., NIELSEN, M. T., PAULSEN, G. M., and HEYNE, E. G. 1981. Preharvest sprouting and α-amylase activity in hard red and hard white winter wheat cultivars. Cereal Chem. 58:424-428.

MICKA, J. 1955. Bacterial aspects of soda cracker fermentation. Cereal Chem. 32:125-131.

MILLER, B. S., and TRIMBO, H. B. 1970. Factors affecting the quality of pie crust. Bakers Dig. 44(1):46-55.

MILLER, B. S., TRIMBO, H. B., and POWELL, K. R. 1967a. Effects of flour granulation and starch damage on the cake making quality of soft wheat flour. Cereal Sci. Today 12:245-247, 250-252.

MILLER, B. S., TRIMBO, H. B., and SANDSTEDT, R. M. 1967b. The development of gummy layers in cakes. Food Tech. 21:377-385.

MILLER, B. S., AFEWORK, S., POMERANZ, Y., and BOLTE, L. 1981. Wheat hardness: Time required to grind wheat with a Brabender automatic micro hardness tester. J. Food Sci. 46:1863-1865.

MILLER, B. S., POMERANZ, Y., and AFEWORK, S. 1984. Hardness (texture) of hard red winter wheat grown in a soft wheat area and of soft red winter wheat grown in a hard wheat area. Cereal Chem. 61:201-203.

MONTZHEIMER, J. W. 1931. A study of methods for testing cake flour. Cereal Chem. 8:510-517.

MOORE, W. R., and HOSENEY, R. C. 1985. The leavening of bread dough. Cereal Foods World 30:791-792.

NAGAO, S., IMAI, S., SATO, T., KANEKO, Y., and OTSUBO, H. 1976. Quality characteristics of soft wheats and their use in Japan. I. Methods of assessing wheat suitability for Japanese products. Cereal Chem. 53:988-997.

NEEL, D. V., and HOSENEY, R. C. 1984a. Sieving characteristics of soft and hard wheat flours. Cereal Chem. 61:259-261.

NEEL, D. V., and HOSENEY, R. C. 1984b. Factors affecting flowability of hard and soft wheat flours. Cereal Chem. 61:262-266.

NGO, W., HOSENEY, R. C., and MOORE, W. R. 1985. Dynamic rheological properties of cake batters made from chlorine-treated and untreated flours. J. Food Sci. 50:1338-1341.

OLCOTT, H. S., and MECHAM, D. K. 1947. Characterization of wheat gluten. I. Protein-lipid complex formation during doughing of flours. Lipoprotein nature of the glutenin fraction. Cereal Chem. 24:407-414.

PARKS, J. R., HANDLEMAN, A. R., BARNETT, J. C., and WRIGHT, F. H. 1960. Methods for measuring reactivity of chemical leavening systems. I. Dough rate of reaction. Cereal Chem. 37:503-518.

PIEPER, W. E. 1971. Basic principles of saltine production. Baking Ind. 135(5):69-73.

PIZZINATTO, A., and HOSENEY, R. C. 1980. Rheological changes in cracker sponges during fermentation. Cereal Chem. 57:185-188.

POMERANZ, Y., AFEWORK, S., and LAI, F. S. 1985. Determination of hardness in mixtures of hard red winter and soft red winter wheats. I. Bulk samples. Cereal Chem. 62:41-46.

PRATT, D. B., Jr. 1963. Soft wheat flour. Air classification as a building block approach. Bakers Dig. 37(4):40-42.

PRITCHARD, P. E., and STEVENS, D. J. 1973. The influence of processing variables on the properties of wafer sheets. Food Trade Rev. 43(5):11-16.

PRITCHARD, P. E., and STEVENS, D. J. 1974. The influence of ingredients on the properties of wafer sheets. Food Trade Rev. 44(6):12-18.

PRITCHARD, P. E., and WADE, P. 1972. Development of a test baking procedure for wafer sheets. Food Trade Rev. 42(8):9-15.

REES, R. F. 1971. Importance of special flours in recipes. Baking Ind. J. 4(6):27-33.

REIMAN, H. M. 1977. Chemical leavening systems. Bakers Dig. 51(4):33-36, 42.

RUSSO, J. V., and DOE, C. A. 1970. Heat treatment of flour as an alternative to chlorination. J. Food Technol. 5:363-374.

SAVAGE, H. L., and OSMAN, E. M. 1978. Effects of certain sugars and sugar alcohols on the swelling of cornstarch granules. Cereal Chem. 55:447-454.

SEEBORG, E. F. 1953. Evaluation of wheat milling characteristics by laboratory methods. Trans. Am. Assoc. Cereal Chem. 11:1-5.

SEGUCHI, M. 1984. Oil-binding capacity of prime starch from chlorinated wheat flour.

Cereal Chem. 61:241-244.

SEGUCHI, M., and MATSUKI, J. 1977. Studies on pan-cake baking. I. Effect of chlorination of flour on pan-cake qualities. Cereal Chem. 54:287-299.

SEIBEL, W., MENGER, A., LUDEWIG, H.-G., and BRETSCHNEIDER, F. 1976. Standardization of a baking test for short sweet biscuits. Getreide Mehl Brot 30(12):235-330.

SIMMONDS, D. H. 1971. Morphological and molecular aspects of wheat quality. Wallerstein Lab. Commun. 34:17-34.

SIMMONDS, D. H. 1974. Chemical basis of hardness and vitreosity in the wheat kernel. Bakers Dig. 48(5):16-18, 20, 22, 24, 26-29, 63.

SIMMONDS, D. H., BARLOW, K. K., and WRIGLEY, C. W. 1973. The biochemical basis of grain hardness in wheat. Cereal Chem. 50:553-562.

SMITH, E. E. 1932. Report of the subcommittee on hydrogen-ion concentration with special reference to the effect of flour bleach. Cereal Chem. 9:424-428.

SOLLARS, W. F. 1956. Evaluation of flour fractions for their importance to cookie quality. Cereal Chem. 33:121-128.

SOLLARS, W. F. 1958a. Fractionation and reconstitution procedures for cake flours. Cereal Chem. 35:85-99.

SOLLARS, W. F. 1958b. Cake and cookie flour fractions affected by chlorine bleaching. Cereal Chem. 35:100-110.

SOLLARS, W. F. 1959. Effects of the water-soluble constituents of wheat flour on cookie diameter. Cereal Chem. 36:498-513.

SOLLARS, W. F. 1961. Chloride content of cake flours and flour fractions. Cereal Chem. 38:487-500.

SOLLARS, W. F., and BOWIE, S. M. 1966. Effect of the subfractions of starch tailings on cookie diameter. Cereal Chem. 43:244-260.

SOLLARS, W. F., and RUBENTHALER, G. L. 1971. Performance of wheat and other starches in reconstituted flours. Cereal Chem. 48:397-410.

SPIES, R. D., and HOSENEY, R. C. 1982. Effect of sugars on starch gelatinization. Cereal Chem. 59:128-131.

SPIES, R. D., and KIRLEIS, A. W. 1978. Effect of free flour lipids on cake-baking potential. Cereal Chem. 55:699-704.

STEELE, I. W. 1977. The search for consistency in biscuit doughs. Baking Ind. J. 9(9):21, 23-24, 33.

STENVERT, N. L., and KINGSWOOD, K. 1977. The influence of the physical structure of the protein matrix on wheat hardness. J. Sci. Food Agric. 28:11-19.

SUGIHARA, T. F. 1978a. Microbiology of the soda cracker process. I. Isolation and identification of microflora. J. Food Prot. 41:977-979.

SUGIHARA, T. F. 1978b. Microbiology of the soda cracker process. II. Pure culture fermentation studies. J. Food Prot. 41:980-982.

SYMES, K. J. 1965. The inheritance of grain hardness in wheat as measured by particle size index. Aust. J. Agric. Res. 16:113-123.

SYMES, K. J. 1969. Influence of gene causing hardness in the milling and baking quality of two wheats. Aust. J. Agric. Res. 20:971-979.

THEWLIS, B. H., and WADE, P. 1974. An investigation into the fate of sulphites added to hard sweet biscuit doughs. J. Sci. Food Agric. 25:99-105.

THOMSON, L. S. 1976. Flour needs for the commercial cracker process. Cereal Foods World 21:642-644.

TRIMBO, H. B., and MILLER, B. S. 1973. The development of tunnels in cakes. Bakers Dig. 47(4):24-26, 71.

TRIMBO, H. B., MA, S., and MILLER, B. S. 1966. Batter flow and ring formation in cake baking. Bakers Dig. 40(1):40-42, 44-45.

TSEN, C. C., KULP, K., and DALY, C. J. 1971. Effects of chlorine on flour proteins, dough properties, and cake quality. Cereal Chem. 48:247-255.

TSEN, C. C., BAUCK, L. J., and HOOVER, W. J. 1975. Using surfactants to improve the quality of cookies made from hard wheat flours. Cereal Chem. 52:629-637.

UCHINO, N., and WHISTLER, R. L. 1962. Oxidation of wheat starch with chlorine. Cereal Chem. 39:477-482.

UPADHYAY, M. P., PAULSEN, G. M., HEYNE, E. G., SEARS, R. G., and HOSENEY, R. C. 1984. Development of hard white winter wheats for a hard red winter wheat region. Euphytica 33:865-874.

UPTON, E. M., and HESTER, E. E. 1966. Nonstarchy polysaccharides and proteins of soft wheat flour tailings. Cereal Chem. 43:156-168.

VAN WAZER, J. R., and ARVAN, P. G. 1954a. Chemistry of leavening. 1. Milling Prod. (Feb.):5-8, 22.

VAN WAZER, J. R., and ARVAN, P. G. 1954b. Chemistry of leavening. 2. Milling Prod. (Mar.):5-7, 17.

VARRIANO-MARSTON, E. 1985. Flour chlorination: New thoughts on an old topic. Cereal Foods World 30:339-343.

WADE, P. 1971a. The effect of flour quality on the properties of semi-sweet biscuits. Food Trade Rev. 41(7):19-25.

WADE, P. 1971b. Technology of biscuit manufacture: Investigation of the process for making semi sweet biscuits. Chem. Ind. pp. 1284-1293.

WADE, P. 1972a. Flour properties and the manufacture of semi sweet biscuits. J. Sci. Food Agric. 23:737-744.

WADE, P. 1972b. Technology of biscuit manufacture: Investigation of the role of fermentation in the manufacture of cream crackers. J. Sci. Food Agric. 23:1021-1034.

WADE, P. 1972c. Flour properties and the manufacture of cream crackers. J. Sci. Food Agric. 23:1221-1228.

WADE, P. 1985. Wheat kernel hardness in relation to biscuit (cookie) making properties of flours. Pages 93-97 in: Analyses as Practical Tools in the Cereal Field—An ICC Symposium. K. M. Fjell, ed. Norwegian Grain Corp., Oslo.

WADE, P., and WATKIN, D. A. 1967. Process control in the biscuit industry. Pages 39-44 in: Symp. Ser. 24. Inst. Chem. Eng., London.

WAINWRIGHT, A. R., COWLEY, K. M., and WADE, P. 1985. Biscuit making properties of flours from hard and soft milling single variety wheats. J. Sci. Food Agric. 36:661-668.

WHISTLER, R. L., MITTAG, T. W., and INGLE, T. R. 1966. Mechanism of starch depolymerization with chlorine. Cereal Chem. 43:362-371.

WICHSER, F. W. 1958. Air-classified flour fractions. Cereal Sci. Today 3:123-126.

WILLIAMS, P. C. 1979. Screening wheat for protein and hardness by near infrared reflectance spectroscopy. Cereal Chem. 56:169-172.

WILSON, J. T., DONELSON, D. H., and SIPES, C. R. 1964. Mechanism of improver action in cake flours. I. The relation between flour specific surface and chlorine distribution. Cereal Chem. 41:260-274.

YAMAZAKI, W. T. 1953. An alkaline water retention capacity test for the evaluation of cookie baking potentialities of soft winter wheat flours. Cereal Chem. 30:242-246.

YAMAZAKI, W. T. 1954. Interrelations among bread dough absorption, cookie diameter, protein content, and alkaline water retention capacity of soft winter wheat flours. Cereal Chem. 31:135-142.

YAMAZAKI, W. T. 1955. The concentration of a factor in soft wheat flours affecting cookie quality. Cereal Chem. 32:26-37.

YAMAZAKI, W. T. 1959a. Flour granularity and cookie quality. I. Effect of wheat variety on sieve fraction properties. Cereal Chem. 36:42-51.

YAMAZAKI, W. T. 1959b. Flour granularity and cookie quality. II. Effects of changes in granularity on cookie characteristics. Cereal Chem. 36:52-59.

YAMAZAKI, W. T. 1959c. The application of heat in the testing of flours for cookie quality. Cereal Chem. 36:59-69.

YAMAZAKI, W. T. 1972. A modified particle-size index test for kernel texture in soft wheat. Crop Sci. 12:116.

YAMAZAKI, W. T., and DONELSON, D. H. 1972. The relationship between flour particle size and cake-volume potential among Eastern soft wheats. Cereal Chem. 49:649-653.

YAMAZAKI, W. T., and DONELSON, D. H. 1973. Evaluating soft wheat quality of early generation progenies. Crop Sci. 13:374-375.

YAMAZAKI, W. T., and DONELSON, J. R. 1983. Kernel hardness of some U.S. wheats. Cereal Chem. 60:344-350.

YAMAZAKI, W. T., and GREENWOOD, C. T., eds. 1981. Soft Wheat: Production, Breeding, Milling, and Uses. Am. Assoc. Cereal Chem., St. Paul, MN.

YAMAZAKI, W. T., and LORD, D. D. 1971. Soft wheat products. Pages 743-776 in: Wheat: Chemistry and Technology, 2nd ed. Y. Pomeranz, ed. Am. Assoc. Cereal Chem., St. Paul, MN.

YOUNGQUIST, R. W., HUGHES, D. H., and SMITH, J. P. 1969. Effect of chlorine on starch-lipid interactions. (Abstr. 5) Cereal Sci. Today 14:90.

CHAPTER 8

FLAT BREADS

HAMED FARIDI
Nabisco Brands, Inc.
RMS Technology Center
East Hanover, New Jersey

I. INTRODUCTION

For thousands of years, cereals have been the staple food for most people. The saving in energy and the increased productivity when cereals are consumed directly are so great that even the Sumerians, 4,000 years ago, recognized that "a given acreage of land put down to wheat or barley filled more stomachs, more quickly and more cheaply than the same land given over to livestock"(Tannahill, 1973).

Although no record exists of where and when bread originated, there is little doubt that flat breads are the most ancient of all breads. Baking cereal gruel into a palatable flat bread was discovered in the Neolithic period (late Stone Age). The gruel was made into a dough and shaped into flat cakes, which were baked on hot stones or directly in a fireplace. By the end of the Neolithic period (about 1800 B.C.), the Egyptians had spread knowledge of breadmaking and fermentation throughout the Mediterranean world. Thus, by the time of the Bronze and Iron ages (1800 B.C.–1 A.D.) flat, sourdough bread had become the main European food. In the Middle Ages, loaves of bread were rounded or semicircular and flat and marked with a cross, a Christian symbol that made it easier to break the loaf (Pomeranz and Shellenberger, 1971). Readers desiring more detailed information could consult the extensive study of Wahren (1959, 1961, 1962) on the history of bread baking in ancient Egypt and other North African countries, including a description of baking techniques of bedouins and nomads of the Egyptian deserts.

At present, flat breads are consumed around the world. Until a few years ago, few people in the United States had heard of flat breads, but the growing success of restaurants with cultural roots in India, Greece, Mexico, Turkey, and various Arab countries and a growing tendency to travel abroad have changed all that. The increasing popularity of flat breads in the United States has resulted in the installation of a number of Middle Eastern bakeries. The products of these commercial bakeries, known as pita, are sold either fresh, refrigerated, or frozen (Rashid, 1983).

Breads of the world may be divided into three groups with respect to their density: 1) those with high specific volume (v/w), such as Western white pan bread; 2) those with medium specific volume, such as French and rye breads; and 3) those with low specific volume, such as pita and *lavosh* breads and flat breads of Northern Europe, the Middle East, and India. Thus, flat breads are dense in texture, mostly crust with little crumb, and usually round but sometimes triangular or rectangular. Their diameter varies from 5–10 cm up to 70 cm; in thickness, they range from paper-thin to 4 cm thick. The crust is thin and light with brown or dark spots. The crumb is small in quantity and coarse and dense. Flat breads have a higher crust-to-crumb ratio than do pan breads.

II. TYPES OF FLAT BREADS

Table I describes the major flat breads produced around the world, some of which are shown in Figs. 1 and 2. Among the numerous types, the most popular

TABLE I
Traditional Flat Breads[a]

Type of Bread	Country or Region	Description
Arepa	Venezuela	Disk-shaped corn bread made from dehulled and cooked maize dough called *masa*. Dough pieces are flattened to the desired thickness and cooked on each side for approximately 2 min on a clay or metal griddle.
Bakoom	Egypt	Disk-shaped, two-layered bread made from white flour, water (60%), and salt (0.1%). All ingredients are mixed, and the dough is shaped immediately and proofed for 90 min prior to baking at 300° C for 40–60 sec.
Balady	Egypt	Two-layered, disk-shaped flat bread made from flour of 85–100% extraction. The water absorption is very high (70–80%). All ingredients are mixed and fermented for 90 min. The dough is then shaped into balls, flattened to disk-shaped loaves, and proofed for an additional 30 min while covered with a thin layer of wheat bran. It is baked at 500° C for 40–60 sec. The bread weighs about 150 g.
Bannock	Scotland	Round, flat pancake, usually unleavened, made of oatmeal, barley flour, or wheat flour.
Barbari	Iran	Usually 70–80 cm long, 25–30 cm wide, and 2.5 cm thick. All ingredients (flour, water, salt, soda, and sour starter) are mixed to the proper consistency and fermented for 2 hr. Dough balls (800–900 g) are flattened into an oval shape and rested for 20 min. Then a teaspoonful of a concentrated, boiled, and cooled mixture of flour, water, and oil is poured on the surface, to make it shiny and brown after baking. The dough is then docked and grooved with fingers, to form five or six 1-cm-deep rows. The final proof time is 15 min. The oven is dome-shaped, made of fire-resistant bricks, and heated by petrol burners. The baking time is 8–12 min at 220° C. The bread weighs 700 g.
Barley *maloug*	Yemen	Very similar to *maloug* but made from barley flour.

(*continued on next page*)

TABLE I (continued)

Type of Bread	Country or Region	Description
Battawi	Egypt	Disk-shaped flat bread made from flour (87% extraction), water (50%), salt (0.1%), and fenugreek seed flour (2.5%). The ingredients are mixed, and the dough is shaped immediately and baked at 300°C for 40–60 sec. No proofing is required.
Bazlama	Turkey	Very popular flat bread made from flour, water, salt, and sour starter. After 2–3 hr of fermentation, 200- to 250-g pieces of dough are cut, rounded, sheeted to a thickness of 4–5 mm, and baked on a hot plate.
Burr	Saudi Arabia	Whole wheat Arabic bread, or *mafrood*.
Caucasian bread	Soviet Union	Made from butter-yeast dough in round, flat pieces (17–18 cm in diameter, 3–3.5 cm thick). It includes butter, sugar, milk, yeast, vodka, salt, saffron, eggs, and flour. The dough is fermented for 1–2 hr, cut into 300-g pieces, rounded, and pressed into round pieces. The surface is covered with eggs. The dough is proofed for 20 min and baked for 8–10 min at 200°C.
Chapati (*roti*)	Pakistan, India, Tibet, Mongolia, China	Flour is mixed with water and kneaded by hand to the desired consistency. The water absorption is usually high, ranging from 65 to 75% and sometimes as high as 80%. The dough is consequently slack. It is allowed to rest for a length of time depending on the convenience of the cook. After resting, the dough is divided, rounded, rested for a couple of minutes, and then sheeted with a rolling pin or by hand to a flat, round shape. The sheeted dough is immediately transferred to a preheated hot plate, where it is turned two or three times during cooking. The total cooking time is 70–100 sec. The bread is usually made at home.
Corn bread	Egypt	Disk-shaped, white flat bread made from whole corn flour, water (60%), and salt (0.1%). All ingredients are mixed, and the dough is shaped and baked at 300°C for 5–8 min. No fermentation is required. The bread weight varies, depending on the baker.
Dosai	India, Sri Lanka	Slightly crisp, round pancake made from a fermented mixture of cereals and black gram. Rice is the preferred cereal.
Fatier	Egypt	Made from the same dough as *mabbatt* and *kabbouri*. The dough is stretched very thin, melted butter is spread over it, it is folded over and stretched again, and more butter is spread over it. The many thin layers of dough coated with butter are placed in a well-buttered pan and baked.
Fiti	Egypt	Unleavened pancake made from wheat flour, salt (1%), and water (125%). It is made from a batter and cooked on an oiled griddle called a *doka*. When sorghum flour is used, the bread is called *senesen*.
Glarus	Switzerland	Flat, round bread.
Gomme	Turkey	Made from a stiff dough of flour and milk. The dough pieces are flattened to disks (35–40 cm in diameter, 2–5 cm thick), placed on a hot stone, and covered with a thin iron plate with hot ashes around and on top.
Hamursuz, Peksimet, or *Halka*	Turkey	*Bazlama* type of flat and fermented breads baked in a "peel oven."

(continued on next page)

TABLE I (continued)

Type of Bread	Country or Region	Description
Idli	India	Pancake type of rice cake, with a coarse particle size, steamed into small cakes.
Injera	Ethiopia	Fermented, round flat bread (50–60 cm in diameter) made from sorghum. The dough is fermented for two or three days, boiled in excess water, cooled, proofed, and steamed for 2–3 min on a clay griddle.
Kabbouri	Egypt	Made from the same dough as *mabbatt*, but with a smaller diameter (20 cm).
Khameri roti	India, Pakistan	Usually made during summer. Sugar and salt are added. Milk, buttermilk, curd, or yogurt partly replaces the water. A sponge made from whole wheat flour, yogurt, salt, and sugar is fermented overnight. A second portion of flour, soda, and water are added the following day. The bread is baked like *roti*.
Kisra	Sudan	Thin bread made from sorghum with a small portion of millet or wheat flour. A thick paste of flour, water, and starter is fermented for 12–16 hr and baked for 30–40 sec on a lightly oiled metal or clay sheet.
Korsan	Saudi Arabia	Flat, circular bread (60 cm in diameter) with a single thin layer, made from whole wheat flour, water (65%), and salt (0.35%). All ingredients are mixed, fermented (30 min), remixed, and fermented again (60 min). Dough balls of 180–200 g are rounded, proofed for 45 min, flattened, and baked at 200°C for 2 min. The bread weighs about 65 g. It has a light brown crust and no crumb.
Lavosh	Iran, Turkey	Paper-thin white bread about 60–70 cm long, 30–40 cm wide, and 2–3 mm thick. Ingredients (flour, water, salt, and sour starter) are mixed to the proper consistency and fermented for 1–3 hr. Dough balls of 300 g are put on a round wooden plate and worked to the proper thickness with rollers of different sizes. The dough is then put on a cushion and stuck to the walls of the oven and baked for only 60–90 sec at 250°C. The oven is barrel-shaped and forms a 45° angle with the floor; it is heated with a petrol burner. The bread weighs about 225 g.
Lefse	Sweden	Flat bread made from a blend of wheat and potato flours.
Mabbatt	Egypt	Semicircular loaf about 40 cm in diameter, with a thin crust and a thin crumb layer, which has large, open cells. The formula consists of a blend of corn and wheat flours (1:3), water (85%), salt (1%), and yeast (1%). It occasionally includes sugar, egg, and honey. After mixing and a short fermentation, dough balls of 750 g are flattened. A layer of bran is sprinkled over the dough. The flattened pieces are each divided into two semicircles, proofed for 30 min, and baked for a few minutes. The bread has a low moisture content and a good shelf life.
Maloug, or Yemeni bread	Yemen	Thin, round bread (35 cm in diameter) made from white flour, water (40%), salt (0.2%), and yeast (0.1%). All ingredients are mixed and fermented for 2 hr. Pieces of dough are stretched out to about 35 cm in diameter and placed by hand against the inner wall of a clay oven, previously heated by burning wood and cardboard. After about 90 sec, the bread is scraped from the oven wall with a long metal spatula.

(*continued on next page*)

TABLE I (continued)

Type of Bread	Country or Region	Description
Matzo	Israel, United States	Baked product consumed during the Jewish feast of Passover to commemorate the exodus from Egypt. It is round or rectangular. Typically, the recipe uses about 100 parts of flour to 38 of water. This mixture is gently rolled together in a mixer to form a crumbly "dough." There is no dough development. The sheeter presses the mix together to form a sheet, which, after reduction, is simply laminated with two to six layers. After further gauging, the sheet becomes clear and strong. It is heavily docked and cut and is then baked for 1 min at 400°C. The high temperature causes much blistering, but the heavy docking keeps the blisters small. The blisters take up color, and the rest of the biscuit remains pale. Some moisture continues to be lost after the oven exit; a final moisture content of about 3% is typical.
Millet bread	Egypt	Disk-shaped or rectangular flat bread made from whole millet flour, water (50%), salt (0.1%), and fenugreek seed flour (2%). The ingredients are mixed, and the dough is immediately shaped and baked at 300°C for 40–60 sec. No proofing is required.
Mlinci	Yugoslavia	Thin, *chapati* type of bread baked in Croatia.
Mon-le-bway	Burma	Crisp, thin sheets, made of fried rice flour batter. They are very light and filled with air bubbles.
Moroccan sourdough bread	Morocco	Round bread (18 cm in diameter, 2 cm thick) made in two stages. A portion of white flour (70 parts), water (45%), and yeast (1%) are mixed and allowed to ferment for 90 min. Then salt (2%), the remaining flour (30 parts), and water (15%) are added and mixed to optimum development. The dough is fermented for an additional 30 min, rounded, flattened (18 cm in diameter, 5 mm thick), and proofed for 45 min. The bread is baked for 20 min at 220°C.
Moroccan whole wheat bread	Morocco	The same size as Moroccan sourdough bread. All ingredients (flour, water, salt, and yeast) are mixed, and the dough is immediately flattened (18 cm in diameter, 5 mm thick), allowed to ferment for 3 hr, docked (holes are punched or cut in it), and baked in a relatively cool oven (180°C) for 55 min.
Naan	India, Pakistan	Fermented flat bread made from flour, yeast (2%), sugar (1%), salt (2%), water (35%), yogurt (25%), and shortening (6%). The dough is mixed, immediately divided into balls, and allowed to ferment while covered with wet cloth. It is then sheeted into thin, oblong, flat pieces and baked at 315°C for 2 min.
Nepalese bread	Nepal, northern India	Flat, triangular fried bread made of whole wheat flour, yeast (1%), water (30%), fresh milk (5%), sugar (1%), salt (0.5%), egg (2%), and small quantities of cinnamon, nutmeg, and cloves. All ingredients are mixed to optimum development. The dough is allowed to rise for 90 min and then is divided into 250-g balls and allowed to rest for an additional 30 min. Each ball is rolled into a circle, 18–20 cm in diameter, on a lightly floured board. Each circle is cut into quarters and deep fried in cooking oil at 350°C.

(continued on next page)

TABLE I (continued)

Type of Bread	Country or Region	Description
Oklablomos	Greece	Round and flat.
Orus	Egypt	Breakfast bread made for special occasions. The formula includes wheat flour, fresh milk (25%), butter (25%), sesame seed (6%), and some olive oil. All ingredients are mixed together. The dough balls are flattened to circles 12 cm in diameter. A small handful of filling (chopped dates, sesame seed, and olive oil) is placed in the center of each piece. The dough is folded over the filling, sealed, and flattened again to a 12-cm circle. The dough pieces are allowed to proof overnight and are baked early in the morning.
Paratha	India	Unleavened bread made from whole wheat flour, salt (2.5%), water (65%), and shortening (40%). The dough is mixed with some of the shortening and sheeted like *chapati* and *puri*, with one side smeared with shortening. Then it is folded, twisted, resheeted, and cooked on a preheated hot plate. When both sides are partly cooked, it is smeared with sufficient shortening on both sides and cooked to completion (4–5 min).
Plank	Wales	Flat bread.
Puri	India	Dough is prepared as for *chapati*, sheeted (with the help of a little shortening), gently lowered into preheated oil, and flipped over once (while being pressed gently with a slotted spatula, to help puffing). The frying time is 1–2 min.
Raghif	Israel	Thick flat bread made by peasants from whole wheat flour and water. It has a soft crust, indented on the bottom from the pebbles on which it is rested while baking.
Ra'rou, or *mashtouah*	Egypt	Thin, crisp bread. It is sprinkled with water and covered with a cloth to freshen it before eating, or it is crushed and eaten in milk or soup. The formula includes wheat and corn flours (1:1), okra flour (10%), salt (1%), water (75–80%), and sourdough. After mixing, the dough is fermented for 1 hr, flattened to a thin layer, and baked for a few minutes.
Rye crisp	Sweden, Soviet Union, Norway, Finland	Thin crisp bread made from whole rye or a blend of rye and wheat, mixed intensely to incorporate air. The dough is flattened, cut, and baked for about 7 min, to dry with little browning.
Sangak	Iran	Sourdough flat bread, 70–80 cm long, 40–50 cm wide, and 3–5 cm thick. Sometimes its surface is sprinkled with sesame or poppy seeds. The ingredients (flour, water, salt, and sour starter) are mixed to the proper consistency and fermented for 2 hr. Water absorption is very high (85%), with the result that a batterlike dough is produced. Balls of 500 g are flattened on a special paddle, docked with fingers, and placed on the floor of a previously heated oven covered with small stones, which are moistened from time to time with a soap solution. The oven is dome-shaped and heated with a petrol burner. The baking time is 3–5 min at 250° C. The bread weighs about 400 g.

(*continued on next page*)

TABLE I (continued)

Type of Bread	Country or Region	Description
Shamey, white Arabic *mafrood*, or pita	Egypt, Syria, Lebanon, Saudi Arabia, Jordan, United States, Canada	Two-layered, disk-shaped flat bread made from white flour (72% extraction). The water absorption is very low (40–50%). All ingredients are mixed and fermented for 90 min. The dough is shaped into balls, flattened to disks, and proofed for 30 min before baking at 500° C for 40 sec. The bread weighs about 120 g.
Shamsy, or sunny bread	Egypt, Sudan	Squared block-shaped or disk-shaped bread with a brownish crust and white, firm crumb, made from a lean formula. Flour (87% extraction), water (50%), and salt (0.1%) are mixed, and the dough is shaped into pieces and left under the intense heat of the sun for 3 hr. It is then baked at 300° C for 15 min. The bread weighs around 300 g.
Sinn bread	Egypt	"All-bran" bread made from bran, white flour (10%), sugar (1%), salt (1%), vegetable oil (0.5%), and water (60%). It is tan and has a sweet, nutty flavor. After mixing, the dough is fermented for 1 hr, scaled, flattened with a wooden roller, and placed on a bran-coated tray for proofing for 1 hr. The loaves are baked for 3–5 min at 500° C.
Steam bread	China	Round, small bread made from white flour, water (65%), sugar (6%), shortening (3%), and yeast (1%). All ingredients are mixed and fermented for 3–4 hr. An additional 30% flour is mixed with the dough, which is then sheeted and molded, cut into small pieces (3–6 cm long, 2–3 cm in diameter), and proofed for 50 min. The pieces are steamed in special pots over boiling water for 10 min.
Taftoon	Iran	Round, sourdough flat bread (40–50 cm in diameter) with small holes on its surface. The ingredients (flour, water, soda, salt, and sour starter) are mixed to the proper consistency and fermented for about 1 hr. Dough balls (350–400 g) are flattened and worked into the proper thickness with rollers of different sizes. The dough is put on a cushion stuck to the walls of the oven and baked for 2 min at 250° C. The oven is barrel-shaped and forms a 45° angle with the floor; it is heated with a petrol burner. The bread weighs about 300 g.
Tamees	Saudi Arabia	Flat, circular bread (35 cm in diameter) made from white flour, water (65%), salt (0.4%), yeast (0.1%), sugar (0.2%), shortening (6%), and baking soda (0.2%). All ingredients are mixed and fermented for 45–60 min. The dough is then divided into 700- to 800-g balls, which are fermented for 2–3 hr and then molded, flattened, and baked at 300° C for 2 min. The bread weighs about 500 g. It has a golden brown crust with dark spots and little white crumb.
Tandouri roti	Pakistan, India	The method of preparation is similar to that for *chapati*, except that salt is sometimes added at the time of kneading, and the dough balls are generally heavier. *Roti* is baked in a *tandoor*, an oval in-ground oven, the walls of which are plastered with clay. It is heated by burning wood or natural gas. The sheeted dough is placed on a cloth pad and, with the help of the pad, is pasted to the heated walls of the *tandoor*.

(*continued on next page*)

TABLE I (continued)

Type of Bread	Country or Region	Description
		Depending on the heat contained in the walls of the *tandoor*, the *roti* may be baked in 60–90 sec. The baked *roti* is recovered with a long iron, the end of which is formed into an L shape.
Tannouri	Saudi Arabia	Round flat bread (24 cm in diameter) made from brown flour, water (50%), salt (0.2%), and yeast (0.4%). All ingredients are mixed to optimum development and fermented for 1 hr. The dough is divided into 250-g balls and proofed for another hour. It is molded, flattened, and baked at 500°C for 1 min. The bread weighs about 175 g. It has a golden brown crust with dark spots, very little crumb, and a pierced upper surface.
Tannouri, or *gubbuz*	Iraq, Jordan, Afghanistan	Thin, round flat bread (35 cm in diameter) made from flour, water, salt, and starter. The baking time is about 20 min in a *tannour* type of oven.
Terabelesi	Tunisia	Round bread (20 cm in diameter) made from white flour, water (60%), salt (2%), and yeast (1%). The ingredients are mixed to optimum dough development and fermented for 30–45 min. The dough is divided into 700-g balls and rolled out to a thickness of 2 cm. Proof time is 45 min. Four cuts are made across the top to form a square before the dough is sent to the oven. The baking time is about 35 min at 220°C. The bread weighs about 650–700 g.
Tortilla	Mexico, Central America, United States	Unfermented, round flat bread prepared from alkali-cooked maize. It may also be fried, to make taco shells. Tortillas are also prepared from white wheat flour or sorghum.
Village bread, *tanok*, or shepherd's bread	Iran, Afghanistan	A common bread baked in rural areas. Almost all kinds of village breads are round (30–40 cm in diameter); they vary from paper-thin to 3.5 cm thick. The ingredients (flour, water, and salt) are mixed by hand and fermented for 30–90 min. Dough balls are rolled to the desired thickness and baked lightly in large flaps on sheet metal over a fire fed by straw or in an oven consisting of an egg-shaped hole in the ground heated by charcoal, wood, or dung in the bottom of the oven. The flattened dough is stuck to the walls of the oven and left for 2–4 min, depending on the oven temperature.
Yufka	Turkey	Thin, round flat bread (40–50 cm in diameter) made from flour, water, and salt kneaded and rested for 30 min. Dough pieces (150–200 g) are rounded, sheeted, and baked.
Zanzibar flat bread	Zanzibar	Made from wheat flour, water, salt, and soda. The sour starter is prepared the previous night, and the bread is baked in an egg-shaped oven.

[a] Data from Adler (1958); Kuzazli et al (1966); Pellet and Shadaverian (1970); Pomeranz and Shellenberger (1971); Maleki (1972); Nawar (1974); Chaudri and Muller (1976); Desikachar (1977); Tabekhia and Toma (1979); Milatovic (1979); Faridi and Finney (1980); Finney et al (1980b); Faridi and Rubenthaler (1983a, 1983b); Sawaya et al (1984); K. R. Davis, A survey of Egyptian village bread, unpublished report, 1985, Dep. Bacteriol. Biochem., Food Res. Cent., Univ. Idaho, Moscow; and Mousa and Al-Mohizea (1987).

Fig. 1. Popular flat breads in the United States.

are Egyptian *balady*, Arabic bread, *barbari, chapati*, Swedish crisp bread, and extruded flat breads.

A. Egyptian *Balady* Bread

Balady bread, produced in Egypt and other Arab countries, is a special type of

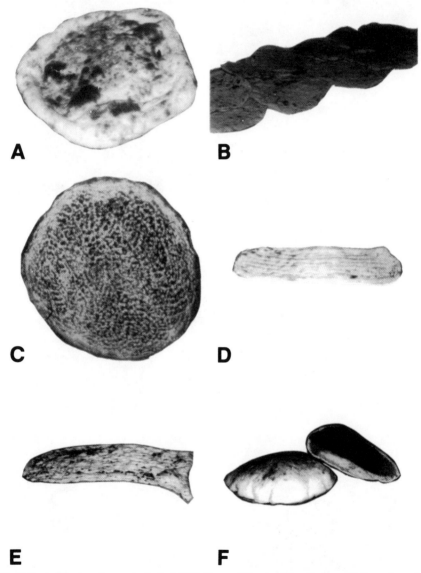

Fig. 2. Typical flat breads popular in the Middle East and North Africa: A, *tannouri*; B, *mashtouah*; C, *tamees*; D, *barbari*; E, *sangak*; F, *mafrood*. (A, C, and F reprinted, with permission, from Mousa and Al-Mohizea, 1987; B used, with permission, from K. R. Davis, A survey of Egyptian village bread, unpublished report, 1985, Dep. Bacteriol. Biochem., Food Res. Cent., Univ. Idaho, Moscow)

product. The formula is simple: flour (80–90% extraction), water, a small amount of salt, and a source of yeast, either regular yeast or a mother dough. The bread has two special characteristics. First, the absorption is very high, usually 75–85% (flour basis), and consequently the dough has the consistency of a batter. Faridi and Rubenthaler (1983b) used a dough with a farinograph water absorption of 200 Brabender units (BU), compared to 400–500 BU in regular white pan bread. The dough of *balady* bread tends to be sticky and is normally scaled by hand. Second, the baking temperature is high, in the range of 500–600°C, and the baking time is only a minute (Fig. 3).

The dough is mixed in the usual manner and permitted to ferment for several hours. It is then scaled, and the dough pieces are deposited on a tray covered with a thin layer of bran. The dough ball is then pressed into a flat pancake shape by hand. After an additional 30 min of fermentation, the excess bran is shaken from the dough (a considerable amount adheres), and the loaf is introduced into the oven. The bread is hearth-baked. The high temperature immediately causes the dough to rise, more from steam formation than from carbon dioxide production. At the end of the baking period, which averages 60 sec, the bread is 7–10 cm high at the center of the loaf. As it cools, it falls to some extent. In packing and shipping, it returns to an essentially flat shape.

When the loaf is cut into two portions, the bread crumb adheres to the top and

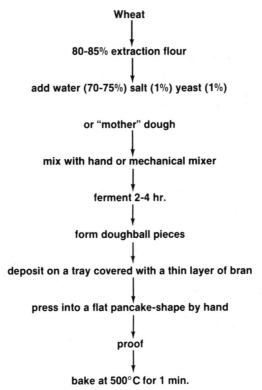

Fig. 3. Traditional procedure for the production of *balady* bread.

bottom crusts, leaving an open space, or pocket, between the top and the bottom. This is convenient to many consumers, because the cavity can be filled with beans, vegetables, or other foods, and the filled bread can be eaten without plates or forks. On the average, the crust and crumb each constitute 50% of the loaf (Rizk et al, 1960a, 1960b; Dalby, 1963, 1969; Doerry, 1982).

B. Arabic Bread

Many varieties of leavened flat breads are consumed in the Middle East and North Africa. One type of pocket bread is very popular. It is made in various regions, with much the same basic formula and processing methodology, and the finished products from different regions have the same general appearance. These products, named *shamey*, Syrian bread, *kohbz, mafrood*, etc., are classified under the general category of Arabic leavened flat bread. Traditionally, the bulk dough is fermented by airborne wild bacteria and yeasts, but baker's yeast or sodium bicarbonate is sometimes included in the formulation. Sodium bicarbonate serves as a leavening agent in addition to airborne microorganisms and imparts a typical soda flavor. The yeasty aroma of baker's yeast, when used as the single source of leavening, is less acceptable in traditional Arabic bread, especially in rural areas.

There are two types of Arabic bread, white and brown, depending on the extraction rate of the flour. White Arabic bread is made from flour of 72% extraction; brown Arabic bread is made from flour of 90–95% extraction.

The bread is made in the following steps: mixing, primary fermentation, dividing (scaling), intermediate proofing, two-dimensional sheeting, final proofing, and baking. A wide range of equipment (Fig. 4) and procedures are used. In most upgraded and automated bakeries, the standard formula includes flour, water (40–55%), salt (1%), yeast (1%), and sugar (0.5%). The dough is rather dry. Abdel-Rahman and Youssef (1978) based the water absorption on a

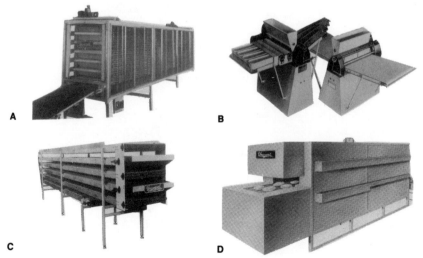

Fig. 4. Equipment used for the production of Arabic bread: A, fermentation cabinet; B, molder and - divider; C, proofer; D, oven. (Courtesy Universal Oven, Westbury, NY)

farinograph consistency of 800 BU. Many bakeries use a little sugar to stimulate fermentation in winter or a little more salt than usual to slow fermentation in summer. In automated bakeries, it is impossible to bake satisfactory two-layered Arabic bread without yeast. Unlike *balady* bread, which is baked at very high water absorption (75–85%), Arabic bread is made with low water absorption, and therefore steam expansion in the very hot oven may not be the primary factor in the ballooning effect that causes the separation into two layers.

The primary fermentation period is about 45 min, depending on the temperature of the added water. The temperature of tap water in winter slows down the fermentation rate; in summer, when the ambient temperature often reaches 35–40°C, the dough may become overfermented. After primary fermentation, dough balls of 125–200 g are prepared, and a 10-min intermediate proof is usually allowed. Some small bakeries do not employ any intermediate proof. However, El-Haramein and co-workers[1,2] found that proof has a significant influence on the crumb texture. The dough balls are then sheeted in two directions, by commercial sheeting rolls. The final proof is 25 min, followed by baking at 450°C for 45 sec (Maleki and Daghir, 1967; Hallab et al, 1974; Faridi and Rubenthaler, 1983b).[1,2]

C. *Chapati*

It is estimated that about 80% of the wheat consumed in India (Sinha, 1964) is in the form of *chapati* or *roti*, a type of round flat bread that has often been called an unleavened pancake. This bread is made from wheat, sorghum, pearl millet, or sometimes maize in India and Pakistan. Wheat is the preferred grain, but in many areas it is not available for *chapati* production.

Chapati making involves the following steps: 1) milling the wheat to prepare *atta*, a coarse flour, 2) mixing the *atta* with water, with or without the addition of salt and fat, to form a dough, 3) resting the dough, and 4) dividing, shaping, and cooking the *chapati* (Chaudhry, 1968). In rural areas, there is no standard procedure for the milling of coarse flours such as *atta* from wheat. The equipment varies from a primitive stone grinder (locally known as a *chakki*), driven by animals, water, or human power, to modern steel roller mills. The user often removes coarse bran particles by passing the flour through a sieve.

Chapati dough consists of flour and water. A small amount of salt, fat, or sodium bicarbonate may also be added. It is much softer than bread dough and is prepared and mixed by hand.

After resting 30–90 min, the dough is divided into balls of suitable size, which are then flattened and shaped, either with the help of a rolling pin or between the palms of the hands. The weight of the dough ball varies from region to region, ranging from 50 g to more than 1 kg. *Chapati* is usually cooked on a plate heated by burning wood or charcoal, an electrical heating element, or gas burners. The

[1] F. J. El-Haramein, P. C. Williams, G. O. Ferrara, and J. P. Srivastava. Factors affecting the experimental baking of two-layered Syrian flat breads. Unpublished report, 1987. Int. Cent. Agric. Res. Dry Areas, P.O. Box 5466, Aleppo, Syria.

[2] F. J. El-Haramein, P. C. Williams, J. P. Srivastava, M. Nachit, and A. Sayegh. Selecting durum wheats on the basis of flat bread quality. Unpublished report, 1987. Int. Cent. Agric. Res. Dry Areas.

time required for cooking is 2–5 min, depending on the temperature of the plate and the size, weight, and thickness of the dough (Aziz and Bhatti, 1960).

When wheat is not available, other cereals are used to make *chapati*. Murty and Subramanian (1982) estimated that 70% of the sorghum grown in India is used to make *roti*. For the production of *chapati* from sorghum, warm water is usually added in increments, and the flour and water are kneaded into a dough that has little cohesiveness. The dough is shaped or rolled into a thin circle and baked. The *roti*, if properly made, should puff during baking. In some households, a part of the flour may be cooked or soaked in water overnight to improve the cohesiveness of the dough. Good sorghum *roti* should be creamy white, with a few slightly burnt spots, and have a soft texture and a bland flavor.

Millet *roti* is traditionally prepared with stone-ground millet flour having an extraction rate of 85–100% (Rashid, 1974; Desikachar, 1977). The meal is mixed with water and sometimes salt to produce a dough of proper consistency. The dough is then rested under a moist cloth, shaped into a ball, rolled or patted into a disk, and baked on a greaseless iron plate. Millet *roti* generally does not puff well, but puffing is encouraged by placing it in the fire for a few seconds or by damping the upper surface of the baking dough with a moistened cloth before placing it in the fire. Cooking time depends on the temperature of the plate and the size of the *roti*. Mason and Hoseney (1980) reported that sorghum and millet *roti* doughs trap gas less efficiently and tear more easily than do wheat *chapati* doughs.

D. *Barbari* Bread

Barbari, an oval leavened bread popular in Iran, is 70–80 cm long, 25–30 cm wide, and approximately 2.5 cm thick. The ingredients include flour (72–77% extraction), water (60%), salt (2%), yeast (1%), and soda (0.35%). The mass is mixed by hand or in a mechanical mixer and then allowed to ferment for 2–3 hr. It is made into balls weighing approximately 900 g, which are flattened into an

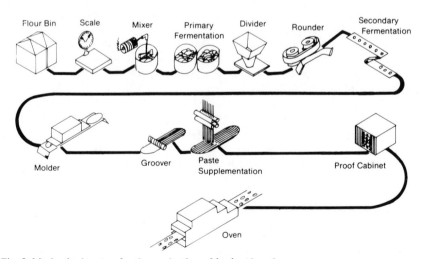

Fig. 5. Mechanized system for the production of *barbari* bread.

oval shape and rested for 20 min. To improve the crust color, a tablespoonful of a flour paste is poured and smeared on the flattened dough, which makes the bread shiny and brown after baking. The paste is prepared by making a flour-water slurry (10%, w/v), bringing it slowly to a boil, while stirring, and letting it cool down to room temperature. The dough is then docked and grooved with the fingers to form five or six 1-cm-deep rows, primarily for decorative purposes. The final proof is 15 min. The oven is dome-shaped, made of fire-resistant bricks, and heated by oil burners. The bread is baked for 8 to 12 min at 220° C, and a loaf weighing approximately 700 g is produced.

The traditional methods for *barbari* bread production lack uniformity and are often unsanitary as well. These problems are overcome in mechanized production (Fig. 5). The ingredients are mixed and fermented for 30–40 min; no soda is included in the formula. Next, the dough is divided into 600-g balls, which are rounded, left to rest for 1 or 2 min, flattened, and shaped. Flour paste is then applied, and the doughs are proofed for 20–40 min. Baking is performed in a conveyor type of oven at 220° C for 10–14 min.

E. Swedish Crisp Bread

Swedish crisp bread, which has become one of Sweden's major exports, is a low-cost, nutritious, and good-tasting snack flat bread. Until the development of modern baking techniques in the nineteenth century and the emergence of automated crisp-bread machinery in this century, this wafer-thin "cracker," made of unfermented dough, was the most popular homemade food in the Scandinavian countries. According to historians, Vikings brought it on their expeditions over 1,000 years ago, and medieval kings used it to carry them through years of famine, because crisp bread, when properly baked and stored, retains its crispness for months, even years.

There are numerous types of crisp breads, using wheat, rye, or a blend of the two (Haenel and Tunger, 1976), and sometimes sprinkled with sesame seed. The most popular in Scandinavia, particularly in Sweden, is brown crisp bread; the most popular worldwide is light crisp bread, or ice bread (Bressler, 1979).

The formula for ice crisp bread contains flour, iced water (130%), and salt (1%). The dough temperature is adjusted to 6° C or less (Karp and Garping, 1966; Manley, 1983). In automated production (Fig. 6) the dough is pumped through a pressure-beater unit, where it is whipped intensely while being cooled and supplied with compressed air, until it is effectively a cold foam. It is pumped to a forming machine, where it is rolled to a precise, uniform thickness. The sheeted dough is dusted, brushed, and cut to the desired dimensions. Next, the cut pieces of dough are forwarded to a conditioning band, to stabilize the dough consistency and to melt the top-dust flour, and then they are baked. After baking, the moisture content and temperature are still elevated. The product is therefore passed through electronic drying and cooling units before it is ready for stacking, sorting, sawing into portion-sized pieces, and packaging (Bressler, 1979).

Brown crisp bread, made from 100% whole rye or wheat-rye flour, is produced by a somewhat different process, involving long and intense kneading. The kneaded dough is fermented for several hours before being formed into a sheet of uniform thickness. It is then dusted and cut to the desired dimensions. After a short proof time, the doughs are baked for 6–10 min, depending on the desired thickness of the bread.

F. Extruded Flat Breads

The application of extrusion cooking to the manufacture of flat breads has led to a great increase in the number of such products in the food market. These products can be categorized according to texture, appearance, crispness, and taste. These characteristic qualities are obtained both by appropriate application of the extrusion technique and by the suitable choice of ingredient mixtures (Breen et al, 1977; Andersson et al, 1981; Meuser and van Lengerich, 1984).

In high-temperature, short-time extrusion cooking, the raw material and water can be fed directly into extruders. Because of vapor expansion, the extrudates acquire the porosity desirable in flat bread without the need for yeast or other leavening agents. The effects of extrusion cooking and of forcing the dough through a narrow die can be seen in a striplike orientation of the totally gelatinized starch mass (Fig. 7A). For comparison, Fig. 7B is a scanning electron micrograph of rye flat bread produced in the traditional way. In extrusion cooking of flat breads, the extent of product expansion depends mainly on the

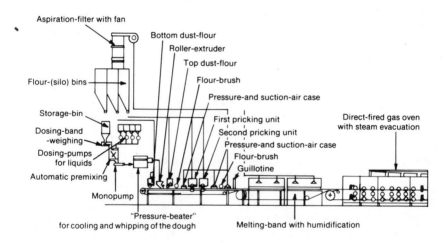

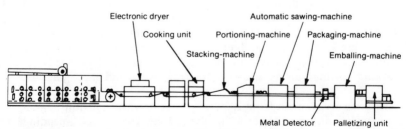

Fig. 6. Automated system for the production of Swedish crisp bread. (Reprinted, with permission, from Bressler, 1979)

feed rate and the amount of water and type of protein in the formula (Antila and Seiler, 1982). Protein enrichment stabilizes the product and improves texture and taste. Dixon (1983) described an automated method for the production of flat breads using a twin-screw extruder in which all ingredients were fed into the extruder simultaneously. Strips (from one to four) of flat breads were extruded very rapidly, and the thickness of the ribbon could be easily adjusted during a run. The ribbons were then roasted at 180°C to increase flavor and remove moisture. Most extruded flat breads look more like thick, airy crackers than like bread (Dixon, 1983). Using wheat varieties with a wide range of baking quality and sprouted wheats of high amylase activity, Linko et al (1982) demonstrated that the baking quality and amylolytic activity are not crucial factors in the production of extruded crisp breads.

In experiments with slightly crushed and whole (unmilled) kernels as the ingredient, the use of whole kernels increased the number of particles in the product that were not completely homogenized during processing, an outcome considered desirable in crisp bread (Antila et al, 1984).

G. Flat Breads Made from Nonwheat Cereals

In many parts of the world where wheat is not widely grown, flat breads are made from other grains, such as maize or sorghum. Because nonwheat grains and their end uses are not within the scope of this monograph, nonwheat flat breads are discussed only briefly.

Arepa is a typical Venezuelan corn bread, traditionally made from mechanically dehulled, partially degermed corn. It resembles a disk or a short cylinder, 8–10 cm in diameter and 1–3 cm high, with a rounded edge and flat or convex faces, depending on the region. The processed grain is cooked and stone-milled to a mass called *masa*. The *arepas* are then formed manually; small portions of the dough are shaped into balls and then flattened to the desired thickness. Final treatment includes cooking for approximately 2 min on each side, on a clay or metal griddle. Occasionally, they are cooked further over a burning charcoal for improved toasted flavor (Cuevas et al, 1985).

Tortillas are unfermented flat breads prepared from alkali-treated and steeped maize. They are a staple food in Mexico and Central America and are

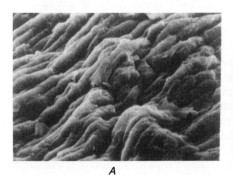 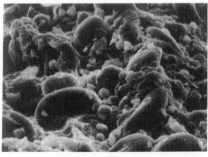

| A | B |

Fig. 7. Scanning electron micrographs of Swedish flat breads made by a high-temperature, short-time extrusion process (A) and a traditional baking method (B). (Reprinted, with permission, from Antila et al, 1984)

very popular in the United States. The traditional method includes cooking the maize in boiling lime (ash) solution for a short time or steeping it overnight. The cooked or steeped maize is washed to remove alkali and then ground into *masa*, which is flattened into thin disks and baked on a hot griddle for 30–60 sec on each side. They puff during baking. Tortillas may also be fried, to make taco shells. They are also prepared from white wheat flour or from sorghum (Rooney and Serna-Saldivar, 1987).

Injera is a fermented flat bread made from sorghum in Ethiopia. It is 50–60 cm in diameter, light to dark brown, slightly shiny, and soft, with uniformly distributed "fisheyes" (air bubbles on the surface) and a smooth lower crust. *Injera* may be made from teff, sorghum, or corn. Its quality is influenced largely by the process, time of fermentation, and cereal used. The process is a lengthy one, requiring two or three days of primary fermentation for the preparation of a batter, which is boiled in excess water, cooled, and fermented for a second time. The batter is then poured in a thin layer on a clay griddle, covered, and steamed for 2–3 min (Rooney et al, 1986).

Kisra, produced from sorghum and a small portion of millet or wheat flour, is a very popular bread in Sudan. A thick paste of flour and water is inoculated with a starter from a previous batch of the bread. After the dough has been fermented for 12–16 hr, it is diluted with water, to form a thin batter. It is then baked, on one side only, for 30–40 sec on a lightly oiled metal or clay sheet. Good-quality *kisra* peels off as a single piece, without breaking, and yields a soft, flexible, paper-thin pancake. Whiteness and a soft, moist (but not spongy) texture are desirable.

Dosai, widely consumed in southern India and Sri Lanka, is a slightly crisp, yet flexible, thin pancake made from a fermented mixture of cereals and black gram. Rice is the preferred cereal, but others are used in some areas. Another product, *idli*, is similar but has a coarse particle size and is steamed into small, white, sour-leavened cakes (Rooney et al, 1986).

III. INGREDIENTS

A. Flour

Wheat is the most widely used cereal in southern Asia, the Middle East, and North Africa (Hoover, 1974). The wheats used are of variable quality. In general, whiteness is desirable in bread, and thus white wheats are preferred over red wheats. Some research has been conducted to study the suitability of U.S. and Australian wheats for the production of flat breads. Mousa et al (1979) determined the influence of U.S. wheat classes, flour extractions, and baking methods on the quality of *balady* bread. Faridi et al (1981) and Faridi and Rubenthaler (1983b, 1984b) tested a number of U.S.-grown wheats for the production of Iranian and North African flat breads and Chinese steam bread and found that soft white wheats were very suitable. Qarooni et al (1987b) tested 33 Australian wheat samples for suitability for the production of Arabic bread and reported that the best-quality bread was made from white flour produced from hard wheat, of intermediate dough strength, with flour protein in the range of 10–12%, and with starch damage above 6%. Qarooni and co-workers (1987b) attributed the suitability of hard wheat flour to its high water-absorption capacity.

In general, flours with extraction rates of 80% or lower are preferred for the production of most types of flat breads, but flours with extraction rates of 90% or higher are also utilized in certain locations, especially in rural areas of the Middle East (Haq and Chaudhry, 1976; Shurpalekar et al, 1976) and in Scandinavian crisp breads. Nazar and Hallab (1973) prepared white Arabic bread from flour of 65% extraction, and Dalby (1963) reported that an acceptable *balady* bread can be prepared from a 75:25 blend of Egyptian wheat flour of 82% extraction and imported wheat flour of 72% extraction. Tobekhi and El-Esley (1982) used a flour of 87.5% extraction for the production of *balady* bread. *Naan*, a type of leavened flat bread, has been prepared from flour of lower extraction, e.g., patent flour (Dutta et al, 1980), but in certain areas of Afghanistan and Pakistan, whole wheat *naan* is also consumed (Rashid, 1983). Williams and Blaschuk (1980) and Bakhshi et al (1979) used various classes of wheat to produce flat breads. Austin (1980) attempted to classify Indian wheat varieties according to their suitability for various bakery products, including flat breads.

Ibrahim et al (1983) studied the effect of flour particle size on the quality of Egyptian *balady* bread. The largest particles had the lowest protein content and produced a loaf of poor volume and appearance; the smallest particles produced bread of the largest volume, with good grain and texture and excellent color.

Flour strength and protein content are important factors in the production of flat breads. Traditionally, flat-bread production in North Africa, the Middle East, southern Asia, and northern Europe was adapted to the strength of the local wheats, mainly soft or durum. Therefore, the use of high-protein, strong U.S. and Canadian wheats may cause problems in small-scale production of flat breads, such as production in neighborhood bakeries. *Balady* dough, for example, must be overmixed, to break down the gluten network, for optimum bread production.

The water absorption of flat breads encompasses a broad range. Unlike Western pan breads, in which the optimum water absorption is 60–65%, flat breads have optima ranging from very low (38%), as in matzo, to very high (85%), as in Iranian *sangak* (Faridi and Rubenthaler, 1983c). Figure 8 shows the absorption and dough consistency required for the production of five Iranian breads. Faridi and Rubenthaler (1984c) studied the effect of various levels of water absorption (optimum plus or minus 10%) on the physical quality of white and whole wheat pita breads. The quality of both breads improved at higher water absorption (optimum plus 10%), probably because the generation of more steam during baking produces finer and more uniform crumb and crust. Qarooni et al (1987a) adjusted the baking absorption of Arabic bread dough to the farinograph consistency of 850 BU. They reported that when the dough consistency was above 850 BU, as a result of the addition of less water, the dough sheeting quality improved, but the baked loaves were dark in crust color and tended to break during rolling and folding tests. The crumb had a dense grain and became denser as the amount of water was decreased. On the other hand, at lower dough consistency, sticky doughs with inferior sheeting quality were formed. The resultant breads were asymmetrical and light in crust color, with blistering of the surface, and the crumb had a woollier appearance, with large air cells, as the baking absorption increased.

Rao et al (1987) developed an objective method for measuring the consistency of *chapati* dough with the Henry Simon Research Water Absorption Meter. Using various levels of water absorption, they demonstrated that this instrument gave more sensitive resolution between consistencies of dough samples than the Brabender Farinograph or the General Foods Texturameter. They also reported that variation in flour particle size, protein content, and damaged starch content accounted for 88% of the variability in the dough consistency of several wheat flours.

Flat breads are to a large extent tolerant of wheat sprout damage, which may result from late-season rainfalls. Finney et al (1980b) made flat breads from flour

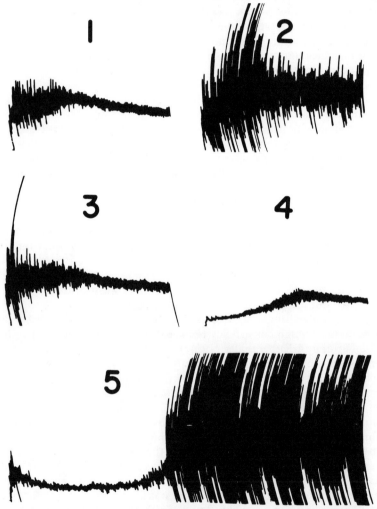

Fig. 8. Ten-gram mixograms of flours used to produce five Iranian flat breads: 1, *barbari* (78% flour extraction) with 60% water absorption; 2, *lavosh* (82% flour extraction) with 45% water absorption; 3, *taftoon* (84% flour extraction) with 60% water absorption; 4, *sangak* (87% flour extraction) with 85% water absorption; 5, village bread (97% flour extraction) with 48% water absorption. (Reprinted, with permission, from Faridi et al, 1982b; © 1982 by Institute of Food Technologists)

milled from highly field-sprouted wheat and reported that seven of the nine tested breads were judged to be equal to breads produced from sound wheat flour. Linko et al (1984) produced crisp flat breads by direct continuous extrusion cooking from mixtures of wheat and rye flours (1:1) of normal and high amylase activity (falling numbers 183 and 61, respectively). No significant differences were observed between products obtained from flours with normal amylase activity and those from high-amylase flours. Orth and Moss (1977) reported that with reasonable protein content and a falling number over 120, *chapati* of adequate quality could be made from a flour of 80–85% extraction, typical of flours used in India. However, variability in the water absorption resulted from the degree of sprout damage, which would be likely to cause manufacturing difficulties. Production of Middle Eastern pocket bread was reported to be more difficult with flour from sprout-damaged wheat. Blisters, poor crumb color, surface cracks, and difficulties in rolling and folding were apparent to varying degrees in the sprout-damaged samples. The samples with the highest protein content tended to be better than those of lower protein but were still inferior to the sound sample.

Many Third World countries in which flat breads are popular do not grow enough wheat to satisfy the demand for bread and thus must import an increasing amount of wheat (Dendy et al, 1970; Tsen et al, 1974; Bond, 1975; El-Saied and El-Farra, 1983). In most cases, this places a strain on the economy, so efforts have been made to use alternative native grains or plant materials to replace part of the wheat flour required for breadmaking. Flat breads, in general, are more tolerant of the addition of nonfunctional supplements than is pan bread, but undesirable effects are still noticed. For example, the dough may become quite sticky and difficult to handle, especially in the scaling stage and when the dough ball is pressed into a flat, pancake form. In the oven, nonfunctional additives such as potato flour may also modify the gluten matrix structure to such an extent that steam generated during baking escapes readily through the surface of the loaf. Consequently, the loaf does not rise. It is improperly baked and generally unsatisfactory.

Some supplements, however, may even improve the quality of flat breads. Prabhavathi et al (1976) reported an improvement in the *chapati*-making quality of Indian wheats when blended with durum wheat. Williams et al (1988) reported that Arabic breads made from durum wheat flour had a softer texture than those made from common wheat flour. The improvement in quality could be due to the hard grain texture of durum wheat; milling causes more starch damage in durum flour than in flour from softer wheats. A high level of damaged starch results in greater absorption of water and higher starch gelatinization and consequently a softer crumb. Morad et al (1984) reported that the replacement of bread flour with up to 30% ground sorghum produced acceptable Egyptian *balady* bread. Awadalla (1974) and Awadalla and Slump (1974) reported using 20% whole or decorticated millet flour with 80% Dutch wheat flour of 100, 80, or 70% extraction in the production of Egyptian flat breads. The baking quality was impaired, as judged by loaf volume and internal and external characteristics, but it was improved by the addition of dough conditioners ($KBrO_3$, fat, and calcium stearoyl lactylate). The sorghum flour had the greatest effect on the quality of the bread with wheat flour of 80% extraction. Blending of sorghum flour (10%) with wheat flour was recommended by Mustafa (1973) for

Sudanese breads.

Hamed et al (1973b, 1973c) reported that adding 10–15 parts of sweet-potato flour to 100 parts of high-extraction wheat flour (93% extraction) produced good *balady* bread, and Elias et al (1977) reported that adding 3–6% potato flour to conventional *balady* dough yielded an acceptable loaf. Maize, a staple crop in North Africa, is used for the production of flat breads, with or without blending with wheat flour.[3] Replacement of bread flour with up to 30% ground whole maize flour produced acceptable *balady* bread (Faridi and Rubenthaler, unpublished data).

The suitability of blending wheat and triticale flour for the production of *chapati* was studied by Rashid and Hawthorn (1974), Sharma et al (1977), and Sekhon et al (1980). A blend of triticale flour with wheat flour (1:3) gave as good a *chapati* as wheat flour alone. Triticale varieties with high protein content (16%) were the most suitable for *chapati* making. When triticale alone was used, the bread was darker and somewhat tougher than that made with wheat flour.

The *chapati*-making qualities of blends of wheat flour with flours of tapioca, *jowar, bajra*, and Bengal gram were studied by Murty and Austin (1963a, 1963b). Wheat flour could be mixed with 15–20% tapioca or other nonwheat flours without adversely affecting its *chapati*-making qualities, such as sweetness and palatability. An exception was Bengal gram flour, which gave a prominent flavor to the *chapati* even at low levels. These flour blends did not lose their good *chapati*-making qualities during storage for up to 30 days after the preparation of the blends.

B. Leavening Agents

The practice of leavening by natural fermentation dates back to early history. Although baker's yeast has now been introduced to the baking industry in most of the world, native or spontaneous fermentation, usually initiated by inoculating with a starter dough, is still the major source of leavening for flat-bread production. The starter may also be called mother dough, *torsh*, or *khameer*, depending on the region. The starter, if prepared fresh, is kept for a variable period of time to obtain proper acidity and aroma. It may be prepared in single or multiple stages to attain a desirable aroma, as evaluated by native bakers.

Yonema (1973), during a visit to Afghanistan, studied the preparation of starters for the production of *naan*. His experimental work included microbiological examination of local starter doughs. This study revealed the presence of two varieties of yeast isolated from traditional cultures, both belonging to the species *Hansenula anomala*.

The microbiology of dough starters for Iranian breads was studied by Tadayon (1976, 1978) and Azar et al (1977). In general, Iranian bakeries use portions of sourdough saved from previous days for leavening. These uncontrolled stock cultures are used over and over again. Except for the village bread, which may sometimes be unleavened, other Iranian breads are leavened.

---

[3] K. R. Davis. A survey of Egyptian village bread. Unpublished report, 1985. Dep. Bacteriol. Biochem., Food Res. Cent., Univ. Idaho, Moscow.

Azar et al (1977) studied 18 samples of sourdough starter collected from 14 bakeries in the Teheran area. Of the total bacteria found in the starter, 77% were *Leuconostoc* and *Lactobacillus* species, which are essential for the development of the desirable flavor and sour taste of *sangak* bread. The small number of yeasts present in the starter were responsible for leavening the dough. Tadayon (1976, 1978) studied 81 yeast cultures isolated from 43 bakeries in Shiraz, Iran. They exhibited morphological and physiological characteristics similar to those of nine species in the genus *Saccharomyces*. Yeast cultures were also examined for their gas production ability; 11 of the cultures had abilities comparable to those of commercial baker's yeast.

Microbiological studies of Egyptian *balady* bread were reported by Abdel-Malek et al (1973, 1974) and Mahmoud et al (1976). Not surprisingly, flours of 90% extraction generally contained larger microbial populations than flours of 72% extraction. The bulk of the flora in fresh dough consisted of, in order of importance, *Lactobacillus brevis, L. fermenti, L. plantarum, L. casei,* and *L. helveticus.* Some of the bacterial spores survived the baking process (Grecz et al, 1985).

The use of baker's yeast for the production of flat breads is increasing. Although the yeasty flavor is not very popular, the ease of handling has encouraged many bakers to substitute yeast for the sour starter. Baker's yeast is used in small quantities (0.1–2%, flour basis), to avoid vigorous fermentation, which in most cases is detrimental to the quality of flat breads. Sodium bicarbonate is very commonly used as a bread-leavening agent in the Middle East, especially Iraq, Iran, and Pakistan (Kouhestani et al, 1969; Adib and Al-Mukhtar, 1975; Faridi et al, 1983b; Rashid et al, 1982). The implications of the use of soda in flat-bread production are discussed later in this chapter.

C. Minor Ingredients

Most flat breads are made from a lean formula comprising flour, water, salt, and a leavening agent. Use of salt at a level of 1–2% (flour basis) is common. Mousa et al (1979) reported that *balady* breads made from unmalted flours were preferred to those made from malted flour. Rubenthaler and Faridi (1981) reported improvement in the quality of five flat breads by the addition of malted barley and ascorbic acid. Faridi and Rubenthaler (1984c) reported that shortening at the level of 1% improved the physical quality of pita breads, but at 2% made the texture excessively soft and fragile. Maleki et al (1981) reported that shortening and sodium stearoyl-2-lactylate (SSL) improved the quality of Iranian *barbari* bread.

IV. PRODUCTION

A. Mixing

Although optimum mixing is required for the production of high-quality breads of any type, flat breads are generally more tolerant of undermixing or overmixing than pan bread is. As mentioned earlier, when the flours are too strong, the dough is overmixed to mechanically degrade the gluten structure;

this is a common practice in many flat-bread-producing countries, such as Egypt.

Mechanical mixers are becoming increasingly popular worldwide. However, mixing by hand is still very common in neighborhood bakeries in North Africa, the Middle East, and southern Asia. Starting at one end of a trough and finishing at the other end, the hand mixing process usually requires 25–50 min, depending on the trough size. Each section of the dough is mixed for only a small part of the entire period. Because the mixing is not done under controlled conditions, it is difficult to determine the mixing requirements of many flat breads. However, under laboratory conditions, optimum flat breads were made when the dough was fully developed (Mousa et al, 1979; Faridi et al, 1981; Faridi and Rubenthaler, 1984a). Mousa et al (1979) reported that sensory scores of *balady* breads made by the straight-dough procedure were higher than those made by the continuous-mix system. Qarooni et al (1987a) studied the effect of mixing time on the quality of Arabic bread. Loaves were prepared from doughs mixed for 2 min (undermixed), 4 min (optimally mixed), and 7 min (overmixed). Sections of the dough samples taken after mixing clearly indicated that the dough mixed for 2 min (Fig. 9a) was poorly developed. The samples mixed for 4 min (Fig. 9b) showed good development, as the protein matrix was composed

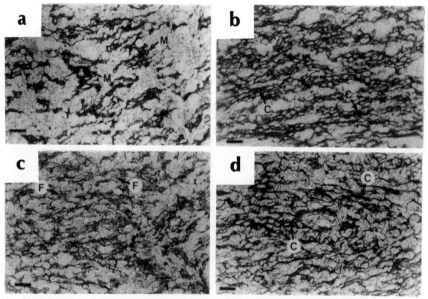

Fig. 9. Photomicrographs of dough sections stained with ponceau 2R to demonstrate protein development. C = continuous protein matrix; F = thin protein film; M = protein mass; bar = 5 μm. a, Dough mixed for 2.0 min. The coarse protein mass and the absence of a continuous protein matrix indicate that the dough is underdeveloped. b, Dough mixed for 4.0 min. The protein mass has been stretched out to form a continuous, intensely stained protein matrix that surrounds nearly all the starch granules. c, Dough mixed for 7.0 min. The protein has been overdeveloped and stretched out to form thin, little-stained films of protein that flow over the surface of many of the starch granules. d, Dough mixed for 2.0 min., after final sheeting. The sheeting has further developed the protein to form a continuous, intensely stained matrix that surrounds the starch granules. (Reprinted, with permission, from Qarooni et al, 1987a)

of extensively interconnected sheets of intensely stained protein. The samples mixed for 7 min (Fig. 9c) contained areas having a marked increase in the amount of veiling protein, resulting in a sticky dough that was difficult to process. When the sample mixed for 2 min was sheeted (Fig. 9d), the sheeting process had a marked effect on the degree of gluten development. The gluten matrix was further developed by the rolling required to sheet the doughs, and the matrix more closely resembled that of the other two samples. However, sheeting had relatively little effect on the samples mixed for 4 and 7 min. Table II presents the quality evaluation of the loaves from these mixing trials. The overmixed doughs produced bread with the best overall loaf scores, having superior internal properties and keeping quality; however, because of a slight blistering and lack of symmetry, it was inferior to bread prepared with the fully developed dough. The undermixed doughs scored lowest in all regards. Thus, undermixing clearly presented the greatest threat to loaf quality. However, the extra development imparted by the sheeting process meant that scores were not as bad as the underdeveloped microstructure might suggest. The overmixed dough, although producing a loaf that was superior to those made from fully developed dough, had the drawback of being excessively sticky (Qarooni et al, 1987a).

B. Fermentation

Many flat breads, such as Indian *chapati*, are unleavened. Therefore, a rest time of 30–90 min is given, for adequate hydration of starch and protein in the dough. Most local bakers and homemakers find the rest time necessary for the production of good *chapati*. In general, flat breads require less fermentation than does white pan bread, mostly because flat breads are made with a weaker type of flour and less dough development, and consequently vigorous fermentation is detrimental to their physical quality.

El-Shimi et al (1977) produced *balady* bread by both the activated dough method (high yeast activity) and the pre-liquid-ferment method. By either method, acceptable *balady* bread was produced from hard and semisoft wheat flour but not from soft flour. Breads produced by the pre-liquid-ferment method had higher loaf volume and specific volume than those produced by the activated dough method (Hamed et al, 1973a).

An interesting variation of the usual bread production is a type known as *Shamsy*, or sunny bread, made in the Sudan and in upper Egypt. After the dough

TABLE II
Score Assessed for Breads Made from Underdeveloped,
Optimum, and Overdeveloped Doughs[a,b]

Quality Factors	Mixing Time		
	2.0 min	4.0 min	7.0 min
External	37.0 ± 0.8	41.8 ± 1.9	41.0 ± 2.1
Internal	38.6 ± 1.8	39.5 ± 1.6	41.2 ± 1.2
Second-day score	24.8 ± 3.2	32.2 ± 3.7	35.5 ± 1.4
Total score	66.9 ± 3.1	75.7 ± 2.3	78.5 ± 2.2

[a] Source: Qarooni et al (1987a); used by permission.
[b] The scores are the average of four batch replicates carried out over four days. They represent percent of maximum score, plus or minus the standard deviation.

pieces are scaled and flattened in pancake form, the trays are exposed to the intense heat of the sun for a period of up to 3 hr (Fig. 10). The heat of the sun produces a characteristic crust in the dough piece, which carries through the oven. It is claimed that bread made in this manner has a distinctive flavor[4] (Dalby, 1966) and that it is much preferred by many consumers. This traditional process is reported to be slowly dying out.

[4]Refer to note 3.

Fig. 10. Sunny bread doughs fermenting while being exposed to the intense heat of the sun for up to 3 hr. (Used, with permission, from K. R. Davis, A survey of Egyptian village bread, unpublished report, 1985, Dep. Bacteriol. Biochem., Food Res. Cent., Univ. Idaho, Moscow)

C. Molding

A crucial step in flat-bread production is the flattening and sheeting and the concomitant degassing of the dough. The optimum dough thickness is 2–10 mm, depending on the particular bread. Each type has a very small range of dough thickness that produces an optimum bread, and small variations change bread quality significantly. This is true in traditional bakeries (Dalby, 1963; Faridi and Finney, 1980) and automated systems (Schnee, 1980), as well as in laboratory production (Maleki and Daghir, 1967; Mousa et al, 1979; Faridi and Rubenthaler, 1983b; Qarooni et al, 1987a). Rubenthaler and Faridi (1982) reported that consistent dough thickness is essential for the evaluation of the baking performance of flours in most flat breads, particularly in those that puff during baking (pocket breads). Figure 11 illustrates the wide variation in bread characteristics of white Arabic bread (pocket development, upper and lower crumb, and crust formation) that result when the dough thickness is varied from 2 to 7 mm. There are also subtle differences (not shown) in crust and crumb color. When the dough thickness was 2 mm, pocket formation did not occur; a dough thickness of 5–7 mm resulted in an atypical crust-crumb ratio. The bread produced from the 3- and 4-mm moldings had all the desirable characteristics. Similar findings were also reported by Qarooni et al (1987a). Quality evaluation of wheats for flat-bread production or investigation of biochemical changes during baking (browning, starch gelatinization, etc.) may thus be influenced by small variations (1 mm or less) in dough thickness.

D. Baking

Proof time for flat breads is usually very short, if they are proofed at all, and they all have different baking requirements. Many of them can be baked in regular ovens, like pan breads, but some require higher temperatures, in the range of 350–550° C, for producing high-quality breads, and special hearth ovens are necessary for such breads. High oven temperatures reduce the baking time and, consequently, increase the moisture content and softness of the bread. Some flat breads are best made when they are baked over a hot plate or are deep fried. Pan breads, on the other hand, are baked at a lower temperature for a longer period. Faridi and Rubenthaler (1984c) reported that white and whole

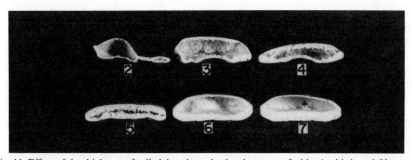

Fig. 11. Effect of the thickness of rolled dough on the development of white Arabic bread. Numerals indicate dough thickness, in millimeters, after sheeting. (Reprinted, with permission, from Rubenthaler and Faridi, 1982)

wheat pita breads baked at 480° C for 90 sec were of significantly better quality than those baked at 260° C for 4–6 min. The longer baking time in the oven and the lower temperature caused rough and dry crumb. Breads baked at a higher temperature for less time had more desirable crusts. The upper and lower crusts were completely separated in all treatments. Similar results were found for Egyptian *balady* bread (Faridi and Rubenthaler, 1984a).

El-Samahy and Tsen (1981) reported that baking at increasing temperatures from 288 to 343° C for 4 min reduced the loaf weight and significantly darkened the top and bottom crusts of *balady* bread. In another study (El-Samahy and Tsen, 1983), they measured the heat penetration inside *balady* during baking. The internal temperature of the loaf increased linearly with the baking time. The internal temperature at the central portion approached the maximum (99° C) after 1.5 min in a 343° C oven. Increasing the dough weight from 125 to 225 g did not have a significant effect on the profile of heat penetration in the central portion of the bread.

P. H. Rao et al (1986) studied the effect of baking temperature on the quality of *chapati* bread. Baking at the relatively low temperature of 150° C increased the baking time and also resulted in a lower-scoring *chapati* with undesirable grayish spots. A baking temperature in the range of 205–232° C was considered optimum. Puffing time was found to be critical, as even a slight variation greatly influenced the *chapati* quality, particularly with respect to color and pliability. The puffed height of the *chapati* depended on the temperature of the oven. As the temperature increased up to 343° C, the puffed height also increased. The puffing was only partial when the temperature was 176° C. The moisture content and the mixing time of the dough also affected the puffed height during baking.

Some of the traditional ovens used for the production of flat breads in the Middle East and North Africa are shown in Fig. 12. Traditionally, people in towns prepared bread doughs at home and brought their molded loaves to a public oven, to be baked for a fee (Adler, 1958). In smaller villages, however, every family had its own oven. In traveling through the Middle East, one may still see what appear to be small huts built of clay with low entrance doors. The "hut" is an oven called a *tarboon*. It consists of a cupola of burnt clay, about 3 ft in diameter, with a hole on the top, which can be closed with a lid. The floor of the interior, or hearth, is covered with a layer of pebbles for better distribution of heat. The fire surrounding the cupola is fueled by any available material that burns—wood, olive oil cake, etc. When the fire has burned down, ashes are heaped up around the outside to keep the heat in, and the oven is ready for baking (Adler, 1958).

A more common type of oven in the Middle East is called an ash *tanour*. It is similar to a *tarboon* but has a fire hole near the bottom and a loading hole, about a foot in diameter, at the top. A small fire is maintained on the hearth. A worker spreads the soft dough on a paddle, which is inserted through the upper hole into the oven, where the dough sticks to the inside of the arched cupola wall and bakes while adhering to it. The thin bread, only a few millimeters thick, is done in less than a minute. It is then removed from the wall of the oven with large L-shaped implements called tongues. Commercial bakeries in towns use procedures different than those used in rural areas. Ovens are made of brick, like Dutch ovens, but with a high arch. With the fire burning on one side of the hearth, dough is continuously put in on peels, inflates immediately, bakes, and is

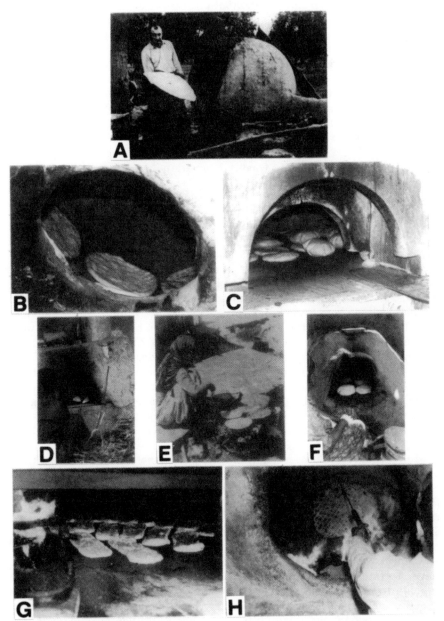

Fig. 12. Typical ovens for traditional production of flat breads: A, Turkish type of farm oven; B, oven for *tamees* bread in Saudi Arabia; C, oven for *burr* bread in Saudi Arabia; D, oven for *balady* bread in Egypt; E, baking on a curved iron plate, called a *sadj*, in Israel; F, baking of sunny bread in Egypt; G, baking of *barbari* bread in Iran; H, baking of *taftoon* bread in Iran. (A and E reprinted, with permission, from Pomeranz and Shellenberger, 1971; B and C reprinted, with permission, from Mousa and Al-Mohizea, 1987; D and F used, with permission, from K. R. Davis, A survey of Egyptian village bread, unpublished report, 1985, Dep. Bacteriol. Biochem., Food Res. Cent., Univ. Idaho, Moscow; H reprinted, with permission, from Faridi and Finney, 1980)

removed from the other side (Adler, 1958).

In recent years, various private and government agencies in most Middle Eastern and southern Asian countries have emphasized the installation of semiautomatic bakery plants for the commercial production of flat breads. Such plants are now in operation throughout these regions (Faridi and Finney, 1980; Doerry, 1982), and several techniques and ovens for the production of flat breads have been patented (Haffa, 1977; Saadia, 1977; Simon, 1979; Athenstadt et al, 1980; Papantoniou and Papantoniou, 1980; Hohn et al, 1981, 1982; Williams and Montaruli, 1983; Goglanian, 1986). Bahadori (1972) designed an oven for continuous automatic production of flat breads.

E. Quality Evaluation

Quantifying the quality of baked products has been a challenge to researchers. Two primary sensory characteristics, i.e., color and texture, of flat breads have been measured by a number of laboratory instruments, such as the Instron Universal Testing Machine (Ebeler and Walker, 1983). Kannur et al (1973) quantitatively determined the color of *chapati* by a colorimetric method. Also, measurements of texture indicators by a number of tests conducted with the Instron have been investigated, and significant correlations between panel scores, hardness, and yield point measurements were found for flat bread samples kept at room temperature for several hours (Abrol, 1972; Rashid and Hawthorn, 1974). Textural properties of Tunisian and Moroccan breads were evaluated using a Fudoh Rheometer fitted with a disk 2.5 cm in diameter and an automatic-stop accessory adjusted to penetrate the bread crumb 3 mm (Faridi and Rubenthaler, 1983b). For *balady* and white Arabic breads, the instrument was fitted with a wire 0.29 mm in diameter, which cut through a 1-cm slice of bread. Khan et al (1972) used the precision Penetrometer for determining the softness of *chapati*. G. V. Rao et al (1986) used a Warner Bratzler shear press and a device called a pliability tester to measure textural changes in *chapati* during storage.

Mousa et al (1979) used a semitrained sensory panel for measuring textural characteristics of Egyptian *balady* bread. A range was given to each characteristic, i.e., crust color, crust character, crumb color, grain and texture, flavor and taste, and chewing. Chewing was specified as tough at one end of the range and weak at the other end. Mann (1970) studied the shelf life of unleavened and leavened flat breads by evaluation of color, aroma, texture, taste, and overall acceptance by a panel of judges. Chaudhry (1968) studied the quality of U.S. wheats for *chapati* production. *Chapati* was evaluated on the basis of color, flavor, texture, and general acceptability. White club and Pakistani wheats were superior to red wheats. Color was the major factor lowering the grade of the experimental samples. Yasin et al (1965) used a different set of criteria to test the quality of *chapati*, including stiffness, staleness, brittleness, and pliability. Austin and Ram (1971) described a number of procedures to determine the palatability of flat breads. Morad et al (1984) evaluated the quality of Egyptian *balady* bread by subjective measurement of the extent of puffing during oven baking. Excellent puffing was defined as complete separation of the two crust layers; satisfactory puffing, as the separation of the two layers except in some parts of the loaf. Bread that did not puff was rated poor. Faridi and Rubenthaler

(1984c) adopted a visual scoring system for evaluation of U.S. pita bread, with a possible overall score of 65 (Table III). Scoring depended on the following desired characteristics: complete separation of the upper and lower crusts so that they are of the same thickness; soft, white, and moist crumb; light and shiny crust with brown spots; and pocket formation.

Recently Williams et al (1988) and Qarooni et al (1987a) developed procedures for evaluating wheat quality for the baking of white Arabic bread. The descriptive procedures allow objective assessment of Arabic bread quality. The scoring system adopted by Qarooni et al (1987a), shown in Table IV, allows a maximum overall score of 150. Breads are categorized as follows: excellent (135–150), good (120–134), satisfactory (105–119), fair (90–104), unsatisfactory (75–89), poor (60–74), and very poor (below 59).

The scoring system of Williams et al (1988) is based on dough-handling properties at dividing and sheeting, crumb color and texture, bitability, flavor, and regularity in the appearance of the final product. Each of these attributes is described below.

Dough Handling. Weak doughs cause serious problems in sheeting. Strong doughs with high elasticity shrink after sheeting and give thick loaves with poor texture and bitability.

Color. Gray indicates poor flour quality and possible mold contamination. Burnt patches usually indicate overfermentation or high alpha-amylase activity. Ideally, Arabic bread should be of an even cream to pale brown color with no patches.

TABLE III
Scoring Factors for Pita Bread Quality[a]

Factor	Score
Crust color	10
Crumb color	5
Upper to lower crumb ratio	20
Pocket formation	20
Crumb texture	10
Bread score (overall)	65

[a] Source: Faridi and Rubenthaler (1984c); used by permission.

TABLE IV
Loaf Scoring for Arabic Pocket Bread[a]

Quality Factor	Maximum Score	Quality Factor	Maximum Score
First-day external		Internal	
Area index[b]	5	Quality separation	16
Crust smoothness	5	Evenness of the layers	5
Shape	7	Grain appearance	5
Crust color	8	Grain uniformity	5
Cracks	7	Crumb texture	7
Blisters	8	Quality of tearing	7
Ability to roll and fold	10	Crumb color	5
		Second day	
		Ability to roll and fold	30
		Quality of tearing	20

[a] Source: Qarooni et al (1987a); used by permission.
[b] Measured on 6-cm-diameter samples.

Texture. The internal crumb should be rich in small cells, 1–2 mm in diameter, evenly distributed about both internal surfaces. If the internal crumb is waxy, there is no cell structure. Poor texture is usually an indication of under- or overfermentation. Texture is also affected by sheeting. If the dough piece is too thick, the cells of the final loaf will be large (coarse), with thick cell walls. This is often the case if the doughs are too strong.

Separation of the two layers should be even, each layer being about 1.5–2.2 mm thick. Frequently one layer will be thicker than the other. This can develop during proofing, if bubbles of gas form closer to one surface than the other. It can also be caused by oven conditions. If the oven is too hot at the floor level, the bottom layer will be thin and may burn. The other layer will be too thick and will have poor crumb texture and bitability.

Aroma. This important component of taste should be judged separately by trained personnel.

Bitability. The test is conducted by actually biting the bread. Arabic bread may be soft, chewy or sticky, tender, tough and leathery, or hard. These conditions reflect flour properties. For example, if the protein content of the flour is high and damaged starch content is low, the resultant bread will probably be tough to bite.

Appearance. Bread should show regularity in roundness. When folded tightly, it should not crack.

All judging factors are arbitrarily scored 0–5; the bread with the highest total score is the best.

F. Experimental Baking (Microbaking)

Standardization has become an important goal in flat-bread-producing countries, where Westernization and an improved standard of living have necessitated mass production of staple and convenience foods. The introduction of various types of flat breads into the United States and Western Europe has also contributed to the need for the standardization of ingredients and large-scale production. Another factor is the dramatic change in international food trade in the 1970s. Many developing countries, striving to establish industrial bases, lost their food-exporting capabilities, for a number of reasons, including mass migration to urban centers of farm villagers seeking better-paying jobs and the amenities of urban life. Consequently, severe food shortages occurred throughout the Third World. On the other hand, the oil boom of the 1970s caused a dramatic increase in per capita income and food consumption in the oil-producing countries, where bread is a major part of the diet. Today, most of the wheat grown in the United States, Canada, Australia, and Argentina is exported overseas. To establish a long and lasting market, efforts in breeding and production should include attention to the suitability of wheat varieties for end uses in the importing countries (Bond and Moss, 1982; Faridi and Rubenthaler, 1983c). Many of the major wheat customers are flat-bread-consuming countries. For example, Egypt alone purchased nearly 40 million bushels of soft white wheat and 1 million tons of flour from the United States in 1983. Standard test baking methods for evaluating the suitability of wheat varieties for flat-bread production have been developed in the United States

(Faridi et al, 1981), Australia (Qarooni et al, 1987a), and Canada (Williams et al, 1988).

One of the first objectives in bread-baking research should be to develop a standard laboratory baking method that simulates large-scale production, the result of which might help the commercial bread producer. Because traditional methods are predominantly used for flat-bread production, the significance of standard methods is magnified. For example, Chaudhry (1968) asked three Indian and Pakistani housewives to prepare doughs from 300 g of *atta* of 95% extraction from Gaines wheat. The prepared doughs were immediately placed in a large farinograph bowl and the instrument switched on. The amounts of water used were 49, 59, and 71.5%, and the dough consistency was recorded as 1,000, 780, and 410 BU, respectively. The cook who used the least amount of water also used some vegetable oil in preparing the dough. It was surprising that the doughs varied to such a large degree and still each could be made into acceptable *chapati*.

Maleki and Daghir (1967) developed a laboratory microbaking method for the production of white Arabic breads, 8 cm in diameter and 7 mm thick. The standard method included mixing, first fermentation (8 min), rounding by hand, second fermentation (10 min), rolling to circular dough sheets (8 cm in diameter and 4 mm thick), and third fermentation (45 min). A small muffle furnace was used as a baking oven. To prevent direct contact between the dough sheets and the hearth of the furnace, a thin, elevated asbestos pad was used as a baking floor. The breads were baked at 450°C for 45 sec, the time-temperature combination most commonly used in Lebanese commercial bakeries.

Chaudhry (1968), Austin and Ram (1971), Shurpalekar and Prabhavathi (1976), and P. H. Rao et al (1986) developed standard baking methods for Indian *chapati*. Rashid et al (1982) reported an experimental baking technique for Pakistani *naan*. In addition, Olewnick et al (1984) developed a laboratory procedure for pearl millet *chapati* (*roti*) and found water absorption to be critical. If the meal was finely ground and optimum absorption was maintained, all pearl millet cultivars tested produced satisfactory *roti*.

A method for laboratory production of Egyptian *balady* bread using sour starter was described by Mousa et al (1979). Faridi et al (1981) developed microbaking techniques for *barbari, taftoon, lavosh*, and *sangak*, to evaluate the suitability of four U.S. wheat classes for Iranian breads. Faridi and Rubenthaler (1983b) also developed experimental laboratory baking techniques for four North African breads: Tunisian *Terabelesi*, Moroccan, white Arabic, and Egyptian *balady*. They reported wide differences in baking performance among the 13 wheat varieties evaluated. They also reported an experimental laboratory method for Chinese steam bread (Faridi and Rubenthaler, 1984b).

V. SHELF LIFE AND STALING

A. Staling

Flat breads (except for crisp breads) in general have a poor shelf life, and as a result households in flat-bread-producing countries traditionally need to bake or purchase the bread twice a day. There is little literature on the subject. In contrast, the literature on the staling of pan loaf breads is extensive. In general, the term *staling* "indicates decreasing consumer acceptance of bakery products caused by changes in the crumb other than those resulting from the action of spoilage organisms" (Ponte, 1971). The effect of various ingredients and additives on the staling of pan bread has been studied extensively (Ponte, 1971). A number of changes occur in bread with age: the crumb firms and its water absorption capacity lessens, enzyme susceptibility of the starch is lowered, etc.

Khan and Bhatti (1968) observed that Pakistani *roti* containing 4% nonfat dry milk and 2% glycerol monostearate (GMS) could be kept for three days at room temperature and for 30 days at refrigeration temperature without any noticeable changes in composition and organoleptic qualities. Hussain and Satti (1968) reported that *roti* prepared with GMS and added vital wheat gluten remained soft and resilient for up to 72 hr at 35°C.

Yasin et al (1965) found that the inclusion of nonfat dry milk improved the quality of stored *chapati*. An increase in storage temperature from the ambient temperature to 37°C increased the rate of moisture loss, which adversely affected the softness of the *chapati*, making it hard, dry, and brittle. The rate of moisture loss corresponded to the rate of freshness loss. The loss of moisture from *chapati* containing GMS and milk was comparatively low.

The effect of baking time and temperature on the staling of Egyptian *balady* bread was studied by Faridi and Rubenthaler (1982, 1984a). *Balady* breads baked at 540°C for 1 min were initially softer and remained softer during 72-hr storage than those baked at 260°C for 6–7 min. They also reported similar results with U.S. pita bread (Faridi and Rubenthaler, 1984c): breads baked at higher temperatures and containing a higher level of water in the formula were initially softer and remained softer during 72-hr storage. Supplementing the shortening in the pita bread formula slowed the staling rate and made the texture excessively soft by the second or third day of storage. The softness is a negative feature, since most pita breads are used in sandwiches and should hold the filler.

Swaranjeet et al (1982) studied the role of leavening and supplements on the keeping quality of *chapati*. Yeast leavening, GMS, SSL, and α-amylase supplements notably improved the texture and flavor of whole-meal *chapati*. Maleki et al (1981) reported that different combinations of emulsifiers, sugar, shortening, and soy flour improved the keeping quality of *barbari* bread; the best result was obtained when 0.5% SSL was added with 3% shortening during a three-day storage period. Faridi and Rubenthaler (1984b) reported that freezing was the best practice for the storage of steamed breads. Crumbs of the steamed breads stored at ambient and refrigerator temperatures became 50 and 100% firmer, respectively, than breads stored in a freezer.

Parihar and Chatterji (1956) used X-ray diffraction crystallography to study the staling of *chapati*. Much less retrogradation of starch was observed in a *chapati* with increased gluten content (1:1 ratio of starch and gluten) than in a *chapati* made of pure starch or wheat flour. The latter *chapati* was determined to be stale; thus the presence of gluten retarded staling. Storage temperature was also an important factor in staling. The retrogradation of starch was greatest at 2°C and decreased with a rise in temperature. In *puri* (fried *chapati*), a fat-amylose complex was formed, and the rate of retrogradation was decelerated. Nath et al (1957) observed a slight increase in the fat acidity and reducing-sugar content of *chapati* during storage.

B. Shelf Life Extension

The flat-bread formula is generally lean; thus the water activity (a_w) is slightly higher than that of pan bread of equal moisture content. Ingredients such as sugar and salt, which may lower the a_w of pan bread, are usually not included in flat-bread formulations. This formulation, plus the tropical and semitropical climates of countries in which flat breads are popular, encourages the growth of molds. The establishment of automated plants to produce and package flat breads in the Middle East has also extended the time interval between baking and consumption from a few hours to several days. Consequently, the effect of preservatives on the keeping quality of flat breads has been studied by numerous workers. Khan and Bhatti (1968) observed that *roti* containing 0.2% sorbic acid could be kept for three days at room temperature and for 30 days at refrigerator temperature without noticeable microbial growth. Hussain and Satti (1970), however, reported that sorbic acid imparted a bitter aftertaste to *roti* and therefore was not recommended as a preservative for flat breads. Rao et al (1964, 1966) reported that *chapati* containing salt and sorbic acid, packaged and heat-sealed in polyethylene pouches, kept satisfactorily and remained acceptable for six months under diverse tropical conditions. Sorbic acid could be replaced either by incorporating citric acid (0.4%), sugar (3%), and salt (2.5%) or by heating the packaged *chapati* in an oven for 2 hr at 90° C (Arya et al, 1977). Free sorbic acid was a better preservative than its potassium salt. The addition of 0.15% propionate to the *chapati* formula was sufficient to retard mold growth for up to 120 hr (Yasin et al, 1965).

Al-Mohizea et al (1987) investigated the microbial spoilage of Arabic bread. They reported that initial microbial loads (immediately after baking) were low, mainly because of the high oven temperature and the thin layer of dough. No coliforms were detected, presumably because they are unable to survive the high temperature. The microbial load of the air and the relative humidity inside the package played a major role in bread spoilage. Molds were the first noticeable sign of spoilage.

VI. ENRICHMENT

Flat breads are the staple food for people in North Africa, the Middle East, and southern Asia (Irani, 1987). Thus enrichment of the basic dough formula with various nutrients is an ideal way to increase the overall nutritional status of indigent populations.

A. Protein

Nutritional studies in the Middle East point to the prevalence of protein-calorie malnutrition. The literature on attempts to supplement various flat breads with animal and vegetable proteins as well as specific amino acids is extensive.

Nutritional improvement of *chapati* supplemented with soy flour was reported by several investigators (Bhat, 1977). Imtiaz (1962) reported that *chapati* prepared from whole wheat flour with 15% medium-fat soybean flour and 10% dry milk gave the best balanced diet. Lindel and Walker (1984) prepared *chapati* with various blends of cereal flours enriched with soy flour. In all cases, the calculated protein efficiency ratios (PERs) were improved; the corn-soy *chapati*, with a PER of 2.15, had the highest calculated ratio, compared to 1.41 for the unblended corn. Shyamala and Kennedy (1962) reported that the PER increased from 1.65 to 3.08 in *chapati* supplemented with 10% defatted soy flour; it increased to 3.07 in *chapati* supplemented with 10% nonfat dry milk. Although these investigators claimed that a soy supplement at that level was acceptable to consumers, Bass and Caul (1972) demonstrated that expert members of a taste panel can easily detect and characterize the flavor of soy in fortified *chapati* flour and bread.

Ali et al (1964) found that *chapati* containing 2% shark meat flour could satisfy the protein requirement of growing children and that wheat flour containing about 3–5% shark meat flour would be suitable for nursing mothers. Arabic and Indian breads were supplemented with fish protein concentrate (FPC) by Nikkila et al (1976). Significant differences in nutritional value were observed in the different types of FPC used. Taste panel tests showed that breads supplemented with 10% FPC were all acceptable. The FPC raised the PER of breads to a level equal to that of casein. The high temperature required for the baking of Arabic bread (500°C) or for cooking *puri* in oil (190°C) did not significantly decrease the protein value of FPC when the heating time was about 1 min. The effects of FPC and cottonseed flour on *chapati* quality were also studied by Archer (1970).

The nutritive value of Arabic bread was improved when it was supplemented with whey protein (Khalil and Hallab, 1975). Abdel-Rahman and Youseff (1978) fortified Arabic bread with soybean flours; 8% defatted and 3% full-fat soy flours were organoleptically acceptable levels.

Fortifying white Arabic bread with various levels of chick-pea flour improved the nutritive value of the bread. Organoleptically, however, only up to 20% supplementation was acceptable (Hallab et al, 1974).

Protein enrichment of Egyptian *balady* bread has been attempted by several investigators. Khalil et al (1976) fortified Egyptian wheat and maize bread with FPC up to a level of 7.5%. Shehata and Fryer (1970) determined that supplementing *balady* bread with 10% chick-pea flour increases the nutritional value of the bread while also improving its organoleptic and physical properties. Finney et al (1980a) reported that supplementing wheat flour with up to 20% faba bean flour did not change the organoleptic characteristics of the bread. Hussein et al (1974) supplemented *balady* bread with broad-bean proteins (raw or stewed), and Dalby (1969) used cottonseed, chick-pea, and dried yeast as protein supplements in the bread. An acceptable *balady* bread supplemented with 5% cottonseed flour was prepared by El-Sayed et al (1978).

Arafah et al (1980) conducted feeding trials with rats to study the effects of fortification of *balady* bread with FPC and green algae. Baking at 400°C resulted in a loss of 12–21% of L-lysine in bread made with FPC. The PER of FPC-supplemented bread was 2.6, compared to a mean value of 1.0 for unsupplemented bread. Diets based on bread containing algae (6% in the composite flour) did not improve the growth rate of rats. Patel and Johnson (1975) and Patel et al (1977) reported marked increases in lysine, histidine, arginine, threonine, and tyrosine in Moroccan flat breads supplemented with a 10% horsebean protein isolate. In breads supplemented with up to 20% of the isolate, the protein quality improved without alteration of the breadlike aroma, flavor, and textural quality. These studies indicate that flat breads, in general, can be fortified to improve the nutritional status of undernourished rural populations.

Several studies were designed to enrich flat breads with synthetic amino acids. Maleki and Djazayeri (1968), in a study of the effects of amino acid supplementation on the protein quality of Arabic bread, showed that the addition of 0.3% L-lysine to flour significantly improved the protein quality of the resulting bread. Supplementation of lysine-fortified flour with 0.62% DL-threonine caused a further improvement of protein quality, but methionine had no effect. Baking did not change the protein quality of bread supplemented with lysine or lysine and threonine. Hussein et al (1973) fortified *balady* bread with enough DL-methionine and L-lysine to supply the equivalent of 10% of the protein in the diet. The PER rose significantly with the amino acid supplement, from 1.28 to 2.19 for the DL-methionine supplement and to 3.32 when L-lysine was also added. *Balady* breads containing added 0.35% lysine, 0.13% DL-threonine, and 0.7% DL-methionine (flour basis) produced the same growth rate in rats as a diet supplemented with 6.6% FPC (Arafah et al, 1980). Iranian village bread was fortified with L-lysine by Hedayat et al (1968, 1971, 1973). Preliminary studies with rats showed significant improvement in animal growth. However, no significant improvement was observed in the nutritional status of a group of Iranian village children who received lysine-fortified bread for 210 days.

B. Vitamins and Iron

Symptoms of vitamin deficiency, such as night blindness, are not uncommon in the Middle East (McLaren, 1969; Shakir and Demarchi, 1971). Vaghefi and Delgosha (1975) studied the fortification of Iranian *sangak* bread with vitamin A and found that about 70% of the added vitamin was recovered after baking. They recommended that yeast be enriched with vitamin A to eliminate operational mistakes by the miller or baker. Eid and Bourisly (1986a, 1986b) fortified flour with thiamine, riboflavin, niacin, and iron in the production of Arabic and Iranian breads. The loss caused by baking was minimal in enriched Arabic bread but was pronounced in the fortified Iranian bread. Maleki and Daghir (1967) reported a significant loss of riboflavin in both white and brown Arabic breads, but the loss of niacin under the same baking conditions was negligible. The retention of riboflavin was higher in vitamin-enriched samples than in unenriched bread. Added niacin was retained completely. The effect of baking on the vitamin content of Iranian village bread was studied by Hedayat et al (1968). Added niacin was remarkably stable, but about 50% of the added riboflavin and vitamin C and 80% of the vitamin A were destroyed during baking.

Enrichment of flour with iron is a practical way to increase the iron intake of people in the Middle East. Nazar (1970) and Nazar and Hallab (1973) reported no adverse effect on dough characteristics and baking quality when up to 50 mg of iron was added per pound of flour, regardless of the type of iron salt used.

VII. NUTRITIONAL STUDIES

Flat breads constitute an important item of food as a source of energy and vital nutrients, such as vitamins, minerals, and proteins, for people in North Africa, the Middle East, and southern Asia. In extreme cases, which occur often in rural and remote areas, flat breads may contribute as much as 90% or more of the total food intake. The consumption of cereals at this high rate and the deficiency or even complete absence in the diet of other foods, such as dairy products and meats, have led to various nutritional deficiency diseases.

A. Effect of Processing

The removal of bran and shorts during milling affects the nutritional value of the resulting flour. Faridi et al (1982a) reported that the PER of breads and doughs gradually increased with an increase in flour extraction rate. Baking lowered the PER of five Iranian breads; the effect of fermentation on PER was insignificant. The effects of the milling extraction of the flour, the type and length of fermentation (yeast-raised, sourdough, and unleavened), and baking conditions (varying time and temperature) on the relative bioavailability of magnesium (Faridi et al, 1983c), zinc (Faridi et al, 1983a), and iron (Ranhotra et al, 1981) were determined in studies with weanling rats. The process of breadmaking did not improve the bioavailability of magnesium, but it did improve the bioavailability of zinc and iron. No direct relationship could be established between the bioavailability of these minerals and the protein, phytate, and dietary fiber contents of the breads.

There are conflicting reports on the influence of baking on protein quality. Eggum and Duggal (1977) reported a 5% loss of net protein utilization value of *chapati* as a result of cooking. On the other hand, Shyamala and Kennedy (1962) found the PER of Indian *chapati* to increase during cooking. Khan and Eggum (1978) reported that the baking of wheat, maize, rice, barley, millet, sorghum, and triticale flours into unleavened Pakistani bread (*roti*) affected the nutritive quality only to a minor extent.

El Tinay et al (1979) found no change in threonine and lysine and an increase in tyrosine and methionine during the fermentation of *kisra*. When the dough was fermented at pH 9.3, cooking had only minor effects on the nutritional quality of *kisra* made with flours from various sorghum varieties (Eggum et al, 1983).

Tabekhia and Mohamed (1971) detected an increase in thiamine (6.7%), riboflavin (14%), and nicotinic acid (13.6%) during the fermentation of *balady* dough. The bread was then baked at 350°C for 1.5 min, resulting in a large reduction in thiamine (24.8%) and riboflavin (25.8%); nicotinic acid was affected little (decreasing by 2.5%). Toasting of the bread (120–130°C) for 10 min after baking further reduced the level of thiamine by 36%; riboflavin and nicotinic acid remained unchanged. Tsen et al (1977) demonstrated that PERs of breads were significantly improved and feed conversion ratios were reduced by substituting steaming for conventional oven baking.

Finally, an important note on the baking of flat breads concerns the fuels used in traditional ovens. The breads are generally exposed to flame and smoke and to various fuels, such as petrol, straw, charcoal, wood, and dung. This exposure increases the chance that smoke residues remain in the breads. There has been much concern that smoke, especially from heavy fuels, could be carcinogenic. Research on this subject is recommended.

B. Nutritional Studies of Flat-Bread Flour

Flat breads, generally, are produced from flours of higher extraction rates than those used in American or European breads and are made with a shorter fermentation. Faridi (1981) reviewed some health advantages of a diet high in carbohydrates, especially in flat breads made from whole wheat flour. Many health problems associated with Western diets, such as diabetes and osteoporosis, are rarely observed or reported in rural areas of developing countries (Groen et al, 1966). Whereas many consumers in affluent societies have reduced their intake of cereals in favor of a high intake of meat and fats in the past 50 years, people in rural areas of the Middle East, by economic necessity, have been forced to rely on high-extraction flour (nearly 100% extraction) as their main and occasionally only source of energy and protein. The beneficial effects of a low-fat diet, however, may be counteracted by a high general occurrence of mineral deficiencies.

Prasad et al (1961) described several cases of iron deficiency in people whose staple food was wheat bread, although the bread contained adequate amounts of iron. One of the most comprehensive nutritional studies of human mineral deficiency was conducted during the 1960s and 1970s in southern Iran by Reinhold and co-workers (Eminian et al, 1967; Kohout et al, 1967; Ronaghy et al, 1968, 1974; Reinhold, 1971, 1972, 1975a, 1975b; Haghshenass et al, 1972; Reinhold et al, 1973, 1974, 1975, 1976a, 1976b; Ismail-Beigi et al, 1977; Faradji et al, 1981). Although intakes of zinc, iron, calcium, magnesium, and phosphorus by villagers exceeded the recommended dietary allowances by a considerable margin, mineral deficiencies were found in the villagers. The problem, therefore, was that the minerals were not available to the body, and the researchers suspected that the village diet was the cause. Clues to the sources of interference with the absorption of these elements from the intestine were found by examination of the village bread *tanok*, consumed in the Shiraz region of Iran. The bread contained high levels of fiber and phytate.

Similar studies were reported by Berlyne et al (1973) on Israeli bedouins, among whom osteomalacia is prevalent. It is suspected that this condition is associated with a high phytic acid content resulting from the intake of unleavened bread. Reinhold (1972) and Ter-Sarkissian et al (1974) reported that daily phytate intake by adults ranged from 2 to 5 g. In a later study, Ismail-Beigi et al (1977) found that the removal of phytate from the village bread did not reduce its capacity to bind metal ions and concluded that the formation of complexes of divalent metals with fiber caused, at least in part, the decreased availability of dietary iron and zinc in whole wheat bread. In support of this viewpoint, Reinhold et al (1976a) found a significant correlation between fecal dry matter and metal excretion. In a similar study in Pakistan, the diet of eight children with late rickets and two women with osteomalacia was changed from *chapati* to bread made with white flour, for seven weeks; levels of blood serum calcium, inorganic phosphorus, and alkaline phosphatase returned to normal (Ford et al, 1972).

Although discussion of the nutritional implications of wheat fiber and phytate is beyond the scope of this chapter, it is clear that when whole wheat unleavened flat breads supply more than 90% of calories and protein, deficiencies in zinc, iron, and calcium caused by fiber and phytate binding may result.

An interesting observation came from Dutta et al (1980), who reported that high consumption of *chapati* may be an important factor in the prevention of folate deficiency in northern India. Similar results in southern Iran were reported by Russell et al (1976).

C. Nutritional Studies of Soda in Flat Breads

Sodium bicarbonate (soda) is commonly used as a leavening agent in flat breads in Iran (Kouhestani et al, 1969; Faridi et al, 1982b), Pakistan (Finney et al, 1980b; Rashid et al, 1982), Iraq (Adib and Al-Mukhtar, 1975), Saudi Arabia (Mousa and Al-Mohizea, 1987), and other Middle Eastern countries. Its popularity as a leavening agent is based on its low cost, lack of toxicity, ease of handling, and relative tastelessness. Phytic acid in whole wheat flour is susceptible to yeast and bacterial activity and is destroyed by phytase during fermentation. As the pH of the dough decreases, the rate of phytase activity increases. Therefore, when soda is included in the bread formula, the dough pH increases and phytase activity is lowered. Faridi et al (1983b) studied the effect of soda leavening on the phytic acid content and physical characteristics of Middle Eastern breads. Their studies on the effect of soda on dough rheological properties revealed why it is so popular with small-shop bakers. As the mixograms in Fig. 13 demonstrate, the soda-supplemented doughs were

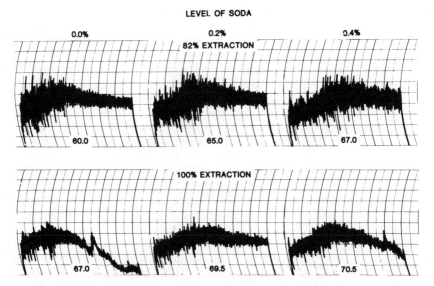

Fig. 13. Effect of 0.2 and 0.4% soda on water absorption (values below the curves) and dough-mixing properties of 82 and 100% extraction flours, measured with a mixograph. (Reprinted, with permission, from Faridi et al, 1983b. ©1983 by Institute of Food Technologists)

distinctly stronger in functionality, yielding doughs with improved handling properties, characteristics that the small-shop and home bakers find desirable.

The effects of various types of leavening on the amount of phytic acid lost during fermentation and baking of *taftoon* and *naan* are shown in Fig. 14. Soda reduces the rate of destruction of phytic acid during fermentation. The destruction of phytic acid was higher for *taftoon* (made with flour of 82%

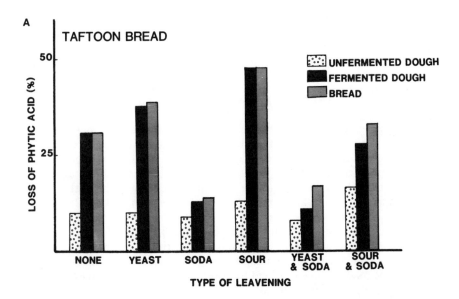

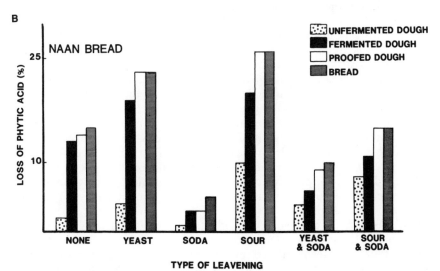

Fig. 14. Effect of various types of leavening on loss of phytic acid during dough fermentation and baking of *taftoon* (A) and *naan* (B). (Reprinted, with permission, from Faridi et al, 1983b. ©1983 by Institute of Food Technologists)

extraction) than for *naan* (made with flour of 100% extraction). These results corroborated the findings of Reinhold (1975b) that phytate in flours of 90–100% extraction, used for making Iranian village breads, was resistant to destruction by phytase. Soda at the level of 0.4% (flour basis) has a retarding effect on the hydrolysis of phytic acid with no apparent advantage to bread quality. In many bakeries of the Middle East, soda is used at 1% (Faridi and Finney, 1980), a level at which its impact would be even more pronounced. In parts of the world where bread constitutes a major part of the diet, a high phytic acid content in the bread is suspected to be an important factor in widespread mineral deficiencies. Consequently, the elimination of soda as a leavening agent could significantly reduce the phytic acid content of breads and increase the availability of some minerals.

VIII. CONCLUSIONS AND FUTURE OUTLOOK

During the past decade, the consumption of flat breads has increased dramatically in the United States and likely will continue to increase in the future. The consumption of a greater amount of flat breads would be desirable for a number of reasons: 1) the formula is simple and the ingredients are few, so that these breads have a lower caloric content on a dry basis than other breads and are more appealing and desirable to diet-conscious people; 2) flours of higher extraction rates are generally used, and flat breads are a good source of dietary fiber; 3) the crust-to-crumb ratio is high: flat breads are less crumbly than pan breads and most of the time do not require toasting; 4) the natural pocket of some flat breads makes them good food carriers, ideal for sandwiches; and 5) flat bread may be made from soft wheat flour (and sometimes durum) and may not require much gluten of high quality.

For the baker, the advantages of flat-bread production include little proofing space, great tolerance of changes in formula or procedure, and simple packaging requirements. The marketing of flat bread could therefore provide bakers with a more equitable return on their investment, by decreasing production and distribution costs.

Much of the research on flat breads conducted at the Western Wheat Quality Laboratory of the Agricultural Research Service of the U.S. Department of Agriculture, in Pullman, Washington; the Bread Research Institute of Australia, in Sydney; and the Grain Research Laboratory of the Canadian Grain Commission, in Winnipeg, has focused on developing practical and reproducible baking methods for the production of various types of flat breads. The wide variety of flat breads makes it mandatory to focus on better understanding of their physical, chemical, and rheological properties and on the development of new products based on traditional characteristics of the breads described in this chapter.

More research is also needed to identify potential hazards to health in traditional methods of flat-bread production. Outstanding among them are the impact of lifelong consumption of breads exposed directly to flame and smoke during baking. The implications of the consumption of breads leavened with soda (and as a result containing high levels of phytic acid) also deserve descriptive in-depth study.

LITERATURE CITED

ABDEL-MALEK, Y., EL-LEITHY, M. A., and AWAD, Y. N. 1973. Microbiological studies on Egyptian "balady" bread-making. I. Microbial content and chemical properties of the flour. Chem. Mikrobiol. Technol. Lebensm. 2:60-62.

ABDEL-MALEK, Y., EL-LEITHY, M. A., and AWAD, Y. N. 1974. Microbiological studies on Egyptian "balady" bread-making. II. Microbial and chemical changes during dough fermentation. Chem. Mikrobiol. Technol. Lebensm. 3:148-153.

ABDEL-RAHMAN, A.-H. Y., and YOUSSEF, S. A. M. 1978. Fortification of some Egyptian foods with soybean. J. Am. Oil Chem. Soc. 55:338A-341A.

ABROL, Y. P. 1972. Studies on chapati quality. I. Discoloration of chapatis. Bull. Grain Technol. (Hapur, India) 10:41-46.

ADIB, N., and AL-MUKHTAR, F. 1975. Guide for Iraqi Cooking and Baghdadi Dishes, 6th ed. Al-Muthann Library Publ., Baghdad, Iraq. (In Arabic)

ADLER, L. 1958. Breadmaking in the land of the Bible. Cereal Sci. Today 3:28-30, 32.

AL-MOHIZEA, I. S., MOUSA, E. I., and FAWZI, N. M. 1987. Microbiological studies on two common types of bread in Saudi Arabia. Cereal Foods World 32:610, 612.

ALI, S., MAVA, A. S., and WAHIF, M. 1964. Protein value of wheat flour supplemented with shark meat flour. Pak. J. Sci. Res. 16(2):41-43.

ANDERSSON, Y., HEDLUND, B., JONSSON, L., and SVENSSON, S. 1981. Extrusion cooking of a high-fiber cereal product with crispbread character. Cereal Chem. 58:370-374.

ANTILA, J., and SEILER, K. 1982. Production of flat bread by HTST—Extrusion cooking with wholemeal of wheat and rye as raw materials. Pages 909-914 in: Proc. World Cereal Bread Congr., 7th.

ANTILA, J., SEILER, K., SEIBEL, W., and LINKO, P. 1984. Production of flat bread by extrusion cooking using different wheat/rye ratios, protein enrichment and grain with poor baking ability. Pages 107-128 in: Extrusion Cooking Technology. R. Jowitt, ed. Elsevier Sci. Publ. Co., New York.

ARAFAH, A., ABASSY, M., MORCOS, S., and HUSSEIN, L. 1980. Nutritive quality of baladi bread supplemented with fish protein concentrate, green algae, or synthetic amino acids. Cereal Chem. 57:35-39.

ARCHER, J. L. 1970. Performance of protein-supplemented flours in chapatis. M.S. thesis, Kansas State Univ., Manhattan.

ARYA, S. S., VIDYASAGAR, K., and

PARIHAR, D. B. 1977. Preservation of chapatis. Food Sci. Technol. 10:208-212.

ATHENSTADT, D., BORMANN, H., DORIAS, B., FRANZ, E., KRETSCHMER, P., and MAACK, E. 1980. Manufacture and baking of flat bakery products. German Democratic Republic patent 142,790.

AUSTIN, A. 1980. Consumer's and processor's preferences choose wheat according to your choice. Indian Miller 10:15-18.

AUSTIN, A., and RAM, A. 1971. Studies on chapati making quality of wheat. Indian Counc. Agric. Res. Tech. Bull. 31-34, New Delhi.

AWADALLA, M. Z. 1974. Native Egyptian millet as supplement of wheat flour in bread. II. Technological studies. Nutr. Rep. Int. 9:69-78.

AWADALLA, M. Z., and SLUMP, P. 1974. Native Egyptian millet as supplement of wheat flour in bread. I. Nutritional studies. Nutr. Rep. Int. 9:59-68.

AZAR, M., TER-SARKISSIAN, N., GHAVIFEKR, H., FERGUSON, T., and GHASSEMI, H. 1977. Microbiological aspects of sangak bread. J. Food Sci. Technol. (Mysore) 14:251-254.

AZIZ, M. A., and BHATTI, H. M. 1960. Quality considerations for chapatis (unleavened pancakes). Agric. Pak. 12:157-163.

BAHADORI, M. N. 1972. New oven designed for thin bread. Food Eng. 44(10):91.

BAKHSHI, A. K., SEKHON, K. S., SEHGALL, K. L., and GILL, K. S. 1979. Chapati-making performance of some promising wheat strains. J. Res., Punjab Agric. Univ. 16:267-271.

BASS, J., and CAUL, J. F. 1972. Laboratory evaluation of three protein sources for use in chapati flour. J. Food Sci. 37:100-104.

BERLYNE, G. M., BEN ARI, J., NORD, E., and SHAINKIN, R. 1973. Bedouin osteo-malacia due to calcium deprivation caused by high phytic acid content of unleavened bread. Am. J. Clin. Nutr. 26:910-911.

BHAT, C. M. 1977. Effect of incorporation of soy flour, peanut flour and cottonseed flour on the acceptability and protein quality of chapatis. Ph.D. thesis, Ohio State Univ., Columbus.

BOND, E. E. 1975. Technical aid for the baking industry in India and Bangladesh. Food Technol. Aust. 27:21-23.

BOND, E. E., and MOSS, H. J. 1982. Test baking—From breeder to baker. Pages 649-652 in: Proc. World Cereal Bread Congr., 7th.

BREEN, M. D., SEYAM, A. A., and BANASIK, O. J. 1977. The effect of mill by-products and soy protein on the physical characteristics of expanded snack foods. Cereal Chem. 54:728-736.

BRESSLER, S. 1979. New opportunities for Swedish crispbread. Food Eng. Int. 4(3):34-35.

CHAUDHRI, A. G., and MULLER, H. G. 1976. Chapatis and chapati flour. Milling 152(2):22-25.

CHAUDHRY, M. S. 1968. Preparation and evaluation of attas and chapatis from U.S. wheats. Ph.D. thesis, Dep. Grain Sci. Ind., Kansas State Univ., Manhattan.

CUEVAS, R., FIGUEIRA, E., and RACCA, E. 1985. The technology of industrial production of precooked corn flour in Venezuela. Cereal Foods World 30:707-708, 710-712.

DALBY, G. 1963. The baking industry in Egypt. Bakers Dig. 36(6):74-77.

DALBY, G. 1966. Bread baking in the Sudan and in Saudi Arabia. Bakers Dig. 37(3):64-66.

DALBY, G. 1969. Protein fortification of bread: Balady bread as an example. Pages 174-180 in: Protein-Enriched Cereal Foods for World Needs. M. Milner, ed. Am. Assoc. Cereal Chem., St. Paul, MN.

DENDY, D. A. V., CLARKE, P. A., and JAMES, A. W. 1970. The use of blends of wheat and non-wheat flours in breadmaking. Trop. Sci. 12:131-143.

DESIKACHAR, H. 1977. Processing of sorghum and millets for versatile food uses in India. Page 41 in: Proc. Symp. Sorghum Millets Hum. Food. Int. Assoc. Cereal Sci. Technol. (ICC), Vienna.

DIXON, J. M. 1983. Unique cooker-extruder. Food Eng. Int. 8(4):41-43.

DOERRY, W. 1983. Baking in Egypt. Cereal Foods World 28:677-679.

DUTTA, S. K., RUSSELL, R. M., and CHOWDHURY, B. 1980. Folate content of North Indian breads. Nutr. Rep. Int. 21:251-256.

EBELER, S. E., and WALKER, C. E. 1983. Wheat and composite flour *chapaties*: Effects of soy flour and sucrose-ester emulsifiers. Cereal Chem. 60:270-275.

EGGUM, B. O., and DUGGAL, S. K. 1977. The protein quality of some Indian dishes prepared from wheat. J. Sci. Food Agric. 28:1052-1055.

EGGUM, B. O., MONOWAR, L., BACH KNUDSEN, K. E., MUNCK, L., and AXTELL, J. 1983. Nutritional quality of sorghum and sorghum foods from Sudan. J. Cereal Sci. 1:127-137.

EID, N., and BOURISLY, N. 1986a. Suggested level for fortification of flour and bread in Kuwait. Nutr. Rep. Int. 33:241-245.

EID, N., and BOURISLY, N. 1986b. Bread consumption in Kuwait. Nutr. Rep. Int.

33:967-971.

EL-SAIED, H. M., and EL-FARRA, A.-H. A. 1983. Utilization of aqueous by-products from starch for improving bread quality. Cereal Chem. 60:131-134.

EL-SAMAHY, S. K., and TSEN, C. C. 1981. Effects of varying baking temperature and time on the quality and nutritive value of balady bread. Cereal Chem. 58:546-548.

EL-SAMAHY, S. K., and TSEN, C. C. 1983. Changes of temperature inside the loaves of balady and pan breads during baking. Chem. Mikrobiol. Technol. Lebensm. 8:15-18.

EL-SAYED, K., SALEM, A. E., and BARY, A. A. 1978. Utilization of cottonseed flour in human foods. III. Bread and biscuit supplemented with cottonseed flour. Alexandria J. Agric. Res. 26:619-631.

EL-SHIMI, N. M., SALEM, A. E., MOHASSEB, Z. S., and ZOUIL, M. E. 1977. New method for making balady bread. Alexandria J. Agric. Res. 25:83-86.

EL TINAY, A. H., ABDEL GADIR, A. M., and EL HIDAI, M. 1979. Sorghum fermented kisra bread. I. Nutritive value of kisra. J. Sci. Food Agric. 30:859-864.

ELIAS, A. N., MORAD, M. M., and EL-SAMAHY, S. K. 1977. Use of potato flour for the production of balady bread in Egypt. Lebensm. Wiss. Technol. 10:42-44.

EMINIAN, J., REINHOLD, J. G., KFOURY, G. A., AMIRHAKIMI, G. H., SHARIF, H., and ZIAI, M. 1967. Zinc nutrition of children in Fars province of Iran. Am. J. Clin. Nutr. 20:734-742.

FARIDI, H. A. 1981. Health advantages of a high bread diet and approaches to U.S.-type flat bread production. Bakers Dig. 55(4):6-9.

FARIDI, H. A., and FINNEY, P. L. 1980. Technical and nutritional aspects of Iranian breads. Bakers Dig. 54(5):14-22.

FARIDI, H. A., and RUBENTHALER, G. L. 1982. Experimental baking method for Egyptian balady bread production and effect of oven baking on bread starch gelatinization and shelf life. Pages 106-121 in: Proc. Egypt. Conf. Bread Res., 1st, Cairo.

FARIDI, H. A., and RUBENTHALER, G. L. 1983a. Breads produced from Pacific Northwest soft white wheat. II. Variety breads. Bakers Dig. 57(1):27-30.

FARIDI, H. A., and RUBENTHALER, G. L. 1983b. Experimental baking techniques for evaluating Pacific Northwest wheats in North African breads. Cereal Chem. 60:74-79.

FARIDI, H. A., and RUBENTHALER, G. L. 1983c. Ancient breads and a new science: Understanding flat breads. Cereal Foods World 28:627-629.

FARIDI, H. A., and RUBENTHALER, G. L.

1984a. Effect of baking time and temperature on bread quality, starch gelatinization, and staling of Egyptian balady bread. Cereal Chem. 61:151-154.

FARIDI, H. A., and RUBENTHALER, G. L. 1984b. Laboratory method for producing Chinese steamed bread and effects of formula, steaming and storage on bread starch gelatinization and freshness. Pages 863-867 in: Proc. Int. Wheat Genet. Symp., 6th, Kyoto, Japan.

FARIDI, H. A., and RUBENTHALER, G. L. 1984c. Effect of various flour extractions, water absorption, baking temperature, and shortening level on the physical quality and staling of pita breads. Cereal Foods World 29:575-576.

FARIDI, H. A., FINNEY, P. L., and RUBENTHALER, G. L. 1981. Micro baking evaluation of some U.S. wheat classes for suitability in Iranian breads. Cereal Chem. 58:428-432.

FARIDI, H. A., RANHOTRA, G. S., FINNEY, P. L., and RUBENTHALER, G. L. 1982a. Protein quality characteristics of Iranian flat breads. J. Food Sci. 47:676-677, 679.

FARIDI, H. A., FINNEY, P. L., RUBENTHALER, G. L, and HUBBARD, J. D. 1982b. Functional (breadmaking) and compositional characteristics of Iranian flat breads. J. Food Sci. 47:926-929, 932.

FARIDI, H. A., FINNEY, P. L., and RUBENTHALER, G. L. 1983a. Iranian flat breads: Relative bioavailability of zinc. J. Food Sci. 48:107-110.

FARIDI, H. A., FINNEY, P. L., and RUBENTHALER, G. L. 1983b. Effect of soda leavening on phytic acid content and physical characteristics of Middle Eastern breads. J. Food Sci. 48:1654-1658.

FARIDI, H. A., RUBENTHALER, G. L., and FINNEY, P. L. 1983c. Iranian flat breads: Relative bioavailability of magnesium. Nutr. Rep. Int. 27:475-483.

FARADJI, B., REINHOLD, J. G., and ABADI, P. 1981. Human studies of iron absorption from fiber-rich Iranian flat breads. Nutr. Rep. Int. 23:267-277.

FINNEY, P. L., MORAD, M. M., and HUBBARD, J. D. 1980a. Germinated and ungerminated faba bean in conventional U.S. breads made with and without sugar and in Egyptian balady breads. Cereal Chem. 57:267-270.

FINNEY, P. L., MORAD, M. M., PATEL, K., CHAUDHRY, S. M., GHIASI, K., RANHOTRA, G., SEITZ, L. M., and SEBTI, S. 1980b. Nine international breads from sound and highly-field-sprouted Pacific Northwest soft white wheat. Bakers Dig.

54(6):22-27.

FORD, J. A., COLHOUN, E. M., McINTOSH, W. B., and DUNNIGAN, M. G. 1972. Biochemical response of late rickets and osteomalacia to a chapati-free diet. Br. Med. J. 3:466-467.

GOGLANIAN, A. 1986. Method of preparing perforated pita bread. U.S. patent 4,597,979.

GRECZ, N., BRANNON, R., JAW, R., AL-HARITHY, R., and HAHN, E. W. 1985. Gamma processing of Arabic bread for immune system-compromised cancer patients. Appl. Environ. Microbiol. 50:1531-1534.

GROEN, J. J., BALOGH, M., and YARON, E. 1966. Effect of the Yemenite diet on the serum cholesterol of healthy non-Yemenites. Isr. J. Med. Sci. 2:196-201.

HAENEL, H., and TUNGER, L. 1976. Nutritional evaluation of crispbreads available in the German Democratic Republic. Ernaehrungsforschung 21:15-18.

HAFFA, W. 1977. Process and apparatus for the production of flat bread from wheat dough. German Federal Republic patent 2,553,752.

HAGHSHENASS, M., MAHLOUDJI, M., REINHOLD, J. G., and MOHAMMADI, N. 1972. Iron-deficiency anemia in an Iranian population associated with high intakes of iron. Am. J. Clin. Nutr. 25:1143-1146.

HALLAB, A. H., KHATCHADOURIAN, H. A., and JABR, I. 1974. The nutritive value and organoleptic properties of white Arabic bread supplemented with soybean and chickpea. Cereal Chem. 51:106-112.

HAMED, M. G. E., FODA, Y. H., ABD-EL-AL, A. J., and EL-SAMAHY, S. K. 1973a. Carbonyl compounds in starter, unfermented and fermented dough and "balady" bread. Egypt. J. Food Sci. 1:195-202.

HAMED, M. G. E., HUSSEIN, M. F., REFAI, F. Y., and EL-SAMAHY, S. K. 1973b. Preparation and chemical composition of sweet potato flour. Cereal Chem. 50:133-139.

HAMED, M. G. E., REFAI, F. Y., HUSSEIN, M. F., and EL-SAMAHY, S. K. 1973c. Effect of adding sweet potato flour to wheat flour on physical dough properties and baking. Cereal Chem. 50:140-146.

HAQ, M. Y. I., and CHAUDHRY, M. S. 1976. Supplementation of roti and naan with chickpea flour. Pak. J. Sci. Ind. Res. 19:66-71.

HEDAYAT, H., SARKISSIAN, N., LANKARANI, S., and DONOSO, G. 1968. The enrichment of whole wheat flour and Iranian bread. Acta Biochim. Iran. 5:1-10, 16.

HEDAYAT, H., SHAHBAZI, H., PAYAN, R., and DONOSO, G. 1971. Lysinfield trials in Iran. Pages 44-51 in: Progress in Human

Nutrition, Vol. I. S. Morgen, ed. Avi Publ. Co., Inc., Westport, CT.

HEDAYAT, H., SHAHBAZI, H., PAYAN, R., AZAR, M., BAVENDI, M., and DONOSO, G. 1973. The effect of lysine fortification of Iranian bread on the nutritional status of school children. Acta Paediatr. Scand. 69:297-303.

HOHN, K., HOHN, O., and HOHN, W. 1981. Method of producing dry flat bread. U.S. patent 4,308,285.

HOHN, K., HOHN, O., and HOHN, W. 1982. Arrangement for baking dry flat bread. U.S. patent 4,311,088.

HOOVER, W. J. 1974. Status of the milling and baking industries in Africa and Asia. Cereal Sci. Today 19:153-156.

HUSSAIN, M., and SATTI, M. H. 1970. Development of long-life roti (bread). Pak. J. Sci. Ind. Res. 12:408-411.

HUSSEIN, L., GABRIEL, G. N., and MORCOS, S. R. 1973. Possibility of improving the protein quality of balady bread by enrichment with synthetic amino acids. Z. Ernaehrungswiss. 12:201-203.

HUSSEIN, L., GABRIEL, G., and MORCOS, S. 1974. Nutritive value of mixtures of balady bread and broad beans. J. Sci. Food Agric. 25:1433-1440.

IBRAHIM, S. S., ELIAS, A. N., and EL-FARRA, A. A. 1983. Flour granularity, its effect upon Egyptian balady bread making quality. Egypt. J. Food Sci. 11:81-88.

IMTIAZ, Z. 1962. Fortification of a Pakistani bread recipe with animal protein and calcium and the determination of its biological value. M.S. thesis, Oklahoma State Univ., Stillwater.

IRANI, P. 1987. Bread and baked products in Islamic countries. Past—present—future. Pages 169-174 in: Proc. ICC Congr., 11th, Hamburg. Int. Assoc. Cereal Sci. Technol., Vienna.

ISMAIL-BEIGI, F., FARADJI, B., and REINHOLD, J. G. 1977. Binding of zinc and iron to wheat bread, wheat bran and their components. Am. J. Clin. Nutr. 30:1721-1725.

KANNUR, S. B., RAMANUJA, M. N., and PARIHAR, D. B. 1973. Studies on the browning of heat processed chapatis and heat allied products in relation to their chemical constituents. J. Food Sci. Technol. (Mysore) 10:64-69.

KARP, D., and GARPING, B. 1966. Experiences in evaluation of rye crisp and other flat breads. Brot Gebaeck 20:169-176.

KHALIL, J. K., and HALLAB, H. A. 1975. Nutritive value and chemical composition of Arabic bread supplemented with cheese-whey. Pak. J. Biochem. 8:57-62.

KHALIL, M. K. M., MOUSTAFA, E. K.,

ZOUIL, M. E., and AMAN, M. E. 1976. Effects of supplementation with FPF (fish protein flour) on the nutritional value of Egyptian corn and wheat breads. Egypt. J. Food Sci. 4:47-53.

KHAN, M., and BHATTI, M. B. 1968. Studies on the storage life of tanouri roti, by chemical preservation and by packaging in various containers. Page 92 in: Proc. All Pak. Sci. Conf., 20th, Dacca.

KHAN, M. A., and EGGUM, B. O. 1978. Effect of baking on the nutritive value of Pakistani bread. J. Sci. Food Agric. 29:1069-1075.

KHAN, M. N., ROBINSON, R. J., WANG, M., and HOOVER, W. J. 1972. Effect of protein-rich flour and sodium stearoyl-2-lactylate (SSL) on chapatti (Indian non-leavened bread). (Abstr.) Cereal Sci. Today 17:263.

KOHOUT, E., DUTZ, W., and PETROSIAN, A. 1967. Laboratory findings in Iranian village and tribal communities, 1964. Am. J. Clin. Nutr. 29:410-414.

KOUHESTANI, A., GHAVIFEKR, H., RAHMANIAN, M., MAYURIAN, H., and SARKISSIAN, N. 1969. Composition and preparation of Iranian breads. J. Am. Diet. Assoc. 55:262-266.

KUZAZLI, M. V., COWAN, J. W., and SABRY, Z. I. 1966. Nutritive value of Middle Eastern foodstuffs. II. Composition of pulses, seeds, nuts and cereal products of Lebanon. J. Sci. Food Agric. 17:82-84.

LINDEL, M. J., and WALKER, C. E. 1984. Soy enrichment of *chapaties* made from wheat and nonwheat flours. Cereal Chem. 61:435-438.

LINKO, P., MATTSON, C., LINKO, Y.-Y., and ANTILA, J. 1984. Production of flat bread by continuous extrusion cooking from high alpha-amylase flours. J. Cereal Sci. 2:43-51.

MAHMOUD, S. A. Z., ABDEL-HAFEZ, A. M., EL-SAWY, M., and ABDEL-HAMID, T. H. A. 1976. Microbiology of dough leavening. III. Microbial flora of dough leavened by different starters. Egypt. J. Microbiol. 11:75-84.

MALEKI, M. 1972. Food consumption and nutritional status of 13-year old village and city schoolboys in Fars province in Iran. Ecol. Food Nutr. 2:39-42.

MALEKI, M., and DAGHIR, S. 1967. Effect of baking on retention of thiamine, riboflavin, and niacin in Arabic bread. Cereal Chem. 44:483-487.

MALEKI, M., and DJAZAYERI, A. 1968. Effect of baking and amino acid supplementation on the protein quality of Arabic bread. J. Sci. Food Agric. 19:449-451.

MALEKI, M., VETTER, J., and HOOVER, W. 1981. The effect of emulsifiers, sugars, shortening and soya flour on the staling of barbari flat bread. J. Sci. Food Agric. 32:1209-1211.

MANLEY, D. J. R. 1983. Technology of Biscuits, Crackers and Cookies. Ellis Horwood Ltd., Chichester, U.K.

MANN, A. R. 1970. Studies on improving the shelf-life of chapatis and naan. M.S. thesis, Kans. State Univ., Manhattan.

MASON, B., and HOSENEY, R. C. 1980. Improvement of pearl millet. Third Annu. Rep., Kansas State Univ., Manhattan.

McLAREN, D. S. 1969. The face of childhood malnutrition in the Near East. Page 613 in: Man, Food, and Agriculture in the Middle East. F. Sarruf, ed. Centennial Publ., American Univ. Beirut, Lebanon.

MEUSER, F., and VAN LENGERICH, B. 1984. Possibilities of quality optimization of industrially extruded flat breads. Pages 180-184 in: Thermal Processing and Quality of Foods. P. Zeuthen, J. C. Cheftel, C. Eriksson, M. Jul, H. Leniger, P. Linko, G. Varela, and G. Vos, eds. Elsevier Sci. Publ. Co., New York.

MILATOVIC, L. 1979. Breadmaking in Iran: Production of "barbari" bread. Ind. Aliment. (Pinerolo, Italy) 18:294-297.

MORAD, M. M., DOHERTY, C. A., and ROONEY, L. W. 1984. Effect of sorghum variety on baking properties of U.S. conventional bread, Egyptian pita balady bread and cookies. J. Food Sci. 49:1070-1074.

MOUSA, E. I., and AL-MOHIZEA, I. S. 1987. Bread baking in Saudi Arabia. Cereal Foods World 32:614, 616, 618, 620.

MOUSA, E. I., IBRAHIM, R. H., SHUEY, W. C., and MANEVAL, R. D. 1979. Influence of wheat classes, flour extractions, and baking methods on Egyptian balady bread. Cereal Chem. 56:563-566.

MURTY, D. S., and SUBRAMANIAN, V. 1982. Sorghum roti. I. Traditional methods of consumption and standard procedures for evaluation. Pages 73-78 in: Proc. Int. Symp. Sorghum Grain Quality. L. W. Rooney and D. S. Murty, eds. Int. Crop Res. Inst. Semi-Arid Tropics, Patancheru, A.P., India.

MURTY, G. S., and AUSTIN, A. 1963a. Studies on the quality characteristic of Indian wheats with reference to chapati making. Food Sci. (Mysore) 12:61-64.

MURTY, G. S., and AUSTIN, A. 1963b. Studies on the quality characteristics of Indian wheats with reference to the mixability of their flours with the flours of other food grains and tubers for making chapatis. Food Sci. (Mysore) 12:64-68.

MUSTAFA, A. M. I. 1973. Production of bread from composite flour. Sudan J. Food Sci. Technol. 5:41-46.

NATH, N., SINGH, S., and NATH, H. P. 1957. Studies on the changes in the water soluble carbohydrates of chapatis during aging. Food Res. 22:25-29.

NAWAR, I. A. 1974. Patterns of food consumption and quality of diets of individuals in an Egyptian village. Alexandria J. Agric. Res. 22:13-17.

NAZAR, M. 1970. Effect of iron enrichment of flour on the baking qualities of Arabic bread. M.S. thesis, Fac. Agric. Sci., American Univ. Beirut, Lebanon.

NAZAR, M., and HALLAB, A. H. 1973. Effect of iron enrichment of flour on dough characteristics and organoleptic qualities of Arabic bread. Pak. J. Sci. Ind. Res. 16:115-121.

NIKKILA, M., CONSTANTINIDES, S. M., and MEADE, T. L. 1976. Supplementation of Arabic and Indian breads with fish protein concentrate. J. Agric. Food Chem. 24:1144-1147.

OLEWNIK, M. C., HOSENEY, R. C., and VARRIANO-MARSTON, E. 1984. A procedure to produce pearl millet *rotis*. Cereal Chem. 61:28-33.

ORTH, R. A., and MOSS, H. J. 1977. The sensitivity of various products to sprouted wheat. Page 167 in: Proc. Int. Symp. Pre-harvest Sprouting Cereals, 4th. D. J. Mans, ed. Westview Press, Boulder, CO.

PAPANTONIOU, A., and PAPANTONIOU, K. 1980. Method of preparing pocket pita bread. U.S. patent 4,202,911.

PARIHAR, D. B., and CHATTERIJI, A. K. 1956. X-ray diffraction studies of chapati during cooking and storage. J. Sci. Ind. Res. (New Delhi) 150:115-117.

PATEL, K. M., and JOHNSON, J. A. 1975. Horsebean protein supplements in breadmaking. II. Effect on physical dough properties, baking quality, and amino acid composition. Cereal Chem. 52:791-800.

PATEL, K. M., CAUL, J. F., and JOHNSON, J. A. 1977. Horsebean as protein supplement in breadmaking. III. Effect of horsebean protein on aroma and flavor profile of Moroccan-type bread. Cereal Chem. 54:379-387.

PELLET, P., and SHADAVERIAN, S. 1970. Food Consumption Tables for Use in the Middle East. American Univ. Beirut, Lebanon.

POMERANZ, Y., and SHELLENBERGER, J. A. 1971. Bread Science and Technology. Avi Publ. Co., Inc., Westport, CT.

PONTE, J. G., Jr. 1971. Bread. Pages 675-735 in: Wheat: Chemistry and Technology.

2nd ed. Y. Pomeranz, ed. Am. Assoc. Cereal Chem., St. Paul, MN.

PRABHAVATHI, C., HARIDAS, R. P., and SHURPALEKAR, S. R. 1976. Bread and chapati making quality of Indian durum wheats. J. Food Sci. Technol. (Mysore) 13:313-317.

PRASAD, A. S., HALSTED, J. A., and NADIMI, M. 1961. Syndrome of iron deficiency anemia, hepatosplenomegaly, hypogonadism, dwarfism and geophagia. Am. J. Med. 31:532-536.

QAROONI, J., ORTH, R. A., and WOOTTON, M. 1987a. A test baking technique for Arabic bread quality. J. Cereal Sci. 6:69-80.

QAROONI, J., MOSS, H. J., ORTH, R. A., and WOOTEN, M. 1987b. The effect of flour properties on the quality of Arabic bread. J. Cereal Sci. 7:95-107.

RANHOTRA, G. S., GELROTH, J. A., TORRENCE, F. A., BOCK, M. A., WINTERRINGER, G. L., FARIDI, H. A., and FINNEY, P. L. 1981. Iranian flat breads: Relative bioavailability of iron. Cereal Chem. 58:471-474.

RAO, G. V., LEELAVATHI, K., RAO, P. H., and SHURPALEKAR, S. R. 1986. Changes in the quality characteristics of chapati during storage. Cereal Chem. 63:131-135.

RAO, K. G., MALATHI, M. A., and VIJAYARAGHAVAN, P. K. 1964. Preservation and packaging of Indian foods. I. Preservation of chapatis. Food Technol. (Mysore) 18:108-112.

RAO, K. G., MALATHI, M. A., and VIJAYARAGHAVAN, P. K. 1966. Preservation and packaging of Indian foods. II. Storage studies on preserved chapatis. Food Technol. (Mysore) 20:94-99.

RAO, P. H., LEELAVATHI, K., and SHURPALEKAR, S. R. 1986. Test baking of chapati—Development of a method. Cereal Chem. 63:297-303.

RAO, P. H., LEELAVATHI, K., and SHURPALEKAR, S. R. 1987. Objective measurement of the consistency of chapati dough using research water absorption meter. J. Texture Stud. 174:401-420.

RASHID, J. 1974. Triticale as a cereal for chapati making. M.S. thesis, Univ. Strathclyde, Glasgow, Scotland.

RASHID, J. 1983. Effect of wheat type and extraction rate on quality of Arabic type leavened flat bread. Ph.D. dissertation, Kansas State Univ., Manhattan.

RASHID, J., and HAWTHORN, J. 1974. Triticale as a cereal for chupatty baking. (Abstr.) Cereal Sci. Today 19:414.

RASHID, J., PONTE, J. G., Jr., and DAVIS, A. B. 1982. Effect of flour type and extraction rate on quality of Arabic-type leavened flat bread (naan). (Abstr.) Cereal Foods World 27:464.

REINHOLD, J. G. 1971. High phytate content of rural Iranian bread: A possible cause of human zinc deficiency. Am. J. Clin. Nutr. 24:1204-1206.

REINHOLD, J. G. 1972. Phytate concentrations of leavened and unleavened Iranian breads. Ecol. Food Nutr. 1:187-192.

REINHOLD, J. G. 1975a. Prevalence of nutritional disease in rural Iran and approaches to its correction. Iran. J. Agric. Res. 3(1):1-9.

REINHOLD, J. G. 1975b. Phytate destruction by yeast fermentation in whole wheat meals. J. Am. Diet. Assoc. 66:38-41.

REINHOLD, J. G., HEDAYATI, H., LAHIMARZADEH, A., and NASR, K. 1973. Zinc, calcium, phosphorus, and nitrogen balances of Iranian villagers following a change from phytate-rich to phytate-poor diets. Ecol. Food Nutr. 2:157-162.

REINHOLD, J. G., PARSA, A., KARIMIAN, N., HARMMICK, J. W., and ISMAIL-BEIGI, F. 1974. Availability of zinc in leavened and unleavened wholemeal wheaten breads as measured by solubility and uptake by rat intestine in vitro. J. Nutr. 104:976-982.

REINHOLD, J. G., ISMAIL-BEIGI, F., and FARADJI, B. 1975. Fibre vs. phytate as determinant of the availability of calcium, zinc and iron of breadstuffs. Nutr. Rep. Int. 12:75-85.

REINHOLD, J. G., FARADJI, B., ABADI, P., and ISMAIL-BEIGI, F. 1976a. Decreased absorption of calcium, magnesium, zinc and phosphorus by humans due to increased fiber and phosphorus consumption as wheat bread. J. Nutr. 106:493-503.

REINHOLD, J. G., FARADJI, B., ABADI, P., and ISMAIL-BEIGI, F. 1976b. Binding of zinc to fiber and other solids of whole-meal bread. Page 163 in: Trace Elements in Human Health and Disease, Vol. I: Zinc and Copper. A. N. Prasad, ed. Academic Press, New York.

RIZK, S. S., SEDKY, A., and MOHAMMED, M. S. 1960a. Studies on Egyptian bread. I. Types and methods of baking commercial bread in Egypt. Alexandria J. Agric. Res. 8:83-87.

RIZK, S. S., SEDKY, A., and MOHAMMED, M. S. 1960b. Studies on Egyptian bread. II. Chemical composition of some types of bread consumed in various localities in Egypt. Alexandria J. Agric. Res. 8:89-97.

RONAGHY, H. A., CAUGHEY, J. E., and HALSTED, J. A. 1968. A study of growth in Iranian village children. Am. J. Clin. Nutr.

21:488-494.

RONAGHY, H. A., REINHOLD, J. G., MAHLOUDJI, M., GHAVAMI, P., SPIVEY FOX, M. R., and HALSTED, J. A. 1974. Zinc supplementation of malnourished schoolboys in Iran. Increased growth and other effects. Am. J. Clin. Nutr. 27:112-121.

ROONEY, L. W., and SERNA-SALDIVAR, S. O. 1987. Food uses of whole corn and dry-milled fractions. Pages 399-429 in: Corn: Chemistry and Technology. S. A. Watson and P. E. Ramstad, eds. Am. Assoc. Cereal Chem., St. Paul, MN.

ROONEY, L. W., KIRLEIS, A. W., and MURTY, D. S. 1986. Traditional foods from sorghum: Their production, evaluation, and nutritional value. Pages 317-353 in: Advances in Cereal Science and Technology, Vol. VIII. Y. Pomeranz, ed. Am. Assoc. Cereal Chem., St. Paul, MN.

RUBENTHALER, G. L., and FARIDI, H. 1981. Breads produced from Pacific Northwest soft white wheat. I. Five international breads. Bakers Dig. 55(6):19-22.

RUBENTHALER, G. L., and FARIDI, H. A. 1982. Laboratory dough molder for flat breads. Cereal Chem. 59:72-73.

RUSSELL, R. M., ISMAIL-BEIGI, F., and REINHOLD, J. G. 1976. Folate content of Iranian breads and the effect of their fiber content on the intestinal absorption of folic acid. Am. J. Clin. Nutr. 29:799-802.

SAADIA, D. 1977. A new oven for flat bread. Israeli patent 46,680.

SAWAYA, W., KHALIL, J., KHATCHADOURIAN, H., and AL-MOHAMMED, M. 1984. Nutritional evaluation of various breads consumed in Saudi Arabia. Nutr. Rep. Int. 29:1161-1170.

SCHNEE, W. 1980. Plant for the automatic production of Arabic flat bread. U.S. patent 4,204,466.

SEKHON, K. S., SAXENA, A. K., RANDHAWA, S. K., and GILL, K. S. 1980. Use of triticale for bread, cookie and chapati making. J. Food Sci. Technol. (Mysore) 17:233-235.

SHAKIR, A., and DEMARCHI, M. 1971. Dietary pattern of rural school children in the environment of Baghdad. Trop. Geogr. Med. 23:258-263.

SHARMA, Y. K., DEODHAR, A. D., and MISHRA, A. 1977. Evaluation of physicochemical characteristics determining the chapati-making qualities of triticale. Indian J. Nutr. Diet. 14:140-143.

SHEHATA, N. A., and FRYER, B. A. 1970. Effect on protein quality of supplementing wheat flour with chickpea flour. Cereal Chem. 47:663-670.

SHURPALEKAR, S. R., and PRABHAVATHI, C. 1976. Brabender Farinograph, Research Extensometer, and Hilliff chapatti press as tools for standardization and objective assessment of chapatti dough. Cereal Chem. 53:457-469.

SHURPALEKAR, S. R., KUMAR, G. V., RAO, G. V., RANGA RAO, G. C. P., VATSALA, C. N., and RAHIM, A. 1976. Physico-chemical, rheological and milling characteristics and bread and chapati making quality of Indian wheat. J. Food. Sci. Technol. (Mysore) 13:79-86.

SHYAMALA, G., and KENNEDY, B. M. 1962. Protein value of chapatis and purees. J. Am. Diet. Assoc. 41:115-118.

SIMON, E. B. 1979. Improvements in baking installations for flat bread. Israeli patent 4,750,612.

SINHA, A. C. 1964. Quality of chapatis from Indian and imported wheat. Bull. Grain Technol. (Hapur, India) 2:53-59.

SWARANJEET, K., MANINDER, K., and BAINS, G. S. 1982. *Chapaties* with leavening and supplements: Changes in texture, residual sugars, and phytic phosphorus. Cereal Chem. 59:367-372.

TABEKHIA, M. M., and MOHAMED, M. S. 1971. Effect of processing and cooking operations on vitamins of some Egyptian national foods (wheat flour, bread and rice). Alexandria J. Agric. Res. 19:279-284.

TABEKHIA, M. M., and TOMA, R. B. 1979. Chemical composition of various types of Egyptian breads. Nutr. Rep. Int. 19:377-382.

TADAYON, R. A. 1976. Characteristics of yeasts isolated from bread doughs of bakeries in Shiraz, Iran. J. Milk Food Technol. 39:539-546.

TADAYON, R. A. 1978. Identification of yeasts isolated from bread dough of bakeries in Shiraz, Iran. J. Food Prot. 41:717-721.

TANNAHILL, R. 1973. Food in History. Stein and Day Publ., New York.

TER-SARKISSIAN, N., AZAR, M., GHAVIFEKR, H., FERGUSON, T., and HEDAYAT, H. 1974. High phytic acid in Iranian breads. J. Am. Diet. Assoc. 65:651-653.

TOBEKHI, M. M., and EL-ESLEY, N. 1982. Effect on various methods of common bread making (native bread) on hydrolysis of phytic acid. Food Sci. Technol. Abstr. 302(2M):14.

TSEN, C. C., MEDINA, G., and HUANG, D. S. 1974. Using indigenous flours and starches as bread supplements in developing countries. Proc. IV Int. Congr. Food Sci. Technol. 5:333-341.

TSEN, C. C., REDDY, P. R. K., and GEHRKE, C. W. 1977. Effects of conventional baking, microwave baking, and steaming on the nutritive value of regular and fortified breads. J. Food Sci. 42:402-406.

VAGHEFI, S. B., and DELGOSHA, M. 1975. Fortification of Persian-type bread with vitamin A. Cereal Chem. 52:753-756.

WAHREN, M. 1959. Cereals and bread in old and new North Africa. Brot Gebaeck 13:21-29.

WAHREN, M. 1961. Types of old Egyptian bread and baked products. Brot Gebaeck 15:1-12, 28-32.

WAHREN, M. 1962. Bread and baked products in life and religion of ancient Egyptians. Brot Gebaeck 16:12-20, 30-38.

WILLIAMS, P. C., and BLASCHUK, W. 1980. Comparison of seven different methods for screening wheat for baking quality of "khobz"-type of flat breads. (Abstr.) Cereal Foods World 25:527.

WILLIAMS, P. C., EL-HARAMEIN, F. J., NELSON, W., and SRIVASTAVA, J. P. 1988. Evaluation of wheat quality by baking Syrian-type two-layered flat breads. J. Cereal Sci. (In press)

WILLIAMS, R. W., and MONTARULI, V. 1983. Radiant oven for baking bread. U.S. patent 4,338,823.

YASIN, M., KHAN, A. H., and KARIMULLAH, H. 1965. Studies of the preservation of chapatis. Pak. J. Sci. Ind. Res. 8(3):126-129.

YONEMA, M. 1973. Yeasts associated with Afghanistan bread "naan," their nature and origin. Rep. Tottori Mycol. Inst. (Japan) 10:619-622.

CHAPTER 9

DURUM WHEAT AND PASTA PRODUCTS

J. W. DICK
Department of Cereal Science and Food Technology
North Dakota State University
Fargo, North Dakota

R. R. MATSUO
Grain Research Laboratory
Agriculture Canada
Winnipeg, Manitoba, Canada

I. INTRODUCTION

Durum wheat is unique in that it is generally considered the hardest of all wheats. This, as well as other characteristics, make it most suitable as a raw material for use in the manufacture of such food products as pasta (i.e., spaghetti, lasagne, elbow macaroni) and couscous (gelatinized, dried particles of dough). These types of foods are commonly referred to as "paste" or "alimentary paste" products since they are made into a dough from a base formula of wheat endosperm and water. *Paste* is defined in the dictionary as "any shaped and dried dough prepared from semolina, farina, or wheat flour, or a mixture of these with water, milk or egg."

The use of paste products dates back many centuries. Although such products as pasta were elevated to popularity by the Italians, most certainly they were a part of the earlier cultures of China and the later cultures of many other countries. Pasta-type products have become an important part of the diet for people of European descent. The paste product couscous was popularized in countries of North Africa. Oriental-style noodles are examples of paste products most widely accepted by people of eastern and southern Asia. The widespread consumption of paste products throughout the world is most likely due to their simple formulation and relative ease of processing, their storability and immense versatility, and their low cost relative to some other foods.

Paste products can be produced from several grain sources such as legumes or other cereals, although more often than not durum wheat is the raw material of choice to manufacture products with the most desirable end-product characteristics. Factors that limit the widespread use of durum in some parts of the world

507

include regional preference for other grains, unavailability of adequate supplies of durum, and durum's market price, which is sometimes volatile and traditionally higher than those of competing grains. Paste products traditionally made from grains other than durum include most Oriental wheat noodles, and noodles made from the flour of rice, barley, buckwheat, mung bean, potato, yam, and cassava or from other raw materials. In recent years, numerous foods have been added as dietary supplements to increase the protein or fiber content of these products.

The bulk of durum wheat that is used for food is consumed in the form of paste products. Durum is also utilized in products such as bread, bulgur, puffed cereal, and parched grain. These products, as well as some nondurum paste products, are discussed briefly in this chapter, although most of the information presented emphasizes the relationships among durum characteristics, semolina milling, pasta processing, and end-product quality. Durum wheat is also commonly used as a source of animal feed, but that subject is not included in this review.

II. DURUM WHEAT

A. Classification

All wheats belong to the genus *Triticum* of the Family Gramineae, the grass family. Botanically, wheats are classified into three groups according to their chromosome makeup. Diploid wheats are those with two seven-chromosome sets (14 chromosomes). Einkorn wheat grown in some countries in Asia Minor belongs to this group. Tetraploid wheats have four seven-chromosome sets (28 chromosomes), which are further classified into two genomes. The most important member of this group today is durum wheat. The common wheats (including bread wheats) are hexaploids, with six seven-chromosome sets.

Durum wheat used to be designated *T. durum Desf.*, but the currently accepted designation is *T. turgidum* var. *durum* (Bowden, 1959, 1966).

B. Origin

The origin of wheat is uncertain, but species of *T. aegilopoides* have been found in excavation sites in Syria dating from 8400 to 7500 B.C. (Nishikawa and Nagao, 1977). Durum wheat was grown in Egypt 4,000 years before the Christian era (Harrison, 1934). It is also claimed that durum wheat was grown in the Ukraine about the same time (Dorofeev and Jakubziner, 1973). It is probably safe to state that durum wheat originated in the Middle East, where it is still produced and consumed in substantial quantities.

Durum wheat was introduced to the New World in 1527, when it was brought to Argentina, but did not become an important crop until the beginning of the present century (Vallega, 1973). Cultivation of durum wheat did not begin in the United States until M. A. Carlton brought in durum cultivars from Russia around 1900, although, according to Dorofeev and Jakubnizer (1973), the cultivar Arnautka had been introduced from Russia in 1864. Production of durum wheat in Canada began after the severe rust epidemic of 1916. Thus, the history of durum wheat in North America, currently the major exporter of durum wheat, is relatively short.

C. Production

Relative to common wheat (*T. aestivum*), durum wheat is a minor crop. In the decade 1975–1984, the average annual production of durum wheat was only 5.1% of that of common wheat. Area sown to durum wheat in the same period was 16,490,000 ha, 7% of that sown to common wheat (International Wheat Council, 1986). The average yield of durum wheat is lower than that of common wheat (1.38 vs 1.91 t/ha).

The world durum wheat production for the decade 1976–1985 is given in Table I. Turkey is the largest producer, with an average 5,150,000 t per year, followed by Italy at 3,330,000 t and the United States at 3,200,000 t. North Africa was an exporter of durum wheat in the 1950s but now is the main importer, in spite of the fact that about 70% of the total wheat area (in Algeria, Libya, Morocco, and Tunisia) is planted to durum wheat (Srivastava, 1984). Yield is very low, averaging 0.50–0.80 t/ha, compared to an average yield of 2.09 t/ha in Western Europe and 1.86 t/ha in North America.

D. Export

Around the turn of this century, Russia was the major producer and exporter of durum wheat. Annual production ranged from 4.1 to 5.4 million tonnes, and much of it was exported to Italy and elsewhere in Europe (Alsberg, 1939). As mentioned, North Africa was a net exporter until the 1950s, and Argentina exported a large proportion of its production until the mid-1970s. In 1973, for example, Argentina exported 75% of its total production of 610,000 t. In recent years, the United States and Canada have been the major exporters of durum wheat, accounting for over 90% of the world trade. Exports of durum by the United States and Canada in the past decade are listed in Table II.

III. DURUM WHEAT PRODUCTS

Foods made from durum wheat can be grouped into the two general

TABLE I
World Durum Wheat Production (thousand tonnes)[a]

Year	Western Europe	North America	South America	Near East Asia	North Africa	Others[b]	World Total
1976	4,180	6,600	500	7,370	3,540	4,530	26,700
1977	2,700	3,520	370	6,050	2,290	4,030	19,000
1978	4,500	6,550	430	6,460	2,900	4,530	25,400
1979	4,330	4,770	270	6,460	2,810	4,530	23,200
1980	5,010	5,060	270	7,380	3,140	4,530	25,400
1981	4,520	8,060	180	6,840	2,480	4,040	26,100
1982	4,460	7,190	250	7,410	3,090	3,540	25,900
1983	4,380	4,710	210	6,800	2,700	3,540	22,300
1984	6,670	5,000	180	6,340	2,540	3,040	23,800
1985	5,540	5,170	130	6,740	3,810	3,040	24,600
10-year mean	4,649	5,663	279	6,785	2,930	3,935	24,240

[a] Source: International Wheat Council (1986); used by permission.
[b] Including estimates for the USSR and other centrally planned countries.

categories of paste and nonpaste products. Paste products are those foods produced by mixing water and either durum semolina, durum flour, or whole ground durum to make an unleavened dough and then forming the dough into various shapes of the desired dimensions. These are consumed fresh or processed further to preserve the product. Examples of paste products are pasta and couscous. The nonpaste products are those foods that either utilize the milled products of durum in a high-moisture dough (e.g., leavened or unleavened bread) or are the results of a cooking, steaming, or scorching process performed on the whole kernel of durum such as bulgur (gelatinized wheat kernels) or *frekeh* (parched immature wheat kernels).

A. Paste Products

PASTA

Pasta is the most commonly consumed paste-product made from durum wheat. The term *pasta* has generally been reserved to describe paste products fitting the "Italian" style of extruded foods such as spaghetti or lasagne, and is usually distinguished from the "Oriental" style of sheeted and cut foods called *noodles*, which are commonly made from wheat other than durum. However, those definitions are confused by a category of pasta that is also called noodles. *Pasta noodles* in the United States are made from durum or nondurum wheat and are required to contain a minimum of 5.5% egg solids. Therefore, although durum wheat semolina is the most common raw material used to produce pasta, pasta is sometimes also made from nondurum wheat flour, or farina. Perhaps the best way to minimize the confusion in terminology of these paste foods is to call all of them *pasta* and differentiate among them by their specific formulation, manufacturing process, shape, and specific name.

Pasta in the Italian tradition can be categorized into four main groups or types: long-goods, short-goods, egg noodles, and specialty items. Long-goods include such products as spaghetti, vermicelli, and linguine; short-goods include elbow macaroni, rigatoni, and ziti; egg noodles consist of pasta made with eggs; and specialty items include lasagne, manicotti, jumbo shells, and stuffed pasta. Many additional products fit into one or more of these categories. The name of a

TABLE II
Durum Wheat Exports (thousand tonnes)

Crop Year	Canada[a]	United States[b]
1976/77	1,695.9	1,078.9
1977/78	1,967.7	1,589.4
1978/79	1,349.9	1,792.9
1979/80	1,947.9	2,158.5
1980/81	2,074.9	1,409.8
1981/82	2,310.7	1,938.5
1982/83	2,687.1	1,383.0
1983/84	2,546.0	1,595.0
1984/85	1,826.4	1,420.0
1985/86	1,385.3	1,341.0

[a] Data from Canadian Grain Commission (1986).
[b] Data from USDA (1977–1986).

specific pasta product usually denotes the product's shape, size, and thickness. In some countries, the most common names are defined by the "standards of identity" for a specific product with regard to its general shape and thickness. The number of sizes and shapes that can be produced is virtually unlimited and depends on the shape of the die from which the product is extruded or the cutter with which it is cut. Over 600 pasta shapes are produced (National Pasta Association, 1984), some of which are shown in Figs. 1 and 2. Among the most popular shapes are spaghetti, elbow macaroni, lasagne, shells, and various noodle shapes.

Although pasta is sometimes served freshly made in the home or in

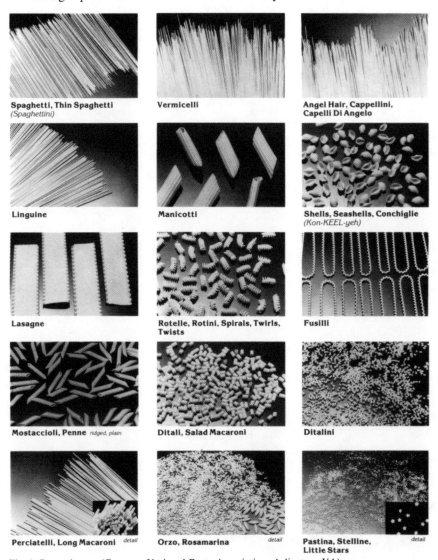

Fig. 1. Pasta shapes. (Courtesy National Pasta Association, Arlington, VA)

restaurants, or as a medium-moisture (28–40%), refrigerated product in supermarkets, most pasta is marketed in three forms—dry and packaged, canned, or frozen. The dry or packaged pasta segment includes individual shapes and pasta dinners (macaroni and cheese) and represents the largest share of the market. Canned pasta includes the basic pasta shapes and stuffed products, usually mixed with tomato sauce. Canned products are preferred by children. In recent years, frozen pasta has become popular, since it can be prepared as an entree or a nearly complete dinner. Another appealing aspect of frozen products is that many of them can be heated by microwave.

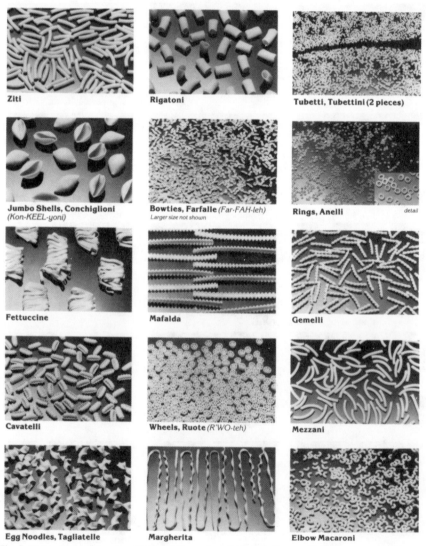

Fig. 2. Pasta shapes. (Courtesy National Pasta Association, Arlington, VA)

The consumption of pasta around the world varies widely (Table III). Whereas consumption has decreased somewhat in some relatively high-consumption countries, such as Italy, Chile, and Switzerland, it has increased significantly in Spain, West Germany, the United States, the United Kingdom, and Japan. These changes in consumption probably are related to economic conditions.

Pasta is gradually losing its "cheap and filling" image, and this makes it an attractive food to more consumers in some markets. This change in image can possibly be attributed to the consumers' realization that pasta has the positive attributes of versatility, convenience, economics, taste, and nutrition. Pasta is versatile in that it is available in numerous shapes and sizes and can be prepared and served with other foods as an appetizer, main dish, side dish, salad, soup, or dessert, thus allowing good menu flexibility for the consumer. It is convenient since it is sold in many forms and at a relatively reasonable price. Because it has a relatively bland flavor, it is complemented by sauces and other foods. Nutritionists consider pasta to be highly digestible. It also provides significant quantities of complex carbohydrates, protein, B-vitamins, and iron and is low in sodium and total fat (Douglass and Matthews, 1982). Durum gliadin proteins are less toxic to gluten-intolerant individuals than are gliadins from other wheats (Costantini, 1985).

COUSCOUS

Couscous, a paste product made from durum, is one of the staple foods in

TABLE III
Approximate Per-Capita Consumption (kg) of Italian-Style Pasta

Country	1971[a]	1986[b]
Italy	29.5	25.0
Argentina	12.3	11.8
Tunisia	10.0	11.8
Libya	12.3	11.4
Venezuela	10.0	11.4
Spain	6.4	10.9
Switzerland	11.8	10.5
Chile	10.9	8.2
Greece	6.4	...[c]
Peru	6.4	6.8
United States	3.6	6.4
France	5.5	6.4
West Germany	...	5.9
Soviet Union	...	5.0
Australia	3.2	4.1
Canada	3.6	4.1
East Germany	...	3.6
United Kingdom	0.5	2.7
Japan	...	1.1
Portugal	6.8	...
Algeria	3.6	...
Mexico	3.6	...

[a] Data from Irvine (1971).
[b] Data from National Pasta Association (1986).
[c] Conflicting figures from various sources range from 6.8 to 31.8 kg.

North Africa. Other grains may be used to produce couscous (e.g., pearl millet in Senegal and maize in Togo), but durum semolina is used in Egypt, Libya, Tunisia, Algeria, and Morocco (Kaup and Walker, 1986).

Couscous is made traditionally by hand in small quantities or commercially by a continuous operation in which several hundred kilograms of product are produced per hour. In either operation, the basic steps include blending of semolina and water, agglomeration to combine semolina particles, shaping to reduce the size and give uniformity to agglomerated particles, steaming to precook, drying to preserve, cooling, grading to separate by size, and storage before consumption or packaging (Fig. 3).

Characteristics of a good-quality couscous include the ability to absorb sauce well, integrity of individual particles during steaming or sauce application, uniform particle size, and absence of particles that stick to one another. These factors affect the mouthfeel of couscous. Like pasta, couscous is a versatile food that can be served in many different ways and with many other foods; it has a natural, bland flavor that is complemented by other foods and sauces.

B. Nonpaste Products

BULGUR

Bulgur (in North America) or *burghul* (in the Middle East and North Africa) are names given to one of the oldest cereal-based foods, which has been consumed for centuries in Turkey, Syria, Jordan, Lebanon, and Egypt (Williams et al, 1984a). Common hard and even soft wheat can be used to

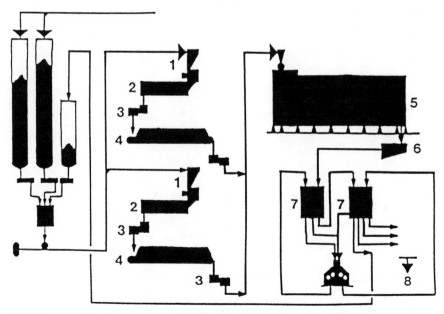

Fig. 3. Commercial couscous process developed by the Buhler Co. 1, blending; 2, agglomeration; 3, shaping; 4, steaming; 5, drying; 6, cooling; 7, grading; 8, storage. (Reprinted, with permission, from Kaup and Walker, 1986)

produce bulgur, although durum wheat is preferred. The high pigment content of durum gives a good color to the product, and durum's natural hardness imparts a firm texture.

The common method of making bulgur in the Middle East, whether it be in a small- or large-scale operation, is to soak the cleaned wheat and cook it to gelatinize the starch, as shown in simplified form in Fig. 4. In a small-scale batch operation, the wheat-water mixture is stirred while heating (Fig. 5) to allow light foreign material to float to the surface for removal. The cooked product is cooled, dried, moistened, peeled (optional), redried, cleaned by winnowing, milled, and sized. Peeling, when done, is to remove the bran. Commercial operations in the United States use continuous processes similar to that shown in Figs. 6 and 7 (Fisher, 1972), in which the dried bulgur is sometimes pearled. Milling reduces the particle size of the product, after which it can be purified or classified into up to four size-grades: coarse, fine, very fine, and flour. Large-scale bulgur operations exist in Syria, India, and the United States.

Preparation of bulgur for consumption depends on the particle size of the product. Coarse bulgur is usually boiled or steamed in the presence of meats or vegetables similar to the manner in which rice or couscous are prepared. Its flavor and texture vary depending on the wheat and the processing method. Fine and very fine bulgur (called *smesma*) are used in several traditional dishes where coarse granulation is not required. Bulgur is a nutritious food that stores well and is relatively simple to prepare (Haley and Pence, 1960). Bulgur flour, which usually contains some foreign material, and the bran obtained in bulgur production are usually fed to poultry.

FREKEH

Frekeh, a wheat-based product, is produced by a smaller industry than that of bulgur. It is eaten in several countries of the Middle East and North Africa (Williams and El-Haramein, 1985). *Frekeh* is made (Table IV) from immature

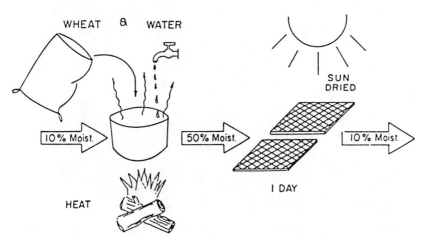

Fig. 4. Basic processing conditions to produce bulgur. (Reprinted, with permission, from Fisher, 1972)

wheat (at about the mid- to late-dough stage). The wheat is cut at the stem and allowed to partially dry in the sun. It is then bunched together (heads laid into a light to medium breeze) and scorched by flames to burn off the awns and leafy material (Fig. 8). Durum is most suitable since the best *frekeh* is made from the largest and hardest kernels. The wheat heads are laid into the wind so that the kernels are not parched excessively. The parching or charring gives the *frekeh* a characteristic flavor. The parched wheat heads are dried in the sun, then separated from the ash, charred awns, and leaves by winnowing in the wind. The grain is threshed by hand or by mechanical means and is separated from the chaff before being bagged or stored in bulk.

The quality of *frekeh* depends on the shape, plumpness, and greenness of the wheat and on the degree of parching, which should range from none to very light. *Frekeh* is a staple like rice, bulgur, and couscous. The ground or chopped grains are often boiled or steamed to be served with the meat of sheep or poultry.

MISCELLANEOUS PRODUCTS

In addition to its use in pasta, couscous, bulgur, and *frekeh*, durum wheat is used to make puffed cereals, bread, hot cereal, desserts, and a filler for pastries. These products, like couscous, bulgur, and *frekeh*, are popular in some regions, unlike pasta, which is consumed around the world.

Puffed wheat is usually produced from selected large kernels of durum and is

Fig. 5. Stirring of wheat and water mixture during the making of bulgur. (Courtesy Grain Research Laboratory Division, Agriculture Canada, Winnipeg)

consumed as a ready-to-eat breakfast cereal. It is most popular in North America. In some countries, durum semolina is sometimes eaten as a hot cereal similar to cereal from nondurum farina.

Although durum is not usually thought of as a bread wheat, its use for bread is quite substantial in the Near East, Middle East, and Italy and occurs in lesser amounts in other countries (Williams et al, 1984b; Williams, 1985). Two-layered, flat breads made from durum flour or blends with flour from other wheats account for the largest bread use for durum. These breads include *khobz* (Syria, Lebanon, and Jordan), *baladi* and *shami* (Egypt), and others. Single-layered flat breads, which are also popular, include *tannour* and *saaj* (Syria and Lebanon), mountain bread and *markouk* (Lebanon), *mehrahrah* (Egypt), and

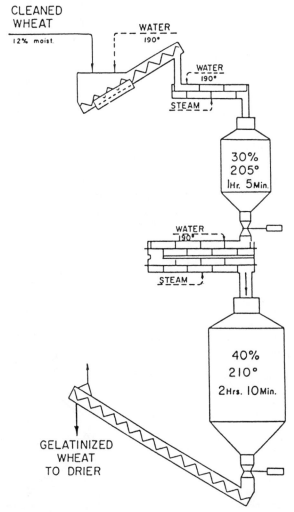

Fig. 6. Commercial process for bulgur. (Reprinted, with permission, from Fisher, 1972)

others. Important factors that must be considered when durum flour is used for bread are gluten strength, starch damage, and pigment content. Durum gluten strength is usually too weak and starch damage and pigment content are usually too high for successful flat-bread baking at the 100% durum level. Therefore, durum flour is often blended with a medium dough-strength, nondurum bakery flour to give bread products with the desired texture and color. (See also Chapter 8 in this volume.)

Foods such as "vegetarian caviar," an appetizer, and "semolina halva," a dessert, are lesser-known products of durum that are eaten in parts of Europe, Asia, and Africa. These are made by boiling the semolina in either vegetable juice, oil, or sugar syrup to give flavor and texture. The boiled, cooled semolina

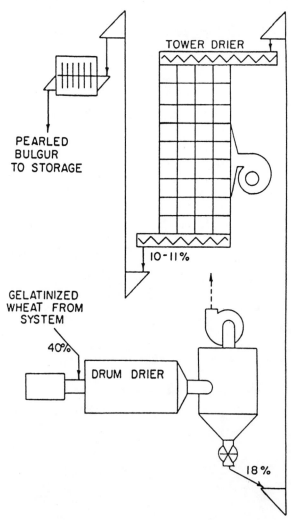

Fig. 7. Bulgur drying and pearling in a commercial process. (Reprinted, with permission, from Fisher, 1972)

is flavored further with spices or the juice of fruits and is garnished with nuts, fruits, or vegetables (T. Chakraborty, personal communication). Another dessert, *Kugel*, is a sweet noodle pudding that originated in Germany but is now marketed in North America.

IV. CHARACTERISTICS DESIRABLE IN DURUM WHEAT FOR SEMOLINA MILLING

A. Test Weight

Test weight is considered an important factor, not only in semolina milling but generally in flour milling. Over a half century ago, Geddes (1934) included test weight among desirable wheat characteristics. He stated that "the higher the test weight the greater is the percentage of endosperm and, in general, the greater the yield of semolina." Millers in most countries agree that high test weight is a requisite for good milling yield.

Test weight has been interpreted as a measure of kernel soundness. Fully mature, plump kernels, undamaged by disease or by the environment, are high in test weight. In Italy, millers buying domestic durum wheat specify a minimum test weight of 82 kg/hl (R. R. Matsuo, personal observation). Such wheat is almost invariably sound and undamaged.

Laboratory tests indicate that semolina yield is positively correlated with test weight. Irvine (1964) reported a correlation coefficient of 0.86 between semolina yield and hectoliter weight for 28 samples of Canada western amber durum wheat grades. In another study of 174 samples, including pure varieties from the Canadian plant-breeding program and grade composite samples from export cargos and from a newly harvested crop, a correlation coefficient of 0.52 was reported (Matsuo and Dexter, 1980a). In another study, in which only pure varieties from the plant-breeding program were examined, however, the correlation between semolina yield and test weight was not significant (Matsuo,

TABLE IV
Sequence of Operations in *Frekeh* Preparation[a]

Operation	Approximate Time of Day
First day	
Cut green wheat stems	6 a.m.–noon
Spread and allow to dry in the sun	
Gather into loose bunches and burn	
Separate burned leafy material	2–7 p.m.
Gather cleaned heads into heaps to dry	
Eighth to twelfth days	
Thresh heads in combine, by donkey power, or by hand	
Clean grains free of chaff and other foreign material	9 a.m.–7 p.m.
Store grains (*frekeh*) in sacks or in granaries for sale	

[a] Source: Williams and El-Haramein (1985); used by permission.

1985). The test weight range was narrow (78–82 kg/hl), and semolina yield was influenced more by varietal characteristics than by differences in test weight.

For samples that are visually sound, i.e., undamaged by disease or by the environment, but that have a large range in test weight, a significant correlation was found between semolina yield and test weight, as shown in Fig. 9 (Dexter et al, 1987). The samples with low test weight had small kernels, so semolina yield was affected by both kernel size and test weight.

In studies involving common wheat, no correlation was found between flour yield and test weight. Hook (1984) reported insignificant correlation with English wheats (soft, hard, spring, winter).

B. Kernel Size

Closely allied with test weight as a milling factor is kernel size, usually reported as weight per 1,000 kernels. Kernel size was reported to be correlated to semolina yield (Matsuo and Dexter, 1980a) in a study in which 174 samples of various origins and of varying degrees of soundness were studied. In another study, of samples selected from a plant-breeding program, the correlation between kernel size and semolina yield was not significant. As these samples were in the late stage of development in the breeding program, the majority of the samples fell in a fairly narrow range of kernel weight (37–43 g/1,000 kernels). Thus, differences in semolina yield were probably due to variety rather than to kernel size.

Fig. 8. Green (immature) wheat being burned in bunches for the making of *frekeh*. (Reprinted, with permission, from Williams and El-Haramein, 1985)

For the earlier study (Matsuo and Dexter, 1980a), the kernel weight ranged from 18 to 54 g/1,000, with a corresponding range in test weight from 72 to 86 kg/hl. Small kernels would be expected to give less semolina since the ratio of endosperm to bran is smaller than for large kernels. This is indeed the case, as small kernels yielded less semolina. Irvine (1979) stated that "1000 kernel weight is the best single index of potential semolina yield." This was based, however, on milling results on only 10 samples.

Samples with low kernel weight tend to give low test weight, and the principal factor contributing to this variation is the packing of kernels in a container. Plump kernels pack more uniformly, giving rise to high test weight, whereas small kernels, usually more elongated, pack more randomly to give lower test weight. Irvine (1961) reported that kernel density remained essentially constant at 1.43 g/cm^3 in hard red spring wheat samples of the same moisture content ranging in test weight from 67 to 84 kg/hl. Therefore, he concluded that test weight is influenced most by packing efficiency, which in turn is affected by the size and shape of the kernel.

Thus, kernel weight, as suggested by Irvine (1979), should be considered an important factor in assessing durum wheat quality.

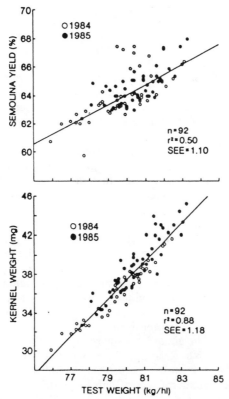

Fig. 9. The relationship of durum wheat test weight to semolina yield and kernel weight. SEE = standard error of estimate. (Reprinted, with permission, from Dexter et al, 1987)

C. Vitreousness

Hard, vitreous kernels are almost synonymous with durum wheat, particularly in Europe where the terms *grano duro* and *blé dur* refer to *T. turgidum* var. *durum* and *grano tenero* and *blé tendre* encompass all classes of *T. aestivum*. Because kernel hardness is desirable for production of semolina, kernels appearing white, chalky (starchy), or opaque are considered undesirable by semolina millers.

Geddes (1934) included vitreousness as one of the desirable characteristics of durum wheat. He stated, "The presence of starchy or piebald kernels reduces the yield of semolina as they pulverize more readily." However, the following year (Geddes, 1935), he expressed the opposite view.

> In regard to starchy kernels, these had no significant influence on semolina yield but the color of the semolina as indicated by carotene content was noticeably affected when starchy kernel content amounted to 30 percent or more.

In 1935, semolina yield obtained in the laboratory was a mere 31–33%. It may be questioned whether differences in semolina yield can be detected at such a low extraction rate.

As experimental milling techniques in the laboratory were refined, results of studies on semolina milling were published. Matveef (1963) reported that for each 10% of nonvitreous kernel content there was a 1% loss in semolina yield in the range of 70–100% vitreous kernel contents. The loss in yield was 1.3% in the 40–70% range and 1.6% under 40%.

Bolling and Zwingelberg (1972) studied the milling characteristics of durum wheats grown in France, the United States, and Canada and reported that semolina yield depended more on the origin than on differences in vitreousness. However, semolina yield for French and Canadian samples varied with the proportion of vitreous kernels. Semolina yield showed little variation, but the amount of flour increased with increased starchiness.

Menger (1973) suggested that the technological relevance of vitreousness had become rather controversial in Europe in light of the fact that the relative importance of vitreousness is greater when coarser semolina is milled and European commercial millers and pasta processors commonly produced and used semolina of finer granulations.

Matsuo and Dexter (1980a) indicated that as starchy kernel content increased, milling yield was not affected but the proportion of flour increased. Protein and yellow pigment contents decreased with increased starchy kernel content. In another study, as starchy kernel content increased, semolina granulation became finer and more flour was produced during milling (Dexter and Matsuo, 1981).

The percentage of vitreous kernels became a grading factor in the Canadian grading system in 1930. The grade schedule required 75% hard vitreous kernels in No. 1 grade, 60% in No. 2, and 30% in No. 3. Currently, the minimum percentages are 80, 60, and 40 for the top three grades, respectively. Because commercial millers are adamant that a high percentage of vitreous kernels is important for high semolina yield, vitreous kernels will continue to be a factor in the grading system.

It is estimated that, on the average, the difference in commercial semolina yield between No. 1 Canada western amber durum and No. 2 can be as much as 3%. As the major difference is normally the percentage of vitreous kernels, the yield decrease can be assumed to be due to nonvitreous kernels.

D. Protein Content

Protein content is generally recognized as an important quality factor in all wheats. In some soft wheat classes, a low level, below 9%, is desirable for use in cakes and cookies, but for wheat used in the pasta industry, a high protein level is desirable. Matveef (1966) showed that a wheat protein content above 13% yielded a satisfactory product, whereas a protein content lower than 11% gave a very poor product. The protein content of commercial durum wheat ranges from 9 to 18% and is higher than that of common wheat (Feillet, 1984).

References to pasta products after the turn of this century indicate that durum wheat was preferred because of its higher protein content. Hunt (1910) claimed that durum wheat "is superior to common wheat because of its higher gluten content and greater density." Bengtson and Griffith (1915) stated that "the large kernels of durum or macaroni wheat are rich in gluten content and are used extensively for the manufacture of macaroni." The first report on the quality of durum wheat produced in Manitoba by the Grain Research Laboratory (Birchard, 1931) included only three factors: test weight, 1,000 kernel weight, and protein content.

A reason for the desirability of high protein content in durum wheat was given by Geddes (1934), who stated:

> Strength is particularly important in macaroni, spaghetti, vermicelli and other long-goods to prevent stretching during the curing or drying process and breakage in the packing and shipping of the dried product. The quantity and quality of the protein is, therefore, important.

The currently accepted view is that protein quantity and quality are important, not for the above reasons but for good pasta cooking quality. A comprehensive review on the role played by proteins and protein components in spaghetti cooking quality was written by Feillet (1984).

E. Ash Content

The mineral constituents in wheat are distributed throughout the kernel but concentrated much more in the aleurone layer than in the starchy endosperm. The distribution of mineral matter varies with the wheat class. For example, normal straight-grade flour milled from red spring wheat to about 75% extraction has an ash content of 0.45–0.50%; flour milled from durum wheat to the same extraction has an ash content of about 0.75%. Ash content in gluten from red spring wheat is 0.25–0.30%, whereas in gluten from durum semolina it varies from 0.85 to 1.00%.

Wheat ash content is influenced more by the environment than by the genetic background. In North Dakota, the average durum wheat ash content in the last five years was 1.65%, in South Dakota 1.84% (Dick et al, 1986), and in western

Canada 1.58% (GRL, 1986). Wheat ash over 1.75%, as in 1983 in western Canada, presents a problem to millers in countries that have a limit on semolina ash content. In Italy, for example, semolina ash may not exceed 0.90% (dry moisture basis). This regulation lowers semolina yield from high-ash wheat since the miller is prevented from adding low-grade, clear flour to the semolina.

Ash content per se has no relationship to semolina color. Wheats from different geographic locations or grown in different growing seasons with a wide variation in ash content show no correlation between these factors and semolina or spaghetti color. If, however, one sample of wheat is milled to increasing semolina extraction with a corresponding increase in ash content, a correlation is apparent between color and ash content (Matsuo and Dexter, 1980a).

In reporting on an international collaborative study on durum wheat quality, Irvine (1979) stated that the main unexpected finding from the study was that "wheat ash is the major single factor determining firmness of cooked spaghetti." This is the only instance in which ash content has been related to spaghetti cooking quality. What makes this finding interesting is that in Italy the "best" pasta, according to several pasta manufacturers, is made from a high-extraction, high-ash semolina (R. R. Matsuo, personal observation). Such a product by law cannot be marketed but apparently is processed for personal consumption.

Since ash content of durum gluten is about three times that of common wheat gluten and since durum semolina produces pasta of superior cooking quality, the mineral constituents may be related to spaghetti cooking quality.

V. DURUM PROCESSING

A. Semolina Milling

The process of milling durum varies, depending on the desired characteristics of the mill products and whether the operation is for experimental or for commercial reasons. Experimental milling is usually done on a much smaller scale than commercial operations.

The objective of experimental durum milling is to produce a milled product of adequate quality to evaluate the potential value of a wheat or a milling technique. In meeting these objectives, research workers should keep in mind the requirements of the commercial miller, although the economics of the experimental milling procedure are not as important as they would be in commercial milling. Limiting factors in meeting the objectives are usually the sample size and the efficiency of the experimental mill. Several experimental milling systems have been described. They include micro- (Vasiljevic et al, 1977), intermediate- (Seyam et al, 1974; Shuey et al, 1977; Dexter et al, 1982), and pilot-scale (Shuey et al, 1980) techniques. An example of an intermediate-scale experimental milling flow is given in Fig. 10. A good rule of thumb is that the greater the size and sophistication of the experimental milling operation, the better it will predict the commercial milling performance or potential. Wheat sample size governs the size of the milling procedure. If the resulting semolina from experimental milling of durum is relatively free of bran particles, it should be suitable for predicting physical, chemical, rheological, and pasta-making characteristics of commercially milled wheat.

The prime objectives of commercial durum milling are maximizing mill yield and profit while satisfying the requirements of the customer, usually the pasta manufacturer. The durum miller must also be cognizant of good wheat selection, handling, and conditioning and cleaning practices, as well as of unique techniques required for grinding and purification (Abercrombie, 1980). Because the durum kernel is very hard (Miller et al, 1982), it yields, when subjected to grinding, relatively high amounts of large particles (semolina). The inherent hardness, which is possibly related to the absence of a specific protein on the surface of the starch granules (Greenwell and Schofield, 1986), makes it necessary to grind durum more vigorously if the desired product is fine flour (Mousa et al, 1983). Economically, it is advantageous to maximize the yield of semolina and minimize flour production since semolina commands a higher selling price. A good commercial mill can be expected to obtain extractions of 60–64% semolina and 8–12% flour (Banasik, 1981).

The mechanics of milling durum wheat differ from those of milling wheat into flour. Meeting the primary objective of maximizing semolina is best accomplished by grinding with corrugated rolls to maintain granularity, then removing small bran particles from the semolina using purifiers, which separate particles based on size and density. For that reason, one finds in a mill designed to grind durum few, if any, smooth rolls and many purifiers. Good purification is required since bran or other dark particles are readily visible in the yellow semolina, and these dark particles or specks, when present in the finished pasta product, are perceived by the consumer to be contaminants. The very hard durum grain must be tempered to a relatively high moisture content

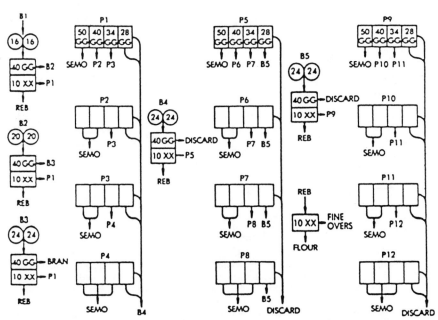

Fig. 10. Milling flow sheet for semolina milling in the Buhler laboratory mill. B = break roll, P = purifier, REB = flour to rebolt sifter, SEMO = semolina. (Reprinted, with permission, from Dexter et al, 1982)

(16.0–16.5%) before grinding. The tempering or conditioning is usually done in two to three stages; the final stage is designed to toughen the bran so that it does not shatter excessively, as this would make it more difficult to separate the bran from the endosperm in the purifier. Figure 11 shows a simplified commercial

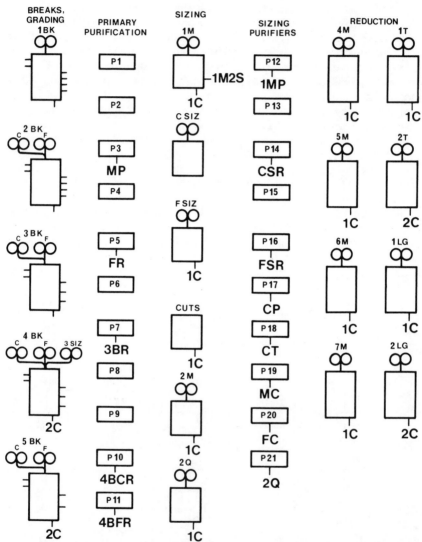

Fig. 11. Simplified schematic flow sheet for a commercial semolina mill. Only those sections involved in the production of collected mill streams are shown. BK = break roll, P = purifier, C = coarse, F = fine, M = middlings, Q = quality, T = tailings, LG = low grade. Semolina streams: MP = medium purifier, FR = fine repurifier, 3BR = third break repurifier, 4BCR = fourth break coarse repurifier, 4BFR = fourth break fine repurifier, 1M2S = first middlings second scalp, siz = sizing, 1MP = first middlings purifier, CSR = coarse sizing repurifier, FSR = fine sizing repurifier, CP = cuts purifier, CT = cuts tailing, MC = medium cuts, FC = fine cuts, 2Q = second quality. Flours: 1C = first clear, 2C = second clear. (Reprinted, with permission, from Matsuo and Dexter, 1980b)

durum milling flow. (Durum milling is also discussed in Chapter 1 of this volume.)

B. Semolina Quality

Although each pasta manufacturer has unique quality requirements for semolina, several common factors are often considered when judging quality (Table V). Factors that are important but not directly related to inherent quality include insect fragment count and total microbial count.

MOISTURE CONTENT
The moisture content is an important factor in processing pasta. Semolina moisture should be as high as is possible without risking the hazards of spoilage or deterioration during storage and the stickiness and poor flow properties associated with excessive moisture. An acceptable level is 13.5–14.5%. Too low a semolina moisture level can cause difficulties with hydration of liquid ingredients in the mixing chamber of continuous pasta extrusion systems, because ingredient retention time in the mixer is limited. Wide ranges in semolina moisture from batch to batch can give an uneven pattern of extrusion of the pasta and therefore should be minimized to maintain uniformity of the finished product (Stehrenberger, 1981).

GRANULATION
Granulation or particle size distribution of semolina is important since it has an effect on the absorption properties of the pasta dough and therefore influences the quality of the finished pasta. Since coarse particles hydrate at a different rate than fine particles, it is undesirable to have too wide a range in particle size, especially in continuous-flow mixers, which have a limitation on mixing time. Fine semolina particles hydrate more rapidly than coarse particles, leaving the coarse particles starved for moisture. These unhydrated particles produce an inferior, unattractive pasta with "white specks." Coarse semolina was most desirable when batch processes were used. The use of large, continuous extrusion systems has caused many manufacturers to prefer a semolina with a

TABLE V
Factors Used to Evaluate Semolina Quality for Pasta Manufacture

Factor	Realistic Level	Preferred Level
Moisture content, %	13.5–15.0	Narrow range, 13.5–14.5
Granulation, μm	100–500	Narrow range, 150–350
Speck count, specks per 64.5 cm^2	30–50	Low
Color score, units[a]	7.5–10.0	High
Grit content, %	0.002–0.005	Low
Ash content, %	0.55–0.75	Low, 0.80 maximum
Protein content, %	11.0–13.0	High, 12.0 minimum
Gluten quality, mixogram classification[b]	3–6	High, 5–8
Amylase activity as falling number, sec	250–500	High, $\geqslant$400
Wheat-class purity, % durum	98–100	100

[a] Method reported by Walsh (1970).
[b] Method reported by Dick (1985).

smaller average-particle size and a more uniform granulation. Semolina particles below 350 μm in diameter are easier to process into a more homogeneous and translucent pasta than coarse semolinas (Manser, 1984). There is some indication, however, that particles of semolina that are relatively small are high in damaged starch, which increases the loss of solids into cooking water during pasta cooking (Matsuo and Dexter, 1980b).

COLOR SCORE

Traditionally, a yellow color in semolina has been associated with good quality. This association was widely accepted in the early days of the pasta industry because durum wheats, which normally had higher pigment contents than nondurum wheats, gave pasta with the best overall quality. The inferior gluten quality in some durum varieties with high concentrations of yellow pigment indicates that yellowness of semolina or pasta does not necessarily parallel the cooking quality. Still, a yellow pasta is considered a mark of quality by many consumers because they assume that the product is made from durum wheat, which generally gives pasta with superior cooking quality compared to that made with nondurum wheat (Dexter et al, 1981b).

The color of semolina is an indication of the color one might expect in the pasta. However, in making that assumption, it is important to remember that color can be influenced by particle size, enzyme content, and processing conditions. A finely granulated semolina appears to to have less of the desirable yellow color than a coarsely ground sample, even though when processed into pasta there might be no perceptive difference between the two. This perception can be minimized when semolina is judged for color by comparing samples of similar particle size. Increases in natural levels of lipoxygenase can reduce the amount of yellow pigments as the semolina is processed into pasta. Although lipoxygenase is concentrated in the germ and bran portions of the durum kernel, cultivars with inherently high concentrations have higher levels of the enzyme in the semolina unless the wheat is steam-conditioned before milling (Irvine, 1971). Mixing semolina with the other ingredients in the presence of oxygen reduces the intensity of the yellow color in the pasta.

The predominant carotenoid pigments in semolina responsible for the yellow color in pasta are xanthophylls and lutein (Sims and Lepage, 1968). Semolina samples can be compared visually using standard samples, photometrically using light reflectance measurements, or chemically by extracting pigments and measuring them spectrophotometrically (Johnston et al, 1980). In addition to yellow pigments, a brownish pigment caused by a copper-protein complex is known to exist (Matsuo and Irvine, 1967). This brownish pigment as well as brownness caused by immature, frozen, or diseased kernels can affect pasta color or brightness. Brown pigments appear to be of greater concern in durum varieties with inherently low yellow-pigment concentrations than in varieties with high levels of yellow pigments.

SPECK COUNT

Specks in semolina are caused by any material with a color that contrasts with the durum endosperm particles that make up the semolina. Speck concentration is usually determined by counting the number of brown or black pieces in a given area on the surface of the semolina. Wheat bran is the most common source of

brown specks, but black specks are usually caused by materials ground in the mill such as discolored or diseased wheat kernels, weed seeds, sunflower seeds, ergot, or dirt. Black specks are most annoying, since they show up so readily. Speck count is best controlled by proper selection, thorough cleaning, and proper conditioning of the wheat before milling. Despite all the precautions, it is impossible to economically produce semolina totally free from specks. The acceptable limit of specks in semolina is established between the buyer and the seller.

GRIT CONTENT

Grit is any hard object (for example metal, stone, or glass) in the semolina that can stick in the pasta extrusion die and cause streaking or tearing of the dough as it passes by. The hard, gritty material might also be extruded and become part of the pasta. Obviously, the presence of grit is undesirable, as it results in "blemished" pieces of pasta that detract from the appearance of the product or in a hazard to the unsuspecting consumer who might bite into a grit-containing product. The grit can be detected in the laboratory by floating semolina in carbon tetrachloride in a separatory funnel; the grit particles, being heavier than semolina, settle to the bottom of the liquid. Thorough cleaning of the wheat, utilizing stoners and magnets, significantly reduces the incidence of grit in the semolina and the pasta.

ASH CONTENT

The ash content in the endosperm of durum is characteristically higher than in the endosperm of other hard wheats. Still, the percentage of ash in durum semolina or flour can be used as a relative measure of bran. Ash content determination is useful for evaluation of durum flour since the finely ground bran is not as visible as the coarse bran in semolina. Durum flour with a high ash content probably contains high amounts of bran material. The ash in commercial durum semolina of about 65% extraction (wheat basis) normally ranges from 0.55 to 0.75% (14% mb) (Irvine, 1971); if the ash content is about 0.5% or lower, the milled product most likely contains some nondurum flour. Ash content values alone are usually not an overriding quality factor in evaluation of semolina, but when combined with other factors they can be useful in explaining peculiarities in pasta appearance or performance.

PROTEIN CONTENT

Durum wheat protein, although not perfect nutritionally, is a valuable source of nutrients for humans. Irrespective of nutritional considerations, however, the protein content of semolina is important because it influences the functional quality of pasta. Adequate amounts of gluten protein are necessary to impart to pasta the desirable attributes of mechanical strength and cooking quality. Correlation coefficients of 0.92–0.97 ($P<0.05$) between total (crude) protein and gluten content of wheat flour have been reported (Kulkarni et al, 1987). The high correlation allows the gluten content to be predicted from the Kjeldahl protein determination, as the latter is more easily and reliably determined on a routine basis.

Semolinas with protein levels of 11.5–13.0% process with little difficulty and can be expected to give satisfactory results (Irvine, 1971). Although lower-protein semolinas can be used, too low a protein is likely to produce pasta with relatively poor mechanical strength in the dried product and less than optimum quality with respect to cooking stability and cooked firmness (Grzybowski and Donnelly, 1979). At the other extreme, protein levels that are too high may result in products that stretch excessively upon extrusion (Irvine, 1971). The practical range of semolina protein content for use in pasta is most narrow for long-goods products, which are placed under considerable weight stress when they hang on the support rod as they pass through the continuous dryer. The quality of gluten proteins is at least as important as the quantity of the proteins.

GLUTEN QUALITY

The functional quality of semolina gluten has been shown to have a direct relationship to pasta quality. Two rheological methods used to determine the relative gluten quality of semolina are the mixograph and the farinograph (Dick, 1985). Other methods such as measurement of elasticity and compressibility of washed gluten, whole-meal sedimentation tests, protein solubility, electrophoresis, chromatography, or specific absorption of protein also contribute useful information about gluten quality (Dick, 1985). Generally, durum semolina proteins with the greatest amounts of insoluble residue protein, high glutenin-gliadin ratios, and low specific absorbance are characteristic of strong mixing gluten and give pasta with superior cooking quality. Polyacrylamide gel electrophoresis studies of gliadin (Joppa et al, 1983; Josephides et al, 1987) have shown that band 45, which is associated with strong gluten, is genetically linked to chromosome 1B. Substitution of chromosome 1B from a strong gluten durum variety into a weak gluten durum variety significantly strengthened the gluten characteristics of the latter as measured by mixograph and cooked spaghetti firmness.

Low-molecular-weight glutenin polypeptides are also highly correlated with gluten strength (du Cros, 1987). High-molecular-weight glutenin polypeptides, although poor indicators of viscoelastic properties, have a positive effect on gluten strength. Durum cultivars possessing gliadin band 42, and producing dough of intermediate strength instead of the weak dough normally associated with gliadin band 42, contain high-molecular-weight glutenin polypeptides not found in cultivars that produce weak doughs.

Statistical evaluation of tests to predict the spaghetti-making quality of durum semolina show that gluten quality, farinograph band width, and mixograph mixing time are significantly correlated with spaghetti cooking quality. Semolina protein content, specific absorption at 280 nm for semolina protein in aqueous urea, sodium dodecyl sulfate sedimentation volume, and farinograph band width have been shown to be the best indicators of cooking quality (Matsuo et al, 1982a).

The characteristics of gluten give an indication of its functional properties. Gluten of good quality should also have a bright yellow color and should be free from shades of brown or grey, which might render the finished pasta dull or dark. Gluten with a bright yellow color will produce pasta with similar color.

AMYLASE ACTIVITY

High levels of α-amylase in semolina influence the quality of the finished pasta. Although the color of the pasta does not appear to be appreciably affected (Dick et al, 1974; Donnelly, 1980; Matsuo et al 1982b), the mechanical strength of dry pasta seems to be weakened (Donnelly, 1980) and pasta cooking quality is affected slightly by high amylolytic activity (Matsuo et al, 1982b); the cooked pasta is slightly softer and sloughs more solids into the cooking water. Commercial pasta manufacturers who use semolina with high amylolytic activity experience uneven dough extrusion across the die, overly soft dough, and long-goods pasta that falls off the drying rods.

Preharvest sprouting of durum is the cause of abnormally high amylolytic activity in semolina. α-Amylase activity is reduced when the wheat is milled into semolina, depending on the wheat, the mill flow, and the milling extraction. Several methods can be used to measure the relative concentration of α-amylase in semolina. The falling number and the amylograph tests are based on changes in starch viscosity of gelatinized starch. In addition, dextrinogenic, nephelometric, and colorimetric tests are available. The falling number test is probably the most widely used because of its relatively high speed, simplicity, and low cost. For semolina, falling number values much below 400 sec are usually indicative of some degree of sprouting in the parent wheat (Donnelly, 1980). A procedure suitable for use outside the laboratory uses a Rapid Visco-Analyzer (Newport Scientific) to provide a "stirring number" for estimating α-amylase activity (Ross et al, 1987).

WHEAT-CLASS PURITY

Durum wheat is considered to be most desirable to produce a granular product for pasta manufacture. Whereas some countries have legal require-ments for the use of durum semolina in pasta, others allow the use of nondurum "semolina," or farina, as it is called in North America. Where allowed, farina is often blended with durum semolina to reduce the cost of ingredients. This practice is prevalent especially during times of low availability of durum, when the price differential between durum and nondurum wheat is high. To monitor wheat-class purity of semolina or pasta because of legal, contractual, or labeling considerations, testing procedures were developed to detect nondurum flour or farina in semolina or pasta. Differences between durum and nondurum samples were noted in sitosteryl palmitate (Hsieh et al, 1980a, 1980b), in protein immunological (Silano, 1975) and electrophoretic characteristics (Burgoon et al, 1985), and in infrared spectra of wheat extracts (Laignelet, 1983). Unfortunately, these differences do not always hold true across wheat classes. Although none of the techniques has met with universal acceptance, they are useful under limited conditions.

Comparisons of pasta made from semolina, farina, and their blends indicate that durum pasta is of better cooking quality than nondurum pasta (Dexter et al, 1981b; Kim et al, 1986). As one might expect, the intensity of the yellow color of nondurum pasta is much lower than that of durum pasta. Both the cooking quality and color of nondurum pasta can be improved with the proper use of higher temperatures in the pasta drying process (Manser, 1980a); the degree of improvement depends on the quality of the starting material and the specific processing conditions.

VI. PASTA PROCESSING

A. Historical Review

The origin of pasta products is not well documented; one can only speculate that the earliest form of wheat consumption was probably in the form of paste products. According to one account (Nori, 1974), wheat cultivation began in Mesopotamia around 6,000–7,000 B.C. From there, wheat spread eastward through India to China and westward to ancient Egypt, Greece, Italy, and the rest of Europe. Around 2,000 B.C., the Egyptians discovered the fermentation process for baking as well as the use of unleavened dough for "noodle products."

Noodles were known in China about 5,000 B.C. (Ohtsuka, 1974). It is assumed that these products were flattened into thin sheets and cut into strips. Details of the development of the extrusion process, in general, and of its application to form flour products are unknown. It appears that paste products utilizing durum wheat were known in Southern Italy in the 12th century (Nori, 1974). The popularity of pasta products eventually led to a crude form of mechanization in the 18th century. This apparently resulted in the development of the mechanical press, initially made of wood, by about 1850 (Hummel, 1966). At the turn of the present century, mixers, kneaders or gramolas, hydraulic extrusion presses, and drying cabinets became available.

About 1934 a simple continuous extruder was developed in France, and about the same time in Italy and Switzerland continuously operated automatic presses made their appearance. In 1946, the first production line that converted semolina into dried pasta in one continuous operation was unveiled by a Swiss firm. Today all large pasta processors use extrusion presses made by only a few equipment manufacturers.

B. Commercial Process

The process of pasta production is simple in that semolina and water are metered into the mixing chamber in a predetermined ratio. The metering rate depends on the output of the press and on dryer capacity. Typical processing lines have production capabilities in the range of 200–1,200 kg/hr. Figure 12 illustrates a typical layout for continuous processing of short-goods and long-goods pasta. The largest press has the capacity of 6,000 kg/hr. At this point, eggs or egg yolk may be added for egg noodles. Vitamins and minerals are commonly dry blended at the durum mill before delivery to the pasta plant.

For mixers open to the atmosphere, the hydrated semolina particles pass via a rotary seal into an evacuated extrusion chamber. In some presses, the mixing chamber operates under vacuum. The process of mixing and extrusion under vacuum minimizes pigment oxidation and removes trapped air bubbles.

In the extrusion chamber, the hydrated semolina pieces are formed into a cohesive plastic mass. The extrusion pressure created by the screw can be as high as 150 atm. The extrusion pressure is a function of the design of the screw—the pitch, the depth, and the width of channel, as well as the rotational speed of the screw. During extrusion, a considerable amount of heat is generated due to friction. The extrusion chamber must be cooled with circulation water to maintain the dough temperature around 50°C.

The dough at the end of the extrusion worm is split into a number of streams to equalize pressures along the long spaghetti die face. If the pressure is uneven, the extrusion pattern becomes irregular, resulting in excessive trimming. In the processing of short-goods, the circular disk must be designed to ensure uniform extrusion; otherwise, the length of the pieces will be nonuniform.

Early dies were machined from a bronze alloy. Bronze, unfortunately, is soft and wears over time, thereby causing a change in the dimensions of the extruded products. At present, the vast majority of dies are fitted with Teflon inserts. Not only does Teflon prolong the life of the die, but it also produces products with a very smooth surface.

In the production of spaghetti, the strands are picked up on rods in the form of a curtain (Fig. 13). As extrusion is never completely uniform, the bottom of the curtain is trimmed, and the trim is chopped into small pieces and fed back into the mixer. The curtains of spaghetti are subjected to a blast of air for surface drying, called "case hardening." The surface drying sets the shape of the strands and prevents the strands from stretching. The curtains are then conveyed into the first of several drying chambers.

Drying is the most critical phase in processing. Moisture must be removed at a uniform rate to prevent moisture gradients within the strands that can cause checking or cracking.

The first stage of drying is the predryer, where pasta moisture is lowered from 30% or so to about 20%. The usual time is approximately 1 hr, with the temperature at 65° C and relative humidity of 65%. In conventional systems, the main dryer is controlled at about 55° C, with relative humidity initially at 95% and decreasing to 70%. The final stage is the conditioning, where the final

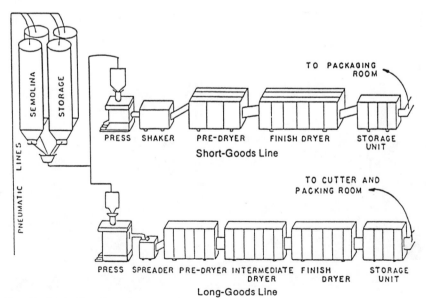

Fig. 12. Production lines for processing short-goods or long-goods pasta. (Reprinted, with permission, from Banasik, 1981)

moisture content of around 12.0% is established and the product is "cured." The total time varies from 15 to 28 hr, depending on the spaghetti diameter and on the drying system.

Many, perhaps the majority, of pasta manufacturers currently use high-temperature (HT) drying—at temperatures over 70° C. Two procedures are commonly used in HT drying. In one scheme, the high temperature is applied immediately at the predryer stage and the initial stage of the main dryer for 2 hr. Thereafter, the temperature is lowered to about 45° C and the relative humidity to about 70% (Pavan, 1979). In the other method, the samples are predried as in the conventional method and then dried at high temperature during the final drying stage. The second method appears to give pasta with improved cooking quality (Manser, 1980b; Dexter et al, 1981a).

In a later study, Manser (1983) reported that samples dried between 60 and 90° C gave pasta of better (firmer and less sticky) cooking quality, especially for samples blended from durum and common wheats. However, pasta color was poor at 80–88° C. Egg products turned brown above 70° C, but the flavor was unaffected. He suggested that 68° C was the optimum drying temperature.

One equipment firm manufactures the *tres haute temperature* (THT, i.e., very high temperature) system for drying pasta at temperatures above 100° C (Anonymous, 1982). For long-goods, the initial phase uses the traditional drying time schedule. When moisture reaches 16%, the spaghetti is stripped, cut, and subjected to the very high temperature. The combination of the traditional method with the THT system reduced drying time from 18–30 hr to three to four hours. However, the above report about the THT system for long-goods seems

Fig. 13. Commercial extrusion of spaghetti. (Courtesy Grain Research Laboratory Division, Agriculture Canada, Winnipeg)

premature. Ollivier (1985a) reported that the THT long-pasta drying system was not operational commercially but an experimental laboratory-scale system. Maximum temperature was reported to be 90°C with a drying time of 3 hr (Ollivier, 1985b).

Abecassis et al (1984) dried spaghetti at 37°C for 30 hr, at 70°C for 10 hr, and at 90°C for 105 min. Color and cooking quality were superior at high temperatures, but the improvement that resulted from increasing the temperature from 70 to 90°C was slight. No discoloration of pasta dried at 90°C was noted.

Korbanov et al (1982) reported a new method for drying pasta utilizing convective hot air, a steam-air mixture, and superheated steam. Extruded pasta is treated with superheated steam, conditioned with saturated steam, and dried with convective heat. The final stage is stabilization with superheated steam for 2 hr. No mention is made of the advantages of this method, i.e., in production time or in pasta quality.

Medvedev et al (1984) studied the effect of heat treatment of dough on pasta drying. When dough is mixed at 60–70°C, the moisture of the extruded product is reduced to 24–25%, thereby eliminating the predrying stage. At 45–50°C drying temperature, drying time is reduced and the percentage of cracked pasta is also reduced.

During drying of short goods, the cut pieces fall from the die onto horizontal shaking screens. Hot air is forced through the screens to reduce surface moisture. As the time on the shakers is short, only a 2–3% drop in moisture occurs. From the shakers, the pieces are spread uniformly onto a screen-belt system that travels back and forth from the top to the bottom in the predryer. The moisture content of the product exiting the predryer is about 21%. During the 12 hr in the main dryer, with temperature and relative humidity controlled continuously, the moisture is reduced to approximately 12.5%. With the conventional drying method, temperatures do not exceed 55°C.

HT dryers for short-goods were available commercially before HT dryers for long-goods. Manser (1980a) stated that the first short-goods drying lines utilizing elevated temperatures were built in the late 1960s. The major impact of HT drying was the reduction in drying time. Because the thickness of most short-goods is less than the diameter of ordinary spaghetti, migration of moisture and establishment of an equilibrium are much faster in short-goods. For systems using temperatures over 100°C, drying time is as short as 105 min (Ollivier, 1985a).

Use of microwave dryers for short-goods was introduced about 15 years ago (Katskee, 1982). A major problem encountered was that a conventional predryer could not be used because at moisture levels of 20% or higher, microwaves tended to cook rather than dry the pasta. Another problem was that case-hardened products could not be dried in a microwave dryer, as the trapped internal moisture caused checking. Technical problems have now been solved so that short-goods can be dried within 1.5 hr by microwave.

C. Laboratory Processing

For centuries, pasta-making was more of an art than a science. The pasta artisan adjusted conditions of processing by "dough feel" and "dough appearance" without reliance on chemical or physical data on the wheat or semolina. As an art, there was little need to apply scientific methods to pasta processing. One can surmise that such a situation existed in Italy, where pasta products as we know them today originated. When domestic production of wheat became insufficient to meet the demand of the pasta makers, wheat had to be imported. Pasta makers could rely on their experience to produce acceptable pasta, but it would have been simpler had they been able to refer to some type of guidelines of desirable quality factors in wheat and semolina.

The first report on a piece of small-scale laboratory equipment for pasta processing was published by Fifield (1934). It consisted of a horizontal pin mixer, a kneader, and an extruder. Extrusion pressure was generated by a motor-driven screw. The quantity of semolina required for processing was not specified, but the dimensions of the mixer indicate that it was about 1 kg. Drying was done in two cabinets. The extruded product was kept in the first cabinet for up to an hour and then transferred to the second cabinet. Humidity was controlled by exhausting moist air through a vent at the bottom and venting fresh air from the top. Drying time was four days.

Binnington and Geddes (1936) described small-scale macaroni processing equipment identical in construction to that described by Fifield (1934) with a couple of modifications. Drying was accomplished in two stages, but unlike that described by Fifield (1934), the main dryer was equipped with a refrigeration unit to remove excess moisture. Drying temperature was 32°C with a drying time of 60 hr.

The first micro-method for making macaroni using 50 g of semolina was reported by Martin et al (1946). The unit consisted of a horizontal pin-type mixer fitted with seals to permit mixing under vacuum, a kneader, a rest press, and an extrusion press. The rest press was a hydraulic press in which the dough was rested under pressure after mixing. According to the authors, "this step makes the dough more translucent and decreases viscosity." Extrusion was accomplished with a hydraulic press. The macaroni was dried in a cabinet maintained at 32°C with a controlled decrease in relative humidity for 48 hr.

Until the 1970s, the advance in laboratory processing was in drying cabinets. The cabinet design requiring a 96-hr drying time (Fifield, 1934) was improved to shorten the time. In the micro-cabinet method of Martin et al (1946), the drying time was shortened to 28 hr. A significant contribution in laboratory dryers for macaroni products was made by Gilles et al (1966). The design incorporated the use of electronic control devices to control temperature and relative humidity independently.

Modification in the micro-method was reported by Matsuo et al (1972). Mixing was done on a 50-g farinograph mixer, which was sealed for mixing under vacuum. The mixing action of the farinograph blades was sufficiently vigorous that the kneading step was unnecessary. For processing pasta on a larger scale using 1.0–2.0 kg of semolina, a laboratory-scale continuous process extrusion press has been used (Walsh et al, 1971; Matsuo et al, 1978). For experimental purposes, continuous operation using the automatic feed system is impractical; thus, the extrusion press is used in a batch process.

European laboratories use macro-equipment of similar design, with a mixing chamber that can be evacuated and an extrusion worm to force the dough through various pasta dies. A small-scale extruder is used in two laboratories in

France. Mixing and extrusion are done under vacuum. A flat piece about 0.8 mm by 4 cm is extruded; disks are stamped out and tests for color and cooking quality performed (Institut Technique des Céréales et des Fourrages, 1983).

Because HT drying of pasta products is very common, HT effects on pasta quality have been studied in the laboratory. Investigations on the effects of HT drying have been reported (Manser, 1980b; Dexter et al, 1981a; Ollivier, 1985b). When applied properly, HT drying intensifies color, increases cooked firmness, and reduces the microbial count of pasta.

It is difficult, if not impossible, to reproduce commercial processing conditions in the laboratory. The processing of pasta at a rate of 500–1,000 kg/hr cannot be scaled down to a laboratory-scale of 10–15 kg/hr. Cubadda (1985) emphasized the differences in processing conditions between the laboratory and the industrial pasta plant. The pasta produced in the laboratory may be very similar in appearance, but very often it is different in certain quality characteristics.

The development of laboratory-scale pasta processing equipment was the first major step that enabled cereal chemists to investigate quality factors in durum wheat and semolina that are important for good quality in pasta. Significant advances have been made in durum wheat technology, and a remarkable improvement in the overall quality of durum wheat is the result.

VII. PASTA QUALITY EVALUATION

Pasta quality is judged by the processor and ultimately by the consumer. The processor of dry pasta desires a finished product that is microbiologically stable, visibly appealing, and elastic and strong enough to resist breakage during cutting, packaging, handling, and shipping. The consumer desires a product that is free from cracks and shattered pieces and that is otherwise appealing with regard to appearance and cooking and eating quality. Processors of canned or frozen pasta want a product that will withstand the rigors of cooking, retorting, hot-fill, or cooling operations so as to retain a high degree of the original physical integrity inherent in the product before the additional processing. In any case, no matter what the form of the pasta, it must meet the expectations of the consumer in order to be accepted.

A. Appearance

Factors that influence the appearance of the pasta product other than its size and shape are its color, uniformity, clarity, and surface texture. Desirable attributes of pasta are that it be translucent, bright yellow, and free from excessive specks, cracks, or checks and that it have a smooth surface.

To the consumer, pasta color is purely subjective, but in a laboratory, where numerous samples must be compared on a day-to-day basis, an objective test is needed since color is perceived differently by different people. Reflectance colorimeters (Walsh et al, 1969; Walsh, 1970) or spectrophotometers (Dexter and Matsuo, 1977) work well for comparing the brightness and intensity of color. Pasta color is influenced by semolina quality, and by the processing conditions of mixing, extrusion, and drying. Semolina of too high a milling extraction or having a high speck count will yield pasta with a less than desirable color. The best color is obtained by extruding a high-quality semolina in a processing system equipped to mix and extrude under vacuum. Whereas pasta color can be improved somewhat by proper manipulation of HT drying cycles, too high a temperature in the pasta dryer can cause excessive darkening of the pasta, making it less appealing.

The uniformity, clarity, and surface texture of pasta can be affected by several factors, including semolina speck count, granulation, grit content, and pasta processing conditions. Black and brown specks in semolina can alter product clarity and surface texture. Semolina with excessively coarse granulation does not hydrate uniformly during extrusion and produces white specks and a rough surface in the pasta. Grit in semolina can tear or chalk the surface of pasta, making it nonuniform or rough. Inadequate mixing time or uneven die extrusion pressure can also adversely affect product uniformity. Improperly dried pasta may develop cracks or checks that affect its appearance, but more importantly, improper drying may reduce mechanical strength and cause immediate or delayed shattering of the product.

B. Mechanical Strength

The mechanical strength of dry pasta is an important consideration since, if the product is fragile, it will not withstand the rigors of cutting, packaging, handling, and shipping. Although raw material quality and the processes of mixing and extrusion have some influence, the drying process is probably the most critical factor in determining pasta strength. Inadequate control of air temperature, relative humidity, flow, or velocity in the dryer can contribute to decreased strength in the finished product. If the drying cycle is tailored to meet the requirements of a specific product, the resulting pasta should be quite hard yet elastic as it exits the dryer. A true test of proper drying is whether these attributes of hardness and elasticity persist in the pasta for several weeks or months after removal from the dryer. Pasta mechanical strength is usually measured subjectively by squeezing or bending the product by hand, but objective comparisons can also be made with instruments such as the Instron Universal Testing Instrument.

C. Cooking Quality

The cooking and eating qualities of pasta are judged by consumers based on their expectations of the products. For example, an experienced consumer would normally expect a softer eating texture from canned pasta than from dried pasta.

The appearance and texture of pasta should be acceptable after cooking, and the pasta should have a pleasing flavor and mouthfeel when eaten. Dried pasta after cooking is generally considered to possess good quality if it has retained some yellow color, has not broken apart or released excess solids into the cooking water, is not sticky or mushy when eaten, and exhibits some firmness to the bite—"al dente," as it is sometimes expressed. These assessments of quality are quite subjective and vary depending on the consumer's judgment.

Although sensory evaluation is probably the most reliable measure of quality, testing laboratories have developed controlled cooking procedures and mechanical objective tests to minimize variables that influence the measurement of pasta quality and to eliminate individual bias. Factors shown to influence the perceived quality of pasta include cooking techniques, cooking time, composition of the cooking water, and handling of the product after cooking. Cooking time varies with the shape and thickness of the product. Hardness and pH of cooking water affect pasta stickiness (Dexter et al, 1983). The relative turbidity of the cooking water is an indication of the amount of breakdown of pasta during cooking and is referred to as cooking loss or cooking-water residue. The residue can be measured by evaporating the cooking water or by freeze-drying (Dexter and Matsuo, 1979).

Several objective procedures used to measure the texture of cooked pasta have been reported (Dick, 1985). Basically, the procedures monitor the response to a force on the cooked pasta of either pushing, pulling, or both to measure such factors as firmness or tenderness, elasticity, compressibility, or stickiness. The combination of these factors gives a good estimation of the relative textural characteristics of the cooked product. Pasta composition with respect to protein content and quality (Grzybowski and Donnelly, 1979), starch to protein ratio (Voisey et al, 1978), and lipid content (Matsuo et al, 1986) have been shown to affect textural properties, the latter two significantly influencing stickiness. The processes of extrusion and drying (i.e., HT drying) also contribute to pasta texture in a significant way (Manser, 1980b). Too high an extrusion temperature (above 50° C) can denature gluten protein, thus adversely affecting pasta texture. The drying process when properly controlled can have a positive effect on pasta texture.

VIII. NONDURUM PASTA PRODUCTS

A. Noodles

The formulations of noodle products made from nondurum wheat tend to be more complex than those used to make durum pasta. Noodle products often contain not only the basic ingredients of water and flour, but other ingredients such as salt, alkaline reagents, artificial coloring agents, and preservatives. Whole eggs from chickens or ducks are sometimes added and can make up as much as 100% of the liquid ingredients of a formulation based on flour weight. Although wheat flour most often makes up the bulk of noodle formulations, other raw materials such as rice, barley, buckwheat, mung bean, potato, yam, or cassava sometimes make up a part of the blend.

Specifications of wheat flour required for nondurum noodles depend on local or regional preferences for noodle color, eating texture, flavor, shelf life, and ease of preparation for consumption. Regional preferences exist for noodle clarity or brightness, yellowness, whiteness, softness, firmness or chewiness, and flavor; the specific preferences depend on eating habits or expectations associated with accepted quality standards for a given noodle shape or type.

Noodle flavor, texture, and appearance depend not only on flour characteristics but also on the specific process used to manufacture the noodle as well as the inclusion of other raw material or chemical additives. Noodles made from low-protein flour generally have a soft bite compared to those made from high-protein flour. Noodle color is intensified by eggs, flour with high protein or high ash, and alkaline reagents. Noodle texture is firmed by steaming the raw noodle and by inclusion of eggs and alkaline salts. The shelf life of noodles varies with moisture content, formulation, processing technique, packaging, handling, and storage conditions. Shelf life requirements in many countries are determined primarily by the proximity of the noodle manufacturer to the marketplace.

B. Noodle Types

Popular noodle types include fresh (raw, Cantonese style), wet or boiled (Hokkien style), dried, steamed and dried (traditional instant), boiled and fried (*yimee*), and steamed and fried (instant, ramen style) (Fig. 14). Production of Chinese noodles is shown in Fig. 15. Each noodle type has at least one unique step in processing that makes it different from the others, although most noodle types share the common processing steps of mixing, kneading, rolling or sheeting, and cutting (Fig. 16).

Raw noodles are sold in the medium-moisture state (approximately 35% moisture), whereas wet noodles are parboiled to approximately 55% moisture, rinsed and cooled with fresh water, and coated with vegetable oil to prevent sticking. Both raw and wet noodles have relatively short shelf lives ranging from a few hours to several days, depending on formulations, packaging, and storage conditions. Refrigeration prolongs noodle shelf life but is not readily available or affordable in many countries where these noodles are consumed.

Low-moisture noodles exhibit relatively long shelf stability (several months) and include dried and fried types. Dried noodles are processed either by drying freshly cut noodles or by steaming and then drying. Steaming toughens the

Fig. 14. Oriental noodles. Top row: various dried noodles; bottom row: left, wet noodles; center two, ramen; right two, packaged wet noodles. (Courtesy P. A. Seib)

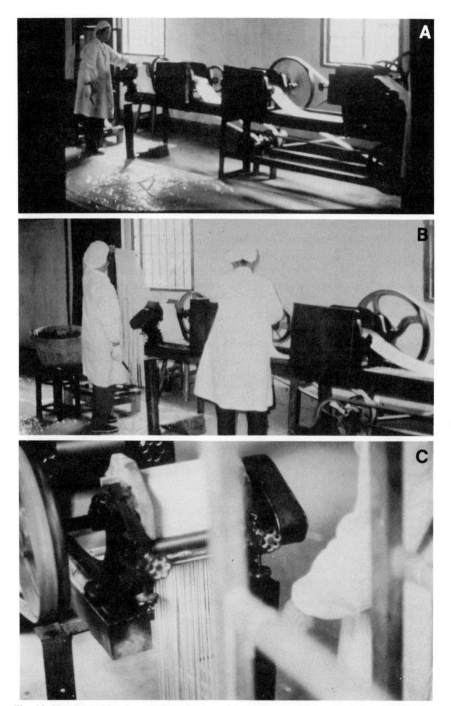

Fig. 15. Noodle-making in a Chinese factory. A, general view of dough mixing and sheeting; B, cutting of dough into noodles; C, detail of noodle cutting. (Courtesy Y. Pomeranz)

noodles and makes them less fragile. Drying is accomplished either by artificial means or naturally by the sun. Processing of instant noodles is done by steaming raw noodles and drying, or by steaming and deep-fat frying noodles before cooling and packaging. Fried noodles are usually boiled and deep-fat fried to an oil content not to exceed 20%. The stability of instant and fried noodles largely depends on their oil content and the type and freshness of the oil used for frying.

Ingredient formulations for each type of noodle vary among and within countries, depending on personal preferences and on the availability and cost of raw materials. For example, in Japan a soft noodle texture usually is preferred and is unique to *udon* noodles (thick noodles made of soft wheat flour) and *soba* noodles (made of buckwheat flour). In China, Korea, and Southeast Asia, noodles with chewy, more elastic texture are preferred. Therefore, the steaming of noodles and the use of noodle formulations including high-protein flour, eggs, or *kan-sui* (alkaline salts) to increase chewiness are common in these countries. Noodles made of 100% rice flour or a blend of rice flour and wheat flour are popular in Taiwan. Noodles made of starch from mung beans, rice, or cassava are also popular throughout Asia, but to a much lesser extent than flour noodles.

To the novice, the terminology of naming noodles can be confusing since noodles of almost identical makeup have different names in various countries. To add to the confusion, noodles of the same formulation within the same country are sometimes named to differentiate noodle thickness or manu-facturing method. For example, *so-men* and *udon* noodles differ only in thickness, and hand-made *so-men* is called *tenobe*, whereas hand-made *udon* is

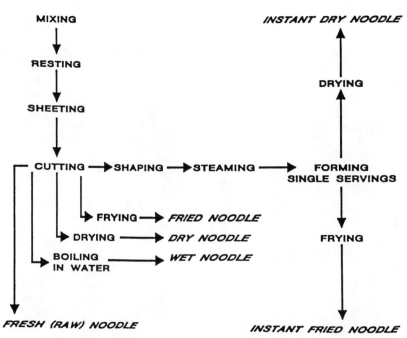

Fig. 16. Types of oriental noodles classified by moisture and degree of precooking. (Reprinted, with permission, from Oh et al, 1983)

called *teuchii* (Nagao, 1981). Because noodles are processed by numerous combinations of processes, formulations, sizes, and shapes, it is important to be specific when comparing their characteristics. Comprehensive coverage of the subject of noodles would require a book in itself.

IX. CONSIDERATIONS FOR THE FUTURE

The demand for durum wheat has the potential to increase as the popularity of pasta rises. Pasta consumption and the demand for durum wheat do not have a correlation of 1.0, yet there is a strong positive relationship. Pasta consumption is expected to continue to rise in developed countries such as Japan, the United Kingdom, and the United States, due not only to increased population but to their increased popularity. This is due, in part, to the consumers' realization that pasta is a nutritious and healthful food. Many underdeveloped countries have expressed the desire to use durum wheat in the future as it becomes affordable, especially some countries in South and Central America that now consume pasta made mostly from nondurum wheat. Countries such as Korea, Taiwan, Singapore, Thailand, and Malaysia, which are traditional users of noodles but not of durum pasta, possess some potential for expanded use of durum.

The challenge to plant breeders and pathologists and cereal chemists lies in developing durum varieties that meet market demands for grain quantity and quality. To accomplish this, they must first meet the needs of farmers who require durum varieties with high grain yield and quality, varieties that will provide the rate of return on investment necessary to compete with other available crops. In addition, grain and end-use quality must be maintained or improved as increases in grain yield are attained. Since protein quantity and quality and soundness of grain influence pasta quality, these factors must be considered in a durum breeding program. Durum varieties possessing some degree of preharvest dormancy are needed to diminish the recurrent problem of grain sprouting, which is accompanied by increased amylase activity that results in diminished market value and end-product quality.

Several advances have been made over the past 20 years with regard to improved resistance to disease, but progress must continue in order to assure disease-resistant varieties for the future. Furthermore, because durum wheat tends to be grown in very localized geographic areas of the world, it is imperative that durum wheat breeders, since they are few, maintain a diverse genetic base in their gene pool to protect against devastating losses due to epidemics. Advances have also been made to improve color, increase gluten strength, and reduce the inherent lipoxygenase activity of durum semolina. Nonetheless, cereal chemists must continue to work closely with plant breeders to guarantee that grain, milling, semolina, and end-use quality are not jeopardized as agronomic improvements are made.

Although considerable research has been reported regarding pasta quality, very little work has been published about the quality requirements of durum wheat for use in bread, couscous, and other products. Perhaps more research will be done in this area as more and better "strong gluten" durum varieties are developed and as the increased potential for durum wheat utilization is realized.

LITERATURE CITED

ABECASSIS, J., ALARY, R. and FEILLET, P. 1984. Influence de la température de séchage sur l'aspect et la qualité de pâtes alimentaire. Ind. Cereal 31:13-18.

ABERCROMBIE, E. 1980. Durum milling. Oper. Millers Tech. Bull. (Mar.–June) pp. 3308–3813.

ALSBERG, C. L. 1939. Durum wheats and their utilization. Wheat Stud. Food Res. Inst. 15(7):336-364.

ANONYMOUS. 1982. Bassano claims victory over +212 degrees F pasta drying. Macaroni J. 64(2):32.

BANASIK, O. J. 1981. Pasta processing. Cereal Foods World 26:166-169.

BENGTSON, N. A., and GRIFFITH, D. A. M. 1915. The Wheat Industry. The MacMillan Co., New York.

BINNINGTON, D. S., and GEDDES, W. F. 1936. Experimental durum milling and macaroni technique. Cereal Chem. 13:497-521.

BIRCHARD, F. J. 1931. Annual Report. Dominion Grain Research Laboratory, Winnipeg .

BOLLING, H., and ZWINGELBERG, H. 1972. Glasigkeit und Griesausbeute bei Durumweizen. Getreide Mehl Brot 26:264-269.

BOWDEN, W. M. 1959. A taxonomy and nomenclature of the wheats, barleys and ryes and their wild relatives. Can. J. Bot. 37:657-684.

BOWDEN, W. M. 1966. Chromosome numbers in seven genera of the tribe tritici. Can. J. Genet. Cytol. 8:130-136.

BURGOON, A. C., IKEDA, H. S., and TANNER, S. N. 1985. A method for detecting adulteration in durum wheat pasta by polyacrylamide gel electrophoresis. Cereal Chem. 62:72-74.

CANADIAN GRAIN COMMISSION 1986. Canadian Grain Exports, Crop Year (1985/86). The Commission, Winnipeg.

COSTANTINI, A. M. 1985. Nutritional and health significance of pasta in modern dietary patterns. Pages 1-8 in: Pasta and Extrusion Cooked Foods: Some Technological and Nutritional Aspects. C. Mercier and C. Cantarelli, eds. Elsevier Applied Science Publ., London.

CUBBADA, R. 1985. Effect of the drying process on the nutritional and organoleptic characteristics of pasta: A review. Pages 79-89 in: Pasta and Extrusion Cooked Foods: Some Technological and Nutritional Aspects. C. Mercier and C. Cantarelli, eds. Elsevier Applied Science Publ., London.

DEXTER, J. E., and MATSUO, R. R. 1977. Influence of protein content on some durum wheat quality parameters. Can. J. Plant Sci. 57:717-727.

DEXTER, J. E., and MATSUO, R. R. 1979. Changes in spaghetti protein solubility during cooking. Cereal Chem. 56:394-398.

DEXTER, J. E., and MATSUO, R. R. 1981. Effect of starchy kernels, immaturity, and shrunken kernels on durum wheat quality. Cereal Chem. 58:395-400.

DEXTER, J. E., MATSUO, R. R., and MORGAN, B. C. 1981a. High temperature drying: Effect on spaghetti properties. J. Food Sci. 46:1741-1746.

DEXTER, J. E., MATSUO, R. R., PRESTON, K. R., and KILBORN, R. H. 1981b. Comparison of gluten strength, mixing properties, baking quality and spaghetti quality of some Canadian durum and common wheats. Can. Inst. Food Sci. Technol. J. 14:108-111.

DEXTER, J. E., BLACK, H. C., and MATSUO, R. R. 1982. An improved method for milling semolina in the Buhler laboratory mill and a comparison to the Allis-Chalmers laboratory mill. Can. Inst. Food Sci. Technol. J. 15:225-228.

DEXTER, J. E., MATSUO, R. R., and MORGAN, B. C. 1983. Spaghetti stickiness: Some factors influencing stickiness and relationship to other cooking characteristics. J. Food Sci. 48:1545-1551, 1559.

DEXTER, J. E., MATSUO, R. R., and MARTIN, D. G. 1987. The relationship of durum wheat test weight to milling performance and spaghetti quality. Cereal Foods World 32:772-777.

DICK, J. W. 1985. Rheology of durum. Pages 219-240 in: Rheology of Wheat Products. H. Faridi, ed. Am. Assoc. Cereal Chem., St. Paul, MN.

DICK, J. W., WALSH, D. E., and GILLES, K. A. 1974. The effect of field sprouting on the quality of durum wheat. Cereal Chem. 51:180-188.

DICK, J. W., D'APPOLONIA, B. L., HOLM, Y. F., and HANSEN, D. 1986. Durum Wheat: 1986 Regional Quality Report. North Dakota Wheat Commission, Bismarck, ND.

DONNELLY, B. J. 1980. Effect of sprout damage on durum wheat quality. Macaroni J. 62(11):8-10, 14.

DOROFEEV, W. F., and JAKUBNIZER, M. M. 1973. The gene pool of Soviet durum wheat (T. durum Desf.). Pages 141-152 in: Proc. Symp. Genet. Breed. Durum Wheat. G. T. Scarascia Mugnozza, ed. Univ. of Bari, Bari, Italy.

DOUGLASS, J. S., and MATTHEWS, R. H.

1982. Nutrient content of pasta products. Cereal Foods World 27:558-561.

DU CROS, D. L. 1987. Glutenin proteins and gluten strength in durum wheat. J. Cereal Sci. 5:3-12.

FEILLET, P. 1984. The biochemical basis of pasta cooking quality; its consequences for durum wheat breeders. Sci. Aliment. 4:551-556.

FIFIELD, C. C. 1934. Experimental equipment for the manufacture of alimentary pastes. Cereal Chem. 11:330-334.

FISHER, G. W. 1972. The technology of bulgur production. Oper. Millers Tech. Bull. (May) pp. 3300-3304.

GEDDES, W. F. 1934. Quality requirements of various industries which use durum wheat. Weekly Market News. Winnipeg, March 28.

GEDDES, W. F. 1935. Annual Report. Grain Research Laboratory, Winnipeg.

GILLES, K. A., SIBBITT, L. D., and SHUEY, W. C. 1966. Automatic laboratory dryer for macaroni products. Cereal Sci. Today 11:322-324.

GREENWELL, P., and SCHOFIELD, J. D. 1986. A starch granule protein associated with endosperm softness in wheat. Cereal Chem. 63:379-380.

GRL 1986. Quality of 1986 Canadian Wheat. Grain Research Laboratory, Winnipeg, Canada.

GRZYBOWSKI, R. A., and DONNELLY, B. J. 1979. Cooking properties of spaghetti: Factors affecting cooking quality. J. Agric. Food Chem. 27:380-384.

HALEY, W. L., and PENCE, J. W. 1960. Bulgor, an ancient wheat food. Cereal Sci. Today 5:203-204, 206-207, 214.

HARRISON, T. J. 1934. History of durum wheat. Weekly Market News, Winnipeg. March 28.

HOOK, S. C. W. 1984. Specific weight and wheat quality. J. Sci. Food Agric. 35:1136-1141.

HOSENEY, R. C. 1986. Principles of Cereal Science and Technology. Am. Assoc. Cereal Chem., St. Paul, MN.

HSIEH, C. C., WATSON, C. A., and McDONALD, C. E. 1980a. Identification of campesteryl palmitate and sitosteryl palmitate in wheat flour. J. Food Sci. 45:523-525.

HSIEH, C. C., WATSON, C. A., and McDONALD, C. E. 1980b. Direct gas chromatographic estimation of saturated steryl esters and acylglycerols in wheat endosperm. Cereal Chem. 58:106-110.

HUMMEL, C. 1966. Macaroni Products. Food Trade Press Ltd., London.

HUNT, T. F. 1910. The Cereals in America. Orange Judd Co., New York; Kegan Paul, Trench Trubner and Co. Ltd., London.

INSTITUT TECHNIQUE DES CÉRÉALES ET DES FOURRAGES. 1983. Qualite des Blés Durs Récolte de France 1983. The Institute, Paris.

INTERNATIONAL WHEAT COUNCIL. 1986. World Wheat Statistics. The Council, London.

IRVINE, G. N. 1961. Annual Report. Grain Research Laboratory, Winnipeg.

IRVINE, G. N. 1964. Uber die Mahleigenschaften von Kanadischen Durumweizen. Getreide Mehl 14:34-39.

IRVINE, G. N. 1971. Durum wheat and paste products. Pages 777-796 in: Wheat: Chemistry and Technology, 2nd ed. Y. Pomeranz, ed. Am. Assoc. Cereal Chem., St. Paul, MN.

IRVINE, G. N. 1979. Durum wheat quality: Comments on the international collaborative test. Proc. Int. Symp.: Matières Premières et Pâtes Alimentaires. G. Fabriani and C. Lintas, eds. I.C.C.-Ist. Naz. Nutrizione, Rome.

JOHNSTON, R. A., QUICK, J. S., and DONNELLY, B. J. 1980. Note on comparison of pigment extraction and reflectance colorimeter methods for evaluating semolina color. Cereal Chem. 57:447-448.

JOPPA, L. R., KHAN, K., and WILLIAMS, N. D. 1983. Chromosomal location of genes for gliadin polypeptides in durum wheat *Triticum turgidum* L. Theor. Appl. Genet. 64:289-293.

JOSEPHIDES, C. M., JOPPA, L. R., and YOUNGS, V. L. 1987. Effect of chromosome 1B on gluten strength and other characteristics of durum wheat. Crop Sci. 27:212-216.

KATSKEE, A. L. 1982. Microwave pasta drying. Macaroni J. 64(4):19-22.

KAUP, S. M., and WALKER, C. E. 1986. Couscous in North Africa. Cereal Foods World 31:179-182.

KIM, H. I., SEIB, P. A., POSNER, E., DEYOE, C. W., and YANG, H. C. 1986. Milling hard red winter wheat to farina: Comparison of cooking quality and color of farina and semolina spaghetti. Cereal Foods World 31:810-819.

KORBANOV, V. T., KALOSHINA, E. N., NAZAROV, N. I., and TSIVTSIVADZE, G. V. 1982. New methods for drying pasta. Izv. Vyssh. Uchebn. Zaved., Pishch. Technol. 5:134-138. (Food Sci. Technol. Abstr. 6M627, 1984)

KULKARNI, R. G., PONTE, J. G., Jr., and KULP, K. 1987. Significance of gluten content as an index of flour quality. Cereal Chem. 64:1-3.

LAIGNELET, B. 1983. Lipids in pasta and pasta processing. Pages 281-283 in: Lipids in Cereal Technology. P. J. Barnes, ed.

Academic Press, New York.

MANSER, J. 1980a. High-temperature drying of pasta products. Buhler Diagram 69:11-12.

MANSER, J. 1980b. Optimal parameters in the production of macaroni products: Long-goods as a case in point. Buhler-Miag Inf. Bull. Buhler Brothers Ltd., Uzwil, Switzerland.

MANSER, J. 1983. High temperature drying of pasta. Molini Ital. 34(3):122-123.

MANSER, J. 1984. Degree of fineness of milled durum products from the viewpoint of pasta manufacture. Buhler-Miag Inf. Bull. Buhler Brothers Ltd., Uzwil, Switzerland.

MARTIN, V. G., IRVINE, G. N., and ANDERSON, J. A. 1946. A micro method for making macaroni. Cereal Chem. 23:568-579.

MATSUO, R. R. 1985. Durum wheat gluten: Non multa sed multum? Pages 365-374 in: Monografie di Genetica Agraria II. Giornate Internazionali sul grano durs. A. Bianchi, ed. Ist. Sperimentale per la Cerealicoltura, Rome.

MATSUO, R. R., and DEXTER, J. E. 1980a. Relationship between some durum wheat physical characteristics and semolina milling properties. Can. J. Plant Sci. 60:49-53.

MATSUO, R. R., and DEXTER, J. E. 1980b. Comparison of experimentally milled durum wheat semolina to semolina produced by some Canadian commercial mills. Cereal Chem. 57:117-122.

MATSUO, R. R., and IRVINE, G. N. 1967. Macaroni brownness. Cereal Chem. 44:78-85.

MATSUO, R. R., BRADLEY, J. W., and IRVINE, G. N. 1972. Effect of protein content on the cooking quality of spaghetti. Cereal Chem. 49:707-711.

MATSUO, R. R., DEXTER, J. E., and DRONZEK, B. L. 1978. Scanning electron microscopy study of spaghetti processing. Cereal Chem. 55:744-753.

MATSUO, R. R., DEXTER, J. E., KOSMOLAK, F. G., and LEISLE, D. 1982a. Statistical evaluation of tests for assessing spaghetti-making quality of durum wheat. Cereal Chem. 59:222-228.

MATSUO, R. R., DEXTER, J. E., and MacGREGOR, A. W. 1982b. Effect of sprout damage on durum wheat and spaghetti quality. Cereal Chem. 59:468-472.

MATSUO, R. R., DEXTER, J. E., BOUDREAU, A., and DAUN, J. K. 1986. The role of lipids in determining spaghetti cooking quality. Cereal Chem. 63:484-489.

MATVEEF, M. 1963. Le mitadinage des bles durs, son evaluation et son influence sur le rendement et al valeur des semoules. Bull. E.F.M. 198:299-386.

MATVEEF, M. 1966. Influence du gluten de blé dur sur la valeur des pâtes alimentaires. Bull. Anc. Eleves Ec. Meun. ENSMIC 213:133-138.

MEDVEDEV, G. M., MALANDEEVA, N. I., KESELEVA, R. F., and SEMKO, V. T. 1984. Drying of pasta from heat-treated dough. Khlebopek. Konditer. Prom. 10:41-42. (Tec. Molitoria 3[1]:7-8, 1985)

MENGER, A. 1973. Problems concerning vitreousness and hardness of kernels as quality factors of durum wheat. Pages 563-571 in: Proc. Symp. Genet. Breed. Durum Wheat. G. T. Scarascia Mugnozza, ed. Univ. Bari, Bari.

MILLER, B. S., AFEWORK, S., POMERANZ, Y., BRUINSMA, B. L., and BOOTH, G. D. 1982. Measuring the hardness of wheat. Cereal Foods World 27:61-64.

MOUSA, I. E., SHUEY, W. C., MANEVAL, R. D., and BANASIK, O. J. 1983. Farina and semolina for pasta production: I. Influence of wheat classes and granular mill streams on pasta quality. Oper. Millers Tech. Bull. (July) pp. 4083-4087.

NAGAO, S. 1981. Soft wheat uses in the Orient. Pages 267-304 in: Soft Wheat: Production, Breeding, Milling, and Uses. W. T. Yamazaki and C. T. Greenwood, eds. Am. Assoc. Cereal Chem., St. Paul, MN

NATIONAL PASTA ASSOCIATION 1984. The Nutrient Profile of Pasta. The Association, Arlington, VA.

NATIONAL PASTA ASSOCIATION 1986. Facts on Pasta. The Association, Arlington, VA.

NISHIKAWA, K., and NAGAO, S. 1977. Komugi no Kanashi (The story of Wheat). Shibata Shoten, Tokyo.

NORI, K. 1974. A brief history of pasta products. Shokuryo No Kagaku 20(Oct.):19-21.

OH, N. H., SEIB, P. A., DEYOE, C. W., and WARD, A. B. 1983. Noodles. I. Measuring the textural characteristics of cooked noodles. Cereal Chem. 60:433-438.

OHTSUKA, K. 1974. The culture of noodle products. Shokuryo No Kagaku 20(Oct):16-18.

OLLIVIER, J. 1985a. Pasta drying at very high temperature: A fact. Pages 90-97 in: Pasta and Extrusion Cooked Foods. Some Technological and Nutritional Aspects. C. Mercier and C. Cantarelli, eds. Elsevier Applied Science Publ., London.

OLLIVIER, J. 1985b. Die Trocknung von Lang-und Kurzteigwaren bei sehr hohen Temperaturen. Getreide Mehl Brot 39:314-317.

PAVAN, G. 1979. High temperature drying process and pasta quality. Report of Officine Maccaniche Pavan S.p.A., Padova, Italy.

ROSS, A. S., WALKER, C. E., BOOTH, R. I., ORTH, R. A., and WRIGLEY, C. W. 1987. The Rapid Visco-Analyzer: A new technique for the estimation of sprout damage. Cereal Foods World 32:827-829.

SEYAM, A., SHUEY, W. C., MANEVAL, R. D., and WALSH, D. E. 1974. Effect of particle size on processing and quality of pasta products. Oper. Millers Tech. Bull. (Dec.) pp. 3497-3499.

SHUEY, W. C., DICK, J. W., and MANEVAL, R. D. 1977. Note on experimental milling evaluation of durum wheat. Cereal Chem. 54:1026-1029.

SHUEY, W. C., MANEVAL, R. D., and DICK, J. W. 1980. Dual-purpose mill for flour and granular products. Cereal Chem. 57:295-300.

SILANO, V. 1975. Determination of soft wheat in semolina and paste foods. Boll. Chim. Unione Ital. Lab. Prov. 1:338-354. (Chem. Abstr. 84[12]:162930q)

SIMS, R. P. A., and LEPAGE, M. 1968. A basis for measuring the intensity of wheat flour pigments. Cereal Chem. 45:605-611.

SRIVASTAVA, J. P. 1984. Durum wheat—Its world status and potential in the Middle East and North Africa. Rachis 3(1):1-8.

STEHRENBERGER, W. 1981. Extrusion. Macaroni J. 63(3):8, 10, 14-15, 18.

USDA 1977–1986. Grain and Feed Market News. U.S. Dep. Agric., Agric. Mark. Serv., Washington, DC.

VALLEGA, J. 1973. Durum varieties cultivated in Argentina and their pastifiable quality related with the Italian macaroni industry. Pages 599-615 in: Proc. Symp. Genet. Breed. Durum Wheat. G. T. Scarascia Mugnozza, ed. Univ. Bari, Bari.

VASILJEVIC, S., BANASIK, O. J., and SHUEY, W. C. 1977. A micro unit for producing durum semolina. Cereal Chem. 54:397-404.

VOISEY, P. W., WASIK, R. J., and LOUGHHEED, T. C. 1978. Measuring the texture of cooked spaghetti. 2. Exploratory work on instrumental assessment of stickiness and its relationship to microstructure. J. Inst. Can. Sci. Technol. Aliment. 11:180-188.

WALSH, D. E., GILLES, K. A., and SHUEY, W. C. 1969. Color determination of spaghetti by the tristimulus method. Cereal Chem. 46:7-13.

WALSH, D. E. 1970. Measurement of spaghetti color. Macaroni J. 52(4):20-22.

WALSH, D. E., EBELING, K. A., and DICK, J. W. 1971. A linear programming approach to spaghetti processing. Cereal Sci. Today 16:385-386, 388-389.

WILLIAMS, P. C. 1985. Survey of wheat flours used in the Near East. Rachis 4(1):17-20.

WILLIAMS, P. C., and EL-HARAMEIN, F. J. 1985. Frekeh making in Syria—A small but significant local industry. Rachis 4(1):25-27.

WILLIAMS, P. C., EL-HARAMEIN, F. J., and ADLEH, B. 1984a. Burghul and its preparation. Rachis 3(2):28-30.

WILLIAMS, P. C., SRIVASTAVA, J. P., NACHIT, M., and EL-HARAMEIN, F. J. 1984b. Durum wheat-quality evaluation at Icarda. Rachis 3(2):30-32.

INDEX